S0-BWY-988

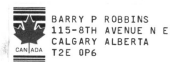
BARRY P ROBBINS
115-8TH AVENUE N E
CALGARY ALBERTA
T2E 0P6

A Publication of

The Society of Economic Paleontologists and Mineralogists

*a division of*

The American Association of Petroleum Geologists

# EUROPEAN

# FOSSIL REEF MODELS

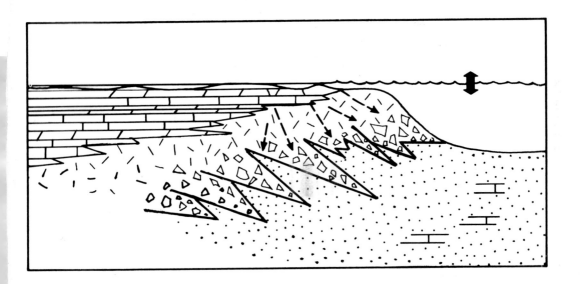

*Edited by*

*Donald Francis Toomey*
*Cities Service Company*
*Midland, Texas*

**SOCIETY OF ECONOMIC PALEONTOLOGISTS AND MINERALOGISTS**

Special Publication No. 30

Tulsa, Oklahoma, U. S. A.

May, 1981

# EUROPEAN
# FOSSIL REEF MODELS

*Edited by Donald Francis Toomey*

## CONTENTS

# EUROPEAN FOSSIL REEF MODELS—AN INTRODUCTION

## DONALD FRANCIS TOOMEY
Cities Service Company (CITCO)
Midland, Texas, USA

## PREFACE

The concept of this volume was initially conceived in June of 1977 at a "Gasthof" situated on the Austrian-Italian border at Nassfeld, Austria. During that Summer I had been invited by the Geologische Bundesanstalt of Vienna, through the courtesy of Dr. Harald Lobitzer, to lecture on Paleozoic organic buildups at the University of Vienna. After a brief stay in Vienna, a small group including Lobitzer, myself, my wife, and Dr. Werner Piller of the University of Vienna, embarked on a field excursion to examine Mesozoic organic buildups throughout Austria. During this excursion we visited many spectacular Mesozoic (mainly Triassic) reef outcrops, and I was impressed by the amount of detailed work that had been, or that was in the process of being, completed on these outcrops by both Geological Survey and academic workers. I was informed that some of this detailed information was already published, albeit in regional journals of rather limited distribution, and of course, written in German. Still, I was unaware of much of this information, especially that related to ongoing work, and even of some of the older work, which until James Lee Wilson's text on *Carbonate Facies in Geologic History* was made available in 1975, had previously not been generally known to an English-reading audience. After visiting exceptional reef exposures at Adnet, I suggested to my host colleagues the possibility of perhaps gathering together a number of Austrian Mesozoic reef studies, translating them into English, and publishing them as a so-called "Reef Volume." We continued to discuss this proposition and considered a number of alternatives until we arrived in the Carnic Alps of southernmost Austria. Here, we were joined by Dr. Erik Flügel, Director of the Paleontological Institute at Erlangen, West Germany, along with a number of his graduate students. During dinner of the first evening we asked Erik Flügel his opinion of our rather preliminary intention to organize a group of Austrian workers for the purpose of publishing an English volume on some of the visited Austrian reef outcrops. His response was immediate and positively enthusiastic, and he further suggested that we broaden the scope of the volume to include "reef models" from all over Europe, concentrating our attention in those areas where new work was planned or in progress. I suggested the possibility of approaching the Society of Economic Paleontologists and Mineralogists with our plans, the ultimate goal being enough material for a Society Special Publication. During the next few days we compiled a list of essential locations and names of potential workers. It was agreed that each of us would write to various colleagues and that Lobitzer would act as Coordinator for our efforts and initially receive any forthcoming manuscripts. Thus the idea for a "European Fossil Reef Models" volume was born.

Upon leaving the Carnic Alps, I continued on a lecture tour which brought me to Germany, Belgium, and England. At most stops I informed interested colleagues of our preliminary plans and solicited their cooperation in this projected endeavor. By the time I had returned to the States, a month later, Lobitzer had already forwarded a list of ten titles with author commitment for this project. Further letter writing by all of us, and the efforts of Dr. David LeMone of the University of Texas at El Paso, on sabbatical leave in Yugoslavia at the time, resulted in the present broad coverage.

Through the efforts of Drs. Charles Ross and Stanley White, succeeding Chairman of the SEPM Publications Committee, I was able to present the outline proposal for the projected "European Fossil Reef Models" volume. This proposal was also presented to the SEPM Council at their annual meeting in San Diego in November 1979. In both cases, the outline proposals were approved, and we were advised to proceed with the endeavor.

The final manuscripts were received in late November, and the editing process which began in December 1979 was completed by June of 1980. The editing procedure was greatly aided by the advice, cooperation, and essential help given by Dr. Orrin H. Pilkey, Jr., of Duke University, the Editor of the *Journal of Sedimentary Petrology,* and his most accomplished editorial assistant, Phyllis Frothingham. Without their generous assistance this volume would not have been possible.

One point that should be mentioned is that of

the twenty-five authors included in this volume, only five acknowledge English as their "mother tongue." This in itself presented some more-than-usual editing problems. I sincerely hope that, as Editor, I have still managed to retain some writer individuality and style, and at the same time, sci-

entific accuracy. Finally, I wish to acknowledge the help of Cities Service Company (CITCO), especially Joseph B. Carl and Michael Cook, for allowing me the necessary cooperation, time, and facilities to complete this task.

## SUMMARY

As an introductory paper to this volume, Longman's "A Process Approach to Recognizing Facies of Reef Complexes" sets the tone for the papers to follow. The authors of most of the papers gathered within this collection, either knowingly or unknowingly, have utilized a process approach to some degree, perhaps some more fully than others.

Longman's study contends that resultant reef morphology we see in Recent reefs is the result of a large set of complex interactions involving both constructive and destructive processes, both of which are superimposed on the cumulative effects of underlying topography and relative sea level changes. Generally, Holocene reef morphology reflects a topography only slightly modified by deposition due to relatively little sea level change during the last several thousand years. Conversely, ancient reef morphology is believed to have evolved over a longer time period; consequently, it may be somewhat more independent of underlying topography and more apt to reflect the nature of the involved biota and their ability to form some sort of a cohesive framework, coupled with marine cementation and a variety of destructive processes.

Study of Recent reefs has provided Longman with evidence of the processes that affect reef facies, specifically in reef complexes possessing a rigid organic framework and generally a steep fore-reef wall. However, Longman contends that the end product we now see reflects the results of only a few thousand years' growth following the latest major Holocene transgression. Only when underlying topography approaches a mature reef profile can well developed reef facies be seen in Holocene reefs.

Longman uses the Belize Barrier Reef complex as his model. Here he finds a predictable sequence of facies which, from a bordering basin to land, consists of distal and proximal talus, reef-slope, reef framework, reef-crest, reef-flat, and back-reef sand. Accordingly, knowledge and understanding of this sequence should aid facies interpretation in all reefs.

In the development of his thesis Longman notes that in a mature reef complex the volume of actual organic framework is very small, whereas the overwhelming sediment consists of reef framework-derived debris in both the fore-reef and

back-reef facies. In petroleum exploration these are the two areas that are most likely to be productive since generally the reef framework facies retains little preserved primary porosity. This is due primarily to the rather common occurrence of marine sedimentation with included sediment infill of voids. The facies patterns we see on reef complexes today are a result of an organic framework composed largely of massive scleractinian corals. Ancient examples of a comparable organic framework can be found in Devonian stromotoporid/tabulate coral buildups, Permian algal/sponge/marine cement structures, and Neogene coral reefs. Different facies do occur in ancient organic buildups where the constructional capability of the dominant organism(s) lacked a rigid framework and marine sedimentation was not too well developed.

Riding's paper, a comprehensive discussion of Silurian reefs and biostromes of northern Europe, follows (significant outcrop localities for this and succeeding papers are plotted in Fig. 1). These reefs occur in cratonic sedimentary sequences, and reef development itself, especially in areas adjacent to the tectonically induced Caledonian Belt, appears to have been restricted by siliciclastic sedimentation, hence, predictable low carbonate and reef proportions. This is well documented in the Welsh Borderlands and Norwegian (vicinity of Oslo) successions. In contrast, marine sequences in the Baltic region (Gotland and Estonia) are relatively thin but are complete and contain high proportions of carbonates and reefs.

Riding identifies and describes various types of Silurian reefs that occur in northern Europe although all are essentially dominated by tabulate corals and/or various stromotoporoids. Accessory reef organisms include rugose corals, calcareous algae, various *problematica,* and bryozoans. His study reinforces the conclusions that reef geometry and biotic composition were controlled through complex interrelationships of a number of environmental parameters and that size and overall organism morphology greatly influenced, and in turn determined, internal reef structure. Silurian bioherms show internal displacement along with differential compaction of adjacent sediments, and this is thought to reflect sediment response both to their own weight and subsequent overburden. Conversely, biostromal sediments behaved more rigidly, and resultant compaction was taken up by stylolitization of adjacent large skeletons.

Burchette's review of European Devonian reefs shows they were widely distributed, but with the exception of the rather familiar reefs of Germany and Belgium, most have not been studied in detail. Earliest reefs occur within the "internal" zone of the Hercynian Orogeny where there was continuous marine sedimentation to at least Middle Devonian time (southeastern Alps, Bohemia, Amorican Massif, and both the Cantabrian and Pyrenees Mountains). In contrast, reef growth did not begin until Middle Devonian time in the "external" zone of the Hercynian Orogeny (southwestern England, the Ardennes, Rhenish Schiefergebirge, Harz Mountains, Poland, and Moravia). In some of these areas Lower and early Middle Devonian sequences are of non-reef facies, and locally, the Lower Devonian may be absent altogether. Irrespective of this early history, reef growth was active during Middle Devonian time and continuous in most areas into the Late Devonian (earliest Famennian) when widespread rise in sea level caused reef extinction and deep water sedimentation became the norm.

Burchette recognizes five broad morphological categories of European Devonian "reefs": banks, biostromal complexes, barrier-reef complexes, isolated reef complexes, and quiet water carbonate buildups. All display characteristic sediment and biotic facies associations and distribution patterns. Biostromal, barrier-reef, and isolated reef complexes exhibit lateral facies zones (e.g., reef-core and back-reef in atolls, barrier-reef and biostromal complexes, and fore-reef in most well developed buildups). Factors which appear to have influenced reef location and development include local tectonics on basin-margin hingelines, synsedimentary faulting and volcanicity, submarine topography, and changes in relative sea level.

Significantly, Burchette notes that following extinction some reefs were buried rather rapidly, whereas others underwent periods of submarine or subaerial exposure and were only buried during latest Devonian or early Carboniferous time. As a result, many reefs developed extensive fissure systems which were filled with fibrous calcite cements and internal sediments. The latter commonly contain biotas somewhat younger than the host reefal carbonates. Early diagenetic carbonate cements appear to be common but show great variability. Early vadose cements are most important in back-reef facies, while synsedimentary submarine fibrous cements predominate in reef-core and fore-reef deposits. In "mud-mounds" stromatactis is an important diagenetic feature.

In the next paper Flügel reports the occurrence of Lower Permian organic buildups in the Trog-

kofel Formation of the Southern Alps (Carnic Alps and Sexten Dolomites), Austria, and Italy. These buildups represent "stratigraphic reefs" that probably formed in a downslope shelf-edge position adjacent to a shallow water carbonate platform. The buildups grew by the interaction of algae (*Tubiphytes* and *Archaeolithoporella*) and various encrusting bryozoans, together with eogenetic carbonate submarine cements, which led to the formation of diagenetic organic boundstones. Flügel contends that both depositional and diagenetic fabrics of the Trogkofel buildups are similar to those of the Permian Reef complex of the southwestern United States.

In the next two papers, Smith describes a barrier-reef complex and bryozoan-algal patch-reefs that occur in the Upper Permian Magnesian Limestone of northeastern England. He contends that a major linear reef, which attained a height in excess of 100 meters, protected the seaward edge of the Magnesian Limestone carbonate shelf in northeastern England. This reef facies is overlain by a widely distributed stromatolite biostrome. Both the barrier-reef and stromatolite biostrome are extensively dolomitized. The foundational substrate of the barrier-reef is a shell coquina, and much of the lower portion of the reef itself is formed of unbedded bryozoan biolithite. Indications of contemporaneous reef rock lithification are found in the presence of biolithite debris in associated reef talus. Middle stages of reef growth are characterized by an increase in proportion of algal rocks and/or laminar organic or inorganic encrustations. Stromatolitic and other laminar rocks became dominant in the latest phases of reef growth. Progressive reef growth and asymmetry appear to have led to a more complete separation of environments both landward and seaward of the barrier-reef; this culminated when the reef approached sea level and formed a lagoon and a starved or semi-starved basin.

Conversely, Lower Magnesian dolomitized patch-reefs (up to 25 meters in length and 8 meters in thickness) are abundant in dolomitized skeletal oolites in Yorkshire, England. They are roughly oval in shape and irregular in section; all are younger than an underlying widespread coquina which served as a stable reef foundation. Most patch-reefs are composed of closely packed "sack-shaped" bodies of ramose bryozoans (*Acanthocladia* dominant, with some *Thamniscus*) along with a low diversity invertebrate assemblage. These patch-reefs formed on a broad, shallow tropical, carbonate marine shelf with their surfaces perhaps less than 2 meters higher than contemporaneous sediment. The main constructional process appears to have been due to mud-trapping and binding by bryozoans, with some encrusting foraminifers and early carbonate sub-

FIG. 1.—Map of Europe delineating locations of reef studies described by the authors of this volume. Numbers plotted on the map correspond to the following: 1. Riding (Silurian); 2. Burchette (Devonian); 3. Flügel (Lower Permian); 4. Smith (Upper Permian, barrier-reef); 5. Smith (Upper Permian, patch-reefs); 6. Bradner and Resch (Middle Triassic); 7. Čar *et al.* (Upper Triassic); 8. Schäfer and Senowbari-Daryan (Upper Triassic); 9. Piller (Upper Triassic); 10. Flügel (Upper Triassic); 11. Turnšek *et al.* (Upper Jurassic); 12. Flügel and Steiger (Upper Jurassic); 13. Masse and Philip (Cretaceous); 14. Carbone and Sirna (Upper Cretaceous); 15. Polšak (Upper Cretaceous); 16. Babić and Zupanič (Paleocene); and 17. Frost (Oligocene).

marine cements hardening the structures and adding bulk. Bryozoa die out in the upper part, and here the reefs are composed of dolomitized algal stromatolites that were probably formed subaqueously in shallower water than the earlier bryozoan portion of the reefs.

The following ten papers are concerned with reefs of Mesozoic age, especially those of the Triassic which are so well represented in the Northern Calcareous Alps of Austria. The first, by Bradner and Resch, is a description of Middle Triassic reefs from the Wetterstein Limestone near Innsbruck, Austria. Here, the edge of the Wetterstein Limestone Platform features two types of reef development: a patch-reef sequence and a massive reef complex (Hafelekar Reef complex). The patch-reefs are relatively small in size and are composed mainly of *Tubiphytes,* calcisponges, and tubular foraminifers. The Hafelekar Reef developed in shallow, well agitated waters in which variable water energies produced zonal patterns that delineate a reef front and a wide reef-flat. A conspicuous zone of skeletal sand shoals shows evidence of emersion during reef development and functions as a barrier that separated the reef from a lagoon. Areas of the reef exposed to high water energy are populated by massive coral forms, calcisponges, and hydrozoans, whereas sheltered reef areas offered refuge to more delicate corals, catenulate sphinctozoan sponges, dendroid *Tubiphytes,* and bushy codiacean algae. Stabilization of reef sediment was accomplished by cementing and encrusting organisms, mainly *Tubiphytes,* as well as by sediments that underwent rapid submarine cementation. Bradner and Resch believe that large scale syngenetic submarine cementation of the reefs was followed by early diagenetic dolomitization. Subsequent sedimentation and cementation produced conspicuous, thick coatings of fibrous spar, and these are interpreted as diagenetic fabrics that formed during an early burial stage.

Čar *et al.* describe circular-shaped coral bioherms of Upper Triassic age from northwestern Yugoslavia. These reefs attain a height of 140 meters and a diameter of up to 180 meters. Two distinctive lithological and faunal associations have been identified. These consist of massive biohermal cores and marginal breccias composed mainly of corals, biolithites, and biopelmicritic limestones, and bedded biopelmicritic limestones between the reefs. The writers believe that the bioherms grew in relatively deep, quiet water. This interpretation is reinforced by the presence of only modest accumulations of marginal breccias and the occurrence of layered pelmicritic limestones between individual bioherms. The reefs are surrounded and interbedded within a sequence of interbedded shales and sandstones.

Schäfer and Senowbari-Daryan concisely document the occurrence, facies development, and paleoecologic zonation of four Upper Triassic patch-reefs from the Northern Calcareous Alps near Salzburg, Austria. One patch-reef grew directly on a carbonate platform in a shallow water setting; two other patch-reef complexes grew out of a basinal setting in two stages, a deep water mud-mound stage and a shallow water reef stage. Another patch-reef, which also developed in a basinal setting, shows only a deep water mud-mound stage. The shallow water stages of the patch-reefs display a lateral facies zonation of five distinctive facies units: coral-sponge facies of the central reef areas, oncolite facies of the upper reef-slope within a zone of high water energy, algal-foraminiferal facies of the lower reef-slope and also the foundation of one of the patch-reefs, reef detritus-mud facies of the deepest part of the reef-slope where reefal sediments interfinger with basinal sediments, and the terrigenous-mud facies of the basin. Organism communities forming the reef biota, and dominated by calcareous sponges and various corals, show distinct distributional patterns, especially calcareous algae, foraminifers, and various *microproblematica,* all of which can be used as facies indicators as well as for differentiating different biotypes within the reef areas.

Piller reinvestigated the Upper Triassic Steinplatte Reef complex of the Northern Calcareous Alps near Salzburg, Austria, and was able to delineate a distinct facies zonation of the carbonate complex. This present facies zonation differs widely from that of other authors in that a distinct reef zone was identified, and the so-called fore-reef breccias are now regarded as back-reef sediments. Additional study of adjacent carbonate platforms demonstrates that these areas, which today are isolated by erosion and tectonics, represent a continuation of a shallow water lagoon of the eastern part of the Steinplatte Platform. In essence, the Steinplatte Reef represents only a minor portion of an Upper Triassic shallow water carbonate platform, some 40 kilometers in width, which was fringed by reefs.

Flügel's paper presents a masterful summary of the facies and paleoecology of all presently-known Upper Triassic reefs from the northern Calcareous Alps of Austria. A work of this scope could only be possible through the combined efforts of the "Erlangen Reef Research Group," of which Professor Flügel is the founder and Director. Of added significance is the fact that all described facies encountered in Upper Triassic reefs of Austria are related to the "Standard Microfacies Types" as defined by Wilson (1975), thus allowing convenient comparison to standardized facies units. In addition, Flügel has provided an informative section in which he compares known Triassic reefs from the European Tethyan region with those from Asia and North America.

Turnšek, Buser, and Ogorelec describe an Upper Jurassic reef complex from northwestern Yugoslavia. This complex is thought to be a barrier-reef that developed along the shelf-margin of an ancient carbonate platform. Preliminary study indicates that this barrier-reef complex can be traced to the southeast as far as Albania. Identified reef facies include a fore-reef area characterized by carbonate breccias and blocks of reef debris and a central reef area containing structures with abundant hydrozoans and corals. The central reef area can be further subdivided by various hydrozoan assemblages into actinostromariid and parastromatoporid zones. The back-reef areas contain locally developed lagoons and patch-reefs composed primarily of the hydrozoan *Cladocoropsis*.

The succeeding paper by Flügel and Steiger focuses on an Upper Jurassic sponge-algal buildup from the Northern Franconian Alps of Bavaria, West Germany. The studied buildup is a single well-developed exposure of a much more widely distributed Upper Jurassic barrier-reef within the Franconian and Schwabian Alps. The principal constructional organisms are siliceous sponges (cup and dish-shaped forms) and cyanophycean algae. The Müllersfelsen buildup developed in several cyclic stages in relatively deep water without strong current and wave action. Three facies types have been recognized; these consist of a sponge-crust boundstone facies characterized by micritic boundstone rich in calcified siliceous sponges, tuberoids, and crusts; a lithoclastic packstone facies in which lithoclasts, tuberoids, and bioclasts are the dominant grain-supported allochems; and a tuberolitic wackestone facies in which mud-supported particles are dominant. Spatial distribution of facies indicates that the sponge-crust boundstone facies is the mound constructional facies, whereas both the tuberolitic wackestone and packstone facies are only developed in areas marginal to the main buildup. The uppermost portion of the Müllersfelsen buildup is dolomitized.

In the next paper Masse and Philip document the occurrence of Cretaceous (Berriasian-Maestrichtian) coral-rudist buildups from three major outcrop areas in France. These are the Paris, Sud-Est, and Aquitaine-Pyrénées Basins. Two types of coral-rudistid formations have been identified, those associated with off-shore "highs" (coral "highs" known only from the Lower Cretaceous, oobioclastic/coral "highs" of the lower Barremian of the Subalpine area, and rudistid banks present only in the Upper Cretaceous) and those associated with carbonate platforms. This latter group can be subdivided into an outer zone with high energy deposits and organic buildups (primarily corals in the Lower Cretaceous and rudistids in the Upper Cretaceous) and an inner zone of quiet to moderate energy deposits characterized by abundant rudistids. The authors note that coral-rudist development appears to be governed by six important physical factors: 1. shallow water conditions, mainly infralittoral, 2. relative basement stability (although in some instances tectonism may create "highs" on which organisms may thrive), 3. eustatic stability or transgression, 4. low terrigenous influx, 5. absence of organism restricting oceanographic conditions, and 6. a warm climate of tropical to subtropical nature.

The Carbone and Sirna paper describes various Upper Cretaceous reef models from central Italy which appear to be related to tectonic evolution of an epioceanic Central Apennine Platform. From early Cenomanian to late Senonian time organic buildups, along with shelf-edge skeletal deposits, were laid down along the present northwestern margin of the Lepini Mountains and the southern portion of the Prenestini Mountains. These sediments overlie rocks of the Cretaceous (Aptian-Albian) restricted platform facies. The Upper Cretaceous shelf-edge deposits are characterized by various sequences of rudistid and coral communities laid down within changing depositional settings.

In the concluding Mesozoic reef paper, Polšak documents an Upper Cretaceous biolithic complex from northern Croatia, Yugoslavia. This complex is regarded as a barrier-reef composed of coral and rudistid bioherms. Associated with the barrier-reef are perireefal breccias that originated from destruction of the reef-front. Basinward, the breccia grades into detrital limestones and, more distally, into basinal sediments with turbidity current structures. The back-reef area contains gastopod biostromes and detrital limestones, and the lagoon contains clastic terrigenous deposits and some small patch-reefs. Polšak notes that biolithic complexes of this type developed on slopes of island arcs during the Upper Cretaceous and lower Paleogene of the Inner Dinarides. These island arcs and corresponding trenches, as well as inter-arc basins, were formed in a subduction zone of the Tethyan oceanic crust beneath the Panonian Plate.

Babić and Zupanič discuss various pore types and fillings that occurred during the developmental history of a Paleocene reef exposed in Croatia, southeast of Zagreb, Yugoslavia. Organic growth and encrustation, internal deposition, and cementation are regarded as constructional processes in pore formation, whereas abrasion, organism borings, dissolution, and fracturing represent destructive processes that produced and modified various types of pore spaces. The authors also include an outline discussion of the paleogeography and depositional history of Paleocene marine carbonates of this region of Yugoslavia.

In the concluding paper of this volume, Frost presents a comprehensive overview of the paleoecologic distribution of Oligocene reef corals in the Vicentin area of northeastern Italy. He identifies coral communities, their different environments, and different stages of ecological successions. As examples, Frost discusses the large Berici barrier-reef/lagoonal complex of the Castelgomberto Limestone. This barrier-reef has a core facies up to 200 meters in thickness, 900 meters in width, and approximately 8 kilometers in length. The lagoonal facies represents most of the complex and extends backward of the barrier-reef for about 30 kilometers. Other buildups in this region include patch-reefs and coppices in the shelf-lagoon. In adjacent areas coral thickets, coppices, and thin fringing-reefs formed along the margin of a large Oligocene volcanic complex. Frost contends that variations of three basic coral assemblages are responsible for the wide spectrum of organic buildups: low diversity "pioneer" or thicket assemblages, high diversity reef-core and flank assemblages, and soft mud substrate assemblages. The environmental factors which appear to have been most important in determining the coral species assemblage, and the community diversity and abundance, are adaptation to water flow, potential to colonize unstable substrates and survive sedimentation, and species-specific dominance relationships. Frost believes that interaction of these factors can explain the observed seral succession of pioneer to intermediate to climax communities in many Tertiary reef structures.

The voluminous amount of information presented in this Special Publication volume not only fills a gap in understanding the European approach to reef studies but also, I believe, provides the necessary data base to allow us (in particular the North American geologist) to incorporate this information in our overall interpretative studies. These studies should serve as an impetus for new investigations and will hopefully broaden our understanding of the complex interrelationships that operate in the reef environment. The data presented in these papers, together with their references, will provide a variety of new ideas and approaches to interpretation and should broaden our own, perhaps somewhat provincial interpretative attitude.

## REFERENCE

WILSON, J. L., 1975, Carbonate Facies in Geologic History: New York, Springer-Verlag, 471 p.

SEPM Special Publication No. 30, p. 9–40, May 1981

# A PROCESS APPROACH TO RECOGNIZING FACIES OF REEF COMPLEXES

## MARK W. LONGMAN
Cities Service Company
Exploration and Production Research
Tulsa, Oklahoma, U.S.A.

## ABSTRACT

Accurate interpretations of facies in ancient reefs requires both proper "classification" of the reef and an understanding of the sedimentologic and biologic processes active during the formation of the reef. Study of modern reefs provides evidence for the processes affecting facies in reef complexes with a rigid organic framework and steep fore-reef wall, but generally reveals only the incipient products of these processes because most Holocene reefs have been growing for only a few thousand years following the Holocene transgression. Only where underlying topography approximates the profile of a mature reef complex can well developed examples of the sedimentologic facies of a mature reef complex be found in modern reefs. Study of these pseudo-mature reef complexes reveals a sequence of sedimentologic facies (from basin toward land) consisting of: 1. distal talus, 2. proximal talus, 3. reef-slope, 4. reef framework, 5. reef-crest, 6. reef-flat, and 7. back-reef sand. Knowledge of the characteristics and distribution of these facies should facilitate facies interpretation in comparable ancient reef complexes.

Reef framework forms only a few percent of the volume of a mature reef complex, whereas the vast majority of the complex consists of debris in fore-reef and back-reef facies; the debris being derived largely from the framework. This has particular significance in hydrocarbon exploration because most wells drilled toward the top of a seismicly defined ancient reef complex would penetrate only the back-reef carbonate sand facies. However, this is as it should be because the reef framework facies generally has little preserved primary porosity due to sediment infilling of cavities and extensive marine cementation.

The facies of modern reef complexes discussed here exist due to the presence of a significant marginal (as opposed to centrally located) shallow water, wave resistant, rigid organic framework composed largely of scleractinian corals. Similar facies may be expected in those fossil reef complexes with a more or less comparable organic framework such as the Devonian stromatoporoid/tabulate coral reefs, the Permian algal/sponge/marine cement reefs, and the Neogene coral reefs. However, different facies must occur in ancient reefs that lacked rigid organic frameworks such as the Paleozoic bryozoan and/or crinoid mounds, Late Paleozoic phylloid algal mounds, Cretaceous rudistid banks, and early Tertiary *Nummulites* banks.

## INTRODUCTION

Reefs have been intensively studied for years because of their importance in carbonate depositional systems, paleontology, paleoecology, hydrocarbon exploration, and biology. However, much remains to be learned about facies and porosity distribution in reefs, mainly because of problems with terminology and the tendency to view all reefs as having similar shapes and facies distributions. Such homogenization of reefs is clearly an oversimplification and has caused major problems among biologists and geologists as shown by the ongoing debate over what is and what is not a reef, and also by the confusion that exists in interpreting characteristics and facies of ancient reefs.

The first step in defining facies of reefs must be to classify reefs so that those with similar facies can be compared. For geologists, the best classifications are based on sedimentologic parameters such as the nature of the framework and geometry of the reef related facies. Most modern coral reefs can be classified as "walled-reef complexes" because they contain a major rigid organic framework, a steep fore-reef wall, and a variety of associated high energy facies, whereas many ancient reefs would be classified as mud-mounds, knoll-reefs, patch-reefs, *etc.* (see Wilson, 1974 and 1975). The objective of this paper is to describe the interaction of depositional processes and biologic activity in reef complexes in order to reveal the characteristic shapes and facies patterns that should exist in both modern and ancient reef complexes. Emphasis is on these reefs as sediment buildups rather than as ecologic units, which means that facies are discussed from a sedimentologic rather than a biologic (paleontologic) point of view. The paper is based on the process approach to sedimentology with the idea that it should eventually be possible to predict the distribution of facies in reefs formed by any type of organism at any time, if processes affecting those organisms, as well as the physical processes acting on the buildup, are understood.

Among geologists, knowledge of facies in ancient reef complexes is limited for three major reasons. First, most studies of modern reefs have involved mainly the attractive zone of active coral growth, while less emphasis was placed on the associated "non-reef" facies which lacked a rigid organic framework. Second, modern reef complexes are relatively young features that have formed during the last few thousand years following the Holocene transgression. Thus, facies tend to be closely tied to underlying topography, and the reefs have not yet developed mature profiles and facies of their own making. Third, and perhaps most important to the interpretation of facies in ancient reefs, the organisms forming reefs have changed through time with resultant effects on reef shapes, facies distribution, and reservoir potential. Thus, study of modern reef complexes provides clues to interpreting ancient reefs, but modern reefs cannot and should not be viewed as direct analogs to ancient reefs, most of which were formed under different environmental conditions by organisms with different ecologic requirements. Geologists should break away from the tradition of interpreting reefs according to the mold established by biologists.

This paper is divided into three parts. First is a discussion of the classification and definitions of reefs to provide a basis for what follows. Second is a section on the physical and biological processes affecting facies in reefs of all types but with emphasis on their effects in reef complexes. The paper concludes with a discussion of facies and facies distribution in reef complexes and the importance of this in the interpretation of ancient reef complexes.

## CLASSIFICATION OF REEFS

Reefs have been defined and redefined, classified and codified, inspected, dissected, investigated, and reinvestigated to the point that it might seem that there is little else that could be done to them. However, it is this same wealth of information, information collected by biologists and paleontologists, ecologists and sedimentologists, ships' captains and naturalists, that has led to all the discussions and disagreement over what is or is not a reef. What is needed, as Stoddart (1969a) so perspicaciously pointed out, is not so much a new theory of reefs, but "standardized procedures to ensure comparability of reef studies and the identification of variations in reefs both on local and regional scales," and, it might be added, through time. The first step to providing this standardization is to recognize the needs of the group involved in the study (needs of geologists differ from those of biologists), and then to classify reefs into comparable assemblages according to parameters important to this group of people.

Classifications of reefs based on organisms (coral reef, rudistid reef, etc.), or shape of the reef in map view (barrier-reef, fringing-reef, faro, atoll, etc.) are of little value to geologists because they reveal almost nothing about the distribution and nature of facies within the reef complex. Of more value are classifications based on shape of the reef in stratigraphic profile and internal facies distribution such as those proposed by Heckel (1974) and Wilson (1974). In Wilson's (1974) classification, shelf-margin reefs are divided into three major types: quiet water mud-mound complexes, shallow water knoll-reef complexes, and walled-reef complexes with a rigid organic framework (Fig. 1). Organisms and facies vary depending on reef type, but this classification is used here as a basis for the topic being discussed, e.g., the characteristics and distribution of facies in walled-reef complexes. Description of the facies in the other types of shelf-margin reefs also needs to be done but is not discussed here.

## DEFINITION OF REEF

Geologists have long been plagued by the disparate definitions of reef which originated because the term was applied by such a wide variety of disciplines to a wide range of organically formed features for such a variety of reasons. Braithwaite (1973) and Heckel (1974) have reviewed the relevant literature and the discussion need not be repeated here. "Reef" is used in this paper for any biologically influenced buildup of carbonate sediment which affected deposition in adjacent areas (and thus differed to some degree from surrounding sediments), and stood topographically higher than surrounding sediments during deposition.

This definition is essentially the same as that proposed by Heckel (1974, p. 96), as well as that proposed by Cumings (1932) for "bioherm," and has two major advantages over the others. First, it is basically the same as that used by most geologists, particularly those in the petroleum industry. Second, it avoids the problem of requiring reefs to have a rigid organic framework as required by the definitions of Lowenstam (1950) and Newell et al. (1953), or potential wave resistance as required by Walker (1974). This is particularly important because 1. types of reef frameworks cover a complete spectrum from non-calcareous mud trapping organisms, such as some algae and sea grass, to firmly inter-cemented hermatypic corals forming massive boundstones, and 2. the framework is only a small part of many reefs and is often destroyed by physical or biological processes before the reef is buried.

The term reef complex is used for the specific type of reef having a significant rigid organic framework which generally forms in high wave energy, shallow water environments, as well as the genetically related facies associated with the framework. Although this use of the term is slightly different from that originally proposed by Hen-

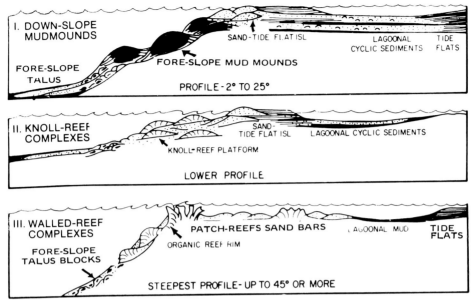

FIG. 1.—Three types of reef complexes described by Wilson (1974). Most modern reefs are walled-reef complexes (Type III).

son (1950), it is consistent with the usage of most recent papers (e.g., Ladd, 1971; James and Ginsburg, *in press*). The term carbonate buildup is used for reefs that apparently lacked a rigid organic framework, although many carbonate buildups had an organic framework formed of uncemented organisms that trapped sediment. Size, shape, facies zonation, and the distribution of organisms are highly variable within both reef complexes and carbonate buildups.

Numerous arguments over what is or is not a reef have resulted from the fact that people perceive reef complexes differently. An analogy with an apple may help clarify this problem. Every apple consists of a core surrounded by the edible fleshy fruit. The fruit could not have formed without this core; indeed, the core is the very reason for the fruit, but the core is generally not what people mention when asked to describe an apple because it is a small and hidden part of the whole. A reef complex is similar in having a core (or framework) that acts as a nucleus to forming surrounding debris deposits. When most petroleum exploration geologists describe a reef, they describe the whole assemblage of framework and debris, while others, such as some biologists, define the reef as being only the organic framework at the center of the associated debris. Both definitions are correct, but the secret to understanding either an apple or a reef complex is to understand the whole entity as a function of its parts.

One just wishes that reefs had "skins" like apples so that we could know exactly where they begin.

Major reef related terms used in this paper are defined in Table 1 in the hope that this list will improve the communication potential of this article.

## CONTROLS ON REEF MORPHOLOGY

A variety of physical and biological factors interact to determine whether or not a reef will form (Fig. 2), and these factors have been extensively discussed in the literature (Wells, 1957; Stanton, 1967; Stoddart, 1969a, 1978, and many others). Indeed, what paper on reefs, other than those that are purely descriptive, has not at least touched upon the genesis of reefs. The reason any confusion still exists as to the controls on reef morphology is because most of these papers describe one or two factors (e.g., underlying topography, nutrients, warm water, *etc.*) as the major control on the reef and ignore the others. Reef morphology is never the result of one or two factors; instead it is the result of many factors interrelated in complex ways as shown in Figure 2. The pyramid shape is intended to show the interrelationships of first order, second order, *etc.* controls on reef growth. To say that a reef forms specifically as a result of this or that factor is pointless. However, it is generally possible to say a reef will not form if a certain factor is missing (e.g., warm

## TABLE 1. DEFINITIONS OF TERMS

| TERMS USED TO DESCRIBE REEF COMPLEXES AND FACIES: | TERMS USED TO DESCRIBE FACIES OF A MATURE REEF COMPLEX: |
|---|---|
| **REEF:** <br> a buildup of carbonate sediment formed by depositional and biological processes with significant topographic relief and influence over deposition in adjacent areas. Definition essentially that of Heckel (1974) | **REEF FRAMEWORK:** <br> rigid organically built framework with potential to resist waves. A fulcrum for the reef complex by acting as both the site for abundant sediment production and a wave baffle to allow reef associated non-framework facies to be deposited in adjacent areas. |
| **REEF COMPLEX:** <br> organic carbonate buildup containing a rigid framework and including those genetically associated facies having major amounts of reef derived material. Buildup must have significant topographic relief and is divisible into many facies including reef framework, reef-crest, back-reef sand, and talus facies. (Definition adapted from Henson, 1950). | **REEF-CREST:** <br> the facies of a reef complex immediately behind the reef framework. Generally composed or coarse rubble ripped from the reef framework by storms. Shallowest facies of a reef complex often extending into intertidal zone. Growth of framework organisms is generally limited by periodic exposure and turbulence. |
| **CARBONATE BUILDUP:** <br> a pile of carbonate sediment formed by depositional and/or biological processes (versus erosional) with significant topographic relief, but lacking a rigid organic framework. May have a framework composed of organisms that trapped sediment but failed to bind to each other. Definition modified from that proposed by Heckel (1974) to exclude reef complexes. | **REEF-FLAT:** <br> relatively flat zone immediately behind reef-crest characterized by water depths of zero to a couple of meters. Scattered framework organisms and associated algae, etc., but generally lacking an extensive organic framework in mature reef complexes. |
| **MATURE REEF COMPLEX:** <br> reef complex in which the profile and facies distribution are relatively independent of underlying "bedrock" topography. Most develop during times of stable or slowly rising sea level so that facies become well differentiated. | **BACK-REEF SAND FACIES:** <br> broad zone of relatively fine-grained skeletal debris derived from the reef framework and mixed with debris from indigenous organisms such as molluscs and foraminifers. Best developed in mature reef complexes. <br> **REEF-SLOPE FACIES:** <br> zone immediately in front of the reef framework and above the talus facies. Generally steeply dipping and below the zone of extensive framework organisms, but delicate and deeper water organisms are common. |
| **IMMATURE REEF COMPLEX:** <br> reef complex in which facies are poorly developed and strongly controlled by underlying topography. Most Holocene reefs are still immature because of the recent rise in eustatic sea level. | **PROXIMAL AND DISTAL TALUS FACIES:** <br> forereef facies composed mainly of skeletal debris derived from the reef. Proximal and distal talus are distinguished by distance from reef framework, grain size, and abundance of planktic foraminifers. |

water, nutrients), or present (e.g., high influx of terrigenous material).

What is meant by reef "morphology"? This question is deceptively simple because reef morphology can be and has been considered at many different scales in both map and cross-sectional view. Morphology of reefs involves features as small as spur and groove structures within a reef to those as large as the distribution of whole reef systems (e.g., in the Australian Great Barrier Reef). For the purpose of this discussion, reef morphology is considered on a relatively large scale with emphasis on the shape of the reef in cross-section. Spur and groove structures and other small scale irregularities such as terraces are not considered.

Reef morphology in this sense is a function of 1. the organisms in the reef and associated deposition, 2. destructive physical and biological processes and 3. inorganic processes such as marine cementation, all superimposed on the effects of the underlying topography and changes in relative sea level. The relative importance of these factors varies with the amount of time that the reef has to grow, or, in other words, the length of time through which sea level is relatively stable

(Fig. 3). Thus, immediately following rapid changes in sea level, reef morphology tends to be strongly controlled by underlying topography, whereas with time the reefs will become more mature (e.g., shape will become more independent of underlying topography), and morphology will become more a function of the physical and biological processes. The major factors affecting reef morphology are discussed below.

### Nature of Reef Building Organisms

Although the study of modern reefs provides clues for interpreting ancient reefs, direct analogies must be made carefully, taking into consideration any changes in the type of the reef building organisms through time. The best known modern reefs have an organic framework composed dominantly of hermatypic scleractinian corals with some coralline red algae and minor associated encrusting foraminifers, other types of algae and coelenterates, sponges, molluscs, *etc.* Hermatypic corals and, to a lesser extent, coralline algae, are capable of rapid growth as well as cementing to other organisms or pre-existing hard substrates, and thus of forming a rigid organic framework of significant size. Because these organisms

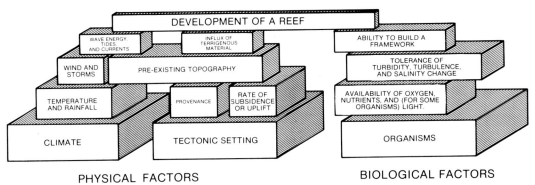

FIG. 2.—Major factors which influence the development of a reef. The pyramid arrangement is intended to show the interrelationships of these factors.

also grow best in areas of low turbidity, good current action, and abundant light, they tend to grow upward to sea level to produce major wave resistant ridges. Growth of the framework forming organisms in the reef generally occurs at a faster rate than deposition in back-reef areas or deep water fore-reef areas, so the reef framework becomes a major topographic barrier between back-reef lagoonal environments and the open ocean. In short, these organisms form walled-reef complexes such as those described in Wilson's (1974) classification.

If this reef framework develops during a prolonged period of stable sea level or slow sea level rise, it develops a cross-sectional profile of its own making; a profile which controls the distribution and relative abundance of associated sedimentologic facies. The combination of rigid framework and progradational growth often produces a nearly vertical to slightly overhanging fore-reef wall with many tens of meters of relief (the height of the wall is a function of underlying topography and/or the interaction of relative sea level rise and amount of progradation). The top of the reef complex tends to be relatively flat and within about a meter of the low tide level. The landward side of the reef complex generally consists of an apron of coralgal skeletal debris carried back from the reef framework. Thus, the general shape of the reef is asymmetric with a nearly vertical front, a flat top, and a back slope of between a few degrees and 20 or 30 degrees (Fig. 4).

Under certain circumstances, reef complexes may form even when organisms capable of interskeletal cementation are not present. A reef formed at the crest of a major fault scarp or shelf break by non-binding organisms such as rudists will assume many of the facies of a true reef complex (Bebout and Loucks, 1974; Bein, 1976). Another way of forming reef complexes is to have extensive and rapid marine cementation occur in

a shelf-margin buildup formed by any type of organisms so that the inorganic cementation serves to bind the reef framework rather than organic interskeletal cementation. According to Mazzullo and Cys (1977) this is the process that led to the formation of the Permian Capitan Reef complex in New Mexico and southeast Texas.

Still a third way of forming a major reef complex with a "rigid" organic framework might be to bind reef forming organisms incapable of cementing to each other with noncalcareous algae. However, this has yet to be documented in the rock record because direct evidence of such algae is hard to obtain after lithification. Such algae (*Archaeolithoporella*?) probably played a role in the formation of the Capitan Reef, but few examples of major reef complexes from other periods in the Paleozoic (except the Silurian-Devonian stromatoporoid/tabulate coral reefs) exist, so binding non-calcareous algae probably played a minor role in forming reef complexes during this time.

Ancient reefs are often assumed to have had the same cross-sectional shape as Recent reef complexes. However, a review of the organisms found in ancient reefs (e.g., Newell, 1972; Heckel, 1974) reveals that most of these ancient reef builders were not capable of forming a rigid organic framework comparable to that of hermatypic corals, and that many thrived in quiet water of relative darkness compared to hermatypic corals. Thus, frameworks of ancient reefs need to be viewed as a spectrum from non-calcareous mud-trapping organisms such as certain algae through delicate branching organisms to massive organic frameworks (Fig. 5). Many ancient carbonate buildups, such as those composed of archaeocyathids, most calcareous algae, calcareous and siliceous sponges, rugose corals, brachiopods, bryozoans, rudistids, and crinoids, apparently lacked a rigid organic framework. Furthermore,

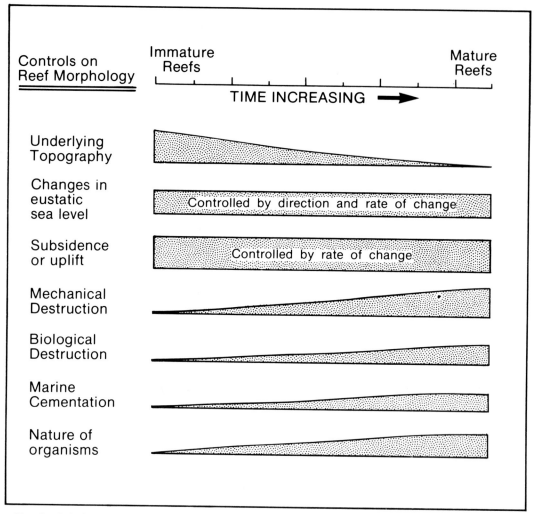

Fɪɢ. 3.—Relative importance of the various processes affecting reef shape (in cross-section) as a function of time.

many of these organisms probably did not have the tolerance of turbulence or the need for abundant light that hermatypic corals possess. Thus, because the parameters controlling the development and shapes of these ancient reefs were different from those controlling shapes of modern reefs, overall shapes and facies distribution were almost certainly different. Numerous studies have shown this empirical observation to be true (Wilson, 1974 and 1975).

The relationship between various Phanerozoic organisms and type of buildup they formed is summarized in Table 2. Information in the table is based on a review of the literature and work by Longman (in prep.). This table is accurate within the limits of the available literature, but as better

classifications and interpretations of reefs become available, it will undoubtedly need to be revised.

The significance of the table is threefold. First, it shows that walled-reef complexes have formed only at certain geologic times when major framework forming organisms were present. Second, it shows an abundance of quiet water mud-mound reefs in Paleozoic rocks. And third, it shows the variety of reef shapes that can be formed by organisms within the same group (e.g., corals).

## Underlying Topography and Changes in Relative Sea Level

Since reefs do not form suspended in air, substrate topography must always control reef morphology to some degree. The problem is to rec-

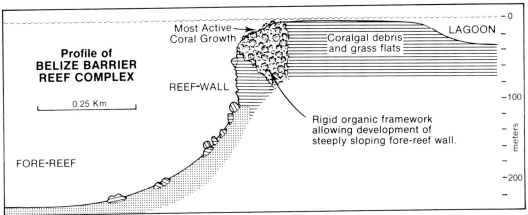

FIG. 4.—Inferred profile for a mature reef complex, based on the profile of the pseudo-mature Belize Barrier Reef complex presented by James and Ginsburg (*in press*). Subsurface geometry of facies is schematic.

ognize the degree to which reef growth has modified the underlying topography (Fig. 6). Many studies have shown that the morphology of Holocene reef complexes, in both map view and cross-sectional shape, is strongly controlled by underlying topography which was shaped during times of lower sea level in the Pleistocene (e.g., Purdy, 1974; Orme *et al.*, 1978b). However, the rapid changes in sea level that occurred during the Pleistocene are likely to have been absent throughout much of the Phanerozoic because glaciers were much smaller or nonexistent (Rutten and Jansonius, 1956; Frost, 1977). Thus, fossil reefs probably had much longer periods of relatively stable sea level in which to grow and develop their own morphology.

Pleistocene lowering of sea level has been well documented in the past two decades with the development of absolute dating techniques based on unstable isotopes (Fairbridge, 1961; Milliman and Emery, 1968; McLean *et al.*, 1978; and many others). During the last Pleistocene lowstand, sea level was about 140 meters below its present level, and this would have resulted in the subaerial exposure of most of the areas where shallow water carbonates are forming today. Following this lowstand, sea level began to rise about 20,000 years ago and at times it rose at a geologically remarkable rate of as much as 15 meters per 1000 years (Milliman and Emery, 1968; Clark *et al.*, 1978) during the Holocene transgression.

The Holocene transgression in any given area was affected by eustatic sea level rise, regional tectonics, and local deformation of the earth's surface by changing ice and water loads (Clark *et al.*, 1978). Thus, the combination of regional transgression curves into a composite curve (e.g., Fairbridge, 1961) must be applied to a given area

with caution. However, a number of studies from different parts of the world (Jamaica—Land, 1974; Australia—McLean *et al.*, 1978; Mexico—Macintyre *et al.*, 1977; the eastern U.S. coast—Milliman and Emery, 1968), along with models developed for a viscoelastic earth based on size of ice sheets (Clark *et al.*, 1978; Clark and Lingle, 1979), suggest that sea level rose rapidly until about 5,000 years B.P. when it reached its present level plus or minus a meter. This means that modern shallow water reefs have had only 5,000 years or so to colonize the surfaces of pre-existing rocks. Thus, although many meters of reef material could be deposited during the Holocene transgression (as much as 33 meters in a part of Alacran Reef according to Macintyre *et al.*, 1977), the cross-sectional profile of most Recent reefs is still strongly controlled by underlying topography. In other words, most reefs are still in an immature stage of development and have not yet developed large scale morphology of their own making.

*Transgression and Regression*

Changes in relative sea level, caused by either eustatic fluctuation or regional uplift and subsidence are a major control on reef shape, not only in the sense of controlling the amount of time a reef has to form as discussed above, but also in controlling the size and shape of the buildup. There are four basic responses of reef growth to changing sea level (Figs. 7 and 8) as have been discussed by Link (1950) and Epting (1978). The first is when a regression (relative lowering of sea level) occurs and causes reefs growing in shallow water to be subaerially exposed. Subaerial exposure, of course, terminates reef growth, but slow regression in combination with suitable environ-

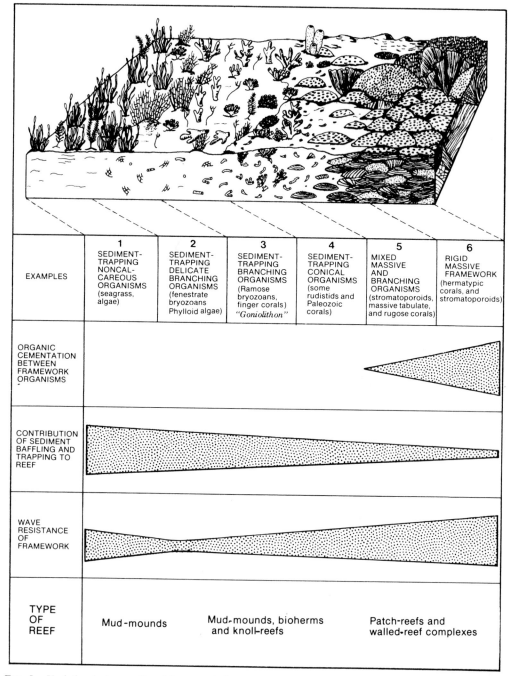

| EXAMPLES | 1 SEDIMENT-TRAPPING NONCAL-CAREOUS ORGANISMS (seagrass, algae) | 2 SEDIMENT-TRAPPING DELICATE BRANCHING ORGANISMS (fenestrate bryozoans Phylloid algae) | 3 SEDIMENT-TRAPPING BRANCHING ORGANISMS (Ramose bryozoans, finger corals) "Goniolithon" | 4 SEDIMENT-TRAPPING CONICAL ORGANISMS (some rudistids and Paleozoic corals) | 5 MIXED MASSIVE AND BRANCHING ORGANISMS (stromatoporoids, massive tabulate, and rugose corals) | 6 RIGID MASSIVE FRAMEWORK (hermatypic corals, and stromatoporoids) |
|---|---|---|---|---|---|---|
| ORGANIC CEMENTATION BETWEEN FRAMEWORK ORGANISMS | | | | | | |
| CONTRIBUTION OF SEDIMENT BAFFLING AND TRAPPING TO REEF | | | | | | |
| WAVE RESISTANCE OF FRAMEWORK | | | | | | |
| TYPE OF REEF | Mud-mounds | Mud-mounds, bioherms and knoll-reefs | | Patch-reefs and walled-reef complexes | | |

FIG. 5.—Variation in types of reef framework from non-calcareous "invisible" frameworks such as sea grass might form to massive organically-bound skeletal biolithites.

mental factors may permit the reef to redevelop at another lower level (Fig. 7A).

The response of a reef complex to a transgression is more complicated and is controlled by the relative rates of the transgression and the rate of growth (deposition) in the reef complex. If the rate of deposition exceeds the rate of transgression, the reef will build upward and basinward

Table 2. Inferred shapes of buildups formed by major groups of organisms throughout the Phanerozoic. Types of buildups are based on Wilson's (1974) classification and are shown in Figure 1. Marine cementation or underlying topography could allow organisms in knoll-reef ramps to form apparent walled-reef complexes. Table compiled from a survey of the literature.

## ORGANISMS IN CARBONATE SHELF-MARGINS THROUGH TIME

| TYPE OF REEF | MUD-MOUNDS | KNOLL-REEF RAMPS | WALLED-REEF COMPLEXES |
|---|---|---|---|
| QUATERNARY | CRINOIDS SPONGES CORALS | SEA GRASS? CORALS | CORALS RED ALGAE |
| TERTIARY | ? | NUMMULITID FORAMS AND SEA GRASS RED ALGAE | CORALS RED ALGAE |
| CRETACEOUS | SPONGES | RUDISTS STROMATOPOROIDS | STROMATOPOROIDS CORALS? |
| JURASSIC | SPONGES ALGAE | CORALS | CORALS RED ALGAE |
| TRIASSIC | SPONGES | CORALS SPONGES RED ALGAE | ? |
| PERMIAN | BRYOZOANS BRACHIOPODS CALCISPONGES CRINOIDS | *TUBIPHYTES* CALCAREOUS ALGAE | SPONGES CALCAREOUS ALGAE |
| CARBONIFEROUS | PHYLLOID ALGAE BRYOZOANS CRINOIDS | CALCAREOUS ALGAE *TUBIPHYTES* CRINOIDS | ? |
| DEVONIAN | CORALS BRYOZOANS CRINOIDS | CORALS STROMATOPOROIDS | STROMATOPOROIDS |
| SILURIAN | CORALS BRYOZOANS CRINOIDS | STROMATOPOROIDS | ? |
| ORDOVICIAN | BRYOZOANS SPONGES CORALS | CRINOIDS? | ? |
| CAMBRIAN | ARCHAEOCYATHIDS *RENALCIS, EPIPHYTON* | ARCHAEOCYATHIDS | ? |
| PRECAMBRIAN | ? | STROMATOLITES | ? |

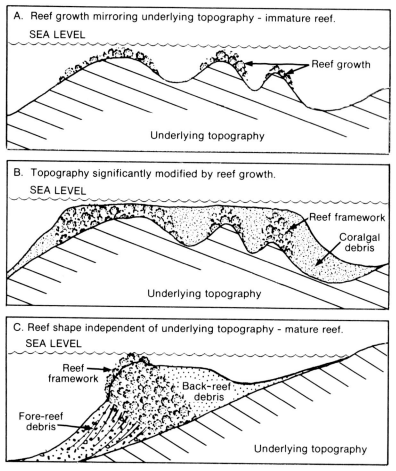

FIG. 6.—Various relationships between reef growth and underlying topography. The length of time available for reef growth is the major control on whether a reef will be immature or mature.

(Fig. 7B and C). This has happened many times in the past and is responsible for many of the large reef buildups in existence (Epting, 1978). The amount of progradation can vary from very great during a slow sea level rise (Fig. 7B) to essentially zero when transgression and deposition are precisely balanced (Fig. 7C).

The response of a reef complex to a transgression that exceeds the rate of deposition is still more complicated. Some people have called on transgressive (landward building) reefs to form under such conditions (Link, 1950; Henson, 1950; Epting, 1978) and in fact some types of reefs will do this (Fig. 8A). However, the profile of a mature reef complex (Fig. 4) is relatively flat topped. When rapid transgression occurs, coral growth will begin in shallower water if possible, but it is clear from the profile that the back-reef area is

not usually significantly shallower than the top of the existing framework facies. Thus, there is no shallower place for the corals to go except a significant distance landward (Fig. 8B). Corals would probably colonize the deeply submerged top of the reef complex after a rapid transgression, but slow rates of deposition there could prevent a significant buildup from forming and the reef could eventually be drowned. This is apparently what is presently happening to the well known Flower Garden Reefs in the Gulf of Mexico (Bright, 1977). Only where a fringing-reef existed on a gently sloping ramp is a transgressive reef complex likely to form. Thus, transgressive reef complexes should be a rare and very minor part of the rock record of reefs.

A second reason that transgressive reef complexes are rare is that the framework facies of a

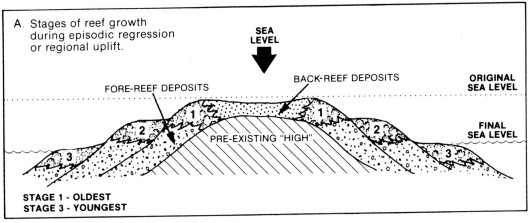

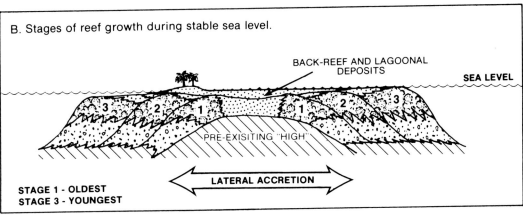

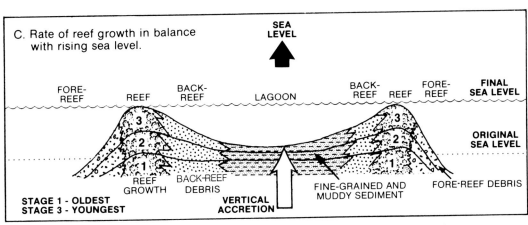

Fig. 7.—Responses of reef growth to regression, stable sea level, and transgression.

reef complex is capable of very rapid growth, even exceeding 6 meters per 1000 years (Macintyre *et al.*, 1977; Adey, 1978). Sea level has apparently risen at a faster rate only during the rapid melting of major glaciers as occurred during the Holocene transgression. Thus, transgressive reef complexes are unlikely to occur in carbonates deposited during non-glacial times.

### Marine Cementation

A very important constructive process in some reef complexes is the formation of marine ce-

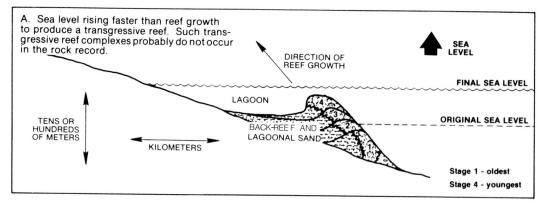

A. Sea level rising faster than reef growth to produce a transgressive reef. Such transgressive reef complexes probably do not occur in the rock record.

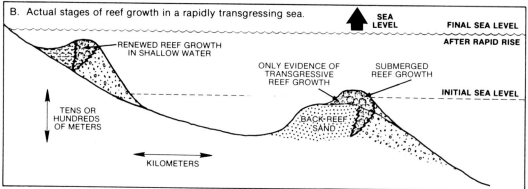

B. Actual stages of reef growth in a rapidly transgressing sea.

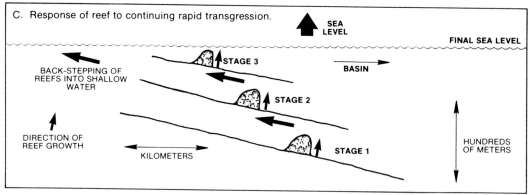

C. Response of reef to continuing rapid transgression.

FIG. 8.—Responses of reef growth when sea level rise exceeds the rate of reef growth.

ments which bind and lithify skeletal material. Marine cements consist of both Mg-calcite and aragonite (Land and Goreau, 1970; Macintyre, 1977) and apparently form most readily where relatively steep carbonate slopes face open oceans so that water is forced through the sediments by waves and currents (Schroeder, 1972; James et al., 1976; James and Ginsburg, in press; Longman, 1980). Cementation may also occur along beaches and at the reef crest where high wave energy is present, but relatively little intergranular cemen-

tation occurs in more deeply submerged or relatively flat parts of reef complexes where less water moves into the sediment, or where movement of grains inhibits cementation.

The importance of marine cementation in the construction of Recent coral reefs has been emphasized by many people (Land and Goreau, 1970; James et al., 1977; James and Ginsburg, in press), and similar marine cements were important in many ancient reefs (Achauer, 1977; Mountjoy and Walls, 1977; James and Kobluk, 1978). Marine cements have played an especially important role in the Permian Capitan Reef of the southwestern United States where they form up to 80 percent of the rock in the so-called "reef framework" facies (Mazzullo and Cys, 1977). The Capitan Reef apparently had a very steep fore-reef wall as indicated by the abundance of large lithified blocks in the fore-reef talus (Newell *et al.*, 1953), but it seems unlikely that the organisms present in the reef (mainly a variety of algae and sponges along with *Tubiphytes* and the problematical *Archaeolithoporella*) could have formed a self-supporting rigid framework with the many tens of meters of vertical relief that existed in the Capitan Reef without this extensive marine cementation (although binding of the framework by non-calcareous algae may have given the reef organisms temporary support until the marine cement formed).

Similar marine cements in other types of carbonate buildups composed of organisms incapable of forming a rigid framework on their own (e.g., rudistids, bryozoans, or phylloid algae) may also have allowed very steep fore-reef slopes to form where they would otherwise have been absent. Such cementation can also occur in relatively deep water reefs as shown by Neumann *et al.* (1977) in their study of buildups in 600–700 meters of water in the Straits of Florida, and marine cements are common in many Lower Carboniferous Waulsortian bryozoan mounds (Washburn *et al.*, 1979).

### Destructive Physical and Biological Processes

All organic buildups, whether true reefs with an organic framework or simply piles of skeletal debris, are the products of both constructional processes involving the organisms in the buildup and marine cementation, and destructive processes such as wave energy, biological erosion, currents, and gravity. Often these destructive processes continue to act even after the death of the organisms in the buildup so that observable characteristics of the buildup reflect the destruction of the reef rather than the biological processes. The nature and intensity of these destructive processes play a major role in determining the distribution and nature of facies in any reef.

Probably the most obvious destructive force acting on a reef is wave action, although this process is active only on those sediments above wave base (usually about 10 meters, but highly variable from reef to reef). Normal day-to-day waves probably do little to affect the shape of a reef through destruction, although anyone who has tried to swim over a reef-crest through average size waves can attest to the intense currents generated. It is the waves and currents induced by

storms that do most of the damage, breaking up the living organisms, and moving large quantities of reef debris around. This destruction has been documented by Cary (1914) who described the huge windrows of gorgonians ripped from a reef in the Tortugas Islands by a hurricane and by Maragos *et al.* (1973) who described the damage inflicted on a reef terrace at depths of 10 to 20 meters (and the resultant movement of coral rubble) by a tropical cyclone on Funafuti Atoll. Several other studies such as those by Perkins and Enos (1968), and Hernandez-Avila *et al.* (1977) have also documented the destructive effects of hurricanes on reefs and the movement of sand to boulder-size rubble that occurs as the result of wave action associated with these storms.

Probably the most thorough documentation of hurricane damage on reefs is that by Hurricane Hattie in 1961 in Belize as described by Stoddart (1962, 1963, 1969b), but his studies were more biological than sedimentological. He found that damage to corals was catastrophic with destruction of about 80 percent of the living corals for an 8 kilometer stretch of the barrier-reef and associated patch-reefs, with lesser damage extending over a zone 50 to 80 kilometers along the reef tract. Much coral debris ("shingle") resulting from this wholesale destruction was carried into the reef-flat or moved into deeper water. From Stoddart's description, it is clear that periodic destruction by such major tropical storms exerts a profound affect on the shapes of reefs both by controlling the distribution of organisms and by transporting vast quantities of skeletal debris.

Bioerosion of carbonate sediments and skeletal debris is another common process in most marine environments and is particularly important in the hard substrates provided by rigid reef frameworks. Biological destruction by boring (siphonodictyid and clionid) sponges is particularly important in modern reefs (Moore and Shedd, 1977). Clionids may produce up to 30 percent of the sediment in some lagoons (Futterer, 1974) and are also important in weakening skeletal material and making it more susceptible to mechanical breakdown (Goreau and Hartman, 1963). Other organisms contributing to the destruction of the reefs include a variety of browsing fish, especially parrot fish (Frydl and Stearn, 1978), echinoids, especially *Diadema* (Stearn and Scoffin, 1977), and a variety of molluscs, sipunculids, polychaetes, and cirripeds (Warme, 1977; MacGeachy and Stearn, 1976). Micritization by boring algae and fungi also plays a significant role in breakdown of carbonate particles (Kobluk and Risk, 1977). With all this extensive biological erosion, it is a bit surprising that any reef material survives to be preserved, and indeed, the work of Land and Moore (1977) and Friedman (1978) suggests that frame-

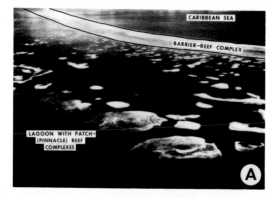

FIG. 9.—Oblique areas photographs of reef complexes. A. Southern part of Belize Barrier Reef and back-reef lagoon with abundant pinnacle-reefs. B. Facies of a pinnacle-reef complex in the Thousand Islands north of Java. C. Facies of the Belize Barrier Reef; reef-crest facies is about 10 meters wide.

works in many ancient reefs may have been obscured by biological destruction.

### FACIES OF MATURE REEF COMPLEXES

Because of the paucity of modern mature reef complexes, any discussion of the facies in such complexes must be based on an understanding of the processes acting on the reef and within the reef complex. Making such a discussion easier is the observation that the underlying topography in a few modern reefs approximates that inferred to have existed in mature reef complexes. Such pseudomature reef complexes may be found in the Belize Barrier Reef (Fig. 9), Australia's Great Barrier Reef system, and Indonesia. However, even with these pseudomature reef complexes, problems still exist in defining reef facies because any division of the reef complex requires the quantification (zonation) of a continuum. Where some will see only a few facies in the reef complex, others will see a hundred.

Stoddart (1969a) recognized a number of zones in windward reefs of the Indo-Pacific atolls including the reef-flat, boulder zone, moat, algal-ridge, and reef front. These terms provide a useful beginning to defining reef facies from a geological point of view, but they are clearly better suited to the biologist rather than a geologist. How does a geologist recognize a moat, for example?

For geological convenience and utility, the reef complex is herein divided into seven facies, mainly because this is the maximum number which a sedimentologist is likely to be able to distinguish in cores from a reef complex. From protected back-reef to open marine fore-reef these seven facies are: back-reef sand, reef-flat, reef-crest, reef framework, reef-slope, proximal talus, and distal talus facies (Figs. 9 and 10), and each of these is defined briefly in Table 3. Boundaries between these zones are gradational. Every facies need not be present in every reef, but each can usually be found, regardless of the shape of the reef in map view (e.g., fringing-reef, barrier-reef, etc.). Only the relative abundance of each facies is likely to be affected by the shape of the reef complex in map view.

Most of the terms used here to describe the sedimentologic facies of reef complexes have been used previously in the literature. Reef-flat is widely used for any relatively flat area behind a "reef" (Stoddart, 1969a), but I propose using it in a more restricted sense to include only the back-reef zone characterized by scattered framework organisms. The term "reef-crest" has generally been used for the zone at the top of the reef framework; e.g., the shallowest zone of the reef complex and it is used here in this sense. This includes Stoddart's (1969a) "boulder zone" and the "rim

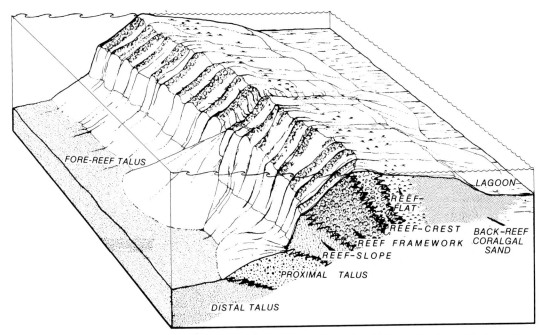

FIG. 10.—Idealized cross-section of facies in a typical mature reef complex similar to that formed by hermatypic corals. The slightly concave nature of the reef framework is artistic, not realistic, and the angle of the reef-slope may be much less than shown.

deposits'' of Flood and Scoffin (1978). The terms reef-flat and back-reef sand (or coralgal sand facies) have been applied to the broad zone of coralgal debris that forms on the backside of some modern mature reef complexes, but the term back-reef sand facies is used here. This facies is poorly developed in many modern reef complexes, but, where present, may become partly emergent to form islands (a vast number of small, low relief South Pacific and Indo-Pacific islands are simply piles of coralgal sand deposited behind reefs). The back-reef sand may grade laterally away from the reef framework into a muddy lagoon.

In the fore-reef area the proximal and distal talus are generally recognized, although an arbitrary boundary exists between the two. Naming the other fore-reef environments is more complex. Both Goreau and Land (1974) and James and Ginsburg (*in press*) have defined several zones on the basis of terraces and slopes, but many of their terms seem to reflect the underlying topography rather than depositional features. Such terms would have little relevance to interpreting facies in ancient reefs. For this reason a simplified approach is used here in which the fore-reef area is divided into the reef framework characterized by

extensive growth of frame building organisms and the deeper water fore-reef slope facies in which the number of frame building organisms is limited for some reason such as light or water circulation.

The characterization of reefs in terms of sedimentologic facies is still in its infancy, especially for mature reefs. Stoddart (1969a) reviewed and summarized the literature on the sedimentologic characteristics of reefs available to that time, but unfortunately the information on reef facies is still sparse. This is an area of active research as shown by the number of recent books dealing with the subject (Maxwell, 1968; James and Ginsburg, *in press*; Frost *et al.*, 1977; The Royal Society's "*The Northern Great Barrier Reef,*" 1978 etc.), and significant advances can be expected in the next few years.

The information presented in the following discussions is based partly on the available literature, but mainly on personal observations and inferences from modern reefs in Belize, the Philippines, Indonesia, the Red Sea, the Caribbean, and Tonga, in combination with information gathered from Tertiary and Pleistocene fossil reefs in many of the same areas. Bear in mind, though, that whereas excellent examples of each facies exist, many intermediate examples also oc-

TABLE 3.—PROCESSES AND CHARACTERISTICS OF FACIES IN MODERN REEF COMPLEXES

| Facies | Process of Sedimentation and Controls on Organisms | Types of Organisms Likely to be Preserved | Grain Size | Sorting | Amount of Framework | Typical Depth (m) | Dominant Rock Type |
|---|---|---|---|---|---|---|---|
| Lagoon | Low energy, much burrowing, sporadic currents and turbidity, possible terrigenous influx | Molluscs, echinoids, miliolids, forams, ostracods | Mud mixed with coarse skeletal debris | Poor | 0% | 5–30 | Wackestone |
| Back-reef sand | Sporadic storms and currents across reef, saltation, gravity sliding | *Halimeda*, miliolids, minor red algae, sparse finger corals | Coarse | Moderate to good | 0% | 1–10 | Grainstone |
| Reef-flat | Sporadic storms, good current circulation, winnowing of mud | Finger corals, red and green algae, larger (benthic) forams, head corals | Coarse–very coarse | Moderate | 0–10% | 1–3 | Grainstone, Scattered corals |
| Reef-crest | High wave energy, constant turbulence, good water circulation | Wave resistant corals and algae | Very coarse | Moderate to good | 0–80% | 0–2 | Grainstone (Minor Boundstone) |
| Reef framework | Good water circulation, high wave energy—sporadic at greater depths | Abundant corals, algae, molluscs, echinoderms, forams | Framework and sand | Poor, Mud in some cavities | 20–80% | 1–30 | Boundstone |
| Reef-slope | Limited light, sporadic turbulence, gravity transport of reef debris | Soft corals, flattened coral plates, sponges | Mixed | Poor | 5–40% | 20–50 | Packstone, Boundstone |
| Proximal talus | Sporadic turbulence, gravity transport, little light, unstable substrate | Few living organisms | Medium to coarse | Poor to good | 0% | 40–100 | Grainstone, Packstone |
| Distal talus | Quiet water, no light, gravity sliding of sediments | Planktic forams | Fine | Moderate to good | 0% | 100–200 | Packstone |

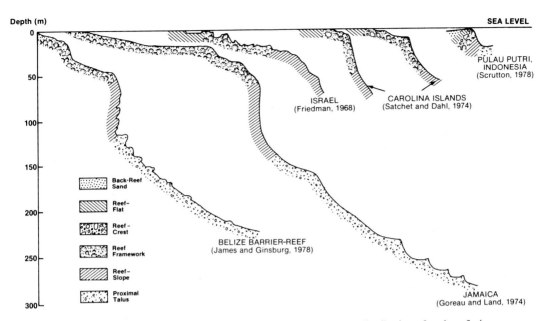

Fig. 11.—Profiles of selected modern reef complexes showing the distribution of various facies.

cur and can be placed into one of the "pigeonholes" only with some difficulty. In addition, local areas of one facies may be found within another facies; for example, coralgal sand often accumulates locally within the reef framework facies. Only by considering relatively large areas of mature reef complexes can good facies interpretations be made.

A final caution is related to the immaturity of Recent reef complexes. As stressed earlier, distribution of facies in immature reef complexes is strongly controlled by irregularities in topography. Considerable simplification of sedimentologic trends on modern immature reef complexes must be done to apply the classification of facies presented here. The fact that Clausade *et al.* (1971) distinguished 68 different morphological zones on a reef near Malagasy, and Battistini *et al.* (1975) defined 125 major morphologic features from reefs in the southern Pacific and Indian Oceans may indicate just how much simplification must be used. While such detailed classifications as those of Battistini *et al.* (1975) may be of use to the biologist interested in making detailed studies of the distribution of organisms, that same detail makes it essentially unusable for the geologist who works with small outcrops containing similar rock types. The simplified classification presented here is designed to be applied to reefs in the rock record, most of which were probably relatively mature with facies belts much better delineated than those found in modern reefs.

### Distribution of Facies in a Reef Complex

Numerous profiles of modern reef complexes are presented in the paper by Satchet and Dahl (1974), and a few idealized profiles of some relatively mature modern reefs from various parts of the world are shown in Figure 11. Of course these reveal only a small sampling of the vast number of possibilities. Relative distribution of the various facies plotted on the profiles reveals that while facies occur in a consistent sequence, significant variations in the widths and depths of each facies exist. These variations are largely due to factors such as pre-existing topography, rainfall and runoff, subaerial exposure, current patterns, wind, and wave energy. Facies should not be considered depth dependent but should be related to the general processes controlling sedimentation and organic activity. Thus, the boundary between reef framework and reef-slope facies apparently approximates the depth below which the major framework organisms do not thrive (due in part to limited light in modern coral reefs); the boundary between reef framework and reef-crest marks the zone of periodic subaerial exposure; and the reef-crest/reef-flat and reef-flat/coralgal sand boundaries may reflect decreasing current action and/or differences in water depth.

All facies of the reef complex are described below and summarized in Table 3. For each facies, the major processes affecting deposition are described along with characteristic types of sedi-

FIG. 12.—Characteristics of living reefs (reef framework facies): A. Spur and groove structures in the framework facies of the Belize Barrier Reef. Dark ellipses are spurs capped by constructional coral growth whereas light areas are channels that funnel sediment off the reef complex. The reef crest facies is present in the foreground and the basin is toward the top of the photo. Spurs are up to 10 meters wide and water depth increases from less than 1 meter at the bottom of the photo to more than 10 meters at the top. B. View of the coral reef framework facies showing prolific growth of *Acropora palmata* and *Millepora* in about 4 meters of water off northern Jamaica. C. Another view of the coral framework in slightly deeper water on a spur off the north coast of Jamaica, note lush coral growth dominated by *Acropora cervicornis* and *Montastraea annularis*, water depth about 10 meters. D. Boundary between coral spur as described in ''C'' showing groove filled with carbonate sand composed of coral, algal, echinoid, mollusc, and foram debris. Picture taken in 12 meters of water off the north coast of Jamaica. E. Reef framework of a pinnacle-reef in the Java Sea showing abundant and diverse coral growth. The dominant corals are *Acropora*, *Stylophora*, and *Fungia*; water depth is 3 meters. F. Fringing-reef coral framework in the Gulf of Aqaba showing steep reef-wall and diverse coral growth; water depth is 2 meters.

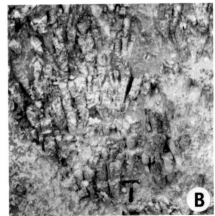

FIG. 13.—Variations in reef framework in Pleistocene reefs: A. Pleistocene reef framework dominated by *Acropora palmata* exposed in a limestone quarry on Barbados. Coral forms about 70 percent of the rock in this part of the framework facies. *Halimeda,* red algae, mollusc fragments, and miscellaneous skeletal debris occur between the coral branches. B. *Montastraea annularis* colonies in slightly deeper water part of the Pleistocene reef framework facies exposed on Barbados.

ments and organisms based on observations in many Holocene and Tertiary coral reefs around the world. Because the reef framework is the key element in the reef complex and controls the formation of both back-reef and fore-reef facies, it is discussed first followed by discussions of the various back-reef facies and, finally, the fore-reef facies. While the descriptions are taken from the study of modern coral reef complexes, application of the processes and sedimentologic characteristics to ancient reef complexes formed by comparable organisms should be possible.

*Reef Framework Facies*

In warm climates, currents rich in nutrients combine with abundant light and good water circulation to make shallow water areas adjacent to deep water an excellent place for growth of those organisms capable of withstanding frequent turbulence. The high level of organic activity may result in the formation of major reef complexes if the colonizing organisms are calcareous. However, this skeletal framework carbonate is often weakened by bioerosion, and is easily broken and moved back over the reef framework by waves and currents or down the front of the reef by gravity. Thus, while much of the reef framework during growth consists largely of *in situ* framework organisms, the framework also contains abundant skeletal debris and is the source of (and support for) the adjacent carbonate facies which are always much larger volumetrically than the framework itself in mature reefs (Ladd, 1950 and 1971). Pictures of selected aspects of the reef framework are shown in Figures 9 and 12.

The key characteristic of the reef framework facies in reef complexes is the presence of a rigid organic framework. The amount of framework in the reef framework facies varies from reef to reef and depends on five factors: 1. nature of the framework organisms, 2. delineation of boundaries with adjacent facies, 3. volume of rock (or sediment) considered, 4. decision to consider the two-dimensional surface of the reef or buried reef rocks, and 5. nature and extent of destructive processes. In a two-dimensional (surface) view of the reef framework, one may find that between 50 and 100 percent of the area is covered with framework organisms (Fig. 12), depending on the type of organisms, slope of the reef, and so on. However, as this extensive framework is buried and incorporated into the subsurface part of the reef complex, destructive physical and biological processes become important and obscure the framework. Thus, a fossil reef framework typically shows between 10 and 70 percent framework with an average amount being 30 percent or less. Land and Moore (1977) and Friedman (1978) might argue that even this low percentage would be too high because they see extensive micritization in reef frameworks caused by both biological and diagenetic processes. Again, the nature of the framework and the intensity of the destructive processes are the critical controls on the percentage of the framework likely to be preserved, and tremendous variation may occur from reef to reef and from time to time within one reef. The above estimates are based on numerous observations of Tertiary and Quaternary reef complexes as seen in outcrop. Within a smaller volume of rock, such as might be available from a cored well, the volume of framework may range from zero percent if the core penetrates a pocket of sand in the framework to as much as 80 or 90 percent as shown in Figure 13.

Within the framework of the reef, skeletal sand composed of fragments of the reef forming and reef associated organisms, such as molluscs and foraminifers, may occur in discrete pockets, in channels, or in lenses up to several meters thick. Mud is sometimes deposited in the reef framework where currents have been baffled. Much sediment within the reef framework is produced locally by boring organisms (Warme, 1977; Moore and Shedd, 1977), and in some parts of the reef, boring may be so extensive that the reef framework is destroyed or obscured (Land and Moore, 1977). Mixing of coarse skeletal fragments with the fine debris produced by these boring organisms typically results in the overall poor sorting of sediment in the reef framework facies.

An important characteristic of many modern reefs is the abundance of penecontemporaneous marine cement. As discussed earlier, the dominant cements are aragonite and Mg-calcite, and both can aid in binding the framework and occluding porosity. Complex histories of cementation, boring, and sedimentation may completely obscure the reef framework locally (Land and Goreau, 1970; Land and Moore, 1977).

Nowhere in the seas is life as abundant and diverse as in the reef framework environment. The shallow water, abundant light and nutrients, and wide variety of ecologic niches combine to create conditions ideally suited to organisms of all types. Hundreds of species of hermatypic corals occur in the Indo-Pacific reefs (Stoddart, 1969a) and algae are at least as diverse. To list all the organisms found in the reef framework is far beyond the scope of this paper, but some of the more common forms having skeletal material likely to be preserved are listed in Table 3.

Topographic relief within the coral framework varies from only a few centimeters in areas where intense wave action restricts growth to low relief encrusting corals and algae (e.g., off Sanur Beach on Bali), to several meters in Belize and the Thousand Islands area in the Java Sea, where well developed spur and groove structures and/or coral pinnacles are present. Lower wave action favors vertical coral growth and the formation of large cavities between branches or colonies. Massive forms may be replaced by more rapidly growing branching and platy forms in such "low energy" reefs.

### Reef-Crest

The shallowest water facies of the reef complex tends to occur at the top of the growing reef and may take either of two distinctly different forms. In the first type, the reef-crest consists of a living coral framework composed mainly of flattened, platy corals or, in lower energy areas, finger corals. Such reef-crests occur in Indonesia and the Philippines. In many ways, this type of reef-crest is similar to the reef framework facies, but coral growth occurs on a nearly horizontal instead of a sloping plane, and diversity of forms is slightly more restricted. The flattened form of the corals may be competitively selected because this shape is an efficient way to use space in the very shallow water of the reef-crest. Periodic exposure during low tides may kill upward growing corals, but nutrients and abundant light are available for organisms that can tolerate the turbulence. Associated organisms are similar to those found in the reef facies.

It is probable that this prolific coral growth on modern reef-crest is the result of the immaturity of the reef complex. If sea level remains stable, continued growth of the corals will build up the reef-crest to the low tide level, and periodic exposure will then kill living corals.

Under such conditions, the reef-crest will probably evolve into a second and more mature type of reef-crest which is characterized by coral rubble and/or rhodoliths. Such a reef-crest occurs today along the Belize Barrier Reef complex and consists of coral fragments ranging in size from sand to large boulders ripped from the reef framework by storms. Coral debris in the reef-crest tends to be extensively bored and heavily encrusted with various types of algae, worms, foraminifers, and a variety of other organisms. Most organisms survive in this high energy environment by living under or between the pieces of coral rubble. Diversity tends to be low relative to that of the reef framework, and living coral in growth position is rare, but certain groups of organisms such as brittle stars, certain foraminifers, algae, etc. flourish.

The reef-crest is one of the narrowest zones of the reef complex and grades into the reef-flat behind the reef. Pictures from this zone are shown in Figure 14.

### Reef-Flat Facies

Between the reef-crest and the back-reef sand facies is a zone of variable width characterized by scattered massive corals, finger corals, much cor-algal debris, and scattered sea grass in relatively shallow water. I propose using the term reef-flat for this zone (others have used reef-flat to include the back-reef sand facies). Compared to the reef framework and reef-crest facies, wave and current energy are relatively low in this zone. This limited circulation combined with the paucity of hard substrates and frequent shifting of the sedi-

FIG. 14.—Characteristics of the reef-crest: A. Prolific coral growth on the reef-crest of the reef complex surrounding Pulau Putri in the Java Sea. Reef-crests in areas ideally suited for coral growth may be covered by living corals, but this is probably a temporary phenomenon; water depth is about 1 meter. B. Reef-crest of the Belize Barrier Reef complex. In this area, coral growth on the reef-crest is inhibited, and the crest is dominated by coral boulder rubble as can be seen breaking the surface of the water; water depth is only a few centimeters at low tide. C. Reef-crest facies of the Belize Barrier Reef showing the characteristic coral rubble. Few corals are present but encrusting red algae, noncalcareous brown algae, and some green algae are common. A variety of other organisms live on and under the algae and boulders; water depth is 1 meter. D. Conglomerate of coral fragments encrusted with red algae to form rhodoliths. Rhodoliths like these are typical of the reef-crest facies where significant wave energy is present. The field of view is about 50 centimeters long, and water depth is 1.5 meters off Sanur Beach, Bali.

ments limits the diversity and abundance of attached filter feeding organisms and corals, although corals may be locally common. Locally these corals may form small patch-reefs. Abundant light facilitates the growth of many organisms including the green alga *Halimeda,* branching and articulated red algae, and many types of non-calcareous algae. Beds of sea grass provide a habitat for a variety of foraminifers, burrowing organisms, and especially molluscs. Holothurians, crustaceans, fish, and other grazing or detritus feeding organisms also tend to be common.

Water depth across the Belize reef-flat is typically one to two meters. Exposure occurs rarely, if ever. Width of the reef-flat varies considerably, depending on pre-existing topography, length of time the reef has been growing, wave energy on the front side of the reef, influx of terrigenous material, and a variety of other factors. In some patch-reefs of Belize and the Java Sea, the reef-flat is only a few meters wide, whereas in other reefs it may exceed a hundred meters in width.

Sediment on the reef-flat is usually moderately sorted. Currents remove most of the mud, but pellets may be associated with the sea grass banks. Large fragments of coral debris are common, and large massive corals may be preserved in growth position. Borings may be present in the massive corals and in the skeletal debris. Individual grains in the reef-flat tend to be angular to subrounded and consist of a complex mixture of fragments of coral, red algae, molluscs, green algae, echinoderms, and foraminifers. Some characteristics of the reef-flat are shown in Figure 15.

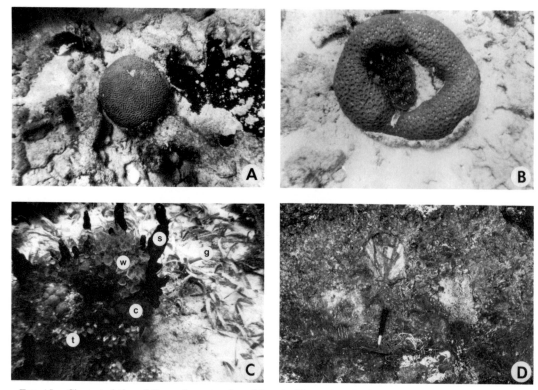

Fig. 15.—Characteristics of the reef-crest and reef-flat: A. Shown is a *Diploria* (brain coral) on the reef-flat of the Belize Barrier Reef complex; coral is about 30 centimeters in diameter. B. Small massive coral 40 centimeters in diameter is an excellent example of a micro-atoll from the reef-flat of Pulau Putri in the Java Sea; *Tridacna* (clam) is nestled in the center of the coral. C. Other organisms in the reef-flat include sponges (s), serpulid worms (w), brown algae (*Turbinaria*, T), sea grass (g), finger corals and branching red algae (not shown), and a variety of algae and molluscs. This cluster of organisms is using a head coral (c) as a substrate and was observed on the Belize Barrier Reef. D. Pleistocene reef-flat with scattered head corals in a skeletal sand matrix. Both *Diploria* and *Montastraea* heads are visible; from Barbados.

## Back-Reef Sand Facies

Wave energy decreases across the reef tract to the point where sediments become so fine-grained and so unstable that attached filter feeding organisms such as corals are no longer able to thrive. At this point, the reef-flat grades laterally into a zone of carbonate sand and mud which is herein called the back-reef sand facies in reference to its position. This facies may be partly exposed as islands or can occur in water as much as 10 meters deep. However, a typical water depth is one to five meters. Some typical characteristics of this facies are shown in Figure 16. These sands have been described by Maxwell *et al.* (1964), Orme *et al.* (1974 and 1978a), and Flood and Scoffin (1978) who used terms such as reef-flat and blanket sands.

Sporadic storms carry material from the seaward parts of the reef complex to this environment where it is mixed with debris from indige-

nous organisms such as molluscs, algae (mainly *Halimeda*), and foraminifers. In modern reefs, the major components of the sand are fragments of corals and calcareous algae, but fragments of echinoderms, molluscs, and foraminifers may also be common. Sorting is usually fair to good and mud is rarely present except where sea grass meadows, which may be very extensive, are present.

The width of the back-reef sand facies is a function of underlying topography, rate of carbonate supply from the reef framework and indigenous organisms, and time. Often it is several tens of meters wide. Prolonged stable sea level can result in back-reef sand flats several kilometers wide, if other parts of the reef complex act as prolific carbonate producers for extended time periods. It is not unusual for the back-reef sand to be piled up by storms to form islands behind barrier-reefs or on pinnacle-reef complexes. These islands, once

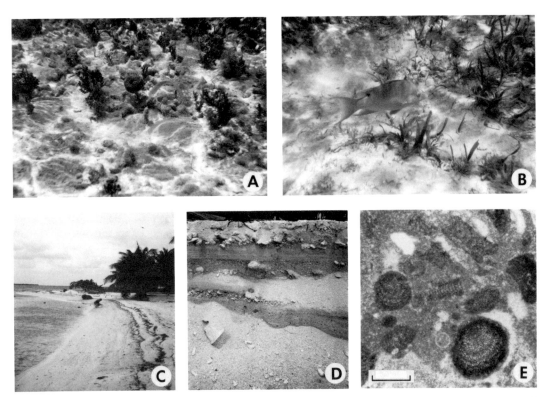

FIG. 16.—Back-reef sand facies: A. Back-reef sand facies of the Belize Barrier Reef complex. The substrate is a mixture of reef derived coralgal debris and grains derived from indigenous organisms such as the *Halimeda* (green algae) shown. *Penicillus* and other algae are also visible; water depth is about 1.5 meters. B. Back-reef sand facies with a sparse cover of *Thalassia*. Foraminifers and other organisms associated with the sea grass contribute skeletal material to the coralgal sand facies, Belize Barrier Reef complex; fish for scale is 20 centimeters long. C. Back-reef sand behind the reefs is often piled up into emergent islands such as Tobacco Cay behind the Belize Barrier Reef shown here. Islands are important pathways through which fresh water may enter reef complexes and cause diagenesis. D. Close-up of back-reef sand facies containing coral sticks and fragments of conch shells. This picture shows a cross-section of the beach sands on Tobacco Cay, Belize; width of field is about 60 centimeters. E. Thin-section photomicrograph of back-reef sand facies from a lower Miocene reef complex in the Philippines; red algae is the dominant grain type.

formed, play an important role in the development of fresh water lenses and diagenesis in carbonates of the reef complex. Characteristics of sediments on these back-reef sand islands have been described by McLean and Stoddart (1978).

Although discussed only briefly here, this coralgal sand facies is one of the most important facies of the reef complex. It is generally very extensive in Tertiary reef complexes, far exceeding the volume of the other shallow water facies of the reef complex. Because of its large volume, high primary porosity (often destroyed during diagenesis), topographically high position, and potential for secondary porosity formed by leaching, this zone offers the best probability as a hydrocarbon reservoir in most Tertiary and older reef complexes around the world.

## Lagoonal Facies

A relatively quiet water environment dominated by muddy sediments may be present within atolls or behind reef complexes. This zone, generally called the lagoon, may or may not be part of the reef complex, depending on the amount of debris derived from the reef framework. The lagoonal sediments in many of the South Pacific islands are part of the reef complex, whereas the Belize lagoon is not because most of the lagoonal sediment is terrestrially derived.

Lagoons typically have water a few meters deep and are characterized by low wave energy and limited water circulation. Carbonate mud and fine sand commonly accumulate in the lagoon and may be extensively burrowed by a variety of organisms such as annelids, echinoids, crustaceans, and molluscs. Skeletal debris produced by these organisms is mixed with the mud and results in generally poor sorting. Little abrasion occurs, but grains are often broken by the burrowers. Terrigenous and organic material may be common, depending on local and regional conditions. Because of the "ugliness" of the muddy lagoonal sediments and the difficulty of studying them statistically, few detailed studies have been made. However, they have been described by Maxwell *et al.* (1964) and are often similar to the inter-reef sediments described by Flood *et al.* (1978).

Lagoonal sediments are relatively easy to recognize by the abundance of mud, poor sorting, types of skeletal material present (particularly molluscs, foraminifers, and *Halimeda*), and general paucity of open water organisms such as planktic foraminifers. However, deep lagoons such as that behind the southern Belize Barrier Reef tend to assume many characteristics of deep water shelf muds including planktic organisms (Scholle and Kling, 1972) and may be difficult to recognize. Although lagoonal sediments are generally of little importance as hydrocarbon reservoirs, they may be important as source rocks.

### Reef-Slope Facies

On the seaward side of the reef framework facies of a mature or pseudomature reef complex is a zone characterized by a steep slope, sparse scleractinian corals, and in modern reefs abundant alcyonarians such as sea whips. Water depth is generally on the order of a few tens of meters, which keeps wave energy relatively low and also limits the amount of light. These factors apparently give the rapidly growing soft corals an ecological advantage over the stony corals, although the stony corals are not completely absent. Those that are present often take the form of flattened plates with corallites only on the upper surface as an adaptation to the limited light conditions. Good sources of information on the reef-slope are the papers of Goreau and Land (1974), Land and Moore (1977), and James and Ginsburg (*in press*).

Defining the upper limit of the reef-slope facies in Recent reefs is often difficult because it grades into the reef framework facies. However, it seems that there is a certain depth more or less related to the availability of light and perhaps also to wave energy below which soft corals are abundant and replace the hard corals. Sclerosponges are important in some reef-slopes, especially where a reef wall is present (Goreau and Land, 1974). In protected areas, such as the Thousand Islands of Indonesia and the Red Sea, this boundary is much shallower than in reefs fronting broad oceans such as in Belize and Jamaica. The average depth of the boundary is about 30 meters but varies markedly depending on the setting of the reef complex.

The lower limit of the reef-slope is generally marked by a distinct decrease in slope and a change from widespread hard substrates to unlithified proximal talus. This depth, too, varies tremendously depending on the setting of the reef complex but is often found at a depth between 100 and 200 meters (Land and Moore, 1977; James and Ginsburg, *in press*). Perhaps the lowering of sea level during the Pleistocene has played a role in forming this boundary in some places, and it would be difficult to say that the boundary occurred at this depth in ancient reef complexes.

Abundant sediment produced in the shallower water facies of the reef complex moves across the reef-slope under the influence of gravity, mainly by sediment creep. Sorting is generally poor to moderate because the coarse material produced in the reef-slope is mixed with a variety of debris coming down from above. Abrasion and erosion, especially by boring organisms, are important processes.

The reef-slope typically dips between 50 and 90 degrees. A variety of organisms such as *Halimeda*, zooantharians, sponges, and sclerosponges are common on the reef-slope and contribute to the sediment (Land and Moore, 1977). Pockets and lenses of sediment become increasingly common toward the base of the slope.

The reef-slope facies is probably the most difficult facies of the reef complex to recognize in ancient rocks because it possesses few diagnostic characteristics. The platy morphology of the corals may be a good indication where platy corals exist, but the best evidence for this facies is its intermediate position between the proximal talus and reef framework facies.

### Proximal Talus

Below the reef-slope is a slope characterized by abundant debris derived from the reef complex and few living calcareous organisms. This is the proximal talus facies. Water depth of the talus is highly variable depending on the type and setting of the reef system. Study of this facies in recent reefs is difficult because it is below the depths ordinarily accessible to SCUBA diving. However, some studies using small submersibles have been done (e.g., Land and Moore, 1977; James and Ginsburg, *in press*), and many ancient examples of reef talus are known (McIlreath and James, 1979).

Wave energy and light are generally low or absent in the talus environment, although sporadic currents may be generated by sediment moving down the slope. Sediment probably moves downslope by gravity both by creep and rapid slumps.

Hard substrates are rare and organisms are sparse.

Reef talus, as the name implies, consists largely of debris derived from the reef complex such as coral fragments, blocks of the reef up to several meters in diameter, *Halimeda,* red algae, and a variety of other biogenic grains. However, there are also talus beds dominated either by debris from the lagoon behind the reef or deep water organisms. Rock types range from mudstone to grainstone with skeletal packstones being the most common. Terrigenous materials may be mixed with reef talus in some areas.

Reef talus associated with Miocene coral reef complexes in Spain contains some beds with fairly good segregation of fossils. Individual beds 10 to 50 centimeters thick may contain abundant fragments of one type of organism such as pectens, rhodoliths, brachiopods, or coral fragments. Fragments are typically abraded. Segregation may be due to the concentration of hydraulically equivalent fossil fragment types (similar size, shape, density, *etc.*), as the debris moved downslope and may prove to be an aid in recognizing proximal talus in the rock record, although storms could probably produce similar segregation in back-reef carbonates.

Sedimentary structures in reef talus include laterally continuous beds and sparse burrows. The beds typically range from a few centimeters to a few meters in thickness and are separated by thin shale partings. This bedding is ubiquitous in talus and can be used with caution to distinguish talus from reef framework and back-reef rocks. The consistent thickness and lateral extent of individual beds are also distinctive. Large scale channels marked by low relief and broad lateral extent also occur in ancient reef talus such as that in the Spanish Miocene (Messinian) reefs.

### Distal Talus

Downslope from the proximal talus grain size becomes finer, and the abundance of planktic organisms increases. Sediments represent a mixture of planktic organisms and material derived from the reef complex. The boundary between distal and proximal talus is arbitrary and gradational. This environment grades into a normal deep water environment beyond the influence of sediments derived from the reef complex.

#### RELATIVE ABUNDANCE OF REEF FACIES

An interesting exercise of some significance is to calculate the relative percent of surface area covered by each of the shallow water facies of the reef complex in a "typical" mature reef. This reveals something about the probability of seeing a certain facies in an outcrop or core. Of course there is tremendous variation from reef complex to reef complex depending on maturity and setting, so it is difficult to pick a typical area, but for the sake of discussion, a small reef complex in the Thousand Islands north of Java is used here. This reef complex contains an island called Pulau Putri and has been described by Scrutton (1978). Both the reef complex and island are more or less circular which makes calculations of relative areas fairly easy, and the reef complex appears to be comparable to many ancient pinnacle-reefs except that it has only about 30 meters of present day relief.

A map view and cross-section of Pulau Putri, drawn essentially to scale, are shown in Figure 17, and the various facies of the reef complex are indicated. Distribution of facies on this shoal was determined by a number of swimming traverses but is only approximate. Assuming that the facies occur in concentric bands (obviously somewhat of an oversimplification) and using the radius of the circle and the width of each facies, it is possible to calculate the area covered by each facies (Table 4). It turns out that for the area of the shallow part of the reef complex the reef framework covers about 10 percent, the reef-crest 5 percent, the reef-flat 12 percent, and the coralgal sand covers 73 percent. No lagoon is present, and the deep water facies in front of the reef complex are not considered in these values. However, if sea level remained constant and the reef framework prograded into deeper water, the percentage of coralgal sand and reef-flat would form an even larger percentage of the reef complex.

Although no similar quantitative estimate of facies abundance has been done for a barrier-reef, aerial reconnaissance and diving on the Belize Barrier Reef indicates that the coralgal sand facies is far more extensive than the reef framework facies (Fig. 10), and that the area covered by each facies would approximate that calculated for the Pulau Putri Reef complex.

### Development of Reef Complexes

Having discussed the factors controlling reef shape and the processes responsible for forming various facies of a reef complex, it is important to consider how the profile of a reef complex changes during prolonged periods of stable sea level. To do this requires knowledge of the rates of accumulation for the various facies of the reef complex. Unfortunately, complete information is not yet available, and even if it was, it would probably show significant variation from reef to reef depending on the nature of the reef building organisms and other environmental factors. Thus, some educated guesses are needed to determine

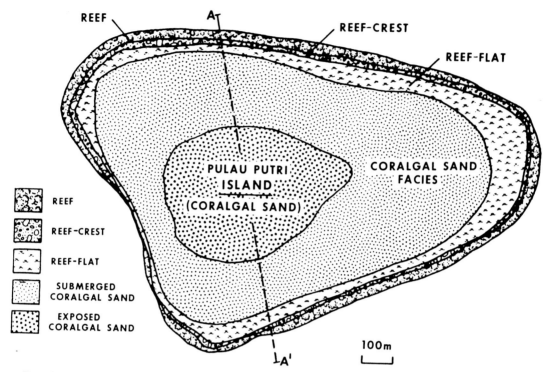

FIG. 17A.—Facies map of Pulau Putri based on reconnaissance swimming traverses and extrapolation; all facies boundaries are gradational.

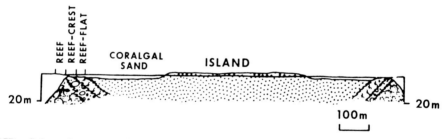

FIG. 17B.—Schematic cross-section of Pulau Putri showing approximate width of facies of the reef complex; subsurface geometry is schematic.

the relative rates of deposition across a reef complex.

There can be little doubt that the rate of growth and accumulation in the reef framework facies exceeds the rate of accumulation of all other facies, nor can there be any doubt that shallower parts of the reef framework above which sporadic subaerial exposure combines with destructive processes induced by storms to equal or exceed the rate of growth or deposition. Thus, on the basinward side of a reef complex, maximum reef growth occurs in shallow water and decreases with increasing depth. Deposition in associated

facies is mainly a function of material carried there from the reef framework but clearly must decrease with distance from the framework, unless sediment bypassing occurs. Hence, proximal talus accumulates more rapidly than distal talus, but sediment bypassing may allow the proximal talus to accumulate more rapidly than the reef slope.

Having established the relative rates of deposition and growth, it is necessary to determine other controls on the reef profile. The following points are accepted as controls:

1. sediment may accumulate up to sea level but

not above (islands can form on the reef complex, but these are of such low topographic relief that they may be ignored in the model);

2. growth of the reef framework facies in combination with marine cementation is capable of forming a vertical or nearly vertical fore-reef wall;

3. non-framework facies consist mainly of loose skeletal debris and can form slopes only as steep as the angle of repose of the sediment; e.g., up to 30–35 degrees in poorly-sorted material and as low as one or two degrees in very fine-grained mud;

4. excess material from skeletal growth in the reef framework zone is transported into back-reef and fore-reef facies; the amount of material carried in each direction can only be estimated and probably varies from reef to reef and place to place within one reef, depending on setting and storm activity.

Using the above guidelines and assuming a stable sea level, stages in the development of the reef can be defined (Fig. 18). Initial deposition on the ramp is fastest at the depth of optimum organic growth where a buildup begins to form (Fig. 18B). With continued growth the buildup reaches nearly to sea level and then begins prograding basinward (Fig. 18C). If sea level remains unchanged, the area between the buildup and shore will fill-in quickly, mostly with skeletal debris of the coralgal facies, and no lagoon will be present (Fig. 18D). The reef, thus formed, would be considered a fringing-reef. Relative subsidence at a rate exceeding deposition in the back-reef sands could result in the formation of a significant lagoon separating a barrier-reef complex from land as Darwin (1842) recognized (Fig. 18E). Of particular interest in this case, however, is the shape of the reef complex with its nearly vertical fore-reef wall ending at the bottom in the fore-reef talus characterized by a gradually decreasing slope. Behind the reef framework is the relatively flat top of the reef complex containing the reef-crest, reef-flat, and coralgal sand facies. Behind the reef complex, a sand slope at the angle of repose may be present, if a significant lagoon exists; otherwise the sand extends all the way to shore.

In Figure 18E, we see the profile of a mature reef complex similar to the Belize Barrier Reef. The same profile is probably comparable to those formed by ancient reef complexes containing a rigid organic framework (e.g., Tertiary hermatypic coral reefs and at least some Devonian stromatoporoid reefs). Delineating the positions of facies in the profile, one can immediately see that most of the volume of reef complexes deposited during slow transgression consists of the coralgal sand and fore-reef talus facies. Width of the coralgal sand facies is a function of rate of sedimentation and time but can be up to several kilometers or more.

TABLE 4.—AREA COVERED BY VARIOUS FACIES OF PULAU PUTRI REEF COMPLEX. DISTRIBUTION OF FACIES IS SHOWN IN FIGURE 9 BUT CALCULATIONS ASSUME CONCENTRIC FACIES BELTS (OBVIOUSLY AN OVERSIMPLIFICATION)

| Facies | Radius Width (m) | Area Covered (m²) | % |
|---|---|---|---|
| Reef | 30 | 107,000 | 10 |
| Reef-Crest | 15 | 51,600 | 5 |
| Reef-Flat | 40 | 131,000 | 12 |
| Coralgal Sand | 500 | 785,000 | 73 |

### SIGNIFICANCE OF THE MATURE REEF PROFILE

Imagine a seismic profile across a mature reef complex such as that in Figure 18E, assuming it had subsequently been covered with a thick shale. In developing a prospect for the reef, most petroleum explorationists would drill for the center of the buildup, and it is clear that the facies most likely to be encountered there is the back-reef coralgal sand facies. Not surprisingly, this is exactly what is seen in cores from the coastal positions of many Tertiary pinnacle-reefs in Indonesia and the Philippines, as well as some Devonian reefs in Canada (Klovan, 1964).

Disagreements between geophysicists and geologists over whether or not something is a "reef" are common, precisely because of this "unexpected" abundance of reef debris without an organic framework at the "top of a reef." However, once the reasons for the disagreement are recognized, the geologists and geophysicists can get back to the business of predicting the reservoir potential of various parts of the reef complex and determining where the best place to drill the next well should be.

### DIAGENESIS AND RESERVOIR POTENTIAL OF REEF COMPLEXES

Reef complexes are especially susceptible to diagenesis because of both their position at shelf-edges and their significant topographic relief. Isopachous fibrous aragonite and micrite-sized or pelletal Mg-calcite are often precipitated on the front side of reef complexes and may completely fill all pores in some parts of the reef framework (James *et al.*, 1977). Perhaps because of relatively limited water circulation, the backsides of reef complexes and patch-reefs behind a barrier are less likely to contain significant amounts of marine cement. However, isolated pinnacles and patch-reefs in front of barriers are vulnerable to marine cementation. The extent of marine cementation is dependent on the nature of the reef framework, time available for cementation, water circulation patterns, and possibly on the presence of certain bacteria.

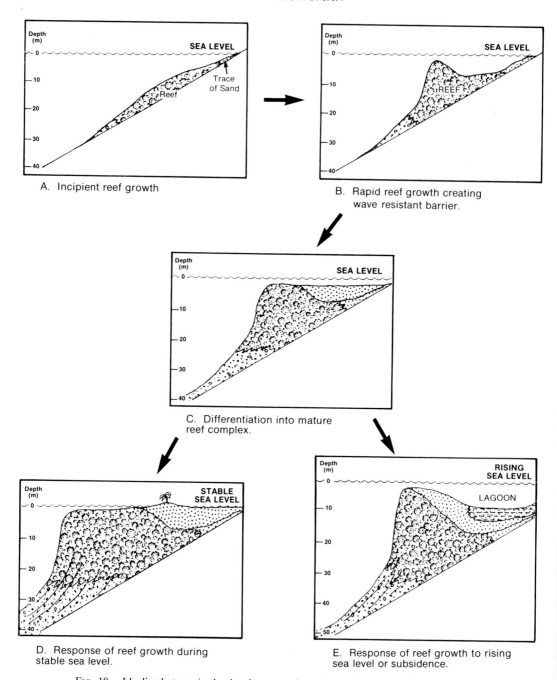

A. Incipient reef growth

B. Rapid reef growth creating wave resistant barrier.

C. Differentiation into mature reef complex.

D. Response of reef growth during stable sea level.

E. Response of reef growth to rising sea level or subsidence.

Fig. 18.—Idealized stages in the development of a reef complex on a gently sloping ramp.

Cementation by calcite and/or leaching occurs in reef complexes during exposure to fresh water. This exposure may be caused by regional uplift or a drop in eustatic sea level. Fresh water (rain) may also enter the reef complex through subaerially exposed islands. This rainwater tends to be undersaturated with respect to calcium carbonate. Significant leaching, and the formation of secondary porosity, particularly above the water table in areas with humid climates, is often the result. Dissolved calcium carbonate may be reprecipitated as equant calcite cement, and cementation may be so extensive that it destroys all porosity within parts of the reef complex. Carbonate diagenesis in the relatively unstable carbonates of an unaltered reef may occur very rapidly, geologically speaking. Significant leaching or cementation can occur in only a few thousand years.

Contrary to dogma, the reef framework is generally not the best reservoir facies of a reef complex. Apparently this idea originated when swimmers on modern reef frameworks observed the large cavities between the corals and speculated that this framework porosity could be preserved and filled with oil. The discovery of the giant Golden Lane Field in a Cretaceous rudistid carbonate buildup in Mexico, in the early 1900s, only strengthened this view because the extensive cavernous porosity (now interpreted as related to early Tertiary karsting by Coogan et al., 1972) was originally interpreted as primary porosity. Blasting and coring of modern reefs reveals that these large framework cavities are quickly filled with skeletal debris, mud, and marine cements, and thus, are rarely, if ever, preserved after burial (Goreau and Land, 1974; Macintyre, 1977). Furthermore, the susceptibility of the reef framework facies to extensive marine cementation, which can fill even the smallest pores inside skeletal debris, makes this facies a poor potential reservoir. As a rule, the reef framework facies can become a good reservoir only if significant leaching occurs during subaerial exposure, or if some other form of secondary porosity is generated (e.g., by dolomitization or fracturing).

Much better potential reservoir facies are the fore-reef talus and back-reef coralgal sand, which tend to be volumetrically much larger than the reef framework and tend to escape most early marine cementation. Primary porosity may be preserved in either facies if exposure to fresh water is inhibited. Alternatively, secondary porosity may form during subaerial exposure. Under certain circumstances, where extensive marine cementation of the reef framework occurs and basinal shales (good source rocks) are present, it seems possible that reef talus may become an excellent reservoir with the cemented reef acting as an updip seal. Under other circumstances, exposure and vadose leaching may produce significant porosity in the topographically high coralgal sand facies making it a good potential reservoir.

## CONCLUSIONS

Reef morphology reflects complex interactions between constructive processes (mainly organic growth and marine cementation) and destructive processes (both mechanical and biological) superimposed on the effects of underlying topography and changes in relative sea level. Shapes of Holocene reefs generally reflect underlying topography which has been modified only slightly by deposition during the last several thousand years when sea level has existed near its present level. Morphology of ancient reefs, which often had much longer to grow, is generally relatively independent of underlying topography and reflects instead the nature of the organisms, their ability to form a rigid framework, the extent of marine cementation, and the intensity and nature of destructive processes.

Underlying topography in a few Recent reefs such as the Belize Barrier Reef approximates that inferred to have existed in mature framework reefs. This permits recognition of a sequence of intergradational, but distinct sedimentologic facies including from the basin toward land: 1. distal talus characterized by common planktic foraminifers and fine-grained coralgal reef debris, 2. proximal talus characterized by coarser coralgal reef debris, 3. fore-reef slope characterized by soft corals, flat platy hard corals, sclerosponges, sponges, and abundant coralgal debris, 4. the true reef consisting of an organic framework and lush growth of corals and algae, 5. reef-crest generally composed of coarse coral rubble, 6. reef-flat with scattered head and finger corals, and small patch-reefs, but no significant framework, 7. coralgal sand facies composed of debris washed back from the reef framework by waves and storms and mixed with debris of indigenous organisms such as *Halimeda*, molluscs, and foraminifers, and 8. the lagoon dominated by burrowed mud with some skeletal debris.

Of these facies, the talus and back-reef coralgal sand facies tend to be volumetrically the most important in most mature reef complexes.

Organisms play a critical role in determining overall shape of a carbonate buildup. Those organisms capable of rapid growth, withstanding high wave energy, and forming a rigid organic framework will form complexes comparable to the coral reef complexes described in this report. Ancient examples of such organisms include stromatoporoids and scleractinian corals. Rapid marine cementation aided in the formation of similar rigid reef structures by calcareous algae and sponges in the Permian and the Triassic (see other papers in this volume).

Delicate or non-framework building organisms such as phylloid algae, rudists, rugose corals, bryozoans, *Nummulites*, and crinoids formed ancient carbonate buildups that are sometimes called reefs, but these structures are very different from modern reef complexes. Organisms such as the above living in shallow water with moderate to high energy typically formed low relief lenticular deposits surrounded by extensive sand facies, whereas those living in quiet water formed mud mounds. The similarity of facies in these buildups to the facies of a scleractinian coral reef needs further study, and any analogies should be made very carefully.

## ACKNOWLEDGMENTS

Like any synthesis, this paper has relied heavily on the work of others whom I gratefully acknowl-

edge. Special thanks to Lynton Land, Gerry Friedman, John McMannus, Charlotte Glenn, and Lyle Baie for reviewing a preliminary version of the manuscript. Cities Service Company provided financial support for field work and permission to publish.

# REFERENCES

ACHAUER, C. A., 1977, Contrasts in Cementation, Dissolution, and Porosity Development Between Two Lower Cretaceous Reefs of Texas, *in* Bebout, D. G., and R. G. Loucks, Eds., Cretaceous Carbonates of Texas and Mexico; Applications to Subsurface Exploration: Bureau Economic Geology, Rept. Investigations No. 89, p. 127–137.

ADEY, W. H., 1978, Coral Reef Morphogenesis: A Multidimensional Model: Science, v. 202, No. 4370, p. 831–837.

BATTISTINI, R., *et al.,* 1975, Elements de Terminologie Recifale Indopacifique: Tethys, v. 7, No. 1, 111 p.

BEBOUT, D. G., AND LOUCKS, R. G., 1974, Stuart City Trend, Lower Cretaceous, South Texas—A Carbonate Shelf-margin Model for Hydrocarbon Exploration: Bureau Economic Geology, Rept. Investigations No. 78, 80 p.

BEIN, A., 1976, Rudistid Fringing Reefs of Cretaceous Shallow Carbonate Platform of Israel: Am. Assoc. Petroleum Geologists Bull., v. 60, p. 258–272.

BRAITHWAITE, C. J. R., 1973, Reefs: Just a Problem of Semantics?: Am. Assoc. Petroleum Geologists Bull., v. 57, p. 1100–1116.

BRIGHT, T. J., 1977, Coral Reefs, Nepheloid Layers, Gas Seeps and Brine Flows on Hard Banks in the Northwestern Gulf of Mexico: *in* Third International Coral Reef Symposium, Proc., v. 1 (Biology), p. 39–46.

CARY, L. R., 1914, Observations Upon the Growth Rate and Ecology of the Gorgonians: Washington, D.C., Carnegie Inst., Tortugas Lab. Papers, v. 5, p. 81–89.

CLARK, J. A., FARRELL, W. E., AND PELTIER, W. R., 1978, Global Changes in Post-glacial Sea Level: A Numerical Calculation: Quaternary Res., v. 9, p. 265–287.

———, AND LINGLE, C. S., 1979, Predicted Relative Sea Level Changes (18,000 Years B.P. to Present) Caused by Late Glacial Retreat of the Antarctic Ice Sheet: Quaternary Res., v. 11, p. 279–298.

CLAUSADE, M., GRAVIER, N., PICARD, J., PICHON, M., ROMAN, M. L., THOMASSIN, B., VASSEUR, P., AND WEYDER, P., 1971, Morphologie des Recifs Coralliens de la Region de Tulear (Madagascar): Elements de Terminologie Recifale: Tethys, Supplement 2, 76 p.

COOGAN, A. H., BEBOUT, D. H., AND MAGGIO, M., 1972, Depositional Environment and Geologic History of Golden Lane and Poza Rica Trend, Mexico, an Alternative View: Am. Assoc. Petroleum Geologists Bull., v. 56, p. 1419–1447.

CUMINGS, E. R., 1932, Reefs or Bioherms?: Geol. Soc. America Bull., v. 43, p. 331–352.

DARWIN, C. R., 1842, The Structure and Distribution of Coral Reefs: London, Smith, Elder and Co., 214 p. (Reprinted 1962, Berkeley, Univ. California Press.)

EPTING, M., 1978, Sedimentology of Miocene Carbonate Buildups, Central Luconia, Offshore Sarawak: Geol. Soc. Malaysia, Second Petrologists Seminar, Kuala Lumpur, 25 p.

FAIRBRIDGE, R. W., 1961, Eustatic Changes in Sea Level: Physics Chem. Earth, v. 4, p. 99–185.

FLOOD, P. G., ORME, G. R., AND SCOFFIN, T. P., 1978, An Analysis of the Textural Variability Displayed by Inter-reef Sediments of the Impure Carbonate Facies in the Vicinity of the Howick Group: Phil. Trans. Royal Soc. London A, v. 291, p. 73–83.

———, AND SCOFFIN, T. P., 1978, Reefal Sediments of the Northern Great Barrier Reef: Phil. Trans. Royal Soc. London A, v. 291, p. 55–71.

FRIEDMAN, G. M., 1968, Geology and geochemistry of reefs, carbonate sediments, and waters, Gulf of Aqaba (Elat), Red Sea: Jour. Sed. Petrology, v. 38, p. 895–919.

———, 1978, Recognition of Post-Paleozoic Reefs: An Experience in Frustration: Tenth International Congress on Sedimentology, Jerusalem, Proc., v. 1, p. 220.

FROST, S. H., 1977, Ecologic Controls of Caribbean and Mediterranean Oligocene Reef Coral Communities: Third International Coral Reef Symposium, Miami, Proc., v. 2, p. 367–373.

———, WEISS, M. P., AND SAUNDERS, J. B., Eds., 1977, Reefs and Related Carbonates—Ecology and Sedimentology: Am. Assoc. Petroleum Geologists, Studies Geology, No. 4, 421 p.

FRYDL, P., AND STEARN, C. W., 1978, Rate of Bioerosion by Parrotfish in Barbados Reef Environments: Jour. Sed. Petrology, v. 48, p. 1149–1158.

FUTTERER, D. K., 1974, Significance of the Boring Sponge *Cliona* for the Origin of Fine Grained Material of Carbonate Sediments: Jour. Sed. Petrology, v. 44, p. 79–84.

GOREAU, T. F., AND HARTMAN, W. D., 1963, Boring Sponges as Controlling Factors in the Formation and Maintenance of Coral Reefs, *in* Mechanisms of Hard Tissue Destruction: Washington, D.C., Am. Assoc. Advancement Science Pub., 75, p. 25–54.

————, AND LAND, L. S., 1974, Fore-Reef Morphology and Depositional Processes, North Jamaica, *in* Laporte, L. F., Ed., Reefs in Time and Space: Soc. of Econ. Paleontologists Mineralogists Spec. Pub., No. 18, p. 77–89.

HECKEL, P. H., 1974, Carbonate Buildups in the Geologic Record: A Review, *in* Laporte, L. F., Ed., Reefs in Time and Space: Soc. Econ. Paleontologists Mineralogists Spec. Pub., No. 18, p. 90–154.

HENSON, F. R. S., 1950, Cretaceous and Tertiary Reef Formations and Associated Sediments in Middle East: Am. Assoc. Petroleum Geologists Bull., v. 34, p. 215–238.

HERNANDEZ-AVILA, M. L., ROBERTS, H. H., AND ROUSE, L. J., 1977, Hurricane Generated Waves and Coastal Boulder Rampart Formation: Third International Coral Reef Symposium, Proc., v. 2 (Geology), p. 71–78.

JAMES, N. P., AND GINSBURG, R. N., The Deep Seaward Margin of Belize Barrier and Atoll Reefs: International Assoc. Sedimentologists Spec. Pub. (in press).

————, ————, MARSZALEK, D. S., AND CHOQUETTE, P. W., 1977, Facies and Fabric Specificity of Early Subsea Cements in Shallow Belize (British Honduras) Reefs: Jour. Sed. Petrology, v. 46, p. 523–544.

————, AND KOBLUK, D. R., 1978, Lower Cambrian Patch Reefs and Associated Sediments: Southern Labrador, Canada: Sedimentology, v. 25, p. 1–35.

KLOVAN, J. E., 1964, Facies Analysis of the Redwater Reef Complex, Alberta, Canada: Bull. Canadian Petroleum Geol., v. 12, p. 1–100.

KOBLUK, D. R., AND RISK, M. J., 1977, Micritization and Carbonate-Grain Binding by Endolithic Algae: Am. Assoc. Petroleum Geologists Bull., v. 61, p. 1069–1082.

LADD, H. S., 1950, Recent Reefs: Am. Assoc. Petroleum Geologists Bull., v. 34, p. 203–214.

————, 1971, Existing Reefs—Geological Aspects: North American Paleontologists Convention, Chicago, 1969, Proc., pt. J, p. 1273–1300.

LAND, L. S., 1974, Growth Rate of a West Indian (Jamaican) Reef: Second International Coral Reef Symposium, 2, Great Barrier Reef Committee, Brisbane, Proc., p. 409–412.

————, AND GOREAU, T. R., 1970, Submarine Lithification of Jamaican Reefs: Jour. Sed. Petrology, v. 40, p. 457–462.

————, AND MOORE, C. H., JR., 1977, Deep Forereef and Upper Island Slope, North Jamaica: Am. Assoc. Petroleum Geologists, Studies Geology, No. 4, p. 53–65.

LINK, T. A., 1950, Theory of Transgressive and Regressive Reef (Bioherm) Development and Origin of Oil: Am. Assoc. Petroleum Geologists Bull., v. 34, p. 263–294.

LONGMAN, M. W., 1980, Carbonate Diagenesis in Nearshore Diagenetic Environments: Am. Assoc. Petroleum Geologists Bull., v. 64 (in press).

LOWENSTAM, H. A., 1950, Niagaran Reefs of the Great Lakes Area: Jour. Geology, v. 58, p. 430–487.

MacGEACHY, J. K., AND STEARN, C. W., 1976, Boring by Macro-organisms in the Coral *Montastrea annularis* on Barbados Reefs: International Rev. Ges. Hydrobiol., v. 61, p. 715–745.

MacILREATH, I. A., AND JAMES, N. P., 1979, Facies Models 13. Carbonate Slopes: Geoscience Canada, v. 5, No. 4, p. 189–199.

MACINTYRE, I. G., 1977, Distribution of Submarine Cements in a Modern Caribbean Fringing Reef, Galeta Point, Panama: Jour. Sed. Petrology, v. 47, p. 503–516.

————, BURKE, R. B., AND STUCKENRATH, R., 1977, Thickest Recorded Holocene Reef Section, Isla Perez Core Hole, Alacran Reef, Mexico: Geology, v. 5, p. 749–754.

MARAGOS, J. E., BAINES, G. B. K., AND BEVERIDGE, P. J., 1973, Tropical Cyclone Creates a New Land Formation of Funafuti Atoll: Science, v. 181, p. 1161–1164.

MAXWELL, W. G. H., 1968, Atlas of the Great Barrier Reef: Amsterdam, Elsevier, 258 p.

————, JELL, J. S., AND McKELLAR, R. G., 1964, Differentiation of Carbonate Sediments in the Heron Island Reef: Jour. Sed. Petrology, v. 34, p. 294–308.

MAZZULLO, S. J., AND CYS, J. M., 1977, Submarine cements in Permian boundstones and reef-associated rocks, Guadalupe Mountains, West Texas and southeastern New Mexico: Soc. Econ. Petrologists Mineralogists, Permian Basin Section, Mineral. Guidebook 77-16, v. 1, p. 151–200.

McLEAN, R. F., AND STODDART, D. R., 1978, Reef Island Sediments of the Northern Great Barrier Reef: Phil. Trans. Royal Soc. London A, v. 291, p. 101–117.

————, ————, HOPLEY, P., AND POLACH, H. A., 1978, Sea Level Change in the Holocene on the Northern Great Barrier Reef, *in* The Northern Great Barrier Reef: The Royal Society, p. 167–186.

MILLIMAN, J. D., AND EMERY, K. O., 1968, Sea Level During the Past 35,000 Years: Science, v. 162, p. 1121–1123.

MOORE, C. H., AND SHEDD, W. W., 1977, Effective Rates of Sponge Bioerosion as a Function of Carbonate Production: Third International Coral Reef Symposium, Proc., v. 2. (Geology), p. 499–505.

MOUNTJOY, E. W., AND WALLS, R. A., 1977, Some Examples of Early Submarine Cements from Devonian Buildups of Alberta: Third International Coral Reef Symposium, Proc., v. 2. (Geology), p. 155–161.

NEUMANN, A. C., J. W. KOFOED, AND G. H. KELLER, 1977, Lithoherms in the Straits of Florida: Geology, v. 5, p. 4–10.

NEWELL, N. D., 1972, The Evolution of Reefs: Scientific American v. 226, No. 6, p. 54–65.

————, RIGBY, J. K., FISCHER, A. G., WHITEMAN, A. J., HICKOX, J. E., AND BRADLEY, J. S., 1953, The Permian Reef Complex of the Guadalupe Mountains Region, Texas and New Mexico: San Francisco, W. H. Freeman, 236 p.

ORME, G. R., FLOOD, P. G., AND EWART, A., 1974, An Investigation of the Sediments and Physiography of Lady

Musgrave Reef—a Preliminary Account: Second International Coral Reef Symposium, Great Barrier Reef Committee, Brisbane, 2, p. 371–386.

———, ———, AND SARGEANT, G. E. G., 1978a, Sedimentation Trends in the Lee of Outer (Ribbon) Reefs, Northern Region of the Great Barrier Reef Province: Phil. Trans. Royal Soc. London A, v. 291, p. 85–99.

———, WEBB, J. P., KELLAND, N. J., AND SARGEANT, G. E. G., 1978b, Aspects of the Geological History and Structure of the Northern Great Barrier Reef: Phil. Trans. Royal Soc. London A, v. 291, p. 23–35.

PERKINS, R. D., AND ENOS, P., 1968, Hurricane Betsy in the Florida-Bahama area—Geologic Effects and Comparison with Hurricane Donna: Jour. Geology, v. 76, p. 710–717.

PURDY, E. G., 1974, Reef Configurations: Cause and Effect, *in* Laporte, L. F., Ed., Reefs in Time and Space: Soc. Econ. Paleontologists Mineralogists Spec. Pub. No. 18, p. 9–76.

RUTTEN, M. G., AND JANSONIUS, J., 1956, The Jurassic Reefs on the Yonne (Southeastern Paris Basin): Am. Jour. Sci., v. 254, p. 363–371.

SACHET, M., AND DAHL, A. L., 1974, Comparative Investigations of Tropical Reef Ecosystem: Background for an Integrated Coral Reef Program: Atoll Res. Bull., No. 172, 77 p.

SCHOLLE, P. A., AND KLING, S. A., 1972, Southern British Honduras: Lagoonal Coccolith Ooze: Jour. Sed. Petrology, v. 42, p. 195–204.

SCHROEDER, J. H., 1972, Fabrics and Sequences of Submarine Carbonate Cements in Holocene Bermuda Cup Reefs: Geol. Rundschau, v. 61, p. 708–730.

SCRUTTON, M. E., 1978, Modern Reefs in the West Java Sea: Indonesia Petroleum Assoc. Spec. Pub., Carbonate Seminar, Jakarta, September 12–19, 1976, Proc., p. 14–36.

STANTON, R. J., 1967, Factors Controlling Shape and Internal Facies Distribution of Organic Carbonate Buildups: Am. Assoc. Petroleum Geologists Bull., v. 51, p. 2462–2467.

STEARN, C. W., AND SCOFFIN, T. P., 1977, Carbonate Budget of a Fringing Reef, Barbados: Third International Coral Reef Symposium, v. 2 (Geology), Proc., p. 471–476.

STODDART, D. R., 1962, Catastrophic Storm Effects on the British Honduras Reefs and Cays: Nature, v. 196, No. 4854, p. 512–515.

———, 1963, Effects of Hurricane Hattie on the British Honduras Reefs and Cays, October 30–31, 1961: Atoll Res. Bull., v. 95, p. 1–142.

———, 1969a, Ecology and Morphology of Recent Coral Reefs: Biol. Rev., v. 44, p. 433–498.

———, 1969b, Post-Hurricane Changes in the British Honduras Reefs and Cays: Atoll Res. Bull., v. 131, p. 1–25.

———, 1978, The Great Barrier Reef and the Great Barrier Reef Expedition 1973: Phil. Trans. Royal Soc. London A, v. 291, p. 5–22.

WALKER, K. R., 1974, Reefs through Time: A Synoptic Review: *in* Principles of Benthic Community Analysis; Notes for a Short Course: Sedimenta IV, Comparative Sedimentology Laboratory, University of Miami, p. 8.1–8.20.

WARME, J. E., 1977, Carbonate Borers—Their Role in Reef Ecology and Preservation: Am. Assoc. Petroleum Geologists, Studies in Geology, No. 4, p. 261–280.

WASHBURN, J. R., JONES, A. V., AND JACKA, A. D., 1979, Deposition and Diagenesis of Chappel Limestone, Fort Worth Basin, Texas: Am. Assoc. Petroleum Geologists Bull., v. 63, p. 548.

WELLS, J. W., 1957, Coral Reefs: Geol. Soc. America Memoir No. 67, v. 1, p. 609–631.

WILSON, J. L., 1974, Characteristics of Carbonate Platform Margins: Am. Assoc. Petroleum Geologists Bull., v. 58, p. 810–824.

———, 1975, Carbonate Facies in Geologic History: New York, Springer-Verlag, 471 p.

SEPM SPECIAL PUBLICATION No. 30, P. 41–83, MAY 1981

# COMPOSITION, STRUCTURE AND ENVIRONMENTAL SETTING OF SILURIAN BIOHERMS AND BIOSTROMES IN NORTHERN EUROPE

ROBERT RIDING
Department of Geology
University College
Cardiff, United Kingdom

## ABSTRACT

Silurian reefs of northern Europe occur in cratonic sedimentary sequences which have been relatively well documented stratigraphically and paleontologically although the reefs have generally been less closely examined than the level bottom communities. Reef development adjacent to the Caledonian Belt was restricted by siliciclastic sedimentation and this is also reflected in the low proportions of carbonate rocks in the Welsh Borderland and Oslo successions. Marine sequences in the Baltic areas of Gotland and Estonia are relatively thin and contain much higher proportions of both carbonates and reefs.

Four main types of reef are recognizable on Gotland. Axelsro (previously termed Upper Visby) and Hoburgen Reefs are essentially tabulate coral and stromatoporoid dominated bioherms of moderate to high diversity. Their dense structure and argillaceous matrix made them locally unstable and prone to marginal collapse and internal displacement. Important accessory reef builders include rugose corals, calcareous algae, *Problematica*, and bryozoans. Similar bioherms, particularly of the smaller tabulate rich Axelsro type, are well developed in the Wenlock Limestone of the Welsh Borderland of England where good examples occur at Wenlock Edge. They are also present in the Oslo Region of Norway, together with tabulate dominated biostromes and *Rothpletzella-Wetheredella* bioherms, and occur at several horizons, sometimes very extensively, in the Llandovery and Wenlock.

In Gotland, Hoburgen reefs are especially widespread at numerous horizons, but a unique feature is the occurrence of Kuppen and Holmhällar type stromatoporoid biostromes which have rigid dense to frame structures and relatively low diversity. They are interpreted as shallow water, high energy linear reefs which developed preferentially in the cratonic interior.

The Estonian Silurian sequence shows close similarities to that in Gotland. Reefs are developed at a number of horizons but are generally little documented in detail.

Reef geometry and organic composition were controlled by environmental factors. The size and morphology of the organisms in turn determined the internal structure of the reefs. The bioherms show internal displacement and differential compaction of adjacent sediments in response to their own weight and to subsequent overburden. The biostromes behaved more rigidly and compaction was taken up mainly by stylolitization of adjacent large skeletons.

## INTRODUCTION

The Silurian successions of Britain, Scandinavia and the Baltic contain organic reefs which are among the best examples of their age in the world. They are constructed primarily by stromatoporoids and tabulate corals, together with calcareous algae, bryozoans and rugose corals, and are well developed in four areas: 1. Welsh Borderland of England, 2. Oslo Region of southern Norway, 3. Swedish island of Gotland in the Baltic Sea, and 4. western Estonia, particularly the islands of Hiiumaa and Saaremaa (Fig. 1). Most of these reefs are biohermal but biostromes also occur, some of which may represent fringing or barrier-reefs.

These areas have attracted geological interest for up to 150 years and all have been the objects of detailed studies: by Murchison (1839) on the Welsh Borderland, Kiaer (1908) on the Oslo Region, Hede (1925–1940) on Gotland, and Kaljo and his co-workers (1970, 1977) on Estonia and the subsurface of Latvia and Lithuania. The occurrence of reef like bodies in these rocks has long been recognized, even though their origin was at first uncertain. As early as 1914 Crosfield and Johnston had noted their similarity at widely separated locations and were comparing examples of them in England, Norway, Gotland and North America.

Despite these studies, and their seminal influence on subsequent research, much still remains to be learned about these bioherms and biostromes and their settings. Significant stratigraphic problems remain in the Oslo and Gotland areas, and the reefs of the Oslo Region have hardly been described. But the intensive stratigraphic and paleontologic documentation which has been carried out, coupled with detailed studies of reefs at Wenlock Edge in England and on the island of Gotland, provide strong incentive for further paleoecologic and sedimentologic analyses of the reefs and their associated facies.

41

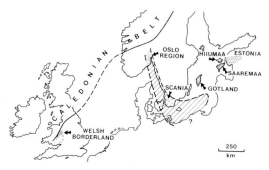

FIG. 1.—Silurian outcrops (stippled) and localities in northern Europe outside the Caledonian mobile belt and the position of the Upper Silurian Oslo-Scania-Baltic Basin (syneclise, cross-hatched).

The Silurian sediments of the Anglo-Baltic area accumulated on the European craton during flooding prior to the conclusion of the Caledonian orogenic cycle. At the continental margin, in western Britain and Norway, deeper water sediments were deposited and this linear belt developed into a zone of more rapid sedimentation and then deformation as the orogeny progressed. The essentially cratonic situation of each of the four main reef bearing areas is reflected by the relative thinness of their rock sequences and their lack of deformation compared to the continental margin, but they also show some distinct differences produced by their situations relative to the craton margin and interior. The reefs of the Welsh Borderland and Oslo regions formed close to the shelf-edge along the southeastern margin of the Iapetus Ocean and their development, particularly in southern Norway, was restricted by siliciclastic sedimentation from adjacent areas of uplift. In contrast, the Baltic area, which was distant from the mobile belt, shows a wider range of reef types over a longer period of development.

All four areas contain a diverse and well preserved shelf biota in the pale buff to green-grey shales and argillaceous limestones commonly associated with the reefs. It is the excellent preservation of these rocks and their enclosed fossils which enabled Roderick Murchison during the 1830's and 1840's to establish within them the classic sections of the Silurian System. Although Murchison noted the presence of the reefs in Britain and Scandinavia during his extensive travels he at first mistook their nature and regarded them as concretionary bodies. It was Charles Lyell, in 1841, who first recognized their organic nature at Wenlock Edge in England, at about the same time that James Hall was comparing the newly discovered Niagaran reefs of the Great Lakes region with Recent coral atolls.

More than a century of subsequent research has refined and extended Murchison's work, establishing the stratigraphic framework of the Silurian of northern Europe and, in the process, documenting the geological setting of the enclosed reefs in great detail. However, this emphasis on stratigraphy and stratigraphic paleontology has not focussed much attention on the reefs themselves. Although a number of valuable studies have elucidated details of some of the reefs, many aspects of their composition, distribution and origin remain unclear. Even since the upswing in interest in carbonates and paleoecology in the early 1960's few studies have specifically dealt with the reefs in detail. Silurian paleoecologic studies of the past 10–15 years have concentrated upon level bottom marine communities and even Manten's (1971) extensive documentation of the Gotland reefs deals primarily with the form and distribution of the reefs rather than with their internal composition and structure.

For these reasons a mere review of the published accounts of the bioherms and biostromes would be of limited value. For this paper I have attempted to go further than previous workers in drawing comparisons between the Welsh Borderland, Oslo and Baltic areas, and also in providing details of reef structure and composition, particularly on Gotland. But the reefs are extensive, the subject complex and most of my contribution is only preliminary to further studies. In Britain and Norway, in particular, I have added little to previous work. Considering its relatively small extent the Wenlock Edge area is the most intensively documented site so far as bioherms are concerned and I have drawn mainly upon the studies by Terence Scoffin and Brian Abbott. The reefs of the Oslo Region are much less well known and I have relied heavily in this account upon information kindly supplied by Nils Martin Hanken, Snorre Olaüssen and David Worsley who are currently working in this area. Gotland contains the finest and most extensive development of Silurian reefs in Europe, and probably in the world, and I have concentrated most attention upon it in this review. Estonia is the one area which I have not visted and I have drawn information solely from the recent volumes edited by Dmitri Kaljo.

The term *reef* continues to convey so many different things to different people that I should say here that I am using it as a general term for any essentially in-place accumulation of carbonate skeletons and associated organically localized material. Much discussion of the nature of reefs has revolved around the problems of interpreting the original environment and strength of the structures, and has proceeded from attempts to discriminate between them on the basis of criteria which are thoroughly subjective. Reefs are com-

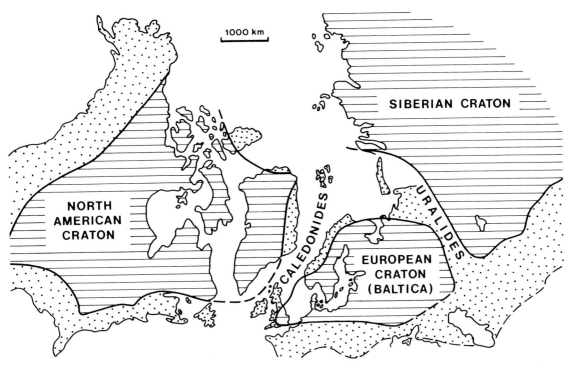

FIG. 2.—Regional setting of the northern European area during the Silurian (after Read and Watson, 1975, fig. 2.1; Ziegler *et al.*, 1977, fig. 2).

positionally and structurally diverse, but they nevertheless share a group identity. Here I attempt to describe the main skeletal components of all the reefs mentioned and to indicate wherever possible their internal structure in simple terms (dense, frame, *etc.*; Riding, 1977). Stromatoporoid morphotypes follow Kershaw and Riding (1978).

### REGIONAL SETTING

Current reconstructions of the northern European region during the Silurian show a triangular continent situated between North America and Siberia and lying close to the equator. It was separated from these continents by two narrow linear seaways to the west and east (Ziegler *et al.*, 1977; fig. 2) whose margins represented the Caledonian and Uralian mobile belts respectively (Fig. 2). This European continent (Baltica of Ziegler *et al.*, 1977) appears to have been a single cratonic unit, although during the latter part of the Silurian a linear basin (the Oslo-Scania-Baltic Syneclise) developed between the Oslo and southern Baltic areas, passing through southern Sweden and possibly linking with the Rheic Ocean to the south (Størmer, 1967, fig. 22).

Late Ordovician regression, perhaps related to glaciation, exposed the continent but it was pro-

gressively flooded during the early Silurian and this produced extensive epicontinental shelf seas, with embayments and islands, which covered much of the western surface (Fig. 1). Land remained to the east, over much of what is now Russia almost as far as the Urals, probably throughout the period (Kaljo, 1972).

A mixed carbonate-siliciclastic sequence accumulated on the shelf and is generally less than 1 kilometer thick, except in the Oslo-southern Baltic Trough where it reaches 2 kilometers. The proportions of limestone, mudstone-shale, and sandstone-conglomerate in the sequence vary with position on the shelf (Fig. 3). The Oslo-southern Baltic Trough succession is dominated by mudstones and shales but in Gotland and Estonia, 300–500 kilometers to the northeast, the Llandovery-Ludlow sequence is only 0.5 kilometer thick and carbonate rocks represent 50–75 percent of the total succession. These Baltic areas were in the cratonic interior, remote from sources of siliciclastic detritus. There are only a few thin sandstone units in the Gotland succession, but the proximity of the area to land at this time is emphasized by the presence of disconformities in the sequence, which often truncate reef bodies and result in a thin sequence which totals only approximately 450 meters from upper Llandovery

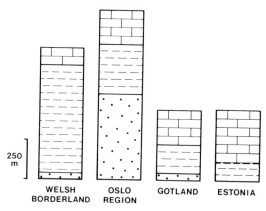

FIG. 3.—Thicknesses of Llandovery-Ludlow sequences in the four areas of reef development showing the proportions of sandstone, shale and limestone/dolostone in each. Note the thinness of the Gotland-Estonia sequences and the relatively high proportions of carbonate rocks they contain.

to upper Ludlow. Both the Gotland and Estonian sequences were deposited in a broad embayment with the Oslo-Baltic Trough to the south and land to the northwest, north and east (Fig. 1). The fineness of the siliciclastic sediment indicates that runoff was low, probably due in part to reduced relief on the land area.

In contrast, the Welsh Borderland and Oslo region have Llandovery-Ludlow sequences approximately 1 kilometer thick in which carbonates represent less than 20 percent of the succession. These western areas were close to the shelf-edge and to the mobile belt and carbonate deposition was constrained by either siliciclastic influx or deep water, or both. The Oslo region experienced influx of coarse siliciclastic sediment reflecting periods of uplift to the northwest (Størmer, 1967, p. 204–207) caused by gradual closure of the Iapetus Ocean. The most important of these were during the Llandovery and the Ludlow. Sandstones and conglomerates comprise approximately 50 percent of the Llandovery-Ludlow sequence which attains a thickness of approximately 1100 meters near Oslo. The Welsh Borderland area was also close to the continental margin but it did not receive coarse sediment influxes from the mobile belt until the Downton. Prior to that a dominantly shale-siltstone sequence about 850 meters thick accumulated in which limestones represent approximately 13 percent and sandstones only 5 percent of the total.

These differences in regional position and sedimentation are directly reflected in the degree and duration of reef development. Reefs in the Welsh Borderland only occur at a single horizon, in the

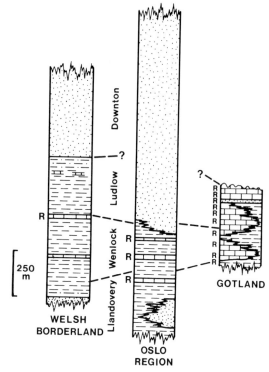

FIG. 4.—Schematic Silurian sequences in northern Europe showing how reef horizons (R) are distributed in time and space relative to lithofacies.

upper Wenlock. In the Oslo region they occur at several levels in the Llandovery and Wenlock, and in Gotland and Estonia there are examples of reefs in the Llandovery, Wenlock and Ludlow (Fig. 4).

In keeping with its intra-cratonic situation the Baltic succession is virtually unaffected structurally. The regional dip on Gotland is only one or two degrees and reef-bearing units can be traced continuously for up to 60 kilometers in coastal cliffs. The Welsh Borderland sequence shows more tectonic influence but is still well preserved and structurally simple. At Wenlock Edge the strata are tilted at 10 degrees to the southeast, and at several other locations in the area reefs occur in small faulted and upfolded areas influenced by both Caledonian (Siluro-Devonian) and Hercynian (Carbo-Permian) deformation. The Oslo Region is the most complex of the areas due to Caledonian folding combined with widespread normal faulting related to extensional rifting of the area during the Late Paleozoic. These effects hamper correlation but the rocks are nonetheless relatively well preserved.

When Murchison visited the Oslo Region in 1841 he assumed that the reef limestones there were of the same age as those in the Welsh Borderland and, similarly, that the non-marine sandstones were of Downton age. Aspects of this error of equating facies similarity with time equivalence have persisted to the present in some areas. Workers in Scandinavia have tended to adhere to long established stratigraphic divisions which combine features of both lithostratigraphic and biostratigraphic units. These divisions, termed topostratigraphic units have been employed in the Oslo and Gotland sequences and possess a broad basic utility. But they necessarily suffer fundamental weaknesses of lithologic heterogeneity on the one hand and diachronism on the other, and it will remain difficult to confidently assess the detailed paleoenvironmental setting of reefs in parts of these successions until they are revised.

Attempts to analyze the patterns of relative sea level movements during the Silurian in northern Europe have confirmed basic trends but have also led to differences in opinion concerning both the timing and cause of smaller scale events. The basic pattern in all four areas is a Llandovery transgression and a regression during the Ludlow. These appear to have been prolonged eustatic changes in sea level probably related to glaciation or global tectonism, or both (McKerrow, 1979, p. 142). In the Welsh Borderland sea level fall during the late Wenlock promoted reef development in the Wenlock Limestone and this was terminated by renewed deepening during the basal Ludlow. Hurst (1975) and McKerrow (1979, fig. 1) recognize a basal Ludlow transgression as a general feature over northern Europe, but Bassett (1976) argues against this view on the grounds that "many sequences in the Anglo-Baltic area suggest widespread regression and not transgression at this level (e.g. Pembrokeshire, Norway, Estonia)" (Bassett, 1979, p. 144). He believes that the deepening at the Wenlock-Ludlow boundary in the Welsh Borderland is a local, tectonically controlled effect. There is no doubt that reef development in all these areas is closely related to water depth and rate of siliciclastic sedimentation and distinct phases of carbonate deposition and reef formation characterize the Baltic sequences. But the degree of correlation between these has still to be established and this is dependent largely upon clarification and correlation of stratigraphic units in Scandinavia and the Baltic.

Northern Europe during the Silurian belonged to the North Silurian faunal realm (Boucot, 1974). Although provincialism gradually increased during the period the Silurian was generally a time of marked faunal homogeneity. Probably this was strongly influenced by global climate and the widespread epeiric seas of the time. Northern Europe (Baltica) occupied an equatorial situation and Ziegler *et al.* (1977, p. 40–41) deduce a trop-

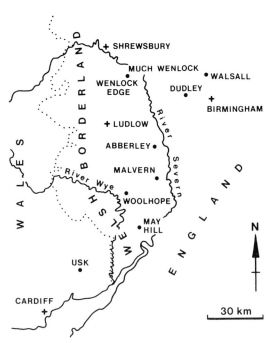

FIG. 5.—Reef localities in the Welsh Borderland.

ical, mainly moist climate for Baltica during the Silurian. Gradual regression towards the end of the period enhanced provincialism and also stimulated reef development in areas remote from sources of siliciclastic sediment. Continental conditions were established over the whole area during the late Silurian.

### Welsh Borderland

The area lying between the English Midlands and Wales is one of the classic regions of British geology. It is dominated by Middle Paleozoic sedimentary rocks and represents the eastern part of the district selected by Murchison (1839) as the type area for the Silurian System. Reefs occur at only one horizon: in a relatively thin carbonate unit, at or near the top of the Wenlock Series, which is known generally as the Wenlock Limestone. Barrier-reefs have been suggested to occur in the area of Wenlock Edge (Symmonds, 1872), but present exposures indicate that only patch-reefs occur. These have been referred to by quarrymen as "ballstones" or "crog-balls" in reference to their massive and irregularly lensoid form, and they have been preferentially mined because of their purity relative to the enclosing argillaceous limestones.

The Wenlock Limestone is never more than 150 meters thick and often, as at Wenlock Edge, it is only 20–30 meters, but it has a wide occurrence over the triangular tract of country between

FIG. 6.—Aerial view to southwest along Wenlock Edge from near Much Wenlock. The line of the escarpment is offset by faults. Photograph from Cambridge University Collection, copyright reserved.

Shrewsbury, Birmingham, and Cardiff and, although it is often very argillaceous, bioherms are known to occur within it at Wenlock, Dudley, Walsall, Abberley, Malvern, Woolhope, May Hill, and Usk (Fig. 5). These localities can all be regarded as being in the Welsh Borderland area except for Dudley and Walsall which are part of the West Midlands conurbation around Birmingham.

The Wenlock Limestone is sandwiched between thick shale-siltstone units; the Wenlock Shale below and the Elton Beds, of Ludlow age, above. These sediments comprise the middle part of a Silurian shelf sequence and were deposited relatively uniformly over the whole region following transgression from the west during the Llandovery (Ziegler, 1970, p. 456–459). The Llandovery-Ludlow sequence on the shelf is approximately 850 meters thick. During the early Wenlock land lay 100–150 kilometers to the southeast and south of the Wenlock Edge area, with basinal facies marking the shelf-edge only 20 kilometers to the west (Bassett, 1974, fig. 7). Siltstones, calcareous shales and thin nodular limestones, constituting the lower part of the Wenlock Shale, were deposited in the northwestern part of the region, including Wenlock Edge, Dudley and Walsall.

Towards the southeast argillaceous and bioclastic non-reef limestones (Woolhope Limestone) up to 60 meters thick were formed. In late Wenlock times limestone deposition with reefs (Wenlock Limestone) replaced argillaceous sedimentation over the entire region east of Usk, Ludlow and Wenlock. The basin margin remained in much the same position as earlier but continued transgression had extended the shelf sea southeastwards. Reef development was greatest between Wenlock and Dudley in the northern part of the area. Fewer, smaller bioherms formed in the Usk-Malvern area which was situated closer to the southern land mass. The Elton Beds, which conformably overlie the Wenlock Limestone at all localities where bioherms occur, mark a return to fine-grained siliciclastic shelf sedimentation. In this context the Wenlock Limestone represents a pause in siliciclastic sedimentation and shallowing of the shelf probably related to a regression (Scoffin, 1971, p. 212), followed by continued deepening during the early Ludlow (Hurst, 1975; Bassett, 1976). Major regression during the late Ludlow was followed by the start of Old Red Sandstone molasse sedimentation during the Downton (Pridoli).

At Wenlock the strata dip 10 degrees to the southeast and the Wenlock Limestone forms an escarpment, Wenlock Edge, between the River Severn and River Onny. The other bioherm localities, which are all to the east and south, occur in small upfolded or upfaulted inliers. Intensive quarrying and mining of the limestone at Dudley, although now discontinued, has left good exposures, particularly at Wren's Nest Hill. Wenlock Edge, 35 kilometers to the west, continues to be the site of extensive quarrying. Since these two localities are also in areas of good reef development they are now the best sites for examining the bioherms, and Wenlock is easily the better of the two, a fact indicated by the concentration of reef studies there in recent years. At present most outcrops further south in the Welsh Borderland are poor exposures in long disused quarries.

*Wenlock Edge.*—The outcrop formed by the Wenlock Limestone near Much Wenlock in Shropshire (where it has been renamed the Much Wenlock Limestone Formation by Bassett *et al.*, 1975) is generally less than 1 kilometer wide and it forms the 30 kilometers long Wenlock Edge (Fig. 6). It thickens northeastwards from 21 meters near the River Onny to 29 meters near Much Wenlock. Bioherm development is restricted to the area northeast of Easthope (Fig. 7) and consists of numerous relatively small patch-reefs (Fig. 8). Murchison (1833) described the succession in this area and noted "beds in Wenlock Edge contain many concretions of very great size and of highly crystalline structure" (p. 475).

In 1839 Murchison published his famous trea-

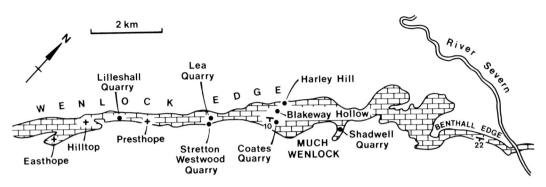

FIG. 7.—Outcrop of the Wenlock Limestone along the northeastern part of Wenlock Edge, England (after Bassett *et al.*, 1975, fig. 8).

tise on "The Silurian System" and added more details about the Wenlock Limestone. He regarded the massive irregular "ballstones," enclosed by argillaceous limestones and shales, as concretions formed by solidification or crystallization and noted that "where these concretions are prevalent, the strata undulate or are contorted." He was impressed by their size: "In one quarry recently opened (called the Yell), the small concretionary beds dip on the one side to the north, on the other to the southeast, and the center of the hill consists of one massive ballstone, which when I visited the spot, was laid bare to a depth of about *eighty feet!* the surface alone being covered with a few thin nodular beds. The exact width of this mass of ballstone had not been proved when I last visited these quarries, but it is doubtless very great, and must be of considerable value on account of its superior quality and accessible position" (p. 211).

Prestwich (1840) supported the concretionary hypothesis but Lyell (1841) noted the corals which they contained and became the first person to interpret them as organic structures. The corals and stromatoporoids of these bioherms were described by Edwards and Haime (1855) and Nicholson (1886–1890) respectively. Crosfield and Johnston (1914) give an excellent account of the "ballstones" at Wenlock Edge. They emphasize that "over 90 percent of the corals and stromatoporoids are found to be in position of growth in ballstone, while only 16 percent are in position of growth in the adjacent stratified rocks" and conclude, "Ballstone rock is the relic in place of large coral and stromatoporoid colonies still in the positions in which they originally grew" (p. 221). Hill (1936) briefly comments on the bioherms, which by then were generally regarded as coral reefs. The geology of the area has been described by Davidson and Maw (1881), Pocock *et al.* (1938), Whittard (1952), Greig *et al.* (1968), Holland *et al.* (1969), Hains (1970), Shergold and Bassett (1970), and Bassett *et al.* (1975). Several unpublished theses have dealt specifically with the paleoecology and sedimentology of the reefs at Wenlock Edge (Colter, 1957; Scoffin, 1965; Abbott, 1974) and have resulted in publications by Scoffin (1971, 1972), and Abbott (1976).

Southwest from Benthall Edge bioherms increase in size and number towards Easthope (Fig. 9). Scoffin (1971, fig. 1) gives an impression of barrier-type development at Hilltop where 80 percent of the upper 17 meters of the Wenlock Limestone consists of reef rock, compared with 25 percent near Much Wenlock and 10 percent at Benthall Edge. Bioherms do occur 1 kilometer southwest of Hilltop (Greig *et al.*, 1968, p. 179) but they are very small, less than one meter across, and there does appear to be a dramatic reduction in reef development near Easthope.

Shergold and Bassett (1970, p. 118–125) and Scoffin (1971, p. 180–183) distinguished a number of off-reef and inter-reef facies in the limestone, most of which are based upon rock types recognized by the quarrymen. They are essentially variations on nodular argillaceous limestones and crinoidal limestones and are illustrated by Shergold and Bassett (1970, figs. 4–13). The inter-reef beds show a coarsening upward sequence through the following lithofacies (Fig. 9):

A. thin-bedded nodular limestone-shale alternations, quarrymen's "Bluestone"; and
B. nodular shaly limestones, quarrymen's "Jack's Soap"; the limestones are argillaceous crinoid wackestones in both these lithofacies. B is a more shaly variety of A;
C. medium-bedded crinoid limestones, quarrymen's "Measures"; crinoid packstone-wackestones;
D. coarse thin-bedded crinoid limestone, quarrymen's "Gingerbread"; crinoid grainstones.

Grain-size, grain-support and sparry calcite increase upwards generally through these inter-reef facies, but lateral and vertical distribution of them is nevertheless quite variable. Facies A shows a

FIG. 8.—Small lensoid bioherm in the Wenlock Limestone, 3 kilometers northeast of Much Wenlock, England.

gradational contact with the underlying Wenlock Shale, it dominates the inter-reef sequence and encloses lentils of facies B. Facies C and D, which are crinoid limestones, only occupy the upper 7 meters of the 29 meters Wenlock Limestone sequence. Reefs occur most commonly between 3 and 10 meters from the top, e.g., in association with facies C and the upper part of A (Scoffin, 1971, p. 183). Bioherms are rare in the lower part of facies A and in facies D (Shergold and Bassett, 1970, p. 121).

Clay content and nodularity of the limestones increase down sequence in the inter-reef and distally in the off-reef. Off-reef facies occur southwest of Hilltop, and Scoffin (1971, p. 182–183) recognizes two main types:

E. thin-bedded fine bioclastic grainstones adjacent to the termination of major reef development near Hilltop and Easthope;

F. thin-bedded limestone-shale alternations, in which the limestones are argillaceous wackestone-mudstones, in the more distal off-reef.

Shergold and Bassett (1970, p. 118–119) subdivide facies F into lower "tabular limestone" and upper "nodular limestone" units (Fig. 9). There is a thin but extensive basinward progradation of facies D (Scoffin, 1971, p. 183) which Shergold and Bassett (1970, p. 119–120) name "crinoidal limestone" at the top of the sequence southwest of Hilltop. Above facies D there is a relatively rapid transition to green-brown siltstones of the Elton Beds.

There are numerous quarry exposures of the upper half of the limestone and the enclosed bioherms along the northeastern part of Wenlock Edge and its extension as Benthall Edge. The most important of these are shown in Figure 7. The disused quarry faces tend to be dirty and those of the working quarries change rapidly, so it is difficult to provide locality details of clean faces which will remain unchanged. Lea, Coates and Shadwell are large working quarries and Lilleshall is one of the largest disused quarries. Facies D, the "gingerbread" bioclastic gravels overlying the reefs, are well seen at Coates Quarry and Blakeway Hollow. The underlying Wenlock Shales (Coalbrookdale Formation) are well exposed in road cuts at Harley Hill, 1.5 kilometers north-northwest of Much Wenlock on the Shrewsbury Road, and the overlying Elton Beds are seen at Shadwell and Stretton Westwood Quarries (see Shergold and Bassett, 1970, fig. 14). The bioherms are described in detail by Scoffin (1971) and much of the following information is drawn from his work.

The biohermal limestone is a tabulate coral-stromatoporoid-algal dense to frame structure with bryozoans, crinoid fragments, and occasion-

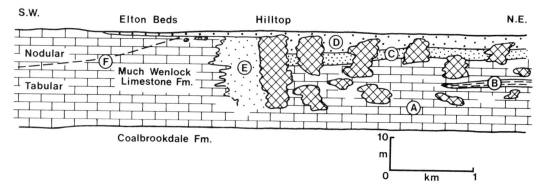

FIG. 9.—Lithofacies in the Much Wenlock Limestone Formation (Wenlock Limestone) around Hilltop on Wenlock Edge, England. A. crinoid biomicrite ("bluestone"); B. biomicrite ("Jack's Soap"); C. crinoid biomicrite ("measures"); D. crinoid biosparite ("gingerbread"); E. biosparite; F. fossiliferous micrite. Bioherms are crosshatched; after Scoffin (1971, fig. 2) and Shergold and Bassett (1970, p. 119–120).

FIG. 10.—Bioherm margin, showing interfingering contact with bedded limestone at Coates Quarry, Wenlock Edge; hammer is 28 centimeters long.

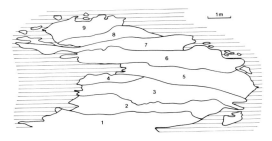

FIG. 11.—Cross-sectional plan of an Axelsro type bioherm at Coates Quarry, Wenlock Edge, showing growth stages marked by clay seams; from Abbott (1976, fig. 6), also see Scoffin (1971, fig. 17).

al pockets of brachiopods and bivalves. It forms highly irregular lensoid shapes in vertical sections, with numerous lateral outgrowths and invaginations (Fig. 10). The bioherms average 12 meters across and 4.5 meters thick and are largest at Hilltop where they reach 100 meters wide and 20 meters thick (Scoffin, 1971, p. 183). The distribution of reefs vertically and laterally shows no clear pattern, other than an increase upwards in the sequence and southwest towards Hilltop, but Scoffin (1971, p. 183–186, fig. 8a) notes a preferred northeast-southwest extension of growth.

Many of the bioherms rest upon basal lenses of coarse skeletal debris consisting of crinoids, bryozoans, brachiopods and corals (Scoffin, 1971, p. 188–189) which forms a well washed grainstone (Abbott, 1976, p. 2120). Once reef development was established growth appears to have been upward and outward from numerous loci, resulting in a succession of irregular overlapping lenses which are often separated by thin layers of green micrite and grey shale (Fig. 11). This coarse and irregular nodularity led to the term "ballstone" and is a characteristic feature of the internal structure of the reefs, referred to by Crosfield and Johnston (1914, p. 199): "the *ballstone* masses are generally ovoid and lenticular, and these are themselves most frequently composed of smaller, lenticular or phacoidal masses, which fit closely together."

Clay sedimentation appears to have profoundly influenced reef development and noticeably affects the lateral profile of bioherms: "generally indentations in the margins of the reefs correlate with clay-rich zones in the adjacent stratified limestone. This is most obvious in the case of bentonites" (Scoffin, 1971, p. 190). This control of reef growth by the rate of fine sediment deposition also extends to the formation of internal subdivisions within them; veneers of shale and micrite locally extinguish growth within a reef and provide new surfaces for lateral colonization. In this way the lenticular "ballstone" structure was built up by loose sediment interfering with organic growth. The resulting clay seams have been used as time planes to reconstruct the sequence of bioherm development (Scoffin, 1971, p. 196; Abbott, 1976, p. 2122).

Contacts with inter-reef facies are sharp and occasional talus wedges of coarse debris extend outwards from the reefs. Rarely crinoid beds are also found thinning away from reefs. The upper surfaces of bioherms are normally convex and their relief appears to have been relatively low, ranging from 0.5 meter in reefs 5 meters across to a maximum of 3 meters in the largest reefs. "The growth surface was horizontal to slightly convex, with irregular hollows, and sloped at a gentle angle (about 15 degrees) down to the reef margins which had low reliefs and were regularly swamped by loose sediment" (Scoffin, 1971, p. 216).

Tabulate corals dominate the reefs, forming 10–20 percent of the volume (Scoffin, 1971, p. 198–202); and Abbott (1976, p. 2120) describes a vertical succession within individual bioherms from *Halysites,* through *Heliolites* and *Favosites* to stromatoporoid dominated communities reflecting a gradual increase in water movement. The principal stromatoporoids in the bioherms are domical *Actinostroma* and *Stromatopora,* and laminar *Stromatopora* and *Labechia.* Bryozoans (*Hallopora, Rhombopora, Fistulipora,* and *Fenestella*) are also abundant and rugose corals (*Entelophyllum*) are common. Laminar tabulates (*Alveolites* and *Thecia*) are also present. Non-skeletal stromatolites, occasionally associated with the problematic micro-organisms *Rothpletzella* and *Wetheredella,* occur as irregular crusts a few millimeters to a few centimeters thick lining reef cavities and veneering skeletons and sediment. Associated organisms are crinoids, brachiopods, ostracodes, foraminifers, gastropods and trilobites.

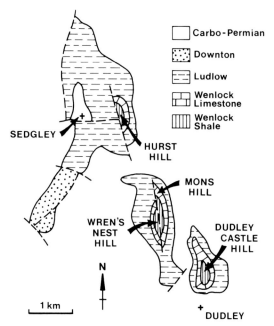

FIG. 12.—Silurian outcrops near Dudley, England.

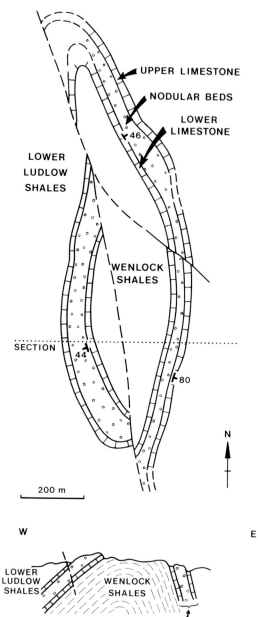

FIG. 13.—Map and section of Wren's Nest Hill pericline near Dudley, England (from Butler, 1939, pl. 3; Hamblin et al., 1978, fig. 1).

The principal organisms: non-laminar tabulates, stromatoporoids, rugosans and bryozoans form an intricate irregular framework in which most upper surfaces are encrusted by stromatolites while bryozoans commonly encrust undersurfaces. A large proportion of original cavity space was largely filled by biomicritic matrix penecontemporaneously with the growth of organisms (Scoffin, 1971, p. 209–210), indicating that the reef organisms had to cope with more or less continuous fine sediment deposition. Skeletal grainstones were unable to filter down into the dense frames and are localized in pockets and lenses in the reefs, probably representing open hollows in the reef surface (Scoffin, 1971, p. 209). The upper parts of cavities unfilled by biomicrite often show a layer of fine micrite, representing a late stage of internal sedimentation, followed by a final sparry calcite cement plug. Original cavities which were subsequently completely filled by particulate sediment as reef development proceeded can be recognized from downward growing encrusting organisms, such as *Fistulipora* bryozoans, beneath framework masses (Scoffin, 1972, fig. 2b). Most of the residual cavity which was ultimately sparite-filled is associated with *Halysites* palisades (Scoffin, 1972, p. 568; Abbott, 1976, p. 2123), but there are also some intra-skeletal voids.

The inherent weakness of these reefs is reflected by post-burial compaction features including: "rotation of the solid masses, the fracturing of skeletons, the squeezing and splitting of micrite

and the injection of clay minerals through new openings in the reef" (Scoffin, 1971, p. 203; also see Abbott, 1976, fig. 7). These enhance the original lenticular components of the reefs as Crosfield and Johnston (1914, p. 223) realized, although they attributed the movements to earth movements rather than differential compaction. Secondary porosity was also produced, on a limited scale, by preferential dissolution of bryozoan and coral skeletons usually on the ceilings of original growth cavities (Scoffin, 1972, p. 575).

Crosfield and Johnston (1914, p. 221) concluded that the reefs at Wenlock Edge grew in shallow but relatively quiet water, an interpretation supported by Scoffin (1971, p. 216) who also recognized evidence for extreme shallowing which ultimately terminated reef growth in the area.

*Dudley.*—Wenlock Limestone with reefs outcrops in three en echelon periclinal inliers between Dudley and Sedgley, 14 kilometers west-northwest of Birmingham (Fig. 12). These form small, but locally prominent, hills: Dudley Castle Hill, Wren's Nest Hill, with Mons Hill just to the north, and Hurst Hill. Intensive mining has created underground caverns at Castle Hill and the Wren's Nest, but these are now in a state of collapse. The area is famous for well preserved fossils from the limestone, especially trilobites such as *Calymene blumenbachi*, the "Dudley Locust."

No specific studies of the reefs have been made here, but the stratigraphy of the limestone is described by Butler (1939), and general details of the local geology are to be found in Murchison (1839, p. 483–487), Jukes (1859), Whitehead and Eastwood (1927), and Whitehead and Pocock (1947).

The Wenlock Limestone is approximately 60 meters thick and includes two relatively pure limestone units, the Lower and Upper Quarried Limestone, separated by shales and nodular limestones termed the Nodular Beds. Butler (1939) recognized the following units:

lower Ludlow Shale

|  |  |
|---|---|
| | 5. Passage Beds (1.3 m), nodular limestones and shales; |
| | 4. Upper Quarried Limestone (7.3 m), bedded bioclastic and stromatoporoid limestone; |
| Wenlock Limestone | 3. Nodular Beds (31.0 m), shales with nodular and thin bioclastic limestones; |
| | 2. Lower Quarried Limestone (12.8), bedded bioclastic stromatoporoid limestone; |
| | 1. Basement Beds (3.4 m), bioclastic limestones and shales. |

Wenlock Shales

At Dudley the areas of more massive and purer limestone were termed "crog-balls" by quarry-men. These occur most commonly in the Nodular Beds where they are generally 3–6 meters wide and 1.5–3 meters high and are reef masses dominated by tabulates, stromatoporoids, rugose corals and bryozoans with pockets rich in brachiopods (Butler, 1939, p. 448). "Crog-balls" also occur in the Lower and Upper Quarried Limestones but they are fewer and flatter and contain mainly crinoid fragments, tabulates, and bryozoans (*Coenites*), with no large rugosans or stromatoporoids. In a comparison between the Wenlock Edge "ballstones" and the Dudley "crog-balls" Croffield and Johnston (1914, p. 210–211) concluded that the Dudley "crogs" were not similar to Wenlock Edge "ballstone." However, they were comparing "ballstone" with "crog" material from the Lower Quarried Limestone. From Butler's description it seems clear that the "crog-balls" of the Nodular Beds are reefs comparable with those of Wenlock Edge, but that the "crog-balls" of the Quarried Limestones are flatter lenses of coarse bioclastic debris.

The best exposures at present are at Wren's Nest Hill which is now a nature reserve (Hamblin et al., 1978). The periclinal fold which brings up the Wenlock Limestone is asymmetric, dipping westwards at 50 degrees and eastwards at 70 degrees (Fig. 13). The Lower and Upper Quarried Limestones have been removed in deep trenches on the east side of the hill, and in a complex series of caverns on the west side where an underground canal was used to ship out the rock.

*Other Localities.*—Bioherms are known in the Wenlock Limestone from Walsall, Abberley and Malvern (Mitchell et al., 1961; Phipps and Reeve, 1967), Woolhope (Squirrell and Tucker, 1960), May Hill (Lawson, 1955), and Usk (Walmsley, 1959) (Fig. 5). At present, exposures at these areas are relatively poor and there has not been much specific emphasis of the bioherms except by Penn (1971) at Malvern.

### Oslo Region

Bioherms and biostromes are widely distributed at several levels in the Llandovery and Wenlock rocks of the Oslo Region (Kiaer, 1908) but have not yet been described in detail. They occur within a variable sequence of marine sediments (Bjørlykke, 1974, p. 24–31) which is influenced by influx of siliciclastic detritus from the north and northwest (Størmer, 1967, p. 204–207) reflecting the proximity of the Caledonian mobile belt. Marine transgression over this shelf area during the lower and middle Llandovery was followed by progradation of sand from the north during the upper Llandovery. Carbonate deposition, including reef development, was established three times, in the late Llandovery, lower Wenlock and middle Wenlock, and alternates with shale-mud-

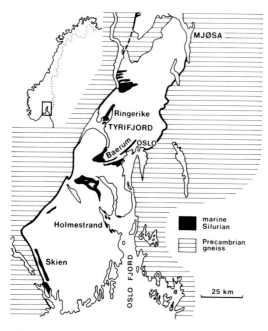

FIG. 14.—Distribution of marine Silurian sediments in the Oslo Region. Other rocks in the Oslo Graben are undifferentiated. Localities with reefs are named; from Dons and Larsen (1978, pl. 1). Details of the Ringerike area are shown in Figure 16.

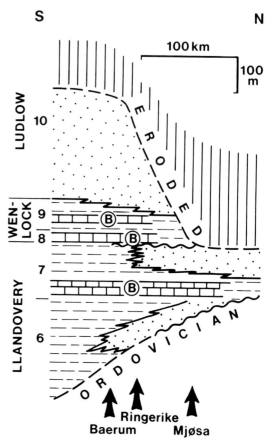

FIG. 15.—Position of bioherms and biostromes (B) in the broad sedimentary facies context of the Silurian of the northern Oslo region; scales very approximate, adapted from Bjørlykke (1974, figs. 11 and 12).

stone-sandstone sedimentation (Fig. 4). Marine deposition during the Llandovery and Wenlock resulted in the formation of 600 meters of rock and was terminated in the late Wenlock by the onset of continental Old Red Sandstone-type conditions.

The Oslo Region was folded fairly intensively during the Caledonian Orogeny but the main character of the area was created by intracratonic rifting associated with volcanicity and subsidence during the Lower Permian (Dons and Larsen, 1978). The result is that Lower Paleozoic sedimentary rocks are now preferentially preserved within the Oslo Graben where they are associated with a diverse suite of intrusive and extrusive Late Paleozoic rocks which cover more than 75 percent of the downfaulted area. The graben extends north-northeast from Oslo to Lake Mjøsa where it joins the Caledonian nappe area, and south-southeast into the Skagerrak extension of the North Sea (Fig. 14). This area is 200 kilometers long and 35–65 kilometers wide and is bounded on the west and east by Precambrian gneisses. Further information on the geology of the region is provided by Holtedahl (1960), Henningsmoen and Spjeldnaes (1960), and Seilacher and Meischner (1964). The Steinsfjord area of Ringerike, where good examples of the reefs occur, is described by Whitacker (1977).

A particular problem of the Oslo Silurian succession concerns lithostratigraphy and correlation. Kjerulf (1855), followed by Kiaer (1908) divided the local Cambro-Silurian sequence into ten units, termed stages. These were changed into series by Strand and Henningsmoen (1960). However, they do not correspond with either formal stages or series; instead they combine both lithostratigraphic and biostratigraphic characters and they exhibit diachronism and time equivalence (Bassett and Rickards, 1974). Broadly, "Stages" 6 and 7 represent the Llandovery, 8 and 9 the Wenlock, and 10 the Ludlow (Fig. 15). These have been applied to the whole area and have provided a convenient interim nomenclatural scheme, but they require revision. A new lithostratigraphical scheme with revised correlations is in preparation.

Silurian rocks with bioherms and biostromes outcrop in patches, usually no more than 15 kilometers across, from Ringerike in the north

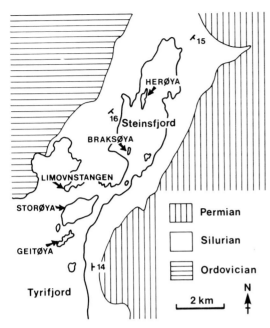

FIG. 16.—Localities, Ringerike district, Norway (after Whitacker, 1977, fig. 1).

FIG. 17.—Upper Llandovery stromatoporoid and tabulate coral dominated bioherm enclosed by bedded bioclastic limestone, Limovnstangen, Ringerike, Norway.

through Baerum, and Holmestrand to Skien in the south (Fig. 14). The following descriptions are based upon Kiaer (1908), with details from Hanken et al. (1978) which utilize a few of my own observations in Tyrifjord. I am indebted to Nils Martin Hanken, Snorre Olaüssen, and David Worsley for making information, on which most of this account depends, available from their work in progress.

Bioherms and biostromes occur at three general horizons in the Llandovery and Wenlock (Fig. 15):

1. upper Llandovery bioherms at Ringerike;
2. lower Wenlock bioherms at Ringerike and Skien;
3a. mid-Wenlock bioherms at Holmestrand and Skien;
  b. late Wenlock biostromes at Ringerike, Baerum, Holmestrand and Skien.

These areas were all in marginal marine situations during much of this time but at present it is difficult to place them more precisely in their environmental setting.

*Llandovery.*—Fronian (lower upper Llandovery) bioherms occur in Stage 7b at Ringerike and Baerum. They are small lensoid masses up to 6 meters across and 2 meters high dominated by halysitids and stromatoporoids with minor favositids, heliolitids and syringoporids. *Girvanella* and bryozoans may have had a binding effect. The matrix is microspar after micrite, with numerous

brachiopods, and rare bumastid trilobites, gastropods and cephalopods. They grow on biosparite banks composed of pentamerid and crinoid debris, and laterally show near-vertical inter-fingering contacts with well-bedded crinoid biosparites. The upper surfaces and overlying sediments suggest that growth stopped at a stage of further diversification as a result of a transgressive episode.

A good example of these bioherms occurs at the base of Stage 7b on Limovnstangen, a peninsula jutting south into Tyrifjord (Fig. 16). It contains laminar and domical stromatoporoids up to 50 cm across with favositids and abundant *Halysites*. There are sharp lateral contacts with bedded bioclastic limestone (Fig. 17).

*Lower Wenlock.*—Sheinwoodian (lower Wenlock) bioherms occur in Stage 8c at Ringerike, particularly on the islands in Steinsfjord and northeastern Tyrifjord. Kiaer (1908, p. 79–82, figs. 19 and 20) describes them at Geitøya and Braksøya (Fig. 16) and also draws attention to a horizon of large stromatoporoids a few meters

FIG. 18.—Stromatoporoid biostrome; lower Wenlock, west side of Geitøya, Ringerike, Norway.

higher in the sequence (p. 89, fig. 19). The bioherms are irregularly dispersed through a 20 meters thick sequence. Individual structures have
maximum diameters of 15 meters and heights of
8 meters, but there are also some very small bodies. Mound-structure appears to be dominated by
encrusting *Girvanella, Rothpletzella, Wetheredella* and *Halysis,* with subordinate halysitids.
Diversity is low and a few brachiopod species are
the only other organisms observed. Radial growth
appears to have taken place from several loci, and
the primary relief was probably low (?0.5 meter).
The upper surfaces are irregularly convex and the
flanks interfinger with bedded limestones. Inter-
reef beds contain a varied fauna of brachiopods
(*Dicoelosia, Skenidioides,* atrypids and rhynchonellids), large solitary corals (*Phaulactis*), and
stromatoporoids. Evaporite pseudomorphs occur
within the bioherms and desiccation cracks occur
in the beds immediately overlying them.

On the west side of the island of Geitøya in
Tyrifjord (Fig. 16) the bioherms are flat-based
coalescive mounds up to 5 meters thick overlying
calcareous siltstones. They have irregular upper
surfaces and are separated from the overlying
stromatoporoid horizon by a few meters of argillaceous limestone. The stromatoporoid bed is a
2 meters thick biostrome of laminar and domical
forms up to 80 centimeters across (Fig. 18).

*Middle-late Wenlock.*—Biostromes and bioherms of medial Wenlock age are developed in
the middle parts of the Steinsfjord Formation
(Worsley, per. comm., 1979) (Stage 9c) of Holmestrand and Skien. In Holmestrand, biostromes
are laterally persistent units 1–4 meters thick,
composed of tabulates (favositids, heliolitids,
"*Thecia*" sp. and syringoporids), rugose corals,
algae, bryozoans (*Coenites* sp.), stromatoporoids, brachiopods, molluscs and ostracodes.
Corals occur both in place and moved (see Kiaer,
1908, fig. 57). Kiaer noted these biostromal units
at three horizons, which he termed Korallenhori-
zonten I, II, and III.

Small bioherms occur locally in the lower part
of Korallenhorizont II and approximately 20–30
meters above Korallenhorizont III. The lower
bioherms are 2–3 meters in diameter and less than
0.5 meter high with relatively flat bases and tops.
Laterally they interfinger with bedded bryozoan
fragment limestones in which *Girvanella* and red
algae encrust the clasts. They are dominated by
the tabulate "*Thecia*" which grows in thin layers
with marginal tongues and sediment inclusions.
The rugosan *Acervularia* is occasionally present.

The upper bioherms occur in a sequence with
high faunal diversity. Kiaer (1908) identified approximately 80 species, not counting algae, *Problematica* and trace fossils. The bioherms are generally only 0.5–0.9 meter wide and 0.4–0.7 meter

high, but one much larger structure 25 meters
across and several meters high also occurs. The
framework consists of stromatoporoids, tabulates
("*Thecia*" and favositids), and colonial rugose
corals. Bryozoans, such as *Coenites,* may have
functioned as binders. The matrix includes both
micrite and sparite. Crinoid debris fills small
channels in the single large bioherm seen.

Biostromes of late Wenlock age occur in the
upper part of the Steinsfjord Formation (Stage 9f)
in the Ringerike, Baerum, Holmestrand and Skien
districts (Fig. 14), a distance of 150 kilometers. In
all exposures they occur approximately 30 meters
below the junction with the overlying red and grey
sandstones of the Ringerike Group. Despite their
wide lateral extent they are only 1 meter thick.
Their persistence is probably a result of a minor
regional transgressive episode. In Baerum, near
Oslo, the biostromes show slight local topography
and are almost biohermal in character. Kiaer
(1908) reports small stromatoporoids, *Favosites*
and *Monticulipora* as major biostrome components at good exposures on the western side of
Herøya in Steinsfjord, Ringerike (Fig. 16). *Amplexopora* also occurs, with brachiopods, cephalopods, gastropods, rugose corals, and *Girvanella*
oncolites, but the overwhelming dominance of
favositids gives the biostrome biota a restricted
aspect. It represents the final accumulation of
stenohaline marine organisms prior to the en-
croachment of non-marine red beds into the region.

## Gotland

The reefs of Gotland are widely known through
the publications of Wiman (1897), Hadding (1950),
Jux (1957), and Manten (1962 and 1971). The geological succession of the island is famous and Gotland, rather than the Silures, almost gave its name
to the system. The sequence is relatively thin,
approximately 450 meters, the dip is very low
and the entire island, 140 kilometers long (including Fårö in the northeast) and 50 kilometers wide,
exposes only Silurian rocks which range from upper Llandovery to upper Ludlow. Although the
succession includes discontinuities it is relatively
complete.

Bioherms or biostromes occur at nearly all
levels in the sequence and their distribution,
stratigraphic setting, size and morphology are
well documented, mainly through Hede's (1925a
and b, 1927a and b, 1928, 1929, 1933, 1936, 1940)
careful mapping of the geology, and Manten's
(1971) extensive examination of them. In contrast
the internal structure, paleoecology and sedimentology of the reefs remain only very generally
understood.

Gotland is situated in the central Baltic (Fig. 1)
on the cratonic Russian Platform. To the west, on

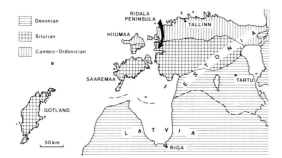

FIG. 19.—Silurian outcrop and localities, Estonia.

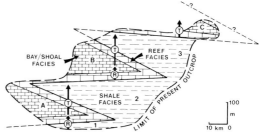

FIG. 20.—Idealized and partially hypothetical diagram of carbonate/shale wedge configuration within the Gotland sequence showing the position of the main reef facies and the alternately shallowing-up/regressive (R) and deepening-up/transgressive (T) nature of the succession. Wedges A (Högklint-Slite), B (Halla-Klinteberg-Hemse), and C (Burgsvik-Hamra-Sundre) are mainly carbonate, with reef facies preferentially developed distally and algal, oolite, bioclastic bay/shoal facies proximally. Wedges 1 (Visby), 2 (Slite-Mulde), and 3 (Hemse-Eke) are essentially shale and argillaceous limestone.

the Swedish mainland, erosion has largely removed Lower Paleozoic sediments, and the Precambrian metamorphic basement is exposed over much of the region forming the Baltic Shield (Størmer, 1967). In Scania, at the southern tip of Sweden, a relatively thick, shale-dominated Silurian succession occurs (Regnéll, 1960, p. 25–31) in the Oslo-Scania-Baltic Syneclise (trough) (Størmer, 1967, fig. 22). In contrast the limestone-shale succession of Gotland formed on a shallower shelf area and has much more in common with the Estonian sequence which is a continuation of the Gotland outcrop northeastwards on the eastern side of the Baltic (Fig. 19).

The Baltic area north of the syneclise is characterized by a thin (less than 1 kilometer) Lower and Middle Paleozoic sedimentary sequence resting unconformably on Precambrian crystalline basement, and boreholes have proved uppermost Precambrian, Cambrian, and Ordovician sediments beneath Gotland. The Caledonide mobile belt was several hundred kilometers to the west and the Baltic sequence is flat-lying and undeformed. Manten (1971, p. 7–32) further summarizes the regional setting of Gotland.

The near horizontality of the beds on Gotland, combined with the rare observation of graptolites in these facies, has presented stratigraphic problems which persist to the present day. The stratigraphic units erected by Hede which are still in use show marked diachronism in the upper part of the succession (Martinsson, 1967; Michael Bassett, pers. comm., 1979). Yet Murchison (1847) quickly recognized the main feature of the Gotland sequence: a series of rock units dipping at a very low angle to the southeast with the oldest strata exposed along the northwestern coastline. The astuteness of this observation is confirmed by the subsequent mistakes made by Lindström (1884) who was misled by the low inclination of the beds (the regional dip is only a fraction of one degree to the southeast) into believing that one series of rocks could be traced all over the island and that *lateral* facies variation alone accounted

for lithological changes. The unlikelihood of this interpretation was becoming clear by the end of the nineteenth century (see Moberg, 1910, p. 40–57) but variants of it continued to be developed (Wedekind and Tripp, 1930) and repeated by workers up to and including Jux (1957) (see Manten, 1971, p. 32–42).

The Gotland sequence is dominated by limestones and shales. Coarser siliciclastic sediments occur at only two horizons, in the Slite Siltstone and the Burgsvik Sandstone, which together represent no more than 10 percent of the total succession (Fig. 3). The limestones, which include a diverse array of bioclastic, reefal, oolitic, oncolitic and argillaceous carbonates, occur as two distinct wedges narrowing to the south overlain by shale wedges narrowing to the north. The top of another shale unit occurs at the base of the exposed succession, and the base of a third limestone unit occurs in the south of the island (Fig. 20). The sequence is relatively thin and disconformities occur at several levels, particularly within the limestone wedges. Those affecting reefs are especially prominent at the top of the Högklint Beds and planing the top of reefs at Kuppen and Holmhällar (see below).

The broad depositional trend throughout the Silurian was shallowing from the north due to influx of fine siliciclastic detritus and the formation of extensive carbonate sediments (Laufeld, 1974a, p. 7). This southern progradation was complicated by transgressive-regressive cycles which produced the pattern of interfingering wedges of shelf carbonates to the north and relatively deeper water shales to the south. By Ludlow time there

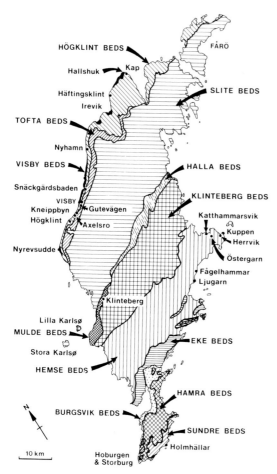

FIG. 21.—Localities and distribution of Hede's stratigraphic units on Gotland.

is evidence that this north-south facies polarity was breaking down to a complex mosaic of very shallow water facies indicating, as everywhere else in northern Europe, the effect of regional regression and the approach of Old Red Sandstone continental sedimentation.

The current stratigraphy is based upon the work of Ernhold Hede who, together with other colleagues, prepared 1:50,000 geologic maps with accompanying memoirs of the whole island. He recognized 13 major stratigraphic units (Fig. 21, Table 1) which are summarized by Hede (1960, p. 44–52) and Laufeld (1974a, p. 7–13). These are topostratigraphic units which combine aspects of both litho- and biostratigraphic units. Consequently, there is not always a correlation between Hede's units and the major lithofacies (Figs. 21 and 22). This is particularly so for the Slite, Klinteberg and Hemse Beds which are all essentially

limestones in the northeastern parts of their outcrops, but pass laterally into shales towards the southwest. Hede's major units are divided into sub-units distinguished either by name or letter (Manten, 1971, p. 277–422; Laufeld, 1974a, p. 7–13). Biostratigraphic correlation with standard European divisions has been made relatively recently by Martinsson (1967) with slight modification by Bassett and Cocks (1974).

Hede (1960) provides a useful guide to Gotland geology and detailed work is greatly helped by Laufeld's (1974b) comprehensive inventory of localities. Manten (1971) describes and illustrates numerous reef outcrops. Stromatoporoids, which are the dominant reef building organisms, have been described by Mori (1968, 1970) and stromatoporoid morphotypes are discussed by Kershaw and Riding (1978). Stel (1978a) describes tabulates, especially favositids, from the Gotland succession. Calcareous algae are described by Rothpletz (1913) and Hadding (1959). Gotland rugosans are described by Wedekind (1927), and Brood (1976) outlines bryozoan paleoecology. Other publications on the paleontology of Gotland are listed by Manten (1971, p. 423–424). The faunal and floral succession through a section of the Visby, Högklint and Tofta Beds near Visby is being documented in detail by a group of specialists.

Murchison, having learned from his original mistaken conception of the Wenlock Edge bioherms as concretions, recognized the reefal character of the Gotland deposits (1847). Studies of the reef limestones were subsequently made by Wiman (1897) and Hedström (1910) near Visby, and by Munthe (1910) in southern Gotland. Both Hadding (1941, p. 79–94) and Manten (1971, p. 56) review work on Gotland reefs. Crosfield and Johnston (1914, p. 212–214) noted the similarity between Wenlock Edge "ballstone" and Gotland reef rock but commented "the longer continuance in time of the reef phase in Gotland . . . has introduced a greater variety into the fauna of the reefs, and also into the lithology and faunas of the associated beds, both at the base of and surrounding the reefs" (p. 214).

Hadding's work (1941 and 1950) represents an important step forward in understanding Gotland reefs. He used thin sections to study the petrography and he plotted out the reef outcrops from Hede's maps to clarify their distribution. Rutten (1958) also surveys the spectrum of reef occurrences on Gotland.

The broad pattern of carbonate and shale wedges which makes up the Gotland sequence was recognized by Wedekind and Tripp (1930), and further emphasized by Jux (1957). Jux's attempts at a general facies analysis of the succes-

TABLE 1.—HEDE'S (1960) STRATIGRAPHIC UNITS FOR GOTLAND WITH APPROXIMATE AGES (AFTER BASSETT AND COCKS, 1974, FIG. 1), MAXIMUM THICKNESSES AND PRINCIPAL LITHOTYPES (AFTER LAUFELD, 1974A)

|  |  |  |  |
|---|---|---|---|
|  | Sundre Limestone | 10 m+ | bioclastic limestone, biostromes |
|  | Hamra Group | 40 m | argillaceous & bioclastic limestone, bioherms |
| LUDLOW | Burgsvik Group | 47 m | oolitic limestone, sandstone, bioherms |
|  | Eke Group | 15 m | shale, argillaceous & bioclastic limestone, bioherms |
|  | Hemse Group | 100 m | bioclastic & argillaceous limestone, shale, biostromes & bioherms |
|  | Klinteberg Group | 64 m | argillaceous & bioclastic limestone, shale, bioherms |
|  | Halla Group/Mulde Marl | 20 m | argillaceous & oolitic limestone, shale, bioherms |
| WENLOCK | Slite Group | 100 m | bioclastic & argillaceous limestone, shale, bioherms |
|  | Tofta Limestone | 8 m | argillaceous limestone |
|  | Högklint Group | 35 m | bioclastic limestones, bioherms |
|  | Upper Visby Marl | 16 m | argillaceous limestone, shale, bioherms |
| LLANDOVERY | Lower Visby Marl | 9 m+ | shale, argillaceous limestone |

sion have been strongly criticized by Manten (1971, p. 39–42) as being idealized and simplistic. Nevertheless, the gross features which Jux stressed do dominate the sequence and his approach should be a stimulus to further refinement of facies patterns within the succession.

Manten (1962 and 1971) made a major contribution by his detailed documentation of the form and distribution of Gotland reefs. He placed little emphasis on sedimentologic or paleoecologic analysis, but he was nevertheless able to recognize three main reef types which he named Upper Visby, Hoburgen and Holmhällar (1971, p. 56–58). In addition, he recognized Stäurnaser and Fanterna types on Stora Karlsø Island off the southwest coast (1971, p. 243). The principal features of the three main reef types, as defined by Manten, are as follows:

*Upper Visby.*—Lensoid to conical reefs with a very marly matrix enclosed by marlstone and argillaceous limestone; typically less than 10 m² in section and with a height:length ratio of between 1:1 and 1:5; dominated by corals and rather laminar stromatoporoids, both of which are relatively small, and lacking calcareous algae and reef detritus; species diversity is moderate and the organic composition is rather variable.

*Hoburgen.*—Very elongate lensoid to conical reefs with a marly matrix, enclosed by limestone and argillaceous limestone; typically about 100 m² in section and with a height:length ratio of between 1:1 and 1:50; dominated by lenticular stromatoporoids and corals, both of which are quite large, together with common calcareous algae; surrounded by crinoidal reef detritus; species diversity is high and the organic composition is generally variable.

*Holmhällar.*—Crescentic reefs with only a little included marl and enclosed by rather pure limestone; typically more than 1000 m² in section, with a height:length ratio of between 1:15 and 1:75; dominated by large round or irregular and high stromatoporoids with calcareous algae very common; surrounded by reef detritus; species diversity is low and the organic composition is rather uniform.

The Upper Visby type of reef mainly occurs in the Upper Visby Beds, near the base of the sequence along the northwest coast, but the Hoburgen type, which is the most common, is found in every unit except the Lower and Upper Visby Beds, the Tofta Beds and the Sundre Beds. The Holmhällar type is found in the Hemse, Hamra and Sundre Beds in the southern part of the island (Manten, 1971, p. 58). The Stäunasar and Fanter-

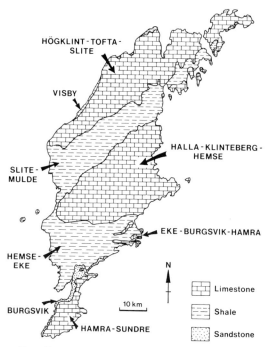

HÖGKLINT-TOFTA-SLITE

VISBY

HALLA-KLINTEBERG-HEMSE

SLITE-MULDE

EKE-BURGSVIK-HAMRA

HEMSE-EKE

N

BURGSVIK

10 km

HAMRA-SUNDRE

Limestone

Shale

Sandstone

FIG. 22.—Lithofacies and their relation to Hede's stratigraphic units on Gotland, compare with Figure 20; from Eriksson and Laufeld (1978, fig. 1).

na reef types are restricted to Karlsø Island and are dominated by corals.

Manten's three main reef types are basically well founded, but they break down in the Hemse reefs of the Östergarn-Ljugarn area, where he admits that variants occur (Manten 1971, p. 357, 374 and 386). These have a greater variety of stromatoporoid morphotypes than either the Hoburgen or Holmhällar types, and a greater density of organisms than the Hoburgen type. They are here named the *Kuppen* type from the locality near Herrvik (Fig. 21). The name Upper Visby, suggested by Manten, is appropriate in the sense that these tabulate dominated structures are common in the Upper Visby Beds, but it also tends to imply a restriction to this unit whereas they also occur in the Lower Visby Beds and similar structures occur in the Klinteberg Beds at Klinteberg (Manten, 1971, fig. 165). In order to avoid direct association between reefs of this type and the Upper Visby Beds, and also to avoid confusion between stratigraphic units and reef types it is here proposed to use the name Axelsro reef type in place of Upper Visby reef type. Good examples occur in cliff sections near Axelsro, 5 kilometers southwest of Visby. Although the Axelsro reefs are mainly tabulate dominated, stromatoporoids also become important in the larger examples

(Ted Nield, pers. comm., 1979). The Hoburgen, Kuppen and Holmhällar types are essentially stromatoporoid reefs. The Axelsro and Hoburgen reefs are bioherms but thin biostromal units are also associated with the Axelsro reefs. The Kuppen and Holmhällar reefs are biostromal so far as is known. Besides these broad compositional and geometric features the various reef types can be compared and contrasted in detailed composition and structure (Table 2). Details, and examples of them, are given below; localities follow Laufeld (1974b).

*Axelsro Bioherms and Biostromes.*—These reefs in argillaceous and bioclastic limestone are mainly small tabulate dominated bioherms which are closely comparable with the Wenlock Edge reefs. Generally they are dense, grading into loose, structures, although locally they are solid. The principal organisms are favositids, heliolitids, halysitids and laminar to low domical stromatoporoids all of which are usually less than 25 centimeters across, although they can be up to 50 centimeters across. There is no distinct boundary between very small examples of these reefs, such as a tabulate overgrown by a stromatoporoid and encrusted by a few bryozoans, and level bottom communities of isolated individuals.

Two Axelsro sub-types can be distinguished: mud-based, usually small structures up to 2 meters across; and larger, gravel-based structures up to 5 meters across (Fig. 23). The presence of gravel lenses beneath many of the larger examples suggests that the reef building organisms preferentially colonized coarse substrates when these were available. The other possible explanation, that the reef itself produced or localized gravel deposition, is weakened by the common occurrence of gravel bands without overlying reefs and the scarcity of gravel bands lateral to reefs. The gravel appears to occur in channels which represent the approach of crinoid shoals, the coarse sediment being swept in from higher energy environments, probably to the northwest.

The Axelsro type are well seen in the Upper Visby Beds along the northwest coast between Nyrevsudde and Hallshuk (Fig. 21). Small examples are well exposed at Halls Huk 3 where a series of them occur at the base of the cliff in the Upper Visby Beds about 10 meters below the base of the Högklint. At this horizon several small structures occur within a lateral distance of 10 meters. The surrounding rock is argillaceous fine to locally coarse bioclastic pale green-gray thin nodularly bedded limestone. The bioherms are irregular lenses 20–100 centimeters wide and 10–40 centimeters high. They are 1–3 meters apart although there are no bioherms for 100 meters or more beside this group. They consist of dense accumulations of favositids and laminar to low dom-

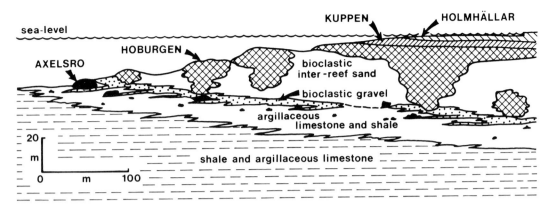

FIG. 23.—Idealized model of Gotland reef development.

ical stromatoporoids with some halysitids and fewer heliolitids. Small patches of bryozoans and *Syringopora* encrust some of the surfaces and there are scattered branched bryozoans, gastropods, and brachiopods (*Leptaena, Atrypa,* rhynchonellids) close to the larger skeletons. The individual favositids and stromatoporoids, which are the principal components, are up to 40 centimeters across, but most specimens are less than 15 centimeters.

An example is shown in Figure 24. This mini-reef occurs 12 meters south of, and 1 meter higher than, the two stromatoporoids figured in Kershaw and Riding (1978, fig. 13). It is composed mainly of halysitids and ragged laminar to low domical

stromatoporoids. The large organism is a stromatoporoid 34 centimeters across. The adjacent limestone contains poorly-sorted bioclastic debris and away from the reef isolated halysitids and stromatoporoids occur, commonly 10 centimeters across. Quite commonly adjacent organisms are overturned indicating moderate current activity sporadically.

Larger examples of these bioherms are well seen in the shore section between Axelsro and Kneippbyn (Fig. 25). These include small structures similar to those at Halls Huk 3 but also larger ones up to 5 meters across. One of the small reefs on the shore due north of Kneippbyn and just outside the military shooting range is unusual

TABLE 2.—FEATURES OF GOTLAND REEF TYPES

|  | AXELSRO | HOBURGEN | KUPPEN | HOLMHÄLLAR |
|---|---|---|---|---|
| Environment (light & energy) | low | moderate | high | very high |
| Dominant organisms | tabulates, domical stromatoporoids | laminar-domical stromatoporoids, tabulates | laminar-domical-bulbous stromatoporoids | cuspate laminar stromatoporoids |
| Diversity | moderate | high | low | very low |
| Matrix | fine, argillaceous | fine-coarse, argillaceous | fine (draped)-very coarse | fine-very coarse |
| Geometry and size | bioherms (small) | bioherms (large)-biostromes | biostromes | biostromes |
| Structure | dense | dense-frame | dense-frame | frame-solid |
| Stability | moderate | low-moderate | high | high |
| Associated sediments | shale, argillaceous limestone, bioclastic gravel | bioclastic sand and gravel | bioclastic gravel, argillaceous limestone, shale | bioclastic gravel |

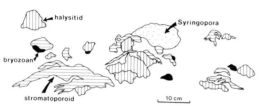

FIG. 24.—Small protobioherm of Axelsro type in the Visby Beds at Halls Huk 3, Gotland; photograph above, and map of organisms below (map courtesy of Ted Nield).

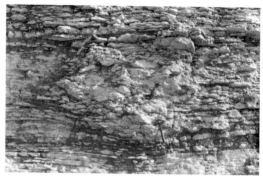

FIG. 25.—Axelsro bioherm showing "ballstone" structure and relationship to surrounding bedded lime-stones and shales; Upper Visby Beds between Axelsro and Kneippbyn, Gotland; hammer 28 centimeters long.

in being composed almost entirely of low domical stromatoporoids and a few favositid colonies (Fig. 26). In this area the Axelsro reefs occur in the Upper Visby Beds with at least one occurring every 50 meters along the section. They mostly range from 1–5 meters across and typically have an irregular lensoid form. Some occur with their basal surface on the argillaceous limestone which makes up most of the Upper Visby Beds, but many, especially the larger ones, rest upon very coarse bioclastic pebbly gravels which channel into the underlying argillaceous limestone. The gravel is made up mainly of large rugose and tabulate skeletons, with some stromatoporoids, up to 25 centimeters across in a bioclastic gravel matrix.

A biostrome formed by halysitids occurs in the Lower Visby Beds at Ireviken 3 (Stel, 1978b, p. 9–12). *Catenipora* colonizes a shale-limestone substrate and results in a *Catenipora, Favosites, Ketophyllum* association forming a bed 1.50 meters thick which is laterally extensive.

Small structures of Axelsro type also occur higher in the Gotland sequence in the Klinteberg Beds at Klinteberg (Manten, 1971, fig. 165).

*Hoburgen Bioherms.*—Hoburgen reefs show the greatest range of internal structure of Gotland reefs. They are essentially stromatoporoid reefs but they contain a diverse biota including rugo-sans, tabulates, bryozoans, brachiopods, calcar-eous algae, and gastropods. The stromatoporoids

range from laminar to bulbous and extended dom-ical forms but most commonly are low to high domes. Nevertheless, stromatoporoid morpholo-gy is less diverse than in the Kuppen reefs; there are fewer laminar and extended domical forms. The Hoburgen reefs have a relatively argillaceous matrix and this may have inhibited laminar forms, which are prone to covering by sediment, and ex-tended domes which perhaps required a firmer and less unstable substrate. The matrix also in-cludes coarse bioclastic limestone.

A feature of particular interest in Hoburgen reefs is the tendency toward a vertical succession from tabulate through stromatoporoid to coral-al-gal dominated sub-units (Nigel Watts, pers. comm., 1978). The tabulate dominated sub-unit resembles the Axelsro reef-type and Axelsro reefs do occur directly beneath Hoburgen reefs in re-gressive sedimentary phases such as the Visby-

FIG. 26.—Very small, stromatoporoid dominated bioherm; Visby Beds, Kneippbyn, Gotland; hammer 28 centimeters long.

FIG. 29.—Large bioherm (right) and adjacent inter-reef sediments (left) in Högklint Beds at Häftingsklint, northwest Gotland.

FIG. 27.—Hoburgen type reefs at Högklint, south of Visby, Gotland. The lower, slightly recessive, part of the cliff is formed by Visby Beds (limestones and shales). The overlying Högklint Beds are here dominated by large bioherms which pass laterally (seen at right) into bedded bioclastic limestones.

Högklint sequence. The coral-algal facies consists of large branched *Solenopora* growths with branched rugosans forming a frame structure.

Good examples of Hoburgen reefs occur in the Högklint, Klinteberg and Hamra Beds and on Lilla Karlsø. They are particularly well seen at many localities in the Högklint Beds along the northwest coast, in the Main Limestone on Lilla Karlsø and in the Hamra Beds at Hoburgen itself.

The abundance of Hoburgen patch-reefs is remarkable. They are common in each of the three major carbonate wedges, and Eriksson and Laufeld (1978, p. 27) counted over 500 in the Högklint Beds alone.

*Högklint Beds.*—The shales and argillaceous limestones of the Visby Beds are overlain by the Högklint limestones which give rise to steep cliffs along the northwest coast. Högklint (literally "high cliff") rises 35 meters above sea level and exposes lensoid Hoburgen reefs enclosed in bio-

clastic sands and gravels. Locally the gravels channel into the underlying Upper Visby sediments. Some reefs of Axelsro type continue from the Upper Visby into the Högklint but the majority of reefs are restricted either to the Visby or Högklint Beds. Those beginning in the Högklint are of Hoburgen type and form lens-shaped masses whose irregularly nodular to lenticular internal structure contrasts with the distinct bedding of the adjacent bioclastic limestone (Fig. 27).

At Högklint the bioherms are up to 50 meters wide and 25 meters thick, but larger individual reefs up to 300 meters across are well seen further north on the coast between Nyhamn and Irevik (Stel, 1978a, p. 115–127). Lower cliffs between these patch-reefs expose bedded inter-reef sediments (Fig. 28). The reefs occur every few hundred meters along this stretch of coast and are often present as tight clusters; at Häftingsklint 1–3 perhaps ten or more lenses make up what superficially is a single large reef (Fig. 29) (Nigel Watts, pers. comm., 1979).

The relationships between reef and off-reef limestone are well seen at Korpklint 1, near Snäckgärdsbaden (Fig. 30) where coalescive reef lenses irregularly overlie well-bedded crinoidal

FIG. 28.—Patch-reefs and intervening bedded limestones; Högklint Beds between Nyhamn and Irevik on Gotland's northwest coast; photo courtesy of Nigel Watts.

FIG. 30.—Hoburgen type bioherms overlying bedded bioclastic gravels in Högklint Beds at Korpklint, Snäckgärdsbaden, north of Visby, Gotland.

FIG. 31.—Bedded bioclastic gravel squeezed up between the bases of two adjacent bioherms by compaction, Högklint Beds, Korpklint 1, Gotland; photo courtesy of Nigel Watts.

FIG. 32.—Concentric lines on the wave-cut platform circling a patch-reef remnant reflect a compactional syncline ("*Philip Structure*") in the Visby Beds beneath the reef. From Jungfrun, near Lickershamn, northwest Gotland; also see top of Figure 33 (photo courtesy of Nigel Watts).

limestones, and compaction has locally squeezed the bioclastic gravels upwards into invaginations between reef lenses (Fig. 31). In turn, the underlying Visby Beds are depressed to produce small compactional synclines. Arne Philip of Visby noticed these circular structures with diameters of 20–70 meters during flights over the wave-cut platform of the northwest coast where erosion has removed the Högklint Beds to reveal the compactional imprints of the patch-reefs upon the Visby Beds (Fig. 32). The occurrence of these "*Philip structures*" has been documented by Eriksson and Laufeld (1978) who realized their value in providing an indication of the spatial pattern of reef development in the Högklint and named them after their discoverer. They also proposed the term "*Cumings structure*" for the updoming of strata above bioherms (Eriksson and Laufeld, 1978, p. 20). Hede's maps give a good general impression of the distribution of the Hoburgen reefs in the Högklint Beds (see Eriksson and Laufeld, 1978, figs. 4–6), but many of the reef areas plotted by him are actually patches of numerous individual reefs (Nigel Watts, pers. comm., 1978). By using Arne Philip's aerial photographs Eriksson and Laufeld were able to map out the pattern of reef occurrences along the 100–200 meters wide cliff and wave-cut platform area of many parts of the northwest coast (Fig. 33) and convincingly demonstrate that many of the reefs are circular and have a non-linear distribution. They conclude that much of the Högklint coastal exposure is in a several kilometer wide patch-reef zone with thousands of individual bioherms (Eriksson and Laufeld, 1978, p. 28).

Although the majority of the Högklint reefs are lensoid in form, they do cluster together (as at Häftingsklint, mentioned above) and locally coalesce to form extensive biostromal masses. This condition is approached at Korpklint 1 and is well

seen at Halls Huk. Between the village of Kap and Halls Huk 3 the cliff continuously exposes 1 kilometer of Högklint reef lenses overlying bioclastic limestones. The Upper Visby Beds are mostly covered by talus except at Halls Huk 1 and Halls Huk 3. For most of its length the cliff is capped by Hoburgen reefs which overlie, and pass laterally into, bedded limestones. The tops of the reefs are only seen where small patch-reefs terminate early in the Högklint sequence, but for the most part the cliff top bevels the reefs off. The bedded limestones range from medium-grained bioclastic thin-bedded limestones, often showing small scale planar cross-bedding and small (less than 2 meters wide) shallow channels, through gravels to thin to medium-bedded coarse pebbly deposits with fragments and whole skeletons of rugosans, tabulates and stromatoporoids up to 30 centimeters across.

The reefs here are bioherms 2–20 meters across which usually coalesce laterally to form a virtually continuous lenticularly based biostrome along the cliff top. The structure is dense to loose in its lower part and consists of large, low to high domical and bulbous stromatoporoids with rugose corals and abundant bryozoans, tabulates and brachiopods. Many of the high domical and bulbous stromatoporoids are on their sides and bulbous forms occasionally occur in coarse, bedded gravels, probably well away from their sites of growth. In between is a talus type deposit of moved, usually small, skeletons.

The upper parts of the reefs are not well seen due to lichen cover and weathering but are probably represented in fallen blocks north of Hallshuk 3 where low to extended domical stromatoporoids up to 35 centimeters across occur with tabulates, stick bryozoans and massive rugosans in a dense, possibly frame, structure.

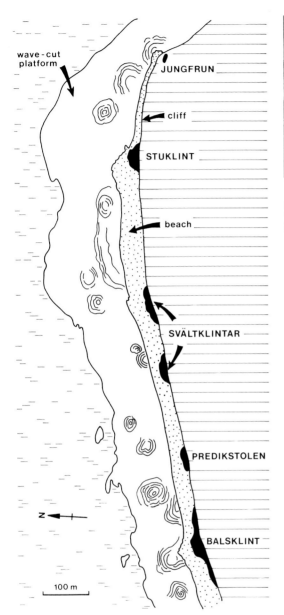

FIG. 34.—Cliffs of biohermal Hamra Beds at Hoburgen, southern Gotland. The hill in the center of the picture is immediately north of Storburg, and is capped by bedded limestones of the Sundre Beds.

FIG. 33.—Distribution of "*Philip structures*" in the Visby Beds on the wave cut platform 25 kilometers north of Visby, Gotland. The remaining reefs in the Högklint Beds along the cliffs are marked in black (from Eriksson and Laufeld, 1978, fig. 15).

There are few very good in place exposures of Hoburgen reef structure in the Högklint, or anywhere else on the island. Weathering and lichen cover obscures the reef structure, even where the immediately adjacent bedded limestones are very clean, and the reef is seen only as an irregularly nodular, rubbly or lenticularly structured mass in which a few large organisms and some of the matrix can be recognized but in which the distribution, composition and mutual relationships of the majority of the components are far from clear.

However, the fallen blocks near Halls Huk 3, together with blocks of laminar stromatoporoid frame at Ireviken 3, suggest that there is a vertical transition from dense to frame structure in the Hoburgen reefs of the Högklint Beds. This is confirmed by exposures in the Gutevägen area (Fig. 21) on the south side of Visby where Högklint reefs are capped by a coral-algal facies of stubbly branched solenoporaceans and branched rugosans forming a tight framework.

Between Gutevägen 3 and Gutevägen 2 reefs overlying bedded limestones are exposed in the upper part of a long low cliff. The underlying limestones range from wavy thin-bedded medium-grained sands with shale partings to coarser gravel units which are locally cross-bedded, and pass upward into medium-bedded coarse to pebbly bioclastic limestones immediately below the reefs. The reefs are up to 4 meters thick and occur as patch-reefs and coalesced patch-reefs forming biostromal units which may extend continuously laterally for up to 100 meters. Individual reefs are never separated laterally by more than 15 meters of bedded limestone.

The reefs consist of mid- to extended domical stromatoporoids up to 30 centimeters across with abundant branched rugosan colonies and scattered coenites. Most of the stromatoporoids are mid- to high domes, many are fallen. The extended domes are often non-enveloping.

For 100 meters or so north of Gutevägen 2 the reefs are overlain by *Solenopora* limestone followed by wavy-bedded laminar stromatoporoid limestone which together are up to 1.5 meters thick. At the road section (Gutevägen 2) the margin of the biostrome where it passes laterally into

FIG. 35.—Hoburgen type bioherms at the "type locality" showing large coalescive lenses passing laterally
into bedded limestones; Hamra Beds at Hoburgen, center of Figure 34.

FIG. 36.—Dipping bedded limestones, possibly representing original depositional slopes, on the south side
of Lilla Karlsö.

coarse-bedded limestone is rubbly and argillaceous with overturned mid-high domical stromatoporoids. On the west side of Gutevägen 4 the
coral-algal facies is overlain disconformably by
laminar stromatoporoid limestone representing
bevelling and recolonization of the reef surface.

*Hoburgen.*—Near the southwesternmost tip of
the island at Hoburgen, Storburg and three small
hills east-northeast of it expose reefs in the Hamra
Beds (Fig. 21). The reef rock itself is mostly lichen
covered, but the extensive exposure shows the
spatial distribution of the reefs and their enclosing
sediment well (Fig. 34). The reefs overlie bioclastic limestone, including oncolitic limestone near
its base, which rests on the Burgsvik Sandstone
and Oolite (see Munthe, 1910, p. 1424–1425).
They are also enclosed laterally and covered by
bioclastic limestone, the overlying layers being
referred to the Sundre Beds (see Laufeld, 1974a,
p. 65; Munthe, 1910, fig. 22). These bioclastic
limestones vary considerably in texture from
coarse sands to conglomeratic gravels, with thin
to thick-bedding, which are mainly composed of
crinoids, coral and stromatoporoid debris. The
seaward side of the first hill north of Storburg
shows at least eight coalescive reef lenses (Fig.
35), each approximately 4 meters high and 8 meters wide, forming a patch-reef cluster 25 meters
long and 16 meters high. Reef structure is best
seen on the south side of this hill (Hoburgen 4,
see Munthe, 1910, fig. 22) where a reef lens 10
meters thick and 25 meters wide, also forming part
of this cluster, is composed of low–mid-domical
stromatoporoids, commonly 30 centimeters across,
with patches of argillaceous and bioclastic matrix.
There is much general disorientation of the large
skeletons and on the northwest side of this bioherm large to low-domical stromatoporoids up to
at least 1 meter across are overturned at the lower

margin of the lens. The enclosing bioclastic limestones form an abrupt contact with the reef.

*Karlsö Islands.*—The Karlsö Islands (Fig. 21),
the largest islands off the western coast of Gotland, are distal southwestward continuations of
the second main limestone wedge of the island,
probably correlating for the most part with the
Slite and Klinteberg Beds. On both Stora and Lilla Karlsö thinly-bedded shale-limestone sequences occur in the northwest and are succeeded
by relatively thick limestone sequences consisting
mainly of bedded bioclastic limestones surrounding patch-reefs. Manten (1971, p. 242–243) distinguished two reef types on Stora Karlsö different
from those of Gotland: the Stäunasar type consists of corals and stromatoporoids, the Fanterna
type of bryozoans and corals. I have not visited
the localities after which these reef types are
named but many of the Karlsö reefs are lenticular
masses enveloped in bedded bioclastic limestones
and they resemble the Hoburgen type. However,
bryozoan biostromal limestones, formed by
branched masses of *Coenites* in arcuate clusters
up to 20 centimeters thick, occur near Smojge 1
on the northern side of Lilla Karlsø and large
branched tabulate colonies, many overturned, are
present south of Sudervagnhus on the west side
of the island. These may resemble the Fanterna
limestones of Stora Karlsö. But both islands are
dominated by steep-sided plateaux of bedded bioclastic limestones enclosing lensoid patch-reefs.
Following Rutten (1958), Manten (1971, p. 242 and
258) believes that these plateaux are cored by
large reefs and he also recognizes "flank reefs"
in the surrounding bedded limestone. It is difficult
to test this idea of large central reefs and it is
possible that the plateaux are constructed internally in much the same way as their margins, with
scattered patch-reefs separated by bedded limestone.

The Karlsö outcrops are particularly interesting

FIG. 37.—Kuppen type biostrome, at the type locality, truncated by a disconformity and overlain by stromatoporoid-rich gravels; Kuppen 2, Hemse Beds, eastern Gotland. Hammer is 28 centimeters long; compare Figure 38.

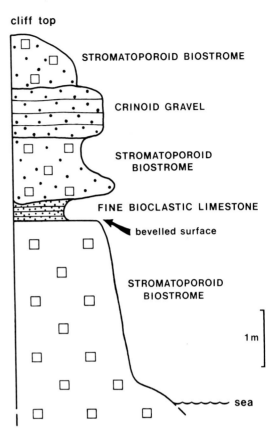

FIG. 38.—Sequence of stromatoporoid biostromes in the Hemse Beds at Kuppen 2, Gotland. The "type" Kuppen biostrome is the lowest unit; also see Figure 41.

because of the steeply dipping margins of the main limestone outcrop in some places, especially on the southern side of Lilla. Here the coarse bioclastic limestones, with a few incorporated patch-reefs, abruptly change from sub-horizontal to a dip of 30 degrees which can be followed through a height of at least 10 meters (Fig. 36). At the northwestern and western sides of the island large blocks of reefal and bioclastic limestone up to 30 meters across have slid into the underlying *Pentamerus gotlandicus* limestone (bioclastic limestone bands separated by shale) crushing and buckling it. Manten (1971, p. 273–275) regards this as probably due to Quaternary rock slides, but Sven Laufeld (pers. comm., 1978) considers it to be of penecontemporaneous Silurian age, and it is quite possible that the Karlsö Islands represent outliers of Klinteberg Limestone which were small carbonate platforms of bedded and reef limestone with steep marginal slopes to deeper water argillaceous sediments down which fragments of partly lithified limestone slipped to form megabreccias which deformed the soft basinal sediments.

### Kuppen Biostromes

Manten (1971, p. 386) recognized that the reefs described here as Kuppen type do not fall neatly into his scheme of reef classification. Those near Herrvik, at Kuppen 1 and 2 (Fig. 21), he regarded as a variant of the Hoburgen type and those at Ljugarn and Fågelhammar he regarded as Holmhällar reefs. But the Kuppen reefs are significantly different from Hoburgen reefs, particularly in the diversity of the stromatoporoid morphotypes which they contain, and the relative paucity of other organisms. Equally, the Ljugarn and Fågelhammar reefs, except for the laminar stromatoporoid frame structure on the south side of Fågelhammar 1, commonly contain high and extended domical stromatoporoids whereas the Holmhällar type is characterized by laminar frame structure.

The Kuppen type is a dense to frame biostromal deposit dominated by large laminar, low to extended domical and, occasionally, bulbous stromatoporoids in a relatively fine-grained grey-green argillaceous or crinoidal limestone matrix. There are occasional stick bryozoans and attached brachiopods, but little else between the stromatoporoids. Branched rugosans and *Syringopora* occur *within* stromatoporoids forming *Caunopora*-type intergrowths.

Kuppen reefs occur in the upper part of the Hemse Beds, particularly unit d of Hede (see Laufeld, 1974a, p. 11–12) in the Katthammarsvik-Ljugarn area of the east coast (Fig. 21). Near Katthammarsvik they occur at several localities, including those around Östergarnsberget (Gannberg, see Manten, 1971, p. 352–359), Grogarns

FIG. 39.—Kuppen biostrome structure showing the juxtaposed laminar and domical stomatoporoids. Close packing is enhanced by compaction, and many of the contacts are stylolitized; Kuppen 2, Hemse Beds, Gotland. Hammer is 28 centimeters long.

FIG. 41.—Sequence at Kuppen 2 showing bevelled upper surface of the biostrome overlain by recessive fine bioclastic limestone.

(Manten, 1971, p. 359–362) and Herrvik and form biostromes 5 meters thick and kilometers in lateral extent which are often dense rather than frame in structure.

*Kuppen.*—At Kuppen, east of Herrvik, a biostrome of this type, taken here as the typical example, is well exposed in sea cliffs (Munthe, 1910, fig. 28; Hede, 1929, p. 40–42). At Kuppen 2 (Fig. 37) the sequence (Fig. 38) is:

cliff top

5. 1.00 m+ stromatoporoid biostrome with crinoidal matrix,
4. 0.80 m bedded crinoid gravel,
3. 1.10 m stromatoporoid biostrome with crinoidal matrix,
2. 0.20–.40 m fine bioclastic limestone with stromatoporoid fragments,

flat bevelled disconformity surface

FIG. 40.—Laminar, domical and bulbous stromatoporoids at Kuppen 2; hammer is 28 centimeters long; detail of Figure 41.

1. 3.40 m+ stromatoporoid biostrome with fine matrix,

sea level.

Unit 1 is the biostrome of Kuppen type. It consists mainly of laminar to extended domical stromatoporoids forming a very dense to frame structure (Figs. 39 and 40). The laminar forms are up to 15 centimeters thick and the extended domes up to 80 centimeters high. The skeletons are separated by draped and laminated fine-medium grained argillaceous limestone patches up to 10 × 20 centimeters in size. Contacts between skeletons are usually stylolitized (Stephen Kershaw, pers. comm., 1976). The topmost 40 centimeters has smaller stromatoporoids and less fine matrix. This unit is truncated upwards by a smooth, flat erosion surface (Fig. 41) overlain by unit 2. Units 3–5 make up a stromatoporoid biostrome similar to unit 1 but with somewhat smaller laminar to extended domical stromatoporoids, abundant coarse crinoidal matrix, and a dense rather than frame structure.

Traced northwest around the coast these units continue for 100 meters virtually unchanged except that units 3–5 merge together. Seventy-five meters northwest of Kuppen 2 unit 1 is 4 meters thick and the base of the cliff exposes 1 meter of grey shale with thin laminar and high domical stromatoporoids. Here the lowest 20–40 centimeters of unit 1 has a coarse crinoidal matrix and the top 1 meter also has a coarser bioclastic matrix than the center of the biostrome which has the normal fine grey-green, draped fill. Beyond 100 meters northwest of Kuppen 2 the disconformity above unit 1 becomes less conspicuous and is replaced near Herrvik village by gravels below and a stromatoporoid bed above.

*Fågelhammar.*—At Fågelhammar 1, 3 kilometers north-northeast of Ljugarn, a rauk (erosional rock remnant) field exposes reef rock which on

FIG. 42.—Small isolated rauk (sea-stack) formed by a single domical stromatoporoid at Fågelhammar 1, Hemse Beds, eastern Gotland; hammer is 28 centimeters long.

FIG. 43.—Toppled high and extended domical stromatoporoids associated with Kuppen type biostromes, Ljugarn, Hemse Beds, eastern Gotland; hammer is 28 centimeters long.

the north side of the locality consists of numerous large mid-extended domical stromatoporoids up to 1.5 meters across, many of which are toppled. One small rauk is a single extended domical stromatoporoid 1 meter high (Fig. 42). As at Kuppen they enclose branched rugose corals and tabulates and include grey-green draped matrix. There is a fair quantity of crinoid debris. On the south side the rauks consist of thin anastomosing, laminar stromatoporoid frame which resembles that seen in fallen blocks at Irevik 3 in the Högklint Beds of the northwest coast, and which also occurs in the Holmhällar reefs.

*Ljugarn.*—Northwards along the coast for 750 meters from the jetty at Ljugarn (Fig. 21) an eroded rauk field exposes reef of Kuppen type. At the northwestern end of the outcrop, 500 meters northwest of Ljugarn 1, shore line exposures show numerous high to extended domical stromatoporoids, often compound and often toppled (Fig. 43) together with laminar forms in a grey-

green fine to coarse bioclastic, often crinoidal, matrix. The structure is less frame, and more dense, than Kuppen itself, due to the lesser abundance of stromatoporoids but they nevertheless constitute 30–40 percent of the rock.

These exposures are mainly in the plane of the bedding and this emphasizes the toppled domes whereas the Kuppen cliff exposures emphasize the laminar and low- to mid-domical forms.

The domical forms are up to 75 centimeters in height and compound specimens are up to 1.50 meters across. Draped fills are lacking. Further to

FIG. 44.—Rauk field (sea-stacks) representing dissected remnants of a Holmhällar type biostrome. The occurrence of enclosed hollows, and ridges parallel to the shore, suggest a karstic solution origin for the rauks with subsequent modification by littoral erosion processes; Holmhällar 1, Sundre Beds, southern Gotland.

FIG. 45.—Ground view of the Holmhällar rauk field shown in Figure 44.

the southeast, in the area up to 400 meters north of Ljugarn 1, the reef retains large domical stromatoporoids but they are less extended, more compound, less toppled and show delicate interleaving with fine matrix which suggests a gradual transition towards Holmhällar type.

### Holmhällar Biostromes

These are stromatoporoid laminar frame structures which appear biostromal in form. Similar structures form parts of other reefs, as in the Kuppen type at Fågelhammar 1 and in the Hoburgen type at Irevik 3, but the Holmhällar type itself is extensively developed in the Sundre Beds, high in the Gotland succession, at a few localities near the southern tip of the island. These occurrences are in raukar fields which expose the limestones well laterally but in which the vertical exposure is limited to only a few meters. Also, the rauks are usually somewhat weathered and lichen covered and the details of the reef structure are not often very clear.

*Holmhällar.*—At Holmhällar 1 the extensive rauk field (Figs. 44 and 45) exposes frame reef rock composed of relatively thin anastomosing laminar stromatoporoids with coarse sand- to gravel-sized bioclastic matrix and occasional small lenses of finer green-grey poorly laminated sediment. The entire exposure, 800 meters long and up to 70 meters wide, is lithologically remarkably homogeneous and, in contrast to the Kuppen reef type, large domical stromatoporoids are very rare; only occasionally are high domes up to 30 centimeters in height seen. Rugose corals occur within some of the stromatoporoids, as at Kuppen, but there are no other conspicuous reef formers and the rock is dominated by laminar stromatoporoids. These are thin, usually 2–10 centimeters in thickness, wavy, and laterally rag-

ged. Their upper surfaces are often very undulose and even cusped and locally rise into miniature high domes up to 15 centimeters in height (Fig. 46). They split laterally and overgrow one another to form a tight frame in which organisms constitute 40–60 percent of the volume. There are no large sparite-filled cavities. Crinoid fragments up to 5 millimeters across make up most of the matrix, but locally much larger crinoids with stems up to 3 centimeters in diameter occur.

Manten (1971, enclosure 2) provides a 1:1000 plan of the Holmhällar rauk field on which he identifies many of the larger rauks by number. They are generally less than 4 meters high. The greatest stratigraphic thickness occurs immediately west of rauk 228 (the reference point for Holmhällar 1 in Laufeld, 1974b, fig. 9) where up to 6 meters of reef is exposed. Even here there is no obvious vertical sequence. The rock shows a crude layering due to the laminar frame structure and dips gently, about 3 degrees, to the southeast. Near the center of the rauk field, in the vicinity of rauk 137, crudely-bedded coarse crinoid gravels with rolled stromatoporoids at their base overlie the reef rock disconformably. At rauk 137B the gravel erosively channels into the reef with up to 1.5 meters of relief.

The reef here is also traversed by a number of narrow (approximately 50 centimeters wide) vertical fissures filled by coarse crinoidal gravel which appear to radiate from a point approximately 250 meters inland. The fill may be of uppermost Silurian (Pridoli) age on the basis of conodonts (Lennart Jeppsson, pers. comm., 1979). At the side of rauk 137B the fissures also appear to cut the coarse gravel overlying the reef.

At Holmhällar the reef deposits, which belong to the Sundre Beds, must almost directly overlie the Hamra reef horizon exposed at Hoburgen 10 kilometers to the west. Manten (1971, p. 189) believed the crescent-shaped outcrop of the Holm-

FIG. 46.—Laminar stromatoporoid frame structure of the Holmhällar biostrome at Holmhällar, Sundre Beds, southern Gotland. Commensal rugosans, crinoid gravel, and lack of diversity in either form or biota are consistent with a high energy environment in which stromatoporoids formed laterally anastomosing flattened or wavy sheets; hammer head is 17 centimeters long.

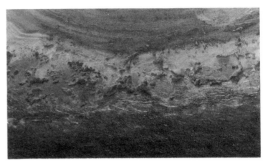

FIG. 47.—Partial aerial view of the Holmhällar rauk field, southern Gotland, showing the crescentric shape with beach-ridges behind.

hällar rauk field (Fig. 47) reflects the original plan of the reef and he regarded Holmhällar reefs, generally, as having this form. This is possible, but the absence of exposures of any lateral off-reef sediments makes it difficult to confirm or deny this interpretation. Rocky headlands on Gotland, as anywhere else, tend to be arcuate and this seems just as likely an explanation of the form of the rauk fields. The Holmhällar field is 500 meters from tip to tip of the crescent and the Hammarshagehällar rauk field 1 kilometer across the bay to the northeast is even larger. The Hammarshagehällar rock has a somewhat lesser volume of in place organisms and more matrix, but it is otherwise similar to that at Holmhällar and is composed essentially of laminar stromatoporoids and crinoid debris. The two outcrops could quite conceivably be parts of the same biostrome, and Holmhällar biostromes, like those of Kuppen type in the Östergarn district, could extend laterally for several kilometers.

*Other Localities.*—Manten (1971) included several rauk fields in the Hemse Beds, such as those at Ljugarn and Fågelhammar, in the Holmhällar reef type. Possibly the development of crescentic rauk fields at these localities encouraged him to do this, but the limestones themselves are only generally similar to Holmhällar itself in being biostromal; the stromatoporoids which dominate them have quite a different form and size. They are here (see above) included in the Kuppen type.

However, there is an element of transition locally between Kuppen and Holmhällar types, as already noted near Ljugarn 1. This is to be expected if the environmental synthesis (presented below) suggesting the Holmhällar type to be a higher energy reef than the Kuppen type is correct

since they are both relatively shallow water biostromal reefs. But in the principal examples exposed on Gotland the two types remain very distinct. Similarly, transitions occur between the Axelsro and Hoburgen reefs and although they are rarely exposed, between the Hoburgen and Kuppen types.

*Model of Gotland Reef Development*

The four main reef types described here from Gotland fit together into an environmental pattern which, ideally, can be seen within one phase of reef growth and which can also be placed in a broad stratigraphic-sedimentologic context allowing comparison with Silurian reefs in other areas (see Discussion). In other words, they are not isolated and mutually unrelated structures but represent responses of sessile organisms in the Silurian Baltic shelf sea to environmental variables which have imposed patterns and trends upon them. This can be seen most simply in vertical facies changes which reflect original lateral variations correlated with depth.

The enclosing sediment, biota, size, and geometry of the reefs described above suggest that the Axelsro, Hoburgen, Kuppen, and Holmhällar types are stages in the response of reef building organisms to progressively shallower and more turbulent conditions (Fig. 48). However, only part of this sequence (the Axelsro-Hoburgen transition of the Visby and Högklint Beds) appears at present to be seen continuously in vertical section, and so the model presented here is interpretive and provisional (Fig. 23).

In the Upper Visby to Högklint Beds sequence of the northwest coast the transition from Axelsro to Hoburgen reefs is clearly seen (Ted Nield and Nigel Watts, pers. comm., 1979), but the replacement of Hoburgen dense reef structure by frame reef in the upper part of this succession is only shown at a few localities, such as Gutevägen 2 and 4, due to both penecontemporaneous and Re-

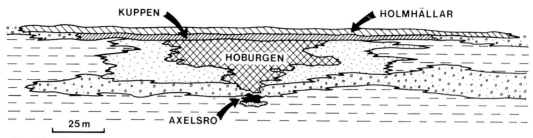

FIG. 48.—Size, geometry, associated sediments and inferred relative positions of bioherm and biostrome types in a shallowing-up sequence. Incidence of laminar organisms, frame structure and rigidity increase upwards.

cent erosion. Similarly, the upper part of Hoburgen reef development is eroded at Hoburgen itself (Storburg 2 and Hoburgen 4), but the Sundre Beds are only approximately 10 meters thick according to Hede (1960, p. 47) so that the equivalent of the Holmhällar reef exposed at Holmhällar 1 must originally have been directly above this horizon at Hoburgen. Although no exposures of Hoburgen reef passing into Kuppen or Holmhällar type are known at present it is suggested here that such a transition is to be expected. The basis for this reasoning lies in the interpretation of the environments of deposition of the reef types. The coarseness of the associated sediments, the tightness of the reef structure and the tendency towards a biostromal form all generally increase from Axelsro to Holmhällar type. In the same sequence there is a trend toward a reduction in biotic diversity and a greater likelihood of erosion of the reef surface. The reef matrix remains fine throughout, but the Kuppen and Holmhällar types are characterized by a reduction in argillaceous matrix and a tendency towards draping of the fill, suggesting filtering of fines down through a relatively tight organic framework. These features are consistent with the suggestion that Axelsro, Hoburgen, Kuppen, and Holmhällar types represent a spectrum of structures grading away from relatively deep, quiet water conditions of reef growth (Axelsro type) towards turbulent, shallow water where rigorous conditions resulted in low diversity, strong, biostromal reefs prone to exposure and erosion (Holmhällar type) (Fig. 23).

The shallow water conditions of formation inferred for the Kuppen and Holmhällar reefs limited upward growth, enhancing biostromal form in contrast to the bioherms of the Axelsro and Hoburgen types, and created the possibility of periodic exposure and resubmergence. The resulting erosion provided bevelled surfaces for subsequent recolonization, as seen at Kuppen 2, and so erosion had the effect not only of removing the upper surfaces of biostromes but also of providing areas for them to become established on. This disrupted trends in vertical reef development

and made it unlikely that a complete, unbroken sequence showing all four phases of reef growth could be preserved. Instead, a sequence with upper biostromes bounded vertically by disconformities (Fig. 48) is more likely to occur.

Organism type and form, biotic diversity, matrix and surrounding sediment composition and texture, as well as reef structure and geometry thus are all thought to reflect gross environmental constraints which are probably basically depth related. Detailed paleoecological considerations are outside the scope of this broad synthesis approach, but it can be noted that tabulates were more successful in quieter, muddier water (Axelsro type), and that stromatoporoid form becomes distinctly laminar in turbulent conditions (Holmhällar type). Highest diversity is shown in Hoburgen reefs, intermediate in the trend deduced here, and is lowest in the Kuppen and, especially, Holmhällar types where rugosans and tabulates were obliged to grow within stromatoporoids, presumably for physical support and protection. Calcareous algae, particularly solenoporaceans, are prominent in the upper levels of some Hoburgen reefs, as at Gutevägan 4. Manten (1971, p. 188) considers algae to be a significant component of the Holmhällar reefs, but he does not identify them more precisely.

Reef distribution can also be viewed in the context of the Gotland sequence as a whole (Fig. 4). Essentially, reefs appear to have developed abundantly at the seaward margins of the carbonate wedges but also occur well within them in association with coarse bioclastic, oolitic, and fine-grained, sometimes oncolitic, limestones. Basinward, argillaceous limestones, calcareous shales and mudstones were deposited. Manten (1971, p. 26) suggests that the coastline was generally to the northwest, trending northeast-southwest. But the configuration of the wedges rather suggests that they are thinning southwards indicating a coast to the north which was roughly trending east-west, although of course it is unlikely that any coast in this situation would have been linear in detail. The overall paleogeographic situation was one in which prograding carbonate wedges built out southwards (or southeastwards) into a shallow partially enclosed basin which deepened towards the Oslo-Baltic Syneclise. Carbonate progradation was accompanied by temporary exposure of the proximal carbonate belt, resulting in subaerial erosion. This is shown in the Högklint-Slite wedge by disconformities associated with the base of the Tofta Beds and indicates the regressive nature of this carbonate progradation. It was followed by a transgressive phase in which facies belts shifted northwards, resulting ultimately in the shale depositional environment moving over the underlying carbonate wedge. This pattern is repeated two or three times so far as the Gotland sequence is concerned and the overall sense of movement was regressive with successive carbonate wedges migrating southwards (Fig. 20). This is consistent with the general trend over the whole of the European craton towards the gradual establishment of terrestrial conditions by the Lower Devonian. Similarly, the distinct north-south polarity shown in the facies belts associated with the first two carbonate wedges (Högklint-Slite and Halla-Klinteberg-Hemse) appears to be breaking down in the upper part of the sequence. This could be due to a shallower, more embayed, mosaic facies pattern developing as the final regression approached, or to increased erosion removing the basal part of the third (Burgsvik-Hamra-Sundre) carbonate wedge. This latter interpretation is based on the possibility that extreme erosion during the regressive phase of carbonate development will locally strip off the entire lower part of the wedge so that subsequent transgression will take place over non-carbonate deposits. Stel and de Coo (1977) have interpreted the upper Burgsvik and lower Hamra-Sundre Beds in southern Gotland as transgressive deposits encroaching over a northeast-southwest trending shoreline (Stel and de Coo, 1977, figs. 27 and 28). It is suggested here that erosion removed preceding carbonate deposits in this area so that the third wedge is substantially thinner than it would otherwise have been. This wedge is thus anomalously thin for Gotland, although it should be emphasized that the preservation of the regressive sequences in the other carbonate wedges is really an unusual feature, peculiar to the Gotland sequence. Progradation of carbonate sediments due to the volume of carbonate production, closely dependent upon extensive reef development (as noted by Laufeld, 1974b, p. 7), seems here to have been at least as important as actual regression in constructing the wedges.

Over much of the shelf during carbonate deposition, patch-reefs of Hoburgen type developed in enormous numbers with inter-reef sediments of surprisingly low biotic diversity between them.

Eriksson and Laufeld (1978, p. 28) estimate that thousands of small Hoburgen bioherms exist in the Högklint Beds and suggest, with Hadding (1956), that larger reefs occupied a zone on the open sea side of them.

Beyond this must have been a zone of Axelsro reefs forming in quieter water and preferentially forming on tongues of crinoidal debris thinning basinward. Nearer shore, probably in a narrower belt than the extensive Hoburgen patch-reefs occupied, Kuppen and Holmhällar biostromes colonized shallow and relatively turbulent environments possibly, if Manten's (1971, p. 189) view is correct, they were crescentic in plan. The Holmhällar reefs in particular suggest a very shallow environment which could have been a fringing-reef or a barrier, but there is no evidence at present of back-reef environments associated with them.

*Estonia*

Bioherms are reported at a number of levels in the Silurian of the eastern Baltic, principally in the lower Llandovery (G1–2), upper Wenlock (J2) and middle Ludlow (K2) (Kaljo, 1970, p. 340–342, fig. 81, 85, and 87). The paleontology of important reef building organisms such as stromatoporoids (Nestor, 1964 and 1966), and tabulates (Klaamann, 1962) has been well documented in the area, but I know of few detailed studies of the bioherms themselves.

Outcrops in Estonia (Fig. 19), together with borehole data from areas to the south in Latvia and Lithuania, indicate a relatively complete Silurian sequence which is wholly in marine or marginal marine facies and is 280–640 meters thick (Kaljo, 1970, p. 340–342). The Baltic coast of the USSR is only 150 kilometers east of Gotland and there are close similarities between the Silurian sequences of the two areas. The eastern Baltic succession accumulated in a broad embayment northeast of the Oslo-Baltic Basin which Nestor and Einasto (1977, fig. 1 and p. 119) regard as a pericontinental sea. As in Gotland the "most general trend in the development of the East Baltic Basin was the gradual regression of the sea, only at times interrupted by relatively short transgressions" (Kaljo and Jürgenson, 1977, p. 148). Kaljo (1970, p. 343) recognizes three major sedimentary cycles in the sequence and emphasizes tectonic control over deposition in the area. Three general episodes of siltstone-shale deposition occur: during the lower Llandovery, upper Llandovery, and upper Ludlow, and can be correlated with transgressions (Kaljo, 1970, p. 330–331). Facies analysis reveals a clear shelf-basin polarity in which bioherm development was located above wave-base in the inner shelf zone (Nestor and Einasto, 1977, fig. 3).

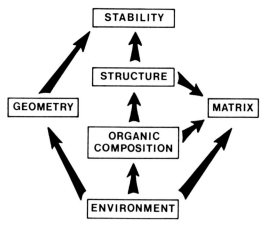

FIG. 49.—Principal interrelations of reef controls and effects.

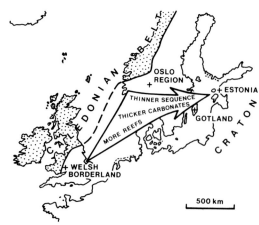

FIG. 50.—Broad trends in the character and distribution of Silurian reef sequences in northwestern Europe.

Carbonate rocks accumulated mainly in the northern and northeastern parts of the area and consequently dominate the Estonian outcrops, while to the south and southwest calcareous and graptolitic shales were deposited (Kaljo, 1970, figs. 80–89; Kaljo and Jürgenson, 1977, figs. 1–8). In contrast to Gotland, dolomites and partly dolomitized limestones are present (Kaljo, 1970, p. 330) and clayey dolomicrites form the near-shore facies belt of the sequence (Nestor and Einasto, 1977, p. 120–121), e.g., in the lower Ludlow of southern Saaremaa (Kaljo, 1970, p. 341, fig. 86). Bioherms occur in the next zone seaward from this, associated with skeletal, oöid and peloid grainstones (Nestor and Einasto, 1977, p. 120).

Aaloe and Nestor (1977) describe bioherms from the lower Llandovery of the Ridala Peninsula and the island of Hiiumaa (Fig. 19) on the western coast of Estonia. They occur at two levels. The lower ones are very small aulocystid-algal structures in the Ridala Member (G1–2R), a discontinuous unit only 1–2 meters thick (Aaloe and Nestor, 1977, figs. 2–4). Less than 10 meters above this, larger bioherms occur in the Hilliste Member (G1–2H) which is up to 7 meters thick. These are relatively complex and variable structures up to 8 meters or more across (Aaloe and Nestor, 1977, fig. 8), composed of tabulates, stromatoporoids, colonial rugosans, algae, bryozoans and heliolitids, and associated with crinoidal limestones. These bioherms appear to be broadly comparable with those of Axelsro type in Gotland.

### DISCUSSION AND SYNTHESIS

Gotland provides the key to understanding the Silurian reefs of western Europe because it contains by far the most complete spectrum of structures, from deeper water tabulate dominated argillaceous bioherms (Axelsro type) through Hoburgen bioherms and Kuppen biostromes to the shallow water, low diversity, stromatoporoid frame biostromes of Holmhällar type. These varieties of in place skeletal accumulation were generated by interaction between the local physical environment and the type of sessile calcareous organisms available at the time. Between them, environment and organisms controlled the structure, geometry and other features of the reefs (Fig. 49).

### Reef Types

The naming of reef varieties after localities, as has been done for Gotland, brings to mind the proliferation of rock terms in igneous petrology and is not a welcome course of action. But in the absence of a comprehensive nomenclature of reef description and classification it is difficult to avoid. It does have the advantage of compressing numerous compositional and structural features into a single name, and provided that the temptation to progressively split types to take account of only minor variations is resisted then the scheme may have long term value.

To the four main Gotland reef types (Axelsro, Hoburgen, Kuppen, Holmhällar) can be added the thin, tabulate dominated biostrome seen at Irevik in the Lower Visby Beds which has counterparts in the Steinsfjord Formation of the Oslo Region. The Fanterna and Stäurnasar types of the Karlsö Islands require further study. Axelsro-type bioherms occur at Wenlock Edge in the Welsh Borderland and the Hilliste bioherms in the lower Llandovery of Estonia are probably also of this type. Most of the Oslo Region bioherms also appear to be of Axelsro type. The lower Wenlock bioherms of Ringerike, such as those on the west side of Geitøya, however, are distinct variants characterized by abundant *Girvanella, Rothpletzella,* and *Wetheredella*. These encrusters are common in Axelsro-Hoburgen reefs but are not usually dominant organisms.

So the picture which emerges is that the great majority of Silurian reefs in northern Europe can be compared directly with those seen on Gotland but that the latter include at least two major types (Kuppen and Holmhällar) not seen elsewhere. Examples in England and Norway can only be compared with Axelsro (plus the Irevik variant), and possibly the dense and less frame parts of Hoburgen reefs. The Estonian reefs are too poorly known to make real comparisons at present. Thus, it seems that reefs outside the Baltic area represent only the deeper, muddier varieties of a spectrum continuing in Gotland into shallower and more turbulent conditions.

### Environment and Distribution

Silurian reefs developed in shallow, probably photic, carbonate environments often with high admixtures of argillaceous material. The organisms were capable of colonizing muddy substrates, but most of them show a preference for gravel bases such as the crinoid lenses below Axelsro reefs at Wenlock Edge and on Gotland and the crinoid limestones flooring the Hoburgen reefs of the Högklint Beds. The on-craton situation meant that regional sea floor topography was subdued. The only evidence of significant slopes associated with reef facies in the area at this time that I know of is on Lilla Karlsø, west of Gotland. Degree of water movement probably ranged from low to moderate around Axelsro reefs to very high over Holmhällar biostromes. The rigidity and stability of the reef structure increased along this gradient.

*Spatial Distribution.*—Enhancement of reef development cratonward, and away from the Caledonian mobile belt, correlates with increase of carbonate sedimentation and thinning of the total marine sequence (Fig. 50). Reef growth was clearly favored by the shallow conditions with low influx of siliciclastic sediment which were maintained for long periods in the Baltic area. This allowed a wide spectrum of reef types, deeper and shallower, patch and linear, to form. Closer to the mobile belt, in England and Norway, deeper water and sediment influx inhibited reef growth and restricted its scale and diversity.

The local lateral distribution of reefs in any one area is best inferred from their vertical distribution in time (see below) but both in the Wenlock Limestone of the Welsh Borderland and in the Högklint Beds of Gotland large areas, probably 1000 square kilometers or more in extent, were covered by patch-reefs and inter-reef sediments. Thin, but extensive biostromes developed both in the Oslo Region and in Gotland. Reef belts were fronted by deeper shale accumulating environments and backed by either low- or high-energy peritidal facies. The Gotland sequences promise to provide detailed records of the spatial distribution of these environments but, to date, facies analysis of them has barely begun and whether the Kuppen and Holmhällar biostromes, which probably were linear reefs, were near- or offshore in position is uncertain.

*Vertical Distribution.*—Reefs are associated with both shallowing-up (regressive) and deepening-up (transgressive) carbonate sequences and are limited mainly to the shallow subtidal portions of them. In Gotland reef development appears to have been concentrated in the outer zones of carbonate wedges adjacent to the outer shelf shales (Fig. 20). The inner parts of the wedges are dominated by peritidal and shoal carbonates and are preferentially associated with disconformable truncations of the sequence which locally bevel reefs.

The idealized reef succession in a regressive sequence presented in the model of Gotland reef development is inferred from disjointed sequences on the island; the most complete of which occurs in the Visby-Högklint Beds of the northwest coast. In England and Norway only the lower parts of this sequence occur, the remainder failing to develop due to adverse environmental conditions. The Gotland model can be used to predict what could have happened in some of these other areas if conditions had been different. For example, at Wenlock Edge the shallowing upward sequence from Wenlock Shale to Wenlock Limestone accompanied by development of Axelsro-type bioherms is directly comparable with the Visby-Lower Högklint Beds sequence on Gotland. The crinoid grainstones (facies D, "gingerbread") at Wenlock have their counterpart in the crinoidal limestones at the base of the Högklint Beds. But whereas, at Wenlock, these gravels appear to terminate reef development, on Gotland they can be seen to represent only a pause in reef formation since they are succeeded directly by the larger and compositionally more complex Hoburgen reefs of the main Högklint sequence. The gravels represent a crinoid shoal belt prograding outwards over the Axelsro reefs and ahead of the Hoburgen bioherms. Consequently, Scoffin's (1971, p. 216) conclusion that "extreme shallowing terminated reef development in the area" at Wenlock Edge needs to be modified in the light of the Gotland succession. The "gingerbread" facies may well have been succeeded by enhanced reef formation if it had not been for the deposition of the silts of

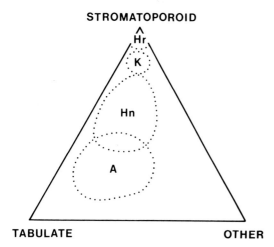

FIG. 51.—Estimated relative organic composition of reef types in terms of main and accessory builders. "Other" includes algae, rugosans, and bryozoans; A, Axelsro; Hn, Hoburgen; K, Kuppen; and Hr, Holmhällar.

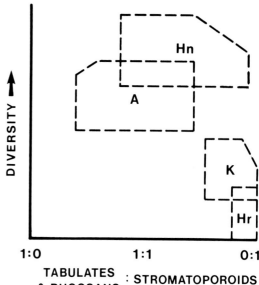

FIG. 52.—Rough estimates of relative diversity and major organism composition in the main reef types; A, Axelsro; Hn, Hoburgen; K, Kuppen; and Hr, Holmhällar.

the Elton Beds. There is no evidence of "extreme" shallowing at Wenlock and reef growth failed to resume because of a regional influx of fine siliciclastic sediment. This latter was the main factor responsible for poor reef formation everywhere adjacent to the mobile belt.

The vertical succession of reef and related facies mirrors their spatial distribution and, where the sequence is relatively complete, as on Gotland, indicates the original lateral variation in composition, structure and geometry of the reef masses. In the shallowing-up Visby-Högklint sequence Axelsro reefs occupied an off-shore belt adjacent to the deeper shelf muds. They passed shorewards into crinoid shoals and then Hoburgen patch-reefs with restricted lagoonal/inter-reef sediments (Fig. 23). Upper coalescence of the Hoburgen bioherms in the Högklint Beds into biostromal masses may reflect the existence of linear reefs of fringing or barrier type separating the patch-reef complex from a peritidal beach or tidal flat environment. The thinness of exposed sequences associated with Kuppen and Holmhällar biostromes and their truncation by erosion surfaces hamper similar analysis, but these biostromes are interpreted to be higher energy varieties of the linear reefs capping the Högklint sequence.

### Composition

*Organisms.*—Tabulates and stromatoporoids are the major groups of reef builders present in the Silurian of northern Europe and they represent two ends of a compositional spectrum (Fig.

51). Axelsro type biostromes and small bioherms are tabulate dominated. Holmhällar biostromes are almost solely built by stromatoporoids. Larger Axelsro bioherms, Hoburgen reefs, and Kuppen biostromes contain different proportions of these organisms. The principal genera involved in reef construction are the stromatoporoids *Actinodictyon, Actinostroma, Clathrodictyon, Densastroma, Labechia, Parallelostroma, Plectostroma, Stromatopora,* and *Syringostroma* (Scoffin, 1971; Mori, 1968, 1970; Steve Kershaw, pers. comm., 1979), and the tabulates *Alveolites, Favosites, Halysites, Heliolites, Syringopora,* and *Thecia.* In addition rugose corals (such as *Acervularia, Amplexopora, Entelophyllum, Phaulactis*), calcareous algae (*Girvanella, Solenopora*), algal Problematica (*Rothpletzella, Wetheredella*), and bryozoans (*Coenites, Fenestella, Fistulipora, Hallopora, Rhombopora, Thamniscus*) are present, usually in lesser quantities. Rugosans and bryozoans are prominent in Axelsro and Hoburgen reefs, calcareous algae are conspicuous in some Hoburgen reefs. *Rothpletzella* and *Wetheredella* form crusts in Axelsro and Hoburgen reefs and are dominant in some of the lower Wenlock bioherms at Ringerike, Norway. Non-skeletal stromatolites described by Scoffin (1971) in the Wenlock Edge reefs may also be important elsewhere but have yet to be reported.

?Spirorbid worms are common, but tiny, additions to the reef mass and free living gastropods such as *Euomphalopterus* and *Euomphalus* (Ted Nield, pers. comm., 1979) are additional inhabitants. Reef associated brachiopods include *Atrypa, Camarotoechia, Cyrtia, Delthyris, Dicoelosia, Eospirifer, Leptaena, Platystrophia, Rhynchotreta* and many more (Manten, 1971; Bassett and Cocks, 1974).

Diversity of preserved organisms increases from moderate in Axelsro bioherms to high in Hoburgen, but then reduces dramatically in Kuppen and Holmhällar biostromes (Fig. 52). Similar trends of increasing diversity with shallowness, followed by reduction in the more rigorous surfzone, are shown by Recent scleractinian reef communities. The stromatoporoids which dominate the Kuppen and Holmhällar biostromes appear to have been the only organisms capable of flourishing in these high energy conditions.

The success of stromatoporoids in shallow reef environments can be attributed mainly to their size, strength and, probably, relatively rapid rates of growth. But they were also apparently tolerant of loose sediment, both coarse and fine, and readily grew laterally to recolonize sediment deposited on the edges of coenostea producing *ragged* margins (Kershaw and Riding, 1978, p. 234). Tabulates appear to have been even better suited to the muddy, and probably somewhat deeper conditions in which Axelsro bioherms and biostromes formed. In this respect major Silurian reef builders were particularly well suited to growth in environments with relatively abundant fine sediment and this equipped them well for colonization of the soft, muddy, level bottom substrates which were widespread over the European craton during the Silurian.

*Matrices and Cavities.*—Mutual arrangement of the organic skeletons (e.g., the structure) in the reefs controlled the degree of volume occupied by loose bioclastic and argillaceous matrix and also the proportion of protected space which could remain unfilled as cavity. In these Silurian reefs the matrix is usually fine and argillaceous, often being a greenish micrite, and the cavity space is slight. Large open spaces in the Axelsro type reefs of Wenlock Edge tended to be infilled penecontemporaneously with reef growth (Scoffin, 1972, p. 566) despite the effect of non-skeletal stromatolites in converting an essentially dense structure into a frame. Of the original cavities only small ones, usually less than 1–2 centimeters in size, between or within skeletons escaped fill (Scoffin, 1972). This applies also even in the tighter frames of the Kuppen and Holmhällar reefs where fines, pumped or settling through the structure, show draped infilling of spaces. The filtering effect of these frameworks is reflected by the absence of coarse material near the centers of the reefs. In

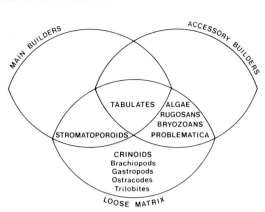

FIG. 53.—Principal sedimentological roles of calcified organisms in Silurian reefs.

more open structures the matrix is less well sorted and includes admixtures of recognizable skeletal debris; mainly crinoid, brachiopod, ostracode and trilobite fragments.

### Structure and Sedimentological Roles

Stromatoporoids and tabulates were the *main* reef builders; *accessory* builders either were attached to these large skeletons (e.g., calcareous algae, bryozoans, *Rothpletzella, Wetheredella,* ?spirorbids), or grew near them (e.g., rugosans); and all these organisms, together with those which spontaneously fragmented after death (crinoids, brachiopods), or which were free living (gastropods, *etc.*), added loose *matrix* which filled hollows of flanked skeletons (Fig. 53). Boring organisms are present in Silurian main reef builders (Stel, 1978a), but they do not appear to have substantially modified the strength of skeletons.

The Silurian reefs of northern Europe may show examples of all the main structural reef types which stromatoporoids were capable of producing. The irregular "ballstone" lenticles within Axelsro and Hoburgen reefs were produced by the amount of fine argillaceous matrix present and the fundamentally dense structure of these bioherms. They had important effects upon both the early and later stability of these accumulations (see Post-depositional Effects). They are absent in the tighter biostromes but what no stromatoporoid appears ever to have developed is an extensively branched form. Consequently, the cavernous frame structure of some scleractinian reefs is lacking in Middle Paleozoic reefs and the closest approach to a real framework is in the mutually encrusting, close packed, and varied morphotypes of the Kuppen biostrome and in the laminar frame of the Holmhällar type.

Calcareous algae, together with the problematic

*Rothpletzella* and *Wetheredella,* constitute the principal accessory encrusters of the reefs. Non-skeletal stromatolites had a similar role in binding and stabilizing the structure.

The coherence of a reef depends fundamentally upon its structure, which is determined by the shape and mutual arrangement of the component skeletons. Encrusting organisms certainly played a part in binding adjacent skeletons together in these reefs, but in most of the examples I have seen they have not been sufficiently abundant to override the effect of the shape of the *main* reef builders. Consequently structures dominated by domical and bulbous stromatoporoids, as is the case with parts of many Hoburgen reefs, were fundamentally unstable and were probably unable to raise themselves much above the local sea floor topography. Flatter skeletal forms were much more stable and the laminar frame Holmhällar reefs show hardly any disorientation of the constituent stromatoporoids.

It is not uncommon in European soft-rock circles to hear "off-the-cuff" remarks to the effect that "Silurian reefs are just debris piles." Overturned skeletons are, indeed, relatively common in Axelsro and Hoburgen reefs, but even in 1914 Crosfield and Johnston (p. 221) emphasized that the great majority of corals and stromatoporoids in the Wenlock Edge reefs are still in place. Since Silurian reefs of Europe range from dense argillaceous to frame structures it is not surprising that those at the "looser" end of this spectrum were susceptible to penecontemporaneous toppling of individual skeletons and local slumping of bioherm margins (see Post-depositional Effects). Local collapse of bioherm margins or toppling of vertically extended skeletons does not, in my view, constitute the difference between a reef and a pile of debris; as Scoffin (1971, p. 191) noted at Wenlock Edge "talus bands are rare," and this is my experience too.

A problem somewhat separate from the question of whether skeletons are in growth position concerns whether bioherms were skeletally constructed at all. In the first case a "bioherm" may be considered really to be a loose pile of moved skeletal material. In the second case, it may be found to lack skeletons altogether! This latter view has been argued for some of the Wenlock Edge bioherms by Abbott (1976). He considers that they lack an adequate skeletal framework and are really mud-banks localized and created through baffling of fine sediment by crinoids. The bioherm which he studied in detail is an Axelsro-type structure in Coates Quarry composed of micrite and *Halysites*. But this tabulate occupies only scattered parts of the bioherm and could only have influenced "a maximum of 23 percent of the structure. This, on its own, would have been inadequate to trap and stabilize the large quantities of lime mud . . ." (Abbott, 1976, p. 2123). Abbott's alternative explanation of crinoid baffling is stimulating, but it too seems to suffer from the same weakness for which *Halysites* is discarded, viz., there is relatively little crinoid debris in the mound. Abbott emphasizes the presence of crinoid material below and lateral to the mound but does not suggest that more than "scattered plates and columnals are present throughout the bioherms" (p. 2125). Bafflers need to be present in quantity to be effective and are very unlikely to be removed more easily from their habitat than the sediment they localize.

Seen in the perspective of the Silurian reef spectrum recognized here, the bioherm Abbott emphasizes is a sparse to dense *Halysites* structure. It is comparable with some Axelsro type reefs in Oslo and Gotland which do grade laterally into level bottom communities. Nevertheless, it does contain an unusually high proportion of micrite and its origin is not clear. But it is not valid to extend the uncertainty caused by this example to all Wenlock Edge bioherms since the majority, together with their counterparts in other areas, contain substantial proportions of tabulates and stromatoporoids. These do not usually contain the frameworks required by Abbott (1976; Abstract) to allow their comparison with "coral reefs," but they do normally possess a dense structure created by the close, in place growth of skeletized organisms.

Watkins (1979) also takes what may be termed an "understated" view of reef building by organisms in the Högklint bioherms near Visby. His detailed measurements are a valuable contribution to knowledge of Högklint reef composition, but it is possible to disagree with his interpretation of them. He (1979, p. 48) regards these Hoburgen type structures as mud-mounds on the grounds that micrite makes up "over half of the volume" (p. 41–42). So far as I know there is no formal definition of a carbonate mud-mound, but both ancient and modern examples have far higher proportions of mud than the 53 percent measured in these mounds. Watkins' (1979, fig. 7) counts show that stromatoporoids alone constitute approximately 15 percent of sediment volume in the bioherm sequence at Korpklint 1, with algae occupying at least a further 7 percent. Rugosans, tabulates and bryozoans are additional components and Watkins' own figures show that around one quarter of the bioherm is composed of skeletons which are in place or reef derived. But he concludes that "organisms did not actually *build* the Högklint bioherms in the area studied. Instead, they successfully exploited an environment of continually accumulating carbonate mud" (p. 42). I think that this exaggerates the importance

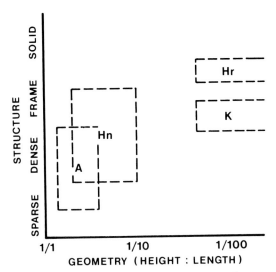

FIG. 54.—Relationship between structure and geometry in Axelsro (A), Hoburgen (Hn), Kuppen (K), and Holmhällar (Hr) bioherms and biostromes.

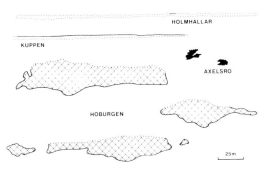

FIG. 55.—Relative cross-sectional shapes and sizes of Silurian reef types; dotted lines indicate erosion or cover.

of physical sedimentation, and significantly underestimates the sedimentological role of the principal skeletal organisms.

### Geometry

The tendency of Hoburgen reefs in shallowing-up sequences (e.g., the Högklint Beds) to flatten and coalesce upwards shows that bioherm and biostrome developments were not wholly separate, and an idealized shallowing sequence shows how these reef geometries may have been related (Fig. 48). Nevertheless, there appears to be a clear distribution between Axelsro-Hoburgen bioherms and Kuppen-Holmhällar biostromes based on presently available exposures both in terms of geometry and structure (Fig. 54). Axelsro and Hoburgen bioherms are irregularly lensoid in cross-section. This becomes converted into an inverted hat-shape where the Hoburgen reefs grade upwards into extended biostromal units (Fig. 55). The basal parts of the bioherms were often rounded in plan view. This is clearly demonstrated by the Hoburgen "*Philip structures*" impressed into the Visby Beds of Gotland's northwest coast (Fig. 33) which are sometimes virtually circular. Oval plans are shown by the Axelsro bioherms of Wenlock Edge (Scoffin, 1971, p. 186).

The biostromes are all very thin bodies with height-length ratios normally exceeding 1:100. Their plan views are uncertain, but may have been irregularly linear. Extremely shallow water is inferred to have been a major control on their overall geometry and their thinness has been enhanced by early post-depositional erosion in littoral-subaerial environments which resulted in bevelling, channelling and crevice formation on their upper surfaces, which are well seen at the Kuppen and Holmhällar localities themselves.

The original local vertical relief of some of the bioherms may have been of the order of 7–10 meters (Nigel Watts, pers. comm., 1979). Kuppen and Holmhällar biostromes probably had the coherence to attain even greater relief, but whether they did is not yet known.

### Post-depositional Effects

Paucity of accessory binding organisms, vertically extended domical or bulbous skeletons, and soft argillaceous substrates, combined in some Hoburgen bioherms, especially those with a dense non-frame structure, to make them inherently unstable. In contrast, Kuppen and Holmhällar frame biostromes appear to have been rigid structures and show no collapse features. These tendencies toward instability or rigidity developed as reef growth progressed and they conditioned later internal responses of the structures to compactional effects. Hoburgen and Axelsro reefs show relative movement of masses within individual bioherms because these sub-units are bounded partly or completely by argillaceous seams which facilitate sliding. The existence of semi-discrete rounded clay-bounded masses within these bioherms is one of the essential features of their structure, and was described early on as "ballstone." It led to incoherence within the reefs which generated lateral spalling (Fig. 56), and which also responded to subsequent overburden by differential movement (Watkins, 1979, p. 36). The relatively rigid frame biostromes took up compaction by intense stylolitization of adjacent skeletons.

Two factors which generate instability in Recent scleractinian reefs, however, were only of minor importance in the Silurian reefs. These are boring of skeletons by organisms and the presence of steep slopes in the reef environments.

TABLE 3.—SEQUENCE OF MEGASCOPIC STRUCTURAL
MODIFICATION EFFECTS IN AXELSRO-HOBURGEN
BIOHERMS

| Syn-depositional | | Post-depositional | |
|---|---|---|---|
| Early | Late | Early | Late |
| lateral spalling ⟶ | | | |
| | ⟵ internal fracture and rotation ⟶ | | |
| | ⟵ sediment injection at base ⟶ | | |
| | compaction of underlying sediment ("*Philip structure*") | ⟶ | |
| | | compaction of ⟶ overlying sediment ("*Cumings structure*") | |

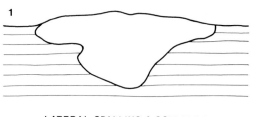

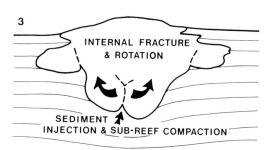

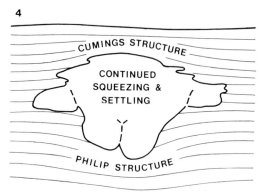

The sequence of development of megascopic structural displacement features in and around Axelsro-Hoburgen bioherms changes from early collapse of reef margins to internal fracture and rotation of masses as the weight of the overlying reef increases (Table 3). The latter effect initiates sub-reef compaction which generates sediment injection. Both compaction and internal movement are increased by subsequent overburden. Consequently, early post-depositional features are principally internal squeezing and rotation of "ballstone" masses and marginal collapse, which give way progressively to external differential compaction effects on the surrounding sediments which create "*Philip*" and "*Cumings structures.*" In contrast, Kuppen and Holmhällar biostromes appear to have been much more resistant to internal deformation due to their rigidity, engendered by a frame-solid structure. But they do show intense internal stylolitization effects, particularly in the Kuppen reefs. Effects on the sediment adjacent to these biostromes are difficult to assess due to lack of exposures.

Apart from Scoffin's (1972) study of cavities in Wenlock Edge bioherms, little work has yet been published on the diagenetic history of Silurian reefs in Europe.

### CONCLUSIONS

The principal reef building organisms during the Silurian in the northern European area were stromatoporoids and tabulate corals with locally significant contributions by rugose corals, calcareous algae, bryozoans and non-skeletal stromatolites. The majority of the reefs were bioherms formed by a mixed assemblage of these organisms and associated with a variety of sur-

FIG. 56.—Stages in the evolution of megascopic displacement features of Axelsro-Hoburgen bioherms caused by penecontemporaneous and early postdepositional differential movement and compaction. 1. and 2. early stages of deposition; 3. late depositional stage; 4. early postdepositional stage. "*Cumings*" and "*Philip structures*" are compactional anticlines and synclines respectively; also see Table 3.

rounding sediments, from calcareous shales and argillaceous limestones to coarse bioclastic grainstones, the latter being a preferred substrate. The internal structure is characteristically a dense to frame accumulation with sub-units outlined by thin shaly layers which impart a general weakness reflected by settling, slumping and differential movement of parts of the bioherms. In Gotland, reefs of these types include the Axelsro and Hoburgen bioherms and this recognition of distinctive varieties of reef can be extended to other areas. The Wenlock Edge reefs, for example, are of Axelsro type as are the Fronian bioherms at Ringerike and Baerum in the Oslo Region. Biostromes are usually less diverse in composition and structure. Relatively quiet water biostromes dominated by tabulates occur in the Steinsfjord Formation (Stage 9c) of the southern Oslo Region and also in the Upper Visby Beds of Gotland at Irevik. Stromatoporoid biostromes representing shallow turbulent environments include the Kuppen and Holmhällar types of Gotland. These may represent fringing-reefs and contrast with the patch-reef developments of Hoburgen type on Gotland which are associated with extensive inter-reef sediments, as for example in the Högklint Beds.

Regional variation in reef development is correlated with position relative to the Caledonian mobile belt and to the craton interior. Reef development appears to have been limited in the Welsh Borderland by fine siliciclastic sedimentation and by the relatively deep water situation. Reefs only occur in the late Wenlock during a temporary shallowing of the sea and reduction of siliciclastic sediment influx. They are of Axelsro type which can be regarded as the deepest of the main reef types.

In the Oslo Region reefs also show signs of the controlling effect of broad sedimentational patterns: bioherms and biostromes are restricted to the upper Llandovery and Wenlock and all show variations on the basic Axelsro type, e.g., they are dominated by tabulate corals and *Rothpletzella-Wetheredella* crusts, and are generally enclosed by relatively fine-grained sediments.

The Gotland sequence is both the thinnest and the most carbonate-rich of the four sequences, and it also contains the most extensive and diverse reef developments. Four distinctive reef types can be recognized: Axelsro, Hoburgen, Kuppen, and Holmhällar, which broadly represent a trend from deeper, tabulate dominated bioherms with a relatively weak internal structure to shallow, stromatoporoid dominated biostromes with a strong rigid framework. Reefs range in age from upper Llandovery to upper Ludlow and can be related to a gross sedimentational pattern of interdigitating carbonate and shale wedges which represent shallowing-up (regressive) and deepening-up (transgressive) phases.

The Estonian sequence is comparable with that in Gotland but the reefs, which also occur at several horizons, are less well documented.

The Axelsro-Hoburgen-Kuppen-Holmhällar range of reef types exhibits contrasting geometric, compositional and structural features which are controlled by environmental factors, on the one hand, and which largely determine their physical stability and responses to post-depositional effects on the other (Table 2). Restriction of the full spectrum of types to the Baltic area reflects the importance of regional controls on reef development and it makes Gotland the prime location for studying the range of in place skeletal accumulations formed in the European area during the Silurian. These reef structures were constructed mainly by stromatoporoids and tabulate corals, and the environmental requirements, growth rates and morphologies of these still problematic organisms determined their essential features.

#### ACKNOWLEDGMENTS

It is a pleasure to acknowledge the help and stimulus of Steve Kershaw, Ted Nield and Nigel Watts at Cardiff who have all collaborated in Silurian reef work on Gotland, and who have provided much information and discussion. For help in understanding reefs in the Oslo Region I am indebted to Snorre Olaüssen, who kindly showed me localities in Tyrifjord, and to Nils Martin Hanken, Snorre Olaüssen and David Worsley who together have generously made available unpublished results of their work which have greatly improved this paper. My stay at the Paleontological Museum, Oslo, was made possible by Gunnar Henningsmoen.

I am particularly grateful to Anders Martinsson and Kent Larsson for providing facilities at the Allekvia Field Station on Gotland through Project Ecostratigraphy. This help has been invaluable in pursuing research on Gotland. Project Ecostratigraphy also provided flying time for aerial photography and I am indebted to Arne Philip for piloting, as well as for kind hospitality and innumerable other aids on the ground on Gotland. Sven Laufeld has helped greatly by showing me localities and providing information and discussion on Gotland geology. I owe Michael Bassett special thanks for his help which has made the drawing together of material for this paper much easier than it might otherwise have been. Rodney Watkins kindly allowed me to see the manuscript of his paper on Högklint bioherms prior to publication.

Michael Bassett, Sven Laufeld, Terry Scoffin, Nigel Watts, and David Worsley critically reviewed various parts or all of the manuscript and their help certainly improved it. This research was supported by funds from the Natural Environment Research Council.

## REFERENCES

AALOE, A., AND NESTOR, H., 1977, Biohermal facies in the Juuru Stage (Lower Llandoverian) in northwest Estonia, in Kaljo, D., Ed., Facies and fauna of the Baltic Silurian: Tallinn, Academy of Sciences, Estonian S.S.R., Institute of Geology, p. 71–88.

ABBOTT, B. M., 1974, Environmental studies of the Wenlock Limestone and comparisons with modern carbonate environments [Ph.D. thesis]: England, Milton Keynes, Open Univ., 286 p.

——, 1976, Origin and evolution of bioherms in Wenlock Limestone (Silurian) of Shropshire, England: Am. Assoc. Petroleum Geologists Bull., v. 60, p. 2117–2127.

BASSETT, M. G., 1974, Review of the stratigraphy of the Wenlock Series in the Welsh Borderland and South Wales: Palaeontology, v. 17, p. 745–777.

——, 1976, A critique of diachronism, community distribution and correlation at the Wenlock–Ludlow boundary: Lethaia, v. 9, p. 207–218.

——, 1979, Discussion of 'Ordovician and Silurian changes in sea level' by McKerrow, W. S.: Jour. Geol. Soc., v. 136, p. 144.

——, AND COCKS, L. R. M., 1974, A review of Silurian brachiopods from Gotland: Fossils Strata, No. 3, 56 p.

——, COCKS, L. R. M., HOLLAND, C. H., RICKARDS, R. B., AND WARREN, P. T., 1975, The type Wenlock Series: Rept. Inst. Geol. Sci., No. 75/13, 19 p.

——, AND RICKARDS, R. B., 1974, Notes on Silurian stratigraphy and correlation in the Oslo district: Norsk Geol. Tidsskr., v. 51, p. 247–260.

BJØRLYKKE, K., 1974, Depositional history and geochemical composition of Lower Paleozoic epicontinental sediments from the Oslo Region: Norges Geol. Unders., v. 305, p. 1–81.

BOUCOT, A. J., 1974, Silurian and Devonian biogeography, in Ross, C. A., Ed., Paleogeographic provinces and provinciality: Tulsa, Oklahoma, Soc. Econ. Paleontologists Mineralogists Spec. Pub. No. 21, p. 165–176.

BROOD, K., 1976, Bryozoan paleoecology in the Late Silurian of Gotland: Palaeogeography, Palaeoclimatology, Palaeoecology, v. 20, p. 187–208.

BUTLER, A. J., 1939, The stratigraphy of the Wenlock Limestone of Dudley: Quart. Jour. Geol. Soc. London, v. 95, p. 37–74.

COLTER, V. S., 1957, The paleoecology of the Wenlock Limestone [Ph. D. thesis]: England, Univ. Cambridge, 311 p.

CROSFIELD, M. C., AND JOHNSTON, M. S., 1914, A study of ballstone and the associated beds in the Wenlock Limestone of Shropshire: Geol. Assoc., Proc., v. 25, p. 193–224.

DAVIDSON, T., AND MAW, G., 1881, Notes on the physical character and thickness of the Upper Silurian rocks in Shropshire, with the Brachiopoda they contain grouped in geological horizons: Geol. Mag., ser. 2, v. 8, p. 100–110.

DONS, J. A., AND LARSEN, B. T., Eds., 1978, The Oslo paleorift, a review and guide to excursions: Norges Geol. Unders., v. 337, 199 p.

EDWARDS, H. M., AND HAIME, J., 1855, A monograph of the British fossil corals: Monogr. Palaeont. Soc., v. 5, p. 245–299.

ERIKSSON, C.-O., AND LAUFELD, S., 1978, Philip structures in the submarine Silurian of northwest Gotland: Sver. Geol. Unders., v. 736, p. 1–30.

GREIG, D. C., WRIGHT, J. E., HAINS, B. A., AND MITCHELL, G. H., 1968, Geology of the country around Church Stretton, Craven Arms, Wenlock Edge and Brown Clee: London, Geol. Survey Great Britain, Memoir, v. 166, 379 p.

HADDING, A., 1941, The pre-Quaternary sedimentary rocks of Sweden, VI, reef limestones: Lunds Univ. Årsskr., N.F., 2, 37, v. 10, 137 p.

——, 1950, Silurian reefs of Gotland: Jour. Geology, v. 58, p. 402–409.

——, 1956, The lithological character of marine shallow water limestones: Kgl. Fysiograf. Sällskap. Lund Förh., v. 26, p. 1–18.

——, 1959, Silurian algal limestones of Gotland: Lunds Univ. Årsskr., N.F., 2, 56, v. 7, p. 1–25.

HAINS, B. A., 1970, The geology of the Wenlock Edge area: London, Inst. Geol. Sci., 61 p.

HAMBLIN, R. J. O., WARWICK, G. T., AND WHITE, D. E., 1978, Geological handbook for the Wrens Nest National Nature Reserve: Newbury, Berkshire, Nature Conservancy Council, 16 p.

HANKEN, N.-M., OLAÜSSEN, S., AND WORSLEY, D., 1978, Silurian bioherms in the Oslo region (Norway): Reef Newsletter 5, p. 33–35.

HEDE, J. E., 1925a, Berggrunden (Silursystemet), in Munthe, H., Hede, J. E., and Von Post, L., Beskrivning till kartbladet Ronehamn: Sver. Geol. Unders., ser. Aa 156, 96 p.

——, 1925b, Berggrunden (Silursystemet), in Munthe, H., Hede, J. E., and Von Post, L., Gotlands geologi. En översikt: Sver. Geol. Unders., ser, C 331, 130 p.

——, 1927a, Berggrunden (Silursystemet), in Munthe, H., Hede, J. E., and Lundqvist, G., Beskrivning till kartbladet Klintehamn: Sver. Geol. Unders., ser. Aa 160, 109 p.

——, 1927b, Berggrunden (Silursystemet), in Munthe, H., Hede, J. E., and Von Post, L., Beskrivning till kartbladet Hemse: Sver. Geol. Unders., ser. Aa 164, 155 p.

——, 1928, Berggrunden (Silursystemet), in Munthe, H., Hede, J. E., and Lundqvist, G., Beskrivning till kartbladet Slite: Sver. Geol. Unders., ser. Aa 169, 130 p.

————, 1929, Berggrunden (Silursystemet), *in* Munthe, H., Hede, J. E., and Lundqvist, G., Beskrivning till kartbladet Katthammarsvik: Sver. Geol. Unders., ser. Aa 170, 120 p.

————, 1933, Berggrunden (Silursystemet), *in* Munthe, H., Hede, J. E., and Lundqvist, G., Beskrivning till kartbladet Kappelshamn: Sver. Geol. Unders., ser Aa 171, 129 p.

————, 1936, Berggrunden, *in* Munthe, H., Hede, J. E., and Lundqvist, G., Beskrivning till kartbladet Fårö: Sver. Geol. Unders., ser. Aa 180, 82 p.

————, 1940, Berggrunden, *in* Lundqvist, G., Hede, J. E., and Sundius, N., Beskrivning till kartbladen Visby and Lummelunda: Sver. Geol. Unders., ser. Aa 183, 167 p.

————, 1960, The Silurian of Gotland, *in* Regnéll, G., and Hede, J. E., The Lower Paleozoic of Scania. The Silurian of Gotland: Twenty First International Geol. Congress, Sess. Norden. Guidebook d, Stockholm, Geol. Survey Sweden, p. 44–87.

HEDSTRÖM, H., 1910, The stratigraphy of the Silurian strata of the Visby district: Geol. Fören. Stockh. Förh., v. 32, p. 1455–1484.

HENNINGSMOEN, G., AND SPJELDNAES, N., 1960, Paleozoic stratigraphy and paleontology of the Oslo region, Eocambrian stratigraphy of the sparagmite region, southern Norway: Twenty First International Geol. Congress, Sess. Norden. Guidebook, Norges Geol. Unders., 30 p.

HILL, D., 1936, Report of "Coral Reef" meeting at Wenlock Edge, the Dudley district and the Oxford district: Geol. Assoc., Proc., v. 47, p. 130–139.

HOLLAND, C. H., RICKARDS, R. B., AND WARREN, P. T., 1969, The Wenlock graptolites of the Ludlow district, Shropshire, and their stratigraphical significance: Palaeontology, v. 12, p. 663–683.

HOLTEDAHL, O., 1960, Geology of Norway: Norges Geol. Unders., v. 208, 540 p.

HURST, J. M., 1975, Wenlock carbonate, level bottom, brachiopod-dominated communities from Wales and the Welsh Borderland: Palaeogeography, Palaeoclimatology, Palaeoecology, v. 17, p. 227–255.

JAANUSSON, V., LAUFELD, S., AND SKOGLUND, R., 1979, Lower Wenlock faunal and floral dynamics, Vattenfallet section, Gotland: Sver. Geol. Unders., ser. C 762 (in press).

JUKES, J. B., 1859, The geology of the South Staffordshire coalfield: Geol. Survey Great Britain Memoir (2nd Ed.).

JUX, U., 1957, Die Riffe Gotlands und ihre angrenzenden Sedimentationsräume: Stockholm Contrib. Geol., v. 1, p. 41–89.

KALJO, D., 1970, The Silurian of Estonia (in Russian with English summary): Tallinn, Academy Sciences, Estonian S.S.R., Institute Geology, 343 p.

————, 1972, Facies control of the faunal distribution in the Silurian of the eastern Baltic region: Twenty Fourth International Geol. Congress, sec. 7 (Paleontology), p. 544–548.

————, 1977, Facies and fauna of the Baltic Silurian (in Russian with Estonian and English summaries): Tallinn, Academy of Sciences, Estonian S.S.R., Institute Geology, 286 p.

————, AND JÜRGENSON, E., 1977, Sedimentary facies of the east Baltic Silurian, *in* Kaljo, D., Ed., Facies and fauna of the Baltic Silurian (in Russian.): Tallinn, Academy Sciences, Estonian S.S.R., Institute Geology, p. 122–148.

KERSHAW, S., AND RIDING, R., 1978, Parameterization of stromatoporoid shape: Lethaia, v. 11, p. 233–242.

KIAER, J., 1908, Das Obersilur im Kristianiagebiete: Vis. Selsk. Skr. Mat.—Nat. Kl. (1906), v. 1, no. 2, 596 p.

KJERULF, TH., 1855, Das Christiania-Silurbecken chemisch-geognostisch untersucht: Universitetsprogram, 68 p.

KLAAMANN, E. R., 1962, Rasprostranenie ordovikskikh i silurijskikh tabuljat Estonii (s opisaniem nekotorykh novykh vidov) i Trudy Inst. Geol. Akad. Nauk Est. S.S.R., v. 10, p. 149–172.

LAUFELD, S., 1974a, Silurian Chitinozoa from Gotland: Fossils Strata, No. 5, 130 p.

————, 1974b, Reference localities for paleontology and geology in the Silurian of Gotland: Sver. Geol. Unders., ser. C 705, 172 p.

LAWSON, J. D., 1955, The geology of the May Hill Inlier: Quart. Jour. Geol. Soc. London, v. 111, p. 85–116.

LINDSTRÖM, G., 1884, On the Silurian Gastropoda and Pteropoda of Gotland: Kongl. Svenska Vetens.-Akad. Handlingar, v. 19, 6, 250 p.

LYELL, C., 1841, Some remarks on the Silurian strata between Aymestry and Wenlock: Geol. Soc., Proc., v. 3, p. 463–465.

MANTEN, A. A., 1962, Some Middle Silurian reefs of Gotland: Sedimentology, v. 1, p. 211–234.

————, 1971, Silurian reefs of Gotland: Amsterdam, Elsevier, 539 p.

MARTINSSON, A., 1967, The succession and correlation of the ostracode faunas in the Silurian of Gotland. Geol. Fören. Stockh. Förh., v. 89, p. 350–386.

McKERROW, W. S., 1979, Ordovician and Silurian changes in sea level: Jour. Geol. Soc. Cond., v. 136, p. 137–145.

MITCHELL, G. H., POCOCK, R. W., AND TAYLOR, J. H., 1961, Geology of the country around Droitwich, Abberley and Kidderminster: Geol. Survey Great Britain Memoir, v. 182, 137 p.

MOBERG, J. C., 1910, Historical-stratigraphical review of the Silurian of Sweden: Sver. Geol. Unders., ser. C 229, 210 p.

MORI, K., 1968, Stromatoporoids from the Silurian of Gotland, 1: Stockholm Contrib. Geol., v. 19, 100 p.

————, 1970, Stromatoporoids from the Silurian of Gotland, 2: Stockholm Contrib. Geol., v. 22, 152 p.

MUNTHE, H., 1910, On the sequence of strata within southern Gotland: Geol. Fören. Stockh. Förh., v. 32, 5, p. 1397–1453.

MURCHISON, R. I., 1833, On the sedimentary deposits which occupy the western parts of Shropshire and Here-
    fordshire, and are prolonged from northeast to southwest through Radnor, Brecknock and Caermarthenshire,
    with descriptions of the accompanying rocks of intrusive or igneous character: Geol. Soc. London, Proc., v.
    1, p. 470–477.
———, 1839, The Silurian System, founded on geological researches in the counties of Salop, Hereford, Radnor,
    Montgomery, Caermarthen, Brecon, Pembroke, Monmouth, Gloucester, Worcester and Stafford: with de-
    scriptions of the coalfields and overlying formations: London, John Murray, 768 p.
———, 1847, On the Silurian and associated rocks in Dalecarlia, and on the succession from Lower to Upper
    Silurian in Smoland, Öland, and Gothland and in Scania: Quart. Jour. Geol. Soc. London, v. 3, p. 1–48.
NESTOR, H., 1964, Stromatoporoids from the Ordovician and Llandovery of Estonia (in Russian): Tallinn, Academy
    Sciences, Estonian S.S.R., Institute Geology, 112 p.
———, 1966, Stromatoporoids from the Wenlock and Ludlow of Estonia (in Russian): Tallinn, Academy Sciences,
    Estonian S.S.R., Institute Geology, 87 p.
———, AND EINASTO, R., 1977, Facies-sedimentary model of the Silurian paleo-Baltic pericontinental basin, in
    Kaljo, D., Ed., Facies and fauna of the Baltic Silurian (in Russian): Tallinn, Academy Sciences, Estonian
    S.S.R., Institute Geology, p. 89–121.
NICHOLSON, H. A., 1886–1890, On some new or imperfectly known species of stromatoporoids: London, Annals
    Mag. Natural History, ser. 5, v. 7 (1886), p. 225–238; v. 18 (1886), p. 8–22; v. 19 (1887), p. 1–17; ser. 6, v. 7
    (1890), p. 309–328.
PENN, J. S. W., 1971, Bioherms in the Wenlock Limestone of the Malvern area (Herefordshire, England): Bur.
    Rech. Géol. Minièr. Mémoire, Colloque Ordovicien-Silurien, Brest, v. 73, p. 129–137.
PHIPPS, C. B., AND REEVE, F. A. E., 1967, Stratigraphy and geological history of the Malvern, Abberley and
    Ledbury Hills: Geol. Jour., v. 5, p. 339–368.
POCOCK, R. W., WHITEHEAD, T. H., WEDD, C. B., AND ROBERTSON, T., 1938, Shrewsbury district including the
    Hanwood coalfield: Geol. Survey Great Britain Memoir, v. 152, 297 p.
PRESTWICH, J., 1840, On the geology of Coalbrook Dale: Trans. Geol. Soc., ser. 2, v. 5, p. 413–495.
READ, H. H., AND WATSON, J., 1975, Introduction to geology, 2. Earth history, II. Later stages of earth history:
    London, Macmillan, 371 p.
REGNÉLL, G., 1960, The Lower Paleozoic of Scania, in Regnéll, G., and Hede, J. E., The Lower Paleozoic of
    Scania. The Silurian of Gotland: Twenty First International Geol. Congress, Sess. Norden. Guidebook d;
    Stockholm, Geol. Survey Sweden, p. 44–87.
RIDING, R., 1977, Reef concepts: Third International Coral Reef Symposium, Miami, Proc., p. 209–213.
ROTHPLETZ, A., 1913, Über die Kalkalgen, Spongiostromen und einige andere Fossilien aus dem Obersilur Got-
    lands: Sver. Geol. Unders., ser. Ca 10, p. 1–57.
RUTTEN, M. G., 1958, Detailuntersuchungen an Gotländischen Riffen: Geol. Rundschau, v. 47, p. 359–384.
SCOFFIN, T. P., 1965, The sedimentology of the Wenlock Limestone [Ph.D. thesis]: Swansea, Univ. Wales,
    208 p.
———, 1971, The conditions of growth of the Wenlock reefs of Shropshire (England): Sedimentology, v. 17, p.
    173–219.
———, 1972, Cavities in the reefs of the Wenlock Limestone (mid-Silurian) of Shropshire, England: Geol. Rund-
    schau, v. 61, p. 565–578.
SEILACHER, A., AND MEISCHNER, D., 1964, Fazies-Analyse im Paläozoikum des Oslo Gebietes: Geol. Rundschau,
    v. 54, p. 596–619.
SHERGOLD, J. H., AND BASSETT, M. G., 1970, Facies and faunas at the Wenlock-Ludlow boundary of Wenlock
    Edge, Shropshire: Lethaia, v. 3, p. 113–142.
SQUIRRELL, H. C., AND TUCKER, E. V., 1960, The geology of the Woolhope Inlier, Herefordshire: Quart. Jour.
    Geol. Soc. London, v. 116, p. 139–185.
STEL, J. H., 1978a, Studies on the paleobiology of favositids (doctoral thesis): Netherlands, Univ. Groningen,
    247 p.
———, 1978b, Environment and quantitative morphology of some Silurian tabulates from Gotland: Scripta Geol.,
    v. 47, p. 1–75.
———, AND DE COO, J. C. M., 1977, The Silurian Upper Burgsvik and Lower Hamra-Sundre Beds, Gotland:
    Scripta Geol., v. 44, p. 1–43.
STØRMER, L., 1967, Some aspects of the Caledonian Geosyncline and foreland west of the Baltic Shield: Quart.
    Jour. Geol. Soc. London, v. 123, p. 183–214.
STRAND, T., AND HENNINGSMOEN, G., 1960, Cambro-Silurian stratigraphy, in Holtedahl, O., Ed., Geology of
    Norway: Norges Geol. Unders., v. 208, p. 128–169.
SYMMONDS, W. S., 1872, Record of the rocks: London, John Murray, 433 p.
WALMSLEY, V. G., 1959, The geology of the Usk Inlier (Monmouthshire): Quart. Jour. Geol. Soc. London, v. 114,
    p. 483–516.
WATKINS, R., 1979, Three Silurian bioherms of the Högklint Beds, Gotland: Geol. Fören. Stockholm Förh., v.
    101, p. 34–48.
WEDEKIND, R., 1927, Die Zoantharia Rugosa von Gotland (bes. Nordgotland). Nebst Bemerkungen zur Biostra-
    tigraphie des Gotlandium: Sver. Geol. Unders., ser. Ca 19, p. 1–94.

————, AND TRIPP, K., 1930, Die Korallenriffe Gotlands. Ein Beitrag zur Lösung des Problems von der Entstehung der Barrierriffe: Centralbl. Mineral. Geol. u. Paläont., B, v. 1, p. 295–304.

WHITACKER, J. H. McD., 1977, A guide to the geology around Steinsfjord, Ringerike: Oslo, Universitets forlaget, 56 p.

WHITEHEAD, T. H., AND EASTWOOD, T., 1927, The geology of the southern part of the South Staffordshire coalfield (south of the Bentley faults): Geol. Survey Great Britain Memoir, 218 p.

————, AND POCOCK, R. W., 1947, Dudley and Bridgenorth: Geol. Survey Great Britain Memoir, 226 p.

WHITTARD, W. F., 1952, A geology of south Shropshire: Geol. Assoc., Proc., v. 63, p. 143–197.

WIMAN, C., 1897, Über Silurische Korallenriffe in Gotland: Bull. Geol. Inst. Univ. Uppsala, v. 3, p. 311–326.

ZIEGLER, A. M., 1970, Geosynclinal development of the British Isles during the Silurian Period: Jour. Geology, v. 78, p. 445–479.

————, HANSEN, K. S., JOHNSON, M. E., KELLY, M. A., SCOTESE, C. R., AND VAN DER VOO, R., 1977, Silurian continental distributions, paleogeography, climatology, and biogeography: Tectonophysics, v. 40, p. 13–51.

SEPM SPECIAL PUBLICATION NO. 30, P. 85–142, MAY 1981

# EUROPEAN DEVONIAN REEFS: A REVIEW OF CURRENT CONCEPTS AND MODELS

TREVOR P. BURCHETTE
Technische Universität Braunschweig
*Braunschweig, West Germany*

## ABSTRACT

Devonian reefs are widespread in Europe, but with the exception of those in Germany and Belgium most have not been widely described in the international geological literature. The present paper aims to amend this situation by providing a general synthesis of published data on all significant European Devonian reef occurrences. Particular attention is paid to their classification and, where possible, to the factors which governed their location and evolution.

The earliest reefs occur in that part of Europe, the "internal" zone of the Hercynian Orogen, which was not strongly affected by the Siluro-Devonian Caledonian earth movements and in which there was continuous marine deposition from the Lower Paleozoic through to at least the Middle Devonian (southeastern Alps, Bohemia, Armorican Massif, and the Cantabrian and Pyreneean Mountains). In the "external" zone (e.g., the "Rhenohercynian" belt, including southwestern England, the Ardennes, Rhenish Schiefergebirge, Harz Mountains, Poland, and Moravia), which experienced uplift and mild deformation on the periphery of the Caledonian Orogen, and where the Devonian usually overlies the Lower Paleozoic unconformably, reef growth did not begin until the Middle Devonian. Here, Lower and early Middle Devonian deposits (and locally the whole Devonian succession) are in a clastic fluviatile or neritic facies. In some cases the Lower Devonian is absent altogether. In most areas, irrespective of this early history, reef growth continued until the late Frasnian or earliest Famennian, when (among other factors) widespread rise in relative sea level caused their extinction, and pelagic sedimentation became the norm.

European Devonian reefs fall into five broad morphological categories: 1. banks, 2. biostromal complexes, 3. barrier-reef complexes, 4. isolated reef complexes (reef-mounds and atolls), and 5. quiet water carbonate buildups ("mud-mounds"). These possess characteristic sedimentary and organic facies associations and distribution patterns. Types 2. to 4. exhibit lateral differentiation into facies-zones (e.g., reef-core and back-reef in atolls, barrier-reef and biostromal complexes, plus fore-reef in most well developed buildups).

Factors which influenced the location and development of reefs in the European Devonian included: 1. local crustal flexures and movement on basin-margin hingelines, 2. synsedimentary faulting, 3. synsedimentary volcanicity, 4. the distribution of small scale elevations on the seafloor (e.g., buried reefs, calcarenite banks), and 5. changes in relative sea level where not clearly attributable to one of the previous factors. Variation in the relationship between reef growth and changes in, or inconsistencies in the rate of change, of relative sea level are expressed in basin-marginal reef complexes as sequences of transgression and regression, and in isolated reef complexes as vertical ecological and facial zonation. Successive minor transgressive pulses within broader trends are recorded in the back-reef facies of Middle and Upper Devonian biostromal and barrier-reef complexes of central Europe, and of Middle and Upper Devonian reefs of the southeastern Alps, as depositional cyclicity on the scale of a few meters.

Time-equivalent deposits of European Devonian reefs are shelf and basinal black shales and pelagic limestones. These commonly contain intercalations of reefal debris as turbiditic or "allodapic" limestones.

Following extinction some reefs were buried relatively rapidly by subsequent sediments, whereas others underwent prolonged periods of submarine, or very locally subaerial exposure, and were buried only in the latest Devonian or early Carboniferous. Concomitant with the late growth and early post-growth stages, many reefs developed extensive systems of fissures which were filled both syngenetically and post-genetically with fibrous carbonate cements and internal sediments. The latter commonly contain faunas which are substantially younger than the host reefal limestones themselves.

Early diagenetic carbonate cements are features in many reefal facies, and vary in character from one of these to another. Early vadose cements (microstalactitic and drusy pore linings) are most important in back-reef facies, while in reef-core and fore-reef deposits synsedimentary submarine fibrous cements predominate. In "mud-mounds" stromatactis is an important early diagenetic structure. Dolomitization is extensive in some areas, but is a difficult feature to qualify; some occurrences, especially in central Europe, may be early stage replacement. Later cementation appears to have followed much the same course in all areas for which data presently exist, though neomorphism and response to tectonic stress of the carbonates was more varied.

To date, no significant hydrocarbon deposits have been discovered in European Devonian reefs, although minor traces do occur in central Europe.

## INTRODUCTION

Many of the Devonian reef limestones in Europe were recognized as such as long ago as the middle of the nineteenth century and since that time they have been the subjects of numerous studies. Over most of this period, however, the emphasis placed on their investigation has been a classical paleontological and stratigraphical one, and in some respects (specifically with regard to paleoecology and sedimentology) less progress has been made towards understanding them than in other parts of the world, such as western Canada, where Devonian reefs have been studied for a much shorter time. A variety of special factors have combined to hinder progress in this field in Europe. Among the most significant of these have been the intricate political constitution and complicated tectonic framework of the continent, slow growth in the popularity of the geosciences in some countries, and the lack, until perhaps recently, of any powerful economic stimulus behind investigations. Only over the last decade has there been any real change in the emphasis or style of research in this area to the kinds of programs which provide more comprehensive and immediately useful interpretative data on the reefs investigated.

Certainly, a general picture of Devonian reefal carbonates in central Europe (Belgium and Germany) has existed for some time, but it is a rather static one derived from the publications of just a small number of communicative workers. For this reason, and because of their highly accessible locations, Devonian reefs of central Europe are perhaps regarded by many as the archetypal pattern for Devonian reef development in Europe as a whole. The Devonian, however, was a period of widespread reef growth, and reef complexes of this age, comparable in scale with those of Belgium and Germany but different in character, also occur in other quarters of the continent. Some of these (e.g., in the southeastern Alps or the Cantabrian Mountains of Spain) have remained comparative strangers to the international geological literature although useful accounts of them do exist, whereas others (e.g., in the Pyrenees) have never been investigated in any more than a broadly stratigraphical fashion, or are only now being studied in detail (e.g., in Poland). In central Europe there is also still a great deal that might be achieved through more detailed studies of sedimentary facies, diagenesis and paleoecology than have hitherto been made.

That an overview comparing and contrasting the Devonian reefal carbonates of Europe has not previously been attempted reflects in part a true lack of useful data, but also probably illustrates one of the major problems deriving from the political and cultural barriers of the region: that is the marked provinciality of research in this field. Because Devonian reefs were so widely distributed, fossil examples occur in nearly every country in Europe (Fig. 1). Data on them tend to be published principally in the local languages (eight of these not including Russian) in a host of jour-

nals with essentially national or sub-national distributions. Marked differences in the research backgrounds of authors also commonly create problems of incompatibility in sets of published data. It can therefore be difficult even for European workers to obtain an accurate comparative impression of Devonian reefs outside their own immediate area.

Those are the human problems. There are also considerable physical obstacles since over the whole region Devonian reef complexes were folded and faulted during the Hercynian Orogeny, and deformation or even metamorphism have been severe enough in some cases to cause acute difficulties in their interpretation (as in southwestern England or parts of the Rhenish Schiefergebirge and the Pyrenees). In southern Europe, within the Alpine Orogen, Paleozoic deposits have experienced *two* major periods of upheaval, and Devonian reefs have been severely disrupted into thrust sheets and fault blocks, and are locally metamorphosed. Extensive dolomitization, or burial beneath younger sediments, are further factors which impede interpretation in some parts of central and eastern Europe.

This paper constitutes an attempt to summarize, compare and contrast published data on European Devonian reefs. It also includes some of the broader conclusions of the author's own unpublished work on Middle and Upper Devonian back-reef facies in Germany and Belgium. The paper has been divided into two parts: in the first part brief stratigraphical and paleogeographical backgrounds of the various European Devonian reefs are presented and in the second part the reefs are classified, and aspects of their growth, paleoecology and sedimentology are considered collectively. The terminology used is discussed in the section on classification.

### STRATIGRAPHICAL, PALEOGEOGRAPHICAL AND PALEOTECTONIC BACKGROUND OF THE MARINE DEVONIAN IN EUROPE

Two opposing schools of thought exist on the nature of the Hercynian Orogeny and the events leading up to it. There are authors who would fit it into the framework of plate tectonics (Burrett, 1972; Laurent, 1972; Nicolas, 1972; Burne, 1973; Johnson, 1973; Riding, 1974), with Europe as a region which incorporated two or more converging cratons or micro-cratons. There are others who maintain that a plate tectonic model has little application in the interpretation of Middle and Late Paleozoic orogeny and paleogeography in this region, and that vertically acting processes, perhaps in response to granitic mantle diapirism, were responsible for Hercynian deformation (Krebs and Wachendorf, 1973 and 1974; Krebs, 1976a and b; Mathews, 1977; Lutzens, 1978; We-

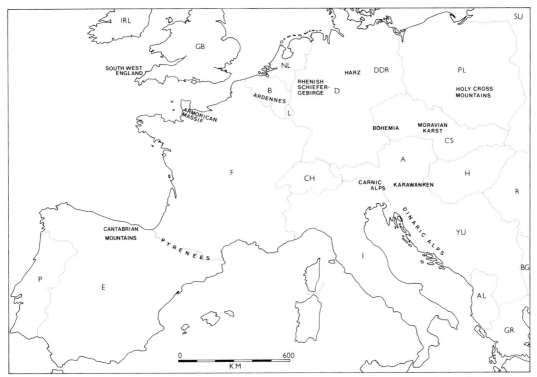

FIG. 1.—Political map of Europe showing areas mentioned in the text in which Devonian reefs occur.

ber, 1978). Certainly, no plate sutures or remnants of oceanic crust have yet been convincingly demonstrated within the orogen, and vertical tectonic movements, commonly quite localized, do appear to have exerted strong influence on sedimentation well before main phase deformation of the Hercynian geosyncline in the Late Carboniferous. Neither view, however, represents a solution which explains all the features of the Hercynian Orogen, and it might be argued that proponents of both have approached the problem somewhat too dogmatically. Some critical areas (e.g., the Late Paleozoic deposits and Hercynian events in southern Europe, and the subsurface in central Europe) are still so poorly understood as to make any attempt to produce an absolute interpretation of the mechanism behind the Hercynian Orogeny in Europe premature. It may be that the hypothesis of granitic diapirism and consequent vertical tectonic movements can eventually be reconciled with a solution within the broader framework of plate tectonics (see e.g., Windley, 1977, p. 201).

Devonian sedimentation in all the areas considered in this paper nevertheless appears to have been *intracratonic* and, although deep water shales and pelagic limestones undoubtedly occur, there is nothing to suggest that these "basinal facies" were anywhere deposited under an abyssal ocean-

ic regime (Krebs and Wachendorf, 1973; Franke *et al.*, 1978). The terms conventionally utilized by authors to denote paleogeographical and paleomorphological features in the European Late Paleozoic context, particularly in central Europe, do not therefore correspond exactly with those which might be applied to the ocean-basin-linked environments where most Quaternary coral reefs occur. "Continent," as in "Old Red Sandstone Continent" (see e.g., Erben and Zagora, 1967; Krebs, 1974, p. 159), does not imply *craton* but a terrigenous source area or stable block, represented in northern Europe principally by the Caledonides; "shelf" similarly does not denote *continental shelf* but shallowly submerged, commonly subsiding ramps or platforms surrounding the stable blocks in all areas; and "basin," as implied above, does not indicate *ocean basin* floored by oceanic crust but is a term applied to the moderately deep water, relatively stable intracratonic troughs.

European authors have commonly distinguished between *external* (Rhenohercynian) and *internal* paleogeographical and paleotectonic zones within the area affected by the Hercynian Orogeny (Fig. 2; see also Krebs, 1974 and 1976b; Krebs and Mountjoy, 1972; Franke *et al.*, 1978), which correspond with the northern and southern

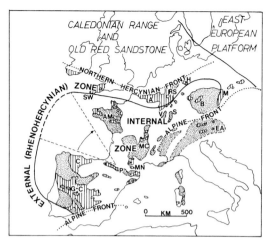

FIG. 2.—Sketch map of Europe showing the main divisions of the Hercynian Orogen, and major outcrops of Paleozoic sediments (vertical lines) and Hercynian igneous and metamorphic rocks (stippled). Modified after Franke et al. (1978, fig. 1); A = Ardennes, AM = Armorican Massif, B = Bohemia, C = Cantabrian zone, EA = Eastern Alpine zone, G-C = Galicia-Castille zone, H = Harz, M = Moravia, MC = Massif Central, MN = Montagne Noire, P = Pyrenees, RS = Rhenish Schiefergebirge, S = Schwarzwald, and SW = southwest England, V = Vosges.

Hercynides respectively. The internal zone in this sense represents the inner portion of the orogen formed by the Saxothuringian and southernmost Rhenohercynian troughs and their lateral extensions, and the external zone the relatively narrow, predominantly neritic "shelf" tract surrounding them in the north and west, stretching from Poland, through central Europe and southwest England, to Portugal (see Franke et al., 1978). The essential stratigraphical and paleotectonic features of these two elements are: 1. that the tectonically relatively stable internal zones, which were not involved in the Late Silurian Caledonian orogeny and subsided only slowly, are characterized by substantially continuous, but comparatively thin (one or two thousands of meters) successions of pelagic deposits ranging from the Lower Paleozoic to at least the Middle Devonian, and in many places into the Lower Carboniferous; whereas 2. the northern external zone, which lay on the fringe of the Caledonian Orogen, experienced mild deformation, uplift and erosion between the Silurian and the Devonian followed by strong subsidence and the deposition of thick Lower Devonian post-Caledonian molasse (the Old Red Sandstone), and, in response to the continued outward migration of the zone of maximum subsidence, marked trangression during the rest of the Devonian and Lower Carboniferous (Brou-

wer, 1967; Erben and Zagora, 1967; Franke et al., 1978). The total thickness of the Devonian sedimentary succession reaches 5000 meters in the external zones (e.g., southwest England and the Rhenish Schiefergebirge).

European authors have also traditionally recognized two Devonian "magnafacies" corresponding to the areas outlined above (see discussion by Brouwer, 1967), the Rhenish magnafacies in the external portion of the orogen, and the Hercynian magnafacies in the internal part, within some areas (e.g., Iberia, southern Rhenish Schiefergebirge, and the Harz Mountains) intercalation of the two (Schmidt, 1935; Erben, 1962 and 1964; Erben and Zagora, 1967). The Rhenish magnafacies denotes the chiefly terrigenous sediments of the northern "shelf" environment where benthonic faunas dominated, the Hercynian magnafacies a basinal assemblage of limestones and shales with a predominantly pelagic fauna, but very little coarse clastic material (Erben and Zagora, 1967; Krebs, 1974). In the southern and eastern Hercynides (the Alps, Bohemia, the Pyrenees and Cantabrians) the central European definitions of these magnafacies break down to some extent since the "Hercynian facies" is developed locally as reefal carbonates. Some authors have considered in this respect that the term "Bohemian" might be preferable to "Hercynian" as a label for the collective facies of the internal zones of the orogen (e.g., Brouwer, 1967), but on sedimentological grounds it might equally be argued that actually three "magnafacies" can be recognized in the European Devonian: a clastic neritic Rhenish, a carbonate neritic (reefal) Bohemian, and a shale and carbonate pelagic basinal Hercynian facies. These terms have, however, been utilized inconsistently for a whole range of features by many authors, and are probably no longer of great practical importance. The features which they signify were well defined only in the Lower Devonian, and following a decrease in clastic input to much of the Rhenohercynian zone during the Middle and Upper Devonian a complex mosaic of neritic carbonate and basinal deposition became typical over much of Europe (see e.g., Brouwer, 1967).

Devonian reef complexes thus occur in most parts of Europe, but with substantial variation in age and character from one to the other (Fig. 3), reflecting the complicated Upper Paleozoic tectonic and paleogeographical evolution of the region. Generally considered, reef growth reached an acme in the late Givetian and early Frasnian and tailed-off shortly before the start of, or during, the early Famennian. Lower and early Middle Devonian reefs are prominent in the internal Hercynian zones of central Bohemia, the southeastern Alps, the Armorican Massif, and in the Can-

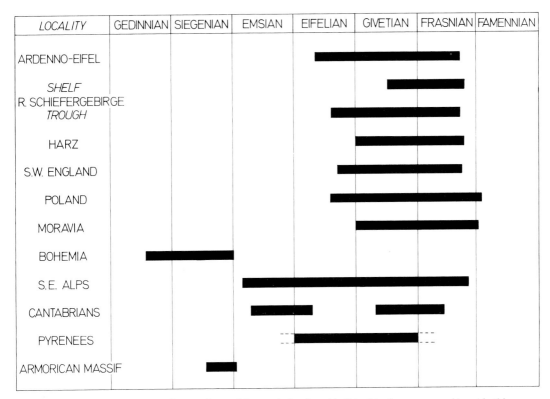

| LOCALITY | GEDINNIAN | SIEGENIAN | EMSIAN | EIFELIAN | GIVETIAN | FRASNIAN | FAMENNIAN |
|---|---|---|---|---|---|---|---|

ARDENNO-EIFEL

*SHELF*
R. SCHIEFERGEBIRGE
*TROUGH*

HARZ

S.W. ENGLAND

POLAND

MORAVIA

BOHEMIA

S.E. ALPS

CANTABRIANS

PYRENEES

ARMORICAN MASSIF

FIG. 3.—Approximate ranges of Devonian reefal growth (bank and buildup) in the areas considered in this paper. Note the difference in age of the points at which reefs first appear in the internal and external Hercynian zones, and that few reefs first appear in the internal and external Hercynian zones, and that few continued to grow beyond the late Frasnian. Data from a number of sources.

tabrians and Pyrenees. Middle and Upper Devonian reefs are characteristic of Poland, central Europe, southwest England, and also of the Cantabrians and southeastern Alps.

All compass directions used in the following descriptions in a paleogeographical sense refer to *present* orientations.

### REGIONAL STRATIGRAPHICAL SUMMARIES

*The Ardenno-Eifelian Tract.*—The area embracing the Ardennes of Belgium and northern France, and their extension into the Eifel of West Germany, is a classical one for Devonian stratigraphy and many of the system's stage names were derived from villages and towns there (Fig. 4). Devonian deposits are preserved in the large east-west-trending Dinant and Namur Synclines, and in the smaller Campine Graben to the north of these, as well as along the northern margin of the Stavelot-Venn Massif (in the Aachen area) and in a swarm of small northeast-southwest-trending synclines in the Eifel (the "Eifelkalkmulden") to the west of the river Rhine (Figs. 4 and 7). All these occurrences belong to a distinct facial,

stratigraphical and tectonic sub-province of the central European Devonian, with an eastern margin roughly delineated by the Rhine, but which is obscured on the western side by younger cover. Isolated outcrops of Devonian shelf limestones do, however, occur as far to the west as the Ferques Inlier, situated between Calais and Boulogne in northern France. Here the *Middle* Devonian rests with unconformity upon Lower Paleozoic sediments (Wallace, 1969).

Devonian sedimentation in the Ardenno-Eifelian tract occurred over the broad, relatively stable shelf which fringed the Wales-Brabant Massif lying to the north (see e.g., Fig. 7) and was bounded to the south by a pelagic shale basin (an extension of the Rhenohercynian Trough). Within the area of the shelf Lower Devonian rests unconformably upon the Lower Paleozoic. The Devonian marine succession was deposited diachronously from south to north over fluviatile lowest Devonian so that the transgression had reached the area of the northern limb of the Dinant Syncline by the Gedinnian (Lecompte, 1970), the northern Eifel by the Emsian (Paulus, 1961),

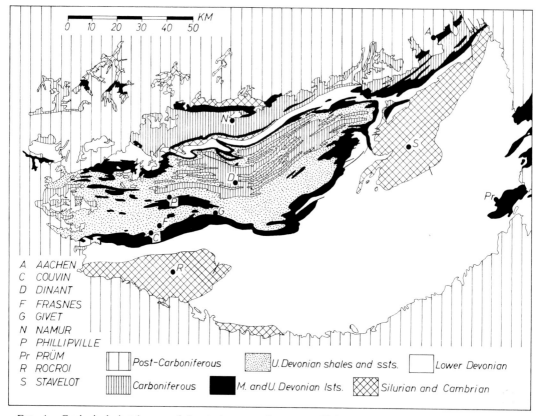

FIG. 4.—Geological sketch map of the Ardennes and western Eifel showing the Devonian outcrop and key localities.

and the northern margin of the Namur Syncline and the Aachen area by the Givetian (Kasig and Neumann-Mahlkau, 1969; Lecompte, 1970); by the early Frasnian it had extended into the incipient Campine Basin to the north of the Brabant Massif (Fig. 7; Lecompte, 1970; Graulich, 1977; Kimpe et al., 1978, fig. 3).

Reefal carbonates were deposited from about the lower Eifelian (Couvinian in Belgium) until the upper Frasnian and constitute a "platform" with a total thickness of 1600–1700 meters on the southern limb of the Dinant Syncline which thins northwards against the Brabant Massif to 300–350 meters in the Namur Syncline (Fig. 5), and to zero in the area of the massif itself which probably remained partially exposed throughout the Devonian (see e.g., Kimpe et al., 1978). A similar relationship exists between the Devonian successions of the Eifel Synclines and the Aachen area.

The Eifelian of the Dinant Syncline and the Eifel is characterized by the widespread development of biostromes, interbedded variably with shales and sandstones, and locally in the higher Eifelian by bioherms of stromatoporoids and cor-

als (Ochs and Wolfart, 1961; Paulus, 1961; Lecompte, 1970; Tsien, 1971). The southern margin of the platform is seen in neither area at this stage, though the appearance of "restricted" limestones in the upper middle Eifelian of the Ardennes suggests a trend towards the polarization of depositional environments over the shelf even at this early stage, with perhaps the development of biostromes at the shelf margin and back-reef facies internally (see Tsien, 1971, p. 146).

Barrier-biostromes probably developed for the first time during the late Middle Devonian since much of the lower and middle Givetian succession in the Ardennes consists of an alternation of restricted ("sublagoonal") and coral-stromatoporoid biostromal facies (Lecompte, 1967 and 1970; Tsien, 1971, p. 128–129). There are also minor transgressive intercalations of deeper water shales. The same is true in the Eifel, where the lower Givetian (Flering and Spickberg Beds) exhibits cyclic alternation of biostromes with back-reef facies and neritic calcareous mudstones (Krebs, 1969a; Burchette, in press). In the Aachen district deposition during the lower Givetian

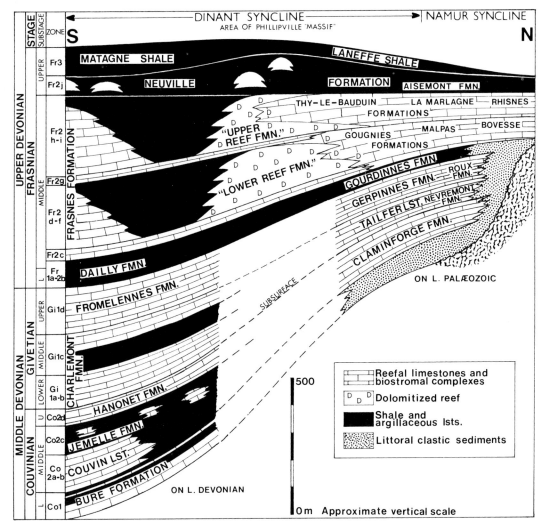

FIG. 5.—Section across the Ardennes showing the principal stratigraphical features of the Middle and Upper Devonian; after Tsien (1975, fig. 1). Note the marked northward thinning and overstep of the succession, and the superposition of barrier-reef complexes and reef mounds in the Frasnian. No horizontal scale intended, though distance north-south about 50 kilometres.

was in a clastic facies, but the middle and late Givetian is represented by open-marine-bank deposits ("Untere Stringocephalenschichten") which pass upwards directly into Frasnian cyclic back-reef deposits (Burchette, in press). The uppermost Givetian in both the Ardennes and the Eifel consists of a thick sequence of biostromal and restricted limestones, but in the Eifel this is completely dolomitized and stratigraphically poorly understood.

At the end of the Givetian, transgression caused stepping back of the margin of the carbonate platform in the direction of the Brabant Massif, thus bringing the organic barrier within the present

Devonian outcrop (seen only in the Dinant Syncline; Fig. 5). Two superposed reefal horizons developed during the early to mid-Frasnian (Fr2d and Fr2h horizons in the lithostratigraphical scheme of Mailleux, 1922 and 1925; and Mailleux and Demanet, 1928), each comprising a barrier-reef complex flanked basinward by stratigraphically equivalent reef-mounds (Lecompte, 1954 and 1970; Tsien, 1971 and 1977). Between the barrier-reef and the clastic shoreline of the Brabant Massif, broad back-reef shelf-lagoons developed. These two reefal sequences can be recognized in both the Ardennes and the Aachen area. Extensive Givetian-Frasnian back-reef facies are also

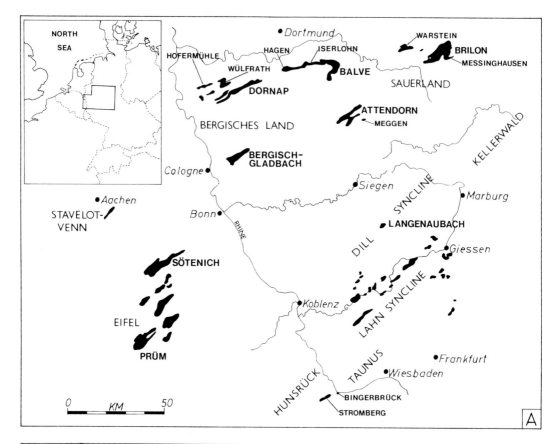

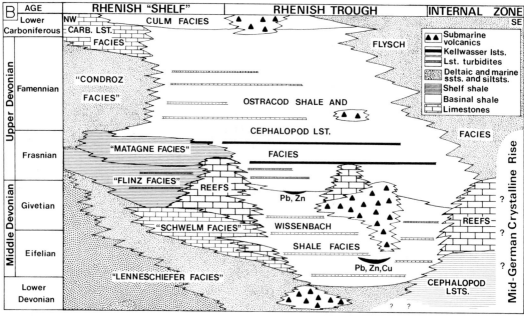

Fig. 6.—A. Sketch map of the Rhenish Schiefergebirge showing outcrops of Devonian buildups (black) and key localities and areas mentioned in the text; based on Krebs (1974, fig. 1); inset gives location within central Europe. B. Highly schematic section through the Devonian of the Rhenish Schiefergebirge showing facies distribution and the stratigraphical context of reefal limestones; modified and simplified after Krebs (1971, fig. 2).

present in the Eifel, but again all but the broadest stratigraphical details have been obscured by the dolomitization.

The isolated Frasnian reef-mounds of the Dinant Syncline are separated laterally from the barrier-reefs, and from each other, by time equivalent ostracode shales, and vertically by thin shales containing limestone bands.

At about the boundary middle-upper Frasnian, transgression over the whole area finally drowned the carbonate platform. This is evident in most areas as an upward transition from back-reef and reefal limestones to dark, commonly nodular shales containing a pelagic fauna ("Neuville" and "Aismont Formations" in Belgium, see Tsien, 1975; also Fig. 5; sequences of "Knollenkalke" and "Plattenkalke" in the Eifel-Aachen tract). The area of the carbonate platform was first covered extensively during the deepening phase by a sheet of argillaceous brachiopod-*Phillipsastrea* limestones 10–30 meters in thickness, and associated with this facies, and the succeeding shales, in the upper Frasnian of the Dinant Syncline are small dome-like bioherms (the Fr2j "mud-mounds"; Fig. 5).

*The Rhenish Schiefergebirge to the East of the Rhine and the Harz Mountains.*—Devonian stratigraphy and reef development in this portion of central Europe (Fig. 6A; summarized by Jux, 1960; Krebs, 1967 and 1974; Krebs and Mountjoy, 1972; Burchette, in prep.) were influenced to a much greater degree by patterns of local tectonism, uplift and volcanicity than in the Ardenno-Eifelian tract. As in the latter area, however, the Devonian stratigraphy exhibits an overall stepwise northward transgressive character over a Lower Paleozoic landmass. Deposition during the Lower and lower Middle Devonian in the Rhenish Schiefergebirge was in a predominantly fluviatile and neritic facies, the "Lenneschiefer facies" (=Rhenish facies), and in a pelagic limestone and shale facies (=Hercynian facies) in the center of the Rhenish Trough which lay to the south (Franke *et al.*, 1978; see Fig. 6B). The southern margin of the trough was determined by a geanticlinal zone (the mid-German crystalline rise) which appears to have remained stationary and relatively stable throughout the Devonian, though periodically it was the source of clastic input to the basin, as well as of flysch sediments during the Carboniferous (Krebs, 1976; Franke *et al.*, 1978). This feature possessed its own reef developments, and in some reconstructions has been considered as the inner "shelf" of the Rhenohercynian Trough (see e.g., Krebs, 1967 and 1974).

The Middle-Upper Devonian Rhenish shelf-edge can be traced from a supposed position along the southern rim of the Dinant Syncline and the Eifel group of synclines, northeastwards through

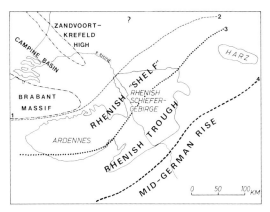

FIG. 7.—Sketch map showing the major paleogeographical divisions of the Rhenohercynian tract; based on data in Krebs (1974) and Kimpe *et al.* (1978). For location see Figure 6A; 1 = Middle and Upper Devonian land masses; 2 = southern margin of Lower Devonian Old Red Sandstone clastic facies; 3 = Middle and Upper Devonian Rhenish shelf margin; 4 = northern margin of the mid-German crystalline rise ("internal shelf" of Krebs, 1974).

the center of the Rhenish Schiefergebirge along a line connecting the southern margins of the Devonian reefal outcrops of Attendorn and Brilon (Krebs, 1967 and 1974; Gwodz, 1972; see Fig. 7). Its course further eastwards probably ran just to the north of the Harz Mountains. The lines followed by successive coastlines of the landmass which lay to the north are, on the other hand, far less clear. Some authors (Jux, 1960; Bartenstein and Teichmüller, 1974; Weggen, 1978, p. 10–11) have assumed that during the Lower and early Middle Devonian this ran in a shallow curve across the North German Plain just to the north of Hanover and into Poland. Newer interpretations suggest that the situation was more complex in the later Devonian, with breakup of the northern part of the area into a number of "massifs" which were at least periodically emergent (see e.g., Krebs and Wachendorf, 1973; Kimpe *et al.*, 1978), and with the late Middle Devonian transgression reaching northwards into the area occupied by the North Sea (see Pennington, 1975; Gotthardt, Meyer and Paproth, 1978; Kimpe *et al.*, 1978).

The earliest reefal carbonates in the area occur in the lowest Middle Devonian Lenneschiefer clastic facies in the Bergisches Land (Spriestersbach, 1942; Jux, 1960). These are biostromal, locally "biohermal" (up to 50 meters), in character, but have been little studied. They are neither as widespread nor as significantly uniform in distribution as the biostromes and "patch-reefs" of Eifelian age in the Ardennes and Eifel.

Large scale carbonate deposition in the Rhenish

Schiefergebirge (the "Massenkalk") did not begin until about the Eifelian-Givetian boundary. Generally considered, the start of reef growth becomes younger from south-to-north, reflecting the progress of the Devonian transgression over the shelf. Growth of reefs in the trough zone began during the latest Eifelian (Stromberg, Meggen, eastern Taunus), except where determined by younger submarine volcanism (e.g., in the Lahn and Dill Synclines, Brilon-Messinghausen, Iberg; see localities in Fig. 6B), whereas in the southern and northeastern Sauerland, at the outer margin of the Rhenish Shelf, it started in about the middle Givetian (e.g., the reefs of Bergisch Gladbach, Brilon and Attendorn). Further to the north in the Bergisches Land, the earliest reefal foundations are nowhere older than about upper Givetian in age. As in the Ardennes, no reef complex in the area to the east of the Rhine continued to grow beyond the upper Frasnian (Krebs, 1974).

Most Devonian reefs in the Rhenish Schiefergebirge were preceded by a bank phase (the "Schwelm Limestone" of Paeckmann, 1922; Lotze, 1928; the "Schwelm facies" of Krebs, 1967 and 1974) comprising undifferentiated complexes of stromatoporoid-coral biostromes and brachiopod and crinoidal limestones, interbedded with bituminous limestones and shales. The Schwelm facies is heterochronous in character and appears several times in the Devonian successions of some districts (e.g., in the Bergisches Land). It is most strongly developed in the middle and upper Givetian, when sediments of this type (reaching 200 meters in thickness) must have been deposited over wide areas of the Rhenish Shelf to form an extensive, more or less continuous carbonate platform upon which buildups (the "Dorp facies" of Krebs, 1967 and 1974) subsequently developed.

In comparison with those of the Ardenno-Eifelian tract, Middle and Upper Devonian reefs of the Rhenish zone were predominantly larger, isolated reef-mounds and atolls, and only in the upper Givetian and lower Frasnian of the Bergisches Land do barrier- or fringing-reefs appear to have developed on a limited scale. In the last area, around Hofermühle and Ratingen, back-reef deposits of this age exhibit small scale cyclicity (Burchette, in press). According to Krebs (1974), three paleogeographically distinct sites of growth can be discerned for the isolated reef complexes of the Rhenish Devonian (Fig. 7):

1. upon submarine volcanic "rises" (Langenaubach, Lahn Syncline, and Iberg in the Harz Mountains) and upfaulted blocks or "Schwellen" (e.g., Stromberg) within the Rhenish Trough;

2. along the margin of the Rhenish Shelf (e.g., the large atolls of Brilon and Attendorn);

3. within the extensive Rhenish Shelf area (e.g., Bergisch Gladbach, Balve, Dornap, Wülfrath, etc.), where they are separated laterally by shale ("Flinz facies") basins.

The reefs which developed on volcanic piles within the trough exhibit a wide range of forms, from minor *in situ* organic skeletal accumulations within the volcanics to atolls several square kilometers in area (Krebs, 1974). Many of these basinal reefs, together with the shelf-margin atolls, are distinguished from those of the more proximal shelf in that they exhibit a common depositional history following termination of growth in the late Frasnian, having been only slowly covered by subsequent pelagic sediments (Krebs, 1974, p. 173). The reefs in other areas were buried rapidly following death by Upper Devonian basinal shales, locally with the development of an intermediate capping biostrome or "mudmound" (the "Iberg facies" of Krebs, 1967 and 1974).

*Southwest England.*—The Devonian shelf of the Ardennes and Rhenish Schiefergebirge runs westwards through northern France, and probably southern England, into southwestern England (Fig. 8A), where Middle and Upper Devonian reefs of significant size also occur. These suffered complex structural dislocation during the Hercynian Orogeny and consequently for a long time were only poorly understood. A number of recent paleontological and sedimentological studies, surpassing earlier investigations in this area (e.g., Taylor, 1951; Dinely, 1961; Braithwaite, 1966 and 1967; House and Selwood, 1966) in terms of scope and detail, have however greatly contributed to their elucidation.

The Lower Devonian in southwest England is developed fairly uniformly in a clastic facies. In southern Devon and northern Cornwall (Fig. 8A) these were of fluviatile aspect until the middle Lower Devonian (Dartmouth Slates), and were

→

FIG. 8.—A. Geological sketch of map of southwest England showing Devonian outcrops and main localities mentioned in the text. Inset 1: outcrops of Devonian reefal limestone in the Tor Bay area (after Scrutton, 1977b, fig. 1). Inset 2: the counties of southwest England. B. Non-palinspastic section through the Devonian of southwest England showing the stratigraphical context and two-dimensional paleogeographical location of the Tor Bay and Plymouth Reef complexes; after House (1975), slightly modified. Note the similarity of the major facial and stratigraphical features to those of the Rhenish Schiefergebirge.

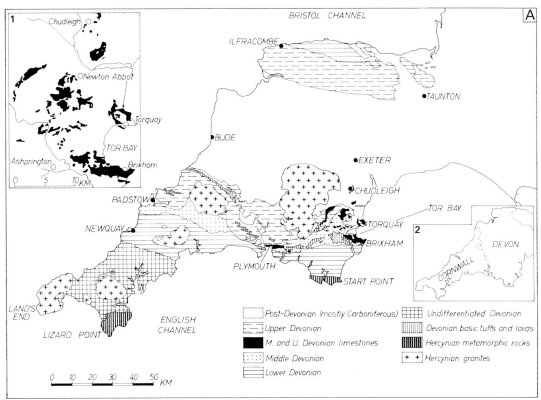

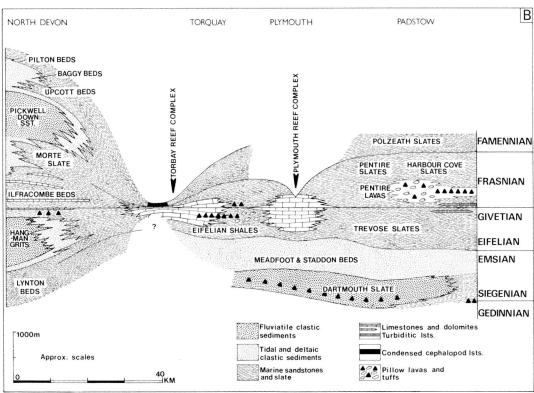

followed in the Emsian by neritic clastic sediments represented by the Meadfoot and Staddon Beds, and in North Devon by the Lynton Beds (Fig. 8B), all exhibiting great similarity to the "Lenneschiefer facies" of the Rhenish Schiefergebirge. During the late Eifelian and the Givetian marked differentiation of depositional environments occurred, with the development in the Torbay-Plymouth district of isolated reefal and volcanic complexes, the deposition of basinal shale facies (Trevose Slates) in northern Cornwall, and of neritic shelf and deltaic clastics, locally with limestone horizons, in the North Devon area (Hangman Grits, Ilfracombe Beds). This is a facies distribution which also closely parallels that in the Rhenish Schiefergebirge (see comparisons by Goldring, 1962; House and Selwood, 1966; House, 1975), and established the pattern for the paleogeographical and depositional evolution of the area during the rest of the Devonian. During the Upper Devonian a thick (over 3000 meters) continental, deltaic and shallow marine clastic succession was deposited in a rapidly subsiding zone around North Devon (see Goldring, 1962; Webby, 1966a and 1966b; House, 1975), whereas the reef complexes in South Devon appear to have developed in an area of high stability which underwent little subsidence during the later Devonian, as indicated by the locally highly attenuated late Frasnian and Famennian limestone sequences associated with them (Fig. 8B). The Upper Devonian of northern Cornwall is represented by further pelagic shales (up to about 1500 meters) and locally pillow lavas. In South Wales and the Bristol-Gloucester district the whole Devonian succession is in a predominantly fluviatile facies (the Old Red Sandstone).

In the areas of both Tor Bay and Plymouth organic carbonate deposition began early in the Eifelian with muddy crinoidal limestones. These become increasingly fossiliferous upwards through the incorporation of stromatoporoid and coral biostromes and can be compared in character with the Schwelm bank facies of the Rhenish Schiefergebirge (see Scrutton, 1977a; Orchard, 1978, p. 925). The development of both complexes reached a peak in the middle and late Givetian,

with their differentiation into reef, fore-reef and back-reef depositional environments, although folding and faulting have complicated the spatial interpretation of these.

Scrutton (1977a and 1977b) considered that the reef in the Torquay area was an isolated atoll-like complex situated at the shelf-margin, and comparable with the large atolls of Brilon and Attendorn in the Rhenish Schiefergebirge. It also probably extended for an uncertain distance towards the northeast, beyond the range of the present Devonian outcrop, and may originally have been as much as 1000 square kilometers in area (Scrutton, 1977a, p. 188). The northern rim of this structure is concealed by Carboniferous sediments, but on its southern margin a southwest-northeast trending belt of "reef-core" facies existed during most of the Givetian and restricted a back-reef lagoon to the north in which dark, thin-bedded and *Amphipora*-biostromal limestones were deposited. The Tor Bay Reef as a whole appears to have ceased accretion by the early Frasnian, although calcarenite shoals probably persisted in the area until about mid-Frasnian (Scrutton, 1977b).

The carbonate complex in the Plymouth district exhibits a similar facies spectrum though back-reef deposition occurred here to the east and southeast of the "reefal" zone (Orchard, 1978, p. 925–926). The reef probably remained as a submarine, possibly locally exposed, topographic high until late in the Devonian when it was covered by basinal shales (Orchard, 1974 and 1977). Scrutton (1977b) has noted similarly that at least a portion of the Tor Bay complex persisted as a submarine rise after extinction, and associated with this in the Late Devonian (Newton Abbot and Chudleigh areas) were carbonate "mud-mounds," closely resembling those of the late Frasnian of the Ardennes.

Both of these reef complexes contain intercalations of lavas and tuffs, and in the Ashprington portion of the Tor Bay area the highest Eifelian, Givetian, and the lowest Frasnian successions consist predominantly of volcanic sediments (the Ashprington Volcanic Group) with only thin limestones. It is possible that the genesis of both these

$\longrightarrow$

FIG. 9.—A. Location map of Poland and Czechoslovakia showing the major geographical features mentioned in the text. B. Vertically exaggerated non-palinspastic section through the "Barrandian" Lower Devonian of Czechoslovakia showing the relationship between the biohermal Koneprusy Limestone and surrounding facies; after Chlupac (1976, fig. 1), slightly modified. V. Lst. = Vinarice Limestone. C. Schematic section through the Devonian of Moravia showing the shelf-to-basin transition and strongly transgressive aspect characteristic of the external Hercynian zones (note the usage of central European stage names in contrast to the Barrandian); after Chlupac (1967, fig. 7), slightly modified. Not to scale, though the total succession in the Brno area has a maximum thickness of about 2000 meters; horizontal distance southeast-northwest 30–40 kilometers.

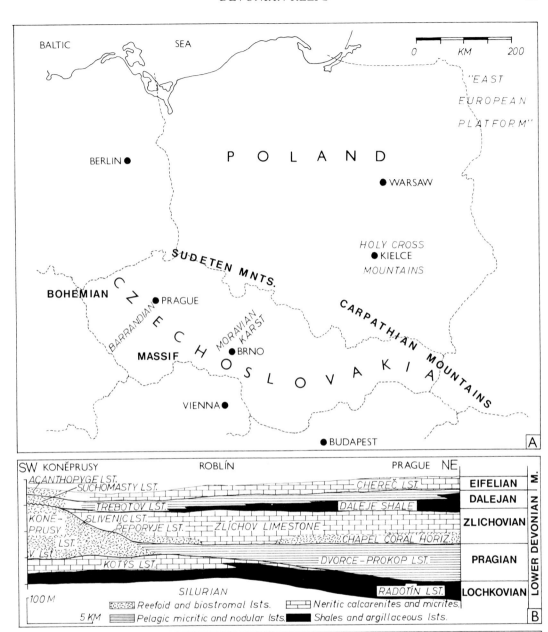

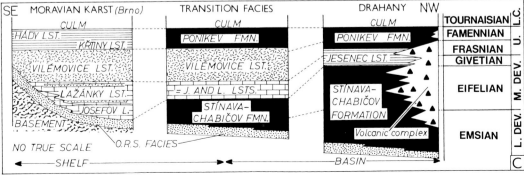

reefs is directly related to the presence of this volcanic center.

*Poland.*—The Polish Devonian represents an eastward extension of the central European Rhenohercynian zone, but paleogeographically presents a more complicated picture. Deposition was restricted to a basin in central and western Poland, bordered to the north and east by the "East European Platform," and to the southeast and southwest by Lower Paleozoic massifs in the Carpathian Foreland and the Sudeten Mountains respectively (Fig. 9A). To the south there was probably continuity with the Devonian basin in Moravia (Dvorak and Freyer, 1965; Chlupac, 1967).

As part of the external Hercynian zone the Devonian in this area also has a strongly transgressive stratigraphy. Lower Devonian marine deposits are in a neritic clastic facies and restricted to westcentral Poland. Elsewhere they are in a transitional or fluviatile Old Red Sandstone facies, or, as over parts of the stable blocks, altogether absent (a feature also of the peripheral parts of the Ardenno-Eifelian tract). In these last areas deposition commenced following marine transgression in the late Emsian (Pajchlova, 1970), and during the Eifelian and Givetian the margins of the surrounding massifs were progressively submerged and covered by shallow marine carbonate sediments.

The Eifelian is characterized by argillaceous limestones, dolomites and shales, locally containing coral-stromatoporoid biostromes (Pajchlova and Stasinska, 1965), and is reminiscent (dolomites excluded) of Eifelian deposits in the Ardennes and Eifel. The higher Eifelian consists of dolomites over almost the whole of Poland (Pajchlova and Stasinska, 1967), and at this stratigraphical level in some marginal areas (e.g., the Carpathian Foreland) anhydrite and gypsum (? sabkha development) are interbedded with dolomites and neritic and peritidal clastic sediments. Similar deposits also occur in the later Middle Devonian in the subsurface of northern Poland (Pajchlova and Stasinska, 1965 and 1967; Chlupac, 1971, p. 135–137).

During the Givetian, widespread development of biostromal shelf limestones formed banks comparable with the "Schwelm-facies" limestones in the Rhenish Schiefergebirge, though in Poland these are also extensively dolomitized which has made interpretation difficult (see e.g., Pajchlova, 1968; Kazmierczak, 1971). Pajchlova and Stasinska (1965 and 1967) recognized the development of "patch-reefs" and concentrations of "massive" stromatoporoids in the middle Givetian of the Holy Cross Mountains.

Devonian carbonates in Poland have not been thoroughly investigated. There are relatively few surface outcrops of these rocks and of the detailed studies made most have been concerned with Upper Devonian reef complexes in rather limited areas. Reefal carbonates of this age occur in the subsurface of Pomerania in northwestern Poland, where they are poorly known only from deep borings, in the area of Lublin in western Poland (Pajchlova and Stasinska, 1967; Neumann, Pozaryska and Vachard, 1975), and in the Holy Cross Mountains and Silesian-Cracovian Plateau in southcentral Poland (Pajchlova, 1970). Best known are the Upper Devonian reefs in the Holy Cross Mountains (Fig. 8A) which have been the subjects of a number of recent studies (Sculczewski, 1968, 1971, and 1973; Kazmierczak, 1971; Racki, 1979, in progress, pers. comm.).

During the latest Givetian and early Frasnian of this area widespread facies differentiation occurred, with the development of massive stromatoporoid-coral reefal limestones in the Kielce district (the "Kielce facies" of Czarnocki, 1928 and 1950), and well-bedded, more argillaceous limestones to the north and south (the "Lysagory facies" of Czarnocki, 1928 and 1950). Szulczewski (1971) considered the reefal limestones to represent a reef-mound situated upon the shelf and surrounded by extensive calcarenitic flank deposits ("detrital facies") which interfingered with the laterally equivalent bedded sediments. The central zone of the structure comprises a complex association of lithofacies ("stromatoporoid-coral facies") which may represent "an extensive reef-fringed bank" (Szulczewski, 1971, p. 111–112). Growth of the reef continued at least up to the Frasnian-Famennian boundary, and possibly into the lower Famennian (Szulczewski, 1971, p. 115), somewhat later than reefs in the central European area. It is capped by a condensed succession of Famennian nodular goniatite limestones. The bedded deposits which surrounded the reef complex throughout the Upper Devonian and eventually covered it were considered by Czarnocki (1950), Pajchlova and Stasinska (1965 and 1967), and Szulczewski (1968 and 1971) to represent deposition in a "basinal" environment. More recent work by Kazmierczak and Goldring (1978), however, suggests that some aspects of the sediments in this facies actually indicate deposition in a much shallower environment; this may have been on an epireefal platform or ramp.

*Bohemia and the Eastern Sudeten Mountains (Moravian Karst).*—Devonian shallow marine carbonates occur in two areas of Czechoslovakia (Fig. 9A), to the southwest of Prague in central Bohemia (the "Barrandian" Syncline), and in the vicinity of Brno in the eastern Sudeten Mountains (Moravian Karst).

The Devonian of the Moravian Karst (Fig. 9C) is the less significant of the two and will not be closely considered. It exhibits facial, faunal and stratigraphical similarities to the external zones of the Hercynian Orogen and probably represents a southeastward extension of the central European Rhenohercynian Trough through Poland into the eastern Sudeten Mountains as a triangular basin situated between the Proterozoic crystalline massifs of Brno and the western Carpathians (Dvorak and Freyer, 1965; Chlupac, 1967 and 1968). The Lower and lowest Middle Devonian of this area is in a fluviatile and neritic clastic facies (up to 400 meters). Associated with southwestward transgression over the Brno Massif at about the Eifelian-Givetian boundary, there was a facies change with differentiation of depositional environments into a basinal one where shales, pelagic limestones and spilitic submarine volcanics were deposited ("Drahany facies"), and a neritic carbonate platform in the vicinity of the stable block ("Moravian Karst facies") which shows many similarities to the Middle Devonian bank and biostromal carbonates of the Ardennes and Rhenish Schiefergebirge (see Dvorak and Freyer, 1965; Chlupac, 1967; Kumpera, 1971). This latter development was deposited as a series of diachronous overstepping carbonate formations containing biostromes and, less commonly, small bioherms of stromatoporoids and corals (Dvorak and Freyer, 1965, p. 406; *Amphipora* and *Stringocephalus* are also common at some horizons (Zukalova, 1961; Dvorak and Havlicek, 1963). Later Upper Devonian deposits consist of pelagic shales and goniatite limestones which pass conformably upwards into Lower Carboniferous Culm deposits (Chlupac, 1967).

Due to differences in sedimentary facies and faunas the Bohemian Lower Devonian has not been easy to correlate with that of central Europe, and the stages of the latter area have been superceded by the local stage names Lochkovian, Pragian, Zlichovian and Dalejan which are based on a combination of the ranges of benthonic and pelagic faunas (Chlupac, 1967, 1968 and 1976; see Fig. 9B), although the Middle Devonian stage names have remained unaltered. More recently correlation has also been facilitated using conodonts (see e.g., Klapper *et al.*, 1978). Deposition was continuous from the Lower Paleozoic through the Lower Devonian.

Within the present Barrandian Lower Devonian outcrop, reefal carbonates occur only in the Pragian (= Siegenian-Emsian) in the vicinity of Koneprusy (the Koneprusy Limestone) at the southwestern end of the syncline. This area probably formed a stable shelf which subsided less rapidly than surrounding "basinal" zones. It was one of consistent shallow marine carbonate deposition throughout the Lower and lowest Middle Devonian, whereas sequences of pelagic platey lime-

stones and shales were deposited in the surrounding areas during this time (Svoboda and Prantl, 1949; Chlupac, 1954, 1956 and 1967).

Three principal elements can be recognized in the Koneprusy Reef complex (Fig. 9B):

1. a pre-reefal bank facies comprising largely *in situ* crinoidal limestones (Vinarice Limestone), which forms a foundation for,

2. a reef facies association (Koneprusy Limestone) containing abundant *in situ* frame building organisms (algae, stromatoporoids, corals) and,

3. a "peri-reefal" facies (Slivenic and Reporyje Limestones) comprising reef talus and calcarenites with a rich benthonic fauna, but few *in situ* frame building organisms.

The last two of these are demonstrably lateral equivalents, although in the past there has been some confusion in their stratigraphical interpretation (see Chlupac, 1954 and 1956, p. 101). The light-colored, unbedded Koneprusy Limestone (up to 200 meters thick) has been interpreted by Chlupac (1954) as a "reef-core" facies which forms dome-shaped bioherms. It interfingers with, and in some areas is overlain by, the calcarenites of the Slivenic facies (up to 90 meters thick) which locally exhibits primary depositional dips (Chlupac, 1968) and also contains an admixture of pelagic faunal elements more typical of surrounding deeper water facies which are absent from the Koneprusy Limestone (Chlupac, 1954 and 1968).

The Koneprusy Reef complex experienced emergence and erosion related to epeirogenic uplift during the late Lower Devonian (Chlupac, 1956, p. 472) and is overlain unconformably (Zlichovian absent) by the Suchomasty Limestone (Dalejan-lowest Eifelian; Fig. 9B). This is a succession (20 meters) of neritic limestones, in part biostromal, which locally has a basal conglomerate and also infills fissures in the Koneprusy Limestone (Chlupac, 1968).

*Southeastern and Eastern Alps.*—Devonian carbonate rocks occur extensively in the "Northern Greywacke Zone" and the area around Graz in the northeastern Alps of Austria, and in the Carnic Alps and Karawanken which take in the area of the southeastern Alps where the borders of Austria, Italy and Yugoslavia meet (Fig. 10A). Only the Devonian reefal carbonates in the southeastern Alps have been investigated in any detail and discussion will therefore be restricted largely to these.

The interpretation of Paleozoic stratigraphy and paleogeography within the Alpine realm is a problem of extreme complexity, and only over the last few years has a picture emerged in any sort of coherent detail, though this is still far from complete (see Flügel, 1970 and 1976; Schönlaub, 1979). Principal reasons for this have been the tectonically highly disturbed and commonly iso-

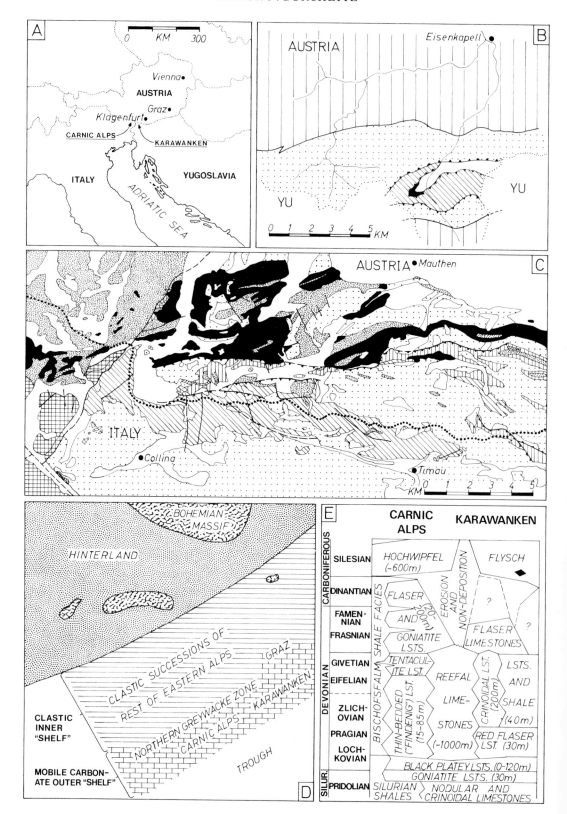

lated character of the Paleozoic outcrops, due to the compound effects of the Hercynian and Alpine Orogenies, and the difficult nature of the terrain. In both the Carnic Alps and Karawanken (Fig. 10B and C), for example, the Devonian reefs were broken up along with the rest of the Paleozoic deposits to form numerous imbricate thrust sheets, and have been dislocated further by major fault systems (Vai, 1971, p. 169; Tessensohn, 1974, p. 92 and 122; Flügel et al., 1977).

As in central Bohemia, deposition appears to have been continuous, though locally condensed, from the Silurian through the Devonian over the whole of the southeastern Alps and, in the Carnic Alps at least, through the Lower Carboniferous as well. The Paleozoic stratigraphy of the Carnic Alps and Karawanken, and probably of the poorly known core zones of the Dinaric Alps (Fig. 1), exhibit a common evolution (Vai, 1975), but correlation, in some respects still controversial, has been hindered by lateral facies variations and differing interpretations of tectonics (see also Clar, 1971; Flügel, 1975 and 1976; Schönlaub, 1979). During the Lower Devonian these areas also possessed a uniform "Hercynian-Bohemian" fauna with strong affinities to the Uralo-Tienshan faunal province of Asia (Flügel, 1964, p. 409; Vai, 1967 and 1975; Schultze, 1968; Erben, 1969; Kodsi, 1971), though it is one which also allows correlation with the central European Rhenish stratigraphy (Flügel et al., 1977). The Bohemian biostratigraphy has consequently also been adopted for the Lower Devonian of the southeastern Alps.

As in other areas, two facies associations have been recognized in the Devonian of the Carnic Alps (Fig. 10E), corresponding in a broad sense with the deposits of 1. a carbonate shelf or platform with reefal limestones and laterally equivalent condensed cephalopod limestones ("Rauchkofel facies"), and 2. a deeper water "basinal" environment ("Bischofalm facies") where pelagic

shale and bedded chert deposition continued from the Lower Paleozoic through the Devonian (Leditsky, 1974, p. 171; Schönlaub, 1979). The carbonate shelf stretched from the Carnic Alps possibly for some 600 kilometers "southeastwards" into the Dinaric Alpine zone (Vai, 1975, p. 294), and the Devonian of the Karawanken represents a continuation on this shelf of the stratigraphy and facies of the Carnic Alps (Flügel, 1964, p. 412; Schönlaub, 1971; Vai, 1971); to the west features have been largely masked by metamorphism (Selli, 1963; Vai, 1975).

In both the Carnic Alps and the Karawanken, the Lochkovian and Pragian are represented in all shelf carbonate successions by fairly uniform deposits. Latest Silurian and early Lochkovian sediments comprise black, platey limestones and shales ("Schwarze Plattenkalke" or "ey-limestone") with intercalations of goniatite- and tentaculite-rich limestones; towards the top of the Lochkovian, though still in the same relatively deep water facies, the deposits acquire a red or yellow coloration ("Rote Flaserkalke"), and crinoidal debris beds become more common (Flügel, 1964; Tessensohn, 1974). Differentiation of depositional environments into reef complexes and epireefal platform occurred between the latest Lochkovian and the early Zlichovian (Schönlaub, 1971; Tessensohn, 1974), although the exact distribution and inter-relationship (and in some respects the interpretation) of these is still problematical. On the epireefal platform, transitional between reef and basin, attenuated sequences of reddish-grey platey ("flaser") limestones, containing autochtonous pelagic fossils and allochthonous reef-derived debris, were deposited (see Deroo, Gauthier and Schmerber, 1967; Schönlaub, 1971; Bandel, 1972 and 1974; Fig. 10E).

In the Carnic Alps and Karawanken, reef growth, and contemporary goniatite limestone deposition, continued until the late Frasnian. Both

←

FIG. 10.—A. Location map for the Carnic Alps and Karawanken. B. Geological sketch map of the Austrian section of the Karawanken illustrating the manner in which Paleozoic deposits are exposed as a series of thrust slices in the "Seeberg window"; after Tessensohn (1974, fig. 1). Blank = Recent; vertical lines = Triassic; Black = overthrust Upper Carboniferous; Stipple = Paleozoic sediments including Devonian reef limestones and associated facies; Cross-hatch = Paleozoic sediments including Devonian marine clastics and banded limestones. C. Geological sketch map of the central Carnic Alps showing outcrops of Devonian reefal limestones and associated facies. After Deroo et al. (1967, fig. 1). Blank = Tertiary and Quarternary; Squares = Permo-Triassic; Coarse stipple = Carboniferous; Black = Devonian reefal limestones; Diagonal lines = undifferentiated reefal Devonian and goniatite limestones; Vertical lines = Devonian and Silurian goniatite limestones; Fine stipple = Ordovician and Silurian. D. Reconstruction of likely paleogeography in the southeastern Alps during the Devonian after "rolling back" the effects of the Hercynian and Alpine Orogenies; after Schönlaub (1979, fig. 75). Not to scale, but the width of the shelf from "shore" to trough may have been around 40–500 kilometers. Note the similarity to the Rhenish shelf paleogeography. E. Schematic representation of Devonian stratigraphy in the "Rauchkofel facies" of the Carnic Alps and Karawanken; after Schönlaub (1979), slightly modified. Black areas indicate volcanics; not to scale.

Deroo *et al.*, (1967) and Bandel (1969 and 1972) have produced "barrier-reef" models for Devonian carbonate sedimentation in the Carnic Alps in which reef-core, fore-reef and back-reef (and off-reef and basinal) facies were recognized, although these authors differed somewhat in their interpretations of the paleogeography. Tessensohn (1974) has produced a similar model for Devonian reefs in the Karawanken, and suggested that these formed an elongate reef zone bordered on both sides by "flaser" limestones and basinal facies, but was unable to orientate the complex. Vertical transitions between reef-core, fore-reef and back-reef facies occur in both areas (Deroo *et al.*, 1967; Bandel, 1972; Tessensohn, 1974), indicating that the complexes retreated and advanced a number of times during their growth.

In the highest Frasnian the reefs were drowned and sedimentation in most places within the area of the former carbonate platform continued in a uniform deeper water goniatite and "flaser" limestone facies, locally highly condensed, which lasted until the start of pre-Hercynian flysch deposition in the mid- to late Viséan (Schönlaub, 1971 and 1979; Tessensohn, 1971 and 1974). In the Karawanken exposure during the late Frasnian or early Famennian resulted in partial erosion of the reef complexes (Tessensohn, 1974 and 1975; Fig. 10E).

In the area around Graz and the Northern Graywacke Zone the Devonian successions also contain extensive platform and "reefal" carbonates, but also exhibit a strong clastic influence. Due to this contrast in facies with the carbonates of the southeastern Alps it was formerly assumed that the Devonian of the northeastern Alps represented deposition in a separate basin (Flügel, 1967 and 1976; Vai, 1971 and 1975). In a recent paleogeographical reconstruction, however, Schönlaub (1979) considered that, after "winding-back" the effects of tectonic deformation, the Devonian depositional regime in the whole eastern Alpine region comprised an inner clastic shelf, with hinterland to the "north," and an outer mobile carbonate shelf, with a basin situated to the "south" (Fig. 10D). In this scheme the Devonian of the Northern Greywacke Zone and the Graz area, with their mixed carbonate-clastic successions, represent deposition in a transitional zone of the shelf, and the predominantly carbonate successions of the Carnic Alps and Karawanken deposition in the aera of the outer shelf. If indeed such a model fits, then the Devonian paleogeography of the eastern Alpine realm exhibits remarkable similarity to that of the Rhenish Schiefergebirge or southwest England. In a paleogeographical situation such as that envisaged by Schönlaub (1979), the reefal complexes of the Carnic Alps and Karawanken might well have resembled in form the large atolls of Brilon and Attendorn in the Rhenish area, rather than barrier-reef complexes as most interpretations seem to imply.

*Spain (the Cantabrian Mountains and the Pyrenees).*—There are two discrete Devonian provinces in Spain: a northern one embracing the Cantabrian, Pyreneean, Iberian, and Catalanian "basins," and a southern one comprising the "Carpentarian basins" of Ciudad Real, Extramadura, and Andalucia. These were linked until separated by epeirogenesis during the Eifelian (Brouwer, 1964a and 1967; Llopis-Llado *et al.*, 1967). In the southern province the Devonian successions consist predominantly of clastic sediments, whereas in the northern there are in most places complete successions from the Silurian to

$\rightarrow$

FIG. 11.—A. Location map for the Cantabrian and Pyreneean Mountains. B. Area of the Cantabrian Mountains showing Devonian paleogeography and major paleotectonic features; after several authors, but mainly from van Adrichem-Boogaert (1967, figs. 58–61). CL = Cordano line; LL = Leon line; PF = Porma fault zone; SGL = Sabero-Gordon line (fault zone); SR = "Santibanez ridge". C. Highly schematic section through the Devonian of the Asturo-Leonese facies belt. After de Coo *et al.* (1971, fig. 1). Not to scale, though north-south distance approximately 20 kilometers, and the total thickness of the succession about 1500 meters on the southern slopes of the Cantabrians (van Adrichem-Boogaert, 1967). Note the manner in which the Upper Devonian unconformity truncates earlier Devonian formations. Reworked fossils from the La Vid and St. Lucia Formations in the Portilla Formation (Reijers, 1973b) indicate that this developed progressively through uplift in the area of the Asturian Geanticline. D. Geological sketch map of the Cantabrian area; after Mendez-Bedia (1976, fig. 1). Note the fashion in which Devonian deposits of the Asturo-Leonese belt are folded around the Asturian Geanticline. E. Geological sketch map of the central and western Pyreneean Mountains; after Mirouse (1967, fig. 1), modified. F. Highly simplified geological sketch map of the eastern Pyrenees, the Mouthoumet Massif, and the Montagne Noire showing outcrops of Hercynian igneous and metamorphic rocks (Stipple), and Paleozoic (cross-hatch) and post-Paleozoic sediments (blank); after Ovtracht (1967, fig. 1), modified. G. Highly schematic section through the Devonian along the axis of the Pyreneean mountain chain from Iraty in the west to Andorra in the east (see this Fig. E), showing very broadly the facies distribution in the area. Modified after Llopis Llado *et al.* (1967, fig. 3), with data from a number of authors.

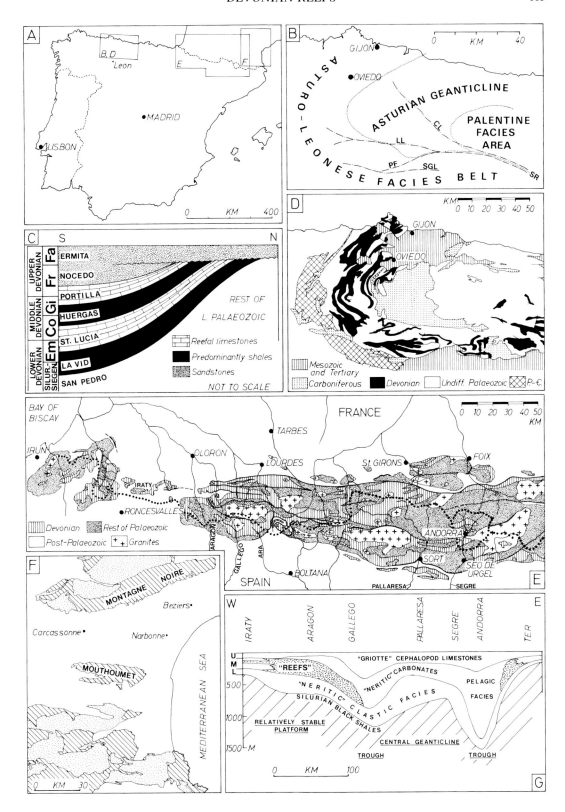

A

B.D
·Leon
·MADRID
·LISBON
0    KM    400

B
GIJON
·OVIEDO
0    KM    40
ASTURO-
ASTURIAN GEANTICLINE
PALENTINE
FACIES
AREA
LEONESE FACIES BELT
LL
CL
PF    SGL
SR

C
S                                    N
UPPER DEVONIAN | Fr | Fa | ERMITA
NOCEDO
MIDDLE DEVONIAN | Gi | Co | PORTILLA
HUERGAS
ST. LUCIA
LOWER DEVONIAN | SILUR. SIEGEN. | Em | LA VID
SAN PEDRO
REST OF
L. PALAEOZOIC
Reefal limestones
Predominantly shales
Sandstones
NOT TO SCALE

D
KM
0 10 20 30 40 50
GIJON
OVIEDO
Mesozoic and Tertiary
Carboniferous
Devonian
Undiff. Palaeozoic
P-E

BAY OF BISCAY
FRANCE
·TARBES
0 10 20 30 40 50
KM
IRUN
·OLORON
·LOURDES
St.GIRONS
·FOIX
·IRATY
·RONCESVALLES
ANDORRA
ARAGON
GALLEGO
ARA
·BOLTANA
SORT
SEO DE URGEL
SPAIN
PALLARESA
SEGRE
Devonian
Rest of Palaeozoic
Post-Palaeozoic
+ + Granites
E

F
MONTAGNE NOIRE
Beziers·
Carcassonne ·
Narbonne·
MOUTHOUMET
MEDITERRANEAN SEA
0    KM    30

W                                                                E
IRATY | ARAGON | GALLEGO | PALLARESA | SEGRE | ANDORRA | TER
U
M
L
"GRIOTTE" CEPHALOPOD LIMESTONES
"REEFS"
"NERITIC" CARBONATES
500
"NERITIC" CLASTIC FACIES
PELAGIC FACIES
SILURIAN BLACK SHALES
1000
RELATIVELY STABLE PLATFORM
CENTRAL GEANTICLINE
1500 M
TROUGH
TROUGH
0    KM    100
G

the late Upper Devonian, and sedimentary facies are more varied (Llopis-Llado *et al.*, 1967). Devonian reef complexes developed in the northern province in the Cantabrians and the Pyrenees, and the many similarities in fauna, stratigraphy and sedimentary facies between the successions of these areas suggest that they originally formed part of the same depositional basin (Brouwer, 1967; Boersma, 1973, p. 343). Certain facial and faunal similarities also exist between the Devonian in northern and eastern Spain and that in Brittany (Brouwer, 1967, p. 43).

In the Cantabrian Mountains, Devonian deposits are restricted to a fold and thrust belt surrounding the "Asturian Geanticline" (Fig. 11D), which was a positive zone throughout most of the Paleozoic, and comprise two complexes of sedimentary facies, those of the Asturo-Leonese and Palentine zones (Brouwer, 1964a, and b; van Adrichem-Boogaert, 1967; see Fig. 11B). The former consists of neritic terrigenous and carbonate sediments, deposited in a crescent bordering the Asturian Geanticline for some 250 kilometers on its southern and western sides. The original western margin of this zone is not seen due to pre-Stephanian uplift and erosion. The Palentine facies, in contrast, was deposited in a low energy, deeper water environment and consists of shales, and argillaceous limestones and sandstones containing a pelagic microfauna and a poorly developed macrofauna of solitary corals and brachiopods which are commonly impoverished and stunted (Kullmann, 1967). This facies is restricted to an area in the southeastern Cantabrians, bounded on its northern margin by the Asturian High, and separated to the east and southeast from the Asturo-Leonese facies belt by the "Santibanez Ridge" (Fig. 11B), part of a Late Paleozoic hingeline though it was itself relatively stable (van Adrichem-Boogaert, 1967). To the east the Devonian is obscured by Mesozoic sediments.

The discrete characters of these two facies belts, which correspond essentially to shelf and basin, were not marked in the Silurian or lowest Devonian, but became more pronounced as tectonic instability in the area increased during the course of the period (see e.g., van Staalduinen, 1973). Their evolution is documented by a gradual polarization in the style of deposition from an originally uniform deep water shaley sedimentary suite (the La Vid Formation and equivalents). A landmass to the northeast, and probably the Asturian Geanticline itself, functioned as intermittent source areas for the clastic sediments present in parts of the Devonian successions in both of these zones (van Adrichem-Boogaert, 1967, p. 173). In the late Upper Devonian, uplift in the region of the Asturian Geanticline caused widespread deposition of siliciclastic sediments

and the development of an intra-Devonian unconformity which cuts across earlier formations, increasing in magnitude towards the massif (see Fig. 11C).

With the exception of a few thin tabulate coral biostromes in the Lower Devonian succession of the Palentine facies zone, with analogues in the basal La Vid Formation of the Asturo-Leonese succession (Brouwer, 1967; Stel, 1975), reefal carbonates occur only in the Asturo-Leonese belt and between the two areas in the vicinity of the "Santibanez Ridge" (Kanis, 1956; Brouwer, 1964, p. 49–50; van Adrichem-Boogaert, 1967). Throughout this zone, the Devonian stratigraphy is remarkably consistent, so that everywhere two horizons of shallow marine carbonate and reef development are distinguishable: the upper Emsian-lower Eifelian St. Lucia Formation (150–200 meters), and the upper Givetian-lower Frasnian Portilla Limestone Formation (60–200 meters). The two are separated by clastic shelf sediments (Huergas Formation; see Fig. 11C).

Carbonate facies in the St. Lucia Formation are arranged in concentric belts paralleling the curved margin of the Asturian Geanticline, with neritic deposits externally to the south and west, and shallower water, restricted ("lagoonal" or back-reef) deposits internally, adjacent to the massif (De Coo, Deelman and van der Baan, 1971; Mendez-Bedia, 1976, p. 19–22). The last comprise chiefly inorganically laminated and fenestral pelsparites and pelmicrites, and locally exhibit desiccation features (Rupke, 1965; Smits, 1965). Towards the Asturian Geanticline these carbonates also grade laterally into red-colored peritidal-neritic clastic deposits (Llopis-Llado, 1960, p. 44–46). The more distal carbonate facies of the St. Lucia Formation is developed principally as a complex of stromatoporoid-coral biostromes, although small, domed bioherms (up to 85 meters in thickness) are present locally in the middle of the formation, and some of these display vertical ecological zonation (Brouwer, 1964; Mendez-Bedia, 1976). On the southern slopes of the Cantabrian Mountains, south of the Asturian Geanticline, scattered "mound-like patch-reefs" also occur at this stratigraphical level (van Adrichem-Boogaert, 1967; Reijers, 1973a, p. 15).

In the Portilla Formation biostromes are also well developed, and there is a very similar distribution of facies to that in the St. Lucia Formation, with a distinct belt of sheltered back-reef facies situated between the Asturian Geanticline and the organic barrier; more distally the biostromes fronted onto an epireefal calcarenite ramp (Reijers, 1972). On the southern slopes of the Cantabrians the Portilla Formation comprises three, locally four, members (Mohanti, 1972; Reijers, 1972 and 1973b), most of which have a biostromal aspect, but at some horizons gently domed reef mounds of corals and stromatoporoids also occur (Reijers, 1972). In both the St. Lucia and Portilla Formations transgressive and regressive tendencies can be recognized.

Bioherms appear to be particularly strongly developed at several horizons of the Devonian in the transition area between the Asturo-Leonese and the Palentine facies belts (along the "Santibanez Ridge"), which probably subsided only slowly during most of the period (Brouwer, 1964b).

The Devonian reefal carbonates of the Pyrenees (Fig. 11E and F) are poorly known and few useful data on them have been published. As in the Cantabrian Mountains, tectonic instability during the Lower and Middle Devonian was reflected in the polarization of depositional environments into marginal shelf and shallow basinal regimes, following on fairly uniform Silurian pelagic shale sedimentation (Mey, 1967). In external zones of the Pyreneean Trough much of the Devonian succession is in shallow marine facies and locally exhibits stratigraphic breaks, whereas in the internal portion (Fig. 11G) deposition was essentially continuous in a limestone and shale pelagic basinal or even neritic facies (see de Sitter and Zwart, 1961; Mey, 1967; Requardt, 1972). The Upper Devonian, as in much of Europe, was characterized by a return to rather uniform pelagic conditions, with the deposition of calcareous mudstones and micritic "griotte" cephalopod limestones in most areas (Llopis-Llado *et al.,* 1967; Mey, 1967; Mirouse, 1967).

Reef growth appears to have followed a pattern similar to that in the Cantabrians, and stromatoporoid-coral biostromes and bioherms developed in the slowly subsiding platform areas of the western, westcentral, and locally in the northern Pyrenees during the Eifelian and the Givetian (Llopis-Llado *et al.,* 1967; Ovtracht, 1967; Boersma, 1973; Joseph, 1975), and in places may have continued until the lower Frasnian (Wesink, 1962).

In the eastern Pyrenees, Montagne Noire, and in the vicinity of the Mouthoumet Massif (Fig. 11F) there are apparently also small Devonian reef occurrences, although little on them has been published. Avais and Joseph (1967, p. 25) have commented on a tendency towards reef building in the Middle Devonian of the Montagne Noire, namely in the Eifelian of the Roquebrun (100 meters of "reefal" carbonates) and the Eifelian and Givetian of the Combe Rolland (30 meters "reefal" carbonates) areas. Biohermal limestones of similar age occur in the area around the Mouthoumet Massif (von Gaertner, 1937; Boyer, 1964) although here the succession is much dolomitized and poorly exposed (Ziegler, 1959).

*Armorican Massif (Northwestern France).*—

Small bioherms and biostromes developed during the Lower Devonian of the northeastern part (Cotentin) of the Armorican Massif (Fig. 1). These reefal carbonates comprise the "Horizon récifal de Baubigny" and the "Biostrome Tetracoralliaires de la Roquelle," members within the "Formation des argillites, grés fines, et grauwackes de décalcification," which is an argillaceous limesone, sandstone and shale sequence (up to 100 meters thick) representing the upper Siegenian in this area (see Bigot, 1904; Poncet, 1972 and 1977). This succeeds a middle Siegenian shallow marine carbonate unit ("Schistes et Calcaires de Néhou") containing stromatolitic and evaporitic facies, and locally thin biostromes (Poncet, 1975).

The Baubigny bioherms are rounded or elongated domed structures which reach 15–20 meters in thickness and exhibit vertical ecological zonation and laterally possess well developed flank deposits (Poncet, 1972 and 1976). Facies peripheral to the bioherms comprise bedded calcarenites, locally containing ooliths and calcareous algal nodules, although in view of the abundance of the remains of such stenohaline organisms as brachiopods, bryozoa and echinoderms (see Poncet, 1976) the depositional environment cannot have been highly restricted. They may represent deposition on an inter-reefal platform sheltered by the stromatoporoid-coral bioherms, but Poncet (1976) considered that these sediments represent back-reef facies (*sic*). The Roquelle biostrome (up to 7 meters in thickness) lies at a slightly older stratigraphical horizon than the bioherms (Poncet, 1977).

## CLASSIFICATION

With regard to morphology and structure, five broad categories of reef can be recognized in the European Devonian (Fig. 12A):

1. biostromes and banks,
2. biostromal complexes,
3. barrier-reef complexes,
4. isolated reef complexes (reef-mounds and atolls),
5. quiet water carbonate buildups ("mud-mounds").

In the following section these are briefly defined, and their distributions discussed. The parameters for classification within these groupings have been broadly used in order to accommodate differences and inconsistencies in the emphasis and quality of the many sources of published data utilized. The terms used, where not otherwise indicated, follow the scheme outlined in a recent comprehensive survey by Heckel (1974). However, without formally proposing modification of the latter, the use in the present paper of the important term *reef* should be qualified. For the sake

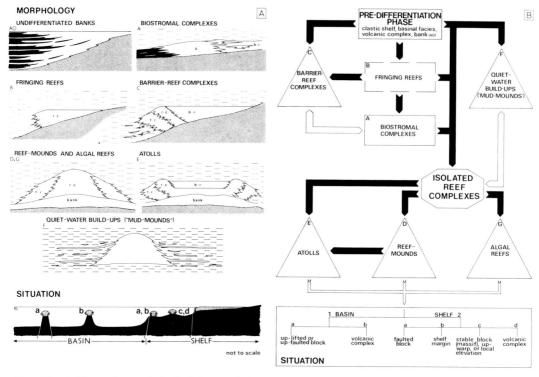

FIG. 12.—A. The main morphological characteristics (above) and paleotectonic locations (below) of reef complexes in the European Devonian. "fringing-reefs" are included here although their occurrence is poorly documented, being possibly present in the Upper Devonian of the northern Ardennes (Tsien, 1977) or the Bergisches Land. Horizontal broken ornament = enveloping sediments; b-r = back-reef facies; f-r = fore-reef facies; r-c = reef-core facies. Other characters refer to section B. B. Flow diagram illustrating the general stratigraphical relationships between reef complexes in the European Devonian; commonly this has a genetic basis. Banks, biostromal complexes and (?) fringing-reefs locally form the foundations for buildup growth, and under correct conditions reef-mounds matured into atolls or showed a trend in this direction. Quiet water buildups in the Ardennes form the foundation phases of some reef-mounds, and themselves succeed reef-mounds and atolls in the Ardennes and southwest England. Though algal reefs have been included here as a separate category most can be considered to be morphologically reef-mounds. Isolated reef complexes show greatest diversity in situation. The whole scheme can be utilized as a notational classification, if required; thus a reef situated on fault block within a basin would be E1a; or a quiet water buildup situated upon a previous shelf reef complex would be F2c. Characters can be added by choice to each section, or be made more specific, to allow for factors not included here.

of simplicity no strict definition has been applied and it is preferred to use this in the European Devonian context as a non-specific master label for all very dense *in situ* accumulations of skeletal carbonate (cf., also Dunham, 1970), even where these did not obviously build a wave-resistant framework; this allows biostromes, banks and biostromal complexes to be considered within the same broad class as the other reef types listed. The term *carbonate complex,* which appears to be the only other convenient alternative, does not necessarily imply the presence of organic remains (unless the prefix *organic* is attached to it, whereby it becomes unwieldy) and is thus considered too general for many applications. On the other

hand, it has been found convenient to use the term *reef complex* to denote the collective facies (e.g., reef-core, back-reef and fore-reef) of any reefal unit. Heckel's (1974) terms *bank* and *buildup* are thus used as more specific subordinates of *reef,* but retain their initial definitions, with the provision that in all cases their formation was strongly influenced by *in situ* organisms.

The overwhelming majority of European Devonian reefs fall into the categories 1.–4. given above. Further division according to their foundation and paleogeographical position is possible and lends itself well to the construction of a convenient notational classification (Fig. 12B). There are definite stratigraphical relationships and sequences of succession between reefs in the European Devonian which are recurrent in both space and time, and these are commonly, though not necessarily, of a genetic nature. Thus, in the evolution of many of the carbonate provinces considered in this paper, banks and biostromal complexes preceded barrier and isolated reef complexes; reef-mounds, through increasing size or lateral growth formed the precursors of some atolls; and mud-mounds, established upon basinal shales or on previous reefs, formed in turn the basal phases of large reef-mounds in the Ardennes. Relationships of this nature apparent between reefs in the European Devonian are documented graphically in Figure 12B.

*Biostromes and Banks.*—Use here of the term *biostrome* is consistent with that of Cumings (1932), and Nelson, Brown and Brineman (1962). Although in the strictest sense *bank* may be applied to a single biostrome, use of the term is preferred here in its wider definition to denote laterally undifferentiated, sheet-like carbonate bodies comprising tiered biostromes or a complex intercalation of biostromes with non-reefal sediments. Except in special cases (e.g., on top of rises in basinal areas) banks possessed little intrinsic relief above surrounding deposits during growth, and therefore cannot be regarded as discrete buildups.

The most extensive example of a bank in the European Devonian is the "Schwelm facies" of the Rhenish Schiefergebirge (Krebs, 1968 and 1974). Deposition of this began in the Middle Devonian (see earlier section) and, although the facies is actually heterochronous, during the late Givetian it must have formed a "carbonate platform several thousand square kilometers extending over the outer [i.e., Rhenish] shelf east of the Rhine River" (Krebs, 1974, p. 170). Essentially similar deposits represent the late Givetian and parts of the Frasnian successions in the Ardennes and the Eifel, in these cases making up the "subturbulent" and "quiet water" depositional zones of Lecompte, 1958 and 1970), and also occur extensively in the late Middle and early Upper Devonian of the other external Hercynian zones such as Poland and the Moravian Karst (see Dvorak and Freyer, 1965; Pajchlova and Stasinska, 1965 and 1967; Chlupac, 1967).

Deposits of Schwelm-facies type are well-bedded, laterally poorly differentiated, and somewhat bituminous. They comprise biostromes mostly of stromatoporoids and corals interbedded variably with dark pelsparites, pelmicrites and shales (Krebs, 1974, p. 174–177).

Considered in isolation, shelf-lagoonal deposits in the Frasnian barrier-reef complexes of the Ardenno-Eifelian tract are, in the broadest sense, banks.

*Biostromal Complexes.*—Eifelian and lower Givetian shelf deposits of the Ardennes and Eifel consist of packets of biostromes and restricted limestones, separated vertically by quasi-basinal shales and argillaceous limestones, and exhibit a much greater degree of lateral differentiation than banks. In these cases the biostromes of stromatoporoids and corals formed broad zones fringing the shelf margin, behind which more restricted shelf-lagoonal deposits accumulated, and which graded distally into deeper off-reef shelf or basinal deposits (Fig. 13); locally small bioherms developed within or above the biostromes. The situation which existed in the St. Lucia and Portilla Formations of the Cantabrian Mountains was also broadly comparable with this model, and the complexes are on a similar scale to those of central Europe.

Reef complexes of this character, which possessed little relief above surrounding depositional environments (and therefore no fore-reef facies) but exhibited marked lateral differentiation into back-reef and biostromal-barrier facies, have been termed *biostromal complexes* in order to distinguish them from the poorly differentiated biostromal banks. Intercalations of non-reefal sediments in the biostromal barrier facies are also less common than in the latter.

*Barrier-Reef Complexes.*—These had the form of a reef belt which fringed a landmass and fronted onto a deep water trough or transitional epireefal platform. Lateral facies differentiation was marked and restricted between the organic barrier, and the shoreline was a broad shelf-lagoon. The whole formed a carbonate "platform" which thinned landwards and had a *steep seaward face* (Fig. 12A); thus fore-reef talus was usually well developed.

Barrier-reef complexes developed at the margins of the Devonian depositional basins in Europe, and locally fringed positive areas on the shelves. Reef complexes which can be assigned to this category occur in the latest Givetian and the Frasnian of the Ardenno-Eifelian tract and the northernmost Bergisches Land, and are possibly represented in the Middle Devonian reef complex associated with the mid-German crystalline rise which was penetrated by the "Saar 1" deep-bore in the southernmost Rhenish Schiefergebirge (Krebs, 1971 and 1974). It is unclear whether or not the Devonian reefs of the southeastern Alps were of this type.

Most of the area of such reef complexes was occupied by the shelf-lagoons. Allowing a 33–50 percent palinspastic correction of Hercynian deformation in the Rhenish Schiefergebirge and Ardennes (see Wunderlich, 1964; Franke *et al.*, 1978, p. 204), the Upper Devonian shelf-lagoons of the Ardenno-Eifelian tract must each have cov-

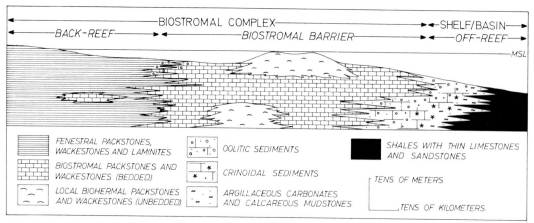

FIG. 13.—Schematic section through a biostromal complex based on the example of the Middle to Upper Devonian Portilla Formation, Cantabrian Mountains; largely after Reijers (1972, fig. 19). Not to scale and horizontal dimensions greatly contracted. Biostromal complexes in the Middle Devonian of the Cantabrians and the Ardenno-Eifelian tract were of broadly similar form, though probably with a less marked tendency towards biohermal development.

ered an area of nearly 10,000 square kilometers (paleogeographical data in Lecompte, 1970), with the back-reef zone reaching possibly 50–60 kilometers in width from reef-to-shore. Elsewhere, however, accurate estimates of spatial relationships and size are not possible, due variously to erosion, obscuration by younger cover, and tectonic disruption. In some cases (e.g., in the Eifel-Aachen tract and the northern Bergisches Land) the shelf-lagoonal deposits are the only portions of the reef complexes which have been preserved, so that the original presence of a restricting reefal rim must be inferred (see Copper, 1966; Kasig and Neumann-Mahlkau, 1969; Krebs, 1974, p. 166).

Although the Frasnian barrier-reefs are preserved in the Ardennes, they are extensively dolomitized (Lecompte, 1970; Tsien, 1977, p. 194), and investigation of their morphologies and facial composition has been largely neglected in favor of the more easily studied reef-mounds at the same stratigraphical levels. Lecompte (1967 and 1970) made scant reference to discrete barrier-reefs, but appears more to have envisaged the seaward margins of the complexes as an area of amalgamation of shelf biostromes, and actually termed the whole complexes "biostromes." Tsien (1971, 1975 and 1977), on the other hand, regarded these features as true barrier-reefs, and in reconstructions (see e.g., Tsien, 1971, fig. 2, 1977, p. 193 and fig. 3) has implied the presence of a wave resistant reef-core zone with a seaward talus slope, but has offered no further sedimentological support for this model. Indeed, many of the successions indicated by these authors as repre-

senting the barrier-reef zone, though dolomitized, are clearly *bedded* (see e.g., Tsien, 1977, p. 194) and therefore probably do not correspond to a resistant core facies. It is, in fact, questionable whether such a facies has yet been satisfactorily demonstrated in the Ardennes. Nevertheless, the similarity which the Frasnian back-reef facies in the Ardennes and Eifel exhibit to those in regions such as the Canning Basin, Western Australia, where Devonian barrier-reef complexes are well exposed (see Read, 1973; Playford, 1976), and also the presence of debris beds in off-reef facies (see e.g., Krebs, 1962 and 1974), would suggest that the organic barrers in these areas were true buildups rather than biostromal.

Tsien (1976 and 1977, p. 199) considered that coral "fringing-reefs" (i.e., wave resistant structures bordering a landmass, but without a back-reef lagoon) developed locally in the northern Ardennes during the late middle Frasnian (Fig. 12A).

*Isolated Reef Complexes.*—These were organic buildups (bioherms) which developed upon submarine elevations within shelf areas and basins and were isolated from direct input of terrigenous sediments. Two broad form categories can be distinguished: 1. reef-mounds without lagoons, and 2. atolls with central lagoons. In both, facies are usually well differentiated into reef-core and fore-reef, and in the atolls also into back-reef; fore-reef facies suggest elevation above the surrounding depositional surface. The foundation is commonly a bank or biostrome (Fig. 12A).

Reef-mounds are dome-shaped in vertical section, circular or somewhat elongated in plan, and exhibit a wide range in size from those several meters or tens of meters in basal diameter (perhaps "knoll-" or "patch-reefs" might be a more useful term here), to complexes over 1 kilometer across. In the latter range they commonly show vertical ecological zonation. Reef-mounds of large size occur in the Middle and Upper Devonian of the Rhenish Schiefergebirge (e.g., Meggen near Attendorn, Bilveringsen near Iserlohn, and in the Lahn Syncline; see Fig. 6A; also Krebs, 1974), and probably in the early Upper Devonian of the Holy Cross Mountains, southern Poland (Szulczewski, 1972). In the Ardennes there are large reef-mounds in the lower Frasnian with basal diameters of over 1 kilometer and amplitudes of 200–300 meters (Lecompte, 1954 and 1970, p. 37 and 41). Smaller reef-mounds with diameters up to several hundred meters are present in all areas (see regional descriptions).

There are a few instances in the Devonian of Europe of what some authors have considered to be "algal" reefs, but whether or not these have been correctly identified as such is debatable since for the most part the published data are not definitive. Morphologically they might all be classified as reef-mounds.

One example is represented by the Lower Devonian Koneprusy Limestone of central Bohemia, in which calcareous algae appear to make up a major proportion of the reef building organisms.

Bandel and Meyer (1975) produced a novel interpretation of a Devonian reef-mound in the Stromberg district of the Hunsrück, southern Rhenish Schiefergebirge (Stromberg Limestone, Waldalgesheim and Bingerbrück Dolomites). This appears to have formed on an uplifted block at least ten kilometers long isolated within the Rhenish trough and is up to 300 meters in thickness. It overlies and is succeeded by crinoidal-coral-stromatoporoid banks up to 60 meters thick. Bandel and Meyer (1975) regarded the Stromberg complex as an algal reef ("Algenriffkalke") which capped the block from the Givetian to the Frasnian.

A third example has been provided by Wallace (1969) who suggested that dolomite mounds (Noces Dolomite) within the mid-Frasnian Beaulieu Shales of the Ferques Inlier, northern France, represent algal mounds, up to 300 meters across and 30 meters in amplitude, which grew in a shallow marine shelf environment.

Atolls have to date been convincingly demonstrated only in the Rhenohercynian area of central Europe, where a large proportion of the reefs are of this type. Some of the reef complexes which can be assigned to this category are those of Brilon, Attendorn, Balve Wülfrath, Dornap, Langen-

aubach, and Bergisch Gladbach in the Rhenish Schiefergebirge, and Iberg in the Harz Mountains (Krebs, 1974). Certain isolated reef complexes in the Frasnian of the Ardennes (e.g., the "Lion Reef" near Frasnes les Couvin) may also have tended towards an atoll-like form (Tsien, 1975, p. 26 and 1977, fig. 3).

The most obvious distinguishing feature of these complexes is the existence of a well developed central lagoonal facies, ringed in by reef-core and fore-reef deposits (Fig. 12A). The atolls are also by far the largest of the isolated buildups. Size estimates for the Brilon and Attendorn complexes, which developed at the Rhenish shelf-margin, suggest that they had areas approaching 100 square kilometers and 70 square kilometers respectively, though others rarely achieved more than 20 square kilometers (e.g., Wülfrath with 17 square kilometers, Iberg with 4 square kilometers; data from Krebs, 1974). Atolls which developed in basinal areas are commonly smaller than those of the shelf, probably because the original bottom environments in the former favorable for the growth of potential reef building organisms were also more restricted in area. The same is true to a less marked extent for reef-mounds. The shape of most European Devonian atolls in horizontal section is particularly difficult to determine, due principally to deformation. In the case of the Attendorn complex, however, palinspastic reconstruction suggests that it possessed a highly indented outline which gave it an original "butterfly"-shaped plan (Gwodz, 1972).

*Quiet Water Carbonate Buildups ("Mud-Mounds").*—In the Ardennes the earliest "mud-mounds" form the basal portions (up to 32 meters) of some mid-Frasnian reef-mounds (e.g., the "Arche Reef" of Lecompte, 1954 and 1970), and further examples developed during the late Frasnian in the area around Phillipville and Couvin, and to the northeast of this between Rochefort and Barvaux (see Lecompte, 1970, fig. 5; Tsien, 1971). Similar mounds developed in the mid- to late Frasnian above the "Tor Bay Reef complex" in the Newton Abbot district of southwest England (Scrutton, 1977a and 1977b).

These buildups are dome-shaped (Fig. 12A), up to 200 meters across and 80 meters in amplitude, and are almost circular in basal section. Lithologically they consist of reddish-colored carbonate mudstone (locally packstones and wackestones) containing large, platey or, rarely, branched colonies of tabulate corals (*Alveolites, Phillipsastrea, Frechastrea*) lying roughly parallel with the successive surfaces of the mounds. Calcareous algae and *Problematica* (*Renalcis, Epiphyton, Rothpletzella,* and *Girvanella*) are also abundant, although stromatoporoids are rare (Lecompte,

1954). These organisms are considered to have played a sediment stabilizing and binding role rather than a frame building one (Lecompte, 1954 and 1970; Scrutton, 1977b). Locally within the mounds there are layers and pockets of brachiopod-echinoderm-bryozoan skeletal debris, and in some cases these are graded, representing in the view of Tsien (1977, p. 196) storm deposits. One of the most distinctive sedimentary structures of the "mud-mounds" in both the Ardennes and southwest England is stromatactis, which may locally make up as much as 60–70 percent of the rock volume.

For the most part the mud-mounds are enclosed in dark shales or argillaceous limestones. Some exhibit extensive lateral interfingering with enveloping sediments, and thin beds of pink micrite may extend from the mounds for several hundred meters into them (see e.g., Tsien, 1977, fig. 12). Talus slopes did not develop around "mud-mounds." It is likely that they grew in a moderately deep water environment, though in view of the abundant calcareous algae probably not below the photic zone, under conditions of low turbulence.

### CONTROLS ON REEF DEVELOPMENT

Devonian reefs in Europe evolved within a complex paleogeographical and tectonic framework, and the factors governing their situation and development were correspondingly diverse. Reefs developed wherever environmental conditions and the relationship between subsidence and sea level stand allowed. Broadly considered, these sites were on the flanks of, or on the shelves peripheral to, the "massifs" of uplifted and deformed pre-Devonian sediments and crystalline rocks, and possibly around localized crustal upwarps. On a more local scale within this framework reef growth was controlled by the distribution of favorable substrates, syndepositional faulting, and submarine volcanicity. Exceptions to the broad patterns of distribution, though not exempt from certain local controls, were the scattered reefs which demonstrably grew within the deep inter-massif troughs.

In the following section the causes of, and controls on, reef growth in the European Devonian are discussed. Most data derive from those few areas, namely central Europe and the Cantabrian Mountains, where Devonian reefs have been most closely investigated in this respect. From others, information is as yet lacking, or is only rudimentary in concept.

*Crustal Flexures and Movement on Basin-Margin Hingelines.*—Lecompte (1967, p. 22–23 and 1970, p. 31) envisaged the controls on Devonian

reef growth in the Ardennes as purely epeirogenic, in that the shelf zone fringing the Brabant Massif subsided generally but also experienced differential subsidence, at least from the Eifelian onwards, giving rise to "arches" and "troughs" with trends oblique to the shelf-margin. The presence of these he inferred from thickness variations in the Devonian succession, and from the localized development of late Middle and Upper Devonian reefs in the vicinities of the supposed rises. Some of these transverse features (e.g., the "massifs" of Givonne, Rocrois, and Stavelot) are also considered to have exerted effective control on depositional facies even during the lowest Devonian, and their influence on reef growth may have persisted into the Lower Carboniferous (Fourmarier, 1963; Lecompte, 1967 and 1970). The restriction of Upper Devonian reef-mounds to these zones, if not an effect of outcrop pattern, suggests that they offered a balance between subsidence and sea level change more favorable to the growth of reefal organisms than the intervening slightly deeper troughs.

The stepwise landward retreat of successive carbonate platforms (Eifelian, Givetian, Frasnian) in the Ardenno-Eifelian area was related by Lecompte (1970, p. 34) to the "northward" migration of a hingeline separating a slowly subsiding ("stable") proximal shelf from a more rapidly subsiding distal shelf and basin. This is illustrated by the north-south thickness variation of the Fr2h horizon (Fig. 5) which at Tailfer (back-reef facies) on the northern margin of the Dinant Syncline is about 160 meters in thickness, at Phillipville 35 kilometers away in the middle of the syncline (barrier-reef and back-reef facies) is 240 meters thick, and in the Frasnes-Couvin district just 15 kilometers further south on the southern margin of the Dinant Syncline (reef-mound) reaches 525 meters in thickness. Lecompte (1970, p. 34) considered the hingeline to have run just to the south of Neuville, the barrier-reef having developed at the edge of the stable shelf and the reef-mounds in the more rapidly subsiding zone seaward of this; during deposition of the underlying F42d reefal horizon the hingeline appears to have run somewhat further south, midway between Couvin and Phillipville.

In the Rhenish Schiefergebirge local vertical tectonic movements controlled by granitic mantle diapirism may also have been superimposed on the overall pattern of basin-margin collapse. According to the model proposed by Krebs and Wachendorf (1973 and 1974) and Krebs (1976a and b) the early stages of such movements were expressed during the Devonian in the development of incipient *rising* domes, flanked by *sub-*

*siding* basins and shelves (see also Paproth, 1976). After Krebs (1976a) examples of such rises in central Europe are represented by Lecompte's Rochefort-Grupont, Phillipville and Rocrois Massifs in the Ardennes, and the Velbert, eastern Sauerland, Ebbe, Bensberg, and Remscheid Anticlines in the Rhenish Schiefergebirge. Krebs also extended the model to include supposedly similar features in other parts of Europe, such as in the Saxothuringian zone, the Armorican Massif, southwest England, and the Iberian Peninsula. Differentiation into this tectonic pattern would have been most marked from the Middle Devonian onwards and in many cases reefs may have grown on the subsiding flanks of these positive structures (Krebs and Mountjoy, in prep.), with conditions having remained widely favorable for reefal development until the mid- to late Frasnian. Over this period several reefs in the Rhenish Schiefergebirge attained thicknesses of around 1000 meters (e.g., Brilon—Bär, 1966; Attendorn—Gwodz, 1972; Balve—Krebs, 1974), while others reached only a few hundred meters (see data in Krebs, 1974, p. 161); this can only be a direct reflection of the local tectonic environments in which the various reef complexes evolved.

In the Cantabrian Mountains synsedimentary movements along the Leon line, a Devonian hingeline bordering the southern margin of the Asturian Geanticline (see de Sitter, 1962; de Sitter and Boschma, 1966), and the associated structural lineaments represented by the present Porma Fault and Sabero-Gordon line (Fig. 11B), appear to have exercised primary control on reef growth and sedimentation in the Asturo-Leonese facies belt (van Adrichem-Boogaert, 1967; van den Bosch, 1969; van Staalduinen, 1973). The Sabero-Gordon line appears to have been active at least from the Middle Devonian until the Tertiary (Rupke, 1965; Evers, 1967, p. 102), and may have been a significant influence on sedimentation substantially earlier. Reijers (1971 and 1973a) considered that the Sabero-Gordon line acted as a facies barrier during deposition of the Portilla Formation and influenced the location and growth of the biostromes and bioherms at this level. Upper Devonian erosion in the region is also restricted to that part of the Asturian Geanticline which lay to the north of the hinge belt (Reijers, 1972, p. 204), and fossils reworked from older Devonian horizons occur in Upper Devonian deposits only to the south of this zone, suggesting uplift or tilting of the hinterland concomitant with continued subsidence in the Asturo-Leonese belt (see e.g., Reijers, 1973a).

*Synsedimentary Faulting.*—If the tectonic model of the European Hercynian Orogen fash-ioned by Krebs and Wachendorf (1973) and Krebs (1976a and 1976b) is accepted as valid, then in many cases faulting and the sort of localized tectonism discussed above may not be readily separable as controls on sedimentation and reef growth since hingelines between the uplifted zones and their subsiding flanks might well have been faulted (Krebs, 1976a, p. 132; Krebs and Mountjoy, in prep.). Such a model, however, is not acknowledged as accurate by all workers in the field. Franke *et al.*, (1978, p. 204) regarded the single most important effect on sedimentation within the Rhenohercynian zone as having been movement on hingelines trending southwest-northeast, parallel to the tectonic strike of the Variscan Hercynian Arc (and therefore to the basin-margin), and considered that the evidence supporting the conclusion by Krebs and Wachendorf that individual *anticlines* in the Rhenohercynian already acted as stable rises during the Devonian is slight (Franke *et al.*, 1978, p. 211). Dvorak (1973), in contrast, has emphasized the significance of major faults trending *normally* to the regional structural grain. It is beyond the scope of this paper to comment on the overall validity of any one of these views.

In the Rhenish Schiefergebirge local synsedimentary faulting has been implicated in the development of the Elberfeld Reef belt (Fig. 14) in the Wuppertal district (Karrenberg, 1965; Krebs, 1975), and in the cases of the atolls of Attendorn (Gwodz, 1972) and Brilon (Paeckelmann, 1936; Bür, 1966); the western margin of the last of these appears to have been governed from the early Givetian to the late Frasnian by syndepositional movement on the northwest-southeast trending (i.e., normally to the shelf-edge) Altenburen Fault line (Krebs, 1974, p. 173). Krebs (1971 and 1974) and Krebs and Mountjoy (1972, in prep.) also expressed the view that the Rhenish Shelf comprised a mosaic of fault blocks which independently tilted and subsided during the Middle and Upper Devonian, with reef complexes developing along the basinward fault scarps of the blocks where subsidence would have been least rapid; inter-reef deposits in the scarp-slope areas (the so-called "inter-reef basins" of Krebs, 1971) would be represented by dark "Flinz" shales. Reef complexes which might be included in this model are those of Brilon, Warstein, Attendorn, Meggen, Balve, and Elberfeld (Krebs, 1971 and 1974; see also Fig. 14). Presumably, the development of such faulted, subsiding blocks is not considered inconsistent with the vertical tectonics model for the central European area.

Krebs and Mountjoy (1972, in prep.) have outlined the following criteria which could indicate

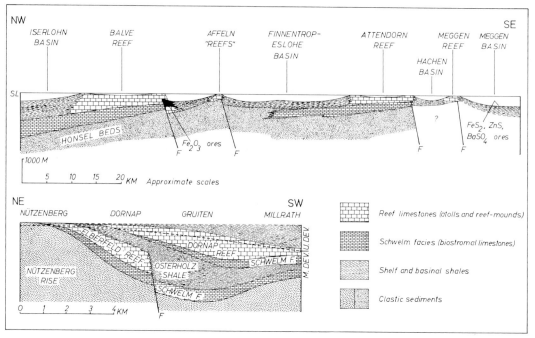

Fig. 14.—Top: Schematic non-palinspastic section across the Rhenish Schiefergebirge during the late Middle or early Upper Devonian showing the manner in which isolated reef complexes may have developed as a result of syndepositional faulting. Note how in this model the reefs are confined to the uptilted (less rapidly subsiding) basinward margins of the fault blocks, the intervening scarp slope troughs forming "Flinz" shale basins; after Krebs (1971, fig. 32), localities given in Figure 6A. Bottom: Non-palinspastic section through the Middle and early Upper Devonian of the Bergisches Land (Rhenish Schiefergebirge, see Fig. 6A) showing a model for the development of the Elberfeld and Dornap Reef complexes marginally to the (rising) Nützenberg Rise. Slightly modified after Krebs (1978, fig. 3). Note the presence of a faulted hingeline between rise and deeper shelf/basin. The reefs have been interpreted by Krebs (1974) as shelf atolls, though the Elberfeld band has the characteristics of a fringing or barrier-reef. No vertical scale intended.

the effect of synsedimentary fault movements in reefal provinces:

1. linear trends of reefs parallel to basin margins;
2. abrupt variations in thickness or facies across suspected fault lines;
3. local association of reefs with off-reef submarine volcanicity, iron oxide mineralization, or more rarely with stratiform lead-zinc mineralization in adjacent basinal shales.

The barrier-reefs of the Ardenno-Eifelian tract, and the large atolls of Brilon and Attendorn (with their small satellite complexes of Warstein and Meggen respectively) in the Rhenish Schiefergebirge, lie along a line which represents the Middle and early Upper Devonian Rhenish shelf-margin (Krebs, 1974), but there are insufficient data to show whether this was also a fault line. Similarly,

in the Asturo-Leonese facies belt of the Cantabrians, there appears to be no evidence as to whether the basin-marginal hingeline was actually faulted; in both cases it must be considered a distinct possibility. No other such trends have as yet been identified in the European Devonian.

Abrupt facies variations characterize most carbonate buildups and this is consequently a difficult criterion to utilize. In Europe, folding and post-burial faulting are commonly so marked as to make this feature of only limited application.

Syndepositional submarine volcanics (basalts and basaltic tuffs) were extruded adjacent to the Balve Atoll in the Sauerland, and may have been related to a fault at the southeastern periphery of this reef complex (Krebs, 1971, p. 77); a similar relationship may have existed during the early stages of the Brilon Reef. Stratiform lead-zinc ore bodies occur in the proximity of the small reef-mound of Meggen (Ehrenberg, Pilger and Schrö-der, 1954; Krebs, 1972), and in the area of Sötern on the souheastern margin (Hunsrück) of the Rhenish Schiefergebirge (Hoffmann, 1966). The latter lay marginally to the (?) barrier-reef com-plex which fringed the mid-German crystalline rise in this area, and may have been due to sub-marine hydrothermal emanations from a fault zone between the basin and the rise (Krebs, 1970).

One further example may have been provided by Szulczewski (1971). On the basis of facies dis-tribution, this author interpreted the Famennian evolution of the Kielce reef-mound in the Holy Cross Mountains of Poland as representing the dissection of the complex into blocks by synsedi-mentary faulting. In other areas of Europe the possibility of a direct relationship between De-vonian reef growth and faulting has not been sat-isfactorily demonstrated.

*Volcanicity.*—The development of Devonian reef complexes in areas of extensive syndeposi-tional volcanicity has been demonstrated only in central Europe and southwest England. Most sig-nificant among the former are those within the area of the Rhenish Trough, namely in the Lahn and Dill Synclines of the southeastern Rhenish Schiefergebirge (see Ahlberg, 1919; Krebs, 1966; Rietschel, 1966), and the Iberg and Elbingerode Reef complexes in the Harz Mountains (Reich-stein, 1959; Franke, 1973; Krebs, 1974). Here, spilitic lavas and tuffs built up submarine rises in the trough, above which water remained consis-tently shallow enough to allow the growth of ree-fal organisms. The depth of the basin may not have been very great in these areas, since volca-nic piles of only a few hundred meters appear to have been adequate to provide such conditions (see Franke *et al.,* 1978, p. 206). In the Lahn and Dill Synclines reef growth reached an acme in the late Givetian and early Frasnian after extensive volcanicity in the area (Krebs, 1974, p. 167). Ex-amples include the Langenaubach Atoll in the Dill Syncline (Krebs, 1966), and the Gaudernbach and Wirbelau Reef complexes in the Lahn Syncline (Krebs, 1960 and 1972; Stibane, 1963). Another characteristic of such reefs is the presence of tuf-fitic layers and other transported volcanic debris within reefal facies, particularly in the back-reef areas (Krebs, 1974).

In southwest England, the Torquay and Plym-outh Reefs developed adjacent to, perhaps fring-ing, a large basic volcanic complex, although the exact lateral facies relationships are still some-what uncertain. Certainly, the Ashprington vol-canic center (see southwest England section) was active periodically from the Eifelian to the Fras-nian (Middleton, 1960, p. 196), and lateral and vertical transitions between detrital and reefal

carbonates and tuffs have been briefly document-ed by Richter (1965) and Scrutton (1977a and b).

*Minor Sea Floor Elevations.*—Included here are cases where relatively small scale positive fea-tures on the sea floor appear by virtue of their elevation, or (?) induration, to have been prefer-entially colonized by reef building organisms. Such features, however, can only realistically be considered valid within the framework set by the broader scale influences on reef growth discussed in this section.

One good example exists in the Frasnian of the Ardennes, where reef-mounds of the "Fr2d" and "Fr2h" horizons (Fig. 5) continued to influence sea floor relief even after burial by basinal shales (Fr2e and Fr2i); this is strongly implied by the manner in which the reef-mounds are commonly stacked one above the other (see Lecompte, 1954; Tsien, 1971). Residual relief over the crests of the older mounds provided shallow water sites which were marginally more favorable for colonization by potential reef building organisms than the sur-rounding areas. Nevertheless, few of the reef-mounds appear to have remained *exposed* on the sea floor for long after their extinction, and so the development of younger complexes directly upon the older ones is rare. It seems likely that differ-ential compaction between the reefs and the sur-rounding shales was responsible for the effect (see Tsien, 1977, p. 199). The "mud-mound" facies in southwest England and the reefal capping lime-stones of the Rhenish Schiefergebirge ("Iberg type"), do however rest directly upon older reefs.

In a number of cases the growth of reef com-plexes appears to have started on an initial bank or mound of skeletal calcarenite. Lower Devonian reef-mounds in both Bohemia and the Armorican Massif, for example, had as foundations localized banks of substantially *in situ* blastoid and crinoid debris and other shelly material (see Chlupac, 1954; Poncet, 1972).

*Reef Growth and Apparent Sea Level Changes.*—Many authors have stressed tectonic mobility as a control on reef growth and sedimen-tation in Europe during the Devonian. Few, how-ever, have considered the possibility that eustatic fluctuations in sea level during the Devonian might also have played a role. It is, of course, notoriously difficult to distinguish on a local scale stratigraphical features which might be attribut-able to eustacy, especially where these could have been overprinted by the effects of local epeiro-genesis. But Devonian marine stratigraphies in most parts of the world are overall transgressive ones, even in areas of apparently relative tectonic stability (clearly the places to look), and it is pos-sible that these include some evidence of eustatic trends. Several black shale horizons in the Upper

Devonian and Lower Carboniferous of central Europe appear to be more or less isochronous and may represent the depositional expression of eustatic sea level rises (Walliser, 1977; Franke *et al.*, 1978). Examples of these are the thin Upper Frasnian "Kellwasserkalk" (Buggisch, 1972; see also Fig. 6B), the Famennian "*Platyclimenia* shale," and the mid-Tournaisian "Liegende Alaunschiefer" (Lower Alum Shale). All have, with their equivalents, a very wide distribution in Europe, and possibly on other continents too (see brief discussion by Franke *et al.*, 1978). The widespread occurrence of late Devonian pelagic limestones in southern and western Europe ("griotte"-type limestones), or the wholesale extinction of reefs in the Frasnian, could conceivably also have been connected with a eustatic sea level rise, but probably reflects as well certain paleomorphological features in these areas. Many regressive-transgressive sequences in Lower Carboniferous successions are also considered to have had a eustatic origin (Ramsbottom, 1973). It is not unreasonable, therefore, to suppose that similar fluctuations may have occurred at an earlier stage of the Devonian, with some black shale incursions onto the shelves or transgressive pulses which resulted in reef initiation or extinction, recording eustatic events. But this is conjecture, and recognition of the direct effects of specifically eustatic sea level changes on reef development remains uncertain for all parts of the European Devonian.

The dynamic relationship between changes in, and changing, sea level in *relative* terms and reefal growth is, in contrast, very clearly discernable. Older reefs were commonly drowned and new ones established during transgressions, decreasing or stable rates of sea level change, or standstills, allowed reefal growth to proceed to a climax, and falls in relative sea level caused exposure of reefs in marginal areas (or those shallowly submerged) and their subsequent erosion and karstification. Three broad conditions in this inter-relationship are illustrated in the European Devonian and can be summarized as follows:

1.  The rate of reef growth exceeded that of sea level rise, causing progradational (advancing) growth of reef complexes, possibly with exposure in marginal areas, or resulting in their extinction.
2.  Equilibrium existed between the rates of reef growth and sea level rise, allowing continuous vertical accretion of reef complexes in (turbulent) shallow water facies, but also possibly with slight progradational growth (e.g., spreading over fore-reef facies).
3.  The rate of sea level rise outstripped that of reefal growth, giving rise to "abortive" reefs (e.g., "mud-mounds"), "cap" facies and retreating biostromes on top of drowned reef complexes, landward retreat of basin-marginal reef complexes, or contributing directly to the extinction of reef complexes by drowning.

These were not static relationships, and reef complexes commonly experienced several phases of interplay of this nature during their growth histories, these being expressed in marginal reef complexes as sequences of transgression and regression and in reef-mounds as vertical ecological and facial zonation. Cases where sea level actually "fell" to expose reefs subaerially will not be considered here. In most examples this can be assumed to have had a tectonic origin, and no connection with reefal growth.

Illustrations of the first and last of the points listed above are provided by the St. Lucia and Portilla Formations in the Cantabrian Mountains. Here, biostromal complexes underwent both "building-out" (advancing) and "building-in" (retreating) phases of growth in response to changes in the relationship between subsidence along the Sabero-Gordon hingeline and the rate of organic carbonate accumulation (T. J. A. Reijers, written comm., 1979). Both formations show an early phase of building-out followed by one of building-in (the latter phase in both also with local bioherm development; see Reijers, 1972; Mendez-Bedia, 1976), and in each case the two phases are more or less symmetrical about a median plane (Fig. 15). In the first of these stages the rate of reefal growth gained on the rate of rise in relative sea level; in the second the reverse was true and a transgressive situation ensued.

The relationships between reef growth and sea level rises in the Ardenno-Eifelian tract were

$\rightarrow$

FIG. 15.—A. Simplified non-palinspastic section through the Emsian-Eifelian St. Lucia Formation, southern Cantabrian Mountains, showing regressive (advancing) and transgressive (retreating) reefal growth phases. Compare also Mendez-Bedia (1976, fig. 18). B. Non-palinspastic section through the Givetian-Frasnian Portilla Formation, Cantabrian Mountains, showing facies distribution and regressive and transgressive episodes of reefal growth. Both sections reproduced with permission of T. J. A. Reijers.

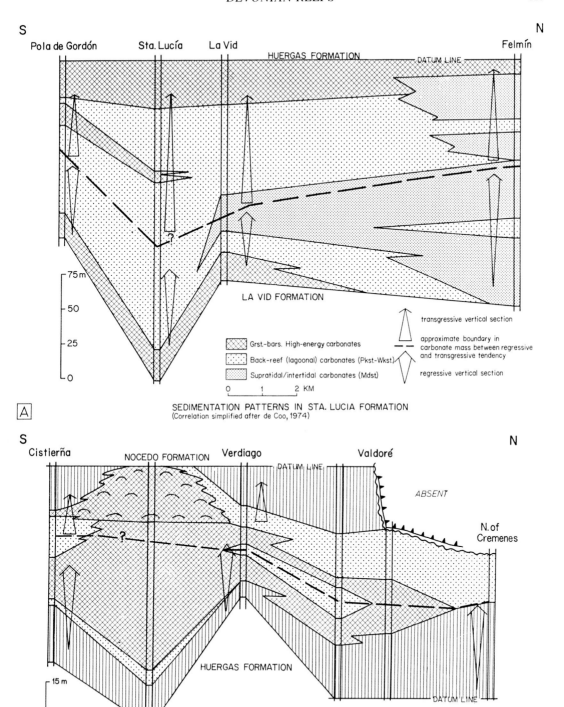

**A**

S

Pola de Gordón    Sta. Lucía    La Vid    Felmín    N

HUERGAS FORMATION    DATUM LINE

75 m
50
25
0

LA VID FORMATION

Grst.-bars. High-energy carbonates

Back-reef (lagoonal) carbonates (Pkst-Wkst.)

Supratidal/intertidal carbonates (Mdst)

0    1    2 KM

transgressive vertical section

approximate boundary in carbonate mass between regressive and transgressive tendency

regressive vertical section

SEDIMENTATION PATTERNS IN STA. LUCIA FORMATION
(Correlation simplified after de Coo, 1974)

**B**

S

Cistierña    NOCEDO FORMATION    Verdiago    Valdoré    N

DATUM LINE

ABSENT

N. of Cremenes

HUERGAS FORMATION

15 m
0

1 KM    6 KM    2 KM    1.3 KM    8 KM

SABERO – GORDON FAULT ZONE

Open marine carbonates

Biohermal    } reef-track carbonates
Biostromal

Back-reef (lagoonal) carbonates

transgressive vertical section

approximate boundary in carbonate mass between regressive and transgressive tendency

regressive vertical section

SEDIMENTATION PATTERNS IN PORTILLA FORMATION
(Correlation simplified after Reijers, 1972)

complex and varied. Devonian successions of the area have a large scale cyclic disposition, in which reefal carbonates alternate with argillaceous limestones and shales of quasi-basinal aspect (see Mailleux and Demanet, 1928; Lecompte, 1970, p. 34). This represents the effects of successive transgressions onto the Ardenno-Eifelian Shelf *within* the broader, first order framework set by the landward retreat of the Middle and Upper Devonian carbonate platforms (see e.g., Fig. 5). In relative terms, periods of equilibrium between the rates of reefal growth and sea level increase alternated with sharp transgressive pulses which drowned reef complexes and gave rise to this well defined megacyclicity. The phenomenon is pervasive, and a subordinate (third order) cyclicity on a scale of just a few meters is expressed in Frasnian barrier-reef complexes of the area as back-reef cyclothems (see later section), and in the "stacking" or cyclic development of biostromal complexes and their associated facies in the Eifelian and Givetian; symmetry among the second and third order cycles is rare.

In certain cases in central Europe, the rate of reef growth also appears to have exceeded, at least temporarily, that of sea level rise, and the complexes affected developed in an advancing fashion. Possible illustrations of this are some of the shelf-atolls of the Rhenish Schiefergebirge, biostromal complexes in the Middle Devonian of the Ardennes and Eifel, and the consecutive Frasnian barrier-reef complexes in the same areas. The megacyclic development of the last of these is expressed in their back-reef successions as overall upward increases in environmental restriction, in which the small-scale depositional cycles become thinner and less well defined (biostromes of stenohaline organisms also decrease in importance) and grade upwards into evaporitic laminites. This suggests seaward advance, with cumulative infilling (the small scale cyclothems being themselves progradational), of the shelf-lagoons in response to vertical development and advancing growth of the restricting organic barrier between the drowning transgressive pulses (Burchette, in press; cf. also Link, 1950; Playford, 1969). A similar facies sequence can be demonstrated in back-reef deposits of the same age in the northern Bergisches Land. It seems likely, therefore, that both the mega and minor cyclicity

in these contexts reflect complex interplay between the rates of sea level changes and sedimentation and organic carbonate accumulation in these broad shelves (see e.g., Pitman, 1978 on the relationship between eustacy, subsidence and sedimentation on passive continental margins), since it is in just such shallow marginal areas that the effects of such processes would be most markedly expressed.

In contrast, reef-mounds which developed basinward of the Frasnian barrier-reef complexes in the Ardennes grew dominantly vertically, with little tendency towards lateral spreading, hence the absence of well developed atolls. Commonly these mounds exhibit vertical ecological and facial zonations (next section) which imply that over their active lives conditions of equilibrium existed between rises in relative sea level and reef growth, but that at times the relationship fluctuated. Considered in combination with thickness variations, this pattern of reef growth does suggest the correctness of Lecompte's (1958 and 1970) view that subsidence was an important control on Devonian reef growth in the Ardennes and Eifel, with differential subsidence between more and less rapidly subsiding parts of the shelf.

These patterns offer no information, however, as to whether the transgressive pulses which apparently repeatedly terminated reefal growth in this area during the Devonian were also due to subsidence. These were undoubtedly rapid in comparison with the gradual rate of rise in relative sea level during the preceding periods of reefal growth, and consequently exceeded the rate at which the reefs could accrete. Isolated reef complexes in the Ardennes appear to have ceased growth at the onset of such transgressions. Deposition in the biostromal and barrier-reef complexes commonly closed with bank or biostromal phases (most distinctly seen in the back-reef zones where they succeed restricted facies) which suggests that in the early stages of transgression the center of reef growth retreated onto the reefal platform. Such features can be demonstrated in the Frasnian reef complexes of the northern Bergisches Land (e.g., Hofermühle, Wulfrath; Fig. 16A), the Ardennes, and the Eifel-Aachen tract (Fig. 16B), and in the Givetian biostromal complexes of the Eifel (e.g., Sötenich; Fig. 17) and the Ardennes (Burchette, in press).

---

→

FIG. 16.—A. Quarry section at the railway station Walheim, near Aachen, looking east. The highest laminated back-reef limestones (A) of the earliest barrier-reef megacycle in the Aachen area are succeeded abruptly by a transgressive stromatoporoid biostromal phase (B), here 4 meters thick though locally thicker, and cross-stratified brachiopod-crinoidal calcarenite (C). The first barrier-reef complex is overlain by a second, and between them is

a 4–5 meter packet of quasi-basinal shale (D), the ''Grenzschiefer,'' with thin argillaceous limestone beds at the base, A–D comprises a full transgressive sequence. Succeeding this is a much thicker (14 meters) advancing reefal phase (E), followed by its own laminated and biostromal back-reef deposits (F). The whole reefal succession in the Aachen area is about 250 meters in thickness. B. Quarry section at Schlupkothen through the Wülfrath Reef showing well-bedded biostromal and *Amphipora* back-reef limestones (Bck) overlain by unbedded transgressive reef-core (Rk). Some tectonic movement at contact. Photograph courtesy of K. Hoenen.

Fig. 17.—Transgressive coral-stromatoporoid bio-strome on top of a lower Givetian biostromal complex at Sötenich, Sötenich Syncline, Eifel. Overlying are the Rodert Beds (pelagic shale and limestones); the biostrome here overlies the back-reef facies of the complex. Note the bulbous morphology of many of the organisms, typical of those in biostromes in the European Devonian. The coral head in the center of the photograph is 22 centimeters across.

**PALEOECOLOGY AND SEDIMENTARY FACIES**

*Reefal Depositional Environments*

Sedimentary facies in Devonian reefs of central Europe have been discussed in detail by Krebs (1974), and these descriptions are valid in principle for all others in Europe; further specific information can be obtained from references given under the regional stratigraphical accounts. In the following section the chief reefal paleoenvironments and sedimentary facies, so far undefined, are only summarized.

*Back-Reef Environment.*—In barrier-reef complexes, atolls and biostromal complexes the organic barrier deposits were much less important, in a volumetric and spatial sense, than those of the back-reef zone (see e.g., Krebs, 1968). Moreover, back-reef sediments offer a wider range of lithofacies types than do the other reefal depositional zones, and these are eminently suitable for study, in many cases allowing unambiguous interpretation of their depositional environments. These sediments have nevertheless been consistently neglected in the course of research into Devonian reefs in Europe. Further aspects of back-reef deposition are considered in a later section, and the sedimentology and paleoecology of Middle and Upper Devonian back-reef deposits in central Europe have also been treated in detail elsewhere (Krebs, 1969a; Burchette, in press).

In general, sediments which accumulated in the back-reef zones of European Devonian reef complexes present a restricted aspect and contain a rather monotonous, low diversity, but commonly high abundance, fauna and flora which reflects the shallowness, predominantly low turbulence, and local high salinity of these depositional environments. The degree of restriction apparent from lithofacies increases internally away from the reefal margin, as does the proportion of fine-grained carbonate to bioclastic debris; so too does the evidence for periodic emergence of the sediments (e.g., fenestrae, desiccation cracks, early vadose cements and cementation-expansion structures, and solution phenomena). Marginal sediments are commonly cryptalgally and inorganically laminated pelsparites and pelmicrites, or *Amphipora* packstones and wackestones, which are locally evaporitic and usually thin-bedded. In near-reef sites the deposits comprise biostromes of branched tabulate corals and branched or bulbous stromatoporoids (see Krebs, 1968 and 1974; Lecompte, 1970). Depositional cyclicity in barrier-reef complexes may modify this pattern on a small scale (see later section), but does not significantly alter the overall facies distribution.

*Reef-Core Environment.*—This represents the highest energy zone in all differentiated reef complexes. It was the tract of maximum accumulation of *in situ* organic skeletons, and probably of greatest faunal diversity. The reef-cores of barrier-reefs and atolls were relatively narrow, only a few hundred meters wide (Krebs, 1974, p. 191), though in biostrome complexes they were substantially wider, being kilometers or tens of kilometers across. The bulks of most reef-mounds appear to consist of reef-core facies.

The growth forms of organisms in the reef-core zones of barrier and isolated reef complexes are usually compact, encrusting ones which were capable of withstanding substantial wave turbulence without being dislodged from the substrate, and in some sites they formed a true framework. Reef-core limestones in these cases are massive and unbedded (Krebs, 1974), and commonly they exhibit evidence of syndepositional cementation, which must also have contributed greatly to their early rigidity. Inter-skeletal cavities are usually filled with coarse, unsorted bioclastic debris. In biostromal complexes the reef-core zone (the term reef-core is not entirely satisfactory in this context, and *reef-tract* or *reef-margin* might be more correctly applied) comprises predominantly bedded (e.g., biostromal) sediments in which the organic growth forms show little adaptation to high turbulence, nor do they obviously form frameworks.

The dominant constructing organisms in the reef-cores of most reef complexes in the European Devonian were stromatoporoids, but locally corals or even algae achieved this status. Pelagic organisms such as conodonts, cephalopods, and tentaculites are rare in the reef-core and back-reef environments (see Lecompte, 1970; Krebs, 1974).

*Fore-Reef Environment.*—Reefs which build up into a shallow, turbulent zone, and maintain moderate relief above the surrounding depositional environment, may develop accumulations of reef derived talus around their flanks (cf. Heckel, 1974, p. 97). Such talus, commonly with primary depositional dip (Fig. 18), is seen in Devonian buildups in many areas of Europe, even in some of the smaller reef-mounds (see e.g., Poncet, 1972). Krebs (1971) recognized two fore-reef subfacies in buildups of the Rhenish Schiefergebirge, corresponding to "upper" and "lower fore-reef flank" zones. The former is characterized by thick-bedded, light-colored, early cemented limestones which largely comprise skeletal fragments and blocks of cemented debris dislodged from the reef-core zone, and which contain a wealth of sedimentary structures (e.g., channels) indicating deposition in a high energy environment. The lower fore-reef deposits consist of dark, thin-bedded limestones containing syndepositional slumps, graded layers and intercalations of basinal shale. This is a pattern applicable in principle to most buildups where fore-reef deposits are developed, and corresponds grossly with Lecompte's (1970) bathymetric zonation (see later section).

Fore-reef deposits carry a diverse authochtonous benthonic fauna of stromatoporoids, corals and echinoderms, with accessory organisms such as bryozoa, brachiopods and trilobites commonly concentrated in nests and pockets (see Chlupac, 1954; Bandel, 1972; Krebs, 1974). The growth forms of *in situ* organisms such as stromatoporoids and corals are predominantly bulbous or laminar and encrusting, the largest examples occurring in the shallow water, high energy, fore-reef deposits.

*Bank Environment.*—Bank deposits are invariably well-bedded and laterally poorly differentiated. The sediments are packstones and wackestones in texture, with fine-grained (peloidal and micritic) matrices, and are somewhat bituminous; they are interbedded variably with argillaceous limestones and shales. The faunas which constructed biostromal banks on Middle and Upper Devonian shelves over wide areas of Europe, especially in the external Hercynian zones, were diverse, and the organisms were specialized for a regime of low or intermittent turbulence. Highly characteristic of such deposits are biostromes dominated either individually or jointly by stromatoporoids and corals with bulbous ("Blockriffe" of Wedekind, 1924; Jux, 1960) or branched ("Rasenriffe") growth forms. Locally *Thamnopora* beds alternate with thin tabular stromatoporoids, suggesting fluctuation in the degree of tur-

FIG. 18.—Quarry section at the flank of the Lion Reef-mound, Frasnes-les-Couvin, Ardennes, showing inclined bedding, regarded by Lecompte (1954) and Tsien (1971) as representing intercalated fore-reef and reef-core deposits.

bulence or turbidity. Layers of crinoidal debris and brachiopods are also common, and in the Givetian there are dense beds of *in situ Stringocephalus*.

Krebs (1974) recognized eight subfacies within the "Schwelm" bank lithofacies association and considered that they represent deposition in a calm, muddy environment. Jux (1960) and Lecompte (1970), in contrast, regarded the biostromes of bulbous stromatoporoids and corals as the products of a high energy intertidal or shallow subtidal regime. It seems likely, in fact, that such deposits formed at or just below fair weather wave base which, within the broad epeiric shelves of the European Devonian, may have been at only a few meters or tens of meters depth. The abundance in Schwelm-type sediments of *Amphipora* sp. (Krebs, 1974), calcareous algae, and calcispheres (see Flügel and Hötzl, 1971 and 1976) also suggests that the depositional environment may have been quite shallow. The fact that essentially similar biostromal deposits also existed intermittently, in a more readily discernable pattern, in the shelf-lagoons of barrier-reef complexes where water depths can seldom have reached more than 10 meters (Burchette, in press) also lends support to this interpretation.

### Principal Reef Building Organisms

Faunal provinciality was marked throughout the world in the Lower Devonian, but progressively lost definition during the course of the period as transgression eliminated many of the "faunal barriers" which had hindered the expansion and diversification of neritic benthonic organisms (see Boucot, Johnson and Talent, 1969; Johnson,

| ORGANISM | | BACK-REEF | REEF | FORE-REEF | OFF-REEF | BIOSTROMES | "MUD-MOUNDS" |
|---|---|---|---|---|---|---|---|
| STROMATOPOROIDS | DOMAL | | | | | | |
| | BULBOUS | | | | | | |
| | LAMINAR | | | | | | |
| | *STACHYODES* | | | | | | |
| | *AMPHIPORA* | | | | | | |
| TABULATE CORALS | DOMAL | | | | | | |
| | BULBOUS | | | | | | |
| | LAMINAR | | | | | | |
| | RAMOSE | | | | | | |
| CALC. ALGAE | *GIRVANELLA* | | | | | | |
| | *RENALCIS* | | | | | | |
| | OTHERS | | | | | | |
| RUGOSE CORALS | | | | | | | |
| ECHINODERMS | | | | | | | |
| BRACHIOPODS | | | | | | | |
| BRYOZOA | | | | | | | |
| MOLLUSCS | | | | | | | ? |
| OSTRACODS | | | | | | | ? |
| CONODONTS | | | | | | | |
| GONIATITES | | | | | | | |
| TENTACULITES | | | | | | | |

ABUNDANT          COMMON          LOCALLY PRESENT          —— RARE

FIG. 19.—Relative abundance of the main groups of reef building and associated organisms in European Devonian reef complexes; data from many sources.

1971; Boucot, 1974). During the Middle and Upper Devonian faunas became more cosmopolitan and reefs of this age are characterized by the global occurrence of a relatively few, but remarkably consistent, associations of dominant genera and sedimentary facies (see Krebs and Mountjoy, 1972, p. 300; discussion by Wilson, 1975).

As well as the more obvious differences in specialization between the facies determined benthonic faunas of the "Rhenish" and "Hercynian-Bohemian" types (one non-reefal, the other reefal) during the Lower Devonian in Europe, there were those derived through relative paleogeographical position which persisted in reefal facies throughout the period. Faunas of the Alpine and Bohemian Devonian, for example, contain a large number of forms characteristic of the Uralian and Tien-shan faunal provinces of Asia (Chlupac, 1954; Flügel, 1964; Vai, 1967), and in Poland too there is some faunal evidence for a connection to the east with the Uralian Basin (Chlupac, 1971, p. 136). Such paleogeographical significance can be attached to these faunal differences in the cases of certain genera and a large number of species, but many genera, including important reef builders, were universal in distribution. On a more local scale (e.g., within northern Europe, or southern Europe) reefal faunas were essentially uniform, the greatest differences being attributable to sedimentary facies variations. In the pelagic regimes faunas were relatively uniform over wide areas (see Erben, 1962 and 1964; Goldring, 1962). The principal reef building groups and their relative roles are considered below.

*Stromatoporoids.*—Variations in the morphology and size of stromatoporoid colonies in Devonian reef complexes reflect to a large extent the turbulence, and probably less directly the turbidity and sedimentation rate, in their environments of growth (see e.g., Lecompte, 1970; Wilson, 1975; Kershaw and Riding, 1978; Kobluk, 1978). Thick laminar and large, low, domed encrusting forms dominated in the reef-core and shallower fore-reef environments where turbulence would have been greatest, and bulbous and delicate dendroid types were prominent in the back-reef and bank environments where turbulence was intermittent and the sediments fine-grained. Deeper, quiet water environments with a low sedimentation rate were characterized by thin laminar growth forms (Fig. 19).

Of the branching stromatoporoids, the robust form *Stachyodes* is abundant in the reef-core, shallow fore-reef and immediate back-reef deposits, but gives way further into the back-reef zone to the delicate dendroid genus *Amphipora* which, as in most Devonian barrier-reef complexes and atolls throughout the world, is an index fossil for low energy, shallow water, possibly restricted, sedimentary facies in the European Devonian. In central Europe *Amphipora* first appeared in the Eifelian (Lecompte, 1970, p. 48). Quite remarkably, neither *Amphipora* nor *Stachyodes* have yet been recorded from the Cantabrian Region (Sleumer, 1969).

Most authors have tended not to make identifications of the other, less-easily distinguished stromatoporoid genera, and have relied principally on the growth forms of these organisms in their interpretations. Notable exceptions to this have been Lecompte (1951 and 1970), Sleumer (1969), Flügel and Hötzl (1971 and 1976), Kazmierczak (1971), and Flügel (1974), though none of these

really represents a detailed study of stromatoporoid genera in a comparative paleoecological sense. Dominant genera in most areas include *Actinostroma, Atelodyction, Clathrodyction, Parallelopora, Stromatopora, Stromatoporella,* and *Syringostroma*.

*Corals.*—Although the role played by corals in European Devonian reefs was secondary to that of the stromatoporoids, in some biostromes in the Portilla Formation, and in the Frasnian "mud-mounds" of the Ardennes and southwest England, they are the dominant constructing organisms and sediment stabilizers. In that part of the Portilla Formation studied by Reijers (1972, fig. 19) corals make up about 70 percent of the fossils in the organic barrier zone, and stromatoporoids account for no more than 5 percent. The predominance of corals rather than stromatoporoids in this formation may be attributable to high sedimentation rates or a greater proportion of fine-grained terrigenous material in the depositional environment than was optimum for stromatoporoids (Brouwer, 1964 and 1968). In the "mud-mounds" the cause may have been water depth, lack of turbulence, or the turbidity of the depositional environment (see e.g., Lecompte, 1970; Tsien, 1971 and 1977; Scrutton, 1977b).

The growth forms of the corals appear to have closely paralleled those of the stromatoporoids within the same depositional environments. The rugose corals (including *Acanthophyllum, Cystiphylloides, Disphyllum, Hexagonaria, Phillipsatrea,* and *Thamnophyllum*) as a group exhibit marked plasticity in form with regard to their environments of growth (Tsien, 1967 and 1971). Simple, small rugose corals characterize argillaceous or fine-grained carbonate sediments deposited in predominantly low energy environments (off-reef, lower fore-reef, bank), but are larger where turbulence was moderate. Compound forms are absent from the lowest energy deposits, but are small or platey where conditions were of low or moderate energy, and are large and compact in the reef-core facies. Rugose corals also exhibit environmentally dictated variations in coenosteal microstructure, and these have been documented for the Middle and Upper Devonian of the Ardennes by Tsien (1967 and 1971).

The tabulates were the most important group of reef building corals during the Devonian (Lecompte, 1970). They exhibit a pattern of morphotypic variation even more similar to that of the stromatoporoids than the rugose corals. Large domal, bulbous and thick laminar forms are most prominent in high energy facies, and small rounded, irregularly branched or platey encrusting types characterize lower-energy and argillaceous facies (Fig. 19).

Particularly common genera among the tabulate

corals are *Alveolites, Favosites, Heliolites,* and *Thamnopora.* The branched tabulate coral genus *Thamnopora* was abundant in all areas of Europe throughout the Devonian. Its importance appears to have been as a sediment producer and baffle rather than a framework constructing organism. In central Europe *Thamnopora* occurs most characteristically in biostromes consisting almost entirely of its finger-like skeletons set in a fine-grained carbonate matrix (though these seldom seem to be in growth position) in banks, biostromal complexes and back-reef deposits. *Thamnopora* is also very common in the St. Lucia Formation in the Cantabrian Mountains (Mendez-Bedia, 1976), and in the Portilla Formation it is one of the most characteristic fossils, being particularly abundant in the biostromal barrier facies (Sleumer, 1969; Mohanti, 1972; Reijers, 1972).

*Calcareous Algae.*—The role and distribution of algae in European Devonian reefs has been little studied. Where present, they have probably not always been recognized as such, and locally where their presence has been suspected the containing rocks have been so affected by diagenesis and subsequent neomorphism or dolomitization that identification has become extremely difficult (see e.g., Wallace, 1969; Bandel and Meyer, 1975; Scruton, 1977a).

Calcareous algae belonging to several groups occur in most shallow water facies, but probably acted most prominently in the reef-core, fore-reef and immediate back-reef environments as encrusters and stabilizers of organic debris (Fig. 19). Only in the core and talus deposits of the Lower Devonian Koneprusy Bioherms of Czechoslovakia do certain calcareous algae (*Solenopora, Sphaerocodium,* and ?*Renalcis*) appear to have achieved the sort of dominance maintained elsewhere by the stromatoporoids and corals (see Chlupac, 1954; Bandel, 1969 and 1972). In central Europe the commonest genera were *Girvanella, Renalcis, Rothpletzella, Solenopora,* and *Sphaerocodium* (see Lecompte, 1970; Tsien, 1971, p. 141–142, and 1977). Tsien and Dricot (1977) recognized two calcareous algal "assemblages" in the Frasnian reef complexes of the Ardennes which may be of environmental significance: back-reef and shelf-lagoonal successions are characterized by a *Girvanella-Ortonella-Rothpletzella* assemblage, and reef-core and fore-reef deposits by an *Epiphyton-Girvanella-Renalcis-Rothpletzella* assemblage.

Although the calcareous (?algal) microfossil *Renalcis* is probably much more widely distributed in European Devonian reefs than most workers have hitherto realized (see e.g., the occurrences documented by Kazmierczak, 1971; Tsien, 1971; Poncet, 1972; Franke, 1973; Krebs, 1974; Tsien and Dricot, 1977; Kazmierczak and Gold-

ring, 1978), it nowhere realized the reef building potential that it did in Devonian reefs of the Canning Basin, Western Australia or of Canada (Playford and Lowry, 1966; Playford, 1969). In Frasnian reef-mounds (particularly the "Lion Reef") and mud-mounds of the Ardennes, however, *Renalcis* is abundant and probably played an important encrusting and stabilizing role on soft or friable substrates (cf. Mountjoy and Jull, 1978).

In bank and back-reef facies, calcispheres (?volvocacean algae) are also a common component of fine-grained carbonate sediments (Flügel and Hötzl, 1971; Krebs, 1974).

*Echinoderms.*—In most areas, the highest concentrations of echinoderm remains occur in bank-type deposits and in the fore-reef facies of buildups, where they are commonly present in significant rock forming quantities. In the southeastern Alps and Bohemia, banks of echinoderm debris formed the foundations for reef growth. Except locally in atolls such as Brilon, Attendorn and Langenaubach (Krebs, 1974, p. 184), echinoderm remains are absent from back-reef facies in central Europe and the Alps, along with many other obviously stenohaline organisms. In the back-reef zones of biostromal complexes in the Ardenno-Eifelian tract and the Cantabrian Mountains, however, where conditions did not become highly restricted from the open marine environment, echinoderm remains make up a significant proportion of the sediments (up to 20 percent), and in the shallow epireefal platform deposits "seaward" of the biostromal complexes are even more abundant (Reijers, 1972).

Crinoids and blastoids were particularly prolific sediment producers in the reef-core and fore-reef facies of the Lower Devonian reef-mounds in the "Barrandian" of Bohemia, in the Armorican Massif, and in the larger reef complexes of the Carnic Alps and Karawanken. According to Chlupac (1968), one of the major components in the biohermal core facies of the Bohemian Koneprusy Limestone is a specialized crinoid fauna characterized by short stems and highly branched holdfasts, while in the peripheral zones of the reef, where turbulence was presumably lower, less specialized crinoid forms predominate.

Echinoderms appear to have made a relatively greater contribution to flank deposits of the European Lower Devonian buildups than to those of the Middle and Upper Devonian, which suggests perhaps that they resembled reefs of the Lower Paleozoic where crinoids were similarly important (cf. Ruhrmann, 1971; Heckel, 1974, p. 106–108).

*Vertical Zonation in Sedimentary Facies and Morphology of Reefal Organisms*

Under normal circumstances turbulence in the marine environment decreases as water depth increases, from a zone of wave breaking at sea level, through one affected by constant wave agitation, to a zone below wave base which is only intermittently or seldom agitated. On epeiric shelves, or in the example of a reef complex where a shelf-lagoon or platform is developed, turbulence decreases laterally, with the highest energy zone to seaward where waves ground on the sea floor/reef, and a decrease in turbulence landward as remaining wave energy is absorbed through bottom friction (cf., models by Shaw, 1964; Irwin, 1965). In the latter case hypersalinity may result in marginal areas, and the greatest effect on water movement in the isolated shelf interior may be the wind. Since these gradients are reflected in Devonian reef complexes by consistent variations in the morphologies of stromatoporoid and coral colonies, where these are *in situ* in the fossil record they can provide accurate independent indications of the water depth or degree of restriction represented by the particular lithofacies in which they occur. This also permits the relationship between reefal growth and changes in relative sea level during the growth history of individual complexes, as discussed in a previous section, to be reconstructed. According to these principles a number of authors have utilized colony shapes and sizes to interpret depositional environments in and around European Devonian reefs. Below, some examples of vertical zonation of this kind are examined, and in the following section vertical and lateral relationships in shelf-lagoons of Frasnian barrier-reef complexes are considered.

Upper Devonian reef mounds in the Ardennes provide a case history which illustrates this well. These complexes exhibit vertical zonation in the types and abundance of reef builders which reflect the turbulence and thus water depth at different stages of their growth. Lecompte (1954 and 1970) recognized five "bathymetric zones" in these reefs and associated sediments, and used them to reconstruct their depositional histories:

5. "turbulent zone" (=breaker zone and immediate subsurface)—large domed and tabular ("massive") encrusting stromatoporoids with subordinate massive tabulate and rugose corals, and common robust branched tabulates and stromatoporoids;
4. "subturbulent zone" (=zone influenced by subsurface wave agitation)—bulbous and laminar stromatoporoids and tabulate corals, accompanied by branched stromatoporoids (*Stachyodes*) and tabulates (*Thamnopora*), and abundant rugose corals, crinoids, and brachiopods;
3. "calm water zone" (=around wave base)—predominantly thin laminar tabulate and ru-

gose corals, with local branched tabulate corals, crinoids, bryozoa, and brachiopods;
2. "still water zone" (=below wave base)—shales containing simple rugose corals, brachiopods, and bryozoa;
1. "deep water zone"—black shales with a pelagic or dwarf fauna.

Lecompte (1970) distinguished between "completely developed" and "incompletely developed" bioherms, according to the bathymetric zone attained by the reef-mound in its growth. A "completely developed" mound which had accreted from a deeper water environment to sea level, and had maintained growth in the turbulent zone, would be expected to exhibit all facies zones in a shallowing-upward succession. Both Lecompte (1970, p. 38) and Tsien (1975 and 1977) have cited the late mid-Frasnian reef-mound at Arche as the type example of such a complex, since it shows well the nature of the vertical transitions. The foundation of this reef is formed by a *Disphyllum* biostrome which lies at the top of a succession of shale and shaley limestones (Daily Formation, Fig. 5) representing bathymetric zones 1. and 2. Deposition of the buildup started with a "mud-mound" (50 meters), representing zone 3., followed by a second story 4. comprising 35 meters of grey-colored grainstones and packstones. The highest zone 5. is one of peloidal grainstones and packstones containing massive stromatoporoids and corals. In most of the contemporary reef-mounds this zone is the thickest, reaching 200–300 meters, and may exhibit several deepening or shallowing episodes, with complementary development of reef talus (Fig. 18) during the shallower phases (Lecompte, 1970, p. 39). This pattern would suggest that following transgression the reef built up into shallow water from a moderately deep water zone faster than any rise in relative sea level, and remained more or less in equilibrium with further rises in relative sea level for the rest of its growth. "Incomplete" bioherms are represented by the late Frasnian "mud-mounds," which were apparently defeated by rising sea level and were unable to accrete into a shallow water depositional zone (Lecompte, 1970, p. 39–41).

Extensive use has been made of this model and terminology by other authors in the European context. A reef-mound in the lower Middle Devonian St. Lucia Formation studied by Mendez-Bedia exhibits comparable vertical zonation, from pre-buildup bank, through a basal zone of laminar and bulbous stromatoporoids and tabulate corals, to one dominated by massive stromatoporoids and corals. Pajchlova and Stasinska (1965 and 1967) on the Polish Devonian, and Poncet (1972 and 1976) on the Lower Devonian of the Armor-

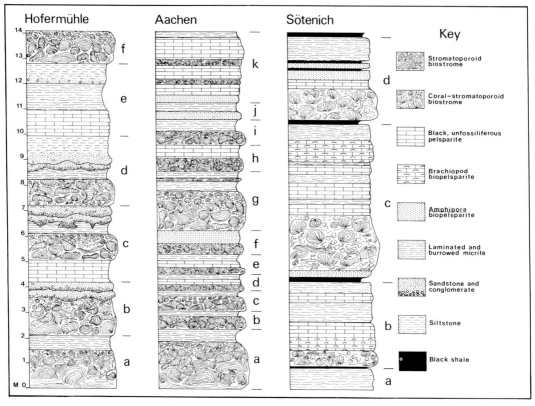

FIG. 20.—Portions of logged sections through cyclic back-reef facies of two Frasnian barrier-reef complexes (Hofermühle and Aachen), and a lower Givetian biostromal complex (Sötenich); all to same scale. Note the marked assymmetry in most cycles; those in the Sötenich section are less restricted in character than the others, and their biostromes contain more corals. At Hofermühle, sandstones and conglomerates make up the highest cycle members, and commonly cut down into the carbonates or overlie a microkarstic solution horizon; from Burchette (in press).

ican Massif, also used Lecompte's scheme in their descriptions.

Only superficial data exist on vertical faunal and facies successions in other European Devonian reefs, and there is still wide scope for further work in this field. Not all variations are consistent with Lecompte's model. Scrutton (1977a, and 1977b, p. 188), for example, considered that sheet-like laminar and tabular stromatoporoids not only predominated in the basal portion of the reef-core facies of the Tor Bay Reef complex, but that they may also have formed part of a reefal flat in the manner demonstrated by Dumestre and Illing (1967) for Devonian reef-mounds in the Spanish Sahara.

### Back-Reef Sedimentation and Sedimentary Cycles

A common feature in the back-reef of Devonian barrier-reef complexes in Europe is the small scale depositional cyclicity mentioned above, which takes the form of thin shoaling-upwards carbonate lithofacies sequences. This feature has been noted in Middle and Upper Devonian successions of the Ardenno-Eifelian tract and the northern Bergisches Land (Burchette, in press), and from the Carnic Alps (Bandel, 1972).

In central Europe there is regional variability in the development of such cycles (Fig. 20). Those in successions to the east of the Rhine River are commonly thicker and more complete than those to the west. Where they are well developed, the former consist of the following sequence:

TOP
6. fluviatile pebbly sandstones (very local),
5. smooth laminated cryptalgal and sedimentary laminites (domal stromatolites very rare),
4. burrowed and fenestral carbonate mudstones,
3. *Amphipora* packstones and wackestones and/or oncolitic and skeletal grainstones,

2. stromatoporoid biostrome (stromatoporoid packstones and wackestones), locally with abundant *Thamnopora*,

1. *Amphipora* packstones and wackestones, or skeletal grainstones.

### BASE

The above succession represents an ideal lithofacies sequence. Compared with this, cycles to the west of the Rhine are attenuated and commonly less well defined. In the late Givetian-Frasnian back-reef successions of the Aachen-Eifel district (e.g., Aachen, and in the Prüm Syncline) and in the Ardennes (e.g., Neuville, Olloy, Tailfer) they commonly comprise just a stromatoporoid biostrome overlain by thin cryptalgal or sedimentary laminites, locally also with an *Amphipora* or calcarenite horizon either sandwiched between the two or preceding the biostrome (Figs. 20 and 21A); less commonly the biostrome may also be absent. Where the biostromal portion of a cycle is fully developed, this commonly exhibits internal vertical zonation of stromatoporoid growth forms (cf., Read, 1973), from domed and tabular at the base, through bulbous forms, to branched (overwhelmingly *Amphipora*) at the top.

The whole cyclic progression represents an upward decrease in turbulence concomitant with shallowing and increasing restriction of the depositional environment. Pseudomorphs after gypsum and anhydrite are abundant in top members of the cycles, suggesting hypersalinity of at least pore waters in marginal sediments. The laminites also commonly contain desiccation structures, vadose cements, and locally (e.g., Hofermühle) the upper bounding surfaces of the carbonate portions of the cycles exhibit microkarstic solution (Burchette, in press; Fig. 21B).

Cycles become thinner and restricted facies more evident in the direction of the paleoshoreline, and thicker, less restricted and less distinct distally. They represent repeated infilling of the shelf lagoons through shoreline progradation following relatively rapid, usually nondepositional transgressive pulses. The shoaling lithofacies successions would have existed as a series of depositional belts mutually interpenetrating at their boundaries, which paralleled the margins of the shelf-lagoons and emergent areas within them. Regional differences in cycle form in central Europe are taken as being a function of differential subsidence, paleoslope, and the widths of the shelf-lagoons (Burchette, in press).

Cyclic deposition in lower Givetian biostromal complexes in the Eifel and the Ardennes was similar in principle to that outlined above, but the cycles themselves are somewhat thicker and sedimentary facies less restricted (Burchette, in press).

Lagoonal deposits in Middle and Upper Devonian atolls of the Rhenish Schiefergebirge are not *obviously* cyclically arranged, but comprise for the greater part monotonous successions of *Amphipora* packstone and wackestone with less predictable and less common intercalations of restricted laminite facies; nor are biostromes of bulbous stromatoporoids common (cf. Krebs, 1969b). This may be because these buildups actually did develop over the tops of faulted blocks and rises in the Rhenish Shelf (Krebs, 1971; Krebs and Mountjoy, 1972) which were more mobile, subsiding more rapidly, than the stable basin-margin shelves to the north and east. In these positions the back-reef lagoons of the atolls would have experienced greater marine influence (less restricted conditions, more coarse reefal debris washed in), though the environment would have remained sheltered. Of interest, however, is the fact that cyclic deposition has been recognized in certain Devonian atolls in western Canada (see e.g., Mountjoy, 1975; Havard and Oldershaw, 1976), and in the light of this further detailed work is required on this aspect of Devonian reef sedimentation in Europe.

In the Carnic Alps, Deroo *et al.* (1967, p. 315) and Bandel (1972, p. 33–40) have noted poorly developed depositional cycles in Lower and Middle Devonian back-reef limestones. Where complete, these are 2.5–8 meters in thickness and exhibit the following shallowing-upwards lithofacies succession:

4. yellow, laminated, finely-crystalline dolomite,

3. grey laminated carbonate mudstone,

2. *Amphipora* (commonly *in situ*) peloidal packstone and grainstone, locally with branched tabulate corals and small bulbous stromatoporoids,

1. intraformational conglomerate of large, angular dolomite intraclasts, commonly imbricated, set in a black peloidal grainstone matrix.

Locally the laminites at the tops of cycles contain desiccation cracks, evaporite pseudomorphs, fenestrae, and small scale tepee structures, and oncolite grainstones are also present in places (Bandel, 1972).

### Off-Reef Deposits and Debris Flows

The influence on sedimentation of many of the reef complexes discussed in previous sections extended beyond their own boundaries into the off-reef (perireefal platform, shelf) and basinal environments. This occurred in the form of sediment laden density currents which moved outwards from elevated portions of the reefs into surround-

ing low energy, deeper water environments, spreading reefal debris over wide areas.

In the Rhenish Schiefergebirge most of these off-reef debris beds ("allodapic limestones") are thin and turbidite-like, commonly graded or laminated, and occur typically intercalated in sequences of dark, pelagic shelf ("Flinz") or basinal shale up to 15 kilometers away from their sources (Krebs, 1974; Engel and Eder, 1975). Megabreccia beds such as those developed locally at the margins of certain reef complexes in western Canada (Cook et al., 1972; Mountjoy et al., 1972) and northwestern Australia (Playford and Lowry, 1966; Playford, 1969) are rare in Europe. Meischner (1964) and Eder (1970 and 1971) made detailed studies of allodapic limestones derived from the vicinity of the Balve Atoll in the Sauerland, and Tucker (1969) has described similar crinoidal limestone turbidites from the upper Givetian-lower Frasnian Marble Cliff Beds, northern Cornwall, which were probably derived from local carbonate-topped submarine rises ("Schwellen"). Allodapic limestones comprising coral and crinoidal reef debris also occur in Upper Devonian basinal shales ("Cos Plattenkalk") in the Prüm Syncline in the southern Eifel, and suggest the presence of a barrier-reef in this area during the Upper Devonian (see Krebs, 1962 and 1974).

Interesting examples of reef derived debris beds have been described from the Upper Devonian of the Holy Cross Mountains (Poland) and the Givetian of the southeastern Alps. In both areas the off-reef environments, considered to have been "basinal" in Poland (Szulczewski, 1968 and 1971) and "abyssal" in the Alps (Bandel, 1972, p. 61; and 1974), were subject to influx of material from adjacent reefs. The debris beds (up to 50 centimeters thick in Poland and 2 meters in the Alps) are coarsely graded and contain mixed faunas and floras, and material derived from several reefal zones; reef-core and fore-reef (large fragments of stromatoporoids, rugose and tabulate corals, and crinoid debris), and back-reef (abundant calcispheres, ostracodes, Amphipora, and calcareous algae) environments are all represented. Most characteristic of such deposits are large, tabular lithoclasts of pelsparite and pelmicrite, reaching the extreme proportions of 60 centimeters across

and 2 centimeters in thickness, which must have been at least partially indurated prior to reworking. In these cases the depositing currents channelled into the underlying sediments, and the sources of the clasts may have been the top layers of these. Both Szulczewski (1968) and Bandel (1972) considered that the beds represent slump sheets or turbiditic incursions into the basinal environment. The Polish examples have, however, subsequently been examined by Kazmierczak and Goldring (1978) who suggested that they were deposited in shallow, subtidal channels not far removed from the source areas of the various sedimentary components. They based their interpretation on the obvious early cementation, lithology and small amount of transport damage exhibited by the large lithoclasts, as well as the multiprovenant fauna and sedimentary structures within the beds. Intensive burrowing of graded beds and local boring of lithoclasts in the examples described by Bandel (1972, p. 57) also suggest that these were deposited at depths substantially shallower than abyssal, even though the associated pelagic flaser limestones locally exhibit solution phenomena and manganiferous crusts (see also Bandel, 1974). Franke et al. (1978) considered that the pelagic flaser limestone facies represents deposition on epireefal or "epicontinental" platforms at depths no greater than a few hundred meters, and probably locally shallower (cf., also the model for Devonian pelagic limestone deposition by Tucker, 1973 and 1974).

### POST-GROWTH DEPOSITIONAL HISTORY

Following extinction, reef complexes came under the influence of non-reefal sedimentary and physical processes. In the European Devonian two broad courses of post-growth evolution are apparent:

1. rapid burial of the reef complex;
2. continued exposure of the complex in a shallow submarine or subaerial environment for a time prior to burial.

*Rapid Burial.*—Many of the European Devonian reef complexes were buried relatively rapidly by subsequent sediments, locally with intervening transgressive reefal phases (e.g., "mud-mounds").

←

FIG. 21.—A. Quarry section at the railway station, Walheim, near Aachen, looking east, showing overturned cyclic back-reef deposits in the upper of the two Frasnian barrier-reef complexes in this area. Two cycles are visible, each consisting of a stromatoporoid biostrome (B) overlain by more thinly-bedded sedimentary and cryptalgal laminites (L). The higher cycle (left) also contains *Amphipora* packstones (A). Central biostrome is 2 meters in thickness. B. Section in back-reef facies of a Frasnian barrier-reef complex at Hofermühle, near Wülfrath, Bergisches Land, showing carbonate cycle (right) overlain by fluviatile sandstones and conglomerates (left). The top surface of the limestone (laminite) is a karstic one, and hollows and cavities in the limestone are infilled with quartzite pebbles and sandstone. Staff graduated in 10 centimeter intervals.

In central Europe, the successive biostromal complexes and banks were drowned during transgressive pulses and covered without break by black shales, giving rise to the gross megacyclic alternation of reefal limestones and basinal shales discussed earlier. In similar fashion Upper Devonian reef complexes in the western and southern Rhenish Schiefergebirge were covered in the later Frasnian by black basinal shales up to 900 meters in thickness (Krebs, 1974).

In the Cantabrian Mountains both the St. Lucia and Portilla Reef developments were succeeded by terrigenous shelf deposits, the influxes of which were probably related to episodes of uplift of the Asturian Hinterland (Mohanti, 1972, p. 154; Reijers, 1973a). The small Lower Devonian bioherms and biostromes in the Armorican Massif were also covered at an early post-growth stage by shallow marine siliciclastic sediments (Poncet, 1972).

*Prolonged Submarine or Subaerial Exposure.*— In some areas reefs were succeeded by condensed or incomplete sequences of pelagic goniatite limestones, while in the surrounding basins complete successions of black shale up to 700 meters in thickness were deposited (Krebs, 1974, p. 173). This suggests that the extinct reefs retained submarine relief and acted as rises ("Schwellen") within deep water areas. In southwest England, growth of the Tor Bay Reef complexes ceased by about mid-Frasnian and it is overlain in the Chudleigh area by a highly condensed succession (50 meters) of shales and goniatite limestones, representing deposition from the late Frasnian through the Famennian (Tucker and van Straaten, 1970; House and Butcher, 1973). At the margin of the complex, however, basinal ostracode shales directly overlie the reef limestones (Scrutton, 1977b; see Fig. 8B).

Following extinction of the Upper Devonian reef in the Kielce district of the Holy Cross Mountains goniatite limestones ("*Manticoceras*" and "*Cheiloceras* limestones") were widely deposited on its flanks (Szulczewski, 1971, p. 117), although oolitic "shoals" persisted in areas of highest relief until the latest Famennian. The complex was eventually covered by "basinal" sediments. Comparable post-growth histories can be demonstrated for several of the isolated reef complexes in the central and eastern Rhenish Schiefergebirge, including those of Attendorn (Gwodz, 1971), Brilon (Bär, 1968), and Langenaubach (Krebs, 1966 and 1972); these were reefs which developed at the Rhenish shelf-margin and trough of this area (see Krebs, 1974).

Some reefs became subaerially exposed and suffered erosion following their extinction. There is, for example, a local hiatus between the Lower Devonian Koneprusy Bioherm of Bohemia and the overlying neritic limestones, which in places possess a basal conglomerate and even infill fissures in the bioherm (Chlupac, 1954 and 1967). In adjacent areas sedimentation was continuous. Submarine exposure and fissuring between reef extinction (in the late Frasnian) and burial (in the Lower Carboniferous) have been demonstrated for the Iberg Atoll in the northern Harz Mountains (Franke, 1973), and for the Langenaubach complex in the Rhenish Schiefergebirge (Krebs, 1963 and 1966). There are also local stratigraphic gaps, though perhaps not erosional, above the Brilon (Bär, 1968) and Balve (Schäfer, 1975) reef complexes.

In the Carnic Alps there appears to have been an overall trend during the Frasnian towards pelagic sedimentation, even in reefal areas, whereas in the Karawanken there is an angular unconformity with transgression conglomerate between Givetian and Frasnian reef limestones and the overlying Lower Carboniferous flysch deposits (Schönlaub, 1979); Lower Carboniferous sediments also infill karstic fissures in the reefal limestones (Tessensohn, 1974).

Uplift caused exposure and led to intense karstification of late Middle and Upper Devonian reef limestones during the latest Devonian and lowest Carboniferous in the vicinity of the Brabant Massif (Pirlet, 1967; Kimpe et al., 1978), though away from this zone the Devonian-Carboniferous transition is continuous.

## ASPECTS OF DIAGENESIS

Diagnesis has been much neglected in most studies of Devonian reefs in Europe and there are large gaps in the presently available data. Most information stems from a few investigations made in central Europe and in the Cantabrians, and in many cases this must necessarily also be considered valid for complexes elsewhere. There is wide scope for further work in this field. The most important aspects of diagenesis in European Devonian reef complexes are considered below.

*Evaporites.*—Laminated sediments in shelf-lagoon successions of central Europe commonly contain calcite pseudomorphs after gypsum and anhydrite (Burchette, in press), but these minerals are themselves no longer present in surface exposure, and in most cases were probably replaced during early diagenesis. Minor deposits of anhydrite mineral have been encountered, however, in several deep boreholes in the Ardennes (van Tassel, 1969; Graulich, 1977), and in the Polish Lowlands (Pajchlova and Stasinska, 1965 and 1967; Chlupac, 1971).

*Early Cementation.*—Intense early cementation, causing local reductions in primary porosity of up to 60 percent (Schneider, 1977; Krebs, 1978), was a feature of all reef complexes. All depositional environments were affected, although there are marked differences between them in the characters of the cements precipitated.

In marginal back-reef areas subaerial exposure of carbonate sediments was widespread, and early vadose cements abound. Fenestral laminites commonly contain well developed microstalactitic iron-free calcite cements, particularly where they form the top members of sedimentary cycles (Burchette, in press). Elsewhere, vadose cements occur as drusy pore linings and cavity fills of iron-free calcite. Early submarine cementation was probably also a widespread feature of these Devonian back-reef environments, though in this setting it is much more difficult to recognize. Lithoclasts derived from underlying pelmicrite beds are common at the bases of Frasnian back-reef cycles in central Europe, even where there is no evidence for subaerial exposure prior to reworking; these may have been indurated penecontemporaneously by micrite cements. Diagenesis in back-reef deposits of the Rhenish Schiefergebirge and the Ardennes will be dealt with in detail elsewhere (Burchette, in prep.).

The most characteristic and visually striking early diagenetic features in reef-core and fore-reef deposits are isopachous fibrous and radiaxial-fibrous cavity filling cements. These are present in many of the reef complexes in areas as widely dispersed as central Europe (Schwarz, 1927; Krebs, 1969b; Schneider, 1977), southwest England (Scrutton, 1977a, and 1977b), Bohemia (see Chlupac, 1954, plates 3 and 5), and the southeastern Alps (Bandel, 1972). They represent penecontemporaneous cementation of reefal material in subtidal environments while interparticular voids and cavities in the sediments were still open to influx of normal sea water (Krebs, 1969b). Associated with the cements, commonly interstratified with them, are multiple generations of internal sediments, consisting of calcarenites, pelsparites, pelmicrites and micrites containing small fossils. Walls of cavities were also commonly encrusted prior to, and locally during, growth of the fibrous layers by laminated micrite and pelmicrite rinds which may be of algal origin (Krebs, 1969b; Orchard, 1974). In some buildups, fragments of early fibrous cements and cavity fillings were incorporated as debris in the fore-reef deposits (Krebs, 1974).

The fibrous cements in fore-reef and reef-core facies of European Devonian reefs closely resemble in character and occurrence examples described from other Devonian reefs in western Canada (Schmidt, 1971; McGillavray and Mountjoy, 1975; Mountjoy and Walls, 1977) and northwestern Australia (Playford and Lowry, 1966). The radiaxial-fibrous habit of many of these sug-

Fig. 22.—Convoluted bedding planes (?tectonic cause) picked out by stromatactis and local laminar corals in a quarry section through an upper Frasnian mudmound, Croissettes Quarry, Vodecée, Ardennes. Height of quarry face 25–30 meters.

gests that they in part represent early replacement fabrics of precursor radial-fibrous carbonate cements (cf. Kendall and Tucker, 1973). They may consequently be analogous with the syndepositional fibrous and botryoidal aragonite cements known from the seaward facing margins (reef-wall and talus) of present coral reefs in the Bahamas and off Belize (see Ginsburg, Schroeder, and Shinn, 1971; Schroeder, 1974; Ginsburg and James, 1976).

*Stromatactis.*—A highly distinctive feature of Frasnian "mud-mounds" in the Ardennes and southwest England is the poorly understood structure "stromatactis" (Dupont, 1881), which may locally account for 60–70 percent of the rock volume (Fig. 22). There has been wide misuse of the term with regard to the European Devonian, however, and it has been applied rather indiscriminately to a whole range of fibrous calcite replacements and cavity infillings, including those considered above. Stromatactis proper appears to be restricted to the "mud-mounds" and mud-mound appearing facies.

In vertical section, stromatactis forms laminar to "domal" or broadly branching "cavities" in the host limestones, which have flat or undulating bases and highly digitate grooves (Fig. 23A and B). In horizontal section, stromatactis forms extensive interconnected networks and sheets. The cavities are filled (usually) with isopachous layers of fibrous or scalenohedral calcite, and in some places they have been enlarged by solution. Locally, fossil fragments appear to "float" in the cement, and rarely, remnants of calcareous algae also occur. Such observations led Lecompte (1937, 1954, and 1970) to propose that stromatactis represents the infilled molds of dissolved or decomposed carbonate skeletons or soft bodied organisms. Tsien (1971 and 1977) has suggested further that these structures exhibit ecologically determined morphological variations as do the associated corals. The occurrence and forms of the mud-mound stromatactis, however, makes an early diagenetic interpretation more convincing (see also the comments of Wilson, 1975), though it seems equally unlikely that the simple dewatering hypothesis proposed by Heckel (1972) can be widely applied in these Frasnian examples. It is possible that stromatactis in these cases might represent the result of penecontemporaneous cementation of surface sediment layers on the mud-mounds ("crustification") and subsequent removal, through scouring, or collapse and compaction (due to seismic shock?), of underlying unconsolidated material (Bathurst, 1978 pers. comm., in press). Neumann, Kofoed, and Keller (1977) have speculated on a similar origin for some stromatactis on the basis of observations made on deep water "lithoherms" in the Straits of Florida and at the foot of the Little Bahama Bank, biohermlike structures which appear to offer at least a partial Recent analogue for the mud-mounds of the European Devonian.

*Fissuring.*—A characteristic process during the late growth and early post-growth stages of many reef complexes in the European Devonian, even those which did not obviously undergo subaerial weathering, was the development of cavities and cross-cutting vertical and horizontal fissures. These were filled both syngenetically and postgenetically with internal sediments and fibrous (now radiaxial-fibrous) calcium carbonate cements. Fissures of this type seem to be rare in back-reef facies, and are thus restricted largely to the reef-core and fore-reef deposits.

In some cases authors have offered a tectonic interpretation (Chlupac, 1954, p. 118; Szulczewski, 1971; Franke, 1973), whereas others have preferred a solution/karstification origin (Braithwaite, 1967; Tsien, 1977), or a combination of both (e.g., Orchard, 1974).

The fissures are commonly filled with fibrous cements and grey or red, laminated peloidal and micritic sediment, clearly mechanically deposited (Fig. 24A and B), which closely resemble the early cements and internal sediments described above. In many cases laminations can be correlated between adjacent cavities. The macroscopic

appearance of such internal sediments in reef complexes as widely distributed as southwest England, the Frasnian reef-mounds of the Ardennes, the complexes of Wülfrath, Brilon, Langenaubach, and others, in the Rhenish Schiefergebirge, and Iberg in the Harz Mountains (Schneider, 1972) is remarkably consistent.

Commonly the fissure filling internal sediments contain ostracodes, conodonts and other small fossils which may be older than the host rocks. Conodonts in such deposits in the lower Frasnian Plymouth Limestone, for example, are of Famennian age (Orchard, 1974). Vertical fissures (up to 60 meters deep) in the Lower Devonian Koneprusy Bioherm of Bohemia (Chlupac, 1954), and others in the Upper Devonian reef complex in the Kielce district of Poland (Szulczewski, 1971 and 1973; Osmolska, 1973; Stasinska, 1973), have preserved multiple generations of internal sediments and breccias, indicating several opening phases, and highly characteristic younger faunas of corals and trilobites.

*Later Cementation, Neomorphism and Compaction.*—In all reefal facies the porosity remaining after early diagenesis was sealed more or less completely by second generation phreatic cements consisting predominantly of drusy ferroan calcite (see e.g., Reijers, 1974; Schneider, 1977; Krebs, 1978). In most of the Givetian and Frasnian reef complexes this stage can be set in age as Late Devonian to Mid-Carboniferous (Schneider, 1977). Subsequent neomorphism consisted generally of aggrading recrystallization of calcilutites and calcisiltites to micro- and pseudospars, and aggressive enlargement of cement crystals.

Of importance in European Devonian reefs, since the effects of Hercynian deformation occur in all areas (and in some of the Alpine as well), is recrystallization of carbonates in response to tectonic stress and metamorphism. The latter is particularly marked in the western part of the Carnic Alps (Vai, 1975; Schönlaub, 1979) and in the southernmost Rhenish Schiefergebirge (Bandel and Meyer, 1975). Schneider (1977, p. 55) noted the presence in areas of deformation of fractures and kink bands in cement crystals, as well as recrystallization and cataclasis. On a larger scale solution planes and vertical stylolites are further common tectonic effects, but are regionally variable in importance.

$\rightarrow$

FIG. 23.—Examples of stromatactis from an upper Frasnian mud-mound exposed in Croissettes Quarry, Vodecée, Ardennes. Both photographs of loose blocks, orientated. A. Note isopachous fibrous calcite and internal sediments (S) infilling the stromatactis cavities. B. Note the great variability in form of the stromatactis cavities, with a vertical change from laminar to highly digitate. Note also the presence of laminar corals (C), some fractured by compaction, and shaley interlayers. Hammer handle for scale (both photographs) is 37 centimeters long.

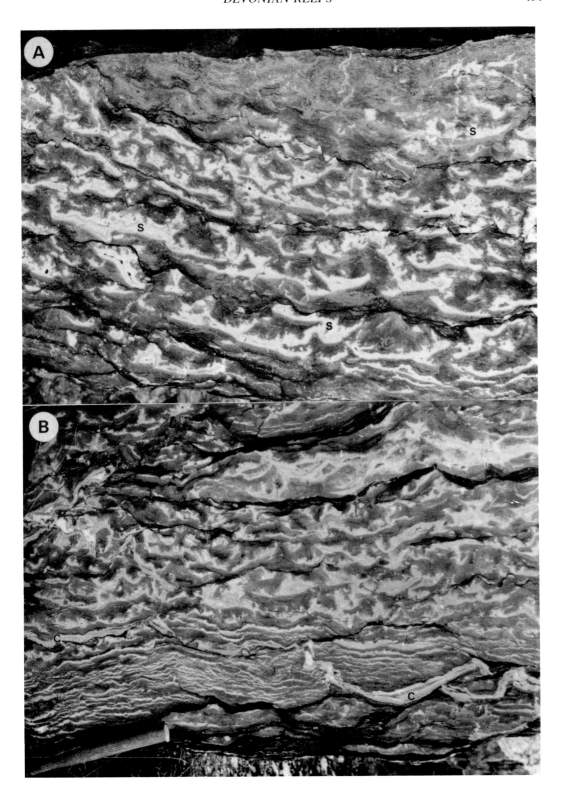

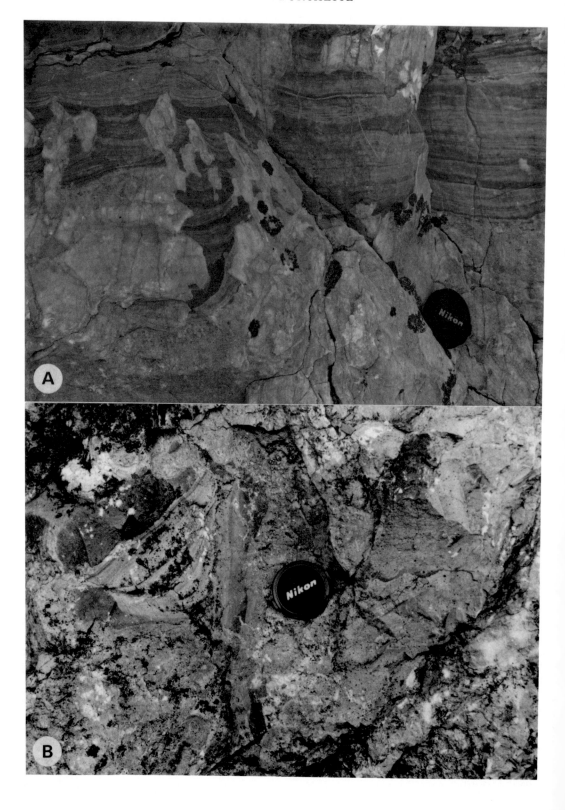

In some areas compaction appears to have been marked, producing stylolitized bedding planes, diminution in bed thickness, and grain-to-grain suturing of bioclasts. The last is especially common among bulbous stromatoporoids in bituminous or argillaceous limestones such as those of the Schwelm-type bank facies and some back-reef biostromes in central Europe. Suturing probably occurred prior to precipitation of the final indurating cement phase (cf. Bathurst, 1975, p. 465), which suggests that such deposits remained essentially unconsolidated or relatively plastic until covered by substantial overburden.

Replacement of carbonates by chert and idiomorphic authigenic quartz is common in some reef complexes and in most cases probably represents a late diagenetic process (Schneider, 1973a and b and 1977). Reijers (1972, p. 194 and 1974) and Sleumer (1969, p. 24), in contrast, considered authigenic quartz and chert which replaced fossils in the Portilla Formation to be the result of a relatively early diagenetic process.

*Dolomitization.*—Extensive dolomitization is encountered in the Givetian and Frasnian of the Ardenno-Eifelian tract and in the Eifelian and Givetian of Poland and the Carnic Alps. Elsewhere, as for example in the Rhenish Schiefergebirge to the east of the Rhine and the Cantabrian Mountains, dolomitization is patchy and closely associated with faults, stylolites, and Permo-Triassic paleomorphology (Reijers, 1972; Krebs, 1974; Schneider, 1977); neither is dolomitization of isolated reef complexes extensive.

Fine-grained unfossiliferous dolomite makes up much of the attenuated succession in the most marginal zones of Frasnian shelf-lagoons in the Ardennes (i.e., on the northern limb of the Namur Syncline; Lecompte, 1970, p. 42). Comparable thin, fine-grained black dolomite horizons also occur at the tops of some sedimentary cycles in Frasnian back-reef successions in the Aachen area. In both cases these may represent penecontemporaneous replacement of the original carbonate minerals (Burchette, in press). In the central Ardennes, however, dolomitization is most severe in the critical area at the margin of the late Givetian-Frasnian carbonate platform, where limestones of the barrier-reef "core" and immediate back-reef facies have been replaced by medium to coarsely crystalline dolomite. Lecompte

(1970) considered this to be largely an early diagenetic feature, and that the greater primary porosity of the reefal deposits predisposed them to dolomitization. The same dolomite phase, up to several hundred meters in thickness, occurs in Givetian and Frasnian shelf and back-reef limestones in the Eifel synclines (the "Muldenkerndolomit"), where it has long been a stratigraphical problem. Richter (1974) interpreted the Eifel dolomites as the result of a combination of both early and later diagenetic processes, and proposed a model involving production of dolomitizing brines in a marginal lagoonal environment and its subsequent transport through the host body by squeezing due to increasing overburden pressure. Why this process should have been so paleogeographically restricted when similar depositional and diagenetic conditions must have existed over most of this Upper Devonian carbonate platform is not clear. The origin of these dolomites is a problem which clearly requires further investigation on a wider scale. Their position at the platform-margin in a zone of mixing of normal marine and fresh or hypersaline pore waters may be significant.

*Hydrocarbons.*—To date no significant deposits of hydrocarbons have been discovered in European Devonian reefs, although some exploratory work towards this end has been carried out, particularly in West Germany (see e.g., Bartenstein and Teichmüller, 1974; Krebs, 1975; Nowak and Keshav, 1977; Weggen, 1978). In the Rhenish Schiefergebirge and the Harz Mountains pyrobitumen (impsonite) is locally present infilling residual porosity in reef limestones (Jux, 1960; Krebs, 1974). Krebs (1978) has demonstrated that in the Iberg complex in the Harz, liquid hydrocarbons were introduced prior to the conclusion of late diagenesis (i.e., sealing of the porosity) and later thermally altered to impsonite. The time of mobilization of hydrocarbons from potential source rocks (black shales?) would have varied somewhat over the area of the Rhenish Schiefergebirge in response to differential rates of overburden accumulation; migration in the north would have begun at some stage in the early Upper Carboniferous, and in the west and south at about the start of Hercynian tectonism in the Late Carboniferous (Krebs, 1978).

←

Fig. 24.—Fissure-fillings in Devonian reefal limestones. A. Lower Frasnian Plymouth Limestone, southwest England. Red and grey-colored peloidal and micritic limestones, associated with fibrous crusts on cavity walls, infilling probable solution cavities in (?) fore-reef limestones. Note how the bands can be correlated between cavity branches. b. Mid-Frasnian Lion Reef-mound, Frasnes-les-Couvin, Ardennes. Red and grey-colored, laminated peloidal and micritic limestones and fibrous calcite crusts infilling (?) karstic pockets and fissures in reef-core limestones. Lens-cap for scale is 5 centimeters across.

## CONCLUSIONS

Five broad classes of reef can be recognized in the European Devonian. These are:

1. banks,
2. biostromal complexes,
3. barrier-reef complexes,
4. isolated reef complexes,
5. quiet water carbonate buildups or mud-mounds.

These different reef types are characterized by distinctive sedimentary facies, facies distributions, and constituent organic communities. Types 2.–4. exhibit lateral differentiation of sedimentary and organic facies.

Reef position and growth were determined principally by the distribution of favorable environments and by changes in relative sea level controlled by the following factors:

1. crustal flexures and movement on basin-margin hingelines,
2. synsedimentary faulting,
3. the development of volcanic islands and local submarine rises,
4. the distribution of minor sea floor elevations (e.g., differential compaction around buried reefs, calcarenite banks, *etc.*),
5. changes in relative sea level, or its rate of change, where these cannot clearly be attributed to one of the above.

Many European Devonian reefs (buildups) were sources of carbonate debris, as calciturbidites, for surrounding deeper water environments.

Most Devonian reefs in Europe exhibit evidence of early cementation, including microstalactitic and drusy non-ferroan calcite cements in back-reef facies, and radial-fibrous (now radiaxial-fibrous) submarine cements in the reef-core, fore-reef and mud-mound (stromatactis) facies. Many also developed fissures during the late growth and early post-growth stages, which were infilled with red internal sediments and fibrous carbonate cements.

Following extinction some reefs were quickly buried by basinal or shelf deposits, whereas others remained exposed on the sea floor, forming sites for the deposition of condensed sequences of goniatite limestones, or were even exposed to subaerial erosion.

No commercial hydrocarbon finds have yet been made in European Devonian reefs.

## ACKNOWLEDGMENTS

I am indebted to Wolfgang Krebs and Peter Carls (both Braunschweig) for supplying literature during this study and for friendly discussion on aspects of European Devonian paleogeography and reef growth. Sincere thanks are also due to Wolfgang Krebs and Klaus Hoenen (Braunschweig) who demonstrated Devonian reefs in the Rhenish Schiefergebirge and Harz Mountains to me in the field; to Hsien Ho Tsien (Louvain-la-Neuve) who introduced me to Devonian reefs in the Ardennes; and to Tony Buglass (Newcastle-upon-Tyne) who provided certain additional data on reef limestones in the Torquay area of southwest England. Thomas Reijers (Shell Research, Rijswijk) kindly permitted the use of unpublished information and diagrams deriving from his studies on Devonian reefs in the Cantabrian Mountains, Spain. I am also grateful to Robert Riding (Cardiff) for discussion.

Part of the work for this paper was carried out under tenure of a Royal Society postdoctoral fellowship at the Technische Universität Braunschweig, West Germany.

## REFERENCES

ADRICHEM-BOOGAERT, H. A. VAN, 1967, Devonian and Lower Carboniferous conodonts of the Cantabrian Mountains (Spain) and their stratigraphic application: Leiden, Leidse geol. Meded., v. 39, p. 129–192.

AHLBURG, J., 1919, Über die Verbreitung des Silurs, Hercyns und rheinischen Devons und ihre Beziehungen zum geologischen Bau im östlichen Rheinischen Schiefergebirge: Berlin, Jahrb. preuss. geol. Landesanstalt, v. 40, p. 1–82.

BANDEL, K., 1969, Feinstratigraphische und biofazielle Untersuchungen unterdevonischer Kalke am Fuss der Seewarte: Vienna, Jahrb. geol. Bundesanstalt, v. 112, p. 197–234.

———, 1972, Paläoökologie und Paläogeographie im Devon und Unterkarbon der zentralen Karnischen Alpen: Stuttgart, Palaeontographica, ser. A, v. 141, 117 p.

———, 1974, Deep water limestones from the Devonian-Carboniferous of the Carnic Alps, Austria, *in* Hsü, K. J., and Jenkyns, H. C., Eds., Pelagic sediments: on land and under the sea: International Assoc. Sedimentologists Spec. Publ. No. 1, p. 93–115.

———, AND MEYER, D. E., Algenriffkalke, allochtone Riffbedecke und autochtone Beckenkalke im Südteil der rheinischen Eugeosynclinale: Mainzer geowiss. Mitteilungen, v. 4, p. 5–65.

BAR, P., 1966, Stratigraphie, Fazies und Tektonik am Briloner Massenkalk-Sattel [unpub. Ph.D. thesis]: Univ. Giessen, West Germany.

————, 1968, Die oberdevonische/unterkarbonische Schichtlücke über dem Massenkalk des Briloner und Messinghäuser Sattels (Ost-Sauerland): Stuttgart, Neues Jahrb. Geol. Paläont. Abh., v. 131, p. 263–288.

BARTENSTEIN, H., AND TEICHMÜLLER, R., 1974, Inkohlungsuntersuchungen, ein Schlüssel zur Prospektierung von paläozoischen Kohlenwasserstoff-Lagerstätten: Krefeld, Fortschr. Geol. Rheinl. Westfalen, v. 24, p. 129–160.

BATHURST, R. G. C., 1975, Carbonate sediments and their diagenesis: Developments in Sedimentology: Amsterdam, Elsevier, v. 12, 658 p.

BIGOT, A., 1904, Réunion extraordinaire de la Société géologique de France en Basse-Normandie en 1904: Paris, Bull. Soc. Géol. France, v. 4, p. 861–908.

BOERSMA, M., 1973, Devonian and Lower Carboniferous conodont biostratigraphy, Spanish central Pyrenees: Leiden, Leidse Geol. Meded., v. 49, p. 303–377.

BOSCH, W. J. VAN DEN, 1969, The relationship between orogenesis and sedimentation in the southwestern part of the Cantabrian Mountains (Spain): Leiden, Leidse Geol. Meded., v. 44, p. 227–233.

BOUCOT, A. J., 1974, Silurian and Devonian biogeography, in Ross, C. A. (Ed.), Paleogeographic provinces and provinciality: Soc. Econ. Paleontologists Mineralogists Spec. Pub. No. 21, p. 165–176.

————, JOHNSON, J. G., AND TALENT, J. A., 1969, Early Devonian brachiopod zoogeography: Geol. Soc. America Spec. Pub., v. 119, p. 1–113.

BOYER, F., 1964, Observations stratigraphiques et structurales sur le Dévonien de la région de Caunes-Minervois: Paris, Bull. Carte Géol. France, v. 60, p. 106–122.

BRAITHWAITE, C. J. R., 1966, The petrology of Middle Devonian limestones in South Devon, England: Jour. Sed. Petrology, v. 36, p. 176–192.

————, 1967, Carbonate environments in the Middle Devonian of South Devon, England: Sed. Geology, v. 1, p. 283–320.

BROUWER, A., 1964a, Deux facies dans le Dévonien des Montagnes Cantabriques méridionales: Oviedo, Breviora Geol. Asturica, v. 8, p. 3–10.

————, 1964b, Devonian biostromes and bioherms of the southern Cantabrian Mountains, in van Straaten, L. M. J. U., Ed., Deltaic and shallow marine deposits: Developments in Sedimentology: Amsterdam, Elsevier, v. 1, p. 48–53.

————, 1967, Devonian of the Cantabrian Mountains, northwestern Spain, in Oswald, D. H., Ed.: Alberta Soc. Petroleum Geologists, International Symp. Devonian System, Calgary, v. 2, p. 37–46.

BUGGISCH, W., 1972, Zur Geologie und Geochemie der Kellwasserkalke und ihre begleitenden Sedimente (Unteres Oberdevon): Wiesbaden, Abh. hess. Landesamt f. Bodenforschung, v. 62, 68 p.

BURCHETTE, T. P., in press, Back-reef sediments and cyclic deposition in the Middle and Upper Devonian of Germany and Belgium.

BURNE, R. V., 1973, Paleogeography of southwest England and Hercynian continental collision: Nature Phys. Sci., v. 241, p. 129–131.

BURRETT, C. F., 1972, Plate tectonics and the Hercynian Orogens: Nature, v. 239, p. 155–157.

CHLUPAC, I., 1954, Stratigraphic study of the oldest Devonian beds of the Barrandian (Czechoslovakian with an English summary p. 204–244): Prague, Sbor. Ustř. úst. geol., v. 21, p. 91–224.

————, 1956, Facial development and biostratigraphy of the Lower Devonian of central Bohemia (Czechoslovakian with an English summary p. 468–485): Prague, Sbor. Ústř. úst. geol., v. 23, p. 369–485.

————, 1967, Devonian of Czechoslovakia, in Oswald, D. H., Ed.: Alberta Soc. Petroleum Geologists, International Symp. Devonian System, Calgary, v. 1, p. 109–126.

————, 1968, Early Paleozoic of the Bohemian Massif: International geol. Congr., 23rd, Prague, 1968, Excursion guide 11 AC, 43 p.

————, 1971, Übersicht über neuere Forschungsarbeiten im Devon Polens (1960–1970): Stuttgart, Zentralbl. Geol. Paläont., Sect. 1, v. (1971), p. 131–145.

————, 1976, The Bohemian Lower Devonian stages and remarks on the Lower-Middle Devonian boundary: Stuttgart, Newsletter Strat., v. 5, p. 168–189.

CLAR, E., 1971, Bemerkungen fur eine Rekonstuktion des variszischen Gebirges in den Ostalpen: Hanover, Zeitschr. Deutsche Geol. Gesellsch., v. 122, p. 161–167.

COO, J. M. C. DE, DEELMAN, J. C., AND VAN DER BAAN, V., 1971, Carbonate facies of the Santa Lucia Formation (Emsian-Couvinian) in Leon and Asturias, Spain: Culemborg (Netherlands), Geol. en Mijnvouw, v. 50, p. 359–366.

COOK, H. E., McDANIEL, P. N., MOUNTJOY, E. W., AND PRAY, L. C., 1972, Allochtonous carbonate debris flows at Devonian bank ("reef") margins, Alberta, Canada: Bull. Canadian Petroleum Geol., v. 20, p. 439–497.

COPPER, P., 1966, Ecological distribution of Devonian atrypid brachiopods: Palaeogeography, Palaeoclimatology, Palaeoecology, v. 2, p. 245–256.

CUMINGS, E. R., 1932, Reefs or bioherms?: Geol. Soc. America Bull., v. 43, p. 331–352.

CZARNOCKI, J., 1928, Aperçu de la stratigraphie du Famennien et du Carbonifère inférieur (culm) dans les parties occidentale et centrale du Massif de Ste. Croix (in Polish with French summary) Warsaw, Pos. Nauk. Państw. Inst. Geol., v. 21, p. 18–21.

————, 1950, Geology of the Lysa Gora Region, Holy Cross Mountains, in connection with the problem of the iron ores at Rudki (in Polish with English summary): Warsaw, Prace Państw. Inst. Geol., v. 1, p. 1–404.

DEROO, G., GAUTHER, J., AND SCHMERBER, G., 1967, Etudes d'environments carbonatés à propos du Dévonien des Alpes Carniques, in Oswald, D. H., Ed.: Alberta Soc. Petroleum Geologists, International Symp. Devonian System, Calgary, v. 2, p. 307–323.

DINELY, D. L., 1961, The Devonian System in South Devonshire: Field Studies, v. 1, p. 121–140.

DUMESTRE, A., AND ILLING, L. V., 1967, Middle Devonian reefs in Spanish Sahara, *in* Oswald, D. H., Ed.: Alberta Soc. Petroleum Geologists, International Symp. Devonian System, Calgary, v. 2, p. 333–350.

DUNHAM, R. J., 1970, Stratigraphic reefs versus ecologic reefs: Am. Assoc. Petroleum Geologists Bull., v. 54, p. 1931–1932.

DUPONT, E., 1881, Sur l'origine des cálcaires Dévoniens de la Belgique: Brussels, Bull. Acad. Roy. Belgique, class sci., ser. 3, v. 2, p. 264–280.

DVORAK, J., 1973, Die Quergliederung des Rheinischen Schiefergebirges und die tektogenese des Siegener Antiklinoriums: Stuttgart, Neues Jahrb. Geol. Paläont. Abh., v. 143, p. 133–152.

———, AND FREYER, G., 1965, Der heutige Stand der Stratigraphie und Paläogeographie des Devons und Unterkarbons (Dinant) im südlichen Teil der Drahaner Höhe (Mähren): Berlin, Geologie, v. 14, p. 404–419.

———, AND HAVLICEK, V., 1963, Brachiopoden der Stringocephalenkalke in Mähren: Prague, Sbor. Ustř. Úst. Geol., ser. Paleontology, v. 28, p. 85–99.

EDER, W., Genese Riff-naher Detritus-Kalke bei Balve im Rheinischen Schiefergebirge (Garbecker Kalke): Vienna, Verhandl. Geol. Bundesanstalt, v. (1970), p. 551–569.

———, 1971, Riff-nahe Detritus-Kalke bei Balve im Rheinischen Schiefergebirge: Göttinger Arb. Geol. Paläont., v. 10, p. 2–65.

EHRENBERG, H., PILGER, A., AND SCHROEDER, F., 1954, Das Schwefelkies-Zinkblende-Schwerspat-Lager on Meggen (Westfalen): Hanover, Beih. Geol. Jahrb., v. 12, 352 p.

ENGEL, W., AND EDER, W., 1975, Limestone turbidites off the Middle Devonian shelf-margin (Padberger Kalk and Flinz), Rheinisches Schiefergebirge, West Germany: International Symp. Fossil Algae, Erlangen, Abs., p. 8.

ERBEN, H. K., 1962, Zur Analyse und Interpretation der Rheinischen und Hercynischen Magnafazies des Devons: International Symp. Silurian-Devonian boundary, 2nd, Bonn-Brussels, 1960, Proc. p. 42–61.

———, 1964, Facies developments in the marine Devonian of the Old World: Redruth (U. K.), Proc. Ussher Soc., v. 1, p. 92–118.

———, 1969, Faunenprovinzielle Beziehungen zwischen unterdevonischen Trilobiten der Karnsichen Alpen und Zentralasiens: Carinthia II, Symp. Paläontologie Stratigraphie Karnischen Alpen, Graz, Spec. Pub. No. 27, p. 19–20.

———, AND ZAGORA, K., 1967, Devonian of Germany, *in* Oswald, D. H., Ed.: Alberta Soc. Petroleum Geologists, International Symp., Devonian System, Calgary, v. 2, p. 53–67.

EVERS, H. J., 1967, Geology of the Leonides between the Bernesga and Portilla Rivers, Cantabrian Mountains, northwestern Spain: Leiden, Leidse Geol. Meded., v. 41, p. 83–151.

FLÜGEL, E., 1974, Stromatoporen aus dem Schwelmer Kalk (Givet) des Sauerlandes: Stuttgart, Paläont. Zeitschr., v. 48, p. 149–187.

———, AND HOTZL, H., 1971, Foraminiferen, Calcisphaeren und Kalkalgen aus dem Schwelmer Kalk (Givet) von Letmathe im Sauerland: Stuttgart, Neues Jahrb. Geol. Paläont., Abh., v. 137, p. 358–395.

———, AND ———, 1976, Paläokologische und statistische Untersuchungen in mitteldevonischen Schelf-Kalken (Schwelmer Kalk, Givet, Rheinisches Schiefergebirge): Bayerische Akad. Wissenschaft, Mathematische-Naturwissensch. Klasse, Abh., v. H156, p. 1–70.

FLÜGEL, H. W., 1964, Das Paläozoikum in Österreich: Vienna, Mitt. Geol. Gesellsch. Wien, v. 56, p. 401–443.

———, 1967, Devonian of Austria, *in* Oswald, D. H., Ed.: Alberta Soc. Petroleum Geologists, International Symp. Devonian System, Calgary, v. 1, p. 99–107.

———, 1970, Fortschritte in der Stratigraphie des ostalpinen Paläozoikums: Stuttgart, Zentralbl. Geol. Paläont., Sect. 1, v. (1970), p. 661–687.

———, 1975, Einige Probleme des Variszikums von Neo-Europa: Stuttgart, Geol. Rundschau, v. 64, p. 1–62.

———, 1976, Fortschritte in der Stratigraphie des ostalpinen Paläozoikums (1970–1975): Stuttgart, Zentralbl. Geol. Paläont., Sect. 1, v. (1975), p. 656–684.

———, JAEGER, H., SCHÖNLAUB, H. P., AND VAI, G. B., 1977, Carnic Alps, *in* The Silurian-Devonian boundary: Stuttgart, I. U. G. S. Pub., ser. A, no. 5, p. 126–142.

FOURMARIER, P., 1963, Les variations du niveau stratigraphique du front supérieur de schistosité dans l'Ouest du synclinorium de Dinant: Paris, Comptes Rendus Acad. Sci., v. 257, 2933–2937.

FRANKE, W., 1973, Fazies, Bau und Entwicklungsgeschichte des Iberger Riffes: Hanover, Geol. Jahrb., ser. A, v. 11, p. 1–127.

———, EDER, W., ENGEL, W., AND LANGENSTRASSEN, F., 1978, Main aspects of geosynclinal sedimentation in the Rhenohercynian zone: Hanover, Zeitschr. deutsche geol. Gesellsch., v. 129, p. 201–216.

GAERTNER, H. R. VON, 1937, Montagne Noire und Massiv von Mouthoumet als Teile des sudwesteuropaischen Variszikums: Abh. Gesellsch. Wissenschaften Göttingen, Mathematische-Physische Klasse, v. 3, 260 p.

GINSBURG, R. N., AND JAMES, N. P., 1976, Submarine botryoidal aragonite in Holocene reef limestones, Belize: Geology, v. 4, p. 431–436.

———, SCHROEDER, J. H., AND SHINN, E. A., 1971, Recent synsedimentary cementation in subtidal Bermuda reefs, *in* Bricker, O. P., Ed., Carbonate cements: Johns Hopkins Univ., Studies in Geology, No. 19, p. 54–59.

GOLDRING, R., 1962, The bathyal lull: Upper Devonian and Lower Carboniferous sedimentation in the Variscan Geosyncline, *in* Coe, K., Ed., Some aspects of the Variscan fold belt: Manchester Univ. Press, p. 75–91.

GOTTHARDT, R., MEYER, O., AND PAPROTH, E., 1978, Gibt es Massenkalke im tiefen Untergrund NW-Deutsch-lands, und können sie Kohlenwasserstoffe führen?: Stuttgart, Neues Jahrb. Geol. Paläont., Mh., v. (1978), p. 13–24.

GRAULICH, J. M., 1977, Le sondage de Soumagne: Brussels, Serv. Géol. de Belgique, Professional Paper No. 139, 55 p.

GWODZ, W., 1972, Stratigraphie, Fazies, und Paläogeographie des Oberdevons und Unterkarbons im Bereich des Attendorn-Elsper Riffkomplexes (Sauerland, Rheinisches Schiefergebirge): Hanover, Geol. Jahrb., ser. A, v. 2, 71 p.

HAVARD, C., AND OLDERSHAW, A., 1976, Early diagenesis in back-reef sedimentary cycles, Snipe Lake Reef complex, Alberta: Bull. Canadian Petroleum Geology, v. 24, p. 27–69.

HECKEL, P. H., 1972, Possible inorganic origin for stromatactis in calcilutite mounds in the Tully Limestone, Devonian of New York: Jour. Sed. Petrology, v. 42, p. 7–18.

———, 1974, Carbonate buildups in the geologic record: a review, *in* Laporte, L., Ed., Reefs in time and space: Soc. Econ. Paleontologists Mineralogists Spec. Pub. No. 18, p. 90–154.

HOFMANN, R., 1966, Lagerstättenkundliche Untersuchungen im Bereich der Schwerspat-Grube Eisen (südwestliche Hunsrück): Stuttgart, Neues Jahrb. Geol. Paläont., Mh., v. (1966), p. 1–35.

HOUSE, M. R., 1975, Facies and time in Devonian tropical areas: Yorks. Geol. Soc., Proc., v. 40, p. 233–288.

———, AND BUTCHER, N. E., 1973, Excavations in the Upper Devonian and Carboniferous rocks near Chudleigh, South Devon: Penzance (U. K.). Royal Geol. Soc. Cornwall, Trans., v. 20, p. 199–220.

———, AND SELWOOD, E. B., 1966, Paleozoic paleontology in Devon and Cornwall, *in* Hosking, K. F. G., and Shrimpton, G. J., Eds., Present views on some aspects of the geology of Cornwall and Devon: Penzance (U. K.), Royal Geol. Soc. Cornwall, p. 45–86.

IRWIN, M. L., 1965, General theory of epeiric clear water sedimentation: Am. Assoc. Petroleum Geologists Bull., v. 49, p. 445–459.

JOHNSON, G. A. L., 1973, Closing of the Carboniferous sea in western Europe, *in* Tarling, D. H., and Runcorn, S. K., Eds., Implications of continental drift to the Earth Sciences: London, Academic Press, p. 845–850.

JOHNSON, J. C., 1971, A quantitative approach to faunal province analysis: Am. Jour. Science, v. 270, p. 257–280.

JOSEPH, J., 1975, Récifs Mesodévoniens des Pyréneés Ossaloises: données paleontologiques comparées: Réunion Annu. Sci. Terre, 11ième, Montpellier Univ. Sci. Tech., Languedoc, Abs., p. 208.

JUX, U., 1960, Die devonischen Riffe im Rheinischen Schiefergebirge: Stuttgart, Neues Jahrb. Geol. Paläont. Abh., v. 110, p. 186–258.

KANIS, J., 1956, Geology of the eastern zone of the Sierra del Brezo (Palencia, Spain): Leiden, Leidse Geol. Meded., v. 21, p. 377–446.

KARRENBERG, H., 1965, Das Alter der Massenkalke im Bergischen Land und ihre fazielle Vertretung: Krefeld, Fortschr. Geol. Rheinl. Westfalen, v. 9, p. 695–722.

KASIG, W., AND NEUMANN-MAHLKAU, P., 1969, Die Entwicklung des Eifeliums in Old-Red-Fazies zur Riff-Fazies im Givetium und unteren Frasnium am Nordrand des Hohen Venns (Belgien-Deutschland): Aachen, Geologische Mitteilungen, v. 8, p. 327–388.

KAZMIERCZAK, J., 1971, Morphologenesis and systematics of the Devonian Stromatoporoidea from the Holy Cross Mountains, Poland: Warsaw, Palaeontologica Polonica, v. 26, p. 1–144.

———, AND GOLDRING, R., 1978, Subtidal flat-pebble conglomerate from the Upper Devonian of Poland: a multiprovenant high energy product: Geol. Mag., v. 115, p. 359–366.

KENDALL, A. C., AND TUCKER, M. E., 1973, Radiaxial-fibrous calcite: a replacement after acicular carbonate: Sedimentology, v. 20, p. 365–389.

KERSHAW, S., AND RIDING, R., 1978, Parameterization of stromatoporoid shape: Lethaia, v. 11, p. 233–242.

KIMPE, W. F. M., BLESS, M. J. M., BOUCKAERT, J., *et al.,* 1978, Paleozoic deposits east of the Brabant Massif in Belgium and the Netherlands: Brussels, Meded. Rijks Geol. Dienst, v. 30, p. 37–103.

KLAPPER, G., ZIEGLER, W., AND MASHKOVA, T. V., 1978, Conodonts and correlations of Lower-Middle Devonian boundary beds in the Barrandian of Czechoslovakia: Marburg, Geologica et Palaeontologica, v. 12; p. 103–116.

KOBLUK, D. R., 1978, Reef stromatoporoid morphologies as dynamic populations: application of field data to a model and the reconstruction of an Upper Devonian reef: Bull. Canadian Petroleum Geology, v. 26, p. 218–236.

KODSI, M. G., 1971, Korallen aus dem Unterdevon der Karnischen Alpen: Vienna, Verhandl. Geol. Bundesanstalt, v. (1971), p. 576–607.

KREBS, W., 1960, Stratigraphie, Vulkanismus und Fazies des Oberdevons zwischen Donsbach und Hirzenheim (Rheinisches Schiefergebirge, Dill Mulde): Wiesbaden, Abh. hess. Landesamt f. Bodenforschung, v. 33, p. 1–119.

———, 1962, Das Oberdevon der Prümer Mulde, Eifel: Wiesbaden, Notizbl. hess. Landesamt f. Bodenforschung, v. 90, p. 211–232.

———, 1963, Oberdevonische Conodonten im Unterkarbon des Rheinischen Schiefergebirges und des Harzes: Hanover, Zeitschr. Deutsche Geol. Gesellsch., v. 114, p. 57–84.

———, 1966, Der Bau des Oberdevonischen Langenaubach-Breitscheider Riffes und seine weitere Entwicklung im Unterkarbon (Rheinisches Schiefergebirge): Frankfurt-am-Main, Abh. Senckenberg. Naturforsch. Gesellsch., v. 511, p. 1–105.

———, 1967, Reef development in the Devonian of the eastern Rhenish Slate Mountains, Germany, *in* Oswald,

D. H., Ed.: Alberta Soc. Petroleum Geologists, International Symp. Devonian System, Calgary, v. 2, p. 295–306.

——, 1968, Facies types in Devonian back-reef limestones in the eastern Rhenish Schiefergebirge, *in* Muller, G., and Friedman, G. M., Eds., Recent developments in carbonate sedimentology in central Europe: Berlin, Springer-Verlag, p. 186–195.

——, 1969a, Über Schwarzschiefer und bituminose Kalke im mitteleuropaischen Variszikum: Hamburg, Erdöl und Kohle, v. 22, p. 2–6; 62–67.

——, 1969b, Early void filling cementation in Devonian fore-reef limestones (Germany): Sedimentology, v. 12, p. 279–299.

——, 1970, Nachweis von Oberdevon in der Schwerspat-Grube Eisen (Saargebiet) und die Folgerungen für die Paläogeographie und Lagerstättenkunde des linksrheinischen Schiefergebirges: Stuttgart, Neues Jahrb. Geol. Paläont. Mh., v. (1970), p. 465–480.

——, 1971, Devonian reef limestones in the eastern Rhenish Schiefergebirge, *in* Müller, G., Ed., Sedimentology of parts of central Europe: International Sedimentological Congress, 8th, Heidelberg, Guidebook, p. 45–81.

——, 1972, Facies and development of the Meggen Reef (Devonian, West Germany): Stuttgart, Geol. Rundschau, v. 61, p. 647–671.

——, 1974, Devonian carbonate complexes of central Europe, *in* Laporte, L. F., Ed., Reefs in time and space: Soc. Econ. Paleontologists Mineralogists Spec. Pub. No. 18, p. 155–208.

——, 1975, Geologische Aspekte der Tiefenexploration im Paläozoikum Norddeutschlands und der südlichen Nordsee: Hamburg, Erdöl-Erdgas Zeitschrift, v. 91, p. 277–284.

——, 1976a, Widerholte Magmenaufstieg und die Entwicklung variszischer und postvariszischer Strukturen in Mitteleuropa: Halle (DDR), Nova Acta Leopoldina, N. S., No. 224, v. 45, p. 23–36.

——, 1976b, The tectonic evolution of variscan *Meso-Europa, in* Ager, D. V., and Brooks, M., Eds., Europe from crust to core: London, Wiley, p. 119–139.

——, 1978, Aspekte einer potentiellen Kohlenwasserstoff-Führung in den devonischen Riffen Nordwestdeutschlands: Hamburg, Erdöl-Erdgas Zeitschrift, v. 94, p. 15–25.

——, AND MOUNTJOY, E. W., 1972, Comparison of central European and western Canadian reef complexes: International geological congress, 24th, Montreal, Proc., Sect. 6, p. 294–309.

——, AND WACHENDORF, H., 1973, Proterozoic-Paleozoic geosynclinal and orogenic evolution of central Europe: Geol. Soc. America Bull., v. 84, p. 2611–2629.

——, AND ——, 1974, Faltungskerne im mitteleuropäischen Grundgebirge–Abbilder eines orogenen Diapirismus: Stuttgart, Neues Jahrb. Geol. Paläont. Abh., v. 147, p. 30–60.

KULLMANN, J., 1967, Associations of rugose corals and cephalopods in the Devonian of the Cantabrian Mountains (northern Spain), *in* Oswald, D. H., Ed.: Alberta Soc. Petroleum Geologists, International Symp. Devonain System, Calgary, v. 2, p. 771–776.

KUMPERA, O., 1971, Das Paläozoikum des mährisch-schlesischen Gebietes des Böhmischen Masse: Hanover, Zeitschr. Deutsche Geol. Gesellsch., v. 122, p. 173–184.

LAURENT, R., 1972, The Hercynides of South Europe—a model: International geological congress, 24th, Montreal, Proc., Sect. 3, p. 363–370.

LECOMPTE, M., 1937, Contribution à la connaissance des "récifs" du Dévonien de l'Ardenne: sur la présence de structures conservées dans des efflorescences cristalline du type "stromatactis": Brussels, Bull. Mus. Royal Hist. Nat. Belgique, v. 13, 14p.

——, 1954, Quelques donneés relatives à la genèse et aux caractères des "récifs" du Frasnien de l'Ardenne, *in* Naturelles de Belgique, Inst. royal Sci. nat. Belgique (translated 1959 by P. F. Moore, Internat. Geol. Review, v. 1, p. 1–23).: Brussels, v. Jubilaire 1 (V. van Straelen), p. 153–194.

——, 1958, Les récifs Paléozoiques en Belgique: Stuttgart, Geol. Rundschau, v. 147, p. 384–401.

——, 1967, Le Dévonien de la Belgique et le nord de la France, *in* Oswald, D. H., Ed.: Alberta Soc. Petroleum Geologists, International Symp. Devonian System. Calgary, v. 2, p. 15–50.

——, 1970, Die Riffe im Devon der Ardennen und ihre Bildungsbedingungen: Marburg, Geologica et Palaeontologica, v. 4, p. 25–71.

LEDITZKY, P., 1974, Die stratigraphische Gliederung des Gebietes zwischen Zollnerhöhe und Zollnersee in den Karnischen Alpen (Österreich): Klagenfurt, Carinthia II, v. 163, p. 169–177.

LINK, T. A., 1950, Theory of transgressive and regressive reef (bioherm) development and origin of oil: Am. Assoc. Petroleum Geologists Bull., v. 34, p. 263–294.

LLOPIS-LLADO, N., 1960, Estudio geológico de las Sierras de la Coruxera Mostayal y Monsacro: Oviedo, Breviora Geol. Asturica, v. 4, p. 3–132.

——, VILLALTA, J. F. DE, CABANAS, R., PELAEZ-PRUNEDA, J. R., AND VILAS, L., 1967, Le Dévonien de l'Espagne, *in* Oswald, D. H., Ed.: Alberta Soc. Petroleum Geologists, International Symp. Devonian System, Calgary, v. 1, p. 171–187.

LOTZE, F., 1928, Das Mitteldevon des Wennetales nördlich der Elsper Mulde: Berlin, Abh. Preuss. Geol. Landesanstalt, v. 104, p. 1–104.

LOWENSTAM, H. A., 1950, Niagran reefs of the Great Lakes area: Jour. Geology, v. 58, p. 430–487.

LUTZENS, H., 1978, Zur tektogenetischen Entwicklung und geotektonischen Gliederung des Harzvariszikums unter besonderer Berücksichtigung der Olistostrom- und Gleitdeckenbildung: Leipzig, Hallesches Jahrbuch für Geowissenschaften, v. 3, p. 81–94.

MACNEIL, F. S., 1954, Organic reefs and banks and associated detrital sediments: Am. Jour. Sci., v. 252, p. 384–401.

MAILLEUX, E., 1922, Traverseé centrale de la Belgique par la valleé de la Meuse et ses affluents de la rive gauche. 1 partie: le Dévonien du bord méridional du synclinal de Dinant: International geological congress, 13th, Brussels, Excursion guidebook A2, 32 p.

———, 1925, Étude du Dévonien du bord sud du bassin de Dinant. Le Dévonien des environs de Couvin: Bull. Soc. Géol. Miner. Bretagne, v. 6, p. 128–163.

———, AND DEMANET, F., 1928, L'échelle stratigraphiques des terraines primaires de la Belgique: Bull. Soc. Géol. Belgique, v. 38, p. 124–131.

MATTHEWS, S. C., 1977, The Variscan foldbelt in southwest England: Stuttgart, Neues Jahrb. Geol. Paläont. Abh., v. 154, p. 94–127.

MCGILLAVRAY, J. G., AND MOUNTJOY, E. W., 1975, Facies and related reservoir characteristics, Golden Spike Reef complex, Alberta: Bull. Canadian Petroleum Geology, v. 23, p. 753–809.

MEISCHNER, D., 1964, Allodapische Kalke, Turbidite in Riff-nahen Sedimentations-Becken, in Bouma, A. H., and Brouwer, A., Eds., Turbidites: Developments in sedimentology: Amsterdam, Elsevier, v. 3, p. 156–191.

MENDEZ-BEDIA, I., 1976, Biofacies y litofacies de la formacion Monjello-St. Lucia (Devonico de la Cordillera Cantabria, N.W. de España): Oviedo, Trab. Geol., No. 9, 93 p.

MEY, P. H. W., 1967, Evolution of the Pyreneean basins during the Late Paleozoic, in Oswald, D. H., Ed.: Alberta Soc. Petroleum Geologists, International Symp. Devonian System, Calgary, v. 2, p. 1157–1166.

MIDDLETON, G. V., 1960, Spilitic rocks in southeast Devonshire: Geol. Mag., v. 97, p. 192–207.

MIROUSE, R., 1967, Le Dévonien des Pyrénées occidentales et centrales (France), in Oswald, D. H., Ed.: Alberta Soc. Petroleum Geologists, International Symp. Devonian System, Calgary, v. 1, p. 153–170.

MOHANTI, M., 1972, The Portilla Formation (Middle Devonian) of the Alba Syncline, Cantabrian Mountains, Prov. Leon, N.W. Spain. Carbonate facies and Rhynconellid Paleontology: Leiden, Leidse Geol. Meded., v. 48, p. 135–205.

MOUNTJOY, E. W., 1975, Intertidal and supratidal deposits within isolated Devonian buildups, Alberta, in Ginsburg, R. N., Ed., Tidal deposits: New York, Springer-Verlag, p. 387–395.

———, COOK, H. E., PRAY, L. C., AND MCDANIEL, P. N., 1972, Allochthonous carbonate debris flows—worldwide indicators of reef complexes, banks or shelf-margins: International geological congress, 24th, Montreal, Section 6, p. 172–189.

———, AND JULL, R. K., 1978, Fore-reef carbonate mud bioherms and associated reef-margin, Upper Devonian, Ancient Wall Reef complex, Alberta: Canadian Jour. Earth Sci., v. 15, p. 1304–1325.

———, AND WALLS, R. A., 1977, Some examples of early submarine cements from Devonian buildups of Alberta: International coral reef symposium, 3rd, Proc., Rosenstiel School Marine Atmospheric Sci. and Univ. Miami, Florida, v. 2 (Geology), p. 155–161.

NELSON, H. F., BROWN, C. W., AND BRINEMAN, J. H., 1962, Skeletal limestone classification: Am. Assoc. Petroleum Geologists Memoir 1, p. 224–252.

NEUMANN, A. C., KOFOED, J. W., AND KELLER, G. H., 1977, Lithoherms in the Straits of Florida: Geology, v. 5, p. 4–10.

NEUMANN, M., POZARYSKA, K., AND VACHARD, D., 1975, Remarques sur les microfacies du Dévonien de Lublin (Pologne): Paris, Rev. Micropal., v. 18, p. 38–52.

NICOLAS, A., 1972, Was the Hercynian orogenic belt of Europe of the Andean type?: Nature, v. 236, p. 221–223.

NOWAK, H. J. AND KESHAV, N. C., 1977, Zur Erdgas- und Erdöl-Exploration in der Bundesrepublik: Hamburg, Erdöl-Erdgas Zeitschrift.

OCHS, G. AND WOLFART, R., 1961, Geologie der Blankenheimer Mulde (Devon, Eifel): Frankfurt-am-Main, Abh. Senckenberg. Naturforsch. Gesellsch., v. 501, p. 1–100.

ORCHARD, M. J., 1974, Famennian conodonts and cavity infills in the Plymouth Limestone (South Devon): Redruth (U. K.), Proc. Ussher Soc., v. 3, p. 49–54.

———, 1978, The conodont biostratigraphy of the Devonian Plymouth Limestone, South Devon: Paleontology, v. 21, p. 907–995.

OSMOLSKA, H., 1974, Tournaisian trilobites from Dalnia in the Holy Cross Mountains: Warsaw, Acta Geologica Polonica, v. 23, p. 60–79.

OVTRACHT, A., 1967, Le Dévonien du domaine nord-Pyrénéen oriental, in Oswald, D. H., Ed.: Alberta Soc. Petroleum Geologists, International Symp. Devonian System, Calgary, v. 2, p. 27–35.

PAECKELMANN, W., 1922, Der mitteldevonische Massenkalk des Bergischen Landes: Berlin, Abh. Preuss. Geol. Landesanstalt, v. 91, p. 1–112.

———, 1936, Erläuterung zur geol. Karte von Preussen, no. 341: Berlin, Blatt Brilon (no. 4617).

PAJCHLOVA, M., 1968, Zur Stratigraphie und faziellen Entwicklung des Devons in Polen, in Prager Arbeitstagung über die Stratigraphie des Silurs und des Devons: Prague, p. 393–405.

———, 1970, The Devonian, in Sokolowski, S., Ed., Geology of Poland, v. 6 Stratigraphy: Pre-Cambrian and Palaeozoic: Warsaw. Geol. Inst. Poland, p. 321–370.

———, AND STASINSKA, A., 1965, Formations récifales du Dévonien des Montes de Sainte-Croix (Pologne): Warsaw, Acta Palaeontologica Polonica, v. 9, p. 249–260.

———, AND ———, 1967, Formation récifales du Dévonien de la Pologne, in Oswald, D. H., Ed.: Alberta Soc. Petroleum Geologists, International Devonian System, Calgary, v. 2, p. 325–330.

Paproth, E., 1976, Zur Folge und Entwicklung der Tröge und Vortiefen im Gebiet des Rheinischen Schiefergebirges und seiner Vorländer vom Gedinne (Unterdevon) bis zum Namur (Silesium): Halle (DDR), Nova Acta Leopoldina, N. S., No. 224, v. 45, p. 45–58.

Paulus, B., 1961, Das Urfttal-Profil in der Sötenicher Eifelkalkmulde (Devon), in Lieber, W., (Ed.), Mineralogische und geologische Streifzug durch die nördliche Eifel: Heidelberg, Jahrestagung Ver. Freunde Mineral. Geol., Der Aufschluss, Spec. Publ. 10, p. 26–41.

Pennington, J. J., 1975, The geology of the Argyl Field, in Woodland, G., Ed., Petroleum and the continental shelf of northwest Europe: Barking (U. K.), Applied Science Publishers, Geology, v. 1, p. 285–291.

Pirlet, H., 1967, Mouvements épeirogiques Dévono-Caronifère dans la région de Visé; la carrière de "La Folie" à Bombaye: Brussels, Ann. Soc. Géol. Belgique, v. 90, p. B103–B117.

Pitman, W. C., 1978, Relationship between eustacy and stratigraphic sequences of passive margins: Geol. Soc. America Bull., v. 89, p. 1389–1403.

Playford, P. E., 1969, Devonian carbonate complexes of Alberta and Western Australia: a comparative study: Perth, W. A., Geol. Survey W. Australia, Rept. 1, 43 p.,

———, 1976, Devonian reef complexes of the Canning Basin, Western Australia: International Geological Congress, 25th, Canberra, Excursion guide no. 38A, 39 p.

———, AND Lowry, D. C., 1966, Devonian reef complexes of the Canning Basin, Western Australia: Perth, W. A., Geol. Survey W. Australia, Bull. 118, 150 p.

Poncet, J., 1972, Les biohermes Éodévonien de l'horizon récifale de Baubigny (Manche). Étude d'un paléomilieu: Paris, Bull. B. R. G. M. France, ser. 2, sect. 4, No. 3, p. 43–65.

———, 1975, Sédimentation recifale Éodévonien dans la Cotentin, in Larsonneur, L., and Dore, F., Eds., International sedimentological congress, 9th, Nice 1975, Excursion guidebook 1: Normandy, Bay of St. Michel and Armorican Massif, p. 98–104.

———, 1976, Faciès carbonatés d'arrière-recif dans l'Éodévonien du Nord-Est du Massif Armoricain (Cotentin): Paris, Bull. B. R. G. M. France, ser. 2, sect. 1, No. 1, p. 49–68.

———, 1977, Le biostrome Éodévonien à Tétracoralliares coloniaux de la Roquelle (Manche): étude d'un paléomilieu: Paris, B. R. G. M. France, No. 89, p. 116–124.

Ramsbottom, W. H. C., 1973, Transgressions and regressions in the Dinantian: a new synthesis of British Dinantian stratigraphy: Yorks. Geol. Soc., Proc., v. 39, p. 567–607.

Read, J. F., 1973, Paleoenvironments and paleogeography in the Pillara Formation (Devonian), Western Australia: Bull. Canadian Petroleum Geology, v. 21, p. 344–395.

Reichstein, M., 1959, Die fazielle Sonderentwicklung im Elbingerode Raum des Harzes: Berlin, Geologie, v. 8, p. 13–46.

Reijers, T. J. A., 1972, Facies and diagenesis of the Devonian Portilla Limestone Formation between the river Esla and the Embalse de la Luna, Cantabrian Mountains, Spain: Leiden, Leidse Geol. Meded., v. 47, p. 163–249.

———, 1973a, Stratigraphy, sedimentology and paleogeography of Eifelian, Givetian and Frasnian strata between the river Porma and the Embalse de la Luna, Cantabrian Mountains, Spain: Culemborg (Netherlands), Geol. en Mijnbouw, v. 52, p. 115–124.

———, 1973b, Hinge movements influencing deposition during the Upper Devonian in the Esla area of the Cantabrian Mountains, Spain: Culemborg (Netherlands), Geol. en Mijnbouw, v. 53, p. 13–21.

———, 1974, Diagenesis in the reefal facies of the Middle to Upper Devonian Portilla Limestone Formation of N. W. Spain: Oviedo, Breviora Geol. Asturica, v. 18, p. 33–48.

Requardt, H., 1972, Zur Stratigraphie und Fazies des Unter- und Mitteldevons in den spanischen Westpyrenaeen: Clausthaler Geol. Abh., No. 13, 76 p.

Richter, D. K., 1974, Entstehung und Diagenese der devonischen und permotriassichen Dolomite in der Eifel: Stuttgart, Contributions to Sedimentology, v. 2, 101 p.

Richter, M. D., 1965, Stratigraphy, igneous rocks and structural developments of the Torquay area: Trans. Devonsh. Assoc. Advancement Sci., v. 97, p. 57–70.

Riding, R., 1974, Model of the Hercynian foldbelt: Earth Planet. Sci. Letters, v. 24, p. 125–135.

Rietschel, S., 1966, Die Geologie des mittleren Lahnstroges: Frankfurt-am-Main, Abh. Senckenberg. Naturforsch. Gesellsch., v. 509, p. 1–58.

Ruhrmann, G., 1971, Riff-nahe Sedimentation paläozoischer Krinoiden-Fragmente: Stuttgart, Neues Jahrb. Geol. Paläont. Abh., v. 138, p. 56–100.

Rupke, J., 1965, The Elsa Nappe, Cantabrian Mountains, Spain: Leiden, Leidse Geol. Meded., v. 32, p. 1–74.

Schäfer, W., 1975, Eine oberdevonische-unterkarbonische Schichtlücke im Bereich des Balver Riffgebietes (Rheinisches Schiefergebirge): Stuttgart, Neues Jahrb. Geol. Paläont. Mh., v. (1975), p. 228–241.

Schmidt, H., 1935, Die bionomische Einteilung der Meeresböden: Geologie und Paläontologie Fortschr., v. 12, p. 1–154.

Schmidt, V., 1971, Early carbonate cementation in Middle Devonian bioherms, Rainbow Lake, Alberta, in Bricker, O. P., Ed., Carbonate cements: Johns Hopkins Univ., Studies in Geology, No. 19, p. 209–216.

Schneider, W., 1972, Zur Genese einiger Rotpelite in den devonischen Massenkalken des ostrheinischen Schiefergebirges: Stuttgart, Neues Jahrb. Geol. Palönt. Mh., v. (1972), p. 415–426.

———, 1973a, Einige Beobachtungen zur Diagenese in den devonischen Karbonatkomplexes des ostrheinischen Schiefergebirges unter besonderer Berücksichtigung der Quartzbildung: Stuttgart, Neues Jahrb. Geol. Paläont. Mh., v. (1973), p. 231–257.

————, 1973b, Ein weiterer Beitrag zur Entstehung verschiedener Generationen authigener Quartze in der Umgebung von Suttrop (Sauerland): Heidelberg, Der Aufschluss, v. 24, p. 33–38.

————, 1977, Diagenese devonischer Karbonatkomplexe Mitteleuropas: Hanover, Geol. Jahrb., ser. D, v. 21, 107 p.

SCHÖNLAUB, H. P., 1971, Stratigraphische und lithologische Untersuchungen im Devon und Unterkarbon der Karawanken (Jugoslawischer Anteil): Stuttgart, Neues Jahrb. Geol. Paläont. Abh., v. 138, p. 157–168.

————, 1979, Das Paläozoikum in Österreich: Vienna, Abh. Geol. Bundesanstalt, v. 33, 124 p.

SCHULTZE, R., 1968, Die Conodonten aus dem Paläozoikum der mitteleren Karawanken (Seeberggebiet): Stuttgart, Neues Jahrb. Geol. Paläont. Abh., v. 130, p. 133–245.

SCHWARZ, A., 1927, Wachstum, Absterben und Diagenese eines paläozoischen Korallenriffes: Frankfurt-am-Main, Senckenbergiana, v. 9, p. 49–64.

SCRUTTON, C., 1977a, Facies variations in the Devonian limestones of eastern South Devon: Geol. Mag., v. 114, p. 165–248.

————, 1977b, Reef facies in the Devonian of eastern South Devon, England: Paris, B. R. G. M. France, Mem. No. 89, p. 124–135.

SELLI, R., 1963, Schema geologico delle Alpi Carniche e Giulie occidentali: Bologna, Giorn. Geol., v. 30, p. 1–136.

SHAW, A. B., 1964, Time in stratigraphy: New York, McGraw-Hill, 365 p.

SITTER, L. U. DE, 1962, The structure of the southern slope of the Cantabrian Mountains (Spain): Leiden. Leidse Geol. Meded., v. 26, p. 255–264.

————, AND BOSCHMA, D., 1966, Explanation geological map of the Paleozoic of the southern Cantabrian Mountains 1:50,000, sheet 1 Pisuerga: Leiden, Leidse Geol. Meded., v. 31, p. 191–238.

————, AND ZWART, H. J., 1961, Excursion to the central Pyrenees: Leiden, Leidse Geol. Meded., v. 26, p. 1–49.

SLEUMER, B. H. G., 1969, Devonian stromatoporoids of the Cantabrian Mountains, Spain: Leiden, Leidse Geol. Meded., v. 44, p. 1–136.

SMITS, B. J., 1965, The Caldas Formation, a new Devonian unit in Leon (Spain): Leiden, Leidse Geol. Meded., v. 31, p. 179–187.

SPRIESTERSBACH, J., 1942, Lenneschiefer (Stratigraphie, Fazies und Fauna): Berlin, Abh. Preuss. Geol. Landesanstalt, v. 203, p. 1–219.

STAALDUINEN, C. J. VAN, 1973, Geology of the area between the Luna and Torio Rivers, southern Cantabrian Mountains, N.W. Spain: Leiden, Leidse Geol. Meded., v. 49, p. 167–205.

STASINSKA, A., 1973, Tabulate corals from Dalnia in the Holy Cross Mountains: Warsaw, Acta Geologica Polonica, v. 23, p. 84–111.

STEL, J. H., 1975, The influence of hurricanes on the quiet depositional conditions in the lower Emsian La Vid Shales of Colle (N.W. Spain): Leiden, Leidse Geol. Meded., v. 49, p. 475–486.

STIBANE, F., 1963, Stratigraphie und Magmatismus des Mittel- und Oberdevons bei Werdorf-Berghausen (Lahn-Mulde, Rheinisches Schiefergebirge): Wiesbaden, Notizbl. Hess. Landesamt f. Bodenforsch., v. 91, p. 119–142.

SVOBODA, J., (Ed.)., 1966, Regional geology of Czechoslovakia, v. 1, The Bohemian Massif, Prague, 688 p.

————, AND PRANTL, F., 1949, The stratigraphy and tectonics of the Devonian area of Koneprusy (Bohemia) (Czechoslovakian with an English summary): Prague, Sbor. Ústu. úst. Geol., v. 16, p. 5–92.

SZULCZEWSKI, M., 1968, Slump structures and turbidites in the Upper Devonian limestones of the Holy Cross Mountains: Warsaw, Acta Geologica Polonica, v. 18, p. 303–324.

————, 1971, Upper Devonian conodonts, stratigraphy and facial development in the Holy Cross Mountains: Warsaw, Acta Geologica Polonica, v. 21, p. 1–129.

————, 1973, Famennian-Tournaisian neptunian dykes and their conodont faunas from Dalnia in the Holy Cross Mountains: Warsaw, Acta Geologica Polonica, v. 23, p. 15–59.

TASSEL, R. VAN, 1969, Anhydrite, célestine et barytine du Givetian au sondage du Tournai: Brussels, Bull. Soc. Belge Géol. Paléont. Hydrol., v. 69, p. 351–361.

TAYLOR, P. W., 1951, The Plymouth Limestone: Penzance (U.K.), Trans. Royal Geol. Soc. Cornwall, v. 18, p. 146–214.

TESSENSOHN, F., 1971, Der Flysch-Trog und seine Randbereiche im Karbon der Karawanken: Stuttgart, Neues Jahrb. Geol. Paläont. Abh., v. 138, p. 169–220.

————, 1974, Zur Fazies paläozoischer Kalke in den Karawanken (Karawankenkalke II): Vienna, Verhandl. Geol. Bundesanstalt, v. (1974), p. 89–130.

————, 1975, Schichtlücken und Mischfaunen in Paläozoischen Kalken der Karawanken: Klagenfurt, Carinthia II, v. 84, p. 137–160.

TSIEN, H. H., 1967, Distribution of rugose corals in the Middle and Upper Devonian (Frasnian) reef complexes of Belgium, *in* Oswald, D. H., Ed.: Alberta Soc. Petroleum Geologists, International Symp. Devonian System, Calgary, v. 2, p. 273–293.

————, 1971, The Middle and Upper Devonian reef complexes of Belgium: Petroleum Geology of Taiwan, v. 8, p. 119–173.

————, 1975, Introduction to Devonian reef development in Belgium: International symposium on fossil corals and reefs, 2nd, Paris 1975, Guidebook, Excursion C, North of France and Belgium, Serv. Geol. Belg., p. 3–43.

————, 1976, L'activité récifale au cours du Dévonien moyen et du Frasnien en Europe occidentale et ses particularités en Belgique: Lille, Ann. Soc. Geol. Nord, v. 97, p. 57–66.

————, 1977, Morphology and development of Devonian reefs and reef complexes in Belgium: Inernational Coral Reef Symp., 3rd, Rosenstiel School of Marine and Atmospheric Sci. and Univ. Miami, Florida, Proc., v. 2 (Geology), p. 191–200.

————, AND DRICOT, E., 1977, Devonian calcareous algae from the Dinant and Namur Basins, Belgium, in Flügel, E., Ed., Fossil Algae: Berlin, Springer-Verlag, p. 344–350.

TUCKER, M. E., 1969, Crinoidal turbidites from the Devonian of Cornwall and their paleogeographic significance: Sedimentology, v. 13, p. 281–290.

————, 1973, Sedimentology and diagenesis of Devonian pelagic limestones (Cephalopodenkalke) and associated sediments of the Rhenohercynian Geosyncline, West Germany: Stuttgart, Neues Jahrb. Geol. Paläont. Abh., v. 142, p. 320–350.

————, 1974, Sedimentology of Paleozoic pelagic limestones: the Devonian griotte (southern France) and Cephalopodenkalk (Germany), in Hsü, K. J., and Jenkyns, H. C., Eds., Pelagic sediments: on land and under the sea: International Assoc. Sedimentologists Spec. Pub. No. 1, p. 71–92.

————, AND STRAATEN, P. VAN, 1970, Conodonts and facies on the Chudleigh Schwelle: Proc. Ussher Soc., v. 2, p. 160–170, Redruth (U.K.).

VAI, G. B.,1967, Le Dévonien inférieur biohermal des Alpes Carniques centrales: Paris, B. R. G. M. France, Mém. No. 33, p. 285–300.

————, 1971, Diskussionsbeitrag zu den Vortragen über das "Variszikum der Ostalpen": Hanover, Zeitschr. Deutsche Geol. Gesellsch., v. 122, p. 169–172.

————, 1975, Hercynian basin evolution of the Southern Alps, in Squyres, C. H., Ed., Geology of Italy: Tripoli, Earth Sci. Soc. Libyan Arab Republic, v.1, p. 293–298.

WALLACE, P., 1969, The sedimentology and paleoecology of the Devonian of the Ferques Inlier, northern France: Quart. Jour. Geol. Soc. London, v. 125, p. 83–124.

WALLISER, O. H., 1977, Probleme der geotektonischen Einordnung der Variskiden: Berlin, Zeitschrift fur angewandte Geologie, v. 23, p. 459–463.

WEBER, K., 1978, Das Bewegungsbild im Rhenoherzynikum-Abbild einer subvaristischen Subfluenz: Hanover, Zeitschr. Deutsche Geol. Gesellsch., v. 124, p. 249–281.

WEBBY, B. D., 1966a, Middle-Upper Devonian paleogeography of North Devon and West Somerset, England: Paleogeography, Palaeoclimatology, Palaeoecology, v. 2, p. 27–46.

————, 1966b, The Middle Devonian marine transgression in North Devon and West Somerset: Geol. Mag., v. 102, p. 478–488.

WEDEKIND, R., 1924, Das Mitteldevon der Eifel. Eine biostratigraphische Studie, 1 Die Tetracorallen des unteren Mitteldevons: Marburg, Schr. bes. Beförd. Naturwiss. Marburg, v. 14, no. 3, p. 1–93.

WEGGEN, K., 1978, Tiefenaufschluss in der Bundesrepublik Deutschland: Hamburg, Erdöl-Erdgas Zeitschrift, v. 94, p. 8–15.

WESINK, H., 1962, Paleozoic of the Upper Gallego and Ara Valleys, Huesca Province, Spanish Pyrenees: Madrid, Estudios de Geología, v. 18, p. 1–74.

WILSON, J. L., 1975, Carbonate facies in geologic history: New York, Springer-Verlag, 471 p.

WINDLEY, B. F., 1977, The evolving continents: England, New York, Wiley, 385 p.

WUNDERLICH, H. G., 1964, Mass, Ablauf und Ursachen orogener Einengung am Beispiel des Rheinischen Schiefergebirges, Rhurkarbons und Harzes: Stuttgart, Geol. Rundschau, v. 54, p. 861–882.

ZIEGLER, W., 1959, Conodonten aus Devon und Karbon Sudwesteuropas (Montagne Noire, Massif von Mouthoumet, Spanische Pyrenäen) und Bermerkungen zur bretonischen Faltung: Stuttgart, Neues Jahrb. Geol. Paläont. Mh., v. (1959), p. 289–309.

ZUKALOVA, V., 1961, On the problem of the boundary between the Middle and Upper Devonian limestones of the Moravian Karst (Czechoslovakian with English summary): Prague, Vestník Ústr. úst. Geol., v. 36, p. 461–463.

SEPM Special Publication No. 30, p. 143–160, May 1981

# LOWER PERMIAN *TUBIPHYTES/ARCHAEOLITHOPORELLA* BUILDUPS
# IN THE SOUTHERN ALPS (AUSTRIA AND ITALY)

ERIK FLÜGEL
Institut für Paläontologie
Universität Erlangen-Nürnberg,
Erlangen, West Germany

## ABSTRACT

Carbonate buildups representing "stratigraphic reefs" were formed during the Lower Permian (Sakmarian) and lower Artinskian at the margins of carbonate platforms in southern Austria and northern Italy. These buildups were formed by the interaction of encrusting algae (*Tubiphytes* and *Archaeolithoporella*), bryozoans, and syndepositional eogenetic radial-fibrous cements.

## INTRODUCTION

Two types of reefs formed in the Southern Alps during the Lower Permian: 1. phylloid algal buildups within the Lower and the Upper *Pseudoschwagerina* Formations of the Rattendorf Stage (see Table 1), and 2. *Tubiphytes/Archaeolithoporella* buildups within the Trogkofel Formation. The Trogkofel buildups represent reefs similar in their ecologic and diagenetic development to the Middle to Late Permian Capitan Reef complex of West Texas and New Mexico.

The objective of this paper is to document biota, microfacies, and diagenetic features of the *Tubiphytes/Archaeolithoporella* reefs in the Sexten Dolomites and the Carnic Alps (Fig. 1).

The interpretations of paleoenvironment and paleogeography of the limestones of the Trogkofel Stage are based on detailed field studies, including investigation of paleontologic and petrographic data in thin-section, geochemical studies (insoluble residues and rare elements), and SEM studies of diagenetic features.

## REGIONAL SETTING

The Sexten Dolomites and the Carnic Alps are located in the Southern Alps, one of the major structural and paleogeographic units of the Alpine chain (Fig. 1).

In the Sexten Dolomites, *Tubiphytes/Archaeolithoporella* limestones of the Trogkofel Stage are only represented as clasts of the "Tarvis Breccia," whereas in the Carnic Alps, autochthonous as well as allochthonous occurrences are known. Autochthonous development can be seen in the "Trogkofel Limestone" and in the somewhat younger "Goggau Limestone," whereas allochthonous occurrences are found in the Tarvis Breccia and in the "Tressdorf Limestone" of the Nassfeld area in the Carnic Alps.

Localities studied in the Sexten Dolomites are situated east of Innichen/S. Candido (Langbühl), and east and southeast of Sexten/Sesto (Villgrater; northwestern slope of Seikofel/Monte Covolo; Kreuzberg Pass/Passo Monte Croce di Comelico; and Col della Croce). At these localities, polymict breccias are exposed and overlie monomict quartzphyllite breccias and Lower Paleozoic quartzphyllites. An exception can be seen in the Seikofel area; here polymict breccia is developed on top of Permian quartzporphyry. The polymict breccia is composed of limestone pebbles, quartzphyllite and quartz pebbles as well as clasts of quartzporphyry, cherts, and sandstones. As can be seen by microfacies criteria, most of the limestone clasts have been derived from biopelsparites, intrabiosparites, and sometimes even biomicrites with *Tubiphytes/Archaeolithoporella*. In addition, other microfacies types including those indicating the existence of a carbonate platform environment (see below), and also rare occurrences of stratigraphically older limestones (Upper *Pseudoschwagerina* Limestone and even Late Carboniferous fusulinid limestones) are present. Except for the Villgrater locality, all breccias have a calcareous matrix thought to be of lacustrine origin (see Flügel, 1980).

The sections investigated in the Carnic Alps (Fig. 1) are located southeast of Forni Avoltri (locality Col Mezzodi), in the Trogkofel and Gartnerkofel area at the Austrian/Italian boundary (localities Straniger Alm; Trogkofel; Tressdorf Alm; Reppwand; Kühweg Alm; Garnitzen Gorge), and between Goggau/Coccau and Tarvis/Tarvisio near the Italian/Austrian border station. The section as seen in Forni Avoltri represents platform development of the Trogkofel Stage (see below). *Tubiphytes/Archaeolithoporella* buildups are chiefly found at the type locality of the Trogkofel Lime-

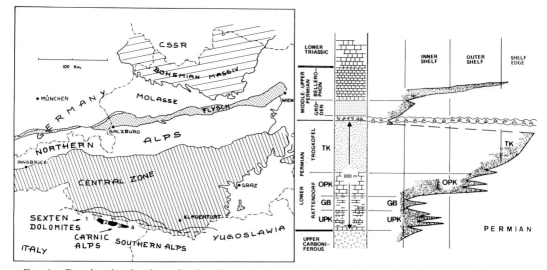

FIG. 1.—Drawing showing the regional setting and paleogeography of the Lower Permian sequence in the Southern Alps: 1. Sexten Dolomites; and 2.–4. Carnic Alps (2. Col Mezzodi near Forni Avoltri, 3. Trogkofel/Gartnerkofel area near the Nassfeld Pass, and 4. Tarvis-Goggau).

stone, in an area east of Nassfeld Pass, and below the Tarvis Breccia in Tarvis. Here the reef facies occur within the Goggau Limestone. The thickness of bedded Trogkofel Limestone is about 180 meters in Forni Avoltri. Thickness of the more massive reef limestone reaches more than 300 meters in the Trogkofel area, and more than 150 meters at Goggau-Tarvis.

The proportions of *Tubiphytes/Archaeolithoporella* limestone in clasts of the Tarvis Breccia seems to be similar in the Trogkofel of the Reppwand and Tarvis areas, but there are striking differences between limestone microfacies below the breccias, and microfacies of the limestone pebbles. The Tarvis Breccia at the top of Trogkofel Mountain consists predominantly of reworked *Tubiphytes/Archaeolithoporella* limestones, and dolomites that are exposed below the breccia. In the type locality of the Tarvis Breccia, up to seven breccia units can be differentiated on the basis of matrix type, modal composition, and type of clay minerals found within the limestone pebbles (Buggisch, 1980). Only the lower breccias reflect the usual Goggau Limestone facies (cyanophycean algal bindstones); most breccias contain up to 70 percent clasts of *Tubiphytes/Archaeolithoporella* limestones which can be rarely recognized in outcrops of the Goggau Limestone. This may indicate diachronous erosion of nearby reefs at the end of Lower Permian time.

A detailed geological and lithological description of the above localities is found in Buggisch and Flügel (1980).

## LOWER PERMIAN STRATIGRAPHY

### General Survey

The basement of the Permian rocks in the Dolomites are Lower Paleozoic quartzphyllites. In the Carnic Alps, post-Variscan sedimentation began during Kasimovian time (approximately equivalent to Late Carboniferous Missouri time), with deposition of the Auernig Formation which is characterized by rhythmic sedimentation patterns of alternating clastic and calcareous members. This 700 meters thick formation is overlain by a Lower Permian sequence of about 900 meters in thickness. This section consists of several formations (see Table 1), which can be assigned to the Asselian, Sakmarian and Artinskian subdivisions according to fusulinid faunas (Kahler, 1974; Kahler and Kahler, 1980). With respect to local stratigraphy, as developed by Heritsch *et al.* (1933), the lower portion of the Lower Permian section corresponds to the "Rattendorf Stage" and the upper part to the "Trogkofel Stage." These stratigraphic units have been used in describing the Lower Permian sedimentary series of the Southern Alps and of Yugoslavia; however, it should be noted that the Trogkofel Stage is not completely documented by autochthonous rock units (Kahler and Kahler, 1980). Fossil groups other than fusulinids, e.g., corals, brachiopods, or calcareous algae, cannot be used as index fossils because of their broad stratigraphic range, their endemic character, and because of the absence of revised taxonomic descriptions of the faunas.

TABLE 1.—STRATIGRAPHY OF THE PERMIAN IN THE SOUTHERN ALPS; UNCERTAIN CORRELATIONS INDICATED BY DASHED LINES. THE FUSULINIDS *SCHWAGERINA GLOMEROSA* AND *SCHWAGERINA MOELLERI* ARE IDENTIFIED AND DEFINED AS IN CURRENT RUSSIAN LITERATURE.

| | USSR | MIDDLE AND EAST ASIA | NORTH AMERICA | SOUTHERN ALPS | | |
|---|---|---|---|---|---|---|
| | UNITS | FUSULINID ZONES | UNITS | FORMATIONS | STAGES | |
| UPPER PERMIAN | TATARIAN | PALAEO-FUSULINA | OCHOAN | BELLEROPHON FM. | BELLERO-PHON | |
| | | CODONO-FUSIELLA | | | | |
| | | LEPIDOLINA + YABEINA | | | | |
| MIDDLE PERMIAN | KASANIAN | NEO-SCHWAGERINA | CAPITANIAN | GRODEN FM. | GRODEN | |
| | UFIMSKIAN | | WORDIAN | | | |
| | KUNGURIAN | CANCELLINA | | | | |
| LOWER PERMIAN | CISJANSKIAN | MISELLINA | LEONARDIAN | TARVIS BRECCIA | TROGKOFEL | |
| | ARTINSKIAN | PSEUDO-FUSULINA VULGARIS | | GOGGAU FM. | | TUBIPHYTES/ ARCHAEO-LITHOPORELLA BUILDUPS |
| | | | | TRESSDORF FM. | | |
| | SAKMARIAN | PSEUDO-SCHWAGERINA SCHELLWIENI | WOLF-CAMPIAN | TROGKOFEL FM. | | |
| | ASSELIAN | SCHWAGERINA GLOMEROSA | | UPPER PSEUDOSCHWAGERINA FM. | RATTEN-DORF | |
| | | SCHWAGERINA MOELLERI | | GRENZLAND FM. | | |
| | | OCCIDENTO-SCHWAGERINA ALPINA | | LOWER PSEUDOSCHWAGERINA FM. | | |

### Lower Pseudoschwagerina Formation

This formation consists of four depositional cycles consisting of clastics and overlying limestones (Homann, 1969), reflecting cyclical deposition in a near coastal inner-shelf environment during alternating regressive and transgressive phases. The lower portion of the cycles is characterized by parallel and cross-bedded sandstones and siltstones as well as quartz conglomerates; these clastic rocks are irregularly interbedded with a few limestone beds. The upper cyclic portion consists of thin- to medium-bedded limestones (predominantly biomicrites with calcareous algae, foraminifers, gastropods, and echinoderms), and thick-bedded to massive carbonates which represent algal buildups formed by various phylloid algae (Flügel, 1977a). The low diversity invertebrate communities, and primarily monomict algal associations, indicate relatively restricted shallow marine subtidal biotopes. The total thickness of this formation is about 160 meters.

### Grenzland Formation

During deposition of the uppermost parts of the Lower *Pseudoschwagerina* Limestone Formation, the source area of terrigenous clastics changed. The erosion of metamorphic as well as acid volcanic rocks, and increasing sedimentation of clastics within a near-shore high energy environment with intertidal to subtidal conditions, appear to be responsible for deposition of the sandstones and silty shales of so-called Grenzland Formation ("Grenzland" means borderland; the type section is located at the Austrian/Italian border). An additional source area for the upper part of this formation is suggested by heavy mineral analysis (Tietz, 1975). Calcareous algae and fusulinids are found in the clastic rocks, and in some of the limestone intercalations (biomicrites with large oncoids, echinoderms, and various groups of algae). These biotas point to the existence of marine but restricted depositional conditions, further substantiated by the occurrence of the trace fossils *Zoophycos, Chondrites,* and *Laevicyclus.* The total thickness of this formation is 125 meters.

### Upper Pseudoschwagerina Formation

This formation is characterized by fine- to medium-bedded limestones with very few clastic intercalations. A medium to high diversity biota consisting of calcareous algae (predominantly dasycladaceans, phylloid algae, and oncoids formed by blue-green or by red algae), smaller foraminifers, fusulinids, corals, bryozoa, brachiopods, gastropods, pelecypods, and echinoderms characterizes this formation. A regular recapitulation of microfacies sequences in various outcrop sections is taken as evidence of a repeated shifting of ecologic zones, from very shallow water areas to off-shore environments in an open-marine shelf lagoon with normal water circulation. The amount of insoluble residue derived from the limestones shows large scaled cyclic distributional patterns of terrigenous influx within the carbonate sequence, indicating sedimentation on a bathymetrically differentiated subtidal outer-shelf platform (Flügel *et al.,* 1971; Flügel, 1971). The total thickness of the formation is 175 meters.

Geochemical data indicate that the percents of strontium carbonate and manganese are similar for the limestones of both the Lower *Pseudoschwagerina* Limestone Formation and the Grenzland Formation, but different for the Upper *Pseudoschwagerina* Limestones, reflecting near coast or off-shore sedimentation as well as rather uniform diagenetic processes within the Upper *Pseudoschwagerina* Limestones.

### Trogkofel Formation

This formation contains massive, often dolomitic limestones, as well as well-bedded limestones which are similar to the Upper *Pseudoschwagerina* Limestones in microfacies types and in lithology. These bedded limestones indicate a continuation of the outer-shelf platform facies during lower Sakmarian and lower Artinskian time (Flügel, 1980). The massive limestones and dolomites exposed at the type locality in the Carnic Alps developed from a coarse-bedded sequence at the base which differs from the Upper *Pseudoschwagerina* Limestones in microfacies (intrabiomicrites). Most of the massive white, or variously colored limestones correspond to *Tubiphytes/Archaeolithoporella* biosparites or biomicrites which can be rich in bryozoans and echi-

noderms. This boundstone facies occurs in the Carnic Alps and the Sexten Dolomites and seems to have been formed at various times upon the carbonate platform and at the shelf-edge (see below).

### Goggau Formation

This unit has been defined by Kahler (1971, p. 15), in order to differentiate bedded or massive white limestones similar in facies to the Trogkofel Limestone, but differing in age (upper Artinskian). The microfacies of these limestones is characterized by an abundance of filamentous cyanophycean algae (*Ortonella* and *Girvanella*), aggregate grains, and a unique foraminiferal association consisting of sessile forms like *Tuberitina* and some miliolids. The matrix is micrite in limestones with *Girvanella subparallela,* or orthosparite in limestones with *Ortonella goggauensis.* Based on the low biotic diversity, the occurrence of abundant algal lumps and ostracodes, together with miliolid foraminifers, the paleoenvironment of the bedded Goggau Limestone seems to have been within protected areas of a shallow shelf-lagoon. As can be shown by fusulinids, *Tubiphytes/Archaeolithoporella* buildups appear to have formed simultaneously within the bedded Goggau Limestones in the Carnic Alps (Forni Avoltri), and in the Sexten Dolomites. In contrast to the bedded Goggau Limestone, this massive reefoid limestone is known only from limestone pebbles within the Tarvis Breccia.

### Tressdorf Formation

Limestones of upper Artinskian age are represented in units of the Tressdorf Limestone which correspond to a polymict stylobreccia occurring as very small erosional remnants northwest of the Nassfeld Pass in the Carnic Alps. The breccia consists of angular and rounded limestone clasts which exhibit various microfacies types. Along with biosparites and biopelmicrites with *Tubiphytes,* and *Archaeolithoporella,* there are bryozoan clasts of densely packed biomicrites with spores, moderately-sorted oosparites, biomicrites with dasycladaceans, quartz-bearing micrites, and compound breccias consisting of biomicrites with ostracodes and sponge spicules. Postulating time equivalency for all the limestone clasts within this broad spectrum of microfacies types, suggests the existence of platform carbonates and carbonate buildups during Artinskian time.

### Tarvis Breccia

Overlying the Trogkofel and Goggau Limestones (Carnic Alps), Lower Paleozoic quartz-phyllites representing Lower Permian quartz-porphyry (Sexten Dolomites), or the Clastic Trogkofel Formation (Karawanken Mountains at the Austrian/Yugoslavian border), is a polymict breccia with a thickness of from 10 meters to 200 meters. This breccia consists of limestones, quartzphyllites, quartz pebbles, sandstones, and some cherts, but limestone clasts predominate. Most breccias have a carbonate matrix (partly a lacustrine micrite and intramicrite), but a siliceous matrix can be found in some locations in the upper part of the breccia sequence. The limestone pebbles have been derived from the substratum, or from nearby exposed sections (Buggisch and Flügel, 1980; Flügel, 1980). Microfacies types of the platform or reef facies of the Trogkofel and Goggau Limestones are represented in the clasts, together with very few well-rounded limestone pebbles containing biotas of Late Carboniferous and Asselian age. The time of formation of the Tarvis Breccias probably corresponds with the *Misellina* (Kahler and Kahler, 1980); these breccias may have formed in connection with orogenic movements at the Lower/Middle Permian boundary ("Saalian phase").

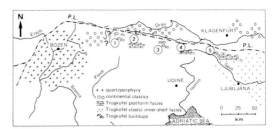

FIG. 2.—Paleogeography of the Trogkofel Stage during late Lower Permian time in the Southern Alps. 1. Sexten Dolomites, 2. Forni Avoltri, 3. Trogkofel, 4. Tavois, 5. Karawanken Mountains. P.L. = Periadriatic Lineament.

### Clastic Trogkofel Formation
### (Kosna Formation)

This formation was recognized by Ramovš (1963) in Slovenia (Ortnek, Carniola; Southern Karawanken Mountains), but may also be developed in the Carnic Alps. The sequence is characterized by very thick rock units (500 to 2000 meters), predominantly clastic rocks (quartz sandstones, quartz conglomerates, and shales), and by intercalations of thin limestone units (bedded limestones and limestone breccias), showing strongly diverging facies types (coral reef limestones, brachiopod limestones, fusulinid limestones, and crinoid limestones). In order to account for the generally low faunal diversity, it is believed that the synchronous existence of soft and hard bottom substrates and the types of lithology and depositional environments must have corresponded to a near coast inner-shelf depositional platform. The age of the clastic Trogkofel Formation is Sakmarian and Artinskian, but the limestone breccias yield pebbles from Asselian and Upper Carboniferous beds (Kochansky-Devidé et al., 1973).

The Lower Permian formations, especially the Tarvis Breccia, are unconformably overlain by Middle Permian red bed facies of the Gröden Formation (Buggisch, 1978).

### TROGKOFEL REEF FACIES *VERSUS* TROGKOFEL
### PLATFORM FACIES

The beginnings of Trogkofel Stage carbonate sedimentation can be seen in sections exposed in the Trogkofel/Gartnerkofel area, and near Forni Avoltri in the Carnic Alps. In the area west and east of Nassfeld, the Trogkofel Reef facies is developed on outer-shelf platform carbonates of the Upper *Pseudoschwagerina* Limestones. Concurrently, an open marine carbonate platform was formed in the area of Forni Avoltri; later, the depositional environment changed to one within a protected inner-shelf lagoon. A similar environment may be assumed for parts of the Goggau Formation with alternating regressive and transgressive phases.

Assuming that most occurrences of the Tarvis Breccia can be regarded as parautochthonous, the Trogkofel Reef facies is found at the following localities: Langbühl, Villgrater, Seikofel, Kreuzberg Pass, Col della Croce, Forni Avoltri (only as a thin intercalation within platform carbonates, and as clasts of Tarvis Breccia), Trogkofel, Reppwand/Gartnerkofel, Kühweg Alm, Garnitzen Gorge, Goggau-Tarvis, and also in the Tarvis Breccia of the Karawanken Mountains in Slovenia (Kosutnik locality, with Tarvis Breccia). With respect to the paleogeographic setting (Buggisch et al., 1976), a shelf-edge position of most Trogkofel *Tubiphytes/Archaeolithoporella* buildups can be postulated (Fig. 2).

Trogkofel platform carbonate microfacies were studied in detail at the Forni Avoltri section (Flügel, 1980). Here approximately 2 percent of the total samples (145 large thin-sections) are mudstones (lacustrine micrites and intramicrites), 36 percent are skeletal wackestones (biomicrites with abundant *Atractyliopsis* and gastropods; biomicrites with abundant *Epimastopora*; biomicrites with abundant sessile foraminifers and calcisponge spicules; biomicrites with abundant phylloid algae forming small mud-mounds; and biomicrites with ostracodes and sponge spicules), 3 percent are packstones (densely-packed echinoderm biomicrites and fusulinid biomicrites), 46 percent are packstones/grainstones (biosparites with *Atractyliopsis* and abundant algal spores; highly diverse biosparites with abundant *Epimastopora*; biosparites with *Epimastopora* and gastropods; biosparites with well-rounded extraclasts

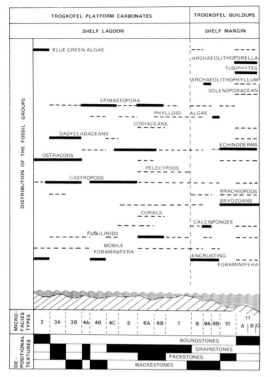

FIG. 3.—Diagram showing distribution of habitats of the common fossil groups and dominant depositional fabrics of the Trogkofel platform carbonates and Trogkofel buildups of the Southern Alps.

and intraclasts, together with oncoids; poorly sorted, highly diverse biosparites with fusulinids, echinoderms, and many palaeotextulariid foraminifers; poorly sorted bio-oncosparites with large oncoids formed by *Girvanella*; ooid-bearing biosparites or oosparites; and biointrasparites with many dasycladacean algae and aggregate grains), 3 percent are grainstones/boundstones (biopelsparites with *Tubiphytes/Archaeolithoporella*), and 10 percent are boundstones (characterized by abundant cyanophycean algae, or by micritic limestones with *Archaeolithophyllum lamellosum* and calcisponges). The cyanophycean algal bindstones seem to have been formed within a protected inner shelf-lagoon (see "Goggau Limestone"); in contrast, the bindstones and framestones with *Archaeolithophyllum lamellosum*, calcisponges, and many epizoans as well as bryozoans, seem to represent small organic mudmounds formed at the margins of carbonate platforms.

As can be seen from Figure 3, the various facies types of Trogkofel platform carbonates and Trogkofel Reefs, are distinguished by various fossil

distributional patterns as well as by dominant limestone fabrics.

## TROGKOFEL BUILDUPS

Because of a generally strong late diagenetic dolomitization of the massive Trogkofel Limestones in the Trogkofel area, it is difficult to quantitatively describe the shape and the size of the *Tubiphytes/Archaeolithoporella* buildups. The different thicknesses of massive Trogkofel Limestones at the Trogkofel area vary between 170 and 330 meters within a distance of less than 1 kilometer. At the Reppwand/Gartnerkofel area they are lens-shaped and seem to be composed of densely packed small mound-like organic buildups. A striking feature of the Trogkofel Stage limestones is the difference in rock colors (grey, dark grey, and red), which has been explained as due to different amounts of terra rossa (Ramovš, 1963), but which may also be due to an interaction of pressure-solution, dolomitization, and an enrichment of clay minerals with iron hydroxides (Flügel and Agiorgitis, 1970).

### Biota

The fauna of the Trogkofel Limestones is taxonomically diverse (Gortani, 1906; Heritsch, 1938; Homann, 1970; Kahler and Kahler, 1941; and Schellwien, 1900), but most of the fusulinids, corals, brachiopods, gastropods, pelecypods, echinoderms, as well as a few trilobites and ammonoids, have been found in the bedded Trogkofel Limestones representing the platform facies of the Trogkofel Stage. More than 80 species have been identified from the bedded shelf facies adjacent to Trogkofel buildups. This is in strong contrast to the relatively low diversity biota of the massive Trogkofel Reef Limestones which consists of only about 25 species of smaller foraminifers, fusulinids, calcisponges, bryozoans, brachiopods, gastropods, pelecypods, echinoderms, and ostracodes. A quantitative survey based on 60 large thin-sections of *Tubiphytes/Archaeolithoporella* boundstones, shows bryozoans and echinoderms (mostly crinoids) as the most abundant invertebrates in Trogkofel buildups (Table 2).

These differences in taxonomic diversity can also be seen in the calcareous algal flora of the Trogkofel buildups and the Trogkofel platform carbonates (Flügel and Flügel-Kahler, 1980). However distinctions between both facies regions follow a more complicated pattern than for the invertebrate fauna. The total algal flora of the Trogkofel and Goggau Limestones consists of 46 species. Within the *Tubiphytes/Archaeolithoporella* microfacies, 16 algal species have been recognized. This diversity is similar to that of microfacies types occurring in protected and in open

TABLE 2.—MAJOR TAXA IN THE *TUBIPHYTES/ARCHAEOLITHOPORELLA* BUILDUPS OF THE TROGKOFEL STAGE IN THE SOUTHERN ALPS (BASED ON 60 LARGE THIN-SECTIONS)

| Taxa | Major Groupings | Frequency (in the total sample, %) | Trophic Group |
|---|---|---|---|
| *Tubiphytes* | Encrusting alga | 100 | Autotroph |
| Bryozoans | Encrusting fenestrate cryptostomes, encrusting or upright cyclostomes | 65 | High-filterer |
| Echinoderms | Crinoids | 60 | High-filterer |
| *Archaeolithoporella* | Encrusting alga | 50 | Autotroph |
| *Tuberitina* | Encrusting calcareous foraminifer | 22 | High-filterer |
| Fusulinids | Mobile calcareous foraminifers | 20 | High-filterer |
| Ostracodes | Mobile arthropods | 20 | Collector |
| Gastropods | Pleurotomarids, grazing forms | 13.3 | Collector |
| *Connexia* | Dasycladacean alga, perhaps allochthonous | 13.3 | Autotroph |
| Hollow open ended tubules | Encrusting blue-green alga? | 11.6 | Autotroph? |
| Palaeotextulariids | Mobile calcareous foraminifers | 11.6 | High-filterer |
| Calcisponges | Segmented Sphinctozoa and non-segmented Wewokellida | 6.6 | Low/high filterer |
| *Tetrataxis* | Encrusting calcareous foraminifer | 3.3 | High-filterer |
| Nodosariids | Mobile calcareous foraminifers | 3.3 | High-filterer |
| Pelecypods | Only fragments | 3.3 | Filterer |
| Small burrow deposit feeder | Unknown | 3.3 | Swallower |
| *Archaeolithophyllum* | Encrusting red alga | 3.3 | Autotroph |
| *Eugonophyllum* | Codiacean green-alga | 1.6 | Autotroph |
| *Neoanchicodium* | Codiacean green-alga | 1.6 | Autotroph |

shelf-lagoonal areas (poorly-sorted grainstones with dasycladaceans). Other microfacies types that characterize the shelf-lagoon environment show a low algal diversity (grainstones with *Epimastopora* and gastropods, with fusulinids, or with ooids), or a moderate algal diversity (wackestones with *Epimastopora*, with *Atractyliopsis*, and bindstones with *Girvanella goggauensis*).

The algal flora of the Trogkofel buildups consists of codiacean and dasycladacean green algae, problematical dasyclads (*Epimastopora*), red algae (*Archaeolithophyllum delicatium* Johnson, *Archaeolithophyllum lamellosum* Wray, *Cuneiphycus aliquantulus* Johnson, *Solenopora* cf. *centurionis* Pia), problematical red algae (*Archaeolithoporella hidensis* Endo), blue-green algae (*Ortonella*), and problematical blue-green algae (*Tubiphytes*). This flora is dominated by encrusting forms such as *Tubiphytes* and *Archaeolithoporella*; dasyclad remains (*Connexia carniapulchra*), as well as fragments of phylloid algae (*Eugonophyllum, Neoanchicodium*) which are rare to very rare and may be interpreted as allochthonous.

Most of the associations identified in thin-sections of the *Tubiphytes/Archaeolithoporella* limestones consist of *Tubiphytes,* bryozoans, echinoderms, *Archaeolithoporella, Tuberitina,* and ostracodes. Classifying these organisms into feeding groups, in order of their abundance (Table 2), indicates the existence of communities dominated

by algae and suspension feeders. Deposit feeders, herbivores, and swallowers are rare. The dominance of suspension feeding sessile invertebrates suggests a firm substrate, a water column rich in nutrients, and rather low turbidity (Babcock, 1977; Walker, 1972).

Based on the trophic guidelines formulated by Turpaeva (1957), general agreement is seen in Trogkofel buildup communities with respect to dominance by one major trophic group (autotrophs), various trophic behaviour by the dominant groups (autotrophs and differently structured high-filterers), the dominance of one species within each trophic group of the community (*Tubiphytes obscurus* or fenestrate bryozoans), and the resultant probability of minimal feeding competition. The *Tubiphytes/Archaeolithoporella*-bryozoan community is different in its trophic structure as compared with the phylloid algal community (Toomey, 1976), or with the archaeolithophyllid alga-sponge community (Toomey, 1979), and this is probably due to the effect of rapid synsedimentary inorganic cementation (see below).

*Tubiphytes* and *Archaeolithoporella*—. *Tubiphytes* Maslov is represented by two species, *Tubiphytes obscurus* Maslov, 1956, and *Tubiphytes carinthiacus* (Flügel, 1966). Both species are encrusting organisms growing upon micritic sediment, carbonate cements, *Archaeolithophyllum,* bryozoans, or upon each other. *Tubiphytes*

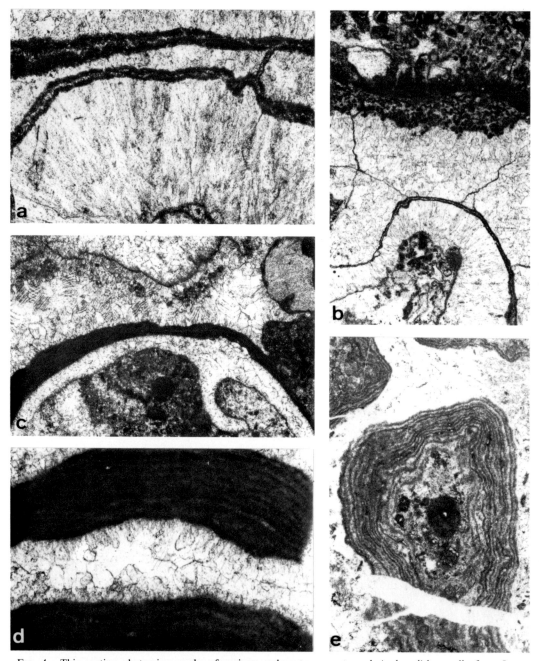

FIG. 4.—Thin-section photomicrographs of various carbonate cements and *Archaeolithoporella* from Lower Permian Trogkofel buildups of the Southern Alps. A. Radial-fibrous calcite crystals orientated normal to thin *Archaeolithoporella* crusts; Tarvis, ×25. B. Partly recrystallized radial-fibrous calcite crystals which formed the substrate of a pelsparitic sediment; Tarvis, ×10. C. *Archaeolithoporella* crust growing on recrystallized fragments of *Archaeolithophyllum*; Seikofel, Sexten Dolomites, ×15. D. *Archaeolithoporella hidensis* Endo, separated by sparry calcite with relicts of fibrous crystals; Seikofel, Sexten Dolomites, ×50. E. *Archaeolithoporella* oncoid circumscribing *Tubiphytes obscurus* Maslov; Tresdorf Limestone, Carnic Alps, ×5.

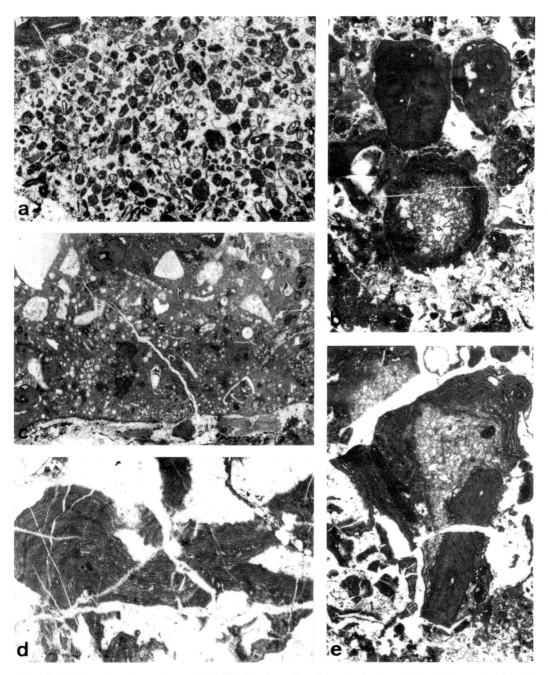

FIG. 5.—Thin-section photomicrographs (×10) of various algae from the Southern Alps. A. Intrapelsparite with small fragments of *Tubiphytes*; Seikofel, Sexten Dolomites. B. *Tubiphytes obscurus* Maslov overgrown by foraminifers (top) and *Tubiphytes carinthiacus* (Flügel) and circumscribed by *Archaeolithoporella hidensis* Endo (center); Kühweg Alm, Carnic Alps. C. Reddish biomicrite with echinoderms, nodosariid foraminifers and *Tubiphytes*; Seikofel. D. *Archaeolithoporella* crusts; Trogkofel, Carnic Alps. E. Association of *Tubiphytes carinthiacus* (Flügel), *Tubiphytes obscurus* Maslov, and *Archaeolithoporella hidensis* Endo; Microfacies B, Kühweg Alm, Carnic Alps.

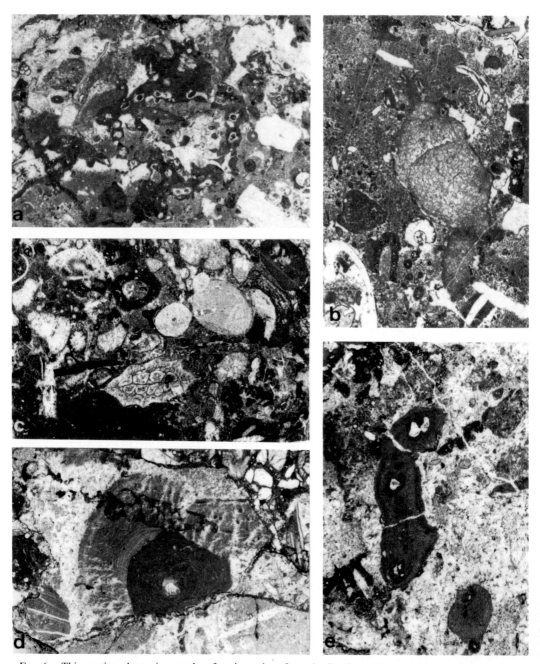

FIG. 6.—Thin-section photomicrographs of various algae from the Southern Alps. A. Micritic algal crusts intergrown with bryozoans; Kühweg Alm, Carnic Alps, Microfacies C, ×10. B. Biopelmicrite with *Tubiphytes carinthiacus*; Kühweg Alm, Carnic Alps, Microfacies C, ×8. C. Biomicrite with bryozoans, echinoderms and *Tubiphytes*; Seikofel, Sexten Dolomites, ×15. D. *Archaeolithoporella* oncoid, partly recrystallized; Sexten Dolomites, ×10. E. *Tubiphytes obscurus* Maslov together with pelmicritic intraclasts; Microfacies A, Sexten Dolomites, ×15.

*obscurus* is cylindrical to ovoid in shape and from 0.3 to 2.5 millimeters in diameter, and several millimeters in length. It consists of very dark to black-colored micrite, which according to SEM studies (Fig. 8) is composed of anhedral calcite crystals 0.8 to about 3 microns in size. Emphasized by a slight difference in the color, these concentrically arranged layers may be seen in thin-section; within these layers are sometimes thin (maximum 5 microns) black concentric "lines." These lines can not be distinguished unequivocally in SEM photographs, but in some micrographs there is an indication of a faint concentric orientation of the calcite crystals. The micritic mass envelopes a subcylindrical hollow cavity which generally is filled with variously sized (10 to 80 microns) anhedral sparite crystals (Fig. 8D). In some samples the periphery of this spar-filled "tube" is bordered by a thin rim consisting of microspar-sized elongated crystals (Fig. 8F) which may reflect internal isopachous rim cement within the "tubes." The diameters of the "tubes" vary between 50 and 200 microns, and some are bifurcated.

*Tubiphytes carinthiacus* (Figs. 5B, 5E, 6B) is characterized by a relatively open meshwork within the micritic zone around the central "tube" ($\phi$ up to 250 microns). *T. carinthiacus* is generally larger than *T. obscurus* (several millimeters in diameter and up to 10 millimeters in length). The diagonal diameter of the mesh structure is 50 to 100 microns. Both species may be found isolated, or together with *T. obscurus* overgrowing *T. carinthiacus*. This relationship and the differentiation of the micritic zone, as well as the differences in the dimensions support the view that *T. carinthiacus* is a separate species rather than a preservational stage.

The biological affinities of *Tubiphytes* are controversial (calcareous alga or hydrozoan). Maslov (1956) argued for interpretation as a blue-green alga because of the similarity between the thin dark threads and very thin trichomes of blue-green algae. The tubes are interpreted as "vanished stems, traces of which remain in the form of a tube filled with secondary calcite," and overgrown by the algae. Babcock (1977, p. 18) has discounted the blue-green algal hypothesis because the organism is complexly organized and has a well defined structural development, and because the organism does not possess filaments, cells, or other diagnostic blue-green algal features. These arguments are not valid if we consider the interwoven threads seen in *Tubiphytes carinthiacus* as remains of trichomes. In addition, the structural organization is thought to correspond to growth layering, which is a common feature of spongiostromate blue-green algal thalli. Another feature common to calcareous blue-green algae is the ultrastructure of the micrite zone around the tubes (see Fig. 9C, 9D) which is similar to that of filamentous porostromate blue-green algae (Flügel, 1977a). It should be noted that a micritic calcification around and between the threads has recently been described from Recent cyanophycean algae, and also from Recent chlorophycean algae similar to the Chaetophoraceae (Flajs, 1977). The spar-filled "tubes" may belong

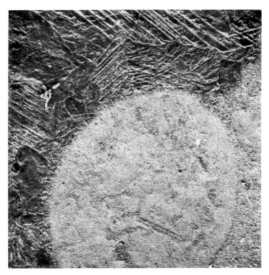

Fig. 7.—SEM micrograph ($\times 60$) of *Tubiphytes obscurus* Maslov showing a thin rim of fibrous calcite crystals and recrystallized blocky carbonate cement; Kühweg Alm, Carnic Alps.

to another alga, which could have been circumscribed by *Tubiphytes*. Associations of epiphytes consisting of filamentous blue-green algae and filamentous green algae are known from modern environments (Fott, 1971, p. 498).

*Tubiphytes* Maslov is known from the Lower Carboniferous to the Upper Jurassic. It is a very important encrusting organism in Late Carboniferous (Pennsylvanian) and Permian organic buildups (Toomey, 1969; Babcock, 1977; Malek-Aslani, 1970), in Middle Triassic sponge reefs of the Alps (Ott, 1967; Kraus and Ott, 1968; Brandner and Resch, this volume), and perhaps in some Upper Jurassic algal-sponge buildups of Germany (Flügel and Steiger, this volume).

Another *microproblematicum*, most probably an alga, is *Archaeolithoporella hidensis* Endo, 1959 (see Fig. 9A, 9B). The name refers to encrustations consisting of alternating dark and light colored calcite. The thickness of lamina ranges from 20 to 30 microns. The dark laminae consist of micritic crystals less than 3 microns; whereas the light laminae are composed of granular, sometimes faintly fibrous crystals, up to 10 microns in size. Generally, not more than 5 to 10 dark laminae can be seen within crusts, which may reach a thickness between 0.2 and 1.2 millimeters. Often the crusts are composed only of 2 to 4 laminae, growing upon *Tubiphytes*. Other substrates are *Archaeolithophyllum*, fenestrate bryozoans, or fibrous calcite crystals. *Archaeolithoporella* also may form oncoids which generally develop around micritic or pelmicritic lithoclasts; these suggest rapid synsedimentary lithification. There are also relatively rare occurrences of crusts up

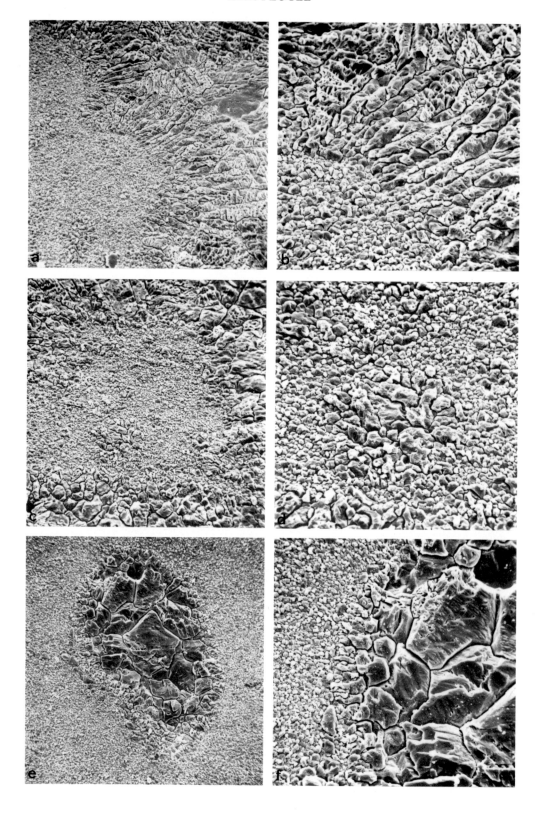

to 20 millimeters thick, with dome-shaped laminae resembling stromatolites (Fig. 5D). Another clue to suggest rapid lithification of *Archaeolithoporella* is the abundance of broken, angular fragments which occur together with regular encrustations.

Babcock (1977, p. 16) has presented conclusive arguments for an algal (probably red alga) interpretation of *Archaeolithoporella*. However, still open to question is whether the light laminae represent the remains of algae (Mazzullo and Cys, 1977, p. 168), or are couplets composed of light and dark algal laminae (compare Babcock, 1977, p. 17). Since the base of the *Archaeolithoporella* crusts is always represented by dark laminae it seems reasonable to assume that both laminae belong to the alga.

### Microfacies

Thin-section studies reveal the existence of three microfacies types which are defined according to depositional and diagenetic fabrics:

*Type A (Fig. 4A–E).*—Biopelsparite with varying amounts of *Tubiphytes* and *Archaeolithoporella*, and irregularly distributed, poorly-sorted peloids of various sizes. Up to 60 percent of the thin-section area consists of sparite, which in places has well developed radial-fibrous fabric (see below).

*Type B (Fig. 5E).*—Intrabiosparite with fragments of *Archaeolithoporella* and angular pelsparitic intraclasts. The intraclasts and fragments of *Archaeolithoporella* are circumscribed by isopachous calcite rims composed of short fibrous crystals. Interparticle voids are filled with neomorphic sparite. Irregularly defined peloids (probably algal peloids) are common. Open spaces may be filled with pelmicritic sediment, which often shows reverse graded bedding.

*Type C (Figs. 5C, 6A–C).*—Reddish biomicrites with frequent bryozoans, *Tubiphytes*, and rare *Archaeolithoporella*. Bryozoans are generally found as encrusting forms as well as fragments.

Biopelsparites and biomicrites of types A and C microfacies are similar in diversity, which is lower than in type B. Type A and type C microfacies are characterized by *Tubiphytes obscurus*, *T. carinthiacus*, varying amounts of *Archaeolithoporella*, thin-shelled ostracodes, and bryozoans. In addition to these organisms, a few calcisponges and gastropods, common echinoderms, and a few dasycladacean fragments and encrusting foraminifers (*Tuberitina*), are found in microfacies type B.

Biopelsparites and intrabiosparites are widely distributed in the Trogkofel Limestones of the Carnic Alps (Trogkofel, Kühweg Alm, Garnitzen Gorge); both types form the dominant limestone pebbles of the Tarvis Breccia in the Sexten Dolomites and at Tarvis-Goggau. *Tubiphytes*-bryozoan biomicrites are known from the Kühweg Alm and Trogkofel (Carnic Alps) and Seikofel (Sexten Dolomites).

### Carbonate Cements

Limestones of microfacies type A exhibit relicts of original carbonate cement fabrics which indicate submarine, syndepositional lithification of these *Tubiphytes/Archaeolithoporella* carbonates (Figs. 4A–B, 8A–B).

The cements consist of radial-fibrous calcite crystals with an average length and width of 700 microns and 60 microns, respectively. The c-axial directions are normal to *Archaeolithoporella* and *Tubiphytes* substrates. SEM micrographs (Fig. 4A) show that fibrous crystals seen in thin-sections are composed of smaller, elongated straight-edged crystals with lengths between 30 to 40 microns and widths between 4 and 8 microns. Some of these smaller crystals may correspond to groups of diverging fibers, which in a few samples are arranged in fan druses similar to those examples described by Mazzullo and Cys (1977, p. 157). The radial-fibrous cement of the Trogkofel boundstones strongly resembles the "ray crystal" type which has been interpreted as inverted aragonite cement. These radial-fibrous cements are generally found on *Tubiphytes* and on *Archaeolithoporella*, sometimes alternating with *Archaeolithoporella* crusts. The syndepositional character of the cements is evident because of the existence of multiple cement generations (200–400 microns thick) encrusted by *Archaeolithoporella* (Fig. 4 A–B), presence of internal sedimentation of skeletal debris and peloids between various layers of cements, and because of coatings of isopachous fringes of short fibrous crystals on broken *Archaeolithoporella* crusts.

Differences appear to have been brought about during diagenesis of the biopelsparites (type A)

←

FIG. 8.—SEM micrographs of *Tubiphytes obscurus* Maslov, Trogkofel Limestone, Carnic Alps. A. Radial-fibrous carbonate cement crystals growing upon fine-grained *Tubiphytes*, ×220. B. Detail of Fig. 8A, ×480. C. Cross-section of *Tubiphytes* surrounded by radially arranged calcite crystals, ×140. D. Detail of Fig. 8C, showing the spar-filled tube of *Tubiphytes*, ×410. E. Tube of *Tubiphytes* exhibiting a well defined boundary, ×180. F. Detail of Fig. 8E, ×450.

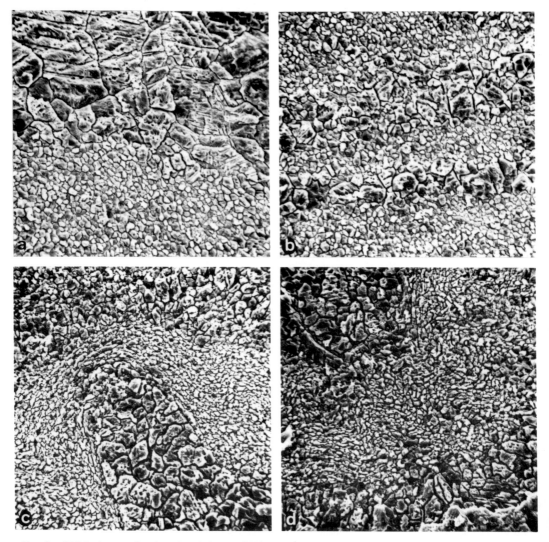

FIG. 9.—SEM micrographs of *Archaeolithoporella hidensis* Endo (A and B) and *Tubiphytes carinthiacus* (Flügel) (C and D); Trogkofel Limestones, Carnic Alps. A. Boundary between micritic *Archaeolithoporella* crust and sparite, ×465. B. Alternating micritic and sparitic layers of *Archaeolithoporella*, ×380. C. Densely laminated and crinkled "laminae" surrounding a spar-filled tube of *Tubiphytes carinthiacus*, ×330. D. Densely folded "laminae" of *Tubiphytes carinthiacus*, ×400.

and the intrabiosparites (type B). Whereas the *Tubiphytes/Archaeolithoporella* boundstones of type A may have been lithified by syndepositional aragonitic submarine cementation, the interparticle porosity of intrabiosparitic grainstones probably was reduced by several generations of cement, which possibly included formation of submarine cement A and blocky cement B, both influenced by freshwater. A freshwater environment has been proved for the carbonate matrix of the basal Tarvis Breccia overlying the Trogkofel Limestones (Flügel, 1980), therefore the paramorphic inversion of aragonite in freshwater

(Folk and Assereto, 1976) during subsurface diagenesis does indeed seem possible.

### Bindstones versus Framestones

The Trogkofel Limestones have generally been considered to be "reef limestones" by geologists working in the Carnic Alps (e.g., Geyer, 1903; Heritsch, 1936; Kahler and Prey, 1963; Selli, 1963) due to the massive outcrop character, and an incorrect interpretation of the monomict breccias intercalated within the Trogkofel Limestones of the Trogkofel (regarded as reef breccias).

The genetic interpretation of the Trogkofel

TABLE 3.—COMPARISON OF THE SOUTHERN ALPINE TROGKOFEL BUILDUPS AND THE PERMIAN REEF COMPLEX OF THE GUADALUPE MOUNTAINS, USA

| | Trogkofel Buildups | Permian Reef Complex |
|---|---|---|
| Major biotic elements | *Tubiphytes, Archaeolithoporella,* encrusting bryozoans, echinoderms | Calcareous sponges, *Tubiphytes, Archaeolithoporella,* bryozoans, echinoderms |
| Dominating growth forms | Laminar and low-encrusting | Upright and laminar |
| Trophic groups | Predominantly autotrophs and suspension feeders; deposit feeders, herbivores and swallowers very rare | Predominantly suspension feeders and autotrophs; very few deposit feeders, predators or herbivores |
| Calcareous algae | *Tubiphytes* and *Archaeolithoporella* abundant, *Archaeolithophyllum* relatively common; phylloid algae, solenoporaceans and dasyclads rare to very rare; filamentous blue-green algae very rare | *Tubiphytes, Archaeolithoporella, Colenella* and phylloid algae abundant; *Archaeolithophyllum* and solenoporacean algae common; filamentous blue-green algae rare; dasyclads absent |
| Major fabric types | Bindstones and grainstones, both strongly recrystallized; subordinate wackestones | Sparry boundstones (frame- and bindstones), extensively recrystallized |
| Carbonate cements | Relicts of syndepositional radial-fibrous fabrics; isopachous calcite rims and granular cements | Syndepositional radial-fibrous fabrics, with isochronous calcite rims |

boundstones depends on whether or not the massive Trogkofel Limestones can be regarded as framestones or bindstones (Embry and Klovan, 1972). Framestones are defined by the existence of *in situ* sessile organisms which can construct a rigid framework in conjunction with the binding effect of encrusting organisms. Bindstones are formed by tabular and lamellar organisms which are able to bind and encrust a large amount of matrix, but where a self-supporting organic fabric is lacking.

Frame building organisms are extremely rare in the Trogkofel Limestones (see Table 2), since calcisponges are rare, and bryozoans belong to encrusting growth forms. Whereas, low laminar encrusting forms are represented by *Tubiphytes*, most of the bryozoans, *Archaeolithoporella*, encrusting calcareous foraminifers (*Tuberitina*), and by *Archaeolithophyllum lamellosum, Tubiphytes* and *Archaeolithoporella* and can be regarded as binding agents (see Cys et al., 1977, p. 245) but not as framebuilders. Since these organisms are found as encrustations upon other encrusters such as bryozoans, and upon synsedimentary formed radial-fibrous cements, these can comprise up to 60 percent of the rock volume. The Trogkofel "reefs" are carbonate buildups which differ in microfacies from equivalent platform carbonates, and from overlying rocks. These buildups are typically thicker than equivalent carbonates.

Carbonate buildups may be categorized into organic or ecologic reefs (buildups consisting of

an organically constructed *in situ* wave resistant framework), stratigraphic reefs (Dunham, 1970), buildups of largely inorganically bound carbonate rock, or banks (organic buildups that originated by *in situ* sediment binding, sediment baffling, local gregarious growth of organisms, or the mechanical accumulation of transported bioclastic debris). Following these definitions and emphasizing the predominance of diagenetic/organic bindstones, the lack of framebuilders, and the absence of coarse-grained breccias, which could be compared with fore-reef sediments, the Trogkofel buildups are interpreted as stratigraphic reefs formed primarily by the combined effects of a syndepositional eogenetic submarine cementation and the binding activities of encrusting algae and bryozoans. This interpretation primarily refers to the *Tubiphytes/Archaeolithoporella* boundstones described as microfacies A, but is probably also applicable to the intrabiosparites of microfacies B, which may indicate the existence of water currents within a downslope environment. Microfacies C indicates a quiet water environment which may have been somewhat deeper or more protected as compared with the well cemented boundstones and grainstones.

### COMPARISONS

Many criteria of the Trogkofel buildups can be compared with those of the Permian Reef Complex of New Mexico and West Texas (Table 3). The biota seems to be similar, with the exception of the abundance of calcareous sponges in the Permian Reef complex (Girty, 1908; Newell et al., 1953). In addition, organism trophic structures are identical, indicating similar biotopes. Some differences can be seen in the composition of the algal flora (Babcock, 1974; Johnson, 1942), in particular, absence of *Collenella* and the rarity of phylloid algae in the Trogkofel buildups.

The major fabric types are similar with respect to predominance of diagenetic/organic boundstones, which exhibit comparable syndepositional radial-fibrous carbonate cement types (Mazzullo, 1977; Mazzullo and Cys, 1977; Schmidt, 1977; Yurewicz, 1977). As a consequence of stronger recrystallization and late diagenetic dolomitization of the Trogkofel Limestones, only relicts of various primary fabrics are preserved within the Lower Permian Trogkofel buildups. Slight differences also exist in the dimensions of "ray or fan crystals," which according to SEM studies are strikingly smaller when observed in thin-section.

## CONCLUSIONS

The Lower Permian (Sakmarian and lower Artinskian, representing upper Wolfcampian and Leonardian time) Trogkofel buildups of the Southern Alps (Carnic Alps and Sexten Dolomites) represent "stratigraphic reefs," which probably formed in a downslope shelf-edge position adjacent to shallow water carbonate platforms (bedded Trogkofel Limestones, and parts of the Goggau Limestone). The major biotic constituents of the reef limestones (predominantly biopelsparites and intrabiosparites) are the algae *Tubiphytes obscurus* Maslov, *Tubiphytes carinthiacus* (Flügel), *Archaeolithoporella hidensis* Endo, and bryozoans, which together with eogenetic submarine carbonate cements contributed to the formation of diagenetic/organic bindstones. Depositional and diagenetic fabrics of the Trogkofel buildups are similar to those of the Permian Reef complex of West Texas and New Mexico.

## ACKNOWLEDGMENTS

This study was supported by a grant from the Deutsche Forschungsgemeinschaft, project F1 41/21. Technical assistance by Christel Sporn, Gerlinde Neufert, and Helmut Keupp is gratefully acknowledged.

## REFERENCES

BABCOCK, J. A., 1974, The role of algae in the formation of the Capitan Limestone (Permian, Guadalupian), Guadalupe Mountains, West Texas and New Mexico, Ph.D. dissertation: Madison, Univ. Wisconsin, 241 p.

————, 1977, Calcareous algae, organic boundstones, and the genesis of the Upper Capitan Limestone (Permian, Guadalupian), Guadalupe Mts., West Texas and New Mexico, in Field Conf. Guidebook, Permian Basin Sect.: Soc. Econ. Paleontologists Mineralogists Pub. 77-16, p. 3–44.

BRANDNER, R., AND RESCH, W., 1981, Reef development in the Ladinian and Cordevolian of the Northern Limestone Alps near Innsbruck, Austria: this volume.

BUGGISCH, W., 1978, Die Grödener Schichten (Perm, Südalpen). Sedimentologisch und geochemische Untersuchungen zur Unterscheidung mariner und kontinentaler Sedimente: Geol. Rundschau, v. 67, p. 149–180.

————, 1980, Die Geochemie der Kalke in den Trogkofel-Schichten der Karnischen Alpen: Carinthia II, Sonderheft 36, p. 101–111.

————, AND FLÜGEL, E., 1980, Die Trogkofel-Schichten der Karnischen Alpen. Verbreitung, geologische Situation und Geländebefund: Carinthia II, Sonderheft 36, p. 13–99.

————, ————, LEITZ, F., AND TIETZ, G.-F., 1976, Die fazielle und paläogeographische Entwicklung im Perm der Karnischen Alpen und in den Randgebieten: Geol. Rundschau, v. 65, p. 649–690.

CYS, J. M., TOOMEY, D. F., BREZINA, J. L., GREENWOOD, E., GROVES, D. B., KLEMENT, K. W., KULLMANN, J. D., MCMILLAN, T. L., SCHMIDT, V., SNEED, E. D., AND WAGNER, L. H., 1977, Capitan Reef—Evolution of a Concept, in Field Conf. Guidebook, Permian Basin Sect: Soc. Econ. Paleontologists Mineralogists Pub. 77-16, p. 201–322.

DUNHAM, R. J., 1970, Stratigraphic reefs versus ecologic reefs: Am. Assoc. Petroleum Geologists Bull., v. 54, p. 1931–1932.

————, 1972, Capitan reef, New Mexico and Texas: Facts and questions to aid interpretation and group discussion: Permian Basin Sec.—Soc. Econ. Paleontologists and Mineralogists Pub. 72-14, no consecutive page numbers.

EMBRY, A. F., AND KLOVAN, E. J., 1972, Absolute water depth limits of Late Devonian paleoecological zones: Geol. Rundschau, v. 61, p. 672–686.

ENDO, R., 1959, Stratigraphical and Paleontological Studies of the Late Paleozoic Calcareous Algae in Japan. XIV. Fossil Algae from the Nyugawa Valley in the Hida Massif: Sci. Rep. Saitama Univ., ser. B, v. 3, p. 177–208.

FLAJS, G., 1977, Die Ultrastrukturen des Kalkalgenskelettes: Palaeontographica, Abteilung B, v. 160, p. 69–128.

FLÜGEL, E., 1966, Algen aus dem Perm der Karnischen Alpen: Carinthia II, Sonderheft 25, p. 3–76.

————, 1971, Palökologische Interpretation des Zottachkopf-Profiles mit Hilfe von Kleinforaminiferen (Oberer Pseudoschwagerinen-Kalk, unteres Perm: Karnische Alpen): Carinthia II, Sonderheft 28, p. 61–96.

————, 1977a, Environmental models for Upper Paleozoic benthic calcareous algal communities, in Flügel, E., Ed., Fossil Algae: New York, Springer Verlag, p. 314–343.

————, 1977b, Verkalkungsmuster porostromater Algen aus dem Malm der Südlichen Frankenalb: Geol. Blätter No-Bayern, v. 27, p. 131–140.

————, 1980, Die Mikrofazies der Kalke in den Trogkofel-Schichten der Karnischen Alpen: Carinthia II, Sonderheft 36, p. 51–99.

————, AND AGIORGITIS, G., 1970, Rotsedimentation im Trogkofel-Kalk (höheres Unter-Perm) der Karnischen Alpen: Anzeiger Math.-Naturwiss. Klasse Österr. Akad. Wiss., Jahrgang 1970, p. 173–178.

————, AND FLÜGEL-KAHLER, E., 1980, Algen aus den Kalken der Trogkofel-Schichten der Karnischen Alpen: Carinthia II, Sonderheft 36, p. 113–182.

————, HOMANN, W., AND TIETZ, G.-F., 1971, Litho- und Biofazies eines Detail-profils in den Oberen Pseudo-schwagerinen-Schichten (Unter-Perm) der Karnischen Alpen: Verhandlungen Geol. Bundesanstalt Wien, Jahrgang 1971, p. 10–42.

————, AND STEIGER, T., 1981, Upper Jurassic Algal-Sponge Buildups of the Northern Frankenalb, Germany: this volume.

FOLK, R. L., AND ASSERETO, R., 1976, Comparative fabrics of length-slow and length-fast calcite and calcitized aragonite in a Holocene speleothem, Carlsbad Caverns, New Mexico: Jour. Sed. Petrology, v. 46, p. 486–496.

FOTT, B., 1971, Algenkunde: Stuttgart, Fischer Verlag, 581 p.

GEYER, G., 1903, Exkursion in die Karnischen Alpen: Exkursionsführer IX. Internation. Geologenkongreß, Wien.

GIRTY, G. H., 1908, The Guadalupian fauna: U.S. Geol. Survey Professional Paper, v. 58, 651 p.

GORTANI, M., 1906, Contribuzioni allo studio de Paleozoico Carnico. I. La fauna permocarbonifera del Col di Mezzodi presso Forni Avoltri: Palaeontographica Italica, v. 12, p. 1–84.

HERITSCH, F., 1936, Die Karnischen Alpen. Monographie einer Gebirgsgruppe der Ostalpen mit variszischem und alpidischem Bau: Geol. Institut Univ. Graz, 205 p.

————, 1938, Die stratigraphische Stellung des Trogkofelkalkes: Neues Jahrbuch Min. Geol. Paläont., Abteilung B, v. 79, p. 63–186.

————, KAHLER, F., AND METZ, K., 1933, Die Stratigraphie von Oberkarbon und Perm in den Karnischen Alpen: Mitt. Geol. Ges. Wien, v. 26, p. 162–190.

HOMANN, W., 1969, Fazielle Gliederung der Unteren Pseudoschwagerinenkalke (Unter-Perm) der Karnischen Alpen: Neues Jahrbuch Geol. Paläont., Monatshefte, Jahrgang 1969, p. 265–280.

————, 1971, Korallen aus dem Unter- und Mittelperm der Karnischen Alpen: Carinthia II, Sonderheft 28, p. 97–143.

JOHNSON, J. H., 1942, Permian lime-secreting algae from the Guadalupe Mountains, New Mexico: Geol. Soc. America Bull., v. 53, p. 195–216.

KAHLER, F., 1971, Karbon und Perm der Ostalpen in Österreich (Kärnten), Italien und Jugoslawien: marines Unterkarbon (Visé), limnisches und marines Oberkarbon, marines Perm: 7. Internation. Kongress Strat. Geol. Karbon, Krefeld, Exkursion V, 32 p.

————, 1974, Fusuliniden aus T'ien-schan und Tibet. Mit Gedanken zur Geschichte der Fusuliniden-Meere im Perm: Rep. Sci. Exped. Northwestern Prov. China, Sino-Swedish Exped., Pub. 52, Invertebrate Paleontology, v. 4, 147 p.

————, AND KAHLER, G., 1941, Beiträge zur Kenntnis der Fusuliniden in den Ostalpen. II. Die Gattung *Pseudoschwagerina* und ihre Vertreter im unteren Schwagerinenkalk und im Trogkofelkalk: Palaeontographica, Abteilung A, v. 92, p. 59–98.

————, 1980, Die Fusuliniden-Fauna der Trogkofel-Stufe in den Karnischen Alpen: Carinthia II, Sonderheft 36, p. 183–254.

————, AND PREY, S., 1963, Erläuterungen zur Geologischen Karte des Nassfeld-Gartnerkofel-Gebietes in den Karnischen Alpen: Geol. Bundesanstalt Wien, 116 p.

KOCHANSKY-DEVIDÉ, V., BUSER, ST., CAJHEN, J., AND RAMOVŠ, A., 1973, Detailliertes Profil durch die Trog-kofelschichten am Bache Kosutnik in den Karawanken: Razprave Diss. Slov. Akad., v. 16, p. 171–185.

KRAUS, O., AND OTT, E., 1968, Eine ladinische Riff-Fauna im Dobratsch-Gipfelkalk (Kärnten, Österreich) und Bemerkungen zum Faziesvergleich von Nordalpen und Drauzug: Mitteilungen Bayer. Staatssammlung Paläont. Hist. Geol. München, v. 8, p. 263–290.

MALEK-ASLANI, M., 1970, Lower Wolfcampian Reef in Kemnitz Field, Lea County, New Mexico: Am. Assoc. Petroleum Geologists Bull., v. 54, p. 2317–2335.

MASLOV, V. P., 1956, Iskopaemye izvestkovye vodorosli SSSR: Akad. Nauk SSSR, Trudy Instituta Geol. Nauk, v. 160, 301 p.

MAZZULLO, S. J., 1977, Synsedimentary Diagenesis of Reefs, *in* Field Conf. Guidebook, Permian Basin Sec.: Soc. Econ. Paleontologists Mineralogists Pub. 77-16, p. 323–349.

————, AND CYS, J. M., 1977, Submarine cements in Permian boundstones and reef associated rocks, Guadalupe Mountains, West Texas and Southeastern New Mexico, *in* Field Conf. Guidebook, Permian Basin Sec.: Soc. Econ. Paleontologists Mineralogists Pub. 77-16, p. 151–200.

NEWELL, N. D., 1955, Depositional fabric in Permian reef limestones: Jour. Geology, v. 63, p. 301–309.

————, RIGBY, J. K., FISCHER, A. G., WHITEMAN, J. E., HICKOX, J. E., AND BRADLEY, J. S., 1953, The Permian reef complex of the Guadalupe Mountains Region, Texas and New Mexico. A study in paleoecology: San Francisco, W. H. Freeman and Co., 236 p.

OTT, E., 1967, Segmentierte Kalkschwämme (Sphinctozoa) aus der alpinen Mitteltrias und ihre Bedeutung als Riffbildner im Wettersteinkalk: Abhandlungen Bayer. Akad. Wiss., Math.-Naturwiss. Klasse, v. 131, p. 1–96.

RAMOVŠ, A., 1963, Biostratigraphie der Trogkofel-Stufe in Jugoslawien: Neues Jahrbuch Geol. Paläont., Monatshefte, Jahrgang 1963, p. 382–388.

———, 1968, Biostratigraphie der klastischen Entwicklung der Trogkofelstufe in den Karawanken und Nachbargebieten: Neues Jahrb. Geol. Paläont. Abhandlungen, v. 131, p. 72–77.

SCHELLWIEN, E., 1900, Die Fauna der Trogkofelschichten in den Karnischen Alpen und der Karawanken: Abhandlungen Geol. Reichsanstalt, v. 16, p. 1–122.

SCHMIDT, V., 1977, Inorganic and organic reef growth and subsequent diagenesis in the Permian Capitan Reef Complex, Guadalupe Mountains, Texas, New Mexico, in Field Conf. Guidebook, Permian Basin Sect. Soc. Econ. Paleontologists Mineralogists Pub. 77-16, p. 93–131.

SELLI, R., 1963, Schema geologico delle Alpi Carniche e Giulie occidentali: Soc. Geol. Italiana, LXII Adunanza Estiva, Bologna, 115 p.

TIETZ, G.-F., 1975, Die Schwermineralgehalte in den Grenzlandbanken (Unterperm der Karnischen Alpen, Standardprofil Rattendorfer Sattel): Carinthia II, v. 164/84, p. 115–124.

TOOMEY, D. F., 1969, The biota of the Pennsylvanian (Virgilian) Leavenworth Limestone, Midcontinent region. Part 2: Distribution of the algae: Jour. Paleontology, v. 43, p. 1313–1330.

———, 1976, Paleosynecology of a Permian plant dominated marine community: Neues Jahrbuch Geol. Paläont. Abhandlungen, v. 152, p. 1–18.

———, 1979, Role of Archaeolithophyllid algae within a Late Carboniferous algal-sponge community, Southwestern United States: Bull. Cent. Rech. Explor.-Prod. Elf-Aquitaine, v. 3, p. 843–853.

———, AND CYS, J. M., 1977, Rock/biotic relationships of the Permian Tansill-Capitan Facies exposed on the north side of the entrance to Dark Canyon, Guadalupe Mountains, Southeastern New Mexico, in Field Conf. Guidebook, Permian Basin Sect.: Soc. Econ. Paleontologists Mineralogists Pub. 77-16, p. 133–150.

———, 1979, Community succession in small bioherms of algae and sponges in the Lower Permian of New Mexico: Lethaia, v. 12, p. 65–74.

TURPAEVA, E. P., 1957, Food interrelationships of dominant species in marine benthic biocoenoses, in Nikitin, B. N., Ed., Transa, Inst. Oceanol. Marine Biol.: USSR Acad. Sci. Press, v. 20, p. 137–148.

WALKER, K. R., 1972, Trophic analysis: a method for studying the function of Ancient communities: Jour. Paleontology, v. 46, p. 82–93.

YUREWICZ, D. F., 1977, Origin of the massive facies of the Lower and Middle Capitan Limestone (Permian), Guadalupe Mountains, New Mexico and West Texas, in Field Conf. Guidebook, Permian Basin Sect.: Soc. Econ. Paleontologists Mineralogists Pub. 77-16, p. 45–92.

SEPM Special Publication No. 30, p. 161–186, May 1981

# THE MAGNESIAN LIMESTONE (UPPER PERMIAN) REEF COMPLEX OF NORTHEASTERN ENGLAND[1]

**DENYS B. SMITH**
Institute of Geological Sciences
London, England

## ABSTRACT

A major linear reef ultimately more than 100 meters high protected the seaward edge of the carbonate shelf of the Upper Permian Middle Magnesian Limestone in northeastern England, and is overlain by an extensive stromatolite biostrome. Rocks of both structures are almost completely dolomitized. The reef is founded on a patchy lenticular coquina, and much of its lower parts is formed of typically unbedded bryozoan biolithite which appears to have formed subaqueously and grew mainly upwards. Some contemporaneous lithification and rigidity is indicated by the presence of biolithite debris in associated talus. Middle stages of reef growth are characterized by a progressive increase in the proportion of algal rocks and laminar organic or inorganic encrustations at the expense of the bryozoa that dominated in early stages, and a tendency towards bedding may indicate shallowing towards the end of this phase. Stromatolitic and other laminar rocks became dominant in latest phases of reef growth, where the evidence of active contemporaneous erosion, roughly horizontal bedding and lateral rather than upwards growth is thought to indicate proximity of the top of the reef to sea level. The growth and increasing asymmetry of the reef led to progressively more complete separation of environments to landward and seaward of the reef, culminating when the reef approached sea level in the formation of a lagoon and a starved or semi-starved basin. Scattered small patch-reefs occur locally in lagoonal beds in the north of the area and considerably larger masses of reef rock in the same area are probably also patch-reefs but could be outliers of a much widened main reef.

The stromatolite biostrome is a relatively uniform tabular body up to 30 meters thick formed of finely-laminated subtidal algal stromatolites on the flat top of the Middle Magnesian Limestone reef. Algal growth forms are diverse only at the lagoonal and basinal margins of the biostrome, the basinal margin also being varied by the presence, in its lower part, of conglomerates composed of rolled cobbles and boulders of biolithite possibly derived from the underlying reef.

## INTRODUCTION AND GENERAL STRATIGRAPHY

A progradational series of three wedges of Upper Permian carbonate rocks in northeast England was formed around the western margin of the English Zechstein Basin, one of a number of major arms of the Zechstein Sea. Each wedge comprises the early and marginal deposits of a major cyclic sequence culminating in evaporites, and those of the first and second cycles feature a diverse biota and lithology indicative of a wide range of shelf and slope environments. The generalized relationships of the various lithological units are shown in Figure 1, and a fuller stratigraphical classification is presented in Table 1.

It is the purpose of this paper to describe and discuss the younger rocks formed at the edge of the carbonate shelf in the first cycle (e.g., in the Middle Magnesian Limestone of County Durham and adjoining areas), and also to describe immediately overlying algal stromatolitic rocks of uncertain age that are typically developed in Hesleden Dene in eastcentral Durham. The Middle

Magnesian Limestone rocks are regarded by the writer as an organic framework reef in the sense of Heckel (1974), or an ecologic reef as defined by Dunham (1970), while the overlying rocks comprise a body of stratiform algal laminates perhaps best classified as a stromatolite biostrome. Reef and biostrome will be described separately. Rocks of both have been almost completely dolomitized, with considerable loss of fine detail. In addition, local dedolomitization has taken place in eastern parts of the Middle Magnesian Limestone reef.

### THE MIDDLE MAGNESIAN LIMESTONE REEF

*History of Research*

The abundant and varied fauna of the Middle Magnesian Limestone attracted the attention of geologists from the early 1800s and diligent collecting and study by Sedgwick (1829), Howse (1848, 1857, and 1890), King (1850), Kirkby (1858 and 1860), and others resulted in the discovery of almost all the invertebrate fauna more than 110 years ago. Subsequent collecting by Trechmann (1913, 1932, and 1945), the first author to suggest that the invertebrates inhabited and built an organic reef, led to the discovery of a few more

---

[1] Published by permission of the Director, Institute of Geological Sciences.

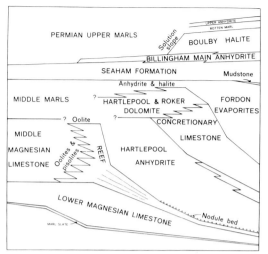

FIG. 1.—Diagrammatic cross-section of the Magnesian Limestone sequence in northeast England, showing the relationships of the main lithostratigraphical units and areas of greatest uncertainity. For thicknesses see Table 1. Reproduced from Smith, Northumbria Nat. Hist. Trans. (in press).

invertebrate genera. Additional work by Pattison (in Smith and Francis, 1967, and unpub. UK Geological Survey reports) has led to many refinements in the paleontology of this important unit. For a full faunal list see Pattison in Smith (1970a).

The early workers had noted that the most prolific sources of shelly fossils were in the lower part of the Middle Magnesian Limestone, but the sinking of Blackhall Colliery shafts through the reef in 1910–12 enabled Trechmann (1913) to make large collections, from which he chronicled a marked upwards faunal impoverishment. Later Trechmann (1925) divided this impoverishment into four main phases, each distinguished by the predominance of certain species which either became smaller and less numerous in the succeeding phase, or died out altogether. Trechmann's view that progressively increasing salinity was the main cause of the impoverishment is probably correct, but environmental changes such as decreasing water depth and greater agitation were almost certainly important contributory factors in younger parts of the reef. Autogenic faunal succession of the type recognized by Walker and Alberstadt (1975) in many other ancient organic communities is a further likely cause of the progressive changes in the Durham reef biota.

The ecology and detailed structure of the Middle Magnesian Limestone reef has received little study although Trechmann (1913) identified bryozoa as the main reef builders, and Woolacott

(1919) noted that the reef consisted of 'remains of invertebrate life that flourished under the protection of the bryozoa, together with an infilling of calcareous or dolomitic sediment.' Woolacott believed that despite the apparently luxuriant growth in parts of the reef, the small size of many individuals probably indicates that conditions for growth were never ideal. Trechmann (1945) briefly discussed the evolving assemblages of bryozoans during the construction of the reef, and concluded that they acted mainly as sediment binders. The important role of submarine laminar encrustations and cements, some organic, others possibly inorganic, which coated and bound together a great diversity of sessile organisms and detritus, especially in upper parts of the reef, also contributed greatly to its bulk, was emphasized by Smith (1958).

In common with many reef rocks of all ages, diagenesis, including almost ubiquitous dolomitization and local dedolomitization, has obliterated much of the evidence by which the primary fabric and ecology of the Middle Magnesian Limestone reef might otherwise have been interpreted. Enough pointers to the distribution, type and role of potential frame and other elements have survived for some impressions of the pattern of reef composition to be inferred. These impressions in turn provide a basis for comparison with less altered reefs in which the interrelationship of biotic and other constituents is better understood. It is fitting to recall that despite intensive study of the essentially undolomitized Capitan Reef of New Mexico by many authors (see Cys et al., 1977), there are still deep fundamental differences of opinion on its mode and environment of origin and on the part played by inorganic processes (e.g., see papers by Yurewicz (1977) and Schmidt (1977). Presumably more detailed study of the Middle Magnesian Limestone reef fabric would reveal equally sharp differences of opinion. I believe, nevertheless, that even if 90 percent of the reef were shown to have been formed by inorganically precipitated carbonate, it would still be reasonable to regard it as an organic structure if without the 10 percent of organisms the reef would not have been formed.

## Distribution

The Middle Magnesian Limestone reef is a linear structure known from surface exposures and boreholes to extend for more than 30 kilometers. Its southerly continuation underground beyond Hartlepool is unknown, although it seems probable that it extended at least for some kilometers and possibly along much of the basin-margin. It is also likely that it extended farther north than its present limit but has been removed by erosion. An extensive drift cover precludes accurate delin-

TABLE 1.—CLASSIFICATION OF UPPER PERMIAN MARINE STRATA AND EVAPORITES IN NORTHEASTERN ENGLAND (TYPICAL THICKNESS IN METERS) AND APPROXIMATE EQUIVALENTS IN THE GERMAN ZECHSTEIN SEQUENCE

| Groups | Formations (NE England) | Cycles | German Zechstein Equivalents |
|---|---|---|---|
| Teesside | Seaham Formation (30) | EZ3 | Plattendolomit |
| Aislaby | Seaham residue (3) (of Fordon evaporites) Hartlepool & Roker Dolomite (60) Concretionary limestone (80) | EZ2 | Stassfurt Evaporites Hauptdolomit |
| Don | Hartlepool Anhydrite (110) | EZ1 | Werraanhydrit & Werradolomit |
| | Middle Magnesian Limestone (100) Lower Magnesian Limestone (50) Marl Slate (1) | | Zechsteinkalk Kupferschiefer |

eation of the outcrop of the reef except in a few places, although the basinward slope commonly gives rise to a marked topographic step. The map showing outcrop distribution (Fig. 2) has been compiled from a range of surface and borehole data. These data, plus the apparent absence of reef-channel delta fans, seem to show that the reef is an almost continuous body, at least in its lower parts, rather than a chain of knolls as envisaged by Trechmann (1913, 1925, *et seq.*), and Woolacott (1919). Reef-channels may have occurred in the Seaham and Easington areas where the course of the reef is particularly circuitous. The sinuosity of the reef is reminiscent of some modern shelf-edge reefs and contrasts sharply with the remarkably smooth curve of the roughly contemporaneous Permian Capitan Reef. It is generally less than 800 meters wide at outcrop and reaches a thickness of more than 100 meters at Blackhall Colliery (Trechmann, 1913) and at Beacon Hill, Hawthorn. It is highly variable in thickness in the Sunderland area as a consequence of the irregularity of the underlying surface on which it rests with marked overlap. Because of a gentle southerly tilt, lower parts of the reef are best exposed in the north of the area, around Sunderland, and higher parts are present and exposed only in the south. The imposition of an easterly dip of 2 to 3 degrees after formation of the reef means that basinward dips quoted in the text must be reduced by this amount in order to restore primary relationships.

### Reef Morphology and Main Sub-Facies

The Middle Magnesian Limestone reef is nowhere exposed in full section and its transverse profile (Fig. 3A) has been interpreted by the author from the evidence of widely scattered exposures. This procedure has obvious shortcomings, especially in its dependence on an assumption of lateral continuity that may not be justified. Nevertheless, there is a considerable degree of uniformity and mutual consistency in the evidence from the various exposures (for positions see Fig. 3B), and the profile thus constructed is probably a reasonable approximation and an advance on that published earlier (Smith, 1958). Within this near classic envelope a number of major sub-facies are widely distinguishable despite marked lateral variation and gradational contacts, and may be recognized by their lithology, fauna and structure. Their distribution within the reef profile is shown in Figure 3A and their main features are described below together with some of the evidence on which the reef profile is based. The geographical position of places referred to in the text is shown in Figure 2.

*The Base of the Reef and the Basal Coquina.*—Most of the main exposures of the base of the reef are in the Sunderland area, where they display so wide a range of configurations that it is difficult to determine which, if any, are normal. This variability is exemplified by the two largest exposures, Down Hill Quarry (Fig. 4A) and at Claxheugh Rock (Fig. 4B) where reef base relief exceeds 20 meters, and reef base slopes exceed 45 degrees for short distances. Boreholes through the reef at Sunderland prove a similar reef base configuration and only at Humbledon Hill is a full thickness of underlying Lower Magnesian Limestone present, and here early parts of the reef are absent. The variability stems from the marked irregularity of the sub-reef surface, which has been interpreted as a major submarine slide plane (Smith, 1970b). The basal reef surface is also uneven and transgressive in Castle Eden Dene, 20 kilometers south of Sunderland. From this scanty evidence it seems that the base of the reef is generally an uneven surface with sharp local relief in the north of the area where the reef commonly lies disconformably on appreciably older strata, and perhaps with less relief in the south where it lies at or near its normal stratigraphical position. No evidence of marked differential settlement of

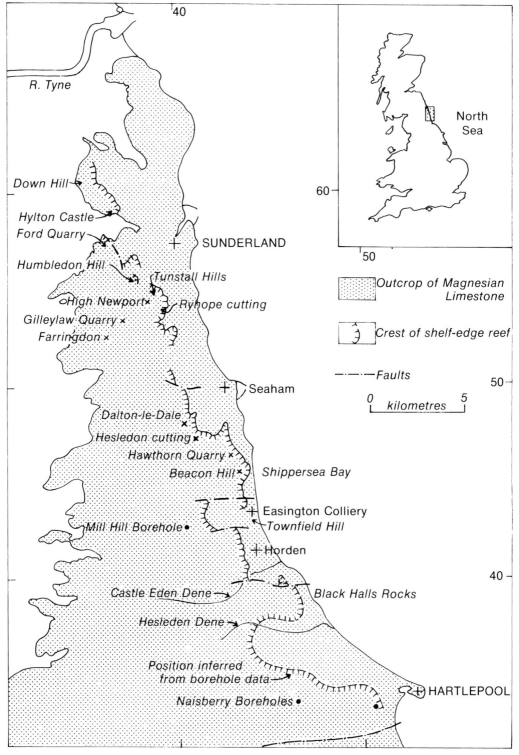

FIG. 2.—Sketch map of part of northeast England showing the outcrop of the Magnesian Limestone, the course of the shoulder or crest of the Middle Magnesian Limestone shelf-edge reef and the location of principal places and exposures referred to in the text.

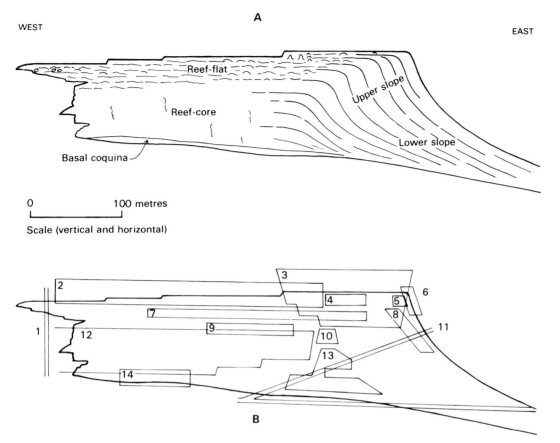

FIG. 3.—A. Generalized profile of the Middle Magnesian Limestone shelf-edge reef and its sub-facies. B. Profile as above, showing approximate position of key exposures referred to in the text. Figures preceded by NZ are the United Kingdom National Grid References.

1. Mill Hill Borehole, Easington. NZ 4122 4248
2. Hesleden Dene. NZ 435 378 to NZ 4715 3705
3. Hawthorn Quarry. NZ 437463
4. Townfield Quarry, Easington Colliery. NZ 4343 4380
5. Quarry at Horden. NZ 4354 4172
6. Quarry, now filled, on east side of Townfield Hill, Easington Colliery. NZ 4355 4373
7. Cold Hesleden railway cutting. NZ 418 473
8. Beacon Hill railway cutting. NZ 442454

9. Humbledon Hill road cutting, Sunderland. NZ 390 553
10. Tustall Hills, north end. NZ 392 545
11. Tunnels at Easington Colliery. Entrance NZ 4355 4418
12. Ford Quarry, Sunderland. NZ 363573
13. Quarry and old railway cutting, Ryhope. NZ 396538
14. Temporary and permanent exposures near Hylton Castle, Sunderland. NZ 359 588

the reef into underlying strata has been detected, nor is there evidence of abnormal pressure solution at the contact. Nowhere does the contact appear to be gradational.

The basal coquina is present only where a full reef sequence is present, commonly in hollows on the underlying surface. It forms a discontinuous belt perhaps 1 kilometer wide and at least 20 kilometers long, and is a poorly layered generally friable cream-colored deposit composed mainly of the remains of bryozoa, brachiopods, gastropods, bivalves, ostracodes, and crinoids. Although in

no sense a biolithite, it is customary to treat the coquina as part of the reef complex because of its faunal integrity and areal coincidence with the reef. In it, shell clasts, many well preserved despite complete dolomitization, are tightly packed in a bioclastic sand matrix. Its fauna is dominated by large brachiopods, especially productids such as *Horridonia horrida* (J. Sowerby), strophalosiids and spiriferids, and is amongst the most varied in the whole reef complex. No evidence of early cementation has been recognized in the basal coquina.

North                                                    South

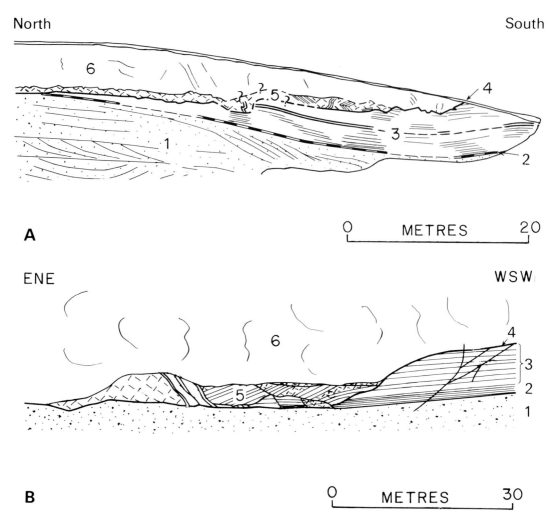

**A**

                                   0        METRES        20

ENE                                                      WSW

**B**

                                   0        METRES        30

FIG. 4.—Sketches showing the uneven base of the Middle Magnesian Limestone Reef and its relationship to underlying Permian strata at Down Hill Quarry (above) and Claxheugh Rock, near Ford Quarry, Sunderland. 1. Basal Permian (Yellow) Sands, 2. Marl Slate, 3. Lower Magnesian Limestone, 4. Inferred slide plane, locally coinciding with the base of the reef, 5. Jumbled slide blocks and breccia, and 6. Reef-core. Modified from Smith (1970b), and published by permission of the Yorkshire Geological Society.

*The Reef-Core and Back-Reef Margin.*—Dense tough slightly calcitic buff and cream-colored dolomite bryozoan biolithite comprises the reef-core, which forms perhaps half of the Middle Magnesian Limestone reef complex. In most places the rock is unbedded and only sparingly jointed. In quarries and natural exposures alike it commonly presents large almost featureless faces (Fig. 5) with little obvious lateral or vertical variation, even in such exposures as Ford Quarry, Sunderland, where almost the whole of the reef-core is visible. Ford Quarry also provides the only exposure of the passage from the reef westwards into equivalent bedded shelf dolomite, the contact being generally vertical for a height of at least 30 meters (Smith 1970a, fig. 17) and displaying only limited interdigitation. In detail, the passage from almost unbedded reef with *in situ* frame builders to granular bedded dolomite is accomplished in a horizontal distance of a few centimeters (Fig. 6). From the spatial relationships it appears that reef and bedded rock are time equivalents and that the top of the reef rock was either at the same level as, or only slightly above, the level of the adjoining shelf floor. Samples taken at close intervals across the contact show that both reef and lagoonal rocks are of similar brown dolomicrite, but that the former is characterized by a mesh of twig-

FIG. 5.—Typically massive reef-core facies, passing (top right) into bedded shelf dolomite; southeast wall of Ford Quarry, Sunderland, scale (left) 1 meter.

FIG. 6.—Reef/back-reef transition, southeast wall of Ford Quarry, Sunderland; scale 1 meter.

like ramose bryozoa mainly in growth position, and the dolomicrite matrix lacks traces of bedding alignment such as are found in the equivalent shelf rock. No washover fans have been noted in back-reef beds at Ford Quarry and reef detritus is minimal. One is driven to conclude that if an observer were to see only the near-reef beds, perhaps in a borehole core, the proximity of a major reef body probably would not be detected. A further feature of significance at the key Ford Quarry and adjoining exposures is the strong evidence of submarine eastwards movement (e.g., towards the reef), and deformation of substantial bodies of bedded shelf sediment while still unconsolidated. This implies that even if the reef rock was fractionally above the level of immediately adjoining back-reef sediment, the overall sedimentary slope was probably eastwards. This has most important implications on the interpretation of contemporaneous reef paleoenvironments and history.

Examination of slabbed specimens, peels, and thin-sections has revealed that the apparent uniformity of the reef-core when viewed from a distance is misleading and that a wide range of fabrics and faunal assemblages is present. It is difficult to estimate the relative proportions of the various constituents because of their intimate admixture, but readily recognizable sessile faunal elements comprise perhaps 10 to 30 percent of about half the core rock. Non-sessile invertebrate remains ranging from large well preserved shells to fine detritus, comprise perhaps 50–60 percent of the rock but generally much less. The remainder is of turbid dolomicrite and dolosparite of which much is apparently structureless, but some contains vague to well defined concentric and nodose laminations. The structureless fine-grained dolomite may be recrystallized lime mud. If this is so, then it must be indigenous to the reef-core because shelf rocks landward of the reef are main-

ly oolitic. Although by analogy with modern reefs, primary voids totalling perhaps 50 to 60 percent of the reef rock probably were present. Readily recognizable infilled initial cavities are uncommon except within the chambers of invertebrates. By contrast, secondary porosity thought to be related to dolomitization and leaching is locally appreciable. In addition, the shape of some cavities suggests that they might once have contained replacive and displacive anhydrite.

From the evidence preserved, the sessile benthos of the reef-core was dominated by cryptostomatous bryozoa, with fenestrate, ramose (twig-like) and stagshorn-like forms all abundantly represented. Many are not specifically identifiable because of the lack of preservation of fine details. Although mixtures of two or more types are common, one of them almost invariably is dominant in any one patch of reef rock, the patches ranging from a few centimeters to a few decimeters across and presumably representing bryozoan colonies. Some patches are in mutual contact both vertically and laterally, whereas others are separated by areas of detritus in which few bryozoa are in growth position and the fauna comprises a varied assemblage of shelly invertebrate remains, including relatively large well preserved bivalves and brachiopods. Crinoid columnals, some forming stems up to 30 centimeters long, are locally abundant near the basinal margin of the core, but are generally only minor components. Burrowing and boring such as characterize most modern reefs, have left few signs in the Durham Permian Reef, although evidence of algal boring could have been obscured during diagenesis and dolomitization. Other notable absentees from the biota are codiacean, dasycladacean and phylloid algae, and sponges.

*Fenestrate Bryozoa.*—Chiefly species of *Fenestella* with some *Synocladia* are the most abundant of the sessile suspension feeders in the reef-

FIG. 7.—Sketch of subspherical encrusted bryozoan biolithite in skeletal rubble, inferred to be upper part of reef-core; road cutting at Humbledon Hill, Sunderland.

FIG. 8.—Derived mass of encrusted bryozoan biolithite in talus of lower reef-slope; rock wall on northeast side of disused railway line, Ryhope Colliery, scale 1 meter.

core. They form dense tangled masses in which their fan-like fronds in growth position locally comprise 20–30 percent of the rock. Spaces between the fronds contain dolomicrite and a limited suite of mainly unabraded complete shells, some of free moving invertebrates, but many of pedunculate and byssate sessile forms that presumably lived near or attached to the fenestellids. Scattered small tubular structures could be the remains of encrusting foraminifers. In the Durham reef only a small proportion of the fenestellids bear laminar or nodose encrustations. Twig-like bryozoans, mainly species of *Acanthocladia* with some *Thamniscus,* are much less abundant than fenestrate forms in early parts of the reef-core, but become increasingly abundant in younger parts where fenestellids progressively decline. In patches of core where twig-like bryozoa are dominant, their carbonate remains generally occupy only a small proportion of the rock, the remainder most commonly being of dolomicrite with only a few invertebrate remains. In places near the base of the reef-core, and increasingly more commonly in higher parts, dolomite associated with the twig-like bryozoa is concentrically laminated and nodular. These laminar encrustations become a volumetrically important and locally a dominant rock forming element. Similarly laminated dolomite is also associated with the stagshorn-like bryozoans, chiefly cf. *Batostomella* which comprise up to 40 percent of some patches of rock in younger parts of the core.

A uniquely instructive exposure of reef-core near the transition to the overlying reef-flat subfacies was revealed temporarily during 1973 in a road cutting at Humbledon Hill, Sunderland. Here about 8 meters of cream-colored reef dolomite, exposed for about 100 meters normal to the reef trend, comprised in its upper parts concentrically-layered subspherical masses of biolithite each 1 to 2 meters in diameter (Fig. 7), surrounded by volumetrically predominant rough-

ly-bedded rubbly shelly and bryozoan detritus. Dense finely-laminated encrustations which comprised much of the biolithite were rare in the detritus, and sample counts by J. Pattison showed marked differences between the fauna of the two rock types. In particular, he found that branching twig-like bryozoans (especially *Acanthocladia* and *Thamniscus*) and the small predunculate brachiopod *Dielasma* virtually comprise the fauna of the biolithite. However, a much more varied fauna of bryozoans, brachiopods, gastropods and bivalves is present in the rubbly detritus. It is not apparent whether the reef rock exposed at Humbledon Hill is typically of the upper part of the reef-core, but biolithite masses similar to those in the cutting are also exposed in a railway cutting at Ryhope both in the reef-core rock of perhaps the same age as that at Humbledon Hill, and also as tumbled blocks in the fore-reef talus (Fig. 8). It is tempting to regard the Humbledon Hill tem-

FIG. 9.—Uneven roughly horizontal bedding in reef-flat dolomite; Townfield Quarry, Easington Colliery, scale 1 meter.

Fig. 10.—Minor dome structure in laminar and bryozoan rocks of the reef-flat sub-facies. A. Castle Eden Dene; scale 15 centimeters; B.–D. Old quarries in Easington Colliery area.

porary exposure as a key to the structure of the reef-core, perhaps there preserved by chance from the pervasive diagenesis which has affected the remainder. The structure has clear echoes in that of the overlying reef-flat sub-facies to which rocks at this exposure may be transitional.

*Reef-Flat Sub-Facies and the Top of the Reef.*—The progressively declining role of bryozoa in the reef-core is accompanied by a similar decline in the number, size and diversity of the larger brachiopods such as *Horridonia, Strophalosia* and *Spirifer* (Trechmann, 1913, 1925, *et seq.*). This trend continues in the rocks of the reef-flat where fenestrate bryozoans are much less abundant and generally smaller than in earlier parts of the reef. Bivalves, gastropods, ostracodes and foraminifers by contrast, although represented in the reef-flat rocks by fewer species than in the core, are relatively much more abundant. Apart from the change in fauna, rocks of this facies differ from those of the core chiefly in being disposed in beds commonly 0.2 meter to 1 meter thick (Fig.

9); internally they differ in their generally higher content of laminar and nodose encrustations. The sub-facies appears to lie across the full width of the reef tract and in places extends onto lagoonal oolites to the west. Its thickness is variable but is probably generally 20 to 40 meters. Subdivisions of the reef belt previously described by the writer (Smith, 1958) apply chiefly to rocks of this sub-facies.

Although extremely uneven in detail, bedding in the reef-flat dolomite is generally horizontal or dips gently eastwards; it is almost certainly a primary feature of the rock. Individual beds are lenticular or vary sharply in thickness from place to place and include irregular masses and mounds of algal-bryozoan biolithite (some rich in *Dielasma*) mainly less than 1 meter high (Fig. 10), surrounded and separated by smaller patches of biolithite and by pockets and lenses of rubbly detritus. The biolithite masses are unevenly spaced, are roughly oval or circular in plan and most are less than 2 meters across. Almost all are circumscribed or

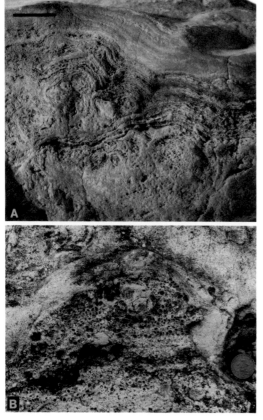

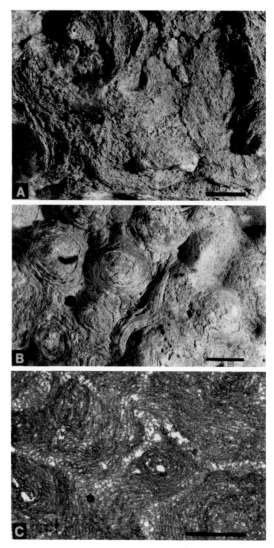

FIG. 11.—A. Dense algal bryozoan biolithite (below) encrusted with concentric to confluent laminae of supposed algal origin; reef-flat sub-facies, old quarry at Yoden, near Horden; scale bar 1 centimeter. B. Mass of tightly packed pedunculate brachiopods (*Dielasma*) encrusted with laminar dolomite; reef-flat sub-facies, railway cutting, Beacon Hill, Hawthorn; the coin is 26 millimeters across.

FIG. 12.—Concentric laminar encrustations in dolomite of reef-flat sub-facies and upper part of reef-core: A. and B. Weathered surface exposure, Cold Hesledon railway cutting; C. Print from thin-section, railway cutting at High Newport, Sunderland; scale bar 1 centimeter.

draped by finely-laminated encrustations up to 0.05 meter thick of fine-grained dolomite; structures within the encrustations range from smoothly flowing to markedly crenulate (Fig. 11). In many places a confused complex of concentric and confluent encrustations makes up almost the entire rock with the encrusted frame, generally a twig-like bryozoan, being barely discernible (Fig. 12). Gross stromatolitic growth forms such as domes and stacked-hemispheres are absent from most of the laminar rocks of the reef-flat, which differ in this respect from those of the succeeding stromatolite biostrome. However, scattered generally small columnar growth forms (Fig. 13) are present in places in basinward parts of the reef-flat. In morphology they resemble colleniform algal stromatolites, but subtle differences in proportions from those of most stromatolites of the

reef-flat suggest closer affinity with a skeletal organism similar to the problematical *Collenella* of comparable parts of the Capitan Limestone of New Mexico (Johnson, 1942; Babcock, 1977).

Dolomite around and between the biolithite masses of the reef-flat is almost infinitely varied in composition and comprises a weakly-bedded chaotic accumulation of abraded derived fragments of biolithite ranging up to 1 meter across (Fig. 14) together with shelly carbonate sands and gravels locally rich in algal debris and oncoids

FIG. 14.—Unevenly bedded laminar and encrusted bryozoan biolithite of reef-flat sub-facies with detached biolithite boulder, north side of Townfield Quarry, Easington Colliery; the hammer is 35 centimeters long.

FIG. 13.—Small scale concentric columnar structures (?algal) in bedded reef-flat dolomite; old quarry at top of Beacon Hill, Hawthorn; scale bar 1 centimeter.

(Fig. 15), and widely bound by a confused complex of laminar sheets and encrustations of fine-grained dolomite. A coarse breccia of abraded blocks of mainly algal biolithite was proved near the landward margin of the reef-flat in the Mill Hill Borehole, Easington. In addition, striking breccio-conglomerates of coarse blocks of algal-bryozoan biolithite are exposed near the seaward margin in Hawthorn Dene and in Townfield Quarry, Easington Colliery. In each place the accumulated blocks are encrusted and presumably bound by anastomosing crenulated sheets of laminated dolomite. Shells and organic remains in these detrital rocks of the reef-flat comprise a biota markedly different from that of the biolithite masses, and dominated by small gastropods and byssate bivalves with pockets of small pedunculate brachiopods (chiefly *Dielasma*), and bryozoan debris.

The relative proportions of biolithite and debris vary across the reef-flat, and biolithite and complex laminar and nodose laminates make up almost the whole rock in a belt adjoining the reef shoulder. The variations are particularly well displayed in a railway cutting at Cold Hesledon, near Seaham, which provides a shallow section across much of the reef-flat. Minor domes of biolithite characterize western (landward) parts of the cutting, and a complex of domes, detritus, algal structures and oncoids is exposed for more than 100 meters in a central stretch. Massive encrusted bryozoan biolithite with curved and steeply dipping laminar sheets predominates for about 70 meters in the eastern part of the cutting. These

eastern exposures are of rock probably close to the contemporary reef shoulder. If this is correct, it suggests that the core and the various reef top belts might rise diachronously eastwards within the reef complex in the manner previously noted in the Capitan Reef (Newell *et al.,* 1953).

The most spectacular and unusual feature of the reef-flat sub-facies is the occurrence of scattered mounds up to 4 meters high and 10 meters across composed of a complex of encrusted bryozoan biolithite and laminar (presumed algal) sheets. They have been noted near the reef shoulder in Hawthorn Quarry (in a part now removed), and in Hesledon Dene where some mounds are simple and others are stacked to a height of 4 meters or

FIG. 15.—Fine conglomerate of oncoids and broken and recemented algal stromatolite chips from old quarry in reef-flat sub-facies, Yoden, near Horden (view of bedding plane from above); scale bar 1 centimeter.

FIG. 16.—Stacked domes mainly of algal laminated dolomite from the reef-flat sub-facies, Hawthorn Dene; the field of view is about 6 meters high, U.K. Geological Survey Photograph L.-145.

so (Fig. 16). Detritus similar to that between the smaller mounds of the reef-flat also surrounds these larger mounds.

Evidence from slabs, peels and thin-sections of dolomite of the reef-flat adds detail to the impressions gained from field examination but does not alter the overall pattern. They show that apart from the faunal changes noted earlier, biolithite masses of this sub-facies are generally similar in lithology to their counterparts in the reef-core. They contain a much higher proportion of nodose and laminar encrustations. They also confirm the field observation that the intervening and surrounding shelly sand-size to boulder-size detritus is almost everywhere coated and bound by laminar alternations of clear and turbid dolomite. In many places, indeed, laminar encrustations make up 80 to 90 percent of the rock, and it is clear from their incorporation of unworn marine invertebrate remains, and broken fragments of both algal-stromatolites and earlier encrustations (Fig. 17), that the laminar structures were formed contemporaneously. In thin-section the oncoids and stromatolite gravels and sheets of the central parts of the reef-flat are distinguished from the more extensive laminar encrustations by their generally finer grain and more regular and finer lamination; but intermediate types exist.

Apart from scattered small grains of quartz, some with pyramidal overgrowths, rocks of the reef-flat appear to contain only indigenous sediments. Ooids, which make up contemporaneous rocks in the lagoon, have not been recognized even in samples from near the lagoonal edge of the reef.

*Reef-Shoulder or Crest and Upper Slope.*—Dolomite rocks of these facies are known only in

FIG. 17.—Contemporaneously fractured multi-generation encrustations in bedded reef-flat sub-facies of possible patch-reef, Gilleylaw Quarry, Silksworth, near Sunderland; scale bar 2 millimeters.

parts of the reef complex corresponding in age to the bedded reef-flat sub-facies and youngest parts of the core. Reef-shoulder and upper slope rock equivalent to the youngest core is exposed at Ford Quarry. Here weak bedding of the incipient reef-flat becomes increasingly rich in nodose and lam-

FIG. 18.—Columnar stromatolites in dolomite of the upper reef-slope a few meters down dip from the local reef-shoulder, northeast side of Ford Quarry, Sunderland; the columns are about 5 centimeters across.

FIG. 19.—Transition from bedded algal dolomite of reef-flat (top left) to steeply dipping massive algal-bryozoan biolithite of upper reef-slope. Figure bottom left provides 1.8 meter scale; north side of Hawthorn Quarry in 1956, since quarried away.

inar encrustations as it slightly thickens and gradually (over about 15 meters) increases in dip to about 30 degrees. Dolomite of this broadly rounded shoulder at Ford is a complex assemblage of distinctively grey weathering dolomite algal laminites, with pockets of small brachiopods, gastropods and bivalves (all commonly coated with laminar encrustations), and with scattered specimens (also coated) of the nautiloid *Peripetoceras freieslebeni* (Geinitz). Bryozoa are uncommon but colleniform algal stromatolites (Fig. 18) are widespread and abundant, the narrow columns extending outwards and upwards from the inclined beds for some meters down the reef-slope. Detached rolled blocks up to 0.3 meter across, mainly of algal biolithites, are scattered unevenly throughout the algal laminites of the uppermost slope, and they too are mostly thickly encrusted.

Reef-shoulder and uppermost slope rocks equivalent to the bedded reef-flat sub-facies overlying the core have been described from Hawthorn Quarry, Easington Colliery, Cold Hesledon, and Horden (Smith and Francis, 1967), but are now exposed only at Cold Hesledon and Horden. Uppermost slope rocks have also been cored in boreholes at Hawthorn Quarry and were cut by tunnels into the reef at Easington Colliery mine. At each exposure of the shoulder the turnover from reef-flat to upper foreslope was much sharp-

er than that in the somewhat older reef rocks at Ford Quarry. The dip of each bed increased to 45 degrees or more in 2 or 3 meters as it merged with generally more massive rocks that then plunged eastwards for at least 15 meters, and probably for more than 30 meters at angles of between 35 and 90 degrees (Figs. 19 and 20). The marked eastwards thickening of individual beds as they turned from flat to slope shows that at this stage the reef-shoulder advanced basinwards generally three to six times faster than it grew upwards. The presence of additional beds on the reef-slope indicates that at times it advanced without any upwards growth. There is a hint from the presence of steps in the adjoining reef top profile at Hawthorn (Smith and Francis, 1967, pl. 9) that youngest bedded parts of the reef-flat might not be represented by rock on the reef-slope, implying phases of upwards growth without advance.

The steeply dipping dolomite rocks of the shoulder and uppermost slope comprise two contrasting types: bryozoan biolithite which forms roughly parallel layers individually up to 3 meters thick, and intervening slightly sinuous laminated sheets or layers commonly 0.1 to 0.3 meter thick. The biolithite is by far the more abundant of the two rock types, and in slabs and thin-sections is seen to be composed of a dense mass of sub-parallel branching zoaria of *Acanthocladia* and cf.

FIG. 20.—Reef-shoulder or crest, showing the transition to bedded reef-flat dolomite (left) and massive steeply dipping bryozoan biolithite of the upper reef-slope, north side of small old quarry, Horden; the hammer is 35 centimeters long.

*Bastostomella.* These make up 15 to 25 percent of much of the rock and locally exceed 40 percent, in a matrix of equant saccharoidal dolomite that contains whole well preserved *Dielasma, Bakevellia* and some gastropods. Sparry calcite is present as partial void fill, mainly within shells, and other calcite fills cruciform and stellate cavities presumably once occupied by anhydrite. Few obvious laminar encrustations are present in most specimens of bryozoan biolithite of the shoulder and upper slope at Horden, but vague concentric structures in some specimens may be relics of former encrustations there. The detailed structure of the laminated layers is less complex than that of the biolithite. They comprise cream-colored slightly calcitic dolomite in which whole shelly fossils are found in layers and pockets parallel with the steeply dipping lamination, and a few fragments of delicate branching bryozoans are similarly disposed. Much of the rock lacks invertebrate remains and the lamination is commonly vague (Fig. 21). In places the laminated layers are characterized by tightly stacked hemispheroidal structures of presumed algal stromatolitic origin. For example, a layer of this latter type 0.6 meter to 0.9 meter thick dips at 35 to 60 degrees down the outermost face of the reef upper slope at Easington Colliery, immediately below the position of the inferred residue of the first cycle (EZ1) Hartlepool Anhydrite (Fig. 22). Formation under at least 10 meters of water is implied. Although the algal origin of these laminar sheets cannot be proved, satisfactory alternative explanations are difficult to envisage. The writer therefore believes that perhaps most of the laminar sheets of the reef upper foreslope are algal and each marks a temporary position of the reef

FIG. 21.—Steeply dipping laminated dolomite of upper reef-slope; borehole core, Hawthorn Quarry, scale bar 3 centimeters.

front. They are not to be confused with bilaterally opposed laminar coatings of the walls of fissures up to 0.6 meter wide in some high eastern parts of the reef-core such as Tunstall Hills. Such coatings are discordant, anastomosing and in some examples display an unfilled or spar-filled central cavity.

*Lower Reef-Slope Facies.*—The bryozoan biolithite and laminar sheets of the upper part of the reef-slope are succeeded downslope by a wedge-shaped deposit interpreted as reef talus. The junction between the two facies is ill defined because of poor exposure, and precise mutual relationships are uncertain. The talus fans thin sharply into the basin and in places as little as 5 kilometers east of the reef-shoulder there appear to be no deposits of the same age as the reef.

Dolomite of the lower part of the reef-slope has been cored in a number of boreholes but is exposed at only three localities and it is not known if these are typical. One of these localities is in tunnels at Easington Colliery mine where lower slope deposits of youngest and possibly intermediate reef age are seen. Others are at Ryhope where talus of the middle phases of reef growth is exposed. Details of the lower slope rocks seen in the tunnels were given by Smith and Francis (1967, p. 137–138 and p. 169–170) who record 9 meters to 15 meters of rubbly shelly dolomite dipping away from the reef at 5 to 10 degrees, overlain by collapse brecciated Cycle 2 carbonates. These beds lie entirely basinward of the reef upper foreslope and probably equate with the full reef sequence. They display an upwards faunal impoverishment similar to that seen in the main mass of the reef. About 400 meters farther west some 20 meters of cream-colored detrital shelly dolomite appear from faunal evidence and spatial relationships to be equivalent to the basal coquina and lower parts of the reef-core.

The main exposure at Ryhope in an old railway cutting, where about 25 meters of generally coarse unbound reef talus dips away from the reef at 20 to 35 degrees (Fig. 23), is probably equivalent in age to much of the later reef-core. The rock is a poorly-sorted rubbly accumulation of derived reef detritus containing the remains of organisms that lived in and on both the reef and the fore-reef slopes. This grades westwards into less rubbly rock that may be autochthonous reef. At this single exposure it is difficult to distinguish the indigenous slope biota from the derived biota but crinoid columnals are relatively much more abundant than in most other parts of the reef complex, and encrustations are relatively less abundant and occur mainly in derived clasts. The reef detritus is mainly of sand-size to cobble-size, and consists of a variety of rock types in which bryozoa and small pedunculate brachiopods are abundant, together with shell debris in which the most common fossils are small gastropods and bivalves. Additionally the talus contains scattered angular to subspherical masses of dense algal-bryozoan biolithite up to 2 meters across (Fig. 8) and contorted and broken slabs of roughly laminated dolomite. Both these rock types can be matched

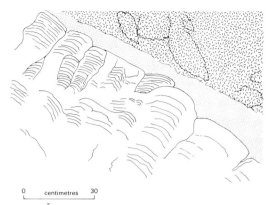

0    centimetres    30

Fig. 22.—Columnar algal stromatolites on outermost face of reef, overlain by mushy dolomite interpreted as an evaporite solution residue (stipple) and limestone breccia interpreted as a dedolomitized collapse breccia. The reef-shoulder is about 10 meters from the top left corner of this view, the intervening dip steepening to 60 degrees; old quarry on east side of Townfield Hill (now filled).

with the two major components of the upper part of the reef-slope from which they were no doubt derived. No unambiguous evidence of contemporaneous cementation of the talus has been recognized and it appears to have accumulated piece-by-piece on a slope nowhere exceeding the angle of rest, and probably generally at depths exceeding 30 meters.

The second exposure at Ryhope is in the bed of Ryhope Dene, 2.5 kilometers south-southeast of the railway cutting, where rubbly shelly reef detritus is notable for its varied ostracode and foraminiferal fauna. Nodosariid foraminifers are particularly abundant here, as they are in fore-reef talus of Zechstein reefs in the Sudeten area of Poland (Peryt and Peryt, 1977).

The interpretation as fore-reef detritus of 46 meters of breccia in a cored borehole at Seaham (Magraw, 1978) is almost certainly erroneous. Careful examination of the cores by the present writer revealed neither fossils nor fragments of reef rock, and showed that the breccia is composed entirely of distinctive rock types found only in post-reef strata.

### The Role of Bryozoa and Laminar Encrustations in the Building of the Middle Magnesian Limestone Reef

The conditions favored by modern bryozoa were reviewed by Cuffey (1970) who stated that while they inhabit seas at all latitudes and environments ranging from intertidal to depths exceeding 600 meters, they flourish principally between 10 and 70 meters on continental shelves in oxygenated waters of normal salinity, clarity,

FIG. 23.—Rubbly dolomitic reef talus of lower reef-slope from rock wall on northeast side of disused railway line, Ryhope Colliery, figure (bending) bottom right).

FIG. 24.—*Acanthocladia* branches arranged roughly concentrically in bryozoan biolithite. Photograph of block in wall of Beacon House, Beacon Hill, Hawthorn, probably from nearby quarry in reef-flat sub-facies; the coin is 26 millimeters across.

moderate agitation and between 20 and 28 degrees Celcius. Some species can adapt to variations of salinity and temperature, whereas others are relatively inflexible in their requirements. Apart from a few forms adapted to an intertidal life style, all modern bryozoa live permanently submerged. As there are no modern fenestellids, it cannot be judged if these important contributors to the Durham reefs required similar environmental conditions to modern bryozoa. Still they certainly thrived and inhabited a sub-tropical shelf sea with water temperatures very likely to have been between 20 and 28 degrees Celcius and depths, as suggested by the sedimentary setting, of 20 to 50 meters.

The role of bryozoa in fossil reef communities has also been considered by Cuffey (1974 and 1977) who notes that bryozoa in some reefs, either alone or in combination with other groups, provided the main frame builder by constructing their calcareous colonies upon earlier skeletons still in growth positions. By so doing they created a semi-rigid lattice which other organisms were able to live in and ultimately to fill or partly fill. They also performed complementary roles by baffling and trapping current borne sediments and by encrusting and thus binding any structures that provided a firm substrate.

The faunal evidence from the core of the Middle Magnesian Limestone reef suggests that where their colonies are densest, fenestellids may have filled all these roles, although Cuffey (pers. comm., 1978), having examined similar reefs, suggests that they may not have constructed a truly rigid framework. Trechmann (1945, p. 253) regarded these and other bryozoans chiefly as reef binders (encrusters). Whatever their precise function might have been, such is their preponderance in much of the Durham reef-core that it is difficult

to avoid the conclusion that the reef is there mainly because of the bryozoa. The apparent fragility of the fenestellids may be misleading in that their many fenestrule openings permitted water flow through the colony so that they could survive even strong currents, and the rarity of encrustations on fenestellids in the reef core shows that secondary stiffening was not essential for preservation. Elias and Condra (1957) considered that a lattice of fenestellids would be able to withstand turbulence when stiffened by a partial filling of submarine calcite cement or lime mud. Reefs dominated by fenestrate bryozoa have also been reported from the Mississippian of southwestern USA (Pray, 1958), the Dinantian of Ireland (Philcox, 1971), and from a number of other Upper Paleozoic carbonate complexes.

As with the fenestellids, stagshorn-like bryozoa such as cf. *Batostomella* appear to have acted in a range of capacities, mainly in younger parts of the reef-core where they commonly are encrusted. They are particularly prone to diagenetic alteration and may formerly have been more abundant than their present distribution suggests. They are major rock forming components, locally forming perhaps 40 percent of the whole. The role of twig-like bryozoa such as *Acanthocladia* is more difficult to determine because they generally form such a small proportion of the rock in areas where their colonies occur. In their case I believe that the common association with concentric and laminar encrustations is probably crucial, turning a comparatively delicate but closely packed framework into an altogether more massive rock former. There are hints from the disposition of *Acanthocladia* branches in some biolithite masses that this bryozoan may have built complex cabbage-like bushy colonies up to 30 centimeters in diameter, with their twiggy skeletons disposed roughly parallel with the margins (Fig. 24). Furthermore, such colonial masses appear locally to have acted as a nucleus, around and upon which further *Acanthocladia* colonies became established so as to bind autochthonous masses up to 2 meters across as at Humbledon Hill (Fig. 7).

The noticeable upwards change in the character of the bryozoan faunas, first observed by Trechmann (1913) and elaborated by him in 1945, may partly reflect evolutionary trends in advance of the extinction of fenestellids during the Upper Permian, but seems more likely to be a response to changing environmental conditions. Some bryozoa, fenestellids perhaps among them, are stenohaline and live mainly in sheltered waters below wave base. Their progressively declining proportions in younger parts of the reef may be due to growth of the reef into shallower, more saline and less sheltered water. Here more adaptable and tolerant forms such as, perhaps, *Acanthocladia* and cf. *Batostomella,* were able to survive and compete more effectively.

Frequent mention has been made of laminar encrustations in various parts of the Middle Magnesian Limestone reef and it is pertinent here to consider their origin, significance, and importance. The laminae are almost invisible on most fresh surfaces of reef rock, but are visible in the field on favorably weathered surfaces and may be accentuated on polished slabs by etching in acid. In thin-sections the laminae are seen as alternating thin irregular discontinuous layers of brown turbid dolomite and almost colorless dolomite microspar. These encrust biotic elements and earlier laminated rock alike and most commonly give rise to concentrically layered sheaths up to 5 centimeters across, but commonly about 1.5 centimeters (Fig. 11). They give the rock a nodular appearance in most cross-sections (Fig. 12C), but in three dimensions they are seen to mimic the surface coated. Individual brown laminae generally are of roughly even thickness, are overlapping and bear abundant small protuberances similar to those found in algal stromatolites. Their color may derive from organic matter. Intervening equant microspar layers are much more variable in thickness, in places being thinner than the brown layers and elsewhere being predominant. No clear evidence of a former fibrous structure in the colorless dolospar has been detected by the writer. The incorporation of invertebrate debris within and on laminar encrustations, and the observation by Pray, 1977) of geopetal fossiliferous sediment in crust lined cavities show that the laminae were formed during active reef growth, and are thus primary reef constituents. In addition to coating some elements of the sessile benthos, they also in places freely coat and thus bind shelly detritus. Encrustations on the walls of cavities in such detritus are present in core rock near Hylton Castle, Sunderland. They also form sheets lining fissures in massive reef rock at Tunstall Hills and elsewhere. From the faintness with which many of the encrusting laminae are preserved and their lateral passage into areas of rock in which their remains cannot now be traced, I suspect that encrustations may formerly have been even more widespread in the Durham reef than is now apparent.

The presence of concentric coatings on elements of the Middle Magnesian Limestone reef was first recorded by Trechmann (1942) in cores from a borehole at Hartlepool. Trechmann did not elaborate on these coatings, but in the same paper he noted 'dolomitized algal concretions' high in the reef. Earlier (1932, p. 170) he recorded 'dolomitized algae' in another bore through reef rock at Hartlepool. Clearly he was aware of the presence of these encrusting structures without discussing their role in reef formation relative to that of other components. The prevalence of such structures in upper parts of the reef, led the writer (Smith, 1958) to emphasize their important reef building role there. I interpreted them as probably the remains or product of lime precipitating marine algae, and believe that the balance of evidence reviewed in the present study supports this interpretation but is not unequivocal. Anderson (in Smith, 1958) gave detailed descriptions of structures in samples of laminated dolomite from Gilleylaw Quarry and Beacon Hill, and tentatively identified tubular algal remains resembling *Girvanella, Ortonella, Bevocastria*, and *Aphralysia.*

Elsewhere in the Zechstein the prominent part played by encrusting laminae in the construction of reef rocks near Possneck, Germany, was recognized by Magdefrau (1933). He recorded a progressive relative increase in the proportion of laminar encrustations ('*Stromaria*') to bryozoa in successively higher parts of the 30 to 60 meters thick reef in terms that could apply equally well to the Durham Reef. He too regarded the encrustations as probably the remains of calcareous algae, but was unable to confirm this because of pervasive dolomitization. Magdefrau does not illustrate '*Stromaria,*' but from his descriptions I take it to be similar if not identical to the encrusting laminar structures of the Durham Reef. Peryt (1978, fig. 4) illustrates algal-encrusted bryozoan biolithite from the Zechstein of Poland.

Laminar and nodose encrustations also play a most important part in the structure of the younger parts of the famous Upper Permian Capitan Reef of New Mexico. Their widespread occurrence there was noted by Newell *et al.* (1953), and they have since been studied in detail by the writer (unpubl.), Cronoble (1974), Babcock (1977), Yurewicz (1977), and Mazzullo and Cys (1977). Babcock and Cronoble applied the name

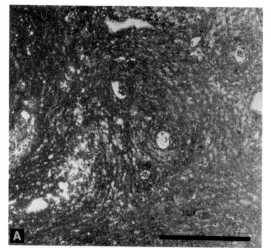

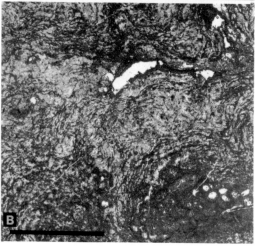

Fig. 25.—A. Concentric laminar encrustations on biotic elements (mainly *Acanthocladia*) in reef rock from High Newport railway cutting, Sunderland; scale bar 1 centimeter. Photo by Professor P. Overlau, University of Namur. B. *Archaeolithoporella* laminar encrustations on biotic elements of the Upper Capitan Reef from the south side of mouth of Walnut Canyon near Carlsbad, New Mexico; scale bar 1 centimeter. (See also Peryt, 1978, fig. 4.)

*Archaeolithoporella* to these laminar structures, which they related to encrusting skeletal red algae, and this view seems to have been accepted by the other writers. Mazzullo and Cys (1977) state that modern equivalents of these algae are weakly aragonite precipitating, and typically inhabit the more dimly lit parts of reef tracts down to depths of 50 or 60 meters. The arguments marshalled by Babcock (1977) favoring an algal origin for *Archaeolithoporella*, though not conclusive, apply equally well to the encrustations in

the Durham Reef. The apparent restriction of *Archaeolithoporella* to Upper Paleozoic reefs is seen as a particularly telling point favoring an organic origin. The alternative view that the encrusting laminae in the Capitan Limestone were constructed mainly by inorganic submarine precipitation has been argued by Schmidt (1977).

Although the encrustations of the Durham shelf-edge reef (and presumably the 'Stromaria' of the Zechstein at Possneck) are similar in morphology and general appearance to *Archaeolithoporella*, and clearly fulfilled an identical function as reef encrusters, cavity fillers and bulk formers. It is not clear, because of their differing modes of preservation, whether they are the same. Careful comparison by the writer of samples of both reveal that some *Archaeolithoporella* crusts are indeed virtually indistinguishable in thin-section from their Durham counterpart (Fig. 25). But most of the Durham encrustations are less irregular than *Archaeolithoporella* and the proportion of turbid laminae to microspar is generally higher. Such differences would perhaps arise from relatively minor differences in local environmental conditions.

### Cementation, Dolomitization and Dedolomitization in the Middle Magnesian Limestone Reef

This study is primarily of reef morphology, biota and environments but some mention must be made of the more obvious diagenetic changes that have affected the Middle Magnesian Limestone reef rocks. In many places, especially in the reef-core, these changes have altered the appearance and mineralogy of the rock so radically that little of the original fabric remains, and its reefoid origin is apparent only from its general context. Detailed study of the diagenesis of the reef rock remains for further study, and the writer's observations are necessarily tentative.

No evidence of early cementation of the basal coquina has been noted, which is still friable in some exposures. But the presence of reworked clasts of biolithite in talus paracontemporaneously derived from the reef-core must indicate some early rigidity in the succeeding reef. It seems most likely in oldest parts of the reef that this rigidity resulted mainly from the tangled network of fenestrate bryozoan fans, each attached to the remains of its predecessors or other firm substrate or organisms. These thus perform both binding and frame building functions as suggested in the preceding section of this paper. Sufficient rigidity was thus probably imparted to allow the early reef to stand slightly higher than its surroundings, and to provide a firm substrate and ecological niches that could be and were colonized by a host of other organisms. In time these lent bulk and strength from which further reef growth could spring. No evidence of early submarine fibrous or other non-laminar cements has been noted by the writer in samples of fenestellid-rich core rock, but the former presence of such cements cannot be ruled out. Cavities in such rock are mainly filled with inorganic debris, apparently structureless fine-grained dolomite or relatively recent sparry calcite. As noted previously, laminar submarine cements in the form of encrustations bind scattered patches of detritus in the early reef rock and thickly coat branching and some fenestrate bryozoans and the walls of intervening cavities so as to create rocks similar to the *Archaeolithoporella* boundstones and crust boundstones of the Capitan Reef. The increasingly high proportion of such submarine encrustations in younger parts of the reef, including fissures, has already been detailed. A degree of early rigidity and cementation in these encrusted rocks of younger parts of the reef may be inferred from the abundance there of broken and reworked biolithite, including earlier generations of encrustations. Further evidence comes from the observation that some bryozoa and small tubular organisms resembling sessile foraminifers appear themselves to be encrusting laminar structures. These were thus presumably moderately firm, so that small pedunculate brachiopods and byssate bivalves appear to have been firmly attached to steeply sloping encrusted surfaces, and were themselves encrusted. It is speculated that this rigidity may have resulted from precipitations of carbonate during formation of the encrustations. The scarcity of clear evidence of early marine void fill other than as encrusting laminae remains a puzzling feature of these rocks.

Dolomitization of the reef appears to have been almost complete, the resulting rock generally being coarser-grained and more porous than its precursor. Only small amounts of calcite, generally less than 5 percent, are present, and these appear to postdate dolomitization and may be comparatively recent. On the assumption that the reef rock was primarily of aragonite and magnesian calcite, and after allowing for the high content of Mg-rich algal deposits, it is clear that large volumes of magnesian ions have been introduced to the reef and calcium ions removed. Even larger scale ionic interchange would have been required to effect the observed dolomitization of the huge volume of time equivalent porous oolitic and other carbonate rocks to landward of the reef. The writer suspects that the dolomitization was accomplished by seepage refluxion on the lines of the model proposed by Adams and Rhodes (1960). The requisite dense magnesium-rich brines were formed either in landward parts of the back-reef shelf/lagoon (now eroded away) or in the broad lagoons and sabkhas of the succeeding Mid-

FIG. 26.—Laminar encrustations on weathered surface of dedolomitized reef-core from old quarry at south end of Tunstall Hills, Ryhope Colliery; the coin is 26 millimeters across.

dle Marls, which presumably at one time lay immediately above the Middle Magnesian Limestone. Dolomitization by landward movement (evaporative pumping?) of marine brines during or after deposition of the Hartlepool Anhydrite might perhaps account for dolomitization of adjacent reef rock, but seems unlikely to have been able to dolomitize equivalent strata extending landwards for 20 kilometers or more.

The presence of coarse spherulites of brown calcite in reef rock at Tunstall Hill and Ryhope was noted by Trechmann (1914 and 1945) and the writer has seen similar calcite in the reef at Horden and Beacon Hill. There are, furthermore, appreciable areas of reef rock composed mainly of coarse-grained roughly equant brown calcite. Each occurrence is in rock relatively near to the shoulder or upper reef fore-slope, and although the calcite locally forms masses many meters across, it comprises only a small proportion of the total reef bulk. Examination of thin-sections shows that despite the coarseness of the calcite crystals, most fossils are moderately well preserved and remain identifiable. Laminar and nodose encrustations are visible on many weathered surfaces (Fig. 26) and in thin-sections are indicated by discontinuous strings of turbid corroded dolomite grains and iron oxides, which most commonly occur within and across individual calcite crystals. Undulose fan-like extinction is noticeable in some of the larger calcites. Manganese dioxide is abundant at Tunstall Hills (Trechmann, 1914).

The patchy occurrence of these coarse-grained calcite rocks and their content of dolomite, other relics and manganese dioxide suggest that they are not primary limestones but are most probably dedolomites. Their content of iron oxides may indicate that the replaced dolomite was ferroan. The timing and cause of the dedolomitization have not

been established, but the limitation of the dedo-
lomite to parts of the reef nearest to the residue
of the Hartlepool Anhydrite is probably more than
coincidence. I infer that the dedolomitization
probably was effected by meteoric ground waters
during gypsification and dissolution of the Hartlepool Anhydrite. This was roughly contemporaneous with the formation of massive partly dedolomitized collapse breccias that now lie against the
reef fore-slope. This dissolution could have taken
place during phases of regional uplift in the Mesozoic, but is more likely to date from major uplift
during the Tertiary (Smith, 1972). No secondary
calcite was reported by Trechmann (1932 and
1942) in boreholes through reef rock at Hartlepool
where thick undissolved anhydrite and gypsum
lies close to or against the reef face.

### MIDDLE MAGNESIAN LIMESTONE REEF ROCKS
### OF UNCERTAIN STATUS

In addition to the semi-continuous reef of the
Middle Magnesian Limestone, a number of isolated masses of generally similar but thinner dolomitized reef rock are present in this formation in
and beyond the southwestern suburbs of Sunderland. They lie 1 to 3 kilometers west of the main
reef mass, and in lithology and fauna are comparable with the upper core and reef-flat sub-facies of the latter. Reef rock forming patches up to
400 meters long and a few meters thick near Silksworth Colliery (Smith, 1958) and High Newport
could be outliers of the main linear reef. If this
were so, then the latter would at one time here
have extended 4 kilometers in width. However
dips in some of the detritus in these reef rocks are
compatible with centrifugal accretion, as if each
mass had its own talus fans. This would suggest
that they were originally discrete patch-reefs.
Much smaller undoubted patch-reefs were met in
temporary excavations in lagoonal dolomite at
Silksworth.

None of the large reef masses are fully exposed
and their relationship to the bivalve and gastropod
rich shallow water lagoonal oolites in which they
appear to lie is generally obscure. Several are
composed mainly of algal-bryozoan biolithite with
a wealth of dense concentric encrustations (Anderson in Smith, 1958, pls. 7 and 8), and contain
pockets and sheets of skeletal rubble rich in bivalves, gastropods and brachiopods. Thick uneven bedding with algal domes up to 1 meter high
characterize the reef rock at Gilleylaw Quarry
(Silksworth Colliery), and that at High Newport
has sack-like masses of algal-bryozoan biolithite
up to 2 meters across, which are similar to those
found in patch-reefs in the Lower Magnesian
Limestone of Yorkshire (Smith, this volume).
Uppermost parts of reef rocks at Silksworth Colliery and near Farringdon are predominantly of

flat-lying and crenulated algal stromatolites,
amongst which oncoids (Smith, 1958, pl. 6) are
locally abundant. Contemporaneous lithification
of these and lower parts of the reef rocks is indicated by abundant evidence of reworking of
biolithite masses. Some tepee-like structures are
also present.

### THE HESLEDEN DENE STROMATOLITE BIOSTROME

*Distribution and Stratigraphical Position*

In the southern part of coastal Durham and presumably originally in the north also, the main reef
complex of the Middle Magnesian Limestone was
succeeded by biostromal algal laminated (stromatolitic) dolomite commonly 15 to 30 meters
thick. Because this rock appears to be conform-
able on the flat top of the reef complex it has been
customary to group the laminites with the reef
despite their different sedimentary style, and this
approach can still be defended. However, the
stromatolite biostrome appears to be considerably
broader than the main reef. Most critically, it appears from exposures at Beacon Point and Shippersea Bay, Hawthorn, that there it may have extended basinward of the underlying reef-shoulder
and subsequently foundered to its present position at least 100 meters below its expected level.
Although the evidence at Hawthorn is inconclusive and alternative explanations cannot be eliminated, it seems possible than an otherwise un-
recognized hiatus separated the formation of the
main reef from that of the biostrome. Probably
within this interval sea level fell and the 100 meters + Hartlepool Anhydrite was built up against
the reef-slope during a succeeding phase of sea
level recovery. Foundering of parts of the biostrome that had extended onto the anhydrite then
accompanied dissolution of the latter. Evidence
apparently supporting this suggestion was formerly exposed in an old quarry at Easington Colliery (see Fig. 22). Here, outermost upper reef-
slope rocks equivalent in age to youngest reef-flat
deposits are in contact with a supposed evaporite
solution residue and associated collapse breccias
without an intervening representative of the stromatolite biostrome. The presence of reportedly
wind abraded quartz sand grains in fissures in the
reef rock at Tunstall Hills, Sunderland (Burton,
1911; Trechmann, 1945) may be seen as evidence
supporting the inferred phase of subaerial exposure of the Middle Magnesian Limestone reef during formation of the Hartlepool Anhydrite.

The age and stratigraphical affinities of the biostromal rocks are uncertain because, on the available evidence, they could be part of the Middle
Magnesian Limestone reef, part of the Hartlepool
Anhydrite of late Cycle EZ1 age, or thirdly a marginal facies of the EZ2 Hartlepool and Roker Dolomite. Their limited fauna has strong links with

that of the main reef complex and contains bryozoa and other elements unknown elsewhere in Cycle 2 in England, favoring an EZ1 age. There are some grounds for believing that almost all the fauna of the biostrome is derived from the main reef. It could also be argued that stratiform algal rocks abound at the Zechstein shelf-margin in Cycle 2 carbonates in Holland, Germany, and Poland, and that a largely algal limestone of this age in England would not be anomalous.

### Morphology and Main Sub-Facies

The stromatolite biostrome is a sheet-like body, exposed from Hawthorn Quarry southwards to Hesleden Dene and encountered also in a few boreholes farther south. No systematic thickness trends may be discerned from the few data available, although the greatest thickness observed, about 30 meters, is at Black Halls Rocks at the basinward side of the biostrome. The rock has been uniformly dolomitized and no evidence of dedolomitization has been reported. Neither the lagoonal nor basinal margins are exposed, but the field evidence suggests that the biostrome comprises a broad uniform median belt perhaps a kilometer wide between narrower and more varied marginal belts. The various belts are thought to be generally parallel with the trend of the underlying reef.

Algal stromatolites of the broad central belt are generally horizontally bedded or dip gently, and comprise saccharoidal cream-colored dolomite of fine sand-size disposed in almost flat to strongly crinkled fine laminae (commonly 40 to 45 per centimeter). No direct evidence of algal filaments has been preserved in any of the laminated rocks examined petrographically, and where laminae are almost flat it is difficult to demonstrate their algal binding. The flat laminae, however, pass laterally into crinkled laminae (Fig. 27) where unmistakably algal hemispheres, individually up to 0.05 meter high, are locally spectacularly stacked. At some levels the pattern of minor growth forms is so distinctive that individual beds may be traced for hundreds of meters, and in one case, for several kilometers. Traces of domes only a few centimeters high but up to several meters across are discernible locally. The laminites lack the disruptions so characteristic of many modern algal bound sediments and exhibit no evidence of early cementation, or of features such as desiccation cracks, fenestral fabric, flat pebble conglomerates, early diagenetic evaporites or paleosol profiles that would indicate periodic emergence. Neither shelly fossils nor evidence of burrowing have been reported from central tracts of the biostrome.

At a number of localities upper parts of the algal laminites are thinly interbedded with unlaminated

Fig. 27.—Algal laminated (stromatolitic) dolomite of the central belt of the Hesleden Dene stromatolite biostrome, suggesting that prevailing current probably came from the left; slabbed borehole core, Hawthorn Quarry, scale bar 3 centimeters.

fine-grained dolarenites, some of which are clearly of altered oolite, and pass up by alternation into oolitic dolomite. A 0.5 meter unit of thinly-bedded dolomite with argillaceous layers marks the top of the laminites at Hawthorn Quarry.

Stromatolite rocks on the lagoonal side of the biostrome are lithologically similar to those of the central belt but are diversified by colleniform, columnar and dome-shaped structures up to 0.6 me-

FIG. 28.—Small stromatolite domes thought to be in western marginal belt of the Hesleden Dene stromatolite biostrome, from west end of Hesleden Dene, photographed in 1954; the domes are about 30 centimeters high.

FIG. 30.—Flat-topped stromatolite domes in eastern marginal belt of Hesleden Dene stromatolite biostrome, Black Halls Rocks; the domes are about 60 centimeters high; U.K. Geological Survey Photograph L.-135.

ter high and 1 meter across (Fig. 28). Interbedded unlaminated and finely cross-laminated dolomite commonly forms a higher proportion of the deposit than in the central more uniform belt, and probably represents intertonguing lagoonal sediment. As in the central belt, no unambiguous evidence of subaerial emersion or of strong wave or

FIG. 29.—Oncoids in isolated pocket between boulders of conglomerate, from eastern marginal belt of Hesleden Dene stromatolite biostrome, Black Halls Rocks; scale bar 1 centimeter.

current activity has been recognized in the laminated dolomite of this western marginal tract, nor have shelly fossils been reported there.

Algal laminites of the inferred basinal fringe of the Hesleden Dene biostrome contain fewer beds of unbound sediment than those of the central and western tracts. They differ also in that lower beds contain wedges up to 8 meters thick of coarse conglomerate and upper beds feature a range of broad low domes. The conglomerates are exposed at Black Halls Rocks and in Hesleden Dene and are composed of a chaotic accumulation of subangular to subrounded blocks of sparingly shelly *Thamniscus*-rich algal biolithite (Smith and Francis, 1967, pl. 12). The matrix consists of smaller mainly angular fragments of biolithite, finely laminated stromatolite debris and, in a few places, subspherical oncoids up to 2 centimeters across (Fig. 29). Virtually all the fauna of the biostrome is found in the biolithite clasts of the conglomerates, and, as Trechmann noted in 1913, virtually every clast regardless of size is coated with many concentric laminae. Evidence of fracturing, reworking and recementation of clasts is abundant.

The source of the biolithite clasts in the conglomerates has not been established as no rocks of comparable character have been reported from the biostrome itself. In lithology and fauna they most resemble rocks of the reef-flat of the Middle Magnesian Limestone, which the conglomerate appears to overlie in Hesleden Dene, and derivation from these is suggested. Crude basinward dipping foresets in the conglomerates are emphasized by irregular sheets and lenses of stromatolitic dolomite, some bearing complex minor hemispherical conical and ripple-like growth forms. A diverse range of stromatolite domes and debris is found in laminites between the breccias. A feature of interest is that thickly coated specimens of *Peripetoceras freieslebeni* are locally present and may be the only contemporaneous fossils in the

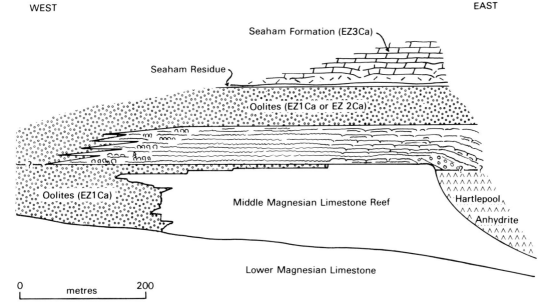

WEST

EAST

Seaham Formation (EZ3Ca)

Seaham Residue

Oolites (EZ1Ca or EZ 2Ca)

Oolites (EZ1Ca)

Middle Magnesian Limestone Reef

Hartlepool

Anhydrite

Lower Magnesian Limestone

0    metres    200

FIG. 31.—Schematic section showing one possible interpretation of the relationship of the Hesleden Dene stromatolite biostrome (unlabeled) to enclosing strata including the Middle Magnesian Limestone reef.

biostrome. Presumably they lived in basinal areas but were washed onto and stranded on the reef by currents and storms.

The domes of higher parts of the biostrome here are less diverse than those associated with the conglomerate. They are commonly about 0.6 meter high (Fig. 30), range to 18 meters across and are concentrated at several separate and readily traceable levels (see Smith and Francis, 1967, for fuller details).

An interpretation of the overall morphology of the Hesleden Dene stromatolite biostrome and its possible relationship with adjacent strata is shown in Figure 31. The main alternative interpretation would place the biostrome entirely landward of the shoulder of the main Middle Magnesian Limestone reef and older than the Hartlepool Anhydrite. Fragments of the biostrome at the base of the cliffs at Shippersea Bay would need to be explained by down reef foundering, or tectonic or other displacement of perhaps 100 meters.

### HISTORY AND CONDITIONS OF GROWTH OF THE MAGNESIAN LIMESTONE REEF COMPLEX AND ITS EFFECT ON THE ENVIRONMENT

The evidence of faunal, structural and environmental changes shows that the Magnesian Limestone reef complex evolved through a series of mutually gradational stages, each genetically related to its predecessor. Reef growth waxed during early and middle phases of this evolution, when conditions were clearly highly favorable and

waned in later stages when decreasingly favorable conditions led to progressive faunal extinction. Reef growth during Middle Magnesian Limestone times may have been terminated by a decline of sea level for which there is some supporting evidence (Smith, 1970a).

Organic reefs throughout Phanerozoic time appear to have been formed wherever an optimal blend of environmental conditions encouraged the growth of a partly sessile benthos with a vigor that exceeded the combined effects of biological, chemical and mechanical erosion. Additional penecontemporaneous rigidity imparted by bio- and physiochemically precipitated calcium carbonate has been and remains an essential factor in the formation and preservation of many reefs. For linear reefs, the optimal coincidence of conditions such as temperature, nutrient supply, oxygenation, light and salinity has been met both inshore and at significant sea bottom slope breaks some distance offshore.

An offshore slope break perhaps 20 to 30 kilometers out seems to have been the favored location for the Middle Magnesian Limestone reef. The close aerial coincidence of the reef and the basal coquina may be genetic in the sense that the presence of the coquina may have altered bottom conditions and configuration so as to favor colonization by reef forming organisms such as bryozoa on the patchy foundation thus provided. Alternatively, it is possible that the fauna contributing to the coquina and the reef biota both found conditions especially favorable there and flourished accordingly. The absence of the coquina from places where younger parts of the reef are missing through onlap shows that a coquinoid foundation was not essential to spreading once the reef was well established. The lack of evidence of emersion or contemporaneous cementation in the basal coquina, its high content of organisms that elsewhere in the Magnesian Limestone are thought to be sub-littoral, and its position relative to offshore facies belts in the underlying Lower Magnesian Limestone all seem to point to subaqueous accumulation. In addition, the excellent preservation of delicate ornament on many of the component shells points to a lower energy environment, probably below wave base.

The dominance of sub-littoral organisms in the lower parts of the reef-core, the lack of evidence of strong contemporaneous erosion except on the seaward side, and the relatively minor role played by algae all suggest that much of the early part of the reef core also was formed under some depth of water. From the evidence of only slight relief at the landward edge of the reef and of reefwards movement of unconsolidated shelf sediments, it appears that during the growth of the core the shelf-edge reef gave rise merely to a gentle step in the basinwards sea floor slope, and did not act as a barrier between shelf and basin waters or biotas. The unmistakable bedding traces in highest parts of the core, the higher proportion there of algal and detrital rock, and the progressive incoming of gastropods and bivalves in place of productid and other brachiopods, combine to suggest shallowing during this slightly later phase of reef growth and perhaps an approach to sea level. The scene I envisage is of a reef tract that formed initially under warm well aerated sea water 20 to 50 meters deep, but progressively built up as an asymmetrical platform with an increasingly steep and high basinward slope along its eastern margin. A small scale relief of a few centimeters seems probable, with scattered thickets of living and dead bryozoan colonies and attached bivalves and gastropods projecting slightly above intervening detritus, and with a few subspherical bryozoan masses up to 2 meters high upon which laminar encrustations formed more profusely than elsewhere. Growth appears to have been predominantly upwards, with little change in the position of the back-reef margin and with only a modest advance of core over equivalent talus fans.

Constraints on the continued upwards growth of the reef are inferred from the generally flat bedded character of the rocks of the reef-flat sub-facies and from the change to predominantly outwards growth, while the abundant evidence of contemporaneous erosion and redeposition of detritus that had been lithified points to exposure

and to phases of high energy. From these inferences, supported by the occurrence of oncoids and stromatolite gravels, and a specialized fauna of gastropods, bivalves and small brachiopods, I believe that the bedded rocks of the reef-flat probably were formed close to sea level and may been peritidal. A reef surface relief of perhaps 1 to 3 meters seems probable, with upstanding rigid or semi-rigid algal and bryozoan-algal masses surrounded by spreads and pockets of clastic reef debris, and invertebrate remains that were themselves undergoing extensive binding by encrusting algae. Early lithification was most probably accomplished mainly by submarine processes, either organically through algal induced carbonate precipitation or inorganically from carbonate saturated slightly hypersaline pore-waters. The apparent absence of fenestral fabric in algal laminites of the reef-flat may be viewed as negative evidence against appreciable emergence.

During this phase of predominantly outward growth the asymmetry of the reef profile became most marked and a diverse biota flourished only on the reef-slope where conditions were most varied. Talus fans that had been formed during core growth and later were rapidly overwhelmed by a seawards advance of the shoulder. The most actively growing part of the reef was the uppermost reef fore-slope which in places at this time approached vertical. During this phase the reef probably became an almost complete barrier, and sediments to the west became truly lagoonal. The roughly contemporaneous growth of patch-reefs in the lagoon and the unmistakable evidence of active erosion around the patch-reefs, and of current movement of oolite grainstones between the reefs, all suggest that water in outer parts of the lagoon, though shallow, was neither tranquil nor hypersaline. The possibility that low energy inshore or that littoral parts of the lagoon were more highly saline, as in some modern partial analogues, cannot be tested since rocks of this inshore belt have now been eroded away from the northern areas.

The uncertainty regarding the course of events after the Middle Magnesian Limestone reef ceased to grow has already been discussed. While the exact course of events and the stratigraphical age of the stromatolite biostrome remain uncertain, this uncertainty bears only slightly on the course of evolution of the reef complex because the biostrome clearly formed on the flat top of the main Middle Magnesian Limestone reef, and inherited a generally similar environment. The virtual absence of a shelly fauna from most of the biostrome, and its greater extent and uniformity compared with the underlying reef suggest a considerable change in the detailed environment. Conditions clearly became much more uniform and remained so for a lengthy period as subsidence continued. The prevalence of low energy levels is indicated by the lack of evidence of active erosion, except along the seaward fringe. There is, furthermore, no evidence of subaerial exposure. On these grounds I infer that the stromatolite biostrome formed essentially subtidally, on a broad shelf generally covered by a few meters of water. I attribute its preservation to the rarity of browsing and burrowing organisms, which itself probably indicates abnormally high salinity at that time. This interpretation conflicts with some earlier interpretations that the algal stromatolites were formed intertidally, but accords with the developing view that some modern and many ancient stromatolitic deposits were formed mainly subtidally (e.g., Achauer and Johnson, 1969; Gebelein, 1969; Playford and Cockbain, 1969; Leeder, 1975; Eriksson, 1977).

The large size and common overturning and fracturing of clasts in the conglomerate wedges, inferred to have been formed along the basinward fringe of the biostrome, implies a degree of contemporaneous lithification and occasional high energy. The wide variation in the growth forms of associated algal stromatolites suggests more varied environmental conditions than those encountered in the central belt; they may at times have been intertidal. Basinward progradation of the several belts is one possible explanation for the upwards progression from conglomerates to more flat-lying stromatolites at Black Halls Rocks. Whether such progradation would have been over carbonates or anhydrite remains unknown, and the cause of the ultimate ending of reef growth is also speculative. The evidence shows only that the algal stromatolites pass upwards by eastwards progradational interdigitation into planar cross-laminated dolomitized oolites with local pisolites (Fig. 31), and that these in turn are overlain by the residue of the EZ2 evaporites.

## ACKNOWLEDGMENTS

While the opinions expressed in this paper are my own, I am greatly indebted to Jack Pattison for his help and advice on general paleontological aspects of the work over many gestative years, and also for reading and suggesting valuable amendments to the text. Roger Cuffey's assistance with the section dealing with bryozoa is gratefully acknowledged and Gordon Smart's and Tony Bazley's comments on the text were most helpful. The specimen photography was capably executed by staff of the IGS Photographic Department, London, under the direction of Martin Pulsford.

## REFERENCES

ACHAUER, C. W., AND JOHNSON, J. H., 1969, Algal stromatolites in the James Reef Complex (Lower Cretaceous), Fairway Field, Texas: Jour. Sed. Petrology, v. 39, p. 1466–1472.

ADAMS, J. E., AND RHODES, MARY L., 1960, Dolomitization by seepage refluxion: Am. Assoc. Petroleum Geologists Bull., v. 44, p. 1912–1920.

BABCOCK, J. A., 1977, Calcareous algae, organic boundstones, and the genesis of the upper Capitan Limestone (Permian, Guadalupian), Guadalupe Mountains, West Texas and New Mexico, *in* Soc. Econ. Paleontologists & Mineralogists, Field Conference Guidebook, Pub. 77-16, p. 3–44.

BURTON, R. C., 1911, Beds of Yellow Sands and marl in the Magnesian Limestone of Durham: Geol. Mag., v. 8, p. 299–306.

CRONOBLE, J. M., 1974, Biotic constituents and origin of facies in Capitan Reef, New Mexico and Texas: Mountain Geologist, v. 11, p. 95–108.

CUFFEY, R. J., 1970, Bryozoan-environment interrelationships—an overview of bryozoan paleoecology and ecology: Earth Min. Sci. Bull., v. 39, p. 41–45, p. 48.

———, 1974, Delineation of bryozoan constructional roles in reefs from comparison of fossil bioherms and living reefs: Int. Coral Reef Symp. Proc., v. 1, p. 357–364.

———, 1977, Bryozoan contributions to reefs and bioherms through geological time: Am. Assoc. Petroleum Geologists, Studies in Geology, No. 4, p. 181–194.

CYS, J. M., TOOMEY, D. F., BREZINA, J. L., GREENWOOD, E., GROVES, D. B., KLEMENT, K. W., KULLMAN, J. D. MCMILLAN, T. L., SCHMIDT, V., SNEED, E. D., AND WAGNER, L. H., 1977, Capitan Reef—Evolution of a concept, *in* Soc. Econ. Paleontologists & Mineralogists, Field Conference Guidebook, Pub. 77-16, p. 201–321.

DUNHAM, R. J., 1970, Stratigraphic reefs versus ecologic reefs: Am. Assoc. Petroleum Geologists Bull., v. 54, p. 1331–1350.

ELIAS, M. K., AND CONDRA, G. E., 1957, *Fenestella* from the Permian of West Texas: Geol. Soc. America Mem. No. 70, 58 p.

ERIKSSON, K. A., 1977, Tidal flat and subtidal sedimentation in the 2250 m.y. Malmani Dolomite, Transvaal, South Africa: Sed. Geology, v. 18, p. 223–244.

GEBELEIN, C. D., 1969, Distribution, morphology and accretion rates of Recent subtidal algal stromatolites, Bermuda: Jour. Sed. Petrology, v. 39, p. 49–69.

HECKEL, P. H., 1974, Carbonate buildups in the geologic record: a review, *in* LaPorte, L. F., Ed., Reefs in Time and Space: Soc. Econ. Paleontologists Mineralogists Spec. Pub. No. 18, p. 90–154.

Howse, R., 1848, A catalogue of the fossils of the Permian System of the counties of Northumberland and Durham: Trans. Tyneside Nat. Fld. Club, v. 1, p. 219–264.

———, 1857, Notes on the Permian System in the counties of Northumberland and Durham: Ann. Mag. Nat. Hist., v. 19, p. 33–52, p. 304–312, p. 463–473.

———, 1890, Catalogue of the local fossils in the Museum of Natural History: Trans. Nat. Hist. Soc. Northumberland, Durham, Newcastle-upon-Tyne, v. 11, p. 227–288.

Johnson, J. H., 1942, Permian lime-secreting algae from the Guadalupe Mountains, New Mexico: Geol. Soc. America Bull., v. 53, p. 195–216.

King, W., 1850, A monograph of the Permian fossils in England: Palaeontogr. Soc. Monograph, 258 p.

Kirkby, J. W., 1858, On some Permian fossils from Durham: Trans. Tyneside Nat. Fld. Club, v. 3, p. 286–297.

———, 1860, On Permian Entromostraca from the Shell-limestone of Durham: Trans. Tyneside Nat. Hist. Club, v. 4, p. 122–171.

Leeder, M., 1975, Lower Border Group (Tournasian) stromatolites from the Northumberland Basin: Scottish Jour. Geol., v. 3, p. 207–226.

Magdefrau, K., 1933, Zur Enstehung der mitteldeutscher Zechstein—Riffe: Centr. f. Miner., Abt. B., No. 11, p. 621–624.

Magraw, D., 1978, New boreholes into Permian beds off Northumberland and Durham: Yorkshire Geol. Soc., Proc., v. 42, p. 157–183.

Mazzullo, S. J., and Cys, J. M., 1977, Submarine cements in Permian boundstones and reef-associated rocks, Guadalupe Mountains, West Texas and southeastern New Mexico, in Soc. Econ. Paleontologists & Mineralogists, Field Conference Guidebook, Pub. 77-16, p. 151–200.

Newell, N. D., Bradley, J. S., Fischer, A. G., Hickox, J. E., Rigby, J. K., and Whiteman, A. J., 1953, The Permian Reef complex of the Guadalupe Mountains region, Texas and New Mexico: San Francisco, Freeman and Co., 236 p.

Peryt, T. M., 1978, Sedimentology and paleoecology of the Zechstein Limestone (Upper Permian) in the Fore-Sudetic area (western Poland): Sed. Geology, v. 20, p. 217–243.

———, and Peryt, D., 1977, Zechstein foraminifera from the Fore-Sudetic area, western Poland, and their paleoecology: Roczn. Polsk. Tow. Geol., Ann. Soc. Geol. Poland, v. 47, p. 301–326.

Philcox, M. E., 1971, A Waulsortian bryozoan reef ('cumulative biostrome') and its off-reef equivalents, Ballybeg, Ireland: Compte Rendue 6e Congr. Int. Strat. Geol. Carbonif., v. IV, p. 1359–1372.

Playford, P. E., and Cockbain, A. E., 1969, Algal stromatolites: deepwater forms in the Devonian of Western Australia: Science, v. 165, p. 1008–1010.

Pray, L. C., 1958, Fenestrate bryozoan core facies, Mississippian bioherms, south-western United States: Jour. Sed. Petrology, v. 28, p. 261–273.

Schmidt, V., 1977, Inorganic and organic growth and subsequent diagenesis in the Permian Capitan reef complex, Guadalupe Mountains, New Mexico, in Soc. Econ. Paleontologists & Mineralogists, Field Conference Guidebook, Pub. 77-16, p. 93–132.

Sedgwick, A., 1829, On the geological relations and internal structure of the Magnesian Limestone: Trans. Geol. Soc. London, v. 2, p. 37–124.

Smith, D. B., 1958, Observations on the Magnesian Limestone reefs of northeastern Durham: Great Britain Geol. Surv. Bull., 15, p. 71–84.

———, 1970a, Permian and Trias, in Geology of Durham County, Hickling, Grace, Ed.: Trans. Nat. Hist. Soc. Northumberland, v. 48, p. 66–91.

———, 1970b, Submarine slumping and sliding in the Lower Magnesian Limestone of Northumberland and Durham: Yorkshire Geol. Soc., Proc., v. 38, p. 1–36.

———, 1972, Foundered strata, collapse breccias and subsidence features of the English Zechstein, in Richter-Bernberg, G., Ed., Geology of saline deposits: Hanover Symp., Proc., 1968, UNESCO, p. 255–269.

———, and Francis, E. A., 1967, The geology of the country between Durham and West Hartlepool: Mem. Geol. Survey Great Britain, London, HMSO, 354 p.

———, in press, The Permian and Triassic Rocks, in Robson, D. A., Ed., Northumbria Nat. Hist. Soc. Trans.

Trechmann, C. T., 1913, On a mass of anhydrite in the Magnesian Limestone at Hartlepool, and on the Permian of south-eastern Durham: Quart. Jour. Geol. Soc. London, v. 69, p. 184–218.

———, 1914, On the lithology and composition of the Durham Magnesian Limestones: Quart. Jour. Geol. Soc. London, v. 70, p. 232–263.

———, 1925, The Permian Formation in Durham: Proc. Geol. Ass., London, v. 36, p. 135–145.

———, 1932, The Permian shell-limestone reef beneath Hartlepool: Geol. Mag., v. 69, p. 166–175.

———, 1942, Borings in the Permian and Coal Measures around Hartlepool: Yorkshire Geol. Soc., Proc., v. 24, p. 313–327.

———, 1945, On some new Permian fossils from the Magnesian Limestone near Sunderland: Quart. Jour. Geol. Soc. London, v. 100, p. 333–354.

Walker, K. R., and Alberstadt, L. P., 1975, Ecological succession as an apsect of structure in fossil communities: Palaeobiology, v. 1, p. 238–257.

Woolacott, D., 1919, The Magnesian Limestone of Durham: Geol. Mag., v. 6, p. 452–465, p. 485–498.

Yurewicz, D. A., 1977, The origin of the massive facies of the lower and middle Capitan Limestone (Permian), Guadalupe Mountains, New Mexico and West Texas, in Soc. Econ. Paleontologists & Mineralogists, Field Conference Guidebook, Pub. 77-16, p. 45–92.

SEPM SPECIAL PUBLICATION NO. 30, P. 187–202, MAY 1981

# BRYOZOAN-ALGAL PATCH-REEFS IN THE UPPER PERMIAN LOWER MAGNESIAN LIMESTONE OF YORKSHIRE, NORTHEAST ENGLAND[1]

DENYS B. SMITH
Institute of Geological Sciences
London, England

## ABSTRACT

Patch-reefs commonly 10 to 25 meters across and 3 to 8 meters thick are abundant in dolomitized skeletal oolite of the Upper Permian Lower Magnesian Limestone in Yorkshire, England, and are themselves dolomitic. They are roughly circular or oval in plan and irregular in section, with a common tendency for a shallow inverted cone to be surmounted by a gentle dome. All are stratigraphically younger than a widespread coquina which lies near the base of the formation and may have provided a stable foundation for reef forming organisms.

Most of the reefs comprise an untidy assemblage of sack-shaped bodies ('saccoliths'), each composed mainly of closely-packed sub-parallel remains of the ramose bryozoa *Acanthocladia* (generally predominant) and *Thamniscus* in a finely crystalline dolomite matrix, which commonly also contains a low diversity community of other invertebrates. It is tentatively suggested that each saccolith is founded on a singly colony of *Acanthocladia anceps* (Schlotheim). The reefs probably were formed entirely under water on a broad shallow tropical carbonate marine shelf, and the tops of most were less than 2 meters higher than surrounding contemporaneous sediment. Mud-trapping and binding by bryozoa appears to have been the main constructional process, with encrusting foraminifers and early submarine cements adding stiffening and bulk. Bryozoa die out in upper parts of the formation, where the reefs are composed largely or wholly of algal stromatolitic dolomite that was also probably formed subaqueously but in shallower water than the earlier bryozoan parts of the reefs.

## INTRODUCTION, GENERAL STRATIGRAPHY AND DISTRIBUTION OF THE REEFS

Roughly circular or oval biohermal structures interpreted as patch-reefs were formed in great numbers on parts of the marginal shelves of the tropical to sub-tropical Upper Permian Zechstein Sea in northeast England. They now form part of the Lower Subdivision of the Lower Magnesian Limestone of Yorkshire and northernmost Derbyshire (Table 1), which is the early carbonate phase of the first of four main rhythmic cycles of the English Zechstein sequence. The purpose of this paper is to describe these reefs and their relationship to surrounding sediments and to discuss their structure and ecology. All have been almost completely dolomitized but their detailed petrology has not been investigated.

The first use of the word 'reef' in connection with these structures in the Lower Magnesian Limestone was by Mitchell (1932) who noted their presence at several places in the Doncaster area of Yorkshire (Fig. 1). However, the presence of rock built in place by a community dominated by bryozoa had previously been recognized by Kirkby (1861). The geographical distribution of the reefs and some aspects of their morphology, fauna

and stratigraphical relationships were briefly chronicled in United Kingdom Geological Survey memoirs describing the outcrop from the vicinity of Leeds to that of Sheffield (Eden *et al.*, 1957; Edwards *et al.*, 1950; Mitchell *et al.*, 1947). A short account of their ecology and structure was given by the writer (Smith, 1974).

Marginal carbonates of the Lower Subdivision of the Lower Magnesian Limestone comprise several sub-facies, the distribution of which is thought to be related to contemporaneous variations of water depth, salinity, bottom configuration, and the levels of wave energy and light. All the reefs seen by the writer lie in a broad (5 kilometers+) belt of dolomitic oolite grainstones which extends discontinuously for some 90 kilometers adjoining and parallel with the inferred shoreline (Fig. 1). Reefs have not been noted between Aberford and Castleford where the shoreline and related shelf facies belts is thought to have swung to the west of the outcrop (Smith, 1974).

In the belt of oolite grainstones there is a widespread and relatively uniform sequence of rock types which typically comprises a basal unit of sandy oolitic dolomite (1 to 3 meters), a 1 to 2 meter coquinoid unit packed with the uncrushed remains of small bivalves and informally known as the '*Bakevellia* Bed' and a 10 to 30 meters unit of sparingly shelly to coquinoid oolitic dolomite

---

[1] Published by permission of the Director, Institute of Geological Sciences.

TABLE 1.—CLASSIFICATION OF PERMIAN STRATA AT
OUTCROP IN YORKSHIRE. THE PATCH-REEFS ALL LIE
IN THE LOWER SUBDIVISION OF THE LOWER
MAGNESIAN LIMESTONE

| | |
|---|---|
| Upper Marls and Evaporites | |
| Upper Magnesian Limestone | |
| Middle Marls and Evaporites | |
| Lower Magnesian Limestone { | Upper Subdivision |
| | Hampole Beds |
| | Lower Subdivision |
| Basal Permian (Yellow) Sands | |

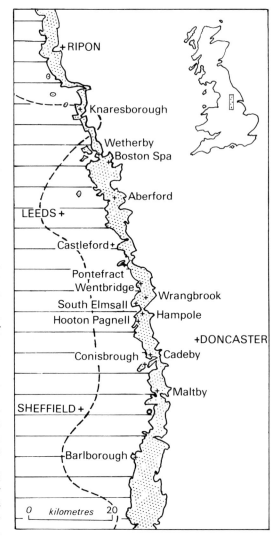

FIG. 1.—Sketch map showing the outcrop of the Lower Magnesian Limestone (stippled) and the position of places mentioned in the text. Carboniferous strata crop out to the west (ruled), and the broken line is an estimated position of the Zechstein shoreline when the reefs were formed.

which, near the top, commonly includes a 0.5 to 1 meter bed of oncoids. Fenestral oolites (0–3 meters and thin multicolored mudstones of the Hampole Beds, overlying a widespread discontinuity (Smith, 1968), complete the sequence. The thick oolitic unit ranges from thin-bedded to massive, generally features evidence of current action in the form of ripples, low-angle planar and trough cross-laminations and minor channels, and is locally bioturbated. It is the prime building stone of the outcrop area. Additionally this unit locally is rich in coated and multiple grains, irregular oncoids up to 1 centimeter across and rounded platy coated clasts up to 15 centimeters across of reworked contemporaneously cemented oolite. Scattered films of dolomite mudstone bear testimony to phases of quiet deposition. Laminae and thin beds of red, green and grey mudstone (which die out against the reefs) are witness to periodic influxes of terrigenous sediment.

From the lithological, sedimentological, and faunal evidence I infer that most of the Lower Magnesian Limestone in this oolite belt was formed in inshore open shelf conditions of moderate energy at depths of perhaps 2 to 10 meters. Except near the top of the oolites, and in the Hampole Beds, there is no unequivocal evidence of phases of subaerial exposure.

The reefs appear to be confined to the thick oolite unit overlying the *Bakevellia* Bed, although some project slightly above the general erosional level of the Hampole Discontinuity and here are surrounded by low cliffs. Where the *Bakevellia* Bed and underlying Zechstein carbonates are absent, through onlap against eminences on the underlying transgression surface, some reefs rest directly on Carboniferous strata. They occur at all levels in the oolite unit, but are most abundant in beds immediately overlying the *Bakevellia* Bed where in places they probably covered half of the local sea floor. Here, with growth, they commonly merged into extensive reef complexes. Merging of adjoining reefs is also a feature in upper parts of the oolite unit. In common with many Paleo-

zoic patch-reefs, reef detritus in adjoining bedded rocks is minimal.

### SIZE, SHAPE AND EXTERNAL RELATIONSHIPS OF THE REEFS

Excluding complexes of coalesced reefs, the reefs in the Lower Magnesian Limestone range from 0.3 meter to perhaps 120 meters across and from 0.15 meter to more than 20 meters in thickness. Most commonly they are 10 to 25 meters across and 3 to 8 meters thick. Many of the larger reefs originate just above the *Bakevellia* Bed or an equivalent horizon.

The smallest and simplest of the reefs are bun-shaped to hemispherical, with almost flat bases (Fig. 2), and lie mainly low in the thick oolite unit. Such reefs are unusual, however, and most of the reefs are compound. In plan, as seen in Cadeby Quarry, Cadeby, where more than twenty reefs have been removed from an area of about a square kilometer, they were clearly shown to be circular or oval. In section in random exposures they assume a wide variety of shapes, a selection of which is shown in Figure 3. Common to many is a theme of early expansion from a central 'root' area, producing a shallow inverted cone shape as successive beds become reefoid, and a gently domed top. The margins of the reefs slope at all angles, recording phases of expansion, migration, standstill and contraction. Interdigitation with the surrounding oolites is common (Fig. 4A). Most of the reefs are roughly symmetrical, and I have recognized no marked differences in their structure and shape between their seaward and landward faces. A few reefs, including several at Cadeby Quarry, are tall, relatively narrow and strongly domed (Figs. 3D and 5), while others are markedly asymmetrical (Figs. 3B and C). None of the latter were seen in plan and their exact trend is therefore unknown, but they appear to mark an eastwards (e.g., basinwards) facing step on the sea floor. Evidence of contemporaneous erosion of the reefs is generally slight or absent.

The sharpness of the margins of the reefs varies with their stratigraphical position in the thick oolite unit. Reefs in the lower two-thirds of this unit are almost all sharply defined (Fig. 4B), reflecting the marked differences in lithology between reef rocks and their bedded equivalents. In the upper part of the oolite unit the lithological contrast between reef and non-reef is less marked than in the lower parts, and reef-margins are commonly gradational. Where reefs extend from low parts of the unit into higher parts, the margins are clearly defined below and less so above.

The relationship of the reefs to enclosing sediments, as clearly displayed in many field exposures, suggests that the upper surface of symmetrical reefs generally was less than 2 meters higher than adjoining oolites and that in many reefs the difference in level was less than half a meter. The top of the largest reefs, such as those at Cadeby Quarry, may have stood as much as 8 meters above the level of the surrounding sea floor if present relationships do not partly reflect differential compaction, for which I see little other evidence. For the asymmetrical reefs, relationships to adjoining bedded rock commonly suggest a step height of 1 to 2 meters. The sediment on the high side is roughly level with the top of

FIG. 2.—Simple hemispherical reef in bedded skeletal oolite; old quarry on western fringe of Conisbrough.

the reef. Dips in beds adjoining the reefs are commonly almost horizontal to within a meter or so of the reef-margin, and quaquaversal dips are generally restricted to sediments overriding the reefs or banked up against their margins.

In size, shape and their relationships to enclosing strata, the patch-reefs of the Lower Magnesian Limestone of Yorkshire are similar to Paleozoic patch-reefs in many other areas including those of the Lower Cambrian of southern Labrador (James and Kobluk, 1978), the Silurian reefs of west-central England (Scoffin, 1971; Riding, this volume) and the Hoburgen-type reefs of the Silurian of Sweden (Manten, 1971). The closeness of the similarity to the Labrador reefs, notwithstanding the faunal differences, is particularly striking and this similarity extends also to much of the internal structure. All these Paleozoic reefs are interpreted as having grown in generally similar open marine shelf environments.

### REEF STRUCTURE AND COMPOSITION

Despite the great variation in the size and shape of the patch-reefs, almost all those in the lowest two-thirds of the thick oolite unit comprise a heterogeneous assemblage of sack-shaped or pillow-shaped masses of bryozoan biolithite. Most of these masses are 0.5 to 1.5 meters across, rarely exceeding 2.5 meters, and they are commonly 0.3 to 1 meter thick. For convenience and brevity in the remainder of this paper these masses will be termed 'saccoliths.'* These are most readily seen in craggy natural outcrops (Fig. 6) where weathering has accentuated mutual contacts, but they are generally discernible even in fresh quarry sec-

---

* Saccolith—a convenient term used here to denote irregular sack-shaped, pillow-shaped or bun-shaped discrete boulder-sized masses of autochthonous reef rock, of whatever composition, within a reef or reef complex.

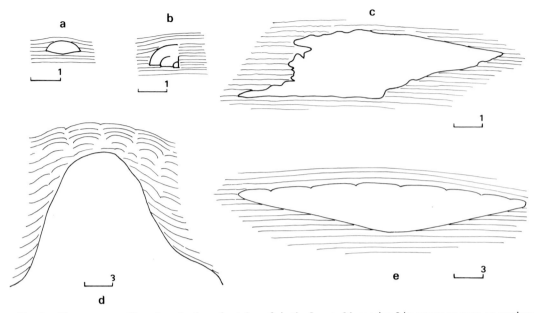

FIG. 3.—Transverse profiles of a selection of patch-reefs in the Lower Magnesian Limestone as seen on random rock faces; scale bars in meters. The examples are from Conisbrough, Aberford, Wentbridge, Cadeby and South Elmsall.

tions. The division of some of the Yorkshire patch-reefs into pillowy masses was first noted by Mitchell *et al.* (1947) at Conisbrough, and reefs at Aberford were subsequently similarly described by Edwards *et al.* (1950). It is possible that the common description of many of the other reefs as 'breccia' also alludes to this distinctive structure, but the size of the component 'clasts' is not recorded in these descriptions. The 'pile of sacks' structure of the early reefs is progressively less distinct both in higher parts of the larger reefs and in reefs originating in the uppermost third of the thick oolite unit. Here an equally distinctive domed stromatolitic structure supervenes (Smith, 1974), and bryozoa are uncommon or absent. Thus the largest reefs are composite structures, composed mainly of masses of bryozoan biolithite in their lower parts and of domed algal stromatolites in their higher parts.

Most of the saccoliths of the early and middle-stage patch-reefs comprise a single relatively uniform unzoned block of biolithite that appears to have grown uninterruptedly to its present size (Fig. 7). The reefs apparently grew by the formation of other similar masses either on top of or beside their predecessors. There is a general tendency for the saccoliths to be elongated roughly horizontally and some are lensoid, giving many reefs a crudely bedded aspect when viewed from a distance (Fig. 6). In other reefs no such order

is apparent, and the saccoliths appear to be disposed randomly. In yet others the saccoliths wrap around each other so as to give rise in places to roughly domal mound-like structures each a few meters across and up to 3 meters high (Fig. 8). A fourth variant, chiefly found in asymmetrical step-like reefs, is for many of the saccoliths to slope in one direction (Fig. 9). Because many of the masses are interlocking and generally tightly packed it seems likely that they lie in their original growth position. Although displacement and detrital accumulation of some of the smaller masses and perhaps toppling and slight redistribution of some of the larger ones cannot be ruled out.

In addition to the predominant unzoned saccoliths described above, some reefs also contain scattered subspherical bodies up to 1 meter across of bryozoan biolithite with a strong concentric structure (Fig. 10) that presumably indicates periodic rather than uninterrupted growth. These bodies resemble masses of *Acanthocladia* biolithite in parts of the slightly younger shelf-edge reef in County Durham (Smith, this volume). They also appear to be somewhat similar to small bioherms reported from the Pennsylvanian rocks of northeast Oklahoma by Bonem (1978).

In several respects it is appropriate to regard the saccoliths in the Yorkshire reefs as individual reefs and the patch-reefs in which they occur as reef agglomerations or complexes. Whatever ter-

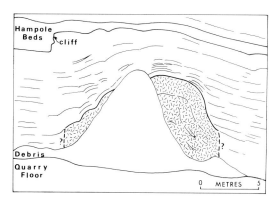

FIG. 5.—Bryozoan reef-core (ornamented) overlain by gently domed algal stromatolitic reef dolomite; north face of Cadeby Quarry. Note the buried cliff (top left) cut into stromatolitic rock by the erosion surface beneath the Hampole Beds.

FIG. 4.—Details of reef-margins at Hooton Pagnell; A. blunt ended thick wedge of skeletal oolite (center right) overlies, underlies and abuts against masses of bryozoan biolithite; B. skeletal oolite (right) abuts against steep face of bryozoan biolithite reef, scale 25 centimeters.

minology is employed, it is clear that the contacts between adjacent saccoliths and between saccoliths and adjoining sediment mark primary rock boundaries. Between saccolith and sediment these boundaries abruptly separate organically supported recrystallized carbonate from grain-supported skeletal oolites, and between adjoining saccoliths they commonly separate areas of rock in which the biotic elements trend in different directions.

The composition of the sack-shaped masses of the Yorkshire patch-reefs is remarkably uniform, both within individual reefs and the reefs in general. Although proportions vary considerably, al-

most all the masses investigated are composed of a complex network of slender ramose (arborescent) bryozoans (commonly 5 to 30 percent, Fig. 11) embedded in a matrix of slightly turbid dolomite microspar (predominant) and dolomicrite, which also contains bryozoan debris (locally abundant) and scattered patches rich in molds of shelly invertebrates. Widely scattered mainly small (less than 5 millimeters) cavities, many elongate, comprise 2 to 8 percent of the volume of most specimens of reef rock. While some are fossil molds or internal cavities, many probably result from volume changes associated with dolomitization. Blocky calcite lines or fills a small proportion of these cavities in most specimens, and a few irregular patches of fine detritus may mark the site of primary voids. The bryozoa in most reefs are overwhelmingly dominated by *Acanthocladia anceps* (Schlotheim) but molds and remains of *Thamniscus dubius* (Schlotheim) are locally predominant, and Kirkby (1861) also recorded *Protoretopora ehrenbergi* Geinitz and *Batostomella* sp. in bryozoa-rich 'beds' (presumably the rocks that are here termed reefs since bryozoa are rare in other parts of the Lower Magnesian Limestone). Shelly invertebrates, where present in the reefs, likewise comprise a limited suite. The biota is dominated by the small supposedly byssate bivalve *Bakevellia binneyi* (Brown) with smaller numbers of the bivalve *B. ceratophaga* (Schlotheim), *Schizodus obscurus* (J. Sowerby), *Liebea squamosa* (J. de C. Sowerby), and of the small pedunculate brachiopod *Dielasma elongatum* (Schlotheim), and small (mainly juvenile) gastropods. Foraminifers and ostracodes are sparingly found, and may formerly

FIG. 6.—Crags of reef dolomite on hillside at Maltby. A. General view; the tree is about 3 meters high. B. Details, showing interlocking masses of bryozoan biolithite; the hammer (center right) is 0.35 meter long.

FIG. 7.—Individual masses of bryozoan biolithite in old quarries 4 kilometers southwest of Conisbrough. A. Small mass; the coin is 26 millimeters in diameter. B. Large mass; the hammer is 0.35 meter long.

have been abundant. Advanced diagenetic alteration of many reefs and of parts of others have virtually obliterated their original fossil content, but their reefoid origin is clearly attested by the almost ubiquitous saccoliths.

The relative proportions of the faunal components of the reefs differ sharply from those of the enclosing bedded oolites. Here accumulations of shelly remains are dominated by *Bakevellia* and *Schizodus* with local concentrations of small gastropods, but with only scattered reworked fragments of bryozoa except near the reefs.

The arrangement of organic remains in biolithite of the reefs, as revealed by the study of weathered surfaces, polished slabs and thin-sections, ranges from disordered to weakly ordered. In the other parts of many (perhaps most) saccoliths there is a noticeable tendency for *Acanthocladia* stems to be either roughly vertical or preferentially oriented, both lengthwise and in the plane of branching roughly parallel with the periphery of the mass of which they form part. This tendency is particularly strong in concentrically zoned saccoliths. The arrangement is closely comparable with that noted in the *Acanthocladia* biolithite

masses of the shelf-edge reef in County Durham. The Yorkshire saccoliths differ mainly in lacking evidence of penecontemporaneous laminar encrustations and in generally being somewhat smaller. Although many of the *Acanthocladia* and other bryozoan stems and branches are fragmentary, the longer specimens (many several centimeters long) are not fully exposed. I suspect them to be part of complexly branching roughly upright bushy colonies in growth position and possibly many centimeters high. In the few places where central parts of the saccoliths could be sampled, roughly vertical branching bryozoan stems predominate in parts while roughly horizontally layered detritus and biolithite predominate in others. Molds of bivalves are unevenly distributed in the masses, locally forming perhaps 30 percent of the rock but in most places being rare or absent. The valves are almost invariably separate and, although disposed at all angles, show some preference to concave upwards, roughly horizontal orientation. The brachiopods (all *Dielasma*) have not been disarticulated, and neither they nor the bivalves have been crushed during lithification.

The incoming of stromatolitic rocks in upper parts of many of the larger reefs and in reefs in the uppermost third of the thick oolite unit is

FIG. 10.—Concentrically layered bryozoan biolithite in large patch-reef; north side of disused railway cutting, 1 kilometer south-southwest of Cadeby.

FIG. 8.—Wrap-around masses of bryozoan biolithite forming dome-like mounds in patch-reefs; sides of abandoned railway cutting 2 kilometers northeast of Wrangbrook, the hammer is 0.35 meter long.

patchy, and apparently at least partly controlled by local factors. In many reefs, layers of stromatolitic laminites are present on the top and sides of the domal mounds built of wrap-around saccoliths and are overlain by further bryozonal reef dolomite. There is a general trend towards an increasingly large stromatolite component in suc-

cessively younger domal mounds and youngest parts of many reefs are entirely stromatolitic. As in the linear reef of County Durham (Smith, this volume), these stromatolitic dolomite rocks are composed of finely crenulated laminae. The laminae are commonly disposed in minor hemispherical or flat-topped domes a few centimeters across which are themselves arranged in complex domes up to several meters across and up to 2 meters high (Figs. 5, 12, and 13). Confluence of the stromatolitic parts of reefs in a number of places, such as Cadeby Quarry and parts of the railway cuttings at Wrangbrook, has resulted in the whole of the uppermost few meters of the rock beneath the Hampole Discontinuity being stromatolitic over substantial areas. Diagenetic changes have oblit-

East                                                                    West

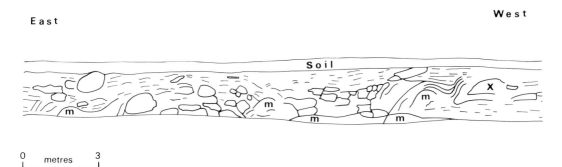

0   metres   3

FIG. 9.—Asymmetrical reef exposed in pipeline trench 1 kilometer east of Wentbridge. The thin-section shown in Figures 11A and 14 is from point X. Areas marked 'm' are of massive bryozoan biolithite similar to that in the saccoliths.

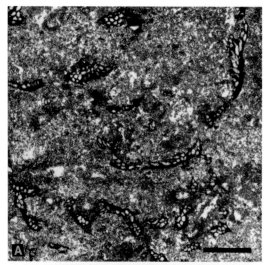

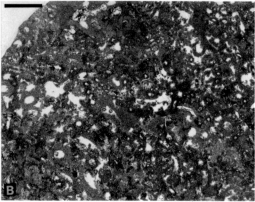

Fig. 11A.—Thin-section of bryozoan biolithite showing bryozoa (probably *Thamniscus dubius*) in a matrix
of dolomite microspar and dolomicrite. Most of the
white areas are voids, some are of blocky calcite, scale
bar 2 millimeters, pipeline trench, Wentbridge, exact
position shown on Figure 9. 11B.—Thin-section of bryozoan biolithite cut normal to the axes of most *Acanthocladia* which appear as doughnut-shaped and elongate
patches of dolomite microspar surrounding cavities,
scale bar 4 millimeters, cutting on disused colliery railway, Cadeby.

erated fine structures from most of the stromatolitic rocks of the Yorkshire reefs. In many
places only the broad outlines of domes attest
their algal origin.

### EVOLUTION OF REEF FAUNAS AND ENVIRONMENTS AND STAGES OF REEF GROWTH

The exceptionally low diversity of the Lower
Magnesian Limestone reef faunas compared with
that of most other patch-reefs, both fossil and
Recent, merits special consideration. With only
one or two species apparently capable of creating
an environment different from that of the surrounding sea bottom, and these slender ramose
bryozoans, rather unusual environmental conditions are indicated.

Clues to these conditions may lie in the nature
of the faunas of the bedded oolitic carbonates surrounding the reefs and also in the faunas and lithology of parts of the Middle Magnesian Limestone reef of County Durham and adjoining areas.
In discussing faunas of first cycle marginal carbonates of the 'Bakevellia Sea' and the contemporaneous Zechstein Sea, Pattison (1970) recognized a faunal diversity gradient which he inferred
declined in the direction of decreasingly favorable
environmental conditions. Amongst bivalves,
Pattison showed that *Bakevellia binneyi* occurs
alone in marginal rocks remote from the open sea.
However this form is joined in turn by *Schizodus*,
*Permophorus*, *Liebea*, *Pseudomonotis* and/or
*Parallelodon* where inferred conditions were less
unfavorable. Other fossils show a similar trend
with euryhaline brachiopods such as *Dielasma*
and tolerant bryozoa such as *Acanthocladia* (both
of which appear only where at least three bivalve
species are present), are joined farther offshore
by a range of stenohaline forms such as *Horridonia* and *Fenestella*. By this rough scale, the
patch-reefs in Yorkshire evidently formed in conditions intermediate between normal marine and
somewhat abnormal. In the context of the Zechstein Sea it seems more likely that this abnor-
mality stemmed from high salinity rather than low
salinity: a figure of perhaps 40 to 45 parts per
thousand might be appropriate. Abnormally high
salinity has also been invoked by Trechmann
(1913 and later) and other workers to explain the
progressive eclipse of reef faunas in younger parts
of the Durham shelf-edge reef. Here a low diversity fauna dominated by *Acanthocladia*, *Dielasma*, and a limited suite of bivalves and gastropods
characterizes the reef-flat sub-facies (Smith, this
volume). In Durham, however, laminar encrustations form most of this younger part of the bulk
of the reef and autogenic community succession
and progressive changes in water depth might be
contributory factors to the low diversity.

From the absence of reefs in the lowest few
meters of the Lower Magnesian Limestone, it
seems probable that conditions were not suitable
for reef formation until some time after the initial
transgression. Bivalves clearly lived here in enormous numbers at the time when the coquinoid
*Bakevellia* Bed was laid down. Here, and throughout the deposition of the overlying thick oolite
unit, neither the bivalves nor the gastropods grew
to their normal sizes except in isolated areas
(Kirkby, 1861). This suggests that even for them
conditions were clearly not altogether favorable.

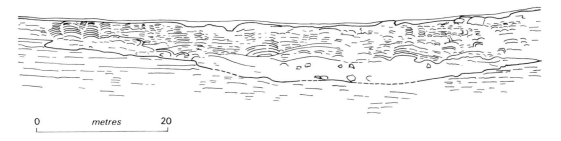

0    *metres*    20

FIG. 12.—Sketch of typical symmetrical reef in middle part of the Lower Magnesian Limestone. Detailed composition has been obscured by diagenesis but the upper part is thought to be mainly stromatolitic and the lower part mainly bryozoan. The stromatolitic rocks extend at least another 30 meters to the right of the area depicted; east wall of old quarry, South Elmsall.

Yet it was in these conditions, perhaps surprisingly, that colonization by bryozoa began shortly after the *Bakevellia* Bed had been formed. This relationship is similar to that in the Durham Reef which also overlies a coquina (Smith and Francis, 1967). One inference is that the coquina modified the substrate by providing a stable surface to which bryozoa could attach themselves. It is no doubt significant that accumulations of shelly detritus have been found by Alberstadt *et al.* (1974) and Walker and Alberstadt (1975) to be a common preliminary to colonization by reef forming organisms.

The evidence shows that, where patch-reefs developed, the sea floor at the site of each initial sack-shaped or concentric mass was colonized by bushy bryozoan growths. The stems of which, occurring in great numbers, became major rock formers. The situation was admirably summarized by Kirkby (1861) who wrote:

"The polyzoan beds also seem to be at first sight an accumulation of drifted materials. This is only a natural surmise on finding so fragmentary an assemblage of remains as these which enter into the composition of these strata. Nevertheless I am disposed to consider them the remains of *Polyzoa* that lived where we find them and that the range of the beds marks the site of an ancient ground or zone altogether (or nearly so) peopled by *Polyzoa*—where they lived and died generation after generation for a long period, the later generations growing on a sea bottom composed of polypidoms that preceded them until at last, owing to reasons unknown, their growth ceased after the accumulation of several thick beds of their remains."

I differ from Kirkby in believing that many of the bryozoan remains are in growth position. Their widely upright orientation would otherwise be difficult to explain. I also disagree with the view

that although they are the most abundant fossil, the bryozoans generally form only a small proportion of the rock. Much of the remainder, now dolomicrite and dolomite microspar, probably originated as early submarine cement and as lime mud. The latter accumulated in the thickets of arborescent and presumably flexible bryozoan stems and branches. Thus they were protected from removal by the currents as evidenced by sedimentary structures in adjoining shelly carbonate grainstones. Parallels could be drawn with the manner in which rough surfaced pebbly areas function as reservoirs of sand in many rocky deserts and protect it from removal by all but the strongest winds, and by the similar protective role of marine grasses on modern carbonate mud-banks. The scattered films of lime mud in grainstones surrounding the reefs may, by selective winnowing, have been the source of the micrite in the patch-reefs. The presence and local abundance of ooids in some reefs bears testimony to the introduction of sediment from outside. Heckel (1974), in seeking to

FIG. 13.—Detail of algal stromatolitic dolomite of South Elmsall Reef, resting on bedded oolite (bottom left).

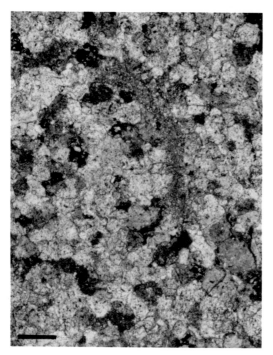

FIG. 14.—Photomicrograph (polarized light) of typical reef rock, showing recrystallized skeletal fragment in matrix of dolomite microspar, scale bar 0.1 millimeter, Maltby, same reef as in Figure 8.

explain the abundance of lime mud in many other Paleozoic reefs where external potential sources of such mud are not evident, suggested that much mud might be derived by the breakdown of indigenous reef organisms including some such as algae. These have left no other trace of their former presence. If this view were correct, the reefs might have acted as lime mud source areas as well as reservoirs. By comparison with other reefs throughout Phanerozoic and Recent time, sediment was probably also contributed to the reefs by boring, burrowing, scavenging and predatory organisms. The visible effects of such activities are scanty in the dolomitized Yorkshire reef rock.

If the views outlined above are correct, it follows that although the reefs are there because of the presence of dense thickets mainly of *Acanthocladia*, the bryozoa probably did not initially form a rigid framework. Instead they formed a lattice of stems that probably acted firstly as nuclei for submarine cements and secondly as baffling and protecting agents, allowing mud to which they may have contributed to accumulate and to provide mutual stiffening, and also acting in a secondary binding capacity. Other organisms such as foraminifers undoubtedly inhabited ecological niches between, within and on the bryo-

zoans, and soft-bodied organisms which have left no recognizable trace of their passing may have been present. It seems likely from the rock structure and bryozoan/matrix relationships that these other organisms played only a minor part in reef construction. A list of references to reef structures in which bryozoa appear to have acted chiefly in a sediment baffling and trapping role is given by Cuffey (1977, p. 189). The apparent widespread absence of laminar encrustations, whether algal or inorganic, is particularly noteworthy in view of their prevalence in *Acanthocladia* biolithite of the Durham shelf-edge reef (Smith, this volume).

As with the bryozoa, there are unresolved doubts on the role of bivalves in the reef communities. Some, such as *Schizodus,* are thought to have been burrowing forms, but I have seen no undoubted burrows of an appropriate size in these reefs. Most of the others, mainly *Bakevellia, Liebea,* and *Permophorus,* are regarded as byssate forms that attached themselves to a firm substrate and might have been attached to the bryozoa. No evidence of such attachment is apparent in the reef rock, and it is clear from the wide distribution of all these bivalves in areas where reefs are absent that they could and did thrive in non-reef environments. In view of this, and in the light of their patchy occurrence, disarticulated state, and weakly preferred orientation in the reefs, I suspect that the bivalves may not have been part of the reef community except incidentally. Most of those now found in the reef rocks were swept in by currents from adjoining areas and trapped (and supported at all angles) by the twig-like bryozoans. In contrast, the small pedunculate brachiopod *Dielasma* is much more common in the reefs than outside, and probably lived there, as did many foraminifers, gastropods and ostracodes.

The total range of species demonstrably indigenous to the Yorkshire patch-reefs is so low that it is doubtful if the stage of diversification widely recognized in reefs by Alberstadt *et al.* (1974) and Walker and Alberstadt (1975) was ever fully achieved. Perhaps the community progressed directly from colonization to dominance.

The progressive upwards diminution in the proportion of bryozoa and other invertebrates in the reefs and the incoming of algal stromatolites is matched by a similar upwards decline in the bivalve-gastropod faunas of the surrounding oolites. Only in their uppermost parts do oolites generally contain a few scattered fragmentary fossils. A similar progressive upwards impoverishment of initially more varied shelly faunas in this formation was noted in somewhat deeper water facies north of the study area by Pattison (1978), and in time equivalent outer-shelf and slope rocks in southern County Durham (Smith and Francis,

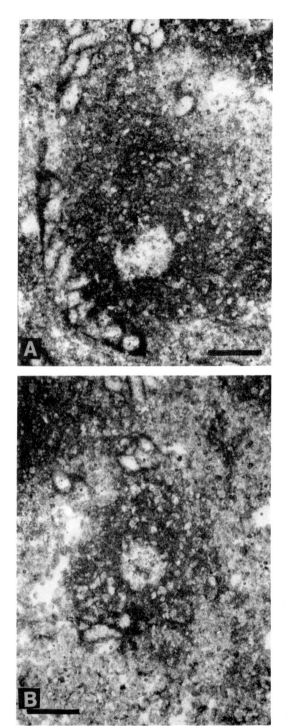

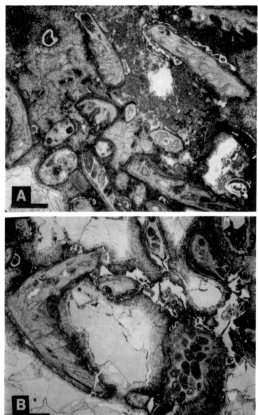

FIG. 16.—Photomicrographs of bryozoan biolithite from patch-reef at Aberford, Yorkshire, showing ramose bryozoa (mostly probably *Thamniscus dubius*) and encrusting foraminifera (*?Calcitornella*, white) with widespread isopachous fringes of radial-fibrous dolomite (grey). The larger white and pale grey patches are of coarse-grained blocky calcite, probably vadose; scale bars 0.5 millimeter.

FIG. 15.—Masses of small tubular structures (?skeletal algae) in bryozoan biolithite. Calcite microspar fills or lines central cavities and also internal cavities in the bryozoa, scale bar 0.4 millimeter; same thin-section in Figure 11A.

1967). Upward growth of the reefs towards the surface of the sea may, by altering environmental conditions, have led to the evolutionary changes in the reef biota. Still it is doubtful if this would explain the changes in the fauna of the surrounding oolites since shells there have been transported from elsewhere. Instead it seems more likely that the faunal attenuation in both reefs and their surroundings is a response to progressively worsening conditions (probably increasing salinity). This eventually exceeded the tolerance of all the invertebrates and left the blue-green stromatolitic cyanophytic algae in charge of the scene. The wide distribution of stromatolitic laminites in upper parts of the thick oolite unit suggests that few browsing organisms existed. The absence of evidence of strong erosion, emer-

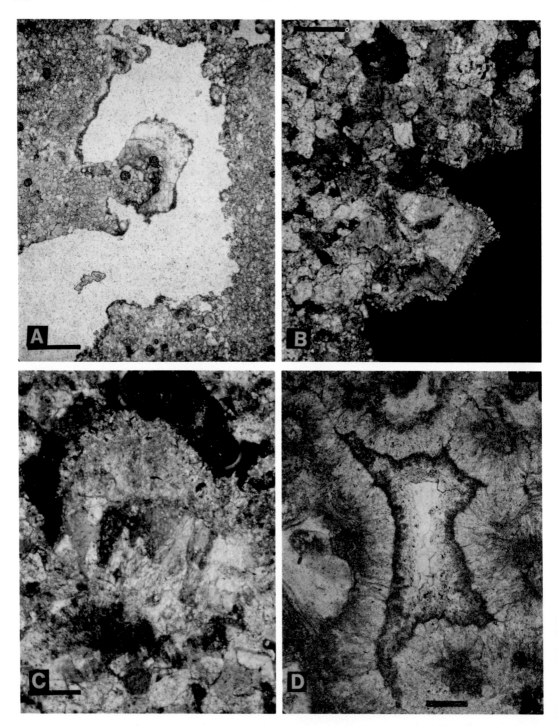

FIG. 17.—Photomicrographs of partial cavity fill in reef rock, Maltby (same reef as Fig. 6B), A to C, and Aberford, D. A—Projecting growth of weakly concentrically layered faintly radial-fibrous dolomite in cavity in dolomite microspar, with narrow coating (dark grey) of pyramidal calcite; scale bar 0.4 millimeter. B—As above, same thin-section, polarized light. Radial extinction, though present in the apparently large crystals (lower center) is very faint; scale bar 0.1 millimeter. C—As B. Fossil fragment (dark-grey, bottom left) is overlain by weakly fibrous columnar dolomite (center, pale) and that by weakly radial-fibrous dolomite (top center, semicircular)

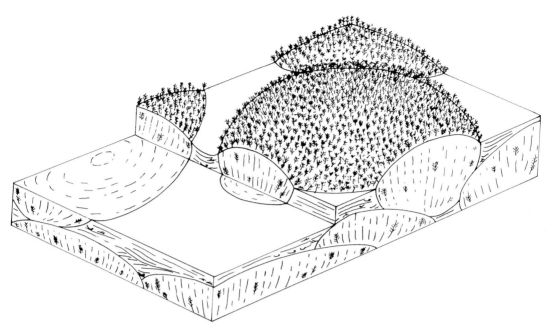

FIG. 18.—Possible sea floor relationships during formation of the patch-reefs in the Lower Magnesian Limestone of Yorkshire, showing saccoliths thought to have formed around colonies of *Acanthocladia anceps*; the area depicted is about 1 × 2 meters. Small gastropods, brachiopods and bivalves, which are locally abundant, are omitted.

gence or even of intertidal conditions in uppermost reef and equivalent rocks leads me to envisage subaqueous formation for most or all of the stromatolitic reef rocks. Similar environmental conditions are postulated to those in which formed the Hesleden Dene stromatolite biostrome of County Durham (Smith, this volume). The tops of the stromatolitic reefs high in the formation were subaerially exposed and locally eroded (see Fig. 5) when relative sea level fell by several meters and the Hampole Discontinuity was cut.

### PETROLOGY AND CEMENTATION OF THE REEF ROCKS

Although no detailed study of these important aspects has been undertaken, some comments are necessary as a foundation for further discussion. The main need is to establish whether cementation was penecontemporaneous or took place after burial.

None of the thin-sections examined provided evidence specifically meeting this need, and no fragments of autochthonous reef rock were found in the scanty talus. Oriented specimens of biolithite from the margins and interior of saccoliths from several reefs almost all proved, on petrological examination, to be less informative than had appeared likely from the appearance of slabbed and weathered surfaces. Dolomite microspar (average size about 0.04 millimeter) forms a high proportion (commonly 50 to 80 percent) of several specimens. Fossils, mainly stems or branches of bryozoans, are completely recrystallized and represented by halo-like aggregates of dolomicrite and fine-grained dolomite microspar (Fig. 14), commonly surrounding cavities or patches of secondary blocky calcite. The dolomite in both matrix and fossils generally forms an equant mosaic of subhedral crystals with abundant anhedra and scattered euhedra. Dolomite in the matrix has totally replaced the original material, presumed to have been cement and carbonate mud. In a few places, diagenetic changes have been less extreme than in general, and these less

←

fringed with small pyramidal crystals of calcite on cavity walls; scale bar 0.1 millimeter. D—Isopachous radial-fibrous dolomite coating bryozoa and partly filling cavity. Much of the remaining cavity is filled with blocky calcite; scale bar 0.1 millimeter.

altered rocks provide valuable clues to the possible primary fabrics of their severely altered counterparts. Dimensions and the main features of the bryozoans are clear in these places, and the matrix contains traces of tubular structures that may be the remains of marine skeletal algae (Fig. 15). Exceptionally well preserved bryozoa, many with encrusting foraminifers, were found in a single specimen of reef dolomite from Aberford (Fig. 16).

Although small cavities are abundant in all the thin-sections examined, none appear to contain internal marine sediment. A search for clear evidence of early marine cements and encrustations proved almost equally fruitless. Only a few small cavities (Fig. 17A through C) have an encrusting fringe of unevenly fan-extinguishing dolomite that may have replaced early radiaxial-fibrous calcite or aragonite. No such encrustations could be recognized on the outsides of most fossils and it is clear that if they or other encrusting marine cements were ever present, few traces of them survived the ensuing neomorphism and dolomitization. In the specimen from Aberford (see above), however, the bryozoa and most other biotic elements bear isopachous fringes and fans of radial-fibrous dolomite (Fig. 16), which presumably had calcite and aragonite precursors. These figures are commonly 0.05 to 0.25 millimeter thick, exceptionally exceeding 0.7 millimeter, and fill or almost fill most of the cavities less than 1.2 millimeters across (Fig. 17D). Locally they comprise 70 percent or more of the rock. Nowhere in this specimen were marine organisms seen to encrust the isopachous fibrous fringes, so that an early marine origin for these cannot be proved. However, subaqueously precipitated, morphologically similar, marine radial-fibrous aragonite has been widely reported coating frame and other elements in modern patch-reefs. A similar early marine origin appears possible for the fibrous fringes in the Yorkshire Permian reefs. Partial micritization of the fibrous dolomite in part of the Aberford specimen hints at a possible origin of the haloes of finely crystalline dolomite surrounding organic elements in many other Yorkshire reef rocks. None of the reefs examined has yielded evidence of former laminar encrustations such as characterize *Acanthocladia*-rich rocks of the Durham shelf-edge reef. This suggests that the organisms presumed to be responsible for such encrustations in Durham were rare or absent in the Yorkshire reefs.

Despite the apparent lack of clear petrographic evidence of early cementation, the lack of crushing of the delicate bryozoan stems and of other invertebrate remains probably indicates that there was little early compaction in the inferred original carbonate mud matrix. Similar evidence in many other former carbonate muds was discussed by Bathurst (1970), who concluded that cement was precipitated in the intergranular pores in sufficient quantity to form a resistant framework before the overburden was thick enough to cause detectable compaction. In the Yorkshire patch-reefs the lack of compaction of early saccoliths by their successors is useful supporting evidence of early cementation, as is the presence of scattered cavities (now filled) too large to have survived in unconsolidated mud. The steep and locally overhanging marginal slopes of many saccoliths are also difficult to explain without such cementation as it appears unlikely that the binding capability of the bryozoa would be adequate. I believe that these gross relationships probably outweigh the lack of unambiguous petrographic evidence, and that early cementation was probably prevalent in the reefs.

### THE ORIGIN OF SACK-SHAPED MASSES (SACCOLITHS) IN THE REEFS

It would be unsatisfactory to leave the subject of these reefs without some discussion of the reasons why they are divided into sack-shaped masses, and why these masses are relatively uniform in size and only rarely exceed 2.5 meters across. Whether the answers to these questions apply also to the origin of somewhat similar masses (but with different organisms), noted in many other patch-reefs such as those in the Lower Cambrian of Labrador (James and Kobluk, 1978), is a matter for interesting speculation.

From the presence of scattered pockets and lenses of detrital sediment between saccoliths, and from the fact that the lower surfaces of sediment interdigitations at the margins of the reefs generally coincide with the tops of masses which continue uninterruptedly into the reef on their inner side, it seems probable that the saccoliths are primary features of the rock. Evidence has already been presented to show that they were probably lithified (and therefore provided a firm substrate) before their successors were formed. Furthermore, the general lack of stratification in most reefs and the common near random arrangement of the saccoliths suggests that at any one time, different saccoliths within the reefs were at different stages of growth. If this inference is correct, it has the important corollary that the saccoliths stopped growing in response to factors acting within or very close to them, rather than because of more general environmental changes.

The suggestion that each saccolith had a finite life governed by indigenous or extremely local factors, and grew to a common size, has clear parallels in many modern organisms and organic communities. This raises the possibility that each mass might have been a single basically organic bush-like entity comprising a closely-packed, essentially monospecific or bispecific assemblage of organisms. The question thus posed by these sack-shaped masses in the patch-reefs of Yorkshire is whether or not the slender cryptostome bryozoan *Acanthocladia anceps* (with or without *Thamniscus dubius*) could have constructed colonies commonly up to 1.5 meters across and 1 meter high, and exceptionally up to 2.5 meters across and 1.5 meters high. Despite the seeming improbability that it could, the clear parallelism of *Acanthocladia* stems and branches in peripheral parts of many masses is compatible with such construction. The massive nature of the biolithite means that few masses are exposed in section and I have been unable to find one in which the implicit trunk and root area is visible. Pending such discovery, single colony construction of the masses remains unproved.

Reference to the literature on bryozoans reveals an extraordinary shortage of information on colony size and attachment of erect Paleozoic species, but it appears that colonies of the large size suggested here for *Acanthocladia anceps* would be exceptional. Nevertheless, the organization of bryozoan colonies is not inconsistent with large size. The presence of carbonate mud or calcareous encrustations around earlier parts of such a colony would not result in death, provided that younger parts of the colony remained clear of the substrate. Death would occur ultimately from natural senility when the colony reached its maximum size. The life span of such a colony would presumably be considerable, an aspect on which the literature on fossil bryozoa is also notably reticent. It seems likely that the death of a colony was followed by even more lengthy exposure on the sea bottom, and by cementation or lithification while other nearby colonies continued to grow, and sediment between the reefs continued to accumulate. In the view that there is some actual and circumstantial supporting evidence and little evidence to the contrary, I therefore advance for debate the proposition that the early patch-reefs in the Lower Magnesian Limestone of Yorkshire are composed of discrete biolithite masses because each mass is founded on a single colony of *Acanthocladia anceps*. These characteristically grew to the sizes now observed. An impression of the sea floor during growth of the reefs is presented in Figure 18.

## CONCLUSIONS

1. Patch-reefs are present in large numbers in the Lower Subdivision of the Lower Magnesian Limestone of Yorkshire, England. They are circular or oval in plan and commonly 10–25 meters across and 3–8 meters thick. In cross-section many of the reefs are roughly symmetrical, with a shallow inverted cone-shaped lower part, rounded interdigitate margins and a gently domed top. Some of the reefs are asymmetrical and many are irregular in cross-section.

2. All the reefs seen by the writer lie in a thick unit of oolitic shelly dolomite grainstone and all are completely dolomitized. They are stratigraphically younger than a widespread coquina which, I believe, provided a substrate suitable for colonization by reef forming organisms. Reefs are generally absent where early beds of the formation are argillaceous.

3. Most of the reefs are formed of an untidy pile of irregular sack-shaped masses (0.5–1.5 meters across) for which I propose the name 'saccoliths.' Most saccoliths contain a dense network of stems and branches mainly of the slender cryptostome bryozoan *Acanthocladia anceps* in a matrix of finely crystalline dolomite. It is suggested that these saccoliths each were formed from a single large *Acanthocladia* colony that grew to a broadly uniform terminal size.

4. Most of the reefs projected only slightly above the sea bed, and talus is uncommon. It seems likely that they were formed mainly of early cement and of carbonate mud trapped and supported by dense thickets of *Acanthocladia* and other bryozoan branches. A small number of other invertebrate species is present and some of these probably lived in the sheltered niches on, between, and within the bryozoans. Current concepts of community progression apply only partly to these reefs, where an exceptionally simple community was only locally diversified, and little specialization seems to have been achieved. Algae may have contributed both to bulk and to cementation.

5. Although the petrographic evidence is inconclusive, there are indications that the saccoliths were cemented contemporaneously or penecontemporaneously on the sea floor, and that others were then formed on top of, beside, or around their predecessors.

6. Towards the top of the oolite unit the fauna is attenuated, probably in response to increased salinity, and reefs at this level (and the upper part of reefs extending from lower levels) are composed mainly or entirely of gently domed dolomitized algal stromatolites.

7. I interpret the bryozoa reefs as having been formed subaqueously some kilometers offshore on a broad shallow marginal carbonate shelf in a tropical or subtropical sea. Uppermost (stromatolitic) parts of the reefs were also probably formed subaqueously though probably nearer to the sea surface as a result of earlier upwards reef growth. Evidence of subaerial emersion and erosion is found only on the top of the youngest reefs which were exposed when sea level fell by several

meters, and the widespread Hampole Discontinuity was formed.

#### ACKNOWLEDGMENTS

I acknowledge with thanks the considerable help
and advice of my colleague Jack Pattison on paleontological aspects of this work and for reading
and commenting on the manuscript, and Pat Cook
and Frank Mckinney for advice on special aspects
of the bryozoa. Dr. J. A. Dickson kindly advised
on some aspects of the petrography. Ramois Gal-
lois, John Kaldi, and Gordon Smith also read the
preliminary draft and made many helpful suggestions. Mr. R. K. Harrison kindly prepared the
photomicrographs of Figures 14, 15, and 16, and
Dr. R. Sanderson those of Figure 17. Remaining
photographic figures were prepared by the photography department of the Institute of Geological
Sciences under the direction of Martin Pulsford.

## REFERENCES

ALBERSTADT, L. P., WALKER, K. R., AND ZURAWSKI, R. P., 1974, Patch-reefs in the Carters Limestone (Middle Ordovician) in Tennessee, and vertical zonation in Ordovician reefs: Geol. Soc. America Bull., v. 85, p. 1171–1182.
BATHURST, R. G. C., 1970, Problems of lithification in carbonate muds: Geol. Assoc., Proc., v. 81, p. 429–440.
BONEM, R. M., 1978, Early Pennsylvanian algal-bryozoan bioherms: developmental phases and associations: Al- cheringa, v. 2, p. 55–64.
CUFFEY, R. J., 1977, Bryozoan contributions to reefs and bioherms through geologic time: Am. Assoc. Petroleum Geologists, Studies in Geology, No. 4, p. 181–194.
EDEN, R. A., STEVENSON, I. P., AND EDWARDS, W. N., 1957, Geology of the country around Sheffield: Mem. Geol. Survey Great Britain, 238 p.
EDWARDS, W. N., MITCHELL, G. H., AND WHITEHEAD, T. H., 1950, Geology of the district north and east of Leeds: Mem. Geol. Survey Great Britain, 93 p.
HECKEL, P. H., 1974, Carbonate buildups in the geologic record: a review, *in* Laporte, L. F., Ed., Reefs in time and space: Soc. Econ. Paleontologists & Mineralogists Spec. Pub. No. 18, p. 90–154.
JAMES, N. P., AND KOBLUK, D. R., 1978, Lower Cambrian patch-reefs and associated sediments: southern Lab- rador, Canada: Sedimentology, v. 25, p. 1–35.
KIRKBY, J. W., 1861, On the Permian rocks of south Yorkshire; and on their paleontological relations: Quart. Jour. Geol. Soc. London, v. 17, p. 287–325.
MANTEN, A. A., 1971, Silurian reefs of Gotland: Developments in Sedimentology: Amsterdam, Elsevier, No. 13, 537 p.
MITCHELL, G. H., 1932, Notes on the Permian rocks of the Doncaster district: Yorkshire Geol. Soc., Proc., v. 22, p. 133–141.
————, STEPHENS, J. V., BROMEHEAD, C. E. N., AND WRAY, D. A., 1947, Geology of the country around Barnsley: Mem. Geol. Survey Great Britain, 182 p.
PATTISON, J., 1970, A review of the marine fossils from the Upper Permian rocks of Northern Ireland and northwest England: Geol. Survey Great Britain Bull., No. 32, p. 123–165.
————, 1978, Upper Permian paleontology of the Aiskew Bank Farm Borehole, north Yorkshire: Inst. Geol. Sci., London, Rept. 78/14, p. 1–6.
SCOFFIN, T. P., 1971, The conditions of growth of the Wenlock reefs of Shropshire (England): Sedimentology, v. 17, p. 173–219.
SMITH, D. B., 1968, The Hampole Beds—a significant marker in the Lower Magnesian Limestone of Yorkshire, Derbyshire and Nottinghamshire: Yorkshire Geol. Soc., Proc., v. 36, p. 463–477.
————, 1974, Permian, *in* Rayner, D. H., and Hemingway, J. E., Eds., The Geology and Mineral Resources of Yorkshire: Leeds, Yorkshire Geol. Soc., p. 115–144.
————, this volume, The Magnesian Limestone (Upper Permian) reef complex of northeastern England.
————, AND FRANCIS, E. A., 1967, The geology of the country between Durham and West Hartlepool: Mem. Geol. Surv. Great Britain, 354 p.
TRECHMANN, C. T., 1913, On a mass of anhydrite in the Magnesian Limestone at Hartlepool, and on the Permian of southeastern Durham: Quart. Jour. Geol. Soc. London, v. 70, p. 232–263.
WALKER, K. R., AND ALBERSTADT, L. P., 1975, Ecological succession as an aspect of structure in fossil com- munities: Palaeobiology, v. 1, p. 238–257.

SEPM Special Publication No. 30, p. 203–231, May 1981

# REEF DEVELOPMENT IN THE MIDDLE TRIASSIC (LADINIAN AND CORDEVOLIAN) OF THE NORTHERN LIMESTONE ALPS NEAR INNSBRUCK, AUSTRIA

RAINER BRADNER AND WERNER RESCH
University of Innsbruck
Austria

## ABSTRACT

The Middle Triassic Wetterstein Limestone platform north of Innsbruck, Austria, forms a gently dipping platform that contains various organic buildups. The upper part of this platform is dominated by a massive reef laid down in shallow water (Hafelekar Reef complex), whereas the lower portion of the Wetterstein Limestone contains conspicuous patch-reefs.

The massive shallow water reef in the upper part of the Wetterstein Limestone shows a distinct biotic zonation. This zonation reflects a rather simple pattern of ecological zones attesting to a smooth transition of biological adaptation by various reef organisms.

Reef growth in the patch-reef sequence appears to have been influenced by subsidence associated with sea level changes, and a gradual increase of skeletal debris derived from actively accreting shallow water reefs. It is thought that a sea level rise in excess of 10 meters would prove sufficient to obstruct patch-reef growth. Sea level rise activated short lived basinal sedimentation adjacent to the reef area. Basinal sediments are characterized by radiolarian-bearing limestones. This same event however, initiated a revival of rapid reef growth on the platform. Here reefs are composed of calcisponges, *Tubiphytes,* and tubular foraminifers. Reef and lagoonal areas are separated by a barrier of skeletal sand shoals.

Large scale syngenetic submarine cementation of the reefs was followed by early diagnetic dolomitization. Subsequent sedimentation and cementation produced conspicuous thick coatings of fibrous spar (?aragonite), and these are interpreted as diagenetic fabrics that formed during an early burial stage.

Synsedimentary tectonics was especially active along the rim of the Wetterstein carbonate platform, as part of a much broader tectonic phase that was active and affected the Middle Triassic Tethys. In general, development of the Wetterstein carbonate platform was a regressive phase frequently disrupted by disturbances caused by continued subsidence in association with block-faulting tectonics.

## INTRODUCTION

Middle Triassic reefs figure prominently in the evolution of reef building organisms, for it was during this time that scleractinian hexacorals originated, and were later to become major contributor to modern shallow marine buildups. Other important reef organisms in Middle Triassic buildups came from an older generation of Paleozoic framebuilders, e.g., *Tubiphytes* and trepostome bryozoans. This transitory phase also saw the development of a variety of calcisponges, specifically as the "large organism" component.

Classical Middle-Triassic buildups of the Dolomites of the Southern Alps, and the reefs of the Austroalpine Upper Triassic (Dachstein Reef complex, Rhaeto-Liassic reefs) have been subjects of extensive research, e.g., Flügel and Flügel-Kahler (1963), Zankl (1969), Lobitzer (1974) and Fabricius (1966). Less attention has been paid to the development of Middle Triassic Wetterstein Limestone reefs of the Northern Limestone Alps, although they too offer opportunity for significant studies on reef growth, sedimentation, erosion, and cementation. In contrast to the Dolomite reefs of the same age, the Wetterstein Limestone reefs show well preserved primary structures, due to a lesser degree of dolomitization.

A major objective of this paper is the re-evaluation of the growth and development of Wetterstein Limestone reefs through analyses of their paleoecology, sedimentology, and diagenesis. The authors also seek to encourage renewed discussion of Schneider's (1964) reef model which has survived without major changes since first proposed. The paper presents the first comprehensive description and analysis of a Wetterstein Limestone platform-margin in its various phases of development.

### Previous Studies

Sedimentological research on the Middle Triassic carbonates of the Northern Limestone Alps was undertaken by Sarnthein in 1965, and this included investigation of Wetterstein Limestone platforms and interspersed basinal areas in the region of Innsbruck. This was followed (Sarnthein, 1967) by a paleogeographical interpretation of the platform and basin assemblages. An important achievement was the discovery of strong platform subsidence whose thickness reached

FIG. 1.—Location map of Wetterstein Limestone reefs investigated by the authors: 1. Nordkette Range, 2. Heiterwand Range, 3. Kaisergebirge of the Northern Limestone Alps, and 4. Dobratsch of the Drauzug Range.

1700 meters as compared to neighboring basins of the same age, but where sediment fillings are only 200–400 meters in thickness.

A series of comprehensive studies by Ott (1967 and 1973) emphasized the dominant role of calcisponges as reef builders. These studies also described and listed a number of newly discovered forms of calcisponges. Furthermore, Ott established that dasycladacean algae, while practically non-existent in reefs, were abundantly represented in lagoonal settings. The distribution of certain genera of dasycladaceans appears to indicate the existence of zonal patterns of lagoonal environments.

Additional contributions to reef paleogeography of the Wetterstein Limestone complexes came from Toschek (1968, Northern Limestone Alps, Kaisergebirge) and Colins (1975, Dobratsch, Drauzug Range). In 1973, Wolff completed a study of the Wendelstein Reef complex in Bavaria, showing evidence of various degrees of complex thicknesses which he attributed to variations of subsidence. A direct interfingering of basinal facies (Partnach beds) and reef facies seems to suggest shallow basinal sedimentation. Micropaleontological research undertaken by Mostler (in Donofrio et al., 1979) has resulted in a biostratigraphic classification of this interfingering relationship west of Innsbruck, Austria.

### Regional Setting and Stratigraphy

The Innsbruck Nordkette Range with its component of Wetterstein Limestone forms part of the Northern Calcareous Alps, one of the main structural units of the Eastern Alps (Fig. 1).

The Northern Limestone Alps belong to the Upper Austroalpine, and they represent an intricately structured pile of nappes, forming an overthrust on top of the Northern Flysch Zone. The internal structure of nappe tectonics is still open to interpretation. The problem, however,

does not affect this study as the entire development of platform-margin took place exclusively within a tectonic unit of the Inn Valley Nappe (Donofrio et al., 1979). The Triassic Period saw the Northern Limestone Alps, as well as the Drauzug Range and the Southern Alps, form a wide marginal area (unstable shelf) which bordered in the southeast (southern Turkey, Cyprus, etc.) newly created oceanic zones. This region of mobile rifting affected the unstable shelf and the course of sedimentation was influenced by varying subsidence, block-faulting, and sea level changes ("labile shelf," *sensu* Bechstädt et al., 1978).

In a general sense, Triassic development of the western part of the Northern Limestone Alps took place in three regressive cycles (Fig. 2). Following a transgression on top of the clastic Permoscythian, each cycle developed in a slow buildup and then followed a rather abrupt ending of the carbonate platforms, e.g., Anisian Steinalm Limestone, Ladinian-Carnian Wetterstein Limestone, and the Upper Triassic Hauptdolomite-Dachstein Limestone. The tops of these carbonate platforms of Wetterstein Limestone, Hauptdolomite, and Dachstein Limestone show evidence of erosion and block-faulting. Platforms alternated with basins of varying depth and these sediments are represented as Reifling beds, Partnach beds, and Hallstatt Limestones. Basinal sediments overlay the tops of drowned platforms (e.g., Reifling beds and Raibl beds).

The Northern Limestone Alps have undergone periodic volcanic activity which produced widespread tuffs and tuffites (pietra verde) during the Aniso-Ladinian Stage, as well as local deposits of Cordevolian basic igneous rock (andesite-porphyrite), found near the village of Lech am Arlberg.

Regressive sedimentation is a typical feature of the Wetterstein Limestone platform. In the Cordevolian, the adjacent basins of argillaceous Partnach beds were superceded by a carbonate platform, and Wetterstein Limestone buildups appear to have been initiated on topographic highs, presumably created by tectonic uplift of the Steinalm Limestone platform. By ignoring alpine tectonics, one arrives at the conclusion that there originally existed a series of longitudinal platforms and corresponding Partnach-type basins, however whether their number was two or three is open to interpretation. One thing is certain, and that is that the platform-margin, as analyzed in this paper, forms the southern border of the southernmost platform along the Innsbruck meridian.

In passing, attention should be called to the existence of lead and zinc deposits of economic significance in the uppermost Wetterstein Limestone.

## Scope of the Present Investigation

This study has focused on the development of the platform-margin as exposed in the Nordkette Range (Fig. 1). It includes facies mapping on a topographic scale of 1:2880, aerial photography, and comparative paleogeographic studies in related mountain ranges, e.g., the Dobratsch, Kaisergebirge, and Heiterwand Ranges (Fig. 1). The authors also prepared 50 polished and etched slabs (20 centimeters in diameter wherever possible) and more than 500 large thin-sections.

### REEF DEVELOPMENT IN THE INNSBRUCK NORDKETTE RANGE

The Hafelekar Reef complex is a prominent feature of the Wetterstein Reef Limestone (Figs. 3 and 4). This reef complex marks the final stage of reef development which originated in the patch-reefs of the lower Wetterstein limestone. The section in the Nordkette Range exposes a sequence of predominantly massive Wetterstein Reef Limestone in the lower part, and well-bedded lagoonal sediments in the upper and northern parts (Fig. 5). Thicknesses were found to be in excess of 1700 meters (Sarnthein, 1965). The patch-reef sequence is clearly visible from the city of Innsbruck, and the reef complex itself can be reached by cable car from Innsbruck.

This carbonate buildup is generally regressive and thus typical of Middle Triassic Tethyan buildups exposed elsewhere in Austria. Carbonate platforms constitute an overgrowth of basins, thus delimiting their area. In the Nordkette Range, such development was modified along the margin of this carbonate platform by synsedimentary tectonics, which probably were coupled with sea level changes.

## Patch-Reef Sequence

The section of the Nordkette Range exposed north of the city of Innsbruck, shows thick-bedded basinal sediments of Reifling Limestone passing laterally into a sequence, some 200 meters thick, of irregularly distributed lenticular carbonate buildups which represent the lower part of the Wetterstein Limestone. These lenticular buildups, up to 40 meters thick, are identified as patch-reefs, an assumption based on the evidence of autochthonous growth of reef building organisms. The areas between these patch-reefs are filled with coarse well-washed reef debris. Significantly, influence of the basinal environment (pale-reddish limestones with ammonites and radiolarians) is a recurrent feature of the patch-reef sequence. Detailed investigation of a single large patch-reef (Fig. 3) has led to the following analysis concerning the relationship between reefal growth and basinal sedimentation. The reef interval (Fig. 6)

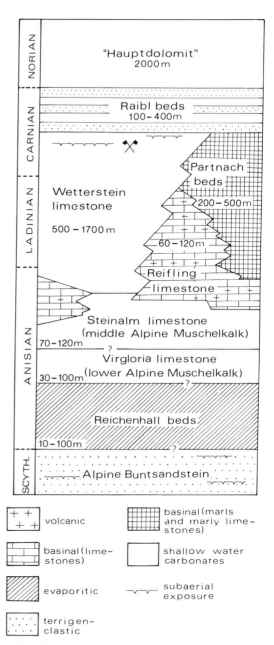

FIG. 2.—Stratigraphic relationships in the western part of the Northern Limestone Alps, modified after Bechstädt *et al.* (1978).

consists of 20 meters of marginal patch-reef displayed in four cycles. The presence of recurrent features (subsequently referred to as "phases") is evident in each cycle (Fig. 7).

*Phase A.*—Early reefal growth was initiated on a stable arenitic substrate (cross-bedded grainstone). This phase clearly shows evidence of ba-

sinal influence by the occurrence of *Daonella,* or fragments of ammonites, radiolarians, conodonts, and some nodosarid foraminifers. Colonization of this environment was pioneered by *Tubiphytes,* primitive dendroid foraminifers (including *Dendrophrya*), and thick-walled sphinctozoan sponges. They formed the foundation for potential reef development. Local trapping and baffling by these organisms resulted in increased accumulation of arenitic biodetritus.

*Phase B.*—Continued development emphasized coarsening-upward sequences which heralded the proximity of reef growth as it expanded into the sedimentation area. A gradual spread of bafflestone created ideal conditions for colonization by dendroid corals (*Thecosmilia* sp.) with encrusting sphinctozoan sponges. This framework was stablized by encrusting *Tubiphytes* (Fig. 16D). Other frame building organisms are: *Ladinella porata* Ott, various bryozoans, and encrusting foraminifers (*Tolypammina* and miliolids). Interstitial space is filled with arenitic detritus consisting of *Dendrophyra,* a few lituolacid foraminifers, *Tubiphytes,* and peloids. Some stromatolitic algae bound fine-grained sediment, and also encrusted the dead framework. Cavities which remained open during this process began to fill up, especially during the early phase of diagenesis, with various generations of fibrous cement ("Grossoolith"). Strong water currents caused a winnowing of siltite and micritic matrix. These currents are a typical factor of Phase B, and are particularly effective along reef-flanks, as well as in the inter-reef areas.

*Phase C.*—This is the phase in which interfingering of previously cemented reef framework and basinal debris occurs. The contact of these sediment units is sharply defined (Fig. 8).

Speedy subsidence, causing "drowning" of the patch-reef abruptly terminated the growth phase and led to a renewed influence of basinal sedimentation. Once again, this change of sedimentation is characterized by filaments, radiolarians, and juvenile ammonites, together with graded layers of arenitic allochthonous reef debris. Shallow water reefs in neighboring areas generally survived the subsidence. In those areas, masses of fine-grained reef debris were swept into suspension during sporadic storms. It is thought that turbidity currents transported debris from shallow waters to the patch-reef areas. The cyclical sequence was completed by an accumulation of lo-cally derived bioarenites, which created conditions similar to those prevailing in Phase A. The curve delineating transgression and regression (Fig. 7) shows slow buildup with sudden termination of a given reef phase. The growth capacity of the reef at this depth proved insufficient, and therefore failed to keep pace with sporadic and sudden subsidences. Hence, reef growth was terminated by a rapid rise of sea level, probably within a range of 5 to 10 meters. A number of factors, such as winnowed grainstones, reef growth, and periodic predominance of pelagic faunal elements, point to the existence of sedimentation just above or below wave base. Consequently, sedimentation is thought to have occurred at a depth of from 15 to 30 meters. The question then remains, how did the alternation of reef buildups, along with the temporary influence of basinal environment on overall sedimentation come to an end? This study has shown that a veneer of debris fills the space between individual reefs, thereby eliminating irregularities in their relief. It is thought that a decrease in subsidence caused shallow water reefs to develop intensively in a direction towards the patch-reef complex, thus eventually burying the patch-reefs under their debris and ending their growth. In general, the decisive factor is to be found in a temporary decrease in subsidence rather than overall acceleration of subsidence. Revival of reef growth was achieved only through a lessening of aggradation. At the depth range of formation of the patch-reefs, this phase of stabilization appears to have been closely related to expanding influence of the basinal environment. The recurrent effects of basinal sedimentation tended to diminish towards the end of the patch-reef sequence, while regressive reef development became the dominant factor. Even at this stage, the sequence maintains features characteristic of the patch-reefs. However, along the gentle slope the strength of currents was barely sufficient to cause a winnowing of fine-grained debris, but insufficient to displace coarse gravel. Apart from this, there is gentle relief descending towards the basin. It is because of this gentle relief that deposits of debris remain in place rather than being carried off by currents. This debris fills the inter-areas between reefs and also interfingers with the patch-reefs. This causes a volumetric imbalance of debris in relation to the growth of patch-reefs.

The Hafelekar Reef complex was formed under

FIG. 3.—Oblique aerial view of the Nordkette Range north of Innsbruck. As shown in the insert scattered patch-reefs prevail in the lower part of the Wetterstein Limestone sequence, whereas the upper part is dominated by massive reefs. The southern escarpment of the Hafelekar summit consists of megabreccias; fissures in the Goetheweg Reef are shown schematically. The arrow points to the patch-reef studied in detail.

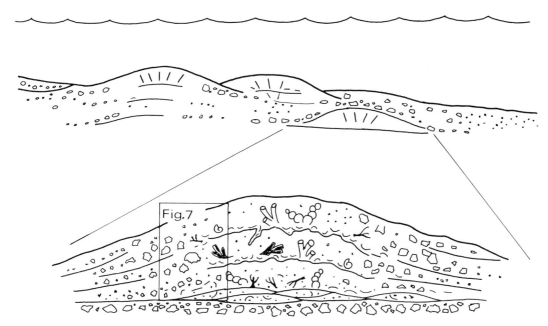

FIG. 6.—Diagrammatic cross-section sketch illustrating facies relationships in the patch-reef sequence. Note recurrent changes of reef growth and basinal sedimentation in the enlarged idealized patch-reef section.

more favorable conditions, e.g., wave agitated shallow water with good circulation.

### Hafelekar Reef Complex

The Hafelekar Reef complex (Hafelekar Riffkorper; Sarnthein, 1965) heralds the first massive reef development in the Nordkette Range, and shows an increasingly regressive tendency. Apart from local complications, due to synsedimentary tectonics along the margin of the platform, the reef shows differentiation in the fore-reef and central reef areas as a result of varying water turbulence and food supply. This conforms with modern reef concepts. One characteristic feature is the occurrence of a reef-flat extending to the north and showing sedimentational characteristics of lagoonal influence (Fig. 10). The main barrier between the reef and the hypersaline lagoon in the north was formed by skeletal sand shoals which show tepee structures and characteristics of vadose diagenesis ("Untere Schollenserie"; Sarnthein, 1965). This barrier, together with frequent emergence of the sand shoals, is seen as one of the main causes for formation of a reef of such magnitude. This agrees with recent descriptions by Ginsburg and Shinn (1964).

Towards the east (Gleirschjöchl, Fig. 4) the reef belt disintegrates; its wedging-out is clearly visible. This break may have been caused by a similar rupture in the sand shoal barrier, causing exit of hypersaline, low-nutritious water from the lagoon, thus causing drastic reduction of reef growth.

*Fore Reef.*—It would be entirely misleading to picture a steep talus slope extending into the basin. Nowhere in the Northern Limestone Alps is there a so-called "Überguss schichtung" comparable to other Triassic Dolomite reefs (Bosellini and Rossi, 1974). The Nordkette fore-reef features a gentle slope of limited extent, suggesting the absence of steep slopes in the area of the reef proper. The angles of inclination of massive beds of breccia, and of the reef itself, show but slight difference. Blocks of rubble are frequently bedded in a matrix of well-sorted grainstone, and cavities within blocks and the matrix are partially filled with graded internal sediment. Tilted geopetal fabric indicates temporary movements of the

←

FIG. 4 (top).—Aerial photograph showing eastern termination of the Goetheweg Reef, with wedging-out clearly visible (arrow). Some large patch-reefs are exposed in the lower part.

FIG. 5 (bottom).—General aerial view of part of the Wetterstein Limestone sequence across the Nordkette Range; left, the earlier reef complex; right, well-bedded lagoonal deposits.

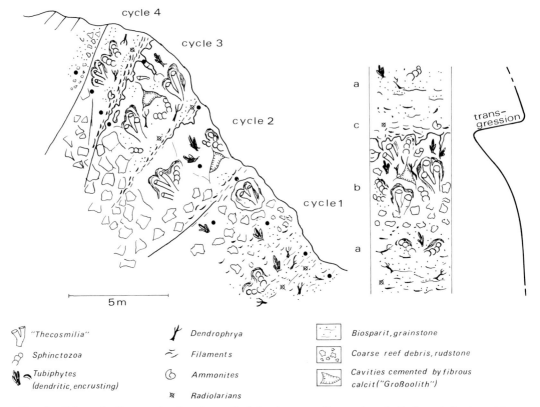

FIG. 7.—Schematic diagram showing the cyclical development of a patch-reef. Each cycle begins with accumulation of arenitic bioclastic debris influenced by basinal environment; sudden subsidence abruptly terminates a growth phase of the reef. The standard cycle can be easily compared with a transgression-regression curve.

talus slope (Fig. 11C), and reef building organisms are not preserved in growth positions. In addition, layers of block rubble may be thoroughly winnowed. Several generations of radiaxial-fibrous calcite (partially dolomite) fill pores of appreciable size. In settings closer to the basin, such as to the west near the town of Zirl (some ten kilometers away), the fore-reef breccia takes on a darker coloring. It is thought that rapid aggradation and a reducing environment have led to enrichment of bitumen within the components of breccia resulting in the darker color. Small foraminifers (miliolids nodosarids) indicate proximity of the basin.

*Megabreccias Caused by Synsedimentary Tectonics.*—A reef-core facies of regressive sediments above the fore-reef, is thought to have been disrupted along the Nordkette by synsedimentary tectonics. Accelerated subsidence caused drowning of the reef body, and this in turn re-created a fore-reef position in the sequence. Thus the Hafelekar Reef complex was divided into two parts with two stages of development—the "Goetheweg Reef," and the younger "Hafelekar Reef." The consequence was an incline (a submarine cliff) whose gradient had been tectonically accentuated. The previously cemented reef limestone disintegrated into blocks several meters in diameter. This tectonic phase also manifested itself with neptunian dikes filled with conodont-bearing filament limestone (Fig. 16A) in lower stratigraphic position. A series of aggradations resulted in large deposits of megabreccias which filled the

$\rightarrow$

FIG. 8 (top).—Thin-section photomicrograph (×3.5) of patch-reef rock with cavity fillings of fine-grained debris and pale-reddish basinal sediments; N 66.

FIG. 9 (bottom).—Stock of slender coralline branches (*Holocoelia* sp.), probably belonging to hydrozoans; Hafelekar Reef, near Hafelekar summit.

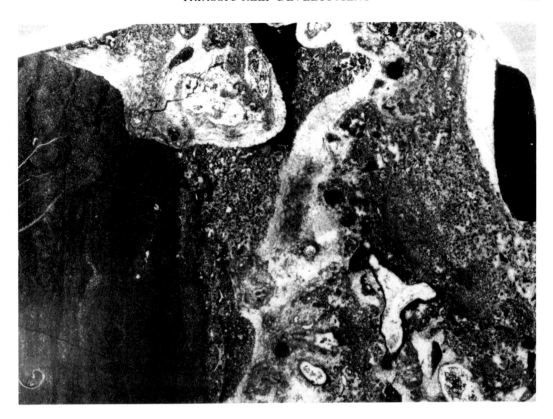

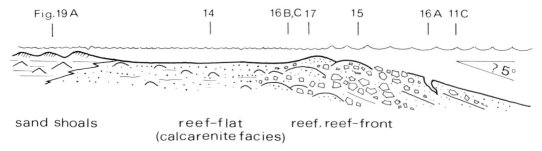

sand shoals                 reef-flat              reef, reef-front
                        (calcarenite facies)

FIG. 10.—Schematic section across the Hafelekar Reef complex; numbered figures refer to the position of typical microfacies and biocoenoses illustrated in this paper.

newly created depressions. Strong water currents prevented settling-out of fine-grained material, thus the self-supporting block framework was originally characterized by conspicuous open pore space and fissures several meters long and decimeters wide. This pore space was later filled by several generations of radiaxial-fibrous calcite (partially dolomite), and finally by blocky calcite (Fig. 11A).

This megabreccia rubble facies was initially studied by Sander in 1936, and originally the "Grossoolithe" had been interpreted as organic structures ("*Evinospongia*"). It is thought that a relief filling of more than 100 meters of rubble was required to allow new reef growth.

*Central Reef Area.*—The early stages of reef growth coincide with a decreased rate in deposition of block rubble. This is a dominant factor in reef development, and also occurs in the earlier patch-reefs. The Hafelekar Reef clearly shows this interrelationship. The formation of arenitic reef debris (Fig. 13) created favorable conditions for initiation of reef growth on top of block rubble. Fine skeletal sand was locally entrapped by scattered, small coral heads, *Tubiphytes*, and calcisponges. In the central reef area, reef structure is remarkably homogenous over long distances, in that the reef structure consists of an organic framework and interstitial grainstone. Significantly, there are no deposits of mud. In essence, areas of coarse rubble are interspersed with homogeneous reef structures.

Early diagenetic dolomitization has emphasized differences in the fabric of both areas. A relatively high degree of dolomitization within the breccia zones may be seen as a result of their permeability, effective over a long period of time, for the transport of Mg-rich groundwater during a low burial stage.

*Reef Framework*

The primary framework consists of corals, calcisponges and *Tubiphytes*. Contrary to earlier views, corals do indeed play a predominant role, mainly in marginal zones, as compared to the role of calcisponges. The prime function of *Tubiphytes* lies in its capacity as a framebuilder in quiet water zones, as well as an encrusting and binding organism in areas of strongly agitated water. Without *Tubiphytes*, many reef building organisms would not have been preserved in their growth position. In addition to this organic cementation, the early cementation by fibrous spar (?aragonite) was of key importance in creating a rigid reef structure.

Among organisms encrusting the primary reef framework are: *Tubiphytes*, calcisponges and, to a lesser degree, various bryozoans, *problematicum Ladinella*, and stromatolitic algae. The reef flora is dominated by intermittent areas covered with tufts of codiaceaen algae, whereas calcareous algae (Solenoporaceae, ?Squamariaceae (Fig. 12), and hemispheroids of Cyanophyceae) play a secondary role. Dasyclad algae are very rare (Ott, 1967). In contrast, spongiostromata algae dominate the reef-flat as a sediment-binding organism which concentrated large quantities of fine-grained sediment (Fig. 14).

Reef cavities and niches were inhabited by foraminifers (e.g., *Dendrophrya*, Fig. 22A), catenu-

→

FIG. 11.—Various types of fore-reef breccias. A. Outcrop photograph taken along the "Goetheweg" showing megabreccias with thick fibrous spar crusts in cavities ("Grossoolith"). B. Thin-section photomicrograph (×2.6) of coarse reef-debris which serves as evidence of repeated reworking of previously lithified particles. Left of center is a hemispheric thallus of *Zonotrichites*; Goetheweg Reef, N 61. C. Thin-section photomicrographs (×3) of tilted geopetal fabrics in fore-reef debris of the Goetheweg Reef, N 153.

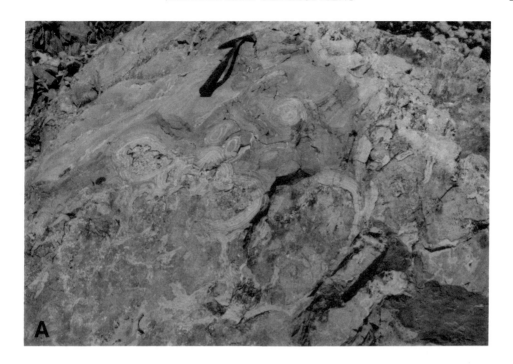

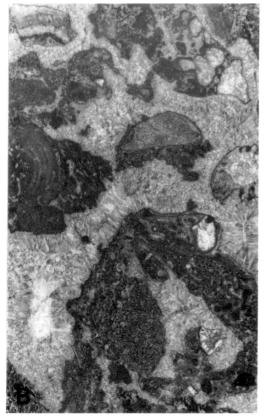

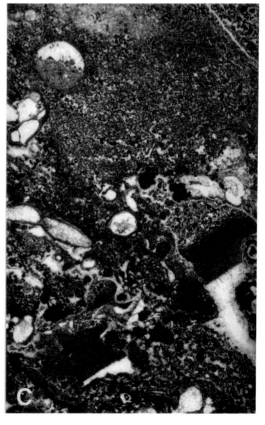

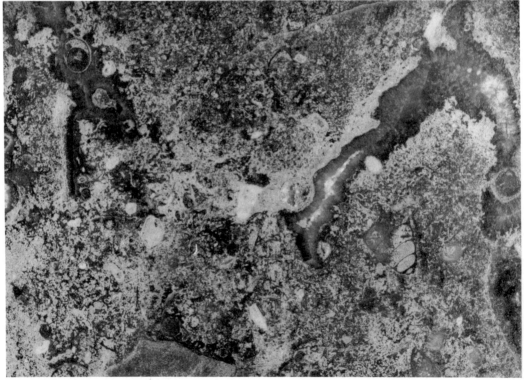

late sphinctozoans (Fig. 18), and various *problematica*. The list of reef dwellers also includes echinoderms, ostracodes, molluscs and brachiopods. Various groups of associated organisms might be termed as biocoenoses scattered in various portions of the central reef area. Although further investigation is required, it can be said that their distribution seems to suggest a zonation pattern (Fig. 10). The central reef area could be interpreted as an area of marginally exposed high energy zones containing highly diverse faunas, and sheltered regions within the reef as well as along the reef-flat. Varying water energy is thought to be the decisive factor in the Hafelekar Reef complex, just as it is the decisive factor in Recent reefs. Table 1 shows the relative abundance of framebuilders present in the Hafelekar Reef complex. The breakdown into various growth forms illustrates their ecological distribution based upon the influence of water energy.

### Reef Biocoenoses

The reef proper accommodates a variety of clearly discernible biocoenoses. Their presence indicates a diversity of ecologies such as prevail on Recent reefs. Biocoenoses listed here serve as examples to illustrate the richness of Middle Triassic reef communities. These organisms are described in the order of functional importance.

*Codiacean Algal Biocoenose.*—This biocoenose seems to favor dwellings in wave-protected reef environments. This assumption appears to be borne out by the existence of fossilized algal tufts which attained 15 centimeters in height, and were fossilized in growth position (Fig. 21). They consist of algae believed to be a hitherto unknown species of the Codiaceae, and possessing branches only one millimeter thick (Fig. 16B). Their matrix is fine-grained. A limited number of cavities left void during sedimentation appear to have been rapidly cemented.

Among secondary framebuilders, only two *problematica* (*Tubiphytes,* and in a general sense also *Ladinella porata* Ott) may be considered as encrusting organisms. Often less important secondary framebuilders are calcareous sponges, remains of echinoderms, small gastropods, pelecypods, brachiopods, rare ostracodes, and small foraminifers (e.g., *Dendrophrya,* and thin-walled Lituoladea, and Duostominidae), as well as small-sized, tube-like *microproblematica*. Codiacean algal tufts do not function as framebuilders, but merely aid in baffling fine-grained sediment.

*Tubiphytes Biocoenose.*—*Tubiphytes obscurus* Maslov, believed to be a problematic blue-green alga, emerges as one of the most important framebuilders and binders from the Permian to the Middle Triassic. It has been found in all biocoenoses listed in this paper. Its main function is binding. On the other hand, there are stocks of dendroid *Tubiphytes* growing in low water energy environments as loosely-jointed tufts up to 50 centimeters in diameter. This type of community inhabits protected reef areas and lagoonal patch-reefs (e.g., Dobratsch, Fig. 16C). *Tubiphytes* is quite often associated with epizoans such as *Ladinella porata,* attached porcelaineous foraminifers, distinct types of small calcareous sponges, thin-walled small foraminifers (species of *Trochammina,* rare specimens of *Ophthalmidium,* and nodosarids), smooth ostracodes, and small gastropods.

*Calcisponge Biocoenoses.*—The Hafelekar Reef housed a variety of organisms which at first sight bear resemblance to inozoa, but are actually ramose corals and hydrozoans (Fig. 9). With their true identity established, the inozoa were designated as two types of calcisponge biocoenoses. One community is based on large cylindrical inozoa (*Peronidella*), as the dominant invertebrate organism that occurs in protected environments. The biotope of this biocoenose is characterized by fine-grained sediment. Other organisms associated with these inozoans are: various molluscs (gastropods and pelecypods), bryozoans, echinoderm debris, bladed as well as cylindrical hydrozoans, rare corals, and sphinctozoan sponges. The epizoans comprise *Tubiphytes,* hemispheroids of Cyanophyceae (some ''*Zonotrichites*''), and *Ladinella porata.* Rare foraminifers include *Dendrophrya,* Lituolacea (particularly *Ammobaculites*), ''*Trochammina*'' *persublima* Kristan-Tollmann, and Involutinidae. The fine-grained sediment of this biocoenose is bound by spongiostromate algae. Strong currents appear to have winnowed cavities clean of fine sediment. These cavities were later cemented by fibrous spar, so-called ''Grossoolith'' structures.

Catenulate sphinctozoan sponges are known to favor as their biotope the protection of sheltered niches in growth cavities, or interparticle cavities

FIG. 12 (Top).—Thin-section photomicrograph ($\times 3.5$) of ?squamariaceans encrusting layers of graded biodetritus; note overgrowth by *Tubiphytes.* Flat cavities are filled with fibrous spar and are believed to be of early diagenetic origin; reef-flat north of Hafelekar summit, N 116a.

FIG. 13. (Bottom).—Photograph of an etched slab ($\times 2.5$) showing typical calcarenite facies at the beginning of reef growth. Intensive bioturbation, a frequent occurrence in this facies, has destroyed or altered much primary lamination; base of Hafelekar Reef, N 128.

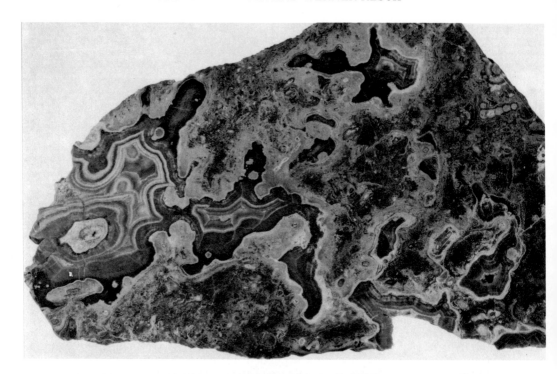

FIG. 14 (top).—An etched slab (×1.4) of typical rock from the reef-flat, showing a framework consisting mainly of catenulate sphinctozoans and *Tubiphytes* encrusted by masses of spongiostromate algae which bind large quantities

TABLE 1.—RELATIVE ABUNDANCES OF THE PRINCIPAL FRAMEBUILDERS PRESENT IN VARIOUS ENVIRONMENTS OF THE HAFELEKAR REEF COMPLEX; IIIII ABUNDANT, ----- COMMON

| ORGANISMS | FORM | CENTRAL REEF AREA | | |
|---|---|---|---|---|
| | | reef flat | protected reef | reef front |
| **CORALS** | | | | |
| Thecosmilia | fasciculate | --------IIIIIIIIIIIIIIIIIIIIIIIIIIIIIIIIIIIIIIIIIIIIIIIIIIIIIIII | | |
| Thamnastrea | massive & platey | | | ---IIIIIIIIIIIIIIIIIIIIIII |
| Montlivaltia | solitary massive | | | ------- IIIIIIIIIIIIIIIIIIIIII |
| **HYDROZOANS** | | | | |
| | bladed | -----IIIIIIIIIIIIIIIIIIIIIIIIIIIIIIIIIIIIIII--------- | | |
| | fasciculate | -------IIIIIIIIIIIIII------- | | |
| **CALCISPONGES** | | | | |
| Inozoa | massive cylindr. | IIIIIIIIIIIIIIIIIIIIIIIIIIIIIIIIIIIIIIIIIIIIIIIIIIIIIIIIIIIIIIIIIIIIIIIIIIIIIIII | | |
| Sphinctozoa | encrusting | ----------------IIIIIIIIIIIIIIIIIIIIIIIIIIIIIIIIIIIIIIIIIIIIIIII | | |
| | catenulate | IIIIIIIIIIIIIIIIIIIIIIIIIIIIIIIIIIIIIIIIIII----------------- | | |
| **TUBIPHYTES** | dendroid | IIIIIIIIIIIIIIIIIIIIIIIIIIIIIIIIIIIIIIIIIIIIIIIII------- | | |
| | encrusting | IIIIIIIIIIIIIIIIIIIIIIIIIIIIIIIIIIIIIIIIIIIIIIIIIIIIIIIIIIIIIIIIIIIIIIIIIIIIIIII | | |
| **SPONGIOSTROMATA** | encrusting & binding | IIIIIIIIIIIIIIIIIIIIIIIII----------------------------- | | |

in block and rubble areas. These particular biotopes also accommodate secondary framebuilders such as encrusting sphinctozoans, *Tubiphytes*, *Ladinella porata*, and rare foraminifers. The process of filling cavities is completed by fibrous spar and graded reef debris (Fig. 18). The entire reef area contains fragment of catenulate sphinctozoan sponges. Those sponges living on the reef-flat are either autochthonous or they are fragments that were wave and current sorted (Figs. 14 and 19B). Earlier research in Triassic calcisponge biocoenoses was by Ott (1967), Senowbari-Daryan (1978), Wolff (1973), and Zankl (1969).

*Thecosmilia-Calcisponge Biocoenose.*—The Wetterstein Limestone is known for its rarity of large coral heads of the kind that occur so prominently in Norian and Rhaetian reefs (Zankl, 1969). In the Wetterstein Limestone reefs, *Thecosmilia* functions as a primary framebuilder (Fig. 16D), but is joined by calcisponges functioning as secondary framebuilders. *Thecosmilia* heads are found in growth position with branches up to 2 centimeters in diameter (Fig. 17). Next to scattered inozoans, calcisponges are represented by encrusting genera such as *Uvanella* and *Annacoelia*, as well as other sphinctozoan sponges whose main distinction is their upright growth position. Dendroid and encrusting *Tubiphytes*, and

←

of fine-grained sediment. Remaining cavities facilitated early diagenetic dolomitization (light-colored zones). Eventually the cavities were filled with alternating layers of fibrous spar and micritic dolomite; N 58.

FIG. 15 (bottom).—Thin-section photomicrograph (×1.7) showing a typical coral-echinoderm–*Tubiphytes* community consisting of rubble with broken and highly diversified framebuilders and reef dwellers cemented by *Tubiphytes*. Note thick crinoid stalk growing on a platey *Thamnasteria*. Large pelecypods are common in this community; reef-front of the Goetheweg Reef, N 111a.

solenoporacean algae should also be added to the list of important framebuilders; additional layers of encrustations and coatings were provided by "*Zonotrichites*" and stromatolitic algae.

There are numerous additional though less important framebuilders and reef dwellers. These include hydrozoans, molluscs (gastropods, pelecypods, and small ammonites), bryozoans, brachiopods, echinoderms, ostracodes, *Ladinella porata,* and spirorbids, all functioning as epizoans. Foraminifers include *Dendrophrya, Ammobaculites,* "*Trochammina*" *persublima, Endothyranella,* Duostominidae, *Tolypammina,* and sessile miliolids. The matrix ranges from silititic to arenitic with local winnowings, although in some places matrix is conspicuously absent. Such a highly diversified biocoenose marks sheltered reef areas of varying depth, an assumption supported by the occurrence of the foraminifers. The *Thecosmilia*–calcisponge biocoenose is a prominent characteristic of Wetterstein Limestone reefs.

*Coral–Echinoderm–Tubiphytes Community (Fig. 15).*—Organisms of this category occur in certain reef-front sediments of the Goetheweg Reef. Although the fossil record of this community is that of a thanatocoenose rather than a biocoenose, connected with transportation and redeposition over short distances. The corals are represented by *Thecosmilia, Montlivaltia,* and massive platy thamnasteriids; echinoderms are represented by crinoids and echinoids, especially spines. In addition to these, the following organisms are often observed in outcrop and in thin-section: attached hemispherical masses of solenoporacean algae, plate-like hydrozoans (Fig. 19D), relatively large pelecypods (up to 1 decimeter in diameter), calcareous songes, e.g., *Annocoelia* (Fig. 19C), and rare catenulate sponges. *Tubiphytes* and *Ladinella porata* as fixosessile *problematica* are autochthonous and important for their cementing and binding function in this environment. Foraminifers are represented in thin-section by *Reophax, Ammobaculites,* and attached specimens (*Tolypammina*). This highly diversified community is typical of an environment of strong wave activity situated on the reef-front.

### Reef Debris

Figure 10 illustrates the similarities in the distribution patterns of reef debris, and the zonation of framebuilding organisms. Water turbulence and reef morphology determine the location of rubble and block rubble, some with diameters of up to decimeters, in the high energy zone mainly along the reef-front.

Arenitic debris consisting of skeletal grains and pellets, and intraclasts occur mainly on the reef-flat (Fig. 13). Most of the fine sediment fraction was winnowed out by strongly agitated water, with the frontal reef area presumably functioning as a barrier which deflected water turbulence. Arenite graded between fine and middle-size fractions was entrapped within the reef framework and contributed to the massive character of the reef. Mud deposits seem to be as scarce as the occurrence of internal reef debris in framework

$\rightarrow$

---

Fig. 16 (right).—Thin-section photomicrographs of various reef carbonates. A. Filament limestone with geopetal deposits of fine-grained reef debris; fissure-filling, N 24, reef biocoenosis (×3.3). B. Codiaceae biocoenosis showing branches of a hitherto unknown species of Codiaceae which aided baffling of fine-grained sediment. Note the characteristic endings of ramifying branches; Goetheweg Reef, N 92a, (×3.3). C. *Tubiphytes* biocoenosis showing detail of a cross-section of stacked masses of dendroid *Tubiphytes obscurus* Maslov. Small-sized sphinctozoans are noted as secondary framebuilders; lagoonal patch-reef, Dobratsch, DO 19a, (×12.2). D. *Thecosmilia*-calcisponge biocoenosis reef framework is built by well preserved, thecosmilians, sphinctozoans, and *Tubiphytes*; patch-reef sequence, N 72, (×3.8).

Fig. 17 (p. 220, top).—Outcrop photograph of a typical *Thecosmilia*-calcisponge biocoenosis. Photograph shows thin and thick-branched thecosmilian species in growth position. Interstitial space is completely filled with fine-grained reef debris. This outcrop is well exposed along the Goetheweg.

Fig. 18 (p. 220, bottom).—Thin-section photomicrograph (×3.9) of catenulate sphinctozoan sponges (*Vesicocaulis* sp.), encrusted by *Tubiphytes* and growing in a reef cavity. The cavity was partially filled by generations of fibrous spar and blocky calcite, Heiterwand Range, HWD 1.

Fig. 19 (p. 221).—Thin-section photomicrographs of various reef rock types. A. Typical example of vadose diagenesis, with microstalactitic cement in grainstones of the sand-shoal facies, north of the Hafelekar summit; N 165a, ×10.2. B. Detail of calcisponge biocoenosis showing sphinctozoans (*Dictyocoelia* sp.) as the primary framebuilders and encrusted by other sphinctozoans, *Tubiphytes,* and *Ladinella.* Fine-grained matrix indicates low water energy of this environment; eastern termination of Goetheweg Reef, N 56, ×2.7. C. Sphinctozoan sponges (*Annacoelia* sp.) encrusting thamnasteroid coral fragment; coral-calcisponge biocoenosis, Goetheweg Reef, N 101, ×2.4. D. Bladed ?hydrozoan and bryozoans (upper left) encrusted by *Tubiphytes.* Reef debris in the interstitial space is well-washed, with only a few remnants of fine-grained matrix preserved through cementing by *Tubiphytes*; reef facies, quarry near Zirl, west of Innsbruck, MW 7c, ×3.2.

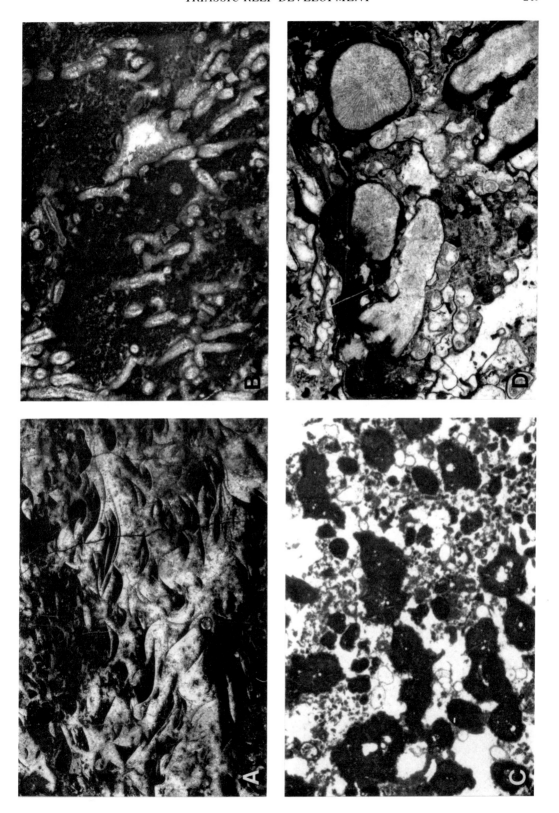

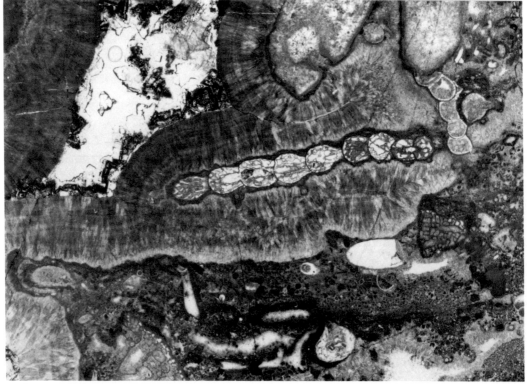

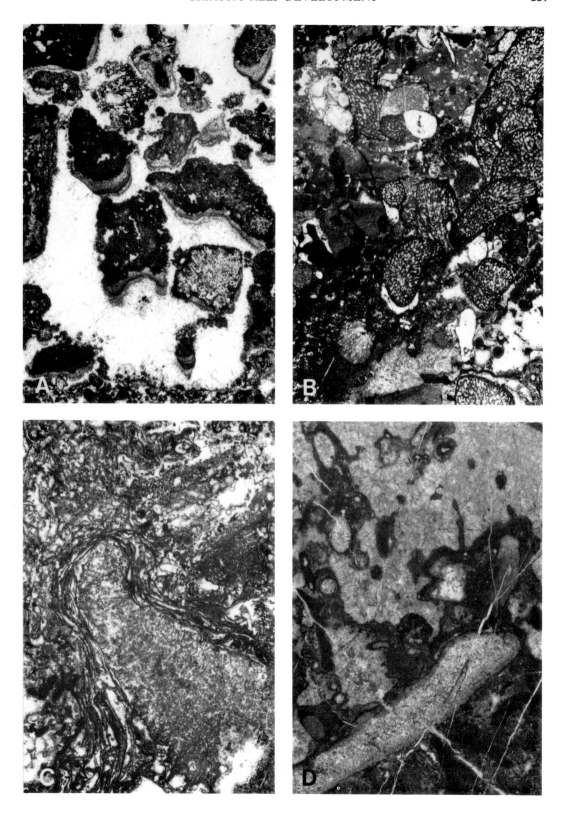

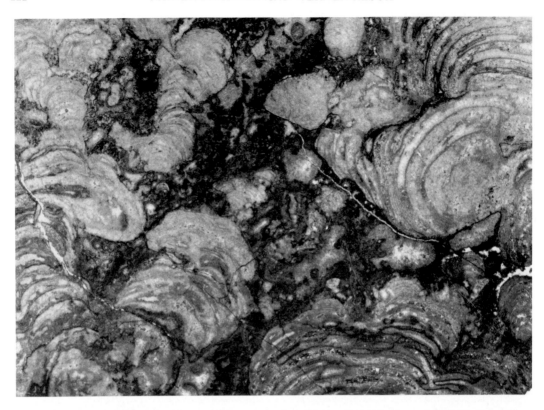

cavities, most cavities were probably closed-off early in the process of rapid cementation. Rather poor bedding is characteristic of the arenites of the reef-flat, and to a lesser extent also in the matrix of the reef framework. Many of these structures were destroyed by intensive bioturbation.

Debris components include intraclasts of cemented reef framework, and lithified fine-grained detritus. It is the composition of the coarse rubble that indicates overall re-working of previously lithified detritus (Fig. 11B), and from this emerges a picture of repeated episodes of reworking followed by rapid cementation.

There are but few signs of bioerosion such as algal borings (Fig. 24I) within reef debris. This shows that as a destructive factor, bioerosion is less effective in Middle Triassic reefs than in Upper Triassic Dachstein reefs, as described by Zankl (1977). Hence, it is thought that the main destructive process was physical abrasion and erosion.

### MICROFOSSILS AND THEIR DISTRIBUTION

The following section is concerned with various microfossils, small-sized remnants of larger invertebrates, and algae as seen in thin-sections. In particular, the foraminifers (Fig. 22) were subjected to detailed examination. Until recently little was known about their paleoecology in the Wetterstein Limestone (Resch, 1979), although it is now believed that their distribution holds promise for future research (Fig. 23).

There is a significant abundance of *Dendrophrya* (Fig. 22A) and other tubular Ammodiscacea in the Wetterstein Limestone. Some of these forms certainly belong to the *microproblematicum Microtubus communis* Flügel, which was thought by Zankl (1969) to belong to the foraminifers. The genus *Dendrophrya* shows a preference for dwelling in cavities, especially amongst larger framebuilders and breccia components. Fragments of this genus are also frequently found, in arenitic sediments adjacent to reefs. Ammodiscidae such as *Ammodiscus, Glomospira,* and *Glomospirella* are found in fine-grained sediments deposited in quiet waters, e.g., in basins and in distal parts of the fore-reef, as well as in deep lagoons, as was shown to be the case in the Kaisergebirge Mountains. The species *Turritellella*

*mesotriasica* Koehn-Zaninetti is less significant in terms of paleoecology. In the central reef areas, genera of the Lituolacea, such as *Reophax* (Fig. 22B and C), *Ammobaculites* (Fig. 22D and E) and, to a certain extent *Trochammina* (Fig. 22G), have been found to develop thick-walled, heavy, structured tests. The opposite seems to be true of the same forms in quiet waters, where thin-walled specimens with fragile structures are dominant. A characteristic species is "*Trochammina*" *persublima* Kristan-Tollmann, whose test wall is extremely thin.

Representatives of *Endothyranella* (Fig. 22F) and *Pseudotaxis* are found in the skeletal sand-shoals, and in the shallow lagoonal environment, whereas different species of the family Duostominidae are common across the entire lagoon, but in a location close to the reefs.

Miliolids are represented by *Agathammina* (Fig. 22K) in lagoons, back-reef areas, and in fine-grained fore-reef debris. *Ophthalmidium* and related genera (*Nodophthalmidium*), and rare miliolids of a more advanced systematic level (e.g., *Karaburunia,* Fig. 22J) migrated from basins in to distal reef-talus, but can also can be found in reef-flat and lagoonal environments.

Involutinidae appear to be represented in the Wetterstein Limestone by the genera *Aulotortus* (Fig. 22H) and *Trocholina*. These are especially abundant in lagoons as well as in back-reef areas and in lagoonal patch-reefs (e.g., Dobratsch). Rare occurrences of *Turrispirillina* specimens were noted in the skeletal sand-shoal deposits.

Adherent species of foraminifers occur more frequently in Upper Triassic and younger reefs than they do the Wetterstein Limestone. We have however observed *Tolypammina*, specimens of Calcivertellinae (Fig. 22I), and rare occurrences of Nubeculariinae. *Tolypammina* may be retrieved intact through acetic acid etching of the carbonate sample.

Attention should be drawn to the existence in the reefs of spirorbids (Fig. 24G) of diverse external ornamentation. These polychaete worms are most frequent in coarse-grained sediments of normal marine environments. Smooth ostracodes are also present; however, they are not discernible in thin-section.

Bryozoans (Fig. 24F) have yet to be thoroughly investigated. They appear as dendritic and en-

---

← 

FIG. 20 (top).—Thin-section photomicrograph (×2.3) of a solenoporacean framestone with *Collenella*-like masses of solenoporaceans up to ten centimeters high. However *Tubiphytes* is also abundant, as in other reef communities; Hafelekar Reef complex, B 3.

FIG. 21 (bottom).—Outcrop photograph showing tufts of thin-branched codiaceans in growth position. Exposure of sheltered reef area as seen along the Goetheweg; coin is two centimeters in diameter.

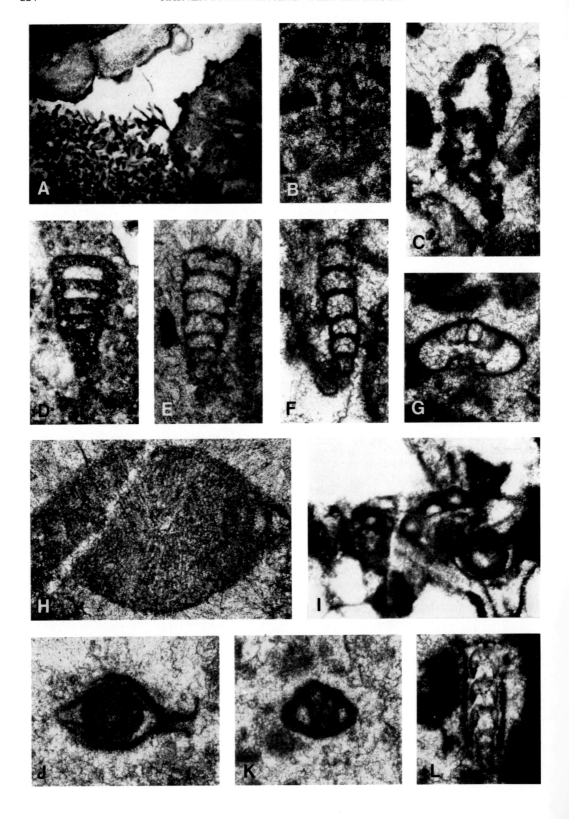

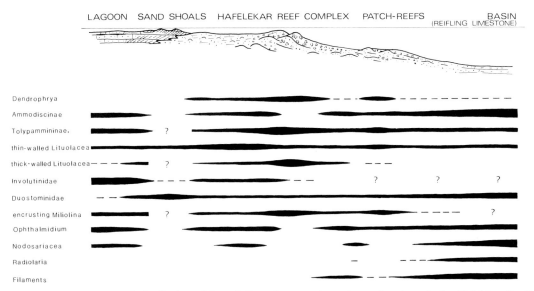

FIG. 23.—Environmental distribution of foraminifers along a schematic section through the Hafelekar Reef complex, used as an example for Wetterstein Limestone reefs. For details of foraminiferal distribution in the reef proper see section on "biocoenoses." Frequency: thick bar = very common, thin bar = rare, dashed line = single occurrence.

crusting growth forms and are probably closely related to Paleozoic Trepostomata. Their contribution to frame building is of secondary importance.

Echinoderm skeletal elements appear frequently in the reef and fore-reef debris. They are mainly derived from crinoids (Fig. 15), and to a lesser degree echinoids and ophiuroids.

Remains of algae, such as stromatolitic algal crusts, *Girvanella,* and patches of filamental Cyanophyceae (Fig. 24H), have a sediment-binding function, and thus a major share in binding and shaping the reefs. Otherwise, their paleoecologic significance is unimportant. Various skeletal segments of *Clypeina* cf. *C. besici* Pantić (Fig. 24C) occur in fore-reef and reef-debris, indicating that they were not confined solely to lagoons. The Wetterstein Limestone lagoon accommodated

various dasyclad algae, and given favorable conditions they became abundant. A hitherto unknown species of Codiaceae (Fig. 16B) was found to have developed dense growths in sheltered reef areas; these forms which grew thin-branched bushes attained a height of 15 centimeters (Fig. 21). Occasional algal borings (Fig. 24I) into various skeletal grains may serve as potential shallow water depth indicators. Vertically stacked patches of Solenoporaceae inhabited high energy reef zones (Fig. 20), and the reef-flats accommodated low accumulations of Squamariaceae which bound arenitic sediment (Fig. 12).

There are numerous occurrences of *microproblematica,* but no new data were elucidated as to their systematic affiliation, although some additional observations may be offered. *Tubiphytes* (Fig. 24D) appears as an encrusting organism in

←

FIG. 22.—Photomicrographs of various foraminifers seen in thin-sections of the Wetterstein Limestone; most are from the Cordovolian of the Dobratsch-Drauzug Range. A. *Dendrophrya* sp. in a spar-filled cavity; Hafelekar Reef, N-116b, ×6.7. B. *Reophax* sp., a small thin-walled form; D-03, ×70. C. *Reophax* sp., a thick-walled specimen; DO-12C, ×30. D. Cf. *Ammobaculites* sp., a thick-walled specimen; DO-12G, ×30. E. Cf. *Ammobaculites* sp., a thin-walled specimen; DO-10C, ×70. F. *Endothyranella wirzi* (Koehn-Zaninetti) from the sand facies of Hafelekar Reef, northern slope of Hafelekar summit; N-165b, ×75. G. *Trochammina jaunennsis* Bronniman and Page, an axial section; DO-19a, ×70. H. *Aulotortus sinuosus* Weynschenk, note radial microstructure; lagoonal patch-reef at Dobratsch, DO-12C, ×70. I. Specimen of Calcivertellinae encrusting a calcareous sponge from the patch-reef sequence in the Nordkette Range; N-72, ×30. J. *Karaburunia* sp. aff. *K. rendeli* Langer; DO-12Aa, ×120. K. *Agathammina* sp.; 12Aa, ×120. L. Specimen of a uniserial nodosarid from the *Tubiphytes* biocoenosis of a lagoonal patch-reef; Dobratsch DO-19a, ×120.

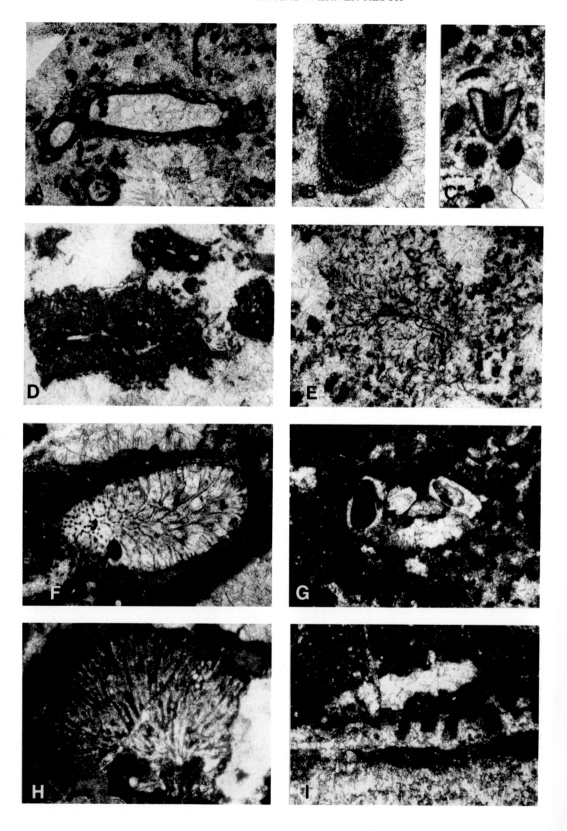

reef and fore-reef environments, as well as in clusters (up to 50 centimeters high) in low water energy biotopes. Their binding function is of especial significance. Some specimens show a low degree of calcification (Fig. 24E), a point not previously observed. Both *Ladinella porata* Ott (Fig. 24B) and *Tubiphytes* occupy only very modest areas in terms of substrate, but both organisms are among the latest settlers on a wide variety of substrates. Another *microproblematicum* of *incertae sedis* is *Lamellitubus cauticus* Ott (Fig. 24A).

Significantly, acetic acid treatment of pelagic carbonates yielded good results. Organism remains found in residues include conodonts, e.g., *Gondolella polygnathiformis* Budurov and Stefanov, *Gladigondolella tethydis* (Huckriede), *Hindeodella (Metaprionidus) pectiniformis* (Huckriede), and *Enantiognathus petraeviridis* (Huckriede), holothurian sclerites, e.g., *Theelia immissorbicula* Mostler, *Kuehnites spiniferporatus* (Zawidzka), and *Calclamnella nuda* Mostler. In addition, these microfossils (identified by A. Donofrio) indicate that these samples have a stratigraphic range from Ladinian to Cordevolian time.

## DIAGENESIS

The entire reef complex has a singular characteristic diagenetic feature, that is, a very early and rapid sediment cementation, presumably submarine. This is evident in the form of reef-debris which appears to have been reworked and recemented. However, the Hafelekar Reef complex lacks signs of vadose diagenesis such as vadose cements, subaerial crusts, and karst features. In this context, sea level changes seem to have been effective only in more shallow water sand-shoals where tepee structures were laid down and where vadose cements were deposited (Fig. 19A).

This study suggests that a dolomitization phase preceded the main cementation. The dolomitization process originated in the highly porous areas of the reef-flat and breccia zones, and spread into adjacent areas. The early stages of this phase of diagenesis appear to have been dependent upon

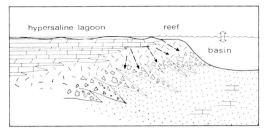

Fig. 25.—Idealized cross-section of a regressive platform-margin as developed in the early burial stage. Permeable reef sediments permit influx of lagoonal saline brines with a high Mg/Ca ratio to enter reef sediments. These percolating pore waters may have caused early dolomitization and subsequent cementation of thick ?aragonite crusts in the main reef body (cross-section vertically exaggerated).

original primary structures (Fig. 14). The question arises as to what factors caused early dolomitization of a reef body that had not even been emergent? First of all, attention should be drawn to the generally regressive character of platform development, for in the course of time, reefs were overlain by lagoonal sediments. Pore waters from temporary hypersaline lagoons, and consequently containing a high Mg/Ca ratio, may have led to the process of dolomitization. The model (Brandner, 1978) as shown in Figure 25, illustrates what appears to be a process of percolating hypersaline ground waters of a higher specific gravity, into areas of permeable reef and reef-debris in the early burial stage. A similar situation seems to apply to subsequent "Grossoolith" cementation. In this instance reef cavities, such as intergranular cavities of the megabreccias, were lined with decimeter-thick isopachous cement crusts. This cavity lining accumulated with several generations of fibrous spar alternating with thin layers of finely crystalline dolomite. The calcite cement attained a thickness of several centimeters and has well developed radiaxial-fibrous structure. It represents secondary formation, probably by inversion of aragonite. This is suggested by the presence of

←

Fig. 24.—Photomicrographs of various microfossils seen in thin-sections of Wetterstein Limestone; if not otherwise noted, all specimens are from the Cordovolian of the Dobratsch-Drauzug Range. A. *Lamellitubus cauticus* Ott from the patch-reef sequence of the Nordkette Range; N-69, ×15.4. B. *Ladinella porata* Ott from the *Tubiphytes* biocoenosis of a lagoonal patch-reef at Dobratsch; DO-19b, ×70. C. Skeletal segment of *Clypeina* sp. cf. *C. besici* Pantié; DO-12C, ×30. D. *Tubiphytes obscurus* Maslov showing stems of porous structure with a central canal and fragile tube-like protuberances; from the *Tubiphytes* biocoenosis; DO-19b, ×17. E. *Tubiphytes* sp. showing a rather low degree of calcification; DO-19b, ×16.5. F. Branch of a trepostome bryozoan; note well preserved autopores and small acanthopores; from reef facies in a quarry near Zirl, west of Innsbruck; MW-7c, ×16.5. G. Cf. *Spirorbis* sp.; DO-13, ×30. H. Blue-green alga *Zonotrichites* sp. with relatively thick filaments; this form is commonly found in lagoonal and skeletal sand-shoal environments; DO-4, ×30. I. Probable algal borings in a skeletal grain from a lagoonal patch-reef at Dobratsch; DO-12B, ×30.

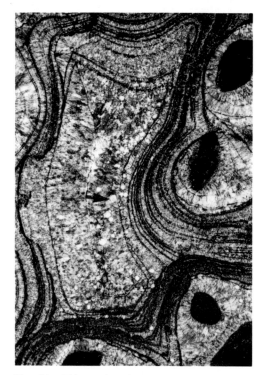

FIG. 26.—Thin-section photomicrograph (×2.8) illustrating phases of cement filling of a reef cavity by alternating layers of radiaxial-fibrous calcite and fine crystalline dolomite. Euhedral quartz crystals (arrow) occur in the central portion of the cavity; Heiterwand Range, HWD 3.

feathered terminations of aragonite crystals, as described elsewhere Folk and Assereto (1976). This assumption is also supported by evidence of celestine mineralization present in fibrous cement crusts. Hence, the strontium may have originated in primary aragonite. Analyses of Recent aragonite cements indicate that strontium is more frequently incorporated in aragonite than in calcite.

Intermittent layers of dolomite may be attributed to primary precipitation of high-magnesium calcite (Germann, 1971). Precipitation of both high-magnesium calcite and aragonite requires, among other factors, an environment of a high Mg/Ca ratio (Folk, 1974). As in the process of dolomitization described above, the regressive reef model permits, especially in the early burial stages, percolation of pore waters from the lagoon, and these contain a high Mg/Ca ratio. The cyclical change of chemistry is orchestrated by outside factors such as periodic influx of fresh water into the lagoon, or by the system of precipitation itself.

Pelites of a reddish color infiltrated the cavity system of the reef during the "Grossoolith" cementation phase, and these pelites are confined to "Grossoolith" cavities within the reef. It may be assumed that they originated in periodically emergent sand-shoals where red pelites are quite frequent.

In addition to calcite and dolomite, there are few rare occurrences of euhedral quartz crystals, generally in the center portions of voids (Fig. 26). The remaining central portions of the voids are filled with phreatic, coarse-blocky calcite crystals. Wherever there is no filling, the voids are preserved as primary voids.

In conclusion, it may be stated that cavities in the Wetterstein Reef limestones played a significant part serving as conduits for various phases of water chemistry diagenesis, especially during the early burial stages.

### GENTLE SLOPE ON THE MARGIN OF THE WETTERSTEIN REEF PLATFORM

The platform-edge gently sloped into the basin, forming the necessary topographic conditions for development of two major reef configurations: patch-reefs along the outer edge but well within the range of wave base, and a massive reef complex, the Hafelekar Reef, in a zone of agitated shallow water (Fig. 27A). Significantly, both configurations show accumulations of an organic framework of coarse debris or previously lithified material. However, both remained largely stationary due to insufficient water energy which failed to carry the debris down the gently dipping slope. Reef growth was most successful in the shallow water zone where strong wave and current action was able to clear reef-shed debris to permit massive buildups. The growth of patch-reefs depends on specific conditions such as:

1. an area that offers protection from assimilation of reef debris; and,
2. development in an area where reef growth is not impeded by periodic rising and falling of sea level.

Consequently, the size of the growth zone depends on the degree of inclination of the slope, with the gentler the slope, the wider the growth zone.

Stabilization of the Middle Triassic platform-margin led to massive carbonate production associated with a relative drop in sea level. This situation normally would have promoted growth of shallow water reefs, however in the Hafelekar setting, this sensitive balance was upset by syn-sedimentary tectonics (Fig. 27B). Such tectonic activities apparently were a regular feature of the Middle Triassic in the Southern Alps (Bechstädt et al., 1978). The sudden destabilization and collapse of the substrate caused an overall collapse

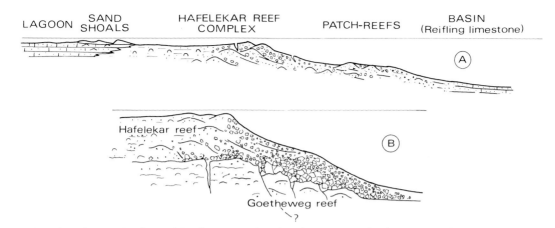

LAGOON   SAND SHOALS   HAFELEKAR REEF COMPLEX   PATCH-REEFS   BASIN (Reifling limestone)

FIG. 27.—Diagram showing reef development at the edge of the Wetterstein Limestone platform near Innsbruck. A. Stratigraphic model showing various reef configurations on a gentle slope, with patch-reefs developed just above or below wave base. Note the gradual transitions of the wide facies belts. B. During Cordevolian time regressive reef development was interrupted by block faulting tectonics, causing a steepening of the slope and a collapse of previously lithified reef rock.

of the lithified Goetheweg Reef rock. This in turn created a depression which was filled by accumulations of megabreccias 100 meters in thickness, and with boulders several meters in diameter. This was followed by a new phase of reef growth (Hafelekar Reef). This tectonic phenomenon might be attributed to a giant submarine rockslide, caused by severe tectonics along the shelf-margin.

### CONCLUSIONS

1. The edge of the Wetterstein Limestone platform near Innsbruck features two distinct types of reef development: one a patch-reef sequence, the other a predominantly massive reef complex (Hafelekar Reef complex).

2. Early generations of patch-reef development were influenced by basinal sedimentation. Minimal sea level changes caused drastic alterations in patch-reef buildup. During the early stages, active reef growth by calcisponges and tubular foraminifers coincided with increased basinal influence. The growth of patch-reefs was obstructed by debris deposits expanding into shallow water reef areas. Permanent subsidence assisted patch-reef growth by insuring equitable ecologic conditions over long periods of time.

3. The Hafelekar Reef complex shows contrasting characteristics of reef development in well agitated shallow waters. Varying water energy produced zonal patterns that delineated a reef-front and wide reef-flat. A conspicuous zone of skeletal sand-shoals shows evidence of emergence, and function as a barrier separating the reef from the lagoon.

4. Patterns of distribution of reef building as-

semblages show dependance upon water energy, and this is reflected in a variety of microfacies. Areas of the reef exposed to high water energy are populated by massive forms of corals, calcisponges, and hydrozoans, whereas sheltered reef zones offer refuge to more delicate corals, catenulate sphinctozoan sponges, dendroid *Tubiphytes,* and to certain types of bushy codiacean algae.

5. Foraminifers of the Wetterstein Limestone are represented by a fairly rich assemblage of species, which is in sharp contrast with foraminiferal assemblages of the Norian-Rhaetian reef limestones. More restrictive environments, especially in lagoonal areas, seem to offer ideal conditions for development of a rich foraminiferal biota of low diversity.

6. Sediment stabilization was achieved by the cementing activities of encrusting organisms, mainly *Tubiphytes,* concurrently as sediments were undergoing rapid submarine cementation.

7. Wave current and sediment reworking and various periods of cementation led to a coarsening of particle size, especially in the reef-flat area, where masses of arenitic debris were bound by spongiostromate algae. Strong currents also prevented formation of mud deposits, and as a result these factors created enormous pore volume, often exceeding 50 percent, which was later filled with fibrous spar.

8. Analysis of the calcareous algal flora, benthic invertebrate fauna, and various types of cement demonstrates that the bank-edge was continuously submerged in well agitated normal marine waters, even up to the edge of the skeletal sand-shoal barrier.

9. In the early burial stages, hypersaline waters seeped into permeable reef areas initiating early dolomitization. Early diagenetic lagoonal influx led to a subsequent cementation phase in which thick ?aragonite coatings accumulated in cavities and alternated with thin dolomite layers.

10. The geometry of the reef complex shows a gentle sloping of the platform-margin. This, in combination with periodic subsidence created two different types of reef development. A gentle slope insured preservation of patch-reefs along the outer-slope area, and protected them from erosion such as affected similar patch-reefs in the Dolomites.

11. The platform-margin appears to have been a zone of indigenous tectonic activity in which block-faulting, neptunian dikes, and megabreccias all formed conspicuous elements during the history of the reef complex.

12. There is no record of Lower Triassic (Scythian) reefs in the alpine Tethys. However, the southern alpine Middle Triassic (Anisian) produced small fringing-reefs of corals and, to a larger extent, calcisponges (Bechstädt and Brandner, 1970). These reefs were associated with dasycladacean algal biostromes (Serla Dolomite correlative with the Steinalm limestone of the Northern Calcareous Alps), and their small size was due to their inability to keep pace with overall subsidence.

13. The biotic composition of these reefs is composed of both older and newer reef building organisms, such as chaetetid, thamnastrean, and calamophyllian corals, new species of calcisponges, and trepostome bryozoans. The *problematicum Tubiphytes*, while an important component in some Permian and Triassic reefs, plays a rather insignficant role in the Anisian reefs.

### ACKNOWLEDGMENTS

Special thanks are given to Professor Dr. F. Purtscheller, who, as pilot of a glider plane, enabled us to produce some rather remarkable aerial photographs. Dr. E. Colins led a field excursion to the Dobratsch Range, Carinthia, Austria, so that we might study the Wetterstein Limestone at this location. Dr. A. Donofrio and Professor Dr. H. Mostler prepared acetic acid residues of some carbonates and identified their contained microfaunas. The writers are also indebted to the following individuals: W. Hanke (thin-section preparation), K. Form (drafting of the illustrations), H. Benedict (assistance in manuscript preparation), and Mrs. B. Prast and Mrs. L. Ortner for typing of the final manuscript draft. Sincere thanks are also extended to the "Alpine Forschungsstelle Obergurgl" for financial contributions.

## REFERENCES

BECHSTÄDT, T., AND BRANDNER, R., 1970, Das Anis zwischen St. Vigil und dem Höhlensteintal (Pragser- und Olanger Dolomiten, Südtirol): Festbd. Geol. Inst. 300-J. Feier Univ. Innsbruck, p. 9–103.

———, ———, MOSTLER, H., AND SCHMIDT, K., 1978, Aborted Rifting in the Triassic of the Eastern and Southern Alps: Neues Jahrb. Paläont. Abh., 156, p. 157–178.

BOSELLINI, A., AND ROSSI, D., 1974, Triassic Carbonate Buildups of the Dolomites, Northern Italy, *in* Laporte, L. F., Ed., Reefs in Time and Space: Soc. Econ. Paleontologists Mineralogists Spec. Pub. No. 18, p. 209–233.

BRANDNER, R., 1978, Tektonisch kontrollierter Sedimentationsablauf im Ladin und Unterkarn der westlichen Nördlichen Kalkalpen: Geol. Paläont. Mitt. Innsbruck, 8. p. 317–354.

COLINS DE TARSIENNE, E. A., 1975, Die tektonische Stellung des Dobratsch unte spezieller Berücksichtigung der Mikrofazies [unpub. Ph.D. thesis]: Innsbruck, Fak. Univ. Innsbruck, p. 1–149.

DONOFRIO, D. A., HEISSEL, W., AND MOSTLER, H., 1979, Zur tektonischen und stratigraphischen Position des Martinsbühels bei Innsbruck: Geol. Paläont. Mitt. Innsbruck, 7, p. 1–43

FABRICIUS, F. H., 1966, Beckensedimentation und Riffbildung an der Wende Trias/Jura in den Bayerisch-Tiroler Kalkalpen: International Sed. Petrograph. Ser., 9, p. 1–143.

FLÜGEL, E., AND FLÜGEL-KAHLER, E., 1963, Mikrofazielle und geochemische Gliederung e nes obertriadischen Riffes der nördlichen Kalkalpen (Sauwand bei Gusswerk, Steiermark, Österreich): Mitt. Mus. Geol. Bergbau Technik, Landesmus, Joanneum, 24, p. 1–128.

FOLK, R. L., 1974, The natural history of crystalline calcareous carbonate: Effect of magnesium content and salinity: Jour. Sed. Petrology, v. 44, p. 40–53.

———, AND ASSERETO, R., 1976, Comparative fabrics of length-slow and length-fast calcite and calcitized aragonite in a Holocene speleothem, Carlsbad Caverns, New Mexico: Jour. Sed. Petrology, v. 46, p. 486–496.

GERMANN, K., 1971, Calcite and dolomite fibrous cements ("Grossoolith") in reef rocks of the Wetterssteinkalk (Ladinian, Middle Trias) Northern Limestone Alps, Bavaria and Tyrol, *in* Bricker, O. P., Ed., Carbonate Cements: John's Hopkins Univ. Studies Geol. No. 19, p. 185–188.

GINSBURG, R. N., AND SHINN, E. A., 1964, Distribution of the Reef Building Community in Florida and the Bahamas (abs.): Am. Assoc. Petroleum Geologists Bull., v. 48, p. 527.

HECKEL, P. H., 1974, Carbonate Buildups in the Geologic Record: a Review, *in* Laporte, L. F., Ed., Reefs in Time and Space: Soc. Econ. Paleontologists Mineralogists Spec. Pub. No. 18, p. 90–154.

LOBITZER, H., 1974, Fazielle Untersuchungen an norischen Karbonat-plattform-Beckengesteinen (Dachsteinkalk-Aflenzer Kalk im südöstlichen Hochschwabgebiet, Nördliche Kalkalpen, Steiermark): Mitt. Geol. Ges. Wien, 66-67, p. 75–92.

OTT, E., 1967, Segmentierte Kalkschwämme (Sphinctozoa) aus der alpinen Mitteltrias und ihre Bedeutung als Riffbildner im Wettersteinkalk: Abh. Bayer. Akad. Wiss., N.F., 131, p. 1–96.

———, 1973, Mitteltriadische Riffe der Nördlichen Kalkalpen und altersgleiche Bildungen aug Karaburun und Chios (Ägäis): Mitt. Ges. Geol. Bergaustud., 21, pt. 1, p. 251–275.

———, AND KRAUS, O., 1968, Eine ladinische Riff-Fauna im Dobratsch-Gipfelkalk (Kärnten, Österreich) und Bemerkungen zum Faziesvergleich von Nordalpen und Drauzug: Mitt. Bayer, Staatssamml. Paläont. Hist. Geol., 8, p. 263–290.

RESCH, W., 1979, Zur Fazies-Abhängigkeit alpiner Trias-Foraminiferen: Jahrb. Geol. B.-A., 122, p. 181–240.

SANDER, B., 1936, Beiträge zur Kenntnis der Anlagerungsgefüge (Rhythmische Kalke und Dolomite aus der Trias): Mineral. Petrogr. Mitt., N.F., 48, p. 27–209.

SARNTHEIN, M., 1965, Sedimentologische Profilreihen aus den mitteltriadischen Karbonatgesteinen der Kalkalpen nördlich und südlich von Innsbruck: Verh. Geol. B.-A., Jg. 1965, p. 119–162.

———, 1967, Versuch einer Rekonstruktion der mitteltriadischen Paläogeograph e um Innsbruck, Österreich: Geol. Rundschau, 56, p. 116–127.

SCHNEIDER, H.-J., 1964, Facies differentiation and controlling factors for the depositional lead-zinc concentration in the Ladinian Geosyncline of the Eastern Alps: Developments in Sedimentology, 2, p. 29–45.

SENOWBARI-DARYAN, B., 1978, Neue Sphinctozoen (segmentierte Kalkschwämme) aus den ''oberrhätischen'' Riffkalken der nördlichen Kalkalpen (Hintersee/Salzburg): Senckenberg, Lethaia, 59, p. 205–227.

TOSCHEK, P. H., 1968, Sedimentological Investigation of the Ladinian ''Wettersteinkalk'' of the ''Kaiser Gebirge'' (Austria), *in* Muller, G. and Friedman, G. M., Eds., Recent Developments in Carbonate Sedimentology in Central Europe: Springer-Verlag, New York, p. 219–227.

WILSON, J. L., 1975, Carbonate Facies in Geologic History: Springer-Verlag, New York, 471 p.

WOLFF, H., 1973, Fazies-Gliederung und Paläogeographie des Ladins in den bayerischen Kalkalpen zwischen Wendelstein und Kampenwand: Neues Jahrb. Paläont. Abh., 143, p. 246–274.

ZANINETTI, L., 1976, Les Foraminifères du Trias. Essai de synthèse et corrélation entre les domaines mésogéens européen et asiatique: Riv. Ital. Paleont., 82, p. 1–258.

ZANKL, H., 1969, Der Hohe Göll. Aufbau und Lebensbild eines Dachsteinkalk-Riffes in der Obertrias der nordlichen Kalkalpen: Abh. Senckenberg. Naturforsch. Ges., 519, p. 1–123.

———, 1977, Quantitative Aspects of Carbonate Production in a Triassic Reef Complex, *in* Taylor, D. L., Ed.: International Coral Reef Symp., 3rd, Proc., 2, p. 379–382.

SEPM SPECIAL PUBLICATION NO. 30, P. 233–240, MAY 1981

# SEDIMENTOLOGICAL CHARACTERISTICS OF UPPER TRIASSIC (CORDEVOLIAN) CIRCULAR QUIET WATER CORAL BIOHERMS IN WESTERN SLOVENIA, NORTHWESTERN YUGOSLAVIA

JOŽE ČAR
Institute For Karst, Postojna, Yugoslavia

DRAGOMIR SKABERNE
University of Ljubljana, Lubljana, Yugoslavia

BOJAN OGORELEC
Geological Survey, Ljubljana, Yugoslavia

DRAGICA TURNŠEK
Institute For Paleontology, Ljubljana, Yugoslavia

LADISLAV PLACER
Geological Survey, Ljubljana, Yugoslavia

## ABSTRACT

Six circular-shaped quiet water coral bioherms of Upper Triassic (Cordevolian) age, ranging from 25 to 140 meters in height and from 50 to 180 meters in diameter, outcrop 20 kilometers north-northwest of Idrija in western Slovenia, northwestern Yugoslavia. Within the biohermal depositional area two distinctive lithological and faunal associations have been identified. These are: 1. the biohermal core and marginal breccias which pass into, 2. surrounding bedded biopelmicritic limestones. The biohermal cores are composed primarily of corals and consist of various biolithites and biopelmicritic limestones. A relatively quiet water depositional environment is suggested for these bioherms. This interpretation is reinforced by the occurrence of only modest accumulations of marginal breccias, and the occurrence of layered pelbiomicritic limestones between individual bioherms. These circular bioherms are surrounded and embedded in a sequence of interbedded sandstones and shales.

## INTRODUCTION

Circular-shaped, quiet water coral bioherms of Upper Triassic age (Cordevolian) outcrop in an area 20 kilometers northwest of Idrija (location of a famous mercury mine), in western Slovenia, northwestern Yugoslavia (Fig. 1A).

Within an outcrop area 500 by 400 meters, six large circular coral bioherms have been mapped (Fig. 1B). At ground surface they appear as regular circular-shaped calcareous structures connected at various levels with either breccias or bedded limestones. The largest bioherm is approximately 180 meters in diameter and 140 meters in height, whereas the smallest bioherm is 50 meters in diameter and about 25 meters in height. The angle of depositional slope averages 65 degrees. Some bioherms exposed on the steep slopes of Kojca Hill stand out conspicuously from the surrounding sediments (Fig. 2). At this location there is very little associated reef talus and we assume that these bioherms are preserved *in toto*.

For this discussion the two largest bioherms (Fig. 1B) and the surrounding sediments have been described in detail, with particular attention being paid to facies relationships and faunal distributions.

## STRATIGRAPHIC SETTING

The first geologist to work in this area was M. V. Lipold (1857). Additional work, mainly of a stratigraphic and tectonic nature, was done by Kossmat (1910) and Berceetal (1959). However, it has only been since 1977 when Placer *et al.*, recognized these structures as coral bioherms, and noted the basic stratigraphic and facies relationships.

The stratigraphic position of these coral bioherms is shown in Figure 3. The bioherms occur in the upper portion of the Upper Triassic Pseudozilian Formation. In this region the Pseudozilian Formation is at least 1000 meters in thickness, and consists of laminated black shales, sandstones and pyroclastics with lenses of layered black limestones, and numerous reef-knolls and patch-reefs. The Pseudozilian Formation is underlain by upper Middle Triassic (Ladinian) igneous rocks (keratophyres and diabases), and associated volcanic tuffs.

In a reef-knoll several meters in diameter, and

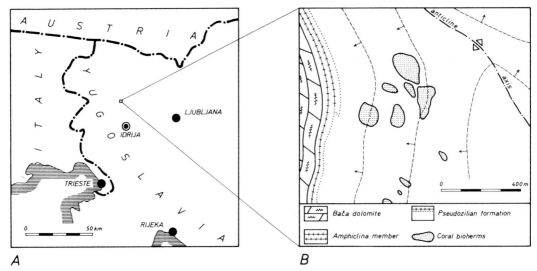

A                                              B

FIG. 1.—Maps locating the studied area in western Slovenia, northwestern Yugoslavia. A. Index map showing area of bioherm outcrops. B. Generalized geological map showing the distribution and size of the studied circular, quiet water coral bioherms.

located in beds below the circular coral bioherms, specimens of the coral *Margarophyllia michaelis* Volz have been identified (Fig. 4A). This coral species is characteristic of the Upper Triassic Carnian Stage (lower Cordevolian). The lower and central portions of the bioherm's reef framework contains corals which are well known from classical Triassic (Cordevolian) beds at St. Cassian in Tyrol (Volz, 1896; Leonardi, 1967). Several species of biohermal corals do have longer stratigraphic ranges, while other biohermal corals such as the bushy form *Hexastrea fritschi* Volz, and the solitary corals *Margarophyllis crenata* Volz (Fig. 4B) and *Myriophyllum dichotomum* (Loretz) (Fig. 4C), are found only in Cordevolian beds (see Table 1).

Beds in the uppermost portion of the bioherms contain coral specimens so strongly recrystallized they cannot be properly identified.

Approximately 70 meters above the highest coral bioherms are a series of limestones and limestone conglomerates designated as the *Amphiclina* Member. Conodonts retrieved from these beds indicate that this interval is of highest Carnian age (Tuvalian), as reported by Placer *et al.* (1977).

On the basis of the above biostratigraphical data we have concluded that growth of the circular coral bioherms began in Cordevolian time and ended in Julian time (Fig. 3).

### REEF ORGANISMS

The primary reef building organisms of the Pseudozilian bioherms are colonial and solitary

corals. Eight coral species have been identified and their stratigraphic ranges shown in Table 1. The skeletal framework of the bioherms is built by these ramose colonial and solitary coral species. All of these corals are considered to have been rather fragile and most suited to live only within quiet water environments.

Other accessory organisms do occur within the bioherms, and these consist of both skeletal and non-skeletal algae, planktic foraminifers, hydrozoans, and molluscs (pelecypods). It should be noted however that most of these are strongly recrystallized and only rarely does the state of preservation allow accurate identification. Undoubtedly, other organisms were more common while

FIG. 2.—Outcrop photograph of the eastern slope of Kojca Hill showing prominent biohermal exposures.

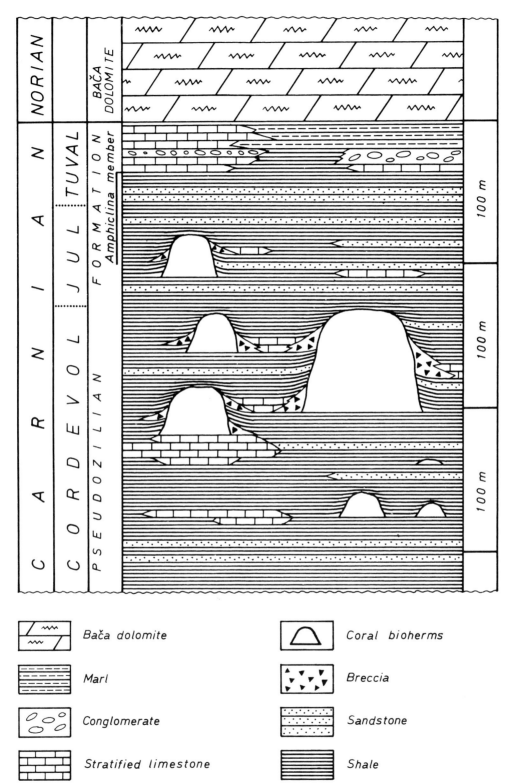

FIG. 3.—Stratigraphic column showing regional time-stratigraphic framework and the intervals of biohermal development.

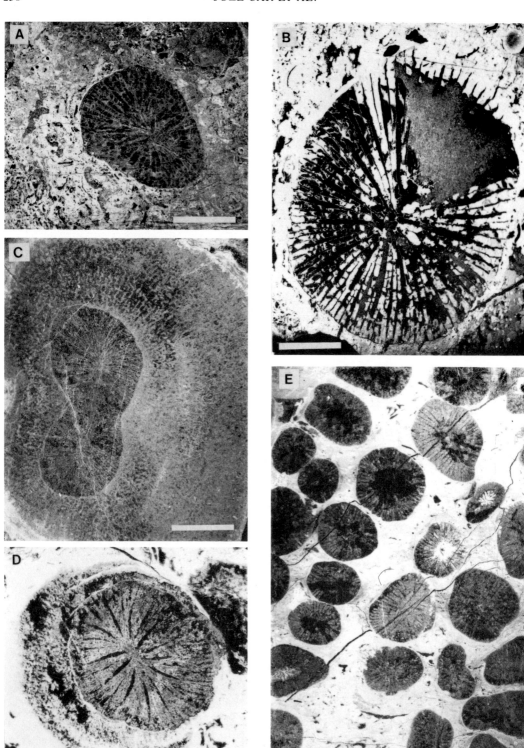

the bioherms were growing, but these cannot be determined after severe diagenetic alterations. A complete listing of accessory species is not yet available.

### LITHOFACIES

Within the Cordevolian biohermal interval two lithological associations have been recognized. They are: 1. rocks that make up the bioherms, and 2. those sediments surrounding the bioherms. Of the biohermal rocks, we recognize both a core facies and a facies of marginal breccias which pass into the surrounding bedded limestones.

### Biohermal Cores

The fundamental characteristic of biohermal core rock is the dominance of *in situ* coral frame builders. However, the abundance of frame building corals varies appreciably. For example, along the marginal portion of the largest bioherm (Fig. 1B), corals may make up to 40 percent of the total biomass, while in other parts of various bioherms, corals may only make up a few percent of the total mass. Actual corallites are relatively narrow and generally attain a height of 40 centimeters, but some corallites up to 80 centimeters in height have been observed.

Among core organogenic limestones both biolithites and bioclastic limestones can be distinguished. The biolithites are composed chiefly of corals and hydrozoans, both of which may be commonly overgrown with non-skeletal algae (Fig. 5A and B). Whereas bioclastic limestones are made up of rich associations of the fragmented remains of organisms that lived on the bioherms. These include debris of corals, hydrozoans, molluscs (gastropods), and echinoderms. Bioclastic fragments are, as a rule, not well-sorted and rounded, but may be encrusted with non-skeletal algae. According to bioclast size, the rocks may be termed either calcirudites or calcarenites. In addition to various size bioclasts, numerous peloids, small oncolites, and rare mud intraclasts may be observed. Rock matrix is usually micrite, and only occasionally has this been diagenetically altered to microspar. According to Dunham's (1962) carbonate classification these rocks would be referred to skeletal/peloidal/intraclastic wackestones and packstones. The rocks of the biohermal cores appear to have been deposited in rela-

TABLE 1.—STRATIGRAPHIC RANGES OF BIOHERMAL CORALS

| Species | Stratigraphic Ranges | | |
|---|---|---|---|
| | Ladinian | Lower Cordevolian | Upper Cordevolian |
| Colonial Corals: | | | |
| *Margarosmilia zieteni* Klipstein | | xxxxxxxxxx | xxxxxxxxxx |
| *Margarosmilia confluens* Münster | xxxxxxx | xxxxxxxxxx | xxxxxxxxxx |
| *Hexastraea fritschi* Volz | | xxxxxxxxxx | |
| *Volzeia badiotica* (Volz) (Fig. 4D) | xxxxxxx | xxxxxxxxxx | |
| Solitary Corals: | | | |
| *Margarophyllia michaelis* Volz (Fig. 4A) | | xxxxxxxxxx | |
| *Margarophyllia capitata* Münster (Fig. 4E) | xxxxxxx | xxxxxxxxxx | xxxxxxxxxx |
| *Margarophyllia crenata* Volz (Fig. 4B) | | xxxxxxxxxx | xxxxxxxxxx |
| *Myriophyllum dichotomum* (Loretz) (Fig. 4C) | | xxxxxxxxxx | xxxxxxxxxx |

tively quiet water environments, especially since their water energy index value does not exceed a maximum of II (Plumley *et al.*, 1962). A higher water energy value is only suggested for those limestones containing oncolites (Fig. 5E). Some oncolites attain a diameter of 4 centimeters.

The micritic matrix of the biohermal cores is often disrupted by irregular burrows up to a centimeter in diameter. Probable soft-bodied sediment ingesting organisms thoroughly bioturbated the biohermal core micrites. These burrows may be filled with sparry calcite cement and micritic sediment, or only filled with coarse mosaics of sparry calcite. Some relatively large vugs, up to

←

FIG. 4.—Thin-section photomicrographs of various coral species from the Upper Triassic (Cordevolian) biohermal facies; bar scale 5 mm in length. A. Transverse section of the corallum of the solitary coral *Margarophyllia michaelis* Volz. B. Transverse section of the corallum of the solitary coral *M. crenata* Volz. C. Transverse section of a bicentric corallum *Myriophyllum dichotomum* (Loretz) encrusted with hydrozoan material. D. Transverse section of corallites of the phaceloid colonial coral *Volzia badiotica* (Volz). E. Transverse section of the corallum of the solitary coral *Margarophyllia capitata* Münster.

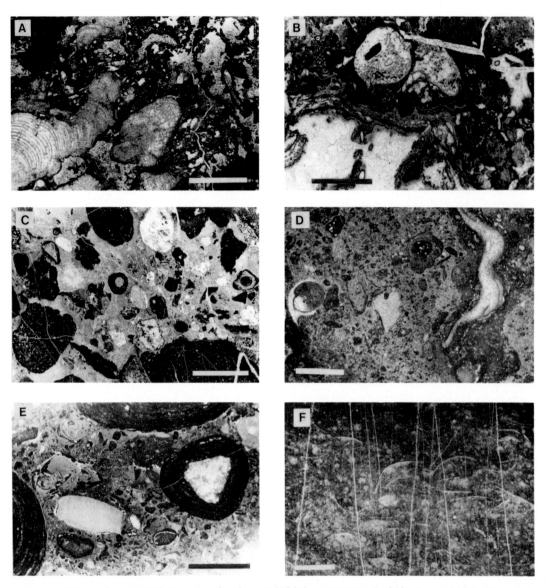

Fig. 5.—Thin-section photomicrographs of various rock fabrics encountered within the coral biohermal facies. A. Typical biolithite showing non-skeletal algal overgrowths on hydrozoans; bar scale 5 mm. B. Recrystallized corallites overgrown with non-skeletal algal coatings, and interspaces filled with sparry calcite; bar scale 5 mm. C. Fine-grained calcirudite composed of a mixture of muddy intraclasts and skeletal fragments; bar scale 1 cm. D. Recrystallized muddy intraclastic biomicrite with some shell debris overgrown with non-skeletal algal coatings; bar scale 2 mm. E. Typical oncolitic structures; rocks contain intraclasts, peloids, and skeletal grains (mainly mollusc fragments); bar scale 1 cm. F. A marly biomicritic limestone from one of the carbonate lenses intercalated within the shales surrounding the bioherms. Fossils consist mainly of thin-shelled molluscan debris along with calcitized radiolarian skeletons; bar scale 1 mm.

several centimeters in length, and closely resembling "stromotactis," are usually filled with geopetal internal sediment.

The carbonates of the biohermal cores also contain trace amounts of authigenic quartz, chert, and pyrite. At a later diagenetic phase these carbonates were recrystallized and lightly and selectively dolomitized.

## Marginal Breccias Transitional to Bedded Limestones

Marginal breccias are only poorly represented on the circular bioherms. They are better represented on the biohermal fore-slopes, and in some cases, in the channels between the bioherms. The breccias are poor-to-moderately-sorted with clasts averaging about 2 centimeters in diameter. The clasts consist of pelmicrites and biomicrites, and large organism fragments. In some instances bioclasts are circumscribed with coatings of non-skeletal algae. Terrigenous grains of quartz and chert up to 4 millimeters in diameter can also be present in these breccias. Cement usually consists of micrite or microspar, clay minerals, and some pyrite. In some occurrences the cement may be slightly dolomitized.

Vugs also occur quite commonly within these breccias, these too are of "stromatactis" type and are usually filled with internal sediment.

Along the margins of these bioherms some clasts may be deformed. Here and in the areas between the bioherms, the breccias pass into bedded limestones up to 2 meters in thickness. These limestones usually pinch out rapidly in all directions. They are classified as slightly-washed biopelmicrites, some contain pelmicritic intraclasts up to 1 centimeter in diameter. Evidence of bioturbation also occurs within these sediments, and among fossils, foraminifers and thin-shelled molluscs are common while calcitized radiolarians are very rare.

## Surrounding Sediments

The circular coral bioherms are surrounded by grey to dark-grey clastic sediments consisting of an interbedded sequence of sand and shale layers. In the lower portion of the formation laminated shales predominate, but interbedded with them are thin beds of fine-grained sandstone. These sandstones may be cross-laminated, and in places are ripple-marked. The amount of sandstone within the section increases upward. Some sandstone beds may even display poor graded-bedding, and these are usually overlain with intervals where bedding may be parallel, inclined, or wavy. Some ripple marks can be found on their upper surfaces. Interbedded with the sandstones are beds of laminated shale. In the upper portion of this sequence the shales are rather conspicuously bioturbated. In places these shales are cut by small erosional channels up to 15 centimeters in depth and 2 meters in width. The channels are filled with coarse-grained sandstone, which in places may be conglomeratic. These conglomerates may contain large intraformational flat pebbles of shale up to 10 centimeters in diameter.

The sandstones are very fine to coarse-grained, and moderately well-sorted. Terrigenous grains consist of poly- and monocrystalline quartz, feldspar, lithic fragments of granitic and extrusive rocks, quartzites, sandstones, schists, and tuffs. These terrigenous grains are bound together with matrix and cement. The major portion of the matrix is apparently not of terrigenous origin, but genetically belongs to the epimatrix (Dickinson, 1970). The epimatrix is predominantly composed of quartz, chlorite, and sericite. Cement is largely quartz, and some illite and feldspars may be present.

The shales are of similar composition as the sandstones, although no lithic grains could be found within them. X-ray diffraction indicates that the coarse-fraction is composed of quartz, muscovite, plagioclase, microcline, chlorite, and calcite. In the clay-size fraction illite and chlorite are dominant.

Shales occur immediately at the base of the circular bioherms, and at their lateral extremities. The contact between the limestones and shales is sharp, and the boundary is indicated by a thin limonitic coating or film. At this marginal boundary are inclusions of quartz and feldspars, cemented and even partially replaced by calcite.

### ENVIRONMENTAL INTERPRETATIONS

Observed features, noted above, of the clastic sedimentary rocks surrounding the coral bioherms indicate that they were deposited in the transitional zone between the shoreline and the shelf (Reineck and Singh, 1975). A prime characteristic for deposition within this transitional zone is the presence of shales interbedded with thin sandy layers. The black shale that occurs at the base of the coral bioherms, and in the area immediately surrounding the bioherms, was deposited in an environment of low water energy. However, the presence of some distinctive interbedded sandstones show characteristics attributed to occasional storm deposited layers (Reineck and Singh, 1975).

A schematic interpretation of the depositional setting of the circular, quiet water coral bioherms, and the adjoining areas, is shown in Figure 6. Field observations tend to suggest that the source of the clastic terrigenous material was situated to the south or southwest of the studied area. Apparently, the clastic material was deposited by continuous currents acting on the fore-slope of the barrier-reef area (Fig. 6). During storms, agitated waters were able to move and sort newly deposited material and carry it into the subsiding transitional zone. During certain intervals, corals lived and developed on the sea bottom and were able to produce small patch-reefs and coral

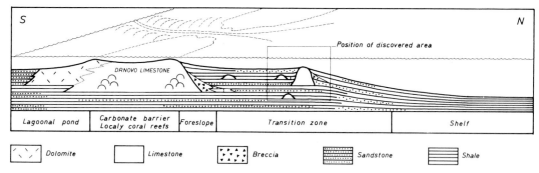

FIG. 6.—Schematic diagram of the interpretation of sedimentary environments in western Slovenia during Upper Triassic (cordevolian) time.

"meadows." During favorable time periods they coalesced and grew into numerous small bioherms and reef-knolls. Most of the smaller coral growths were short-lived and only attained a diameter of a few square meters, and a height of only a few meters. Many of these structures were quickly killed-off by influxes of mud. In areas where clastics were diverted conditions for growth of larger bioherms was satisfactory. Within a uniformly subsiding area, with ideal conditions, some coral bioherms were able to attain a growth height of up to 140 meters. But it should be kept in mind, that this height dimension is the final product of total coral bioherm growth. At any one time the coral bioherms were probably low-relief structures only slightly higher than the surrounding substrate. As a result there is an absence of biotic differentiation, on a vertical scale, of varied coral

biohermal communities. The equilibrium established between coral reef structures and the surrounding sediments is indicated by the contacts of the biohermal cores with the surrounding sediments. However, there does appear to have been time intervals in the growth history of these bioherms when subsidence slowed down and the uppermost portions of the coral structures were exposed to more agitated wave and current conditions. It is at these times when the material for the marginal breccias was derived, and incorporated within the bioherms. The growth of the various intervals of bioherms appears to have ceased when sudden and spasmodic influxes of terrigenous mud engulfed the structures. As shown on the stratigraphic column (Fig. 3), this apparently happened a number of times.

## REFERENCES

BERCE, B., 1959, Poročilo o geološkem kartiranju ozemlja Cerkno-Žiri v letu 1958: Arhiv Geološkega Zavoda v Ljubljani in Rudnika Živega Srebra v Idriji, 82 p.

DICKINSON, W. R., 1970, Interpreting detrital modes of graywacke and arkose: Jour. Sed. Petrology, v. 40, p. 695–707.

DUNHAM, R. M., 1962, Classification of carbonate rocks according to depositional texture: Am. Assoc. Petroleum Geologists Mem. 1, p. 108–121.

KOLOSVARY, G., 1958, Corals from the Upper Anisian of Hungary: Jour. Paleontology, v. 32, no. 3, 636 p.

———, 1966, Angabe zur Kenntnis der Triaskorallen und der begleitenden Fauna der ČSSR, Geol. Práce: Praha, v. 7–8, p. 179–188.

———, 1967, Beiträge zur Kenntnis der Ladin und Liaskorallen von Jugoslawien: Acta Biologica, Szeged, v. 13, p. 159–161.

KOSSMAT, F., 1910, Erläuterungen zur geologischen Karte Bischoflack and Idria: Wien, 101 p.

LEONARDI, P., 1967, Le Dolomite, Geologia die monti tra Isarco e Piave: Trento, 522 p.

LIPOLD, M. V., 1857, Bericht über des geologischen Aufnahmen in OberKrain im Jahre 1856: Jb. Geol. R.A., Wien, v. 8, p. 205–234.

PLACER, L., ČAR, J., OGORELEC, B., OREHEK, S., RAMOVŠ, A., BABIĆ, L., ZUPANIČ, J., ČADEŽ, F., CIGALE, M., AND HINTERLECHNER, A., 1977, Triadna tektonika okolice Cerknega, MVS: Inštitut za geologijo FNT Ljubljana, 58 p.

PLUMLEY, W. J., RISLEY, G. A., GRAVES, R. W., AND KALEY, M. E., 1962, Energy index for limestone interpretation and classification: Am. Assoc. Petroleum Geologists Mem. 1, p. 85–107.

RAMOVŠ, A., 1971, Okamenele korale na Loškem, ozemlju, Loški razgledi: Ljubljana, v. 18, p. 160–163.

REINECK, H. E., AND SINGH, I. B., 1975, Depositional Sedimentary Environments: Berlin, Springer-Verlag, 439 p.

VOLZ, W., 1896, Die Korallen der Schichten von St. Cassian, in Süd-Tirol: Palaeontographica, Stuttgart, v. 43, 124 p.

SEPM SPECIAL PUBLICATION No. 30, P. 241–259, MAY 1981

# FACIES DEVELOPMENT AND PALEOECOLOGIC ZONATION OF FOUR UPPER TRIASSIC PATCH-REEFS, NORTHERN CALCAREOUS ALPS NEAR SALZBURG, AUSTRIA[1]

PRISKA SCHÄFER AND BABA SENOWBARI-DARYAN
Paleontological Institute
University of Erlangen
West Germany

## ABSTRACT

Four Upper Triassic patch-reefs exposed in the Northern Limestone Alps of the Salzburg area of Austria were subjected to detailed study. Especial emphasis was placed on facies development and paleoecologic zonation of the reef complexes.

The Adnet Reef structure grew directly on a carbonate platform in a shallow water carbonate setting; the Rötelwand and Feichtenstein Reef complexes grew out of the Kössen Basin in two distinct stages: 1. a deeper water mud-mound stage and 2. a shallow water reef stage; whereas, the Gruber Reef, which also developed within the Kössen Basin, only shows a deeper water mud-mound stage, since a shallow water phase did not develop at this location.

The shallow water reef stages of the reef complexes display a lateral facies zonation consisting of five different facies units: 1. coral-sponge facies of the central reef areas, 2. oncolitic facies of the exposed upper reef-slopes within zones of highest water energy, 3. algal-foraminiferal facies of the exposed lower reef-slopes, and as the foundation of the Adnet Reef structure, 4. reef detritus-mud facies of the leeward and deepest portions of the reef-slopes, where reefal components interfinger with Kössen basinal sediments, and 5. terrigenous-mud facies of the basin proper. The Rötelwand and Feichtenstein Reef complexes display a linear facies zonation, while the Adnet Reef Structure displays a circular and somewhat patchy facies zonation.

Organism communities forming the reef biota are characterized by the association of various reef framebuilders and their epi- and endobionts. The compositional relationships of the various patch-reefs, their gross morphology, paleoecologic setting, and depositional environments have been studied in detail. This study has demonstrated that calcareous algae, various *microproblematica*, and foraminifers show very distinct distributional patterns within the reef complexes, and can be used as facies indicators as well as differentiating different biotopes within the central reef areas.

## INTRODUCTION

Four Rhaetian patch-reef structures in the Upper Triassic of the Northern Calcareous Alps Austria, have been investigated to determine their developmental history. The aims of this study include analysis of reef growth conditions, vertical reef development, and the zonation of facies units reflecting different sedimentary environments. The relationship between sedimentological-palecological conditions on the one hand, and the biology of reef organisms on the other hand results in the construction of a distinctive reef framework. During this study greatest interest has been devoted to the specific distribution of the frame building organisms, their associations within the reef communities, and their ecologic requirements and adaptation to the biotopes of the reef complexes. The apparent differences of the four reef structures has aided in the interpretation of the local palaeogeographic environmental setting.

This present paper is based on two dissertations by Senowbari-Daryan (1979) and Schafer (1979 under the supervision of Professor Erik Flügel).

The investigated Upper Triassic patch-reefs are located about 15 kilometers south of Salzburg, Austria, on the eastern side of the Salzach River near the town Hallein (Fig. 1A). The Adnet Reef is well exposed in old quarries near the village of Adnet. Here, the polished walls and floors of the quarries give an impressive glimpse into the palecologic zonation of the reef framework within the central reef area. The quarries belong to two different stratigraphic levels and therefore aided in the investigation of the underlaying foundation of the entire reef structure (Fig. 1B). The Rötelwand and the Feichtenstein Reef structures are more suitable to study of facies development of patch-reefs within a deeper water setting (Fig. 1C and D). The Rotelwand Reef lies at the bend of the Mörtelbach Valley, whereas the Feichtenstein Reef is situated directly south of the small village of Hintersee, near Salzburg. The distance between reef structures is about 4 kilometers, and they are separated by a deeper valley containing

[1] Contributions to paleontology and microfacies of Upper Triassic reefs in the Alpine-Mediterranean region, No. 10.

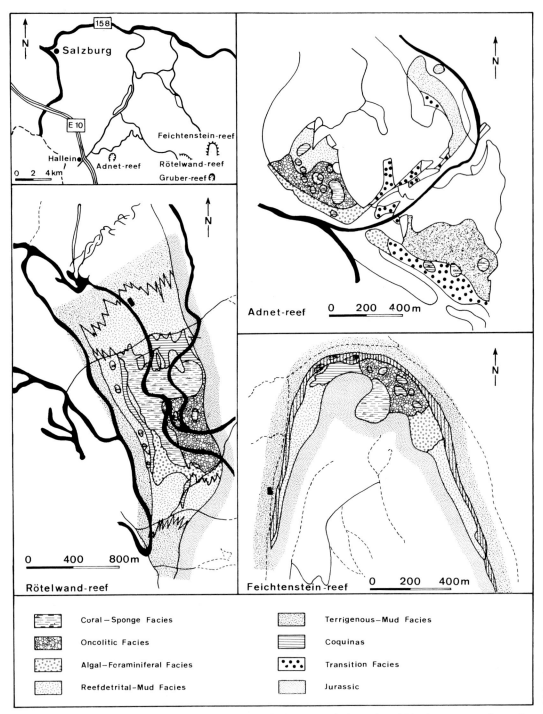

FIG. 1.—Outline maps of the investigated area. A. Distribution of facies units within the Adnet reef complex.
B. The Rötelwand. C. and D. Feichtenstein reef structures.

Kössen basinal sediments. The Gruber Reef, situated about 3 kilometers south-southeast of the
Feichtenstein structure, is characterized by a very
high content of reef framebuilders. This patch-

reef, shows a rather unclear paleoecologic zonation in comparison with the other reef structures,
but because of its abundant biota is more useful
for detailed palaeontological studies.

Tectonically, the investigated area belongs to the eastern portion of the Tyrolian Nappe, which is part of the Oberostalpine. Sedimentary units range from the Upper Triassic (Norian, Hauptdolomite) to the Upper Jurassic (Tithonian, Oberalm beds).

During Upper Triassic time shallow water carbonate deposition predominated. Starting with lowest Jurassic, the depositional conditions changed and shallow water sedimentation was followed by deeper water sedimentation.

According to the environmental scheme presented by Zankl (1971), an extensive platform with shallow water carbonate deposits developed in the Northern Calcareous Alps during the Upper Triassic. During the Norian Stage this platform consisted of a chain of large, but irregularly distributed Dachstein Reef structures. These grew along the southern ramp of the Dachstein Platform, where it graded into the Hallstalt Basin of deeper water sedimentation. Northward, the Dachstein Reefs, and their back-reef sediments interfinger with shallow water deposits of subtidal to intertidal environmental origin. It is possible to distinguish different facies types, such as near back-reef sediments with coarse reef debris and oncolitic far back-reef sediments (Zankl, 1969), shallow subtidal deposits of grapestone facies and mud facies (Piller, 1976), and intertidal Hauptdolomite (Müller-Jungbluth, 1970; Czurda and Nicklas, 1970).

During uppermost Triassic time (Rhaetian) the Dachstein Platform underwent shallow subtidal sedimentation with growth of Dachstein Reefs in the south, while in the adjacent northern area the Kössen Basin developed. This basin resulted from continuous subsidence of the extensive tidal-flat area with Hauptdolomite-type sedimentation. Smaller Rhaetian patch-reefs of up to 500 meters in length and 100 to 200 meters in height developed both on shoals within the Kössen Basin, and on the northern ramp of the Dachstein Platform. These Dachstein Reefs grew under optimal ecologic conditions on the outer platform ramp. The Rhaetian reefs however, separated from the open sea by the Dachstein Platform, developed in more or less protected basinal areas and coeval lagoonal areas with varying restricted ecologic conditions. Nevertheless both show characteristic vertical and horizontal zonation caused by development of different facies units and by a clear biozonation of the central reef area.

In early Jurassic time the carbonate platform was destroyed by tectonic events (e.g., formation of the Penninic Ocean, Dietrich, 1976). It was also separated into isolated blocks, which subsided and were overlain by thin-bedded limestones of deeper water origin. Based on the present relief of the Upper Triassic rocks, Jurassic sedimentation was initiated with red or gray limestones (nodular Adnet Limestones or bedded gray limestones with chert nodules).

The reef structures can be divided into five facies units, which interfinger laterally. They are characterized by type of matrix, content of significant components, and by textural differences. These five facies units delineate six environmental areas as follows:

1. central reef area with patch-reefs separated by furrows filled with reef detrital sands;
2. leeward upper reef-slope with conditions of highest water energy;
3. leeward lower-reef-slope with conditions of reduced water energy;
4. leeward reef-slope and reef tallus;
5. surrounding Kössen Basin;
6. shallow water area with lagoonal conditions.

Subdivision of the facies units according to characteristic biotic components and textural criteria resulted in recognizing 15 microfacies types. Table 1 shows the relation between microfacies types and microfacies criteria based on thin-section analysis.

### Coral-Sponge Facies

Sediments of the coral-sponge facies characterize the central reef area. They are composed of reef framework sediments made up of scleractinian corals, problematical groups of hydrozoans and tabulozoans, calcisponges, calcareous algae, and interstitial reef debris.

The presence of reef framework is indicated:

1. by the relatively high population density of *in situ* organisms in comparison with surrounding sediments;
2. by zonation of the reef builders; and,
3. by definitive biotic depositional fabric.

Two different types of reef framework contributed to the construction of these patch-reefs:

1. framestone to bafflestone with weak biogenetic depositional fabric in which framebuilders are phaceloid and dendroid coral colonies; encrustation of the primary framework is by epizoans and epiphytes: "coral framestone to bafflestone" (Fig. 2B);
2. framestone with strong biogenetic depositional fabric constructed by the framebuilders themselves, and aided by secondary microorganism encrustations: "calcisponge-

TABLE 1.—MICROFACIES CRITERIA OF RECOGNIZED FACIES TYPES IN UPPER TRIASSIC PATCH-REEF STRUCTURES SOUTH OF SALZBURG, AUSTRIA: (x) = RARE; x = COMMON; xx = FREQUENT

| Appearance of microfacies criteria within the facies units | Coral-Sponge Facies | | | | Oncolitic Facies | Algal-foraminiferal Facies | | | Reefdetrital-mud Facies | | | Terrigenous-mud Facies | | | |
|---|---|---|---|---|---|---|---|---|---|---|---|---|---|---|---|
| | biogenous packstone/ wackestone | biogenous mudstone | peloidal grainstone | colored biogenous wackestone | oncolitic grainstone/ rudstone | foraminiferal-cayeuxian packstone grainstone | solenoporacean- foraminiferal- packstone/grainstone | foraminiferal grainstone | detrital wackestone | detrital solenoporacean wackestone | coquina | thecosmilian framestone/wackestone | detrital wackestone | peloidal mudstone/grainstone | coquina/oolites |
| | 1/A | 1/B | 1/C | 1/D | 2 | 3/A | 3/B | 3/C | 4/A | 4/B | 4/C | 5/A | 5/B | 5/C | 5/D |
| micrite | (x) | x | (x) | x | x | x | x | x | x | x | x | x | x | (x) | x |
| sparite | x | | x | | x | x | x | x | | | x | | x | x | x |
| resediments | x | | | | | x | x | | | | | | x | | |
| peloids | x | x | x | | xx | x | x | xx | | x | | x | x | xx | |
| skel. oncolites | | | | | xx | | | | | | | | | | |
| grapestones | | | | | x | | | | | | | | | | |
| micrite envelopes | x | | | xx | xx | x | x | (x) | | x | | | | | |
| ooids | | | | | | | | | | | | | | | x |
| reef debris | x | | (x) | xx | xx | | | | (x) | x | | x | (x) | | |
| reef framework | xx | xx | xx | xx | (x) | | | | | | | x | | | |
| dasycladaceans | (x) | | | x | x | xx | x | (x) | | xx | | | | | |
| solenoporaceans | x | | | x | xx | (x) | x | | | | | | | | |
| porostrom. cyanoph. | | | | x | x | xx | x | | | | | | | | |
| coquinas/echino. | | | | x | | xx | | | x | x | xx | | | | |
| involutinids | x | | | xx | x | xx | x | | x | x | | | | | xx |
| Lithocodium | x | | | xx | xx | | | (x) | | | | x | | | x |
| Alpinophragmium | xx | | | x | (x) | | | | | | | x | (x) | | |
| miliolids | | x | x | | | | | | | | | x | | | |
| Microtubus a.o. | | xx | xx | | | | | | | | | xx | | | |

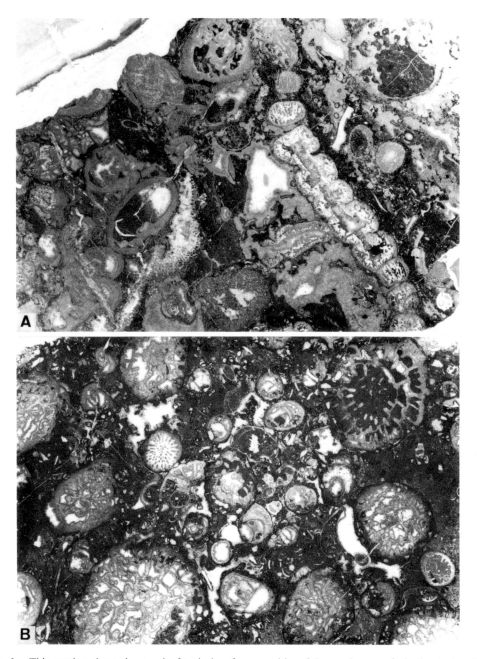

FIG. 2.—Thin-section photomicrograph of typical reef communities of the coral-sponge facies (central reef area). A. Calcisponge community (×2) of the innermost, protected part of reef patches including different calcisponge species (*Paradeningeria weyli* Senowbari-Daryan and Schäfer, *P. gruberensis* Senowbari-Daryan and Schäfer, *P.* sp., *Weltheria* sp. cf. *W. repleta* Vinassa, *Annaecoelia interiecta* Senowbari-Daryan and Schäfer, *Salzburgia variabilis* Senowbari-Daryan and Schäfer) encrusted by tabulozoan and bryozoan colonies, by *Follicatena irregularis* Senowbari-Daryan and Schäfer, *Radiomura cautica* Senowbari-Daryan and Schäfer, *Microtubus communis* Flügel, and by serpulids; Gruber Reef, A/1/2. B. Coral community, with section (×1.5) through coralites of *Stylophyllum polyacanthum* Reuss, *Pinacophyllum* sp., and one coralite of *Stylophyllum paradoxum* Frech. Note over-growths by foraminifer *Alpinophragmium perforatum* Flügel, and the *microproblematicum Lithocodium* sp. Many small gastropods occur within the detrital interstitial sediment of the reef framework; Rötelwand Reef, o/55.

hydrozoan-tabulozoan-solitary coral frame-stone" (Fig. 2A).

Based on the detrital sediment filling the interstices of the reef framework, we can distinguish different microfacies types (Table 1). The type of sediment seems to be related to the association of framebuilders and their position within the central reef area. Four sediment types are recognized, these are:

1. biogenous pack- to grainstones;
2. biogenous mudstones;
3. peloidal grainstones;
4. colored biogenous wackestones.

### Oncolitic Facies

Interstitial reef debris within the reef framework, and in the furrows between the patch-reefs of the central reef area, interfingers with sediments of the oncolitic facies. The oncolitic grain- to rudstone (bio-oncosparite) is composed of reworked framebuilders (mainly coral species of phaceloid-dendroid and massive colonial shape), calcareous algae (nodular thalli of solenoporaceans, porostromate blue-green algae, and dasycladaceans), foraminifers (involutinids, duostominids, "Tetrataxis"), peloids, grapestones (algal lumps), and rare lithoclasts derived from destruction of lithified reef rock (Fig. 3A).

Skeletal oncolites, are constructed by encrustations of microproblematica, e.g., Lithocodium Elliott and Bacinella irregularis Radoicic, which form coatings around reef detrital components of rudite size. Very large oncolites up to 30 centimeters in diameter are restricted to surrounding shallow water environment of the Adnet Reef complex. These sediments shows very poor sorting. Both the coral-sponge facies and the oncolitic facies are thought to have been deposited under shallow water conditions of the highest water energy.

### Algal-Foraminiferal Facies

The algal-foraminiferal facies contains pack- to grainstones (biointrasparites), that are characterized by the presence of dasycladacean algae and of porostromate cayeuxians, grapestones (algal lumps), and grains circumscribed with micrite envelopes; these signify deposits of a shallow water environment. The lack of reef debris in the oncolitic facies indicates a greater distance from the central reef area. These sediments are well-sorted and contain grains dominantly of arenite size (Fig. 3B).

Based on different composition of sedimentary components we can distinguish three microfacies types (Table 1):

1. foraminiferal-cayeuxian packstones/grainstones;
2. solenoporacean-foraminiferal packstones/grainstones; and
3. foraminiferal grainstones.

The predominance of distinct groups of foraminifers and algae, e.g., Involutina, Triasina, dasycladaceans (Diplopora adnetensis Flügel), and "Cayeuxia," within restricted areas, is caused by patchy colonization on the sediment. Still, Involutina, Triasina and "cayeuxians," show a strong affinity to a matrix of sparry calcite, whereas the solenoporaceans and dasycladaceans preferred a micritic environment.

Sediments of the algal-foraminiferal facies, like those of the oncolitic facies, are related to the shallow water sediments of the Dachstein Platform (see Zankl, 1969; and Piller, 1976).

### Reefdetrital-Mud Facies

In addition to coarse reef debris, skeletal oncolites, grapestones, lithoclasts, and calcareous algae are missing from the biogenous wackestones of the reefdetrital-mud facies. The foraminiferal fauna is influenced by the coral-sponge facies (Galeanella, small species of Ophthalmidium), as well as Kössen basinal sediment (Trochammina, and lagenid foraminifers). Sessile miliolid foraminifers encrust shells of pelecypods and crinoidal fragments, which are both very common in this facies. The high content of a primary mud matrix (micrite) indicates more quiet water conditions (Fig. 3C).

This facies unit can be subdivided into three microfacies types:

1. detrital wackestones;
2. detrital solenoporacean wackestones; and,
3. coquinas.

### Terrigenous-Mud Facies

The reefdetrital-mud facies generally interfingers with the terrigenous-mud facies and is characterized by interbedded calcareous mud- to wackestones, marls and shales. They are believed to have been deposited in a basin with water depths up to 100 meters. The sediment contains a restricted microfauna (ostracodes, and foraminifers with lagenids and the typical sessile species Agathammina austroalpina Kristan-Tollmann and Tollmann), and the sediment is strongly bioturbated. Reef biotas occur only as less diversified scleractinian biostromes (Fig. 3D).

We can distinguish four microfacies. These are:

1. thecosmilian biostromes (frame- to wackestones);
2. detrital, bioturbated wackestones;
3. mudstones to grainstones (Fig. 3D); and
4. coquinas and ooidal limestones.

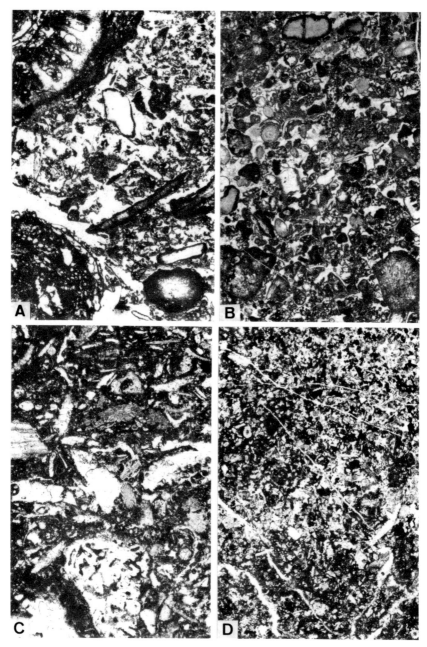

FIG. 3.—Thin-section photomicrographs of different facies units that contributed to the construction of Upper Triassic patch-reefs in the Salzburg area, Austria. A. Oncolitic facies, a shallow water sediment (×3) deposited under high energy environmental conditions. Oncolites range from grain- to rudstone. Detrital biogenous components of rudite size, derived from reworking of reef building organisms (coral fragments, and thalli of solenoporaceans), have been coated by *Lithocodium* sp. and *Bacinella irregularis* Radoicic. In addition to skeletal oncolites, peloids and lithoclasts occur within a matrix of sparry calcite; Rötelwand Reef, r/70. B. Algal-foraminiferal facies; a shallow marine sediment (×6) deposited under less agitated water conditions. Algae (thalli of "*Cayeuxia*" sp., and fragments of dasycladaceans), foraminifers (*Involutina* spp., *Triasina hantkeni* Majzon, and duostominids), grapestones, and peloids all occur within a sparry calcite matrix; a biogenic pack- to grainstone; Rötelwand Reef, x/157. C. Reefdetrital-mud facies; a biogenic wackestone (×2) consisting of detritus of frame building organisms, lamellibranches and echinoderms, within a micritic matrix; Feichtenstein Reef, F/41.5. D. Terrigenous-mud facies (Kössen beds); mainly peloidal mud- or grainstones (×10) with very closely-packed peloids and microfossils (foraminifers and ostracodes) within a sparitic matrix, Feichtenstein Reef, F/74a.

## DISTRIBUTION OF THE FACIES UNITS

Figure 1 shows the distribution patterns of facies units within the reef complexes of Adnet, Rötelwand and Feichtenstein. Both the Feichtenstein and Rötelwand Reef structures have similar facies development (Fig. 9). Characteristic of both reef structures are the strong linear facies zonation and their reversed distribution.

Conversely, the Adnet Reef structure differs not only in biogenic composition of the central reef area, but also in the detritus surrounding the reef structure, and in the distribution of facies units. A more circular facies zonation prevails in the Adnet Reef structure instead of a strong linear facies distribution (Fig. 1 and Fig. 9).

The coral-sponge facies occupies the central reef area of all three reef structures. The biotic composition, distribution of patch-reefs, and the composition of reef communities is described in a following chapter.

The oncolitic facies appears with parautochthonous reefdetrital sediments between the patch-reefs of the central reef area. The main depositional area occurs at the exposed leeward reef-slope, in the immediate vicinity of the coral-sponge facies. In the Rötelwand and the Feichtenstein complexes this facies is restricted to the southwestern and northeastern slopes of the reef structures. It indicates the high energy, shallow water conditions of these reef-slopes, as well as for the coral-sponge facies. In the Adnet Reef complex the reef patches are completely surrounded by sediments of the oncolitic facies. The area of shallow water deposition appears to be more extensive here, than in the Rötelwand and the Feichtenstein complexes.

The algal-foraminiferal facies interfingers with the oncolitic facies on the southwestern side of the Rötelwand Reef complex and on the northeastern side of the Feichtenstein Reef. This distributional pattern intensifies the strong linear, but reversed facies zonation of both reef structures. This facies unit covers an extensive area on nearly all sides of the Adnet Reef complex. It interfingers not only with the oncolitic facies, which surrounds the central reef area, but also forms the foundational unit for initial growth of the reef patches (Fig. 9 and Fig. 10). A transitional facies (detrital foraminiferal grainstones), which contains elements of both shallow water and the basinal environment, is situated between the algal-foraminiferal facies and the terrigenous-mud facies on the northern and eastern side of the reef structure.

The reefdetrital-mud facies suggests location of a protected reef-slope directly adjacent to the coral-sponge facies of the Rötelwand- and Feichtenstein Reefs which grades into the terrigenous-mud facies on the distal reef-slope on all sides of the reef structures. Both reefs are separated by deposits of micritic sediments, of quiet water conditions on the protected reef-slopes. This is in contrast to the deposits of agitated, shallow water conditions, which are limited to the outer exposed reef-slopes. The reefdetrital-mud facies has a limited and rather patchy distribution on the northern and eastern part of the Adnet Reef structure, where it is in close proximity to Kössen basinal sediments.

The terrigenous-mud facies contains varied facies types. Their composition and distribution is dependant on the topography of the sea bottom of the Kössen Basin. The sediments interfinger with reefal deposits on all sides of both the Rötelwand and Feichtenstein Reef structures. In the reef area of Adnet, shallow water platform sediments grade into deposits of the terrigenous-mud facies on the northern and eastern side of the entire complex.

In contrast to the linear facies zonation of the Feichtenstein and Rötelwand Reef structures, and the circular zonation of the Adnet complex, the Gruber Reef shows a very patchy and irregular distribution of sedimentary facies types. The patch-reefs are much more separated from each other than those of the other reef structures, and they are associated with micritic, detrital deposits. Calcisponges, hydrozoans, and both massive and platy coral species dominate the biota and suggest quiet water conditions. Detrital sediments with sparry calcite matrix prevail in the area surrounding the Gruber Reef.

## GENERAL CONSTRUCTION OF THE CENTRAL REEF AREAS

The central reef area appears to be the environment with the most complex sedimentary composition. The center of each reef structure is composed of different, more or less isolated reef patches of 1 to 10 meters in diameter and up to 5 meters in height. These reef patches are separated by channels filled with reefdetrital and oncolitic deposits. The smaller reef patches show a distributional pattern, each of which is characterized by a distinctive assemblage of frame building organisms and associated microbiotas.

Within the Adnet Reef we see a close relationship between the position of the reef patches in the central reef area, their gross morphology, and their internal construction by various framebuilders. The patch-reefs of the Adnet locality are completely separated from each other, and each possesses it's own ecologic zonation. In contrast, the patch-reefs of the Rötelwand and the Feichtenstein locations are more connected with each other and all the reef cores show a general

TABLE 2.—COMPARISON OF FACIES DEVELOPMENT AND PALEOECOLOGY OF FOUR UPPER TRIASSIC PATCH-REEF STRUCTURES SOUTH OF SALZBURG, AUSTRIA

| | Gruber Reef | Feichtenstein Reef | Rötelwand Reef | Adnet Reef |
|---|---|---|---|---|
| vertical development | mud-mound stage lower Kössen beds | reef stage mud-mound stage lower Kössen beds | reef stage mud-mound stage lower Kössen beds | reef stage basement of platform carbonates |
| horizontal facies distribution | lacking | distinct 5 facies units | distinct 5 facies units | indistinct 5 facies units |
| zonation of the facies units | lacking | strong linear zonation | strong linear zonation reversal | weak circular zonation |
| oncolitic facies | lacking | one-sided | one-sided | extended, all-around |
| detrital-mud-facies Kössen beds | all-around | all-around | all-around | reduced, one-sided |
| disconformities of reef growth | lacking | lacking | lacking | present |
| overlaying Liassic sediments | grey facies prevailing, continuous development | grey facies prevailing, continuous development | grey facies prevailing, continuous development | red facies prevailing, discontinuous development |
| reef building organisms | coral/calcisponges hydrozoans calcareous algae | corals/calcisponges hydrozoans calcareous algae | corals/calcisponges hydrozoans calcareous algae | corals calcisponges/hydrozoans calcareous algae |
| degree of diversity | high diversity | high diversity | high diversity | low diversity |
| prevailing extension of reef growth | horizontal | horizontal | horizontal | vertical |
| type of ecologic zonation clearly separated patch-reefs | — | coarse zonation of the whole central reef area | coarse zonation of the whole central reef area | fine zonation of numerous whole central reef area |

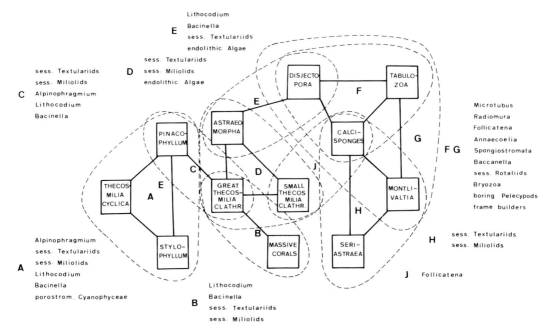

FIG. 4.—Model for the framebuilding organism communities within the reef complexes and their associated epibiota; - - - dashed lines signify those associated framebuilders belonging to one community (marked by - - - lines). Nine communities (A to J) have been distinguished.

ecologic zonation. Since only the deeper water portion of the Gruber Reef is represented, we do not see a clear zonation of the reef building organisms (Table 2). It is apparent that the zonal distribution of reef organisms is caused by different ecologic conditions occurring in the distinct parts of the entire reef complexes, and is related to facies differentiation of the reef environment.

## FAUNAL AND FLORAL INVENTORY

In general, the four reef structures contain a great number of different faunal and floral species. Although the reefs grew in much more protected environments than the Dachstein reefs, they possess a similar diversity of organisms. They also have a much higher organism diversity than the Kössen biostromes (lenses- or bank-like corallian organic buildups).

Each of the reef building organisms shows a typical distribution pattern, so that all parts of the reef area, and each reef community, carries a distinctive qualitative and quantitative biota.

Scleractinian corals play the most important role in the construction of an organic framework. Corals of the families Montlivaltiidae, Stylophylliidae, Thamnasteriidae, Faviidae, Astrocoeniidae, Calamophylliidae, and Procyclolitidae are dominant with about 30 recognized species.

All other reef building groups of organisms are of subordinate importance in regards to reef framework construction, although they too show a rather high species diversity. They occur as associated forms stabilizing the organic framework by encrustation and overgrowth, or by colonizing cavities and protected areas between coral colonies.

Within the central reef area the calcareous sponges are represented with nearly as many species as the scleractinian corals, here they colonize the protected areas within the reef framework. Most of the calcisponges belong to the order of the Sphinctozoans. The families represented are: Celyphiidae, Sebargassiidae, Polytholosiidae, Annaecoeliidae, Verticellitidae, Cryptocoeliidae, and Salzburgiidae, with about 20 recognized species. Among the Inozoans, cylindrical forms like the form Peronidella prevail.

The hydrozoans, a group of problematical systematic position, are represented by three genera: Disjectopora, Spongiomorpha and Stromatomorpha.

The group of bryozoans/tabulozoans, all of them of simple tubular construction, occur as 15 different recognizable forms.

Encrustation of the organic framework by Spongiostromata, questionable algae, problematic microfossils, serpulids, small calcisponges, bryozoans, sessile miliolids, and textulariid foraminifers, is of variable intensity and depends on the type of reef building community present. Stromatolitic algal crusts cover the reef framework and the walls of reef cavities, whereas dasycladaceans, solenoporaceans and porostromate blue-green algae occur more commonly in the surrounding reef detrital facies. These algae are useful indicators for differentiating facies units of the shallow water environment surrounding the reef structures (Schafer and Senowbari-Daryan, 1979).

The organic framework, as well as lithified reef rock, was altered and modified by various boring organisms such as pelecypods, worms, endolithic algae, and a group of organisms of uncertain affinities. Echinoids, holothurians, ophiuroids, gastropods, small ammonites, pelecypods, brachiopods, foraminifers, and ostracodes and other crustaceans are the main reef dwellers. Their fossil remains are concentrated within interstices and cavities of the primary reef framework.

A complete paleontological inventory of the primary reef organisms shows the composition to consist of:

scleractinian corals ............ about 35 species
hydrozoans and/or ischyrosponges ... about 12 species
tabulozoans/bryozoans ........ about 15 species
calcisponges .................... about 30 species
calcareous algae ............... about 25 species
*microproblematica* ........... about 20 species
reef dwelling organisms.... more than 100

## COMPOSITION OF REEF COMMUNITIES

The reef framework of all four reef complexes is composed of a great number of different frame building organisms. Each species shows a unique distributional pattern, and various framebuilders cluster together to form reef communities characterized by typical associations of organisms, and by localization within the central reef area.

Figure 4 shows the inter-relations between different reef framebuilders, their occurrence within distinct reef communities, and their associated epifauna and -flora. Nine reef communities have been distinguished, each of which is significant (Schafer, 1979) and identified:

1. by the association of reef framebuilders;
2. by their associated epi- and endobionts;
3. by the occurrence of significant foraminiferal assemblage; and
4. by the specific type of interstitial sediment present within the cavities of the reef framework.

Accordingly, the following reef communities have been identified as contributing to the construction of the reef framework:

1. *Thecosmilia cyclica–Stylophyllum–Pinacophyllum* community;
2. *Thecosmilia clathrata*–massive coral community;
3. *Thecosmilia clathrata–Pinacophyllum* community;
4. *Thecosmilia clathrata–Astraeomorpha* community;
5. *Astraeomorpha–Spongiomorpha* community;
6. *Disjectopora*–Calcisponge–Tabulozoa community;
7. *Disjectopora*–Calcisponge–Tabulozoa–*Montlivaltia* community;
8. Calcisponge–*Seriastraea–Montlivaltia* community;
9. Calcisponge–*Thecosmilia clathrata* community.

The frame building organisms, occurring in different reef communities, can be reduced to three superordinate morpho-ecologic groups:

1. group of bush-like (dendroid and phaceloid) coral species (communities *A*, B, C and D of Fig. 4);
2. group of finely branched (dendroid) and massive coral species (communities *B*, *D*, E and J); and,
3. group of foliaceous and solitary coral species, calcisponges, hydrozoans (*Disjectopora*) and tabulozoans (communities E, *F*, *G*, *H* and J of Fig. 4).

The coral communities show rare encrustations and low diversity of epi- and endobionts. The *problematica Lithocodium* and *Bacinella* are dominant. Among the relatively high percentage of sessile foraminifers, the form *Alpinophragmium* is a characteristic indicator within coral reef communities.

The biotically diverse calcisponge-hydrozoan-tabulozoan-coral-communities are characterized by pronounced encrustations, caused not only by secondary epi- and endobionts, but by overgrowths of the framebuilders themselves.

The main factor controlling the distribution of reef organisms and communities within the reef center and the patch-reefs seems to be the intensity of water energy. This is because of the localization of communities in relation to the topography of the biotopes, and the type of interstitial sediment within the reef framework. According to the delineation of communities shown in Figure 4, we postulate an increase of water agitation from the right to the left-side of the diagram. This is based on the arrangement of

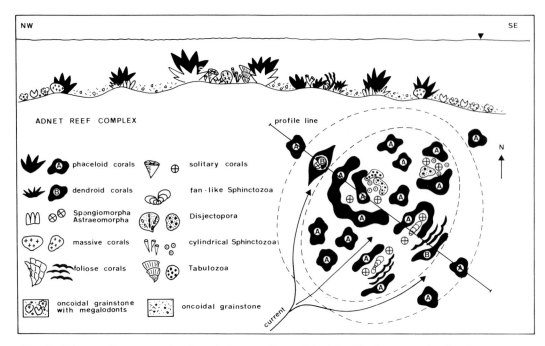

FIG. 5.—Diagram of a cross-section through the central part of the Adnet Reef structure showing the arrangement of the reef patches and their framework composition. This view of the central reef area shows the general morphology and distribution of the different reef patches belonging to a single stratigraphic level.

reef builders and communities within the patch-reefs and the associated increase of interstitial grain sizes. Coral communities can be found in the marginal position of the patch-reefs, whereas coral-hydrozoan- and coral-calcisponge-communities occupied an innermost intermediate area. The protected parts of the patch-reefs appear to have been colonized by calcisponge-hydrozoan-tabulozoan-coral-communities (Fig. 5).

Both homo- and heterotypic reef communities occur within the reef complexes. Most of the homotypic communities, which are built solely by corals, are found in the Adnet Reef, while heterotypic communities prevail in the Rötelwand, Feichtenstein and Gruber Reefs (Table 2).

Each isolated patch-reef within the Adnet complex is built by a dense framework of vertical reef growth caused by the dominance of a large bush-like coral species. Each of the patch-reefs displays a unique biozonation, controlled by horizontal and vertical growth of the reef communities. This biozonation is similar to that of the Rötelwand and Feichtenstein patch-reefs, since similar reef organisms seem to have required similar ecologic conditions on all reef structures. The Rötelwand and the Feichtenstein Reefs show a prevailing lateral extension of reef growth, forming an outer reef that is more biostromal in shape.

The reef centers are solely occupied by one extended reef patch. They contain a rather crude zonation of framebuilders, conditioned by their widely-spaced distribution within the reef area.

During the entire growth history the Gruber Reef remained totally within the deeper water stage of reef development, which is regarded as typical for growth of reef structures in basinal settings (compare Wilson, 1975). The predominance of massive corals, calcisponges, and hydrozoans caused horizontal growth of small isolated patch-reefs.

Figure 5 shows a cross-section through the central portion of the Adnet Reef patches. There appears a close connection between the external shape of the reef patches, their composition by different reef communities, and their distribution within the central reef area. Those reef patches with the highest diversity of framebuilders and the most complex organism communities, occur in the center of the reef area, while those reef patches close to the margin of the reef become less complex. Reef patches in marginal position are generally composed of a vertical succession of homotypic reef communities with relatively sparse populations. The reef patches of the innermost part of the central reef area show a circular and domal morphology, and are character-

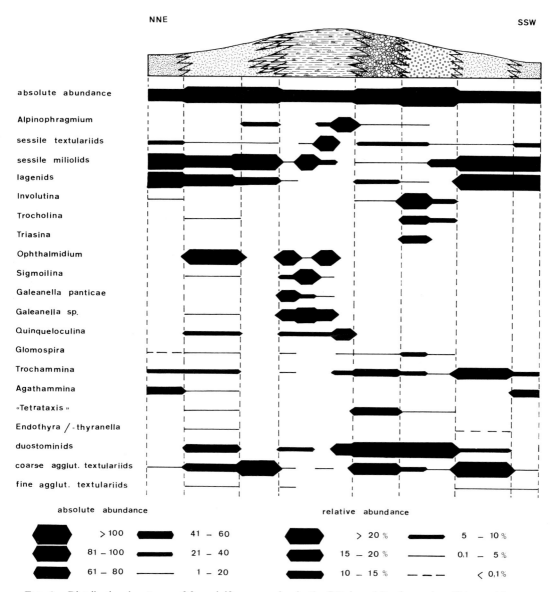

FIG. 6.—Distributional patterns of foraminifers occurring in the Rötelwand Reef complex. This graphic representation shows the absolute abundance of all foraminifers, and the relative abundance of ecologic groups within distinct facies units; see Figure 1 for facies symbols.

ized by a horizontal zonation of reef framebuilders, as well as a vertical community succession.

**DIFFERENT GROUPS OF ORGANISMS AS INDICATORS OF REEF ENVIRONMENTS**

Besides those primary organisms, which play an important role in the construction of the organic reef framework, e.g., corals, hydrozoans and/or ischyrosponges, calcisponges, tabulozoans and bryozoans, there are a group of secondary organisms especially benthonic foraminifers, calcareous algae, and "microproblematical organisms," which appear to be related to specific environmental conditions. These are regarded as useful indicators for different facies units, and are not only restricted to the central reef area, but are present throughout the entire reef complex and into the surrounding deeper water areas. Their distribution show a rather prominent zonation on the various reef structures (Figs. 5–8).

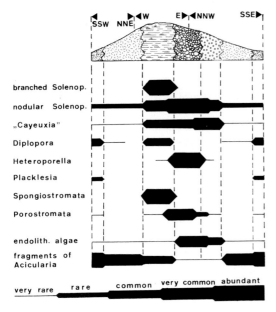

FIG. 7.—Diagram showing the distributional patterns of the main groups of calcareous algae in the Feichteinstein Reef complex.

## Foraminiferal Distribution within the Rötelwand Reef Complex

The quantitative distribution of benthonic foraminifers within these reef complexes (Adnet, Rötelwand and Feichtenstein) has been the subject of a previous publication (Schäfer and Senowbari-Daryan, 1978). It is believed that the distribution of the benthonic foraminifers is related to a number of factors, such as:

1. the morphology of the depositional environment (division of the reef structure into patch-reefs and reef-slope areas and division into areas of shallow water sedimentation and those of basinal development);
2. the condition of the substrate (compact reef frame, sandy or muddy substrate);
3. specific life behaviour of the various groups of foraminifers (forms encrustating reef framework, fixed onto sedimentary particles, forms living on a muddy substrate, or living on or within plants); and,
4. in some cases, to the water depth and the related water chemistry (salinity, and the availability of calcium carbonate as a selecting factor for construction of foraminiferal tests).

Figure 6 shows the quantitative distribution of all foraminiferal groups within the Rötelwand

Reef complex; this distribution pattern is representative for all investigated reef structures in the Salzburg area.

Some organism groups have been found to occur throughout all facies types, and ecologic and facies interpretation is only possible because of overall abundance. Sessile miliolids, textulariids, lagenids, *Trochammina*, and the duostominids belong to this ecologic group. Some other forms show a strong relationship to distinct facies units and various reef associations, and they can be utilized as sedimentary environment indicators without dependence on absolute abundance.

The central reef area is characterized by the abundance of vagile-benthonic miliolids (*Galeanella, Ophthalmidium, Sigmoilina, Quinqueloculina*), and by the sessile form *Alpinophragmium*. *Alpinophragmium perforatum* Flügel is the typical species of the marginal portions of the patch-reefs, where it colonized branches of large, bush-like colonies of the corals *Thecosmilia clathrata, T. cyclica, Stylophyllum*, and *Pinacophyllum*. This foraminiferal species acts contrary to the miliolids, which reach their greatest abundance in the central part of the reef patches within typical pelbiosparites of protected reef cavities (reef associations of single corals, calcisponges, hydrozoans and tabulozoans). This foraminiferal fauna is associated with "*Lituosepta*" and small uni- to biserial Textulariids.

Different forms of the genera *Glomospira* and "*Tetrataxis*" characterize the outer margin of the central reef area and the transition to the leeward reef-slope. Whereas *Glomospira* occurs in fine-grained micritic sediments, "*Tetrataxis*" is more common in coarse-grained reef detrital deposits of the oncolitic facies.

A highly diverse assemblage of involutinids, *Triasina hantkeni* Majzon, and species of *Trocholina* are the "key-foraminifers" of the outer reef-slope of the Rötelwand and Feichtenstein Reefs (Fig. 6), algal-foraminiferal facies and the underlying reef surrounding the shallow water platform facies of the Adnet Reef ("Dachstein facies"). They occur in very high abundances. Involutinids and *Triasina* are not present in the coral-sponge facies.

The protected reef-slopes (reefdetrital-mud facies) show the influence of nearby central reef areas (small miliolids), and of the deeper water terrigenous-mud facies (high content of lagenids). Characteristic of this facies is the high percentage of sessile miliolids which colonize biogenous detritus. This facies unit, as well as the terrigenous-mud facies, is characterized by the agglutinated foraminifer *Agathammina austroalpina*, whose occurrence is restricted to the micritic deeper water facies units.

*Distribution of Calcareous Algae within the*
*Feichtenstein Reef Complex*

The following groups of calcareous algae occur within the four reef complexes (Schäfer and Senowbari-Daryan, 1979):

1. Cyanophyta with Spongiostromata, and Porostromata (*"Cayeuxia"* and *Girvanella*),
2. Chlorophyta with dasycladaceans and probable codiaceans (*Lithocodium* and *Bacinella*, Fig. 8), and
3. Rhodophyta with solenoporaceans and questionable gymnocodiaceans.

Figure 7 shows the distributional pattern of various genera of calcareous algae in the Feichtenstein Reef complex. Their occurrence is restricted to the shallow water environment. Specimens which have been found in deeper water sediments are thought to be redeposited, especially the dasycladaceans and the cayeuxians which occur in the foundational rocks (Dachstein Limestone) of the Adnet Reef structure.

While some algae occur in different ecologic environments, some genera and species characterize distinct reef biotopes. Typical algae of the central reef area are the blue-green Spongiostromata, covering the primary reef framework and the walls of reef cavities, and the massive to ramose red alga *Solenopora alcicornis* Ott, contributing to reef frame construction especially in the base of patch-reefs. Nodular thalli of different solenoporacean species are dominant within the oncolitic facies where reef detrital sediments interfinger with sediments of the coral-sponge facies. The dasycladaceans reach their maximal abundance in the oncolitic facies (*Heteroporella* and *Griphoporella*), and in the algal-foraminiferal facies of the Adnet Reef (*Diplopora*). Transported accumulations of *Diplopora* have been found in pockets of the reef framework, and in the Kössen beds some distance from the reef structure. Fragments of *Acicularia* are predominant in the micritic sediments of the reefdetrital-mud facies. *Lithocodium* and *Bacinella* (of unclear systematic position; probably codiacean algae) are restricted to the oncolitic facies and to the coral communities of the coral-sponge facies. They form skeletal oncolites which are thought to have developed under ecologic conditions of the highest water energy. The heterogenous group of *"Cayeuxia"* reaches maximal abundance in the algal-foraminiferal facies.

Calcareous algae play an essential part in reef frame construction, reef consolidation, reef destruction, and in fixing reef detrital sands surrounding the central patch-reefs (Schafer and Senowbari-Daryan, 1979).

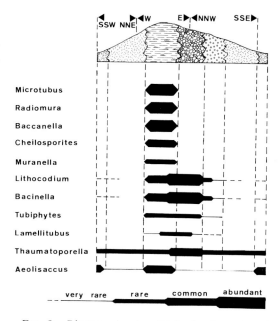

FIG. 8.—Diagram showing distributional pattern of *microproblematica* within various facies units of the Feichtenstein Reef complex.

*Distribution of Microproblematica within the*
*Feichtenstein Reef Complex*

Among the *microproblematica* we can distinguish two groups of organisms which characterize different biotopes of the reef structures (Fig. 8). The first group, consists of the forms *Microtubus communis* Flügel, *Cheilosporites tirolensis* Wähner, *Baccanella floriformis* Pantić, *Radiomura cautica* Senowbari-Daryan and Schäfer, *Lamellitubus cauticus* Ott, and *Muranella sphaerica* Borza, all restricted to the central reef area. These forms occupy biotopes built by calcisponges, solitary corals, hydrozoans, and tabulozoans within the innermost parts of the reef patches. The second group, comprising the genera *Lithocodium* Elliott and *Bacinella* Radoicic have been found only along the marginal portions of patch-reefs, and in the oncolitic facies. They form crusts on the primary reef framework, which is constructed by corals (*Thecosmilia*, *Astraeomorpha*, *Stylophyllum*, and *Pinacophyllum*), but firm encrustations on reef detrital components also may form oncolitic structures 30 centimeters in diameter. The genera *Thaumatoporella* and *Aeolisaccus* are present in all reef structures.

## DEVELOPMENT OF THE REEF STRUCTURES

According to the paleogeographic setting of the Dachstein Platform, the Hallstatt Basin, and the

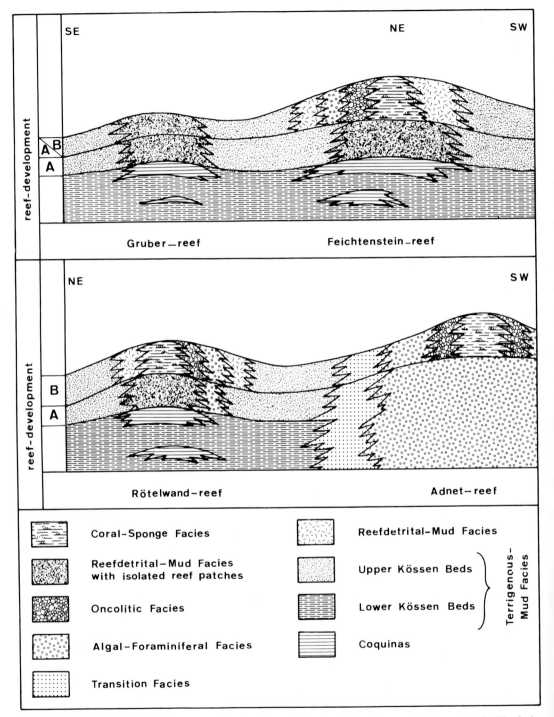

FIG. 9.—Diagram showing development of the four Upper Triassic patch-reefs near Salzburg, Austria. The facies development took place in two distinct depositional stages: A. Deeper water stage with formation of relief on the basin floor and accumulation of an organic mud-mound; B. Shallow water stage with construction of a reef framework. Only the Gruber Reef remained in the deeper water mud-mound stage, while the Feichtenstein and Rötelwand Reefs reached the upper shallow water reef stage. The Adnet Reef is thought to have grown on a foundation of shallow water carbonates; hence a deeper water stage is missing.

Kössen Basin, all four Upper Triassic patch-reef structures of the Salzburg area were initiated simultaneously both in basinal and lagoonal positions north of the extensive carbonate platform. This barrier, upon which shallow water sedimentation took place, separated the smaller patch-reefs from the open sea to the south.

During the Norian Stage the Dachstein Platform was divided into a southern barrier with large but irregularly distributed Dachstein Reefs, and a northern adjacent area with shallow subtidal sedimentation of the "bedded Dachstein Limestone." Further to the north the "Dachstein Limestone" interfingered with the Hauptdolomite, interpreted to be deposits of an extensive tidal flat area (see Zankl, 1971). The Norian Hauptdolomite is well exposed in the Mörtelbach Valley below the Rötelwand Reef structure, and in the Lämmerbach Valley below the Feichtenstein and the Gruber Reefs. Dolomites with algal laminites (PLF-, LLH-C- and LLH-S-textures, after Logan *et al.*, 1964), with millimeter-rhytmites, reworked layers, aphanites, irregular birdseye structures, shrinkage structures, and bioturbation, all indicate supra- to subtidal (mostly intertidal) origin of the Hauptdolomite. In upper Norian time, continuous subsidance of the northern part of the Dachstein Platform (the area with Hauptdolomite sedimentation) led to a change in the overall sedimentary environment. The transitional facies of the bituminous Plattenkalk (biogenous wackestones to mudstones) graded into interbeds of dark fossiliferous limestones (biogenous wackestones), marls and argillaceous shales, which are thought to have been deposited in the deep waters of the Kössen Basin. The southern part of the Dachstein Platform continued with shallow subtidal sedimentation of the Dachstein Reefs, and bedded Dachstein Limestone during Upper Triassic time. In uppermost Triassic time ("Rhaetian"), the Rötelwand, Feichtenstein, and Gruber Reef structures grew out of the Kössen Basin in several distinct stages. The main factors in all reef structures, which appears to have controlled vertical stages of development, are the depth of deposition and related degree of water energy (Fig. 9).

In the development of the Rötelwand and Feichtenstein Reef structures, we can distinguish two depositional stages which have been controlled by the interplay of water energy and rates of sedimentation. Initially, relief was accomplished by mechanical accumulation of pelecypodal/ooidal wackestones or packstones. These deposits were formed by currents of varying intensity on the sea floor of the Kössen Basin, and furnished the foundational unit for the development of reef structures on the sea bottom. This can be seen in the upper part of the Mörtelbach Valley, just below the steep wall of the Rötelwand Reef, and also at the base of the Feichtenstein Reef.

Subsequent reef growth on sporadic hardgrounds (coquinas and crinoidal banks) led to mechanical/biological accumulation of mudmounds through mud-baffling between the reef builders and pelecypod banks. Small patch-reefs (1 to 5 meters in diameter) are completely separated from each other. Their biota consists of calcisponges, hydrozoans, massive platy and fine branched coral colonies, which indicate quiet water environmental conditions. Facies differentiation of the detrital sediments within this mudmound stage is poorly developed. Only the Rötelwand Reef structure shows a lateral facies change in the southern portion with foraminiferal-crinoidal pack- to grainstones (algal-foraminiferal facies) to a northern portion, with biogenous wackestones (reefdetrital mud facies) at the beginning of the mud-mound stage.

Growing actively upward into a level of shallow water with increased water energy, the mud-mound stage of the Rötelwand Reef was first capped by thick pelecypod banks. These coquinas, up to 30 meters in height, comprise nearly one-half of the entire structure. They formed the foundation for settlement of reef building organisms, which were able to construct a highly differentiated patch-reef by active growth. This second reef stage is well shown along the road above the Rötelwand Reef, as well as on the top of the Feichtenstein Reef. The upper reef stage of both the Rötelwand and Feichtenstein Reef structures shows a very distinct lateral zonation of five facies units, reversed in each reef structure. The reef complexes are capped and buried with reefal grainstones and overlain by Jurassic deposits.

The facies development of the Gruber Reef agrees in some instances with that shown for the Feichtenstein and Rötelwand Reefs. However the predominance of marls and shales in the lower Kössen beds indicates a deeper water depositional setting for the Gruber Reef at the initiation of reef development. The mud-mound stage of the Gruber Reef, which is well displayed on the road southwest of the Lämmerbach Bridge, shows a high diversity of reef building organisms. Calcisponges, hydrozoans, and massive and platy corals are clearly dominant over branched coral species, which are more adapted to high energy conditions. In contrast to the Feichtenstein and Rötelwand Reefs, which in the upper reef stage show a zonation of five facies units, this zonation is completely lacking in the Gruber Reef. The deeper water mud-mound stage persisted throughout the time of deposition, and the structure was covered by the thin-bedded red or green nodular limestones of the lowest Jurassic.

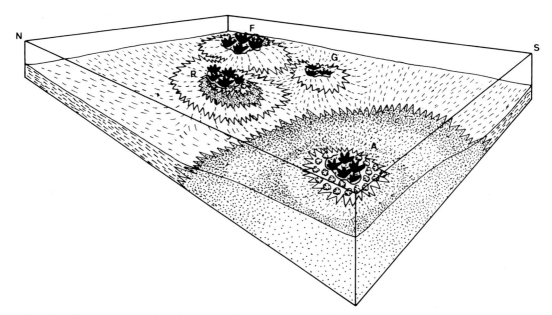

FIG. 10.—Diagram showing the palaeogeographic reconstruction of the Salzburg area during Upper Triassic time. The Adnet Reef structure (A) is positioned on a platform of shallow water carbonates, which grade into the Kössen Basin to the north and the east. The Rötelwand, Feichtenstein, and Gruber Reefs (R, F, and G) have grown on shoals within the Kössen Basin, and reached different stages of reef development.

In comparison with the Rötelwand, Feichtenstein, and Gruber Reefs, the Adnet Reef structure is obviously missing a deeper water stage, the accumulation of a mud-mound. The foundation of this reef structure is well exposed in the Kirchbruch Quarry east of the village of Adnet. Here, the foundation is composed of shallow water pack- to grainstones of the algal-foraminiferal facies, which can be compared to the shallow water bedded Dachstein Limestones of the Dachstein Platform. These deposits interfinger with deeper water Kössen Basin sediments in a transitional facies (Figs. 1 and 9). The entire reef structure is overlain by red nodular Adnet Limestone (lowest Liassic age), which in the center of the structure are green to red nodular cherty limestones. The Adnet Reef complex shows a circular-like distribution of facies units. The biozonation of these patch-reefs, their faunal and floral composition, and their distribution within the central reef area is very different from that of the other three reef structures in a basinal setting.

### PALEOLOGICAL INTERPRETATION OF THE LOCAL SEDIMENTARY ENVIRONMENT

The differences in facies development and distribution within the four patch-reef structures can be explained by the local paleogeographic setting of the depositional area (Fig. 10). These differences, as well as similarities, have been summarized in Table 2.

The Rötelwand and Feichtenstein Reef complexes are thought to have grown in isolated positions within the Kössen Basin. Reef development started with an accumulation of a mud-mound and when this reached a level of higher water energy, a reef framework developed. The reef structures were exposed to currents and wave action which caused a strong linear facies zonation. In the same manner, the Gruber Reef grew out of the Kössen Basin by development of a biogenous mud-mound. However, this structure did not reach wave base with its increased water agitation, a feature which appears to be necessary for construction of a reef framework, as shown in the upper reef stages of both the Feichtenstein and Rötelwand Reefs.

The Adnet Reef structure, beginning with the construction of reef patches directly on a foundation of shallow water sediments, did not have an opportunity to build a deeper water mud-mound stage. Although located close to the water surface at the beginning of reef growth, the absence of distinct current and or preferred direction of wave action, led to a zonation of facies units similar to that of the other reef complexes.

Two other facts reinforce the idea that the Adnet Reef complex was built in more shallow water position than both the Rötelwand, Feichtenstein, and Gruber Reefs. Firstly, disconformities of reef growth with subsolution of the reef surface, and coloring of the sediments, are restricted to the Adnet Reef. Secondly, above the Adnet Reef structure there is a discontinuous development of Liassic sediments, indicating a threshold facies. The continuous facies development, which characterizes the Kössen Basin, prevails at the Rötelwand, Feichtenstein, and Gruber Reef structures. Resedimented Liassic conglomerates from a nearby threshold are restricted to some outcrops between the Adnet and Rötelwand reef areas (see Hudson and Jenkins, 1969), where the northern ramp of the Dachstein Platform plunged into the Kössen Basin.

The local paleogeographic situation results in specific ecologic conditions within the reef structures, and in the differences relating to the construction of the central reef area by reef organisms.

The overall poor diversity of reef building organisms, and the predominance of monotypic reef communities within polymict reef patches, led to differences in sedimentological features to the conclusion, and suggests that the Adnet patch-reefs had to grow under more restricted ecological conditions than those of the other reef structures. Situated on a platform of unknown extent, the Adnet patch-reefs were completely surrounded by shallow water deposition away from the active influence of constant water currents and strongest wave action. Exposure to the water surface intensified restriction of reef growth by lowering the diversity of reef organisms taking part in the actual framework construction.

The Rötelwand and Feichtenstein Reef complexes, although initially developed in deeper water positions, became more exposed to currents and waves in their upper reef stage because of their isolated position within the Kössen Basin. Their unique paleogeographic setting is thought to be the reason for the high diversity of reef building organisms in general, for the high diversity of calcareous sponges, for the predominance of heterotypic reef communities, for the significant zonation of the facies units, and the internal composition of the patch-reefs.

The Gruber Reef, which had been constructed in deeper water basinal conditions, did not grow out of the biogenous mud-mound stage. Therefore a facies zonation is absent and the biota of the patch-reef is characterized by the predominance of massive coral species, calcisponges, hydrozoans, and a subordination of dendroid-phaceloid coral species.

### ACKNOWLEDGMENTS

The scientific program F142/25-27 on ''Mesozoic reefs of the Alpine region'' has been carried out with the generous financial support provided by the Deutsche Forschungsgemeinschaft.

## REFERENCES

DIETRICH, V. J., 1976, Plattentektonik in den Ostalpen. Eine Arbeitshypothese: Geotekt. Forsch, 50, p. 1–84.

HUDSON, J. D., AND JENKINS, H. C., 1969, Conglomerates in the Adnet Limestones of Adnet (Austria) and the origin of the ''Scheck'': Neues Jahrb. Geol. Paläont. Mh., 1969, p. 552–558.

LOGAN, B. W., REZAK, R., AND GINSBURG, R. N., 1964, Classification and environmental significance of algal stromatolites: Jour. Geology, v. 72, p. 63–83.

MÜLLER-JUNGBLUTH, W.-U., 1970, Sedimentologisches Untersuchungen des Hauptdolomits der östlichen Lechtaler Alpen, Tirol: Festb. Geol. Inst. 300-year colloquium, Univ. Innsbruck, p. 255–308.

PILLER, W., 1976, Fazies und Lithostratigraphie des gebankten Dachsteinkalkes (Obertrias) am Nordrand des Toten Gebirges (S Grünau/Almtal, Oberösterreich): Mitt. Ges. Geol. Bergbaustud., 23, p. 113–152.

SCHÄFER, P., 1979, Fazielle Entwicklung und palökologische Zonierung zweier obertriadischer Riffstrukturen in den Nördlichen Kalkalpen (''Oberrhät''-Riffkalke, Salzburg) unpub. Dissertation: Erlengen. Univ. Erlangen-Nürnberg, 263 p. (in press).

——, AND SENOWBARI-DARYAN, B., 1978, Die Häufigkeitsverteilung der Foraminiferen in drei oberrhätischen Riff-Komplexen der Nördlichen Kalkalpen (Salzburg, Österreich): Verh. Geol. B.-A., 1978/2, p. 73–96.

——, AND ——, 1979, Distributional patterns of calcareous algae within Upper Triassic patch-reef structures of the Northern Calcareous Alps (Salzburg): Bull. Centre de Recherches, Pau- SNPA, 3, 2, p. 811–820.

SENOWBARI-DARYAN, B., 1979, Fazielle und paläontologische Untersuchungen an ''oberrhätischen'' Riffen—Feichtenstein- und Gruber-Riff bei Hintersee/Salzburg (Nördliche Kalkalpen) [unpub. Dissertation]: Erlangen Univ. Erlangen-Nürnberg, 352 p. (in press).

WILSON, J. L., 1975, Carbonate Facies in Geologic History: Berlin, Heidelberg, New York, Springer-Verlag, 471 p.

ZANKL, H., 1969, Der Hohe Göll. Aufbau und Lebensbild eines Dachsteinkalk-Riffes in der Obertrias der nördlichen Kalkalpen: Abh. Senckenberg. Naturf. Ges., 519, p. 1–123.

——, 1971, Upper Triassic Carbonate Facies in the Northern Limestone Alps, in Müller, G., Ed., Sedimentology of Parts of Central Europe, Guidebook: Internat. Sedimentol. Congr. 8th. Heidelberg, p. 147–185.

SEPM Special Publication No. 30, p. 261–290, May 1981

# THE STEINPLATTE REEF COMPLEX, PART OF AN UPPER TRIASSIC CARBONATE PLATFORM NEAR SALZBURG, AUSTRIA

## WERNER E. PILLER
University of Vienna, Austria

## ABSTRACT

Reinvestigation of the Upper Triassic (Norian-Rhaetian) Steinplatte "reef" of the Northern Limestone Alps near Salzburg, Austria, revealed a clear facies zonation of this carbonate complex. The upper Kössen Beds are of a basinal facies and are composed of dark bedded limestones, poor in fossils, and partly intercalated with marls. The "Oberrhätkalk," which is the time equivalent of the uppermost part of the Dachstein Limestone, can be subdivided into a fore-reef, reef, and back-reef facies. The fore-reef is characterized by crinoids and by reef derived particles, which were deposited on a slope with dips up to 35 degrees. The upper boundary of the fore-reef is marked by a coquina, built by shells of bivalves and brachiopods. The reef is represented by a narrow zone (less than 100 meters wide) which can be subdivided into: 1. reef-slope, with a diversified organism community of various corals, calcareous sponges, hydrozoans, tabulozoans/bryozoans, and *microproblematica,* embedded in a micritic matrix, and 2. reef-crest, represented as a belt of large phaceolid corals. Facies zonation consists of a back-reef facies, and a patch-reef facies which is characterized by two types of patch-reefs: one dominated by large dendroid corals, similar to the reef-crest, and the other with high organism diversity in a micritic matrix and similar to the reef-slope. Within this facies the most important organisms are megalodontid bivalves. Sediment composition changes rapidly, and the primary particles are oncoids, peloids, "lumps," and various bioclasts. With increasing distance away from the reef, patch-reefs become scarcer and in the eastern lagoon various facies types occur, e.g., grape-stone, foraminiferal-algal, oncolithic, and oolitic. Near the eastern margin of the Steinplatte Platform intertidal algal stromatolites occur, indicating a transition into the Lofer facies of the bedded Dachstein Limestone.

This facies interpretation differs widely from that postulated previously, in that a good reef zone is delineated, and the so called fore-reef breccias are now regarded as part of the back-reef sediments.

Investigations of adjacent carbonate platforms (Loferer, Leoganger Steinberge, and Steinernes Meer) demonstrate that these areas, which today are isolated by erosion and tectonics, represent the continuation of the shallow water lagoon of the eastern part of the Steinplatte Platform. The bedded Dachstein Limestone of all these platforms developed in Lofer facies with cycles of supra-, inter-, and subtidal members. The supratidal member is characterized by green or red marly limestones, the intertidal member by partly dolomitized Loferites (mainly algal stromatolites), and the subtidal member by megalodont limestones with various grain types. In the eastern part of Steinernes Meer the sediments change. Here, limestone was deposited in the subtidal zone and is composed of bioclastic arenites, which are mainly reef derived. The Hochkönig Massif is located eastward of Steinernes Meer and represents a large Dachstein Limestone reef, with a thickness of 700 meters. This reef fringes the carbonate platform on its southeastern edge, with well developed fore-reef breccias towards the basin. Therefore the Steinplatte Reef represents only a small part of an Upper Triassic shallow water carbonate platform, approximately 40 kilometers wide, which was fringed by reefs on the northwestern edge (Steinplatte), and on the southeastern edge (Hochkönig). Whereas the Hochkönig Reef existed throughout almost all of Norian-Rhaetian time, the Steinplatte Reef was initiated only in the uppermost Norian-Rhaetian. During early Norian-Rhaetian time the carbonate platform was connected to the land by widespread tidal-flats of the Hauptdolomite, and only when the Kössen Basin separated this land-connected platform did the Steinplatte Reef develop as a relatively small barrier on the margins of the Kössen Sea.

## INTRODUCTION

Upper Triassic reefs of the Northern Limestone Alps have contributed to our knowledge about reefs in general, and more specifically, towards the evolution of reef building organisms. The Alpine orogenesis caused intensive thrusting and faulting of the Northern Limestone Alps, and since tectonic lines tend to follow facies boundaries, originally connected sedimentation areas were separated by great distances. These disruptive tectonic movements render paleogeographic studies more difficult and therefore our overall knowledge of Upper Triassic reefs is small in comparison with the amount of work undertaken in this area. However, a more favorable situation is discernible with the Steinplatte complex, since this region was spared much tectonic disturbance. Therefore facial and paleogeographic investigations have been encouraged for some time. The reefoid nature of the Steinplatte was first recognized by Mojsisovics (1871, p. 206), and the first faunal list of reef building organisms, as well as a geological map of the area, was presented by Hahn (1910a and b). Initial facies and ecological studies were accomplished by Vortisch (1926 and 1927), and his results formed the basis for future investigations. Vortisch recognized the interfingering of reef limestone with basinal sediments, and he accurately traced the rocks of the basin into reef limestones. The fore-reef area was reported as detrital limestone ("Übergangskalk"), and he measured the dip of the fore-reef slope at about 32 degrees. The reef itself was described by Vortisch as a planoconvex body, which expanded basinwards during growth with an increased angle of slope. Vortisch (1926, p. 51) also mentioned the absence of fore-reef breccias. The beginning of reef growth was initiated by bivalve-coquinas, which acted as hard-bottoms for reef building organisms. Further reef growth was caused by a balance between vertical organism growth and bottom subsidence. In front of the Steinplatte Reef he identified two small reefs in their initial growth stages, and these were described. However, the eastern trend of the reef was not clearly delineated, especially as he postulated a symmetrical shape for the reef.

Investigating another upper Rhaetian reef (Rötelwand, see Schäfer and Senowbari-Daryan, this volume), Sieber (1933, 1934, and 1937), assumed that this reef was built of several patch-reefs ("Riffknospen"), which showed a vertical faunal zonation. He carried these results to the Steinplatte Reef and sought to explain its composition as a series of individual patch-reefs. One of the small reefs in front of the main reef, previously noted by Vortisch, was later interpreted as an individual patch-reef (Sieber, 1937, fig. 4).

Study of Upper Triassic and their facies relationships in the Northern Limestone Alps was given stimulus by A. G. Fischer, who encouraged Ohlen (1959) to reinvestigate the Steinplatte Reef. At the same time Fischer (1964) published his fundamental work concerning the Lofer cyclothems of the bedded Dachstein Limestone. Ohlen (1959) subdivided the Steinplatte complex into 5 facies zones: 1. basin facies, 2. fore-reef facies, 3. reef facies (including a coral facies), 4. a reef calcarenite facies, and 5. a back-reef facies. With respect to vertical development he recognized 5 distinct phases. The "initial phase," characterized by shell beds, built a mound flanked by crinoids. During the "initial phase" reef growth started and B- and C-reefs, the small patch-reefs described by Vortisch (1926), developed. During the "middle and late phases" the Steinplatte Reef rose 100 meters above the basin; the 5 facies zones are best developed in sediments of the "late phase." The basin facies is composed of very fine calcarenites which are made up of transported material derived from the reef. This influence decreases with greater distance from the reef. Ohlen characterized the fore-reef facies by coarse breccias, occurring at the top of the fore-reef slope, but

becoming finer towards the basin. The fore-reef slope attained a dip of 28 degrees. At the reef-front, diversity of the reef biota was conspicuous, e.g., corals, sponge-spongiomorphs, *problematicum* A, foraminifers, and calcareous algae. The organic framework of the reef was built by the coral *Thecosmilia clathrata,* and the reef mud facies with sponge-spongiomorph assemblages and *Thamnasteria* occurring within this framework. Behind the reef, the reef-calcarenite facies appears, and is composed of mud-free calcareous sands populated by corals, megalodonts, dascladacean algae and foraminifers. At a distance of 2 kilometers behind the reef, the back-reef facies commences and is characterized by mud-free fine sands, dominated by foraminiferal-algal associations, and megalodonts, which suggest hypersaline depositional conditions. During the "latest phase" the reef was drowned and a calcareous sand bank was formed, made up largely of crinoids. Therefore a crinoidal bank was the final stage in the development of the Steinplatte Reef. Ohlen also presented paleontological descriptions and ecological interpretations of the biota.

Since the appearance of Ohlen's paper the Steinplatte Reef complex has become widely known and along with the Dachstein Limestone Reef of the Hohe Göll (Zankl, 1969), utilized as an Upper Triassic reef model (e.g., Zankl, 1971; and Heckel, 1975), and as one of the major reef types by Wilson (1975, p. 363), his "framebuilt reef rim."

Twenty years have passed since the investigation of the Steinplatte Reef by Ohlen, and knowledge concerning reefs in general, and Triassic reefs in particular, has increased immeasurably. This was the main impetus for reinvestigation of this reef complex, and this study has not only described and interpreted the Steinplatte Reef, but has also emphasized its paleogeographical position within the Northern Limestone Alps of Austria.

## THE STEINPLATTE REEF COMPLEX

### Geographical and Geological Setting

The Steinplatte complex is situated on the boundary of Salzburg and Tyrol near the West German border. It stretches from north of Waidring (Tyrol) in the west, to Lofer (Salzburg) in the east (Fig. 1). It is a carbonate platform varying in height from 1869 to approximately 1400 meters, and is bordered by a steep wall on the west (Fig. 2) and the south, the latter is called "Sonnenwände." Within the tectonic framework of the Northern Limestone Alps, it belongs to the "Tirolikum" tectonic unit. The upper part, near the eastern boundary, is part of a higher tectonic unit, the "Berchtesgaden unit."

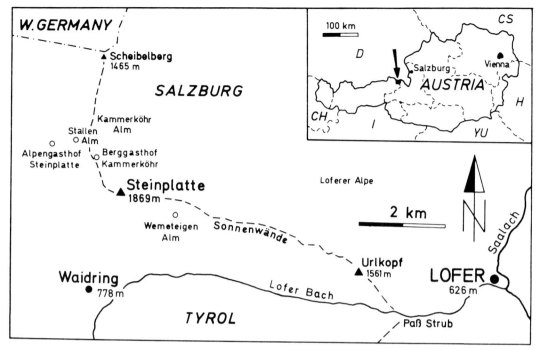

FIG. 1.—Location map of the Steinplatte area.

In the northwest, the Hauptdolomite built the foundation of the Steinplatte, and this interfingers to the east with bedded Dachstein Limestone, both of Upper Triassic (Norian) age. On the western side of the Steinplatte the Hauptdolomite is overlain by Dachstein Limestone, which grades into Kössen Beds. The Kössen Beds thin out to the east. In the western and northwestern parts of the Steinplatte region the Kössen Beds rise up to the Triassic-Liassic boundary, whereas in the central Steinplatte Plateau they are overlain by reef limestone, called "Oberrhätkalk." This reef limestone is actually part of the Dachstein Limestone. The different names are based on older stratigraphy when the Dachstein Limestone was placed into the Norian, and the reef limestone into the uppermost Rhaetian. Later studies have shown interfingering of the uppermost parts of the Dachstein Limestone with the reef limestone, thus demonstrating contemporaneity of both. However, because of the dispute regarding the stratigraphic ranges of the Norian and Rhaetian Stages, and the dispute about the Rhaetian Stage in general (Fabricius, 1974; Wiedmann, 1974; and Tollmann, 1978), we cannot give an exact age for the reef limestones and the contemporaneous Dachstein Limestone. Hence, we use the compromise term "Norian-Rhaetian," but fully recognize that the Steinplatte complex belongs to the highest part of the Upper Triassic.

*Foundation*

The Hauptdolomite built the foundation of the western portion of the Steinplatte complex. This unit is characterized by bedded dolomites, in which many beds show algally laminated sediments attributed to stromatolites, and containing desiccation cracks and shrinkage pores. These structures point to an inter- to supra-tidal origin of these beds. Alternating, but often missing, are dolomites or dolomitic limestones, composed of muddy sediments and carrying a rare bivalve and foraminiferal fauna. There also are bituminous, marly limestones with an bivalve fauna rich in numbers of individuals. This rock type is well exposed in a road cut a few hundred meters below the Steinplatte "Alpengasthaus" (Zankl, 1971, p. 179). Both rock types are thought to reflect a shallow subtidal zone, with the latter exhibiting more restricted environmental conditions. More detailed descriptions of the Hauptdolomite are given by Müller-Jungbluth (1968 and 1970) and Czurda and Nicklas (1970).

Eastwards, below the Sonnenwände, the Hauptdolomite changes to Dachstein Limestone; this facies relationship was recently documented

FIG. 2.—General photographic view of the west side of the Steinplatte complex, Austria. The wooded foreground is built of Hauptdolomite and Dachstein Limestone. The steep meadows below the mounds (A and B) consist of lower Kössen Beds. The visible facies zones of the Steinplatte Reef complex are: basin facies (BF) with "A- and B-mounds" (A, B) and "C-bank" (C) at the base (from "Kammerköhr Gasthaus" (K) to the left edge of the picture); fore-reef facies (from "Kammerköhr Gasthaus" (K) right below the silhouette; above "FR" the fore-reef slope attains its strongest dip); reef facies (R; striking along the background); J ='s Jurassic deposits. The summit of Mt. Steinplatte is not visible; length of outcrop approximately 1.2 kilometers.

by Gökdag (1974). Here, the Dachstein Limestone displays typical Lofer cycles as described by Fischer (1964), with supratidal (A), intertidal (B), and subtidal (C) members. The beds of member C increase in thickness from the west to the east, whereas member B decreases in this direction. The sediments of member C are "pellet-mud facies" and "mud facies" in the west, and "grapestone facies" in the east (Gökdag, 1974, p. 68). The intensity of dolomitization of member C decreases from west to east.

On the western side of the Steinplatte, near Stallen Alm (Fig. 1), the Hauptdolomite is overlain by Dachstein Limestone (Zankl, 1971, p. 179). The transition from Hauptdolomite to Dachstein Limestone reveals a decreasing dolomitic content with the uppermost beds of the Dachstein Limestone containing only dolomitic rhombs that apparently grew during late diagenesis, and member B is generally absent. The most characteristic bed is a cross-bedded oolite (Fig. 3), which is overlain by mottled limestone whose upper surface is a bored hardground.

The transition from the Dachstein Limestone to the Kössen Beds can be observed above these beds. Initially, fossiliferous limestones with bivalves and brachiopods, alternate with marly limestones and marls; their thickness is approximately 12 meters. In the upper portion of these beds the terrigenous content of the limestone and marly intercalations decreases, so that Dachstein Limestone beds (with thicknesses of 3–4 meters) contain corals ("*Thecosmilia*") and megalodonts. These biomicrites are rich in calcareous algae (dasycladaceans, skeletal blue-green algae), and foraminifers (involutinids).

Superimposed on this Dachstein Limestone are the Kössen Beds. They can be subdivided (following Hahn, 1910a, p. 345) into lower and upper Kössen Beds. Whereas the lower Kössen Beds are related to the foundation, the upper Kössen Beds are part of the reef complex. The lower Kössen Beds are poorly exposed and have an estimated thickness of 150–175 meters (Hahn, 1910a). They are principally dark-grey marls or calcareous shales intercalated with fine-grained brownish-grey, marly limestones. Both are poor in fossils and there are only a few beds which carry a rich bivalve fauna.

*Reef Complex*

The reef complex begins above the lower Kössen beds and includes the upper Kössen Beds and the "Oberrhätkalk." This reinvestigation concentrates on their facies zonation and lateral changes and is primarily concerned with the uppermost portion where the reef complex reaches its greatest development.

## Basin (=Upper Kössen Beds)

The rocks that outcrop in the northwestern part of the Steinplatte complex are the upper Kössen basinal sediments, which is agreement with the work of Vortisch (1926) and Ohlen (1959). They are thin-bedded, dark-grey to dark-brown marly limestones with intercalations of marls (Fig. 4), and in the uppermost portions chert layers occur. Thin-sections indicate that these rocks are biopelmicrites to pelmicrites (Fig. 4B) which are bioturbated (Fig. 4C). These sediments have a high amount of pyrite, which often fills cavities of fossils. The pellets are small and thought to be of fecal origin. Brachiopods, bivalves, echinoderm fragments, and gastropods are rare. A poorly represented microfauna of ostracodes, foraminifers, and sponge spicules is present nonetheless.

The foraminiferal assemblage is composed of *Trochammina, Agathammina* and lagenids, especially *Lenticulina,* and thin-walled *Nodosaria* and *Dentalina.* This foraminiferal assemblage, and especially the occurrence of *Lenticulina,* is typical of basinal deposits of the Upper Triassic as known from other localities (e.g., Aflenz Limestone, see Hohenegger and Lobitzer, 1971). These forms are autochthonous and not transported components as thought by Ohlen (1959, p. 86).

To the south and southeast the limestone beds contain more detrital fossil material, mainly crinoid fragments, and grain size increases. On the other hand, marly intercalations decrease and become thinner and rarer. These sediments are biopelmicrites and pelmicrites and their grain size is finely arenitic; cherts are absent. The more coarser-grained sediments are present in the upper beds to the northeast (basinwards).

### Fore-Reef Beds

Further to the south and southeast these sediments show an increase in grain size with the beds becoming thicker and with an absence of marly intercalations. The beds, which lie nearly flat in the basin facies, become more inclined. The sediments are still biopelmicrites or biomicrites (wackestones) with fine to medium arenitic grains (Fig. 5A). The dominant allochems are crinoid fragments, which can be very abundant. These rocks, rich in crinoids, represent the lower part of the fore-reef facies. The boundary between basin and fore-reef cannot be precisely delineated, but is characterized by an increase in abundance and size of crinoid fragments. This rock type is best exposed in the vicinity of "Kammerköhr Gasthaus." From here the dips on the beds increase rapidly and reach a maximum of 35 degrees on the slope between "Kammerköhr Gasthaus" and Mt. Steinplatte; an area not covered by vegetation. The sediments show a continuous increase

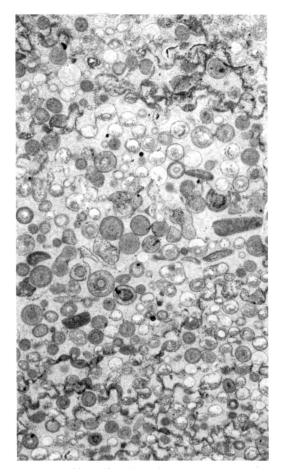

FIG. 3.—Thin-section photomicrograph (×10) of oolites in the uppermost unit of bedded Dachstein Limestone; the foundation of the Steinplatte Reef; SP 78/119.

in grain size and, in addition to crinoid fragments, an increased amount of reefal bioclasts. The color becomes lighter and is medium-grey to medium-brown, and the thickness of the beds increases to a few meters. The bioclasts are composed of corals (especially "*Montlivaltia*"), hydrozoans, solenoporaceans, the problematic alga *Thaumatoporella,* and the encrusting foraminifer *Alpinophragmium.* In addition, small lithoclasts of cemented reef sediment occur. In the lower part of the section the sediments are wackestones, but higher in the section they become packstones with arenitic and fine ruditic components (Fig. 5B). The matrix is micritic and becomes silty, and generally shows geopetal accumulations between larger allochems (Fig. 5B). The autochthonous macrofauna is dominated by crinoids, brachiopods, bivalves, and rare gastropods. The microfauna contains foraminifers, ostracodes, and ho-

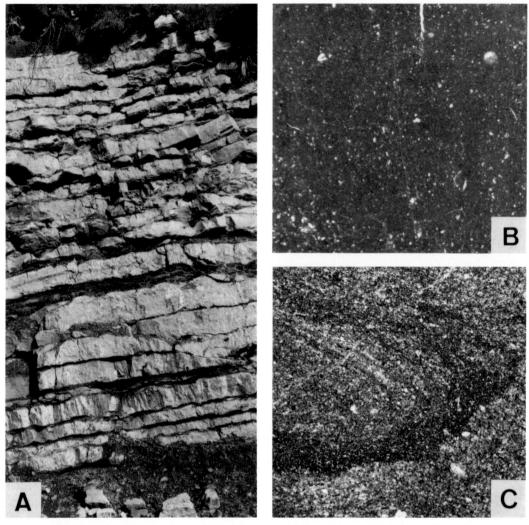

FIG. 4.—Upper Kössen Beds: A. Outcrop photograph of bedded, terrigenous limestones with intercalated marls; road cut near Kammerköhr Alm, height of outcrop approximately 4.5 meters. B. Thin-section photomicrograph (×15) of a wackestone with microbioclasts; ST 77/55. C. Thin-section photomicrograph (×10) of a bioturbated wackestone with small bioclasts; ST 77/68.

lothurians. The foraminiferal fauna reveals a combination of basinal and reef forms, with rare back-reef derived components. The allochthonous forms are mainly fragments of sessile agglutinating forms, *Ophthalmidium* and *Alpinophragmium*-fragments, whereas the autochthonous elements are *Trochammina,* a characteristic species of *Glomospira,* sessile miliolid forms, and various thin-walled lagenids. The most characteristic element of the basinal facies, *Lenticulina,* is very rare, especially in the upper portion of the fore-reef. Here the sediments rapidly change into

a unit dominated by bivalve and brachiopod shells comprising a coquina.

*Coquina.*—This coquina forms a rim which can be traced along the entire exposed strike of the beds from the southwest to the northeast. This rim, only a few meters wide, may be interrupted by crinoid-rich sediments. Usually there is only a small amount of trapped sediment between the bivalve and brachiopod shells, which are generally densely packed (Fig. 6). This sediment is a light-grey, green or reddish mud. The bivalve fauna is of high diversity and contains *Chlamys, Ox-*

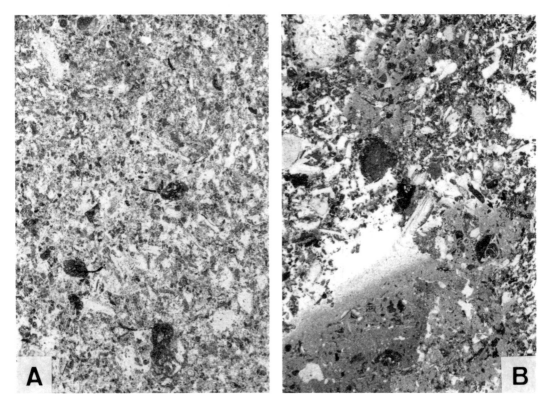

FIG. 5.—Thin-section photomicrographs (×10) of fore-reef sediments. A. Fine-grained bioclastic and lithoclastic packstone of the lower slope; SP 78/3. B. Coarser-grained bioclastic and lithoclastic packstone with micritic and silty cavity fillings; upper slope, SP 78/11.

*ytoma, Lima, Atreta, Rhaetavicula* and *Modiolus*.

### Reef Beds

*Reef-Slope.*—Immediately behind the coquina, and partly covered by it, is a zone characterized by an abundance of reef organisms of high diversity. They consist of several communities which may be zoned. This interval is best developed along the reef-slope on an inclination of 25 to 28 degrees.

Its lowermost part is characterized by an organism community composed of the solitary coral "*Montlivaltia*" *norica* (Frech), calcareous sponges, spongiomorphs, the hydrozoan *Disjectopora*, and the alga *Solenopora alcicornis* Ott. These forms are abundantly encrusted by small sphinctozoans (*Annaecoelia*), tabulozoans/bryozoans, as well as the *microproblematica Microtubus, Baccanella,* and *Radiomura,* and by spongiostromata-crusts.

Behind this initial organism community others follow. One community is dominated by the cylindrical sphinctozoan sponge *Paradeningeria*, another by the glomerate sphinctozoan *Salzburgia* (Fig. 7A), and others by corals. Within the coral communities there are massive forms, like "*Thamnasteria*" and *Astraeomorpha*, and the fragile-dendroid "*Thecosmilia*" *clathrata* (Emmrich), form B of Ohlen, together with calcareous sponges (*Colospongia, Welteria, Peronidella*), and hydrozoans (*Disjectopora, Stromatomorpha, Spongiomorpha*). All are abundantly overgrown by an epifauna of various tabulozoans/bryozoans, sponges like *Annaecoelia* and *Follicatena,* miliolid and agglutinating sessile foraminifers, various *microproblematica* like *Microtubus, Radiomura,* and *Baccanella,* and spongiostromata-crusts (Fig. 7B). Cavities between these organisms are occupied by *Cheilosporites tirolensis* Wähner (Fig. 8A); gastropods may also be abundant.

The sediment between the reef fauna, on the lower part of the reef-slope, is a wackestone or mudstone with fragments of crinoids, reef organisms, molluscs, ostracodes, and foraminifers. The

FIG. 6.—Outcrop photograph of a typical fore-reef coquina consisting of densely packed bivalve shells and brachiopods, marking the upper margin of the fore-reef; eastern ski track of Mt. Steinplatte.

latter are not abundant but contain forms such as *Coronipora* and *Trocholina*, which are generally restricted to the reef-slope. In the upper part of the reef-slope, cavities between the framebuilders are partly washed-out and filled-in by radial-fibrous cements (Fig. 8A).

*Reef-Crest.*—In the uppermost part of the reef-slope the rich biota becomes poorer and there is an abrupt change in faunal composition. The most dominant forms are thick-branched phaceolid corals. The most abundant coral is "*Thecosmilia*" *clathrata* (Emmrich), form B of Ohlen (Fig. 8C), which constructs colonies that attained two meters in height and width. Besides this form, two genera of corals, *Pinacophyllum* and *Stylophyllum* are also common. Other framebuilders, such as sponges (*Molengraafia*), and hydrozoans (*Spongiomorpha, Stromatomorpha*) are of minor importance. The epifauna is also poorer when compared to the reef-slope. Typical and abundant encrusting elements are the sessile foraminifer *Alpinophragmium perforatum* Flügel (Fig. 8B), *Lithocodium* (=*Problematicum* A of Ohlen, 1959), *Bacinella* Pantic (probably part of *Lithocodium*), and crusts of unknown origin (Fig. 8B). Abundant, sessile and miliolid foraminifers

also occur. The characteristic epifauna of the reef-slope, such as tabulozoans/bryozoans, sponges and *microproblematica*, is generally absent. Reef inhabitants are mainly gastropods and echinoids, and in some protected areas, dasycladacean algae.

Sediment between the coral colonies is clearly subdivided into three types. One is characterized by fragmented reef clasts, derived mainly from corals, calcareous algae, *Alpinophragmium* fragments, and lithoclasts. However there are also particles derived from the back-reef. Owing to these influences, the sediment is similar to that of the patch-reef zone, with a high amount of micritized particles, rare oncoids (foraminiferal-*Bacinella-Lithocodium*), and foraminifers (duostominids, *Aulotortus, Auloconus,* and *Pseudotaxis*).

The second type of sediment is restricted to protected areas between coral branches. This is a fine arenitic grainstone, composed of small peloids and abundant foraminifers (Fig. 8C). The matrix is microsparitic and particles often show graded-bedding. The foraminiferal frequency extremely high (compare with Schäfer and Senowbari-Daryan, 1978, p. 93), and the fauna is dominated by the miliolids *Galeanella* and *Ophthalmidium*. A very characteristic form restricted to this facies is "*Lituosepta*."

The third type of sediment is a reddish to greenish wackestone (Fig. 8B). The micritic matrix contains solenoporacean and dasycladacean algal fragments, echinoderm fragments, and foraminifers, especially "*Sigmoilina*" and thick-walled lagenids.

The reef-crest is well exposed on the southern margin of the upper part of the ski trail from Plattenkogel to "Kammerköhr Gasthaus." The width of this zone extends from 30 to 80 meters and strikes southwest to northeast, parallel to the other zones. Continuation of the fore-reef and reef to the northeast is not clear (compare Fig. 12B), due to vegetation cover and tectonic complications.

### Back-Reef Area

A belt with densely-packed, thick-branched corals becomes less densely packed with corals to the east and southeast, and the sediment between the corals changes in that megalodontid bivalves appear. The back-reef area shows differ-

→

FIG. 7.—Polished slab and thin-section photomicrograph of reef-slope sediments. A. Association of various sphinctozoan sponges (*Salzburgia variabilis* Senowbari-Daryan and Schäfer and *Verticillites* sp.), hydrozoans and tabulozoans/bryozoans encrusted by spongiostromata-crusts. Sediment consists of reddish-brown biomicrite; SP 78/22, polished surface, scale in millimeters. B. Sphinctozoan sponge overgrown by tabulozoans/bryozoans and problematic microfossils (e.g., *Radiomura cautica* Senowbari-Daryan and Schafer); SP 78/52 (×4.6).

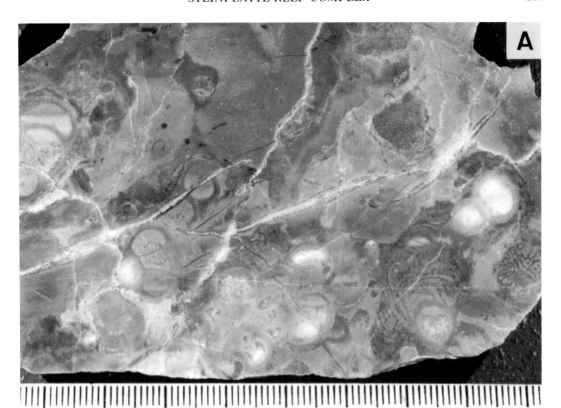

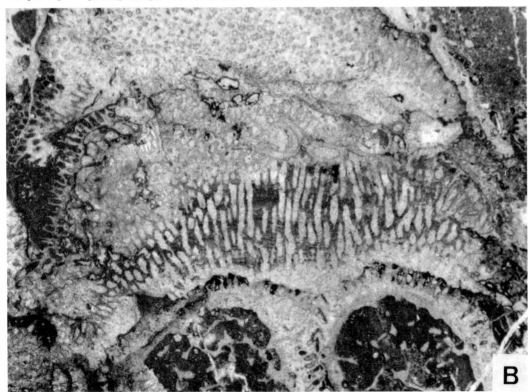

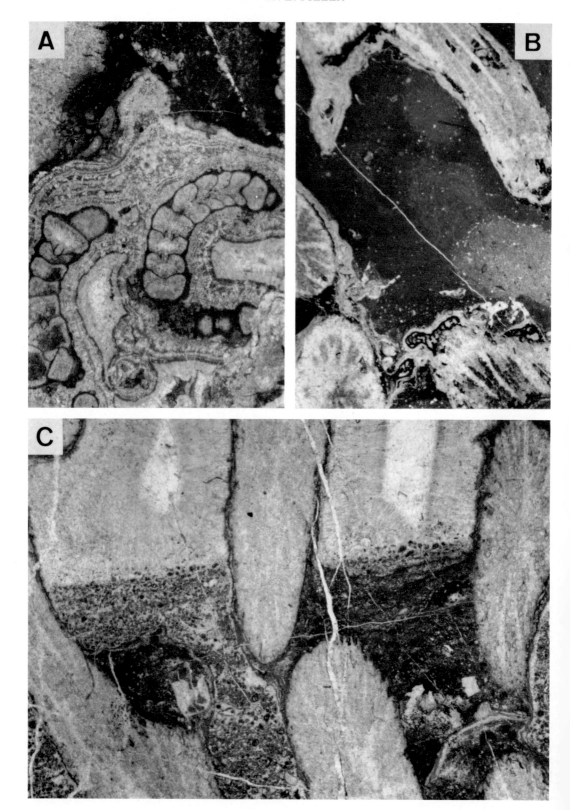

entiation into various facies units, as discussed below.

*Patch-Reef Facies.*—This facies forms a zone more than 500 meters wide located east and southeast of the Steinplatte summit, whereas the boundary between the reef-crest and the patch-reef facies is situated west and northwestward of the Steinplatte summit. The transition appears to be continuous, but is poorly exposed because it is partly overlain by "covering beds" of the Steinplatte complex. The dense coral belt of the reef-crest becomes less dense, and areas free of corals increase. In addition to the corals, mainly "*Thecosmilia*" *clathrata,* form B of Ohlen, the most characteristic macrofossils are thick-shelled megalodontid bivalves (Fig. 9). These bivalves settled both in areas devoid of corals, and between coral heads.

Most patch-reefs are built by "*Thecosmilia*" *clathrata.* Of minor importance are sponges and hydrozoans, with the possible exception of *Lamellata wähneri* Flügel. This problematical hydrozoan seems to be characteristic of those patch-reefs near the reef-crest, or on the backside of the crest.

The sediments of this zone contain several microfacies types which can readily change within short distances. Close behind the reef-crest bioclastic limestones dominate. The bioclasts are derived from the reef-crest and consist mainly of coral and hydrozoan fragments, and molluscs, most with micrite envelopes or encrustations of *Lithocodium* (=*Problematicum* A of Ohlen), and *Bacinella* (Fig. 10A). The grain-size is arenitic to ruditic, and sorting and roundness is highly variable. There are also sediments which consist of poorly-sorted and rounded, medium to coarse arenites. Their most characteristic components are oncoids (Fig. 10B), constructed by algae (*Lithocodium-Bacinella*) and sessile foraminifers, which bind other particles. These oncoids may reach a diameter up to 5 centimeters. Other components are mostly bioclasts, all of which are intensively micritized. Sediments composed of these grains (Fig. 10C) developed from bioclasts which underwent intensive micritization by algal boring. Transitional grains with micrite envelopes of various thicknesses are also discernible. These bahamites

FIG. 9.—Outcrop photograph of megalodontid bivalves on a weathered surface southeast of Steinplatte summit. These organisms together with corals are the most characteristic component of the patch-reef facies.

are frequently inhabited by sessile foraminifers. Besides bahamites and bioclasts, aggregate grains also occur; these are composed of bahamites, bioclasts and foraminifers bound together. The roundness is good and sorting is variable.

Another kind of sediment is represented by fine- to medium-grained arenitic biosparites, with poor rounding but good sorting of the grains. The bioclasts are derived mainly from molluscs, but echinoderm fragments also are common. Bahamite grains and aggregate grains occur abundantly.

Sediments in protected areas close to "*Thecosmilia*" patch-reefs are dominated by corals, echinoderm fragments, molluscs, and dasycladacean algae (Fig. 10D). These components are poorly rounded and sorted. The grain size is arenitic to ruditic and generally a micritic matrix occurs generally filling-in geopetal cavities between coarser grains. These components are always circumscribed by micrite envelopes.

The foraminiferal fauna of the patch-reef facies forms typical associations, but these communities also show transitions similar to sediment types. The dominant foraminifers belong to the Involutinidae (*Aulotortus sinuosus*) Weynschenk, *A. tumidus* (Kristan), and *Auloconus permodiscoides* (Oberhauser), occurring with dasycladaceans and finer-grained sediments, as well as the foraminifers *Glomospirella* and *Pseudotaxis.* In

---

←

FIG. 8.—Thin-section photomicrographs (×7.5) of various sediments and organisms of the reef-slope and reef-crest. A. *Cheilosporites tirolensis* Wähner, inhabiting cavities within the framework of the upper reef-slope; cavities are filled by various generations of cements, SP 78/37. B. Typical reef-crest association of the "*Thecosmilia*" *clathrata* (Emmrich) form A of Ohlen, encrusted by the sessile foraminifer *Alpinophragmium perforatum* Flügel, and thin crusts of unknown origin. The sediment in between is a reddish biomicrite; ST 78/94. C. Fine-grained, partly laminated, sediment that occurs between branches of "*Thecosmilia*" *clathrata* (Emmrich) form B of Ohlen. The sediment consists mainly of small peloids and foraminifers ("*Lituosepta,*" *Galeanella,* and *Ophthalmidium*), embedded in microsparite; SP 78/93.

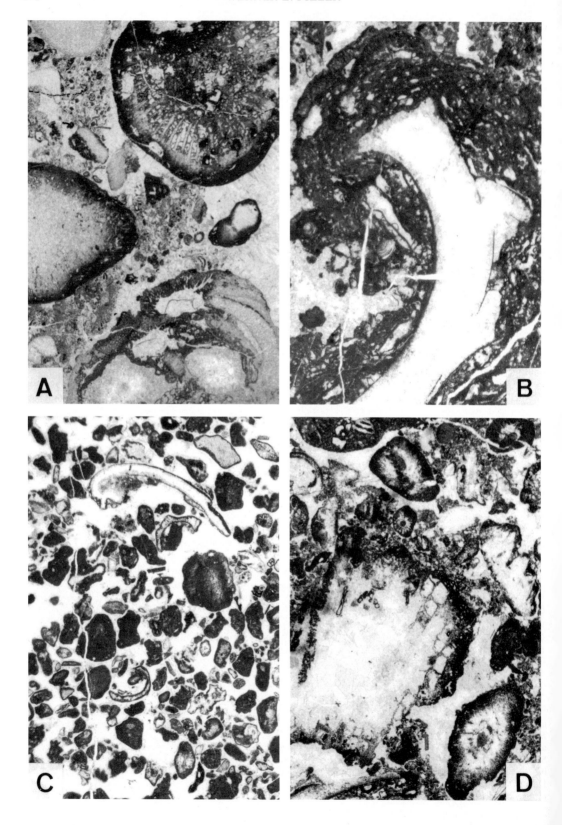

coarser-grained sediments individuals of the family Duostominidae are more numerous, but involutinids still are present and the frequency of both changes. Within coarse-grained biosparites (oncolite and bahamite sediments) sessile foraminifers are very abundant, whereas other foraminifers are rare. All microfacies types contain fragments of *Alpinophragmium,* which generally lives inside "*Thecosmilia*" colonies of the patch-reefs.

The distribution of the sediments has no distinct pattern, but appears to be related to the distance between the reef-crest and the patch-reefs.

In addition to some patch-reefs dominated by "*Thecosmilia,*" others are dominated by assemblages similar to the reef-slope communities. They are constructed by sponges (*Colospongia,* and *Molengraafia*), massive corals ("*Thamnasteria*"), hydrozoans (*Spongiomorpha*), *Cheilosporites,* encrusting *microproblematica* (*Microtubus,* and *Baccanella*), light crusts, and spongiostromata-crusts. Bivalve coquinas are also present. The sediment between these framebuilders is a brownish to reddish micrite (mudstone to wackestone). No zonation was observed within these patch-reefs, but this possibility cannot be excluded. Two of the patch-reefs are situated directly on the path from Steinplatte summit to the Wemeteigen Alm. The spatial distribution of the patch-reefs was not investigated. We can only state that the frequency of patch-reefs decreases to the east, and that the micritic patch-reefs are less abundant.

*Eastern Lagoon.*—The area east of the patch-reef zone was studied only superficially. This area has a distinct facies differentiation, but these facies patterns had not been explained previously. In general, the frequency of patch-reefs decreases near the boundary of the patch-reef facies, and only isolated ones can be seen in the eastern part of the Steinplatte Plateau; near the eastern margin they are completely absent.

In the western part of the Steinplatte Platform the sediments are bioclastic grainstones derived from the patch-reef facies. Fine arenitic material prevails, but the portion of micritic matrix increases. The dominating organisms, except for the megalodontids, are dasycladaceans, skeletal blue-green algae, and foraminifers of the family Involutinidae (*Aulotortus, Triasina, Auloconus*). This organism community belongs to the "algal-foraminiferal facies" which is similar to the "Kalkalgen-Foraminiferen-Detritus Fazies" of the bedded Dachstein Limestone (Hohenegger and Piller, 1975; and Piller, 1976).

In the more eastern parts of the Steinplatte Plateau the sediments belong to the grapestone facies, these contain a high amount of aggregate grains composed of micritized molluscan fragments, peloids, and foraminifers. The grain size of these sediments is highly variable and ranges from coarse to very fine arenites. The sediment contains individual grapestone lumps and algal-foraminiferal lumps (Fig. 11A). In the finer arenitic portions there are less single lumps, and here the main portion is bound together by algal micrite (Fig. 11B). The easternmost part of the Steinplatte (around Urlkopf) also belongs to the grapestone facies, but here there is a higher amount of micritic matrix, and skeletal blue-green algae is also present. These sediments often display cavities which have an irregular or horizontal arrangement, and are thought to have been caused by subtidal algal mats. Laminated algal stromatolites also occur, and these show shrinkage structures and desiccation vugs.

South and southwestward of the Lofer Alm some sediments can be classified with the oncolite and oolitic facies. The oncoids are built by foraminifers and algae and are smaller than those of the patch-reef facies. In general, the sediments are oncosparites, but some micritic matrix occurs. The sediments of the oolitic facies are similar to those of the grapestone facies, but with ooids as the most characteristic grains, although lumps and grapestone lumps are also common (Fig. 11C). Oolites in the strict sense have not been observed.

Also abundant are fine arenitic grain- to wackestones, with a high percentage of fecal pellets and fine-grained bioclasts, which probably could be grouped with the pellet or pellet-mud facies (Fig. 11D).

### Discussion of Facies Interpretation

Since the facies interpretation proposed here differs greatly from interpretations by Ohlen

---

←

FIG. 10.—Thin-section photomicrographs of various sediments present in the patch-reef facies. A. Bioclastic grainstone near the reef-crest; bioclasts derived from corals and sponges and are encrusted by *Lithocodium* and *Bacinella,* SP 78/84 (×5.8). B. Oncolithic grainstone, in which oncoids are built by algae (*Lithocodium-Bacinella*) and sessile foraminifers, which bind other particles and encrust a molluscan shell fragment serving as a nucleus, SP 78/145/2 (×7.4). C. Bahamite grainstone in which the sediment is composed of bahamite grains and micritized bioclasts. Characteristic foraminifers are *Aulotortus sinuosus* Weynschenk and *Pseudotaxis*; SP 78/105 (×10). D. Bioclastic grain- and packstone near a "*Thecosmilia*" patch-reef. Besides coral fragments, dasycladacean algae (*Diplopora*) and foraminifers (*Aulotortus sinuosus*) characterize this microfacies, SP 78/67 (×10).

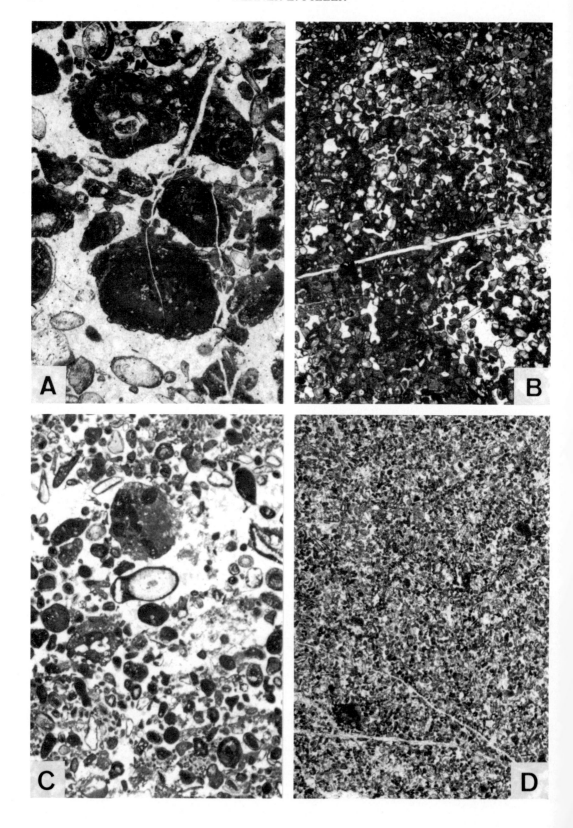

(1959) and Zankl (1971), a short explanatory discussion seems to be in order.

All authors agree on the assignment of the upper Kössen Beds to the basin facies, because of their geological setting, their lithology, and their contained fauna. In addition, there is a continuous transition from basin to the fore-reef, and this is generally accepted. However, differences exist concerning the upper fore-reef, reef, and back-reef facies.

Ohlen (1959) described a fore-reef breccia near the Steinplatte summit, and Zankl (1971, p. 181) reported these breccias "in the karst depression around the summit of Mt. Steinplatte." Conversely, Vortisch (1926, p. 51) noted the absence of fore-reef breccias. Therefore the presence or absence of breccias was of special interest and actively pursued. This investigation confirmed the opinion of Vortisch, since the breccias, correlated by both Ohlen and Zankl to the fore-reef, belong instead to the back-reef sediments (see below).

The second major point of disaggreement concerns the reef itself. Ohlen subdivided the reef facies into a coral facies and into a reef calcarenite facies (Fig. 12A). Both these facies zones are located east of the Steinplatte summit and form a zone more than 2 kilometers wide. The facies interpretation given in this paper (Fig. 12B) shows only a very small reef zone 100 meters in width, and located west of the Steinplatte summit. This zone was subdivided into two sub-zones (reef-slope and reef-crest). Accordingly, the fore-reef breccias and reef facies of Ohlen, are herein placed in the back-reef.

Field observations confirm that the uppermost 30 meters of the slope are densely settled by frame building organisms embedded in a micritic matrix. Ohlen described this organism community (p. 105–106) and attributed it to the "reef-front, or possibly on a fore-reef terrace" (p. 105), but he did not observe a muddy matrix. Instead he reported coarse breccias with components of these organisms. However, if one compares the sediments attributed to the reef-slope, with the fore-reef breccias of Ohlen, which should bear the same organism community (pl. XV, fig. 1–4; pl. XVI, fig. 1–4; pl. XVII, fig. 4), it seems clear, that they differ widely. The rocks now attributed to the reef-slope show frame building organisms in situ, densely encrusted by epizoans, and embedded in a micritic matrix. This organism assemblage and the presence of mud reflects quiet water conditions. On the contrary, the examples of Ohlen represent bio- or intrasparites with large bioclasts with micritic envelopes, and partly encrusted by *Problematicum* A, fragments of *Alpinophragmium,* and a foraminiferal fauna, containing common duostominids. Besides bioclasts, the most abundant components are peloids. The absence of muddy matrix and the fragmentation of the particles point to high water energy. The micritization of the bioclasts, their encrustation by *Problematicum* A, the abundance of peloids, and the dominance of duostominid foraminifers all indicate a back-reef environment. This relationship to the back-reef environment becomes even clearer by examining Ohlen's sample localities which are situated east of the now recognized reef-slope and reef-crest, and which therefore belong to the transitional interval from the reef-crest to the patch-reef zone.

This error originated from several causes: 1. the reef-slope as well as the reef-crest are relatively narrow zones, 2. both zones are located inside an area which is densely covered by dwarf pine woods, 3. Ohlen himself, mentioned that he did not study the reef-front (p. 80), and 4. the rocks of the Steinplatte Reef are not as massive as they appear superficially, but are even-bedded in those areas which superficially appear to be unbedded. Because of the uneven karsted surface one has great difficulty identifying each bed, but this is important, because facies zones are not uniform inside the beds but do show changes. This phenomenon is as yet not solved because of the nearly perpendicular reef walls. Still, this migration of facies zones can be observed on the surface of the west slope of the Steinplatte summit. Here the reef-slope and the reef-crest composing the youngest beds of the Steinplatte complex are located west of the Steinplatte summit (Fig. 12B). If one follows this youngest layer across to the west, it is cut by erosion and the underlying beds are exposed. This erosion exposes the reef-crest and the reef-slope, as well as the coquina, but here these zones are located approximately 200 meters northwest of the equivalent zone in the uppermost layer.

These difficulties may explain why Ohlen (1959) could not recognize the reef zone, and therefore designated the "patch-reef facies" as the reef. This error was discovered mainly due to an in-

←

FIG. 11.—Thin-section photomicrographs of various sediments of the eastern lagoon. A. Grapestone facies with typical aggregates; lumps encrusted by involutinid foraminifers (*Auloconus sinuosus, A. permodiscoides*), SP 78/134L (×10). B. Fine-arenitic grapestone facies in which all components are bound together by algal micrite; SP 78/135L (×7.5). C. Oolitic facies with oolites, aggregates, peloids and lithoclasts; SP 78/131L (×7.5). D. Fine arenitic grainstone with peloids and bioclasts; SP 78/142 (×10).

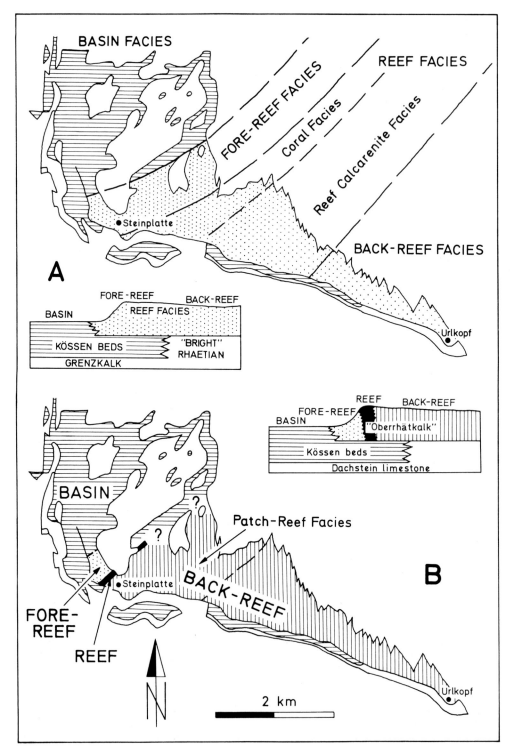

FIG. 12.—Diagram showing facies distribution of the Steinplatte Reef complex. A. Interpretation of Ohlen (1959, pl. II). B. Interpretation given in this paper.

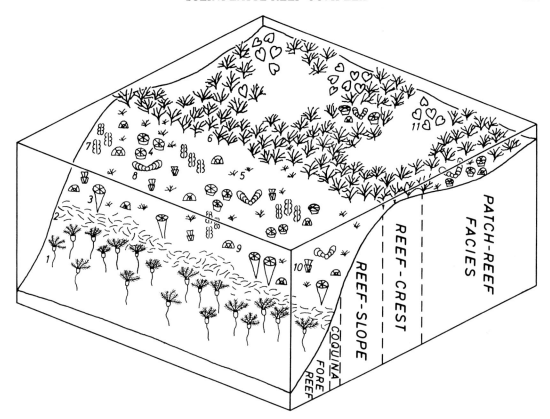

FIG. 13.—Schematic reconstruction of the facies zones and organism communities in the "central" part of the Steinplatte Reef complex. 1. crinoids, 2. bivalves and brachiopod shells, 3. "*Montlivaltia*," 4. massive corals like "*Thamnasteria*," 5. small dendroid corals similar to "*Thecosmilia*" *clathrata* (Emmrich) form A of Ohlen, 6. thick branched corals like "*Thecosmilia*" *clathrata* (Emmrich), form B of Ohlen, 7. *Paradeningeria* and other cylindrical and glomerate calcareous sponges, 8. *Colospongia*, 9. hydrozoans, 10. tabulozoans/bryozoans, and 11. megalodontids.

creased general knowledge of Triassic reefs since the late fifties. Whereas Ohlen did not recognize the "true" reef-zone, he did identify the "patch-reef facies" with its abundance of corals as the only portion of the Steinplatte complex that he could actually classify as a reef facies. But these corals occur together with megalodonts, which are typical lagoonal organisms, frequently occurring in all Upper Triassic lagoons. Furthermore dasycladacean algae, oncoids, peloids, several kinds of aggregates, and a foraminiferal assemblage characterized by involutinids and duostominids are clear indicators of a back-reef environment. Regarding the significance of involutinids for facies interpretation in the Upper Triassic, see Piller, 1978.

*Conclusions Regarding Facies Distribution*

The following facies interpretation can be given (Fig. 12B, and 13) for the Steinplatte Reef complex.

The upper Kössen Beds represent the basin facies, which was laid down in a water depth of approximately 100 meters. There is a continuous transition from the basin to the fore-reef. The dip on the slope reaches 35 degrees, and the detrital components derived from the reef increases towards the upper part of the slope, but no fore-reef breccias are developed. The absence of such breccias may result from water agitation generally too weak to produce large boulders from the reef framework.

The upper boundary of the fore-reef is sharply marked by a coquina. This coquina represents an accumulation of bivalve and brachiopod shells transported for short distances by relatively low water energy. Appreciable transportation can be excluded because some bivalves are still articulated.

Behind the coquina and partly covered by it is the reef-slope with frame building organisms occurring in beds with dips of 20–30 degrees. This

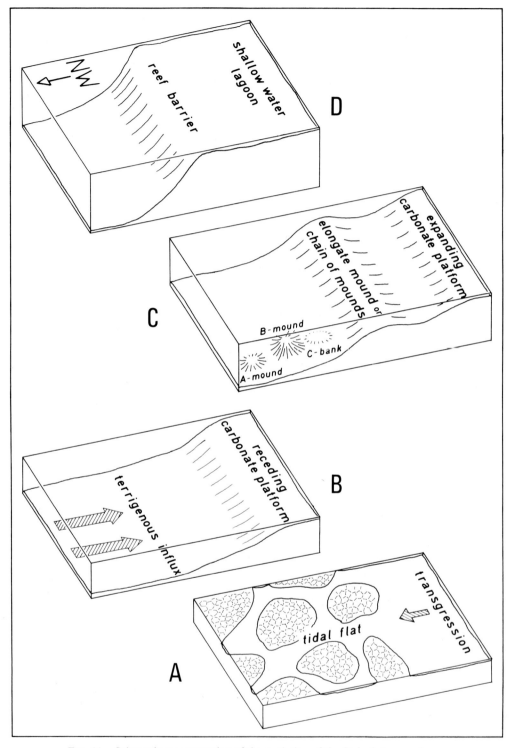

Fɪɢ. 14.—Schematic reconstruction of the evolution of the Steinplatte complex.

reef-slope represents a zone of low energy water turbulence, indicated by its muddy matrix. The reef-slope flattens out in its upper part, and here muddy sediments are absent and the framebuilders consist almost exclusively of thick branched corals. These corals were able to withstand stronger water agitation, and grew in a direction opposite to the water movement. Because of these corals, the restricted faunal associations, and the field relationships, this zone can be identified as the reef-crest, and represents the topographically highest part of the reef that was able to build a small barrier (Fig. 13).

Behind this barrier, water agitation decreased but was strong enough for the growth of patch-reefs built mainly by branched corals. The sediment between the corals are coarse arenites inhabited by thick-shelled megalodontids, calcareous algae, and foraminifers. In protected areas of this environment, such as inside shallow depressions, micritic patch-reefs with high organism diversity occur. These are similar to those that occur in the reef-slope. This area is 1–1.5 kilometers wide and represents the patch-reef facies (Fig. 13).

This zone is connected to a shallow water lagoon with arenitic sand-flats, composed of bioclasts, foraminifers, calcareous algae, and peloids. At a greater distance behind the reef, grapestone, oncolite, and oolitic facies are developed. These suggest shallow water conditions where the sea bottom was stabilized by submarine algal mats. In the easternmost parts of this lagoon, laminated algal mats with shrinkage structures occur, and these suggest inter- and supratidal conditions. Inside this lagoon small intertidal flats developed intermittently.

### Evolution of the Steinplatte Complex

The foundation of the Steinplatte area is the Hauptdolomite which interfingers eastward with bedded Dachstein Limestone (Fig. 14A). Whereas the Hauptdolomite represents widespread inter- and supratidal flats, the sediments of the Dachstein Limestone were laid down under shallow subtidal as well as inter- and supratidal conditions. Eastward the beds of the shallow subtidal environment are dominant. Upsection the boundary between the Dachstein Limestone and Hauptdolomite changes in northwest direction, representing a transgressive phase. This transgression is reflected not only in the migration of the boundary, but also in the type of sediments. The Dachstein Limestone changes from muddy sediments (which are related to pellet-mud and mud facies), to those of the grapestone and oolitic facies.

Over the Dachstein Limestone an inversion of sedimentary conditions was initiated (Fig. 14B), and the supply of terrigenous material from the north increased rapidly causing development of the lower Kössen Beds. Sedimentation varying between carbonates and terrigenous influx is expressed by intercalations of Dachstein Limestone beds with marly limestones and marls of the Kössen Beds. In conjunction with higher content of terrigenous material, the water depth increased several tens of meters. To the east the Kössen Beds interfinger with shallow water limestones, similar to the Dachstein Limestone.

An essential outstanding problem is the reason for subsidence of this part of the carbonate platform. One explanation can be related to tectonics which probably contributed to higher rates of subsidence on the northern margin of the platform. On one hand there is no indication for such a disproportionate subsidence, yet on the other hand the higher subsidence rate could not be the unique reason. Another probable explanation is thought to be the higher supply of terrigenous material, probably related to a climatic change. This terrigenous influx could have reduced carbonate production so that the accumulation of carbonate material was not able to compensate for bottom subsidence, hence the deepening. On the land, terrigenous influence decreased and the carbonate platform continued to develop without incursions of terrigenous sediment.

The lower Kössen Beds are a monotonous sequence of terrigenous sediments, generally poor in fossils, but in the upper part they do become fossiliferous. As the depositional environment shallowed the sea bottom was settled by patchy accumulations of bivalves and brachiopods. These shell banks acted as a hard bottom, and became the substrate for juvenile corals, sponges, and hydrozoans. These organisms were able to build an organic framework in relatively greater water depth, and under quiet water conditions, as suggested by the muddy matrix. Further growth of these organisms gave rise to "mud-mounds." Such "mud-mounds" are represented in "A-, B- and C-reefs," as reported by Ohlen (1959). Ohlen's "C-reef" (Fig. 2) shows a very early stage of development, because it is a flat structure built by corals, hydrozoans, and bivalves embedded in a micritic matrix. This structure is clearly bedded, with layers where the corals and hydrozoans are *in situ*, these alternating with beds composed entirely of fragmented organisms. Hence this structure could be called a bivalve-coral-hydrozoan-bank ("C-bank") according to Ohlen (1959, p. 83).

A more advanced stage is represented in "A- and B-reefs" (Fig. 2). The "B-reef" reflects a "mud-mound" which was raised up approximately 30 meters above the basin floor (Fig. 2, and 15A). The basal part, similar to the "C-bank," contains bedded terrigenous limestones with bi-

WERNER E. PILLER

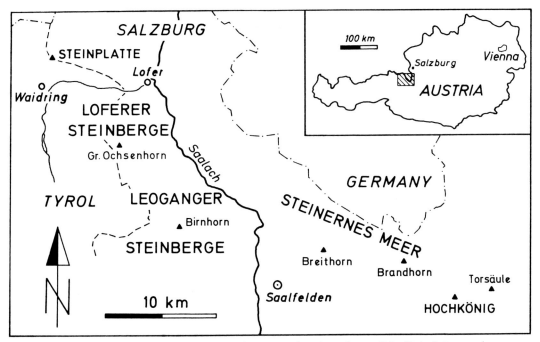

FIG. 16.—Location map of the carbonate plateaus south and southeast of the Steinplatte complex.

valve-coral-hydrozoan assemblages. These colonies are seldom in growth position and most of them are overturned and thought to have been transported a short distance. Such overturning is well represented by a large head of *Spongiomorpha ramosa* Frech (Fig. 15B), also noted by Ohlen (1959, p. 49) located at the base of "B-reef." The branches of these questionable hydrozoans often show intensive borings attributed to bivalves (Fig. 15C). Upwards, the thickness of the beds increase and bedding planes are poorly visible. Because of the almost perpendicular wall, the upper part of the "B-mound" cannot be satisfactorily studied. However from talus samples it does seem clear that there is no great change upward. The content of terrigenous material seems to be less and the material between the framebuilders is carbonate mud. Ohlen (1959, p. 101) reported that this "mud-mound" was covered by *Stromatomorpha rhaetica* Kühn. The sediments of the "mud-mound" slopes are rich in echinoderms, bivalves, and brachiopods, and near the mound coral fragments are abundant.

The latter decrease rapidly with increasing distance away from the mound, but the slope community is poorer and the sediments grade laterally into typical Kössen Beds.

Ohlen's "A-reef" is very similar to "B-mound," but not as massive, and it is tectonically isolated (Fig. 2).

The initial stage of the Steinplatte Reef is similar to that of the "mud-mounds." The basal portion appears to have been a hard ground settled by accumulations of bivalves, followed by frame building organisms, which caused development of a "mud-mound." In contrast to the mounds mentioned above, development continued until the mound extended up into a zone of higher water energy (Fig. 14D). These environments offered excellent conditions for growth to a variety of zoned organism assemblages dependent upon water turbulence: 1. reef-slope with quiet water communities of high diversity, 2. reef-crest with thick branched corals that built a wave resistant barrier, and 3. back-reef lagoon with various facies types and organism assemblages.

←

FIG. 15.—Outcrop photograph and a thin-section photomicrograph of "B-mound." A. General outcrop view of the mound showing the poorly developed bedded limestone at the base and the massive upper portion. The mound is overlain by bedded Kössen Beds. Height of mound is approximately 30 meters. B. A large overturned head of *Spongiomorpha ramosa* Frech from the base of "B-mound." C. Thin-section photomicrograph of *Spongiomorpha ramosa* showing intensive boring by bivalves. Bore holes are inhabited by a multichambered *microproblematicum* (?foraminifer), and are geopetally filled with silty sediment; ST 77/114 (×6).

FIG. 17.—General outcrop view of a portion of the Loferer Steinberge (Mt. Reifhorn, Ochsenhorn) showing typical bedded Dachstein Limestone of the Lofer facies.

The close of reef building is marked by drowning of the entire complex, and later on, the reef, back-reef, and part of the fore-reef were covered by crinoidal biosparites containing small amounts of reworked reef material. At this stage no barrier existed separating the platform from the basin. Thus, small quantities of platform sediments were transported onto the slope and into the basin. These particles are mostly micritized bioclasts. The foraminiferal fauna shows a similar influence, for in addition to autochthonous foraminifers and reworked reef and lagoonal foraminifers, such forms as involutinids (*Aulotortus*) and duostominids are also discernible. These "covering beds" are overlain by Liassic limestones consisting of reddish crinoidal limestones (encrinites) at the base, followed by red and green ammonite-bearing nodular limestones, which were deposited in greater water depths (compare Garrison and Fischer, 1969).

Final questions to be considered are: 1. why the "mud-mounds" developed into "real" reefs? and, 2. why a shallow water lagoon could develop east of the Steinplatte Reef?

The answer to these questions seems to be related to the overall shape of the four structures. The A- and B-mounds have a spherical or subspherical form (Fig. 14C). Both mounds are cut by faults and the dip of the beds suggests that these faults passed near the centers of the mounds. The tops of both mounds seem to be missing. Since both mounds are cut in such a po-

sition we do not believe that both were originally part of an elongated ridge.

The paleogeographical position of the Steinplatte Reef. The overall shape we see today suggests a barrier-reef running from southwest to northeast, and this can be traced for more than one kilometer. Because of the observed shape, we believe that the reef originally had an elongated form, and that during the "mud-mound stage" an elongated ridge, or a chain of "mud-mounds" existed (Fig. 14C). It seems that the location where the Steinplatte Reef originally grew was a preferential site, perhaps because of better water circulation or a higher position within the basin thus causing more rapid growth. This rapid growth probably caused a suppression of "C-bank" (C-reef of Ohlen) by covering it with detrital material because of its close proximity. Both of the "mud-mounds" (A- and B-mounds) would have been only slightly influenced by detrital material from the main reef, but a "mud-mound" ridge or a "mud-mound" chain could have conceivably changed hydrographic conditions so that growth of the framebuilders was suppressed. Because of the limited extent of these "mud-mounds," a slight change in water currents, or a higher influx of terrigenous material could be sufficient to impede growth of the framebuilders. The core of the "mud-mounds" was previously influenced by the surrounding Kössen Beds, and it is thought that the more widespread elongate ridge of the later reef probably had a better opportunity to withstand such changing conditions.

The elongate shape of the reef can also be regarded as a factor in the development of a shallow water carbonate lagoon. At an earlier stage of development the ridge acted as a protecting dam to the influx of terrigenous material to the east and southeast. During the beginning of reef growth, the form of the ridge was nearly symmetrical since sediments of the back-side of the reef are in lower positions, similar to those of the fore-reef slope. These are detrital limestones rich in echinoderms, and geopetally infilled with silty material. The terrigenous content of these sediments clearly decreases upward. Because of the absence of terrigenous influx, the carbonate platform extended farther to the east and southeast during deposition of the Kössen Beds to the north. Owing to the expansion of the platform to the north(west), and the influx of reef debris from the north, the preex-

→

FIG. 18.—Photographs of various structures in Loferite Member B, bedded Dachstein Limestone. A. Loferite with laminated algal mats, sheet- and prism-cracks and shrinkage pores, near Schmidt-Zabirow Hut. B. Weathered surface of Loferite showing polygonal prism-cracks; below Schmidt-Zabirow Hut. C. Polished slab showing reworked loferitic intraclasts at the base, overlain by algal laminated loferites which show sheet-cracks and cavities partly filled with (vadose) silt; near Schmidt-Zabirow Hut, height 9 centimeters.

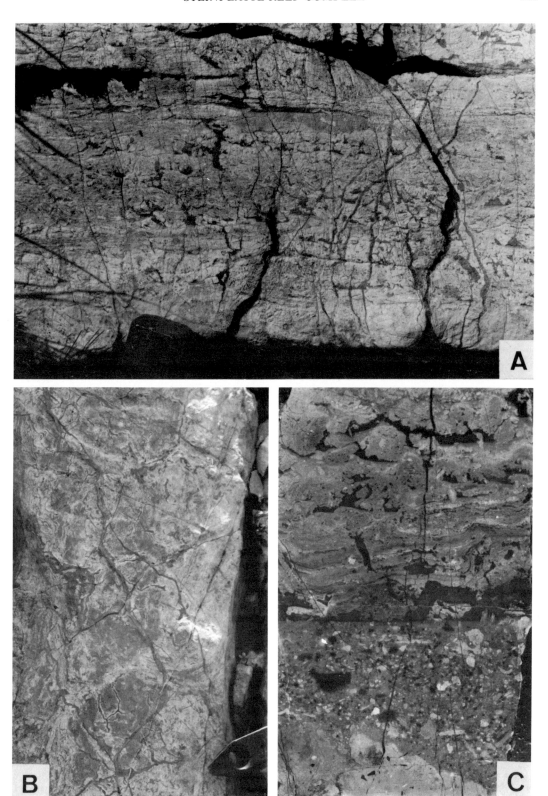

FIG. 19.—General outcrop view looking from the south towards the Leoganger Steinberge (Mt. Birnhorn; height 2634 meters). The carbonate foundation rocks are up to 2000 meters thick and are built of Hauptdolomite; the upper part is bedded Dachstein Limestone.

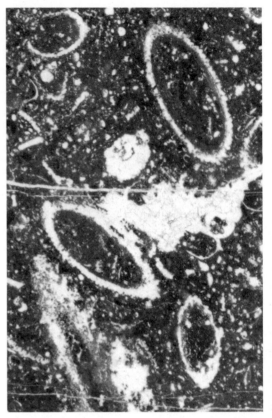

FIG. 20.—Thin-section photomicrograph (×7.5) of a typical dasycladacean micrite (*Griphoporella curvata* (Gümbel); Steinernes Meer, SM 78/8.

isting shallow basin was filled-up, and a continuous carbonate platform developed (Fig. 14D).

## PALEOGEOGRAPHICAL POSITION OF THE STEINPLATTE COMPLEX WITHIN THE NORTHERN LIMESTONE ALPS

The paleogeographical position of the Steinplatte complex within the Northern Limestone Alps is complicated by geologically late intensive alpine tectonics. However, if we look south and southeast of the Steinplatte Plateau, we find three large limestone massifs: Loferer Steinberge, Leoganger Steinberge, and Steinernes Meer (Fig. 16). These massifs are today isolated from each other and from the Steinplatte by faults, but all belong to the "Tirolikum" tectonic phase (Tollmann, 1969). Therefore these massifs are in a relative autochthonous position within the Northern Limestone Alps, a point revealed by their facies distribution and development. Because of the large area covered by these massifs, and the alpine topography, detailed geologic studies have not been forthcoming, however sufficient data is available to demonstrate regional relationships.

### Loferer Steinberge

The massif immediately south of the Steinplatte is the Loferer Steinberge (Fig. 16), which rises to an elevation more than 2500 meters. It is separated from the Steinplatte by a fault system which created the valley of Lofer Bach, connecting the villages of Waidring and Lofer. In the southeastern part of the Steinplatte Plateau the beds dip down to this valley, and the beds of the Loferer Steinberge dip to the northeast. Because of these related dips we can surmise that both massifs were originally connected, even though this transition cannot be traced presently.

The Loferer Steinberge is composed of Hauptdolomite overlain by bedded Dachstein Limestone (Fig. 17). The latter reaches to the top of the massif and is cut by erosion so that younger beds are not preserved. This area has been regarded as a classic locality since Sander (1936) carried out his fundamental studies of depositional environments on these limestones and dolomites. In addition, Fischer (1964) recognized and described the Lofer cyclothems from these Dachstein Limestones, which are now known worldwide (see Wilson, 1975). Therefore the type locality for loferites, which are widespread throughout geologic time (Ginsburg, 1975), is located within this massif. Because of these studies, and additional investigations by Gökdag (1974) on sedimentology and the distribution of oxygen and carbon isotopes, our knowledge of this area is better than for most other regions of the Northern Limestone Alps.

The Hauptdolomite platform foundation is similar to that of the Steinplatte foundation; built mainly of intertidal and supratidal deposits with

FIG. 21.—General view of the southern part of the Hochkönig Massif. The dolomite at the base is overlain by massive Dachstein Reef Limestone.

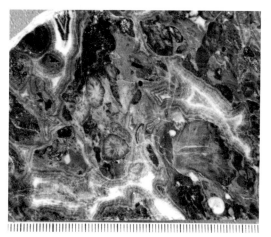

FIG. 22.—Fore-reef breccia of the Hochkönig Reef, the components consist of cemented reef debris bound together by crusts of radial-fibrous cements of various generations; east of Mt. Torsäule; polished surface, scale in millimeters.

algal stromatolites, shrinkage structures and desiccation features. In the uppermost part intercalations of limestone beds become more abundant. This zone represents a continuous transition from Hauptdolomite to Dachstein Limestone.

The Dachstein Limestone is a typically developed Lofer facies as described by Fischer (1964, and 1975). One Lofer cycle, as developed in an idealized manner consists of three members. Member A is a unit composed of red or green argillaceous material, generally following a disconformity, is interpreted as an ancient soil. Member B, can be formed by a variety of rocks, e.g., loferites, laminated loferites, massive loferites, and nonloferitic lutites; all belong to the intertidal zone. Member C represents shallow subtidal conditions and can be expressed by a variety of facies types characterized by different biogenic constituents.

Member A at the Loferer Steinberge consists of green argillaceous limestone. Continuous beds are rarely seen, in most cases the green material fills-in solution cavities within the underlying bed, down to several meters. The sediment is a micrite or microsparite with a fairly high content of dolomite. Biogens are extremely rare and consist of ostracodes and foraminifers. Poorly rounded intraclasts are common, these mainly derived from loferites.

The intertidal member B (Fig. 18) is characterized by the dominance of loferites and other rock types, like homogeneous micrites, which are of minor importance. Member B forms beds from a few decimeters up to two meters in thickness, but most range between 0.5 and 1.0 meter. Laterally these beds often decrease in thickness and may even pinch out. In the upper part of the formation they decrease in thickness. As reported by Gökdag (1974, p. 20) the lower portions of these beds are characterized by fissures and depressions in the underlying layers, and these often contain dense micrites with zebra-structures ("zebra-limestone" of Fischer, 1964, p. 135). These zebra-limestones probably originated by sheet cracks parallel to the bedding planes which were later filled by radial-calcite cements of various generations. The main portion of member B is composed of laminar algal stromatolites (Fig. 18A), which are the most common type of loferite structures. Conglomeratic loferites are also abundant, in which intraclasts of algal crusts occur. These intraclasts, up to several decimeters in length show no rounding, suggesting limited transportation distances. All these loferites show various shrinkage pores (described by Fischer, 1964 and 1975), and cracks (prism- and sheet-cracks). Well developed prism-cracks, producing a polygonal pattern can be seen a few hundred meters below the Schmidt-Zabirow Hut (Fig. 18B). In most cases the loferites show penecontemporaneous dolomitization. Very commonly, cavities are filled with crystalline silt (Fig. 18C) which corresponds to the vadose silt described by Dunham (1969).

The third type of Loferite is represented by member C, which is thought to reflect subtidal conditions, and between beds of members B and C there can either be a disconformity or a continuous transition. Because member C contains abundant megalodontid bivalves, it is called the "megalodont limestone." These limestones are thick-bedded (up to 14 meters) and are related to the pellet-mud and mud facies. In the lower part of member C, reworked algal stromatolites are common. The microfauna is dominated by foraminifers and ostracodes; micrites containing abundant involutinid foraminifers, and micrites with skeletal blue-green algae are also present. The pellets are thought to be of fecal origin. The macrofauna is dominated by megalodonts (most in growth position), but gastropods are also common. In the upper part of the formation, corals ("*Thecosmilia*") are abundant in a few beds, along with oncolitic micrites. Very often solution-cavities occur and are probably related to paleo-karst surfaces. These cavities are geopetally filled with sediments of member A, followed later by radial fibrous calcite cements of several generations.

Algal stromatolites often occur in the uppermost parts of these beds. The transition between typical sediments of member C and these intertidal sediments is continuous. This phenomenon is explained by the fact that the sequence of members A, B and C represent a transgressive development, and that only a small portion in the upper few centimeters form the regressive portion of one cycle. Therefore it is clear that a Lofer cycle consists not only of a transgressive phase but also a regressive phase, and that members need not occur in the same succession. These deviations, briefly mentioned by Fischer, are discussed by Zankl (1971), Gökdag (1974), and Piller (1976).

## Leoganger Steinberge

The Leoganger Steinberge is located south of the Loferer Steinberge (Fig. 16). This massif, rising to a height of more than 2600 meters, shows a similar geological setting as the Loferer Steinberge. However, our knowledge concerning this area is incomplete.

The foundation is also built of Hauptdolomite overlain by bedded Dachstein Limestone, with a transitional zone between. In the south, the Hauptdolomite attains a thickness of approximately 2000 meters (Fig. 19), whereas in the northeast the Dachstein Limestone reaches down into the Saalach Valley, a phenomenon caused by a general tilting of the entire massif.

Here the Dachstein Limestone is developed as the Lofer facies with members A, B, and C. Member A consists of greenish marly limestones, which occur as infillings of solution cavities in the underlying beds. Intertidal member B is dominated by algal stromatolites with various shrinkage structures; member C, rich in megalodontids, and in the upper parts coral-bearing ("*Thecosmilia*"), consists of biopel- or biomicrites. Biopelsparites also occur in the easternmost parts of the massif suggesting higher water energy. The microfauna is similar to that of member C in the Loferer Steinberge, but skeletal blue-green algae seems to be more abundant. Other differences exist in the type of cementation, with microstalactitic ce-

ments occurring in the Leoganger Steinberge, but only very rare in the Loferer Steinberge.

## Steinernes Meer

The beds of the Leoganger Steinberge dip downward in a northeastern direction, into the fault-bounded Saalach Valley. The limestone massif situated north and northeast of the Saalach Valley is the Steinernes Meer. The beds of this massif also dip to the west to this valley, and it is assumed that initially the Steinernes Meer, the Leoganger, and Loferer Steinberge, as well as the Steinplatte, were all interconnected. The Steinernes Meer represents the thickest carbonate platform units in the investigated area, rising up to more than 2600 meters (Fig. 16). The geology is also very similar to the two other massifs, with Hauptdolomite at the base overlain by Dachstein Limestone. Tilting of the entire carbonate platform raised the Hauptdolomite up to a greater height along the southern margin. Our knowledge of this area is somewhat better than of Leoganger Steinberge, primarily because of the studies by Fischer (1964) and Speckman and Zankl (1969).

Whereas in the Loferer and Leoganger Steinberge, the Dachstein Limestone represents a uniform sequence, in the Steinernes Meer there is a distinct facies change from west to east. Typical Lofer cycles are developed in the western part, but in contrast to the Loferer and Leoganger Steinberge, member A consists of reddish limestones. These beds appear to have an intertidal origin, and show a decreased amount of algal stromatolites towards the east. In addition, prism- and sheet-cracks, well developed in the western part (around Mt. Breithorn), decrease towards the east. The color of member B sediments is more pink and white, in contrast to the grey color of these beds in the Loferer Steinberge. Great variations also occur in the thickness of these beds, and in the eastern part there is some pinching-out of beds. The megalodont limestone consists of sediments similar to those of the eastern part of the Leoganger Steinberge where bio(pel)sparites prevail. These biosparites are composed of skeletal blue-green algae (*Cayeuxia*-type), dasycladaceans, involutinids, and algal lumps. The sediments are classified as belonging to the algal-foraminiferal facies; sediments of the grapestone facies can also be observed. Characteristic of these sediments are microstalactitic and meniscus cements. Muddy sediments are of minor importance and decrease in an eastern direction. Only in protected areas do we find dasycladacean micrites, dominated by specimens of the genus *Griphoporella* Pia (Fig. 20).

In the eastern part of the region, members A and B clearly decrease both in thickness and occurrence, and finally disappear in the vicinity of

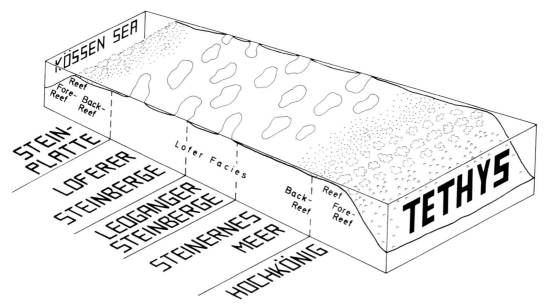

FIG. 23.—Paleogeographic interpretation of the Upper Triassic carbonate platform between Hochkönig and Steinplatte.

Mt. Brandhorn (Fig. 16). Here the Dachstein Limestone consists of beds laid down under subtidal conditions. Between Mt. Brandhorn and Torscharte (the eastern boundary of Steinernes Meer), biosparites and intrabiosparites with increased grain-size are well developed. The components are well-rounded and sorted, and consist of calcareous algae (solenoporaceans, and dasycladaceans), coral fragments, large foraminifers (involutinids and duostominids), and molluscs. Of minor importance are particles derived from sphinctozoan sponges, and hydrozoans. The grain-size is ruditic and arenitic. Vadose cements, common in the western part, are absent here.

### Hochkönig

East of the Torscharte the next massif is Mt. Hochkönig (2491 meters), which is covered by a small glacier. Looking from the eastern margin of Steinernes Meer towards this massif, an increased thickness of beds is clearly visible. Towards the east this bedding becomes indistinct, and the steep wall below Mt. Hochkönig consists solely of massive limestone (Fig. 21). East of Mt. Torsäule, indistinct bedding reappears and the beds show a moderate dip to the east and southeast. Further to the east the rocks are covered by vegetation and cut by erosion and faulting.

Since the investigations of Mojsisovics (1874) and Bittner (1884) the Dachstein Limestone of the Hochkönig massif has been known as a reef limestone. This is important from a stratigraphic point of view because it has been subdivided on the basis of ammonites. Later this reef and its frame building organisms were described by Heissel (1951) and Zapfe (1962) and more recently Zankl (1967, fig. 1), and Tollman (1976, fig. 117) include this reef in a regional paleogeographical reconstruction of the Upper Triassic. However, our knowledge of this reef complex is in fact very limited. Still, we can summarize what is presently known.

The thick-bedded limestones west of Mt. Hochkönig consist of coarse-grained biosparites which contain a high percentage of fragmented framebuilders. These biosparites clearly represent a continuation of sedimentation similar to that present on the eastern part of Steinernes Meer. The unbedded limestones on Mt. Hochkönig represents the reef facies. The original framework is composed of patch-reefs in which the framebuilders are in growth position. The distribution pattern of these patch-reefs seems to be random, and the space between the patch-reefs is filled by reef debris. This debris is the dominant rock type so that the percentage of framebuilders which are *in situ* make up less than 10 percent of the sediment. These framebuilders are similar to those of the Steinplatte Reef, with corals (*"Thecosmilia," "Montlivaltia," Pinacophyllum, Coccophyllum, "Thamnasteria,"* and *Procyclolites*), calcareous sponges (various sphinctozoans and *Peronidella*), hydrozoans, tabulozoans/bryozoans, solenoporaceans, and a great variety of *microproblematica*. The reef debris is cemented together by characteristic thick crusts of radial-fibrous cements of several generations. Presently, we do not know more details about a probable zonation of these patch-reefs, but it does seem that the reef facies itself extends in southwest-northeast direction.

In an eastern direction, around Mt. Torsäule, the framebuilders in growth position decrease, and the massive limestones become thick-bedded. The sediment consists of reef debris, but there are also breccias with clasts up to several decimeters in diameter. The constituents of this breccia are reworked framebuilders which were consolidated before brecciation. The breccias are cemented by thick crusts of radial-fibrous cements similar to reef debris between the patch-reefs of the reef facies (Fig. 22), or the interstitial space is filled by reddish to brownish micrite or silt, which is often millimeter-bedded. This zone can be classified as fore-reef facies. These crusts of radial-fibrous cements of the reef and fore-reef are similar to the "Grossoolithe" reported from the Middle Triassic Wetterstein Limestone reefs of the Northern Limestone Alps (e.g., Germann, 1971).

*Paleogeographical Reconstruction*

As discussed above, we believe that the five massifs (Hochkönig, Steinernes Meer, Leoganger and Loferer Steinberge, and Steinplatte) were all part of one carbonate platform during the Upper Triassic (Fig. 23). This carbonate platform extended approximately 40 kilometers in a northwest to southeast direction. The margin of this platform extended the reef belt to the northwest as well as to the southeast, so that it had a southwestern to northeastern strike.

The southern reef belt at Hochkönig represents a large Dachstein Limestone Reef which attained a thickness of more than 700 meters. This reef developed as a belt, several kilometers wide, between the platform and the Tethyan Ocean. The open marine influence is documented by the occurrence of ammonites within the reef limestone. The fore-reef facies is characterized by well-developed breccias, composed of reef derived fragments and boulders cemented by radial-fibrous cements or by infillings of micritic or silty sediment. The transition to basin equivalents seems to be absent, probably due both to tectonic movements and erosion. Behind the reef facies, in a western and northwestern direction, a sand-flat is exposed, which consists of coarse arenites built of reef gravels. These arenites, thick-bedded near the reef, decrease in grain-size and thickness in a western direction. These arenites extend to a widespread lagoon which reaches from the eastern part of Steinernes Meer to the eastern part of the Steinplatte Plateau. Although these massifs are not situated along a single transect we have projected them to appear as a single section. This large lagoon can be subdivided into two parts on

the basis of varying influence of open ocean waters. The middle part of Steinernes Meer represents a shallow water environment with good ocean water circulation. Therefore abundant growth of calcareous algae (dasycladaceans and skeletal blue-green algae) was possible, and the muddy material was washed away by currents. This led to formation of grainstones, commonly algal sparites. Large areas inside this shallow water lagoon periodically became emergent as they were situated near the lowest water level, this caused the development of vadose cements. Further to the west conditions changed, mainly because of restricted fresh water influx from the ocean. These conditions resulted in the formation of poorly washed sparites (western part of Steinernes Meer and Leoganger Steinberge), and further away, to muddy sediments. This change also caused a decrease of organism diversity. This biota consisted of megalodontids, gastropods, blue-green algae, foraminifers, and ostracodes. Westward lateral pinching out of the beds of member B to non-loferitic micrites (compare Gökdag, 1974, p. 42), reflects the development of intertidal and supratidal flats with shallow water mud-banks in between.

In the lower part of the formation, frequency of intertidal and supratidal flats increases towards the west and northwest as the Dachstein Limestone grades into Hauptdolomite. This unit was laid down under intertidal and supratidal conditions, with intertidal algal-mats, in addition to the formation of gypsum. During deposition of the uppermost part of the formation conditions changed, and the sediments show a lateral sequence similar to that from the Hochkönig to Loferer Steinberge, but in reversed arrangement. The muddy sediments extend to the eastern part of the Steinplatte Plateau, but intertidal sediments clearly decrease. More to the northwest, mud content became less abundant and organism diversity increases, a point which suggests better water circulation. Furthermore, arenitic sediments with patch-reefs occur, and the Steinplatte Reef follows, as was described above. This reef fringes the carbonate platform on its northeastern margin and presents a small barrier towards the Kössen Basin.

The two reefs developed on opposite margins of the platform. The Hochkönig Reef surpasses the Steinplatte Reef in thickness by approximately seven to ten times. Because of the limited data available on the reef facies of the Hochkönig, this segment of the reef complex cannot be compared in detail. The organism communities seem to be similar, but the Steinplatte Reef is built of a dense belt of framebuilding organisms, showing a clear zonation, whereas the reef facies of the Hochkönig Reef consists of randomly spaced patch-reefs with much reef debris in between. Therefore the Hochkönig Reef represents another reef type, similar to the Dachstein Limestone reefs of the Hohne Göll (Zankl, 1969), and the Hochschwab (Lobitzer, 1975), which according to Wilson (1975) can be classified as "knoll reef ramps."

Other differences are discernible as far as cementation is concerned, since thick crusts of radial-fibrous cements so characteristic of the Hochkönig Reef occur only rarely in the Steinplatte Reef. The coarse arenites of the back-reef are also better developed at Hochkönig. Another difference is emphasized by the occurrence of fore-reef breccias at Hochkönig, and their absence at the Steinplatte. This phenomenon does not depend on the dip of the slope, because at Steinplatte the fore-reef slope dips more than 30 degrees towards the Kössen Basin (steeper than the slope of the Hochkönig Reef), and one might expect fore-reef breccias. Instead, the development of such breccias seems to be due primarily to water movement and turbulence. Since this factor is influenced by the position of the reefs relative to facing towards the open ocean, this situation seems to be essentially different. Where-as the Hochkönig Reef is situated on the platform-edge, which is exposed to the Tethyan Ocean, the Steinplatte Reef is located at the opposite edge of the platform and is facing a relatively shallow marginal sea—the Kössen Sea. Owing to these differences we can surmise that not only were hydrographic conditions different, but that meteorologic conditions (e.g., wind direction) were more suitable for reef growth on the southern margin of the platform than on the northern margin.

### ACKNOWLEDGMENTS

I am indebted to Dr. Harry Lobitzer (Geological Survey, Vienna), who aided in the field work, supplied thin-sections and photographs, and critically reviewed the manuscript. Thanks are also given to Professor F. Steininger (Institute of Paleontology, University Vienna) for assistance with the English translation, and to Dr. Priska Schäfer (Institute of Paleontology, University Erlangen) for discussions about the reef building organisms. Preparation of the photomicrographs was done by Ch. Reichel (Institute of Paleontology, University Vienna).

## REFERENCES

BITTNER, A., 1884, Aus den Salzburger Kalkhochalpen. Zur Stellung der Hallstätter Kalke: Verh. kaiserl.-königl. Geol. Reichsanst., 1884, 6, p. 99–113.

CZURDA, K., AND NICKLAS, L., 1970, Zur Mikrofazies und Mikrostratigraphie des Hauptdolomites und Plattenkalk-Niveaus der Klostertaler Alpen und des Rhätikon (Nördliche Kalkalpen, Vorarlberg): Festbd. Geol. Inst., 300-Jahr-Feier Univ. Innsbruck, p. 165–254.

DUNHAM, R. J., 1969, Early vadose silt in Townsend mound (reef), New Mexico, in Friedman, G. M., Ed., Depositional Environments in Carbonate rocks: a Symposium: Soc. Econ. Paleontologists Mineralogists Spec. Pub. No. 14, 139–181.

FABRICIUS, F., 1974, Die stratigraphische Stellung der Rät-Fazies: Österr. Akad. Wiss., Schriftenr. Erdwiss. Kommiss., v. 2, p. 87–92.

FISCHER, A. G., 1964, The Lofer Cyclothems of the Alpine Triassic: Kansas Geol. Survey Bull., v. 169, p. 107–149.

———, 1975, Tidal Deposits, Dachstein Limestone of the North-Alpine Triassic, in Ginsburg, R. N., Ed., Tidal Deposits: p. 235–242, Springer-Verlag, New York.

GARRISON, R. E., AND FISCHER, A. G., 1969, Deep water Limestones and Radiolarites of the Alpine Jurassic, in Friedman, G. M., Ed., Depositional Environments in Carbonate rocks: a Symposium: Soc. Econ. Paleontologists Mineralogists Spec. Pub. No. 14, p. 20–56.

GERMANN, K., 1971, Calcite and Dolomite Fibrous Cements ("Grosoolith") in Reef Rocks of the Wettersteinkalk (Ladinian, Middle Trias), Northern Limestone Alps, Bavaria and Tyrol, in Bricker, O. P., Ed., Carbonate Cements: Studies in Geology, v. 19, p. 185–188.

GINSBURG, R. N., 1975, Tidal Deposits. A Casebook of Recent Examples and Fossil Counterparts: New York, Springer, 428 p.

GÖKDAG, H., 1974, Sedimentpetrographische und isotopenchemische ($O^{18}$, $C^{13}$) Untersuchungen im Dachsteinkalk (Obernor—Rät) der Nördlichen Kalkalpen: Marburg, Diss. Naturwiss. Fak. Univ., 156 p.

HAHN, F. F., 1910a, Geologie der Kammerker—Sonntagshorngruppe. I. Teil: Jahrb. Geol. Reichsanst., v. 60, 2, p. 311–420.

HAHN, F. F., 1910b, Geologie der Kammerker—Sonntagshorngruppe. II. Teil: Jahrb. Geol. Reichsanst., v. 60, 4, p. 637–712.

HECKEL, PH. H., 1974, Carbonate Buildups in the Geologic Record: a Review, in Laporte, L. F., Ed., Reefs in Time and Space. Selected Examples from the Recent and Ancient: Soc. Econ. Paleontologists Mineralogists Spec. Pub. No. 18, p. 90–154.

HEISSEL, W., 1951, Aufnahmen auf den Kartenblättern 124/1 Saalfelden, 124/2 Dienten, 124/3 St. Georgen im Pinzgau, 124/4 Taxenbach, 125/1 Werfen, 125/2 Bischofshofen, 125/3 St. Johann i. P., 125/4 Wagrain der neuen österreichischen Karte 1 : 25.000, früher Blatt St. Johann i. P., 5050 der österreichischen Spezialkarte 1 : 75.000 (Bericht 1950): Verh. Geol. Bundesanst., 1950–51, 2, p. 26–27.

HOHENEGGER, J., AND LOBITZER, H., 1971, Die Foraminiferen—Verteilung in einem obertriadischen Karbonat-plattform—Becken—Komplex der östlich Nördlichen Kalkalpen: Verh. Geol. Bundesanst., 1971, 3, p. 458–485.

———, AND PILLER, W., 1975, Ökologie und systematische Stellung der Foraminiferen im gebankten Dachsteinkalk (Obertrias) des nördlichen Toten Gebirges (Oberösterreich): Palaeogeography, Palaeoclimatology, Palaeo-ecology, v. 18, p. 241–276.

LOBITZER, H., 1975, Fazielle Untersuchungen an norischen Karbonatplattform—Beckengesteinen (Dachsteinkalk—Aflenzer Kalk im südöstlichen Hochschwabgebiet, Nördliche Kalkalpen, Steiermark): Mitt. Geol. Ges. Wien, v. 66/67 (1973/1974), p. 75–91.

MOJSISOVICS, E. v., 1871, Beiträge zur topischen Geologie der Alpen: Jahrb. Kaiserl.-Königl. Geol. Reichsanst., v. 21, 2, p. 189–210.

———, 1874, Faunengebiete und Faciesgebilde der Trias—Periode in den Ost—Alpen: Jahrb. Kaiserl.-Königl. Geol. Reichsanst., v. 24, 1, p. 81–134.

MÜLLER-JUNGBLUTH, W.-U., 1968, Sedimentary Petrologic Investigation of the Upper Triassic "Hauptdolomit" of the Lechtaler Alps, Tyrol, Austria, in Müller, G., and Friedman, G. M., Eds., Recent Developments in carbonate sedimentology in central Europe: Springer, p. 228–239.

———, 1970, Sedimentologische Untersuchungen des Hauptdolomites der östlichen Lechtaler Alpen, Tirol: Festbd. Geol. Inst., 300-Jahr-Feier Univ. Innsbruck, p. 255–308.

OHLEN, H. R., 1959, The Steinplatte Reef Complex of the Alpine Triassic (Rhaetian) of Austria [Ph. D. thesis]: Princeton, New Jersey, Princeton Univ., 123 p.

PILLER, W., 1976, Fazies und Lithostratigraphie des gebankten Dachsteinkalkes (Obertrias) am Nordrand des Toten Gebirges (S Grünau/Almtal, Oberösterreich): Mitt. Ges. Geol. Bergbaustud. Österr., v. 23, p. 113–152.

———, 1978, Involutinacea (Foraminifera) der Trias und des Lias: Beitr. Paläont. Österreich, v. 5, p. 1–164.

SANDER, B., 1936, Beiträge zur Kenntnis der Anlagerungsgefüge. (Rhythmische Kalke und Dolomite aus der Trias). I. und II.: Tschermak Mineral. Petrogr. Mitt., v. 48, p. 27–139, p. 141–209.

SCHÄFER, P., AND SENOWBARI-DARYAN, B., 1978, Die Häufigkeitsverteilung der Foraminiferen in drei oberrhä-tischen Riff—Komplexen der Nördlichen Kalkalpen (Salzburg, Österreich): Verh. Geol. Bundesanst., 1978, 2, p. 73–96.

SIEBER, R., 1933, Paläobiologische Untersuchungen an der Fauna der Rötelwand-Riffmasse in der nördlichen Osterhorngruppe (Salzburg): Anz. Österr. Akad. Wiss., Mathem.-Naturwiss. Kl., 1933.

———, 1934, Weitere Ergebnisse paläobiologischer Untersuchungen an der Fauna der rhätischen Riffkalke der Rötelwand (Osterhorngruppe, Salzburg) und anderer rhätischer Riffgebiete der Nordalpen: Anz. Österr. Akad. Wiss., Mathem.-Naturwiss. Kl., 1934.

———, 1937, Neue Untersuchungen über die Stratigraphie und Ökologie der alpinen Triasfaunen. I. Die Fauna der nordalpinen Rhätriffkalke: Neues Jahrb. Min. Geol. Pal., B, v, 78, p. 123–188.

SPECKMANN, P., AND ZANKL, H., 1969, Das Steinerne Meer—geologischer Zeuge einer tropischen Flachsee: Jb. Österr. Alpenverein, v. 94, p. 85–92.

TOLLMANN, A., 1969, Tektonische Karte der Nördlichen Kalkalpen. 2. Teil: Der Mittelabschnitt: Mitt. Geol. Ges. Wien, v. 61 (1968), p. 124–181.

———, 1976, Analyse des klassischen nordalpinen Mesozoikums: Deuticke, Wien, 580 p.

———, 1978, Bemerkungen zur Frage der Berechtigung der rhätischen Stufe: Österr. Akad. Wiss., Schriftenr. Erdwiss. Kommiss., v. 4, p. 175–177.

VORTISCH, W., 1926, Oberrhätischer Riffkalk und Lias in den nordöstlichen Alpen. I. Teil: Jahrb. Geol. Bunde-sanst., v. 76, p. 1–64.

———, 1927, Oberrhätischer Riffkalk und Lias in den nordöstlichen Alpen. II. Teil: Jahrb. Geol. Bundesanst., v. 77, p. 93–122.

WIEDMANN, J., 1974, Zum Problem der Definition und Abgrenzung von Obernor (Sevat) und Rhät: Österr. Akad. Wiss., Schriftenr. Erdwiss. Kommiss., v. 2, p. 229–235.

WILSON, J. L., 1975, Carbonate Facies in Geologic History: Springer-Verlag, New York, 471 p.

ZANKL, H., 1967, Die Karbonatsedimente der Obertrias in den nördlichen Kalkalpen: Geol. Rundschau, v. 56, p. 128–139.

———, 1969, Der Hohe Göll. Aufbau und Lebensbild eines Dachsteinkalk-Riffes in der Obertrias der nördlichen Kalkalpen: Abh. Senckenberg. Naturforsch. Ges., v. 519, p. 1–123.

———, 1971, Upper Triassic carbonate facies in the Northern Limestone Alps, in Müller, G., Ed., Sedimentology of parts of Central Europe: guidebook, p. 147–185.

ZAPFE, H., 1962, Untersuchungen im obertriadischen Riff des Gosaukammes (Dachsteingebiet, Oberösterreich). IV. Bisher im Riffkalk des Gosaukammes aufgesammelte Makrofossilien (exkl. Riffbildner) und deren strati-graphische Auswertung: Verh. Geol. Bundesanst., 1962, p. 346–361.

SEPM Special Publication No. 30, p. 291–359, May 1981

# PALEOECOLOGY AND FACIES OF UPPER TRIASSIC REEFS IN THE NORTHERN CALCAREOUS ALPS

ERIK FLÜGEL
Institute of Paleontology
University Erlangen-Nürnberg
Erlangen, West Germany

## ABSTRACT

In the Northern Calcareous Alps of Austria and Bavaria, Upper Triassic reefs are known from the Carnian (parts of the Wetterstein Reefs; Tisovec Limestones), and from the Norian and Rhaetian (Dachstein Reef Limestones; "upper Rhaetian" reef limestones; Kössen coral limestones). The Dachstein reefs developed predominantly on the southern exposed platform edges. The upper Rhaetian Reefs were formed upon shoals within the relatively shallow Kössen Basin (Rötelwand, Feichtenstein, Gruberalm), near the inner boundary of the Dachstein Platform (Steinplatte), or upon the Dachstein Platform (Adnet).

Dachstein Reefs and upper Rhaetian Reefs can be compared with respect to their generic and specific composition of the framebuilding biota, but striking differences are evident with regard to constructional types. Upper Rhaetian Reefs seem to have been developed with an initial mud-mound stage followed by a second stage in which an ecological reef formed in the turbulent zone. In the Dachstein Reefs only the second stage appears to be represented, and the distinction between a central reef area (upper reef-crest and reef-flat) with various framebuilding communities, a fore-reef slope with coarse reef breccias, and an extended back-reef area with open and restricted lagoons, is much more pronounced.

The organisms involved in the construction of the primary and secondary framework of Upper Triassic reefs consist of sessile foraminifers, segmented and non-segmented calcisponges, hydrozoans, corals, bryozoans, tabulozoans, calcareous algae, and many *microproblematica*. Reef dwelling organisms are vagile foraminifers, brachiopods, gastropods, lamellibranches, a few ammonites, serpulids, crustaceans, ostracodes, echinoderms, fishes, reptiles, and some algae. Our knowledge of the reef biota is strongly biased with respect to the framebuilding organisms, which have been studied in great detail during the last few years.

The most important framebuilders are corals and calcisponges, followed by hydrozoans and solenoporacean algae. Corals and calcisponges are generally restricted to different parts of the reefs, and in different zones of water energy. A critical review of the reef building organisms from the upper Rhaetian and Dachstein Reefs reveals great difficulties in their systematical treatment, especially corals, "hydrozoans," bryozoans, and "tabulozoans," but also clearly indicates the very strong facies control of the reef biota.

This facies control is expressed by the unique distributional patterns of the foraminifers, calcareous algae, and the *microproblematica*, by which different environments (and facies units) can be recognized. Another hint as to facies control is furnished by a rather regular distribution of the secondary framebuilders, and by the zonal distributional patterns of the various reef communities.

Using distributional patterns and microfacies types of the limestones, 12 "facies units" can be differentiated within the Upper Triassic reef and platform carbonates: these consist of restricted shelf-areas with tidal-flats, open-shelf environments, areas of winnowed edge sands, and reef complexes. The lateral arrangement of these facies units can change depending upon the paleogeographical setting in which the reefs were formed.

In spite of some success in understanding upper Rhaetian Reefs, some important questions still have to be resolved. These include the relationships between the Wetterstein Limestone reefs and the Dachstein Limestone reefs, constructional style of the larger Dachstein Reefs, and the evolution of the reef building communities through time. The latter is one of the topics which is presently being studied by the "Erlanger Reef Research Group," comparing Upper Triassic reefs in the Alps with those in Sicily, Solvenia, and Greece.

## INTRODUCTION

The reefoid character of portions of the Upper Triassic carbonate sequence of the Northern Calcareous Alps was recongized as early as 1850. This is shown by the use of old stratigraphic terms such as "Lithodendron-Kalk," or "Hochgebirgs-korallenkalk," for Upper Triassic limestones rich in dendroid corals. These limestones had been interpreted as ancient reefs by Stur (1871, p. 417), who distinguished a "zone with coral reefs" with-

in the huge Dachstein Limestone masses exposed in the Austrian Salzkammergut area. Even though one important group of reef building organisms, e.g., the corals from the Zlambach Beds and from the Dachstein Limestones, had been described in detail by Frech (1890), investigations of Triassic reefs proceeded slowly beginning with the famous Cordevolian (lowermost Upper Triassic) reefs in Tyrol; these were the subject of the classical monographs by von Richthofen (1860), Mojsisov-

FIG. 1.—Photograph of Gosaukamm Range near Gosau, Upper Austria, showing Upper Triassic (predominantly Norian) Dachsteinkalk Reef Limestones of the central reef area.

ics (1879), and still much later, Bosellini and Rossi (1974).

Studies of the Upper Triassic reefs in the Northern Calcareous Alps began with the monographic work of Wähner (1903) of the Sonnwend Mountains in Tyrol. In this work reef dwelling and reef building organisms from the upper Rhaetian Reef limestones were noted along with sedimentary structures visible in the field, along with distinctive petrographic and characteristics observed in thin-section. Wähner recognized the importance of non-coral organisms like calcareous sponges, hydrozoans, and various *microproblematica*, as framework contributors. Wähner's student Vortisch in 1926 published a very accurate investigation of the upper Rhaetian Steinplatte Reef near Waidring at the Tyrolian-Bavarian border. This reef has become one of the critical locations for facies analysis of Triassic reefs in the Alps (see Ohlen, 1959; Piller and Lobitzer, 1979; Piller, this volume). The first comprehensive study of the biota and construction of an Upper Triassic reef was by Sieber (1937), who described the Rötelwand Reef near Hallein, Salzburg.

After World War II, a new impulse for the study of Triassic reefs was triggered by the work leading to the thesis by Ohlen (1959), who reinvestigated the Steinplatte Reef with current sedimentological and paleontological methods. The Steinplatte Reef model has had a strong influence on the analysis of other upper Rhaetian Reefs both in the Northern Tyrol and Bavaria (Fabricius, 1960, 1966), and on several Dachstein Limestone Reefs (Sauwand near Gusswerk, Styria, E. Flügel and E. Flügel-Kahler, 1963; Hoher Göll, Berchtesgaden Alps; Zankl, 1969). These reefs as well as the

Dachstein Limestone Reefs described by Büchner (1973) from the Gesäuse Mountains in Styria and by H. Lobitzer (1974) from the Hochschwab region, Styria, exhibit distinct zonation patterns representing rather narrow central reef belts, extended back-reef zones and often a tectonically amputated or modified fore-reef slope.

Paleontological investigations of framebuilding organisms and the study of various carbonate microfacies types offered new possibilities for a more detailed interpretation of reef development. Since 1960, fossils from the Dachstein Limestone Reef of the Gosaukamm Range (Fig. 1) have been studied by E. Flügel and by Zapfe (see references). These investigations and the studies by the "Erlangen Reef Research Group" during the last few years has resulted in the recognition of many new species and genera, especially the calcareous sponges, corals, and calcareous algae. The "Reef Research Group" established at the Paleontological Institute of Erlangen-Nürnberg University consists of Erik Flügel, Wolf-Christian Dullo, Massoud Sadati, Priska Schäfer, Baba Senowbari-Daryan, and Detlef Wurm. The overall goals of these group investigations are the detailed descriptions of the fauna and flora of both upper Rhaetian and Dachstein Limestone Reefs, together with the paleoecologic interpretation of the reef complexes based primarily on microfacies analyses (see E. Flügel, 1972, 1978). Reefs which have been investigated, or which are presently being studied, are the Rötelwand and the Adnet Reefs, the Feichtenstein and the Gruberalm Reefs (all near Hallein, Salzburg; upper Rhaetian Reefs), the Gosaukamm Range, Upper Austria, the Kalbling and the Schildmauer Reefs in the Gesäuse Mountains, Styria, and the Hohe Wand Reef southwest of Vienna (cf. Fig. 2). All of these reefs appear to have been developed in marginal zones on Dachstein Limestone Platforms. In the course of these investigations we have utilized the methodical approach, using distributional patterns of foraminifers as indicators of reef zones and ecological niches, and this has been found to be very promising. The value of this method was first demonstrated by Hohenegger and Lobitzer (1971) and Hohenegger (1974) for the Dachstein Limestone Reef of the Hochschwab, and by Hohenegger and Piller (1975a) for the bedded Dachetin Limestones of the Totes Gebirge, Upper Austria.

H. Lobitzer and W. Piller have reinvestigated the Steinplatte Reef (see Piller, this volume), and are presently working on the facies of the Dachstein Limestones in the area of the Loferer Steinberge, Leoganger Steinberge, and the Steinernes Meer, Salzburg. These areas are most important for the interpretation of "Lofer Cyclothems," which are widely known as a result of the work of Fischer (1964).

FIG. 2.—Map showing location of Upper Triassic reefs of the Northern Calcareous Alps studied during the last twenty years; inset photographs are members of the "Erlangen Reef Research Group" (in clockwise direction), Priska Schäfer, Baba Senowbari-Daryan, Erik Flügel, Massoud Sadati, Wolf-Christian Dullo, and Detlef Wurm).

In summary, much information concerning paleontological composition, facies types, and the development of possible reef models has been gathered in the last few years. These data are strongly biased with respect to detailed paleontological investigations. Sedimentological studies, especially of reef diagenesis, apart from the classical investigation of "Anlagerungsgefüge" (depositional fabric) by Sander (1936) have been carried on by Zankl (1969, 1971) and his students (Gökdag, 1974; Mirsal, 1978).

## Scope of the Paper

This paper deals with the construction and development of Upper Triassic reefs in the Northern Calcareous Alps by focusing on the paleoecology and reef building potential of the biota. Difficulties in comparison of different reef complexes arise primarily because of the heterogenous status of available information. This especially is the case for paleontological investigations of the reef biota, which have only been studied systematically in a few upper Rhaetian Reefs (Schäfer and Senowbari-Daryan, this volume) and in a rather small Dachstein Limestone Reef (Dullo, 1979).

Many of the generalizations on the paleoecology of the reef biota are based, therefore, on these rather few studies. Additional information concerning the significance of reef building organisms is given in other papers such as those by Zankl (1969), E. Flügel and E. Flügel-Kahler (1963), and Lobitzer (1974), but the rather broad systematic fossil determinations are difficult to compare with results obtained by studies directed primarily at paleontological and microfacial details. On the other hand, those papers dealing with the Dachsteinkalk Reefs, as well as the work on the other upper Rhaetian Reefs like the Steinplatte, and the buildups in the Tyrolian-Bavarian Alps, are of great value for the analysis of general reef development.

Another difficulty is the still unresolved question of the age of some of the Upper Triassic reefs. Reef development certainly began during the Carnian, whereas most of the Dachstein Limestone Reefs (Göll, Gosaukamm, Gesäuse, and Hochschwab) seem to be Norian in age. The Rhaetian portion of the reefs is open to question, not only because of the scarcity of index fossils (e.g., ammonites or conodonts in the reef facies), but because of the controversial opinions regard-

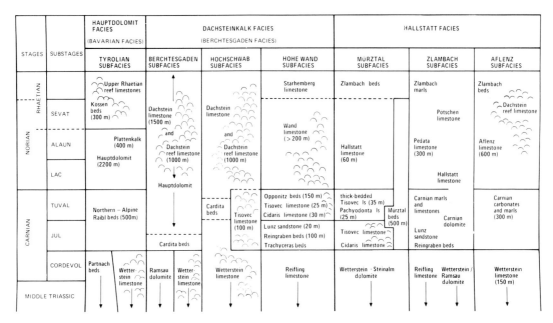

| STAGES | SUBSTAGES | HAUPTDOLOMIT FACIES (BAVARIAN FACIES) | DACHSTEINKALK FACIES (BERCHTESGADEN FACIES) | | | HALLSTATT FACIES | | |
|---|---|---|---|---|---|---|---|---|
| | | TYROLIAN SUBFACIES | BERCHTESGADEN SUBFACIES | HOCHSCHWAB SUBFACIES | HOHE WAND SUBFACIES | MURZTAL SUBFACIES | ZLAMBACH SUBFACIES | AFLENZ SUBFACIES |
| RHAETIAN | | Upper Rhaetian reef limestones | | | Starhemberg limestone | Zlambach beds | Zlambach marls | Zlambach beds |
| | SEVAT | Kossen beds (300 m) | Dachstein limestone (1500 m) and Dachstein reef limestone (1000 m) | Dachstein limestone and Dachstein reef limestone (1000 m) | Wand limestone (> 200 m) | | Potschen limestone | Dachstein reef limestone |
| NORIAN | ALAUN | Plattenkalk (400 m) Hauptdolomit (2200 m) | | | | Hallstatt limestone (60 m) | Pedata limestone (300 m) | Aflenz limestone (600 m) |
| | LAC | | Hauptdolomit | | | | Hallstatt limestone | |
| CARNIAN | TUVAL | Northern – Alpine Raibl beds (500m) | Cardita beds | Cardita beds Tisovec limestone (100 m) | Opponitz beds (150 m) Tisovec limestone (25 m) Cidaris limestone (30 m) Lunz sandstone (20 m) Reingraben beds (100 m) Trachyceras beds | thick-bedded Tisovec ls (35 m) Pachyodonta ls (25 m) Tisovec limestone Cidaris limestone | Murztal beds (500 m) | Carnian marls and limestones Carnian dolomite Lunz sandstone Reingraben beds | Carnian carbonates and marls (300 m) |
| | JUL | | | | | | | |
| | CORDEVOL | Partnach beds Wetterstein limestone | Ramsau dolomite Wetterstein limestone | Wetterstein limestone | Reifling limestone | Wetterstein – Steinalm dolomite | Reifling limestone Wetterstein / Ramsau dolomite | Wetterstein limestone (150 m) |
| | MIDDLE TRIASSIC | | | | | | | |

FIG. 3.—Chart showing accepted Upper Triassic stratigraphy in parts of the Northern Calcareous Alps (simplified after Tollmann, 1976); reef facies noted by symbol.

ing the range of the Rhaetian Stage (Tollmann, 1978; Mostler and Urlichs, 1979).

Important limitations to reconstructing the paleogeographical setting of Upper Triassic reefs are caused by severe Alpine orogenesis during which originally connected geographic areas were tectonically separated. Significant paleogeographical interpretations can only be made in those areas which belong to the same tectonic event. Other large scaled comparisons of the reef structures are rather speculative, but necessary as an impetus for future reef study.

### Regional and Geological Setting

During Alpine orogenesis (mainly Cretaceous and Tertiary time), Mesozoic sediments were subjected to intensive overthrusting. Like other larger tectonic uints, the uppermost unit, the "Oberostalpin," consists of different tectonic sub-units and nappes. In the middle and eastern part of the Northern Calcareous Alps there are three great tectonic sub-units, composed mainly of Mesozoic rocks and consisting from the base to the top, of the Bajuvarikum, Tirolikum, and Juvavikum. All the upper Rhaetian Reefs are found in the Tirolikum. The Dachstein Limestone Reefs belong to the Tirolikum (Göll, Hochkönig, Totes Gebirge), as well as to different nappes of the Juvavikum (Untersberg, Gosaukamm, Gesäuse, Sauwand, Hochschwab, Hohe Wand). Following the concept of Tollmann (1967), the reefs

in the Gesäuse Mountains, the Sauwand, the Hochschwab, and the Hohe Wand Reef, are all situated within the Murztal Nappe of the Juvavikum; the Gosaukamm and the Grimming Reefs belong to the Dachstein Nappe.

The regional distribution of those Upper Triassic reefs which have been studied more intensively is shown in Figure 2. Most of these reefs are situated in Austria, except for some of the upper Rhaetian Reefs in the Bavarian Alps (see Fabricius, 1966) and the Göll Reef at the border of Bavaria and Salzburg (Zankl, 1969).

### Stratigraphy

The Triassic System of the Tethyan region has been subdivided into several stages and substages (Fig. 3), which are defined according to the stratigraphic range of various ammonite species. The boundaries of the substages are still a matter of controversy. The same is true for the age determination of many stratigraphical units which have been proposed for the Alpine Triassic. A thorough review of the characteristics of these units is given by Tollmann (1976), and Figure 3 follows the classification and correlations proposed by this author. The generally accepted differentiation of the Alpine Triassic into three facies zones (Hauptdolomite and Dachsteinkalk facies both characterized by Upper Triassic platform carbonates with reefs, and the Hallstatt facies dominated by subtidal and basinal sediments) can be refined with respect to the development of the entire Triassic section. In contrast to the three main facies zones, which may attain dimensions of several hundred kilometers in length, the subfacies zones are relatively small. Originally, these subfacies zones were arranged in an east-west direction. In Figure 3, only those subfacies zones have been considered which show reefoid development or which are important for speculation concerning the relationships between the shallow water platform and reef carbonates, and basinal sediments.

The following rock/time-units are of importance in the discussion of Upper Triassic reefs in the Northern Calcareous Alps:

1. *Wetterstein Limestone.*—Mainly Middle Triassic (Ladinian) in age, but also developed in the lowermost Carnian (Cordevolian). The reefs of this formation are described by Brandner and Resch (this volume).

2. *Cidaris Beds.*—In the Dachsteinkalk and Hallstatt facies of the eastern part of the Northern Calcareous Alps, a sequence consisting of well-bedded limestones, argillaceous schists, and marls of Carnian age is known. The beds are generally rich in echinodermal fragments. In contrast to the "Northern Alpine Raibl beds" of the Hauptdolomite facies, the terrigenous content is not very important in these sediments. The *Cidaris* beds seem to have been deposited in rather small, shallow, subtidal basins within a shallow water carbonate platform (E. Flügel *et al.*, 1978). Sections rich in calcareous sponges, hydrozoans, algae, and *microproblematica* have been studied in the Gosaukamm Range, and in the Mürztal Alps, Styria. Fossils are found as bioclasts in thin-bedded limestones. The fauna seems to have had no framebuilding components, and consists chiefly of species known from the reef facies of the Wetterstein Limestone.

3. *Tisovec Limestone.*—Kollarova-Andrusova (1960 and 1967) proposed the name "Tisovec Limestone" for massive reef limestones similar in lithofacies to the Wetterstein Limestone, but Carnian in age. Comparable limestones are known from the Dachsteinkalk and Hallstatt facies of the Northern Calcareous Alps as well as from the Western Carpates (Tollmann, 1972 and 1976; Lein and Zapfe, 1971; Dullo, 1980). In the Gesäuse Mountains, the Tisovec Limestone is intercalated between Dachstein dolomite and Dachstein Reef Limestones containing abundant dasycladacean algae. In other localities, especially in the Styrian Mürztal Alps and the Gosaukamm, the Tisovec Limestone is represented as a grainstone with abundant dasycladacean and solenoporacean bioclasts, or with large algal oncoids. Corals and sponges are known, but are not well documented. The overlying beds are Hallstatt Limestones, especially in those sections situated in the Mürztal

subfacies zone. The base of the Tisovec Limestone consists of *Cidaris* beds, or of Carnian arenaceous shales (Schöllnberger, 1974). The age of the Tisovec Limestone is generally regarded as Tuvalian and Julian.

4. *Aflenz Limestone.*—These thin-bedded, dark-colored limestones with cherts, are found in the region south of the Hochschwab Mountains, Styria. Here they are interpreted as a basinal facies which interfingers with the fore-reef zone of the Hochschwab Dachstein Limestone Reef. Lobitzer (1974) noted small patch-reefs in these limestones. Framebuilders are chiefly platy calcisponges and bryozoans; solenoporacean algae and solitary corals are relatively rare. This fauna as well as the microfauna (foraminifers) are similar to that of the central reef areas in the Dachstein Limestone Reef. The age of these small reefs is Norian.

5. *Dachstein Limestone.*—During Norian and Rhaetian times an extensive carbonate platform with peripheral reef zones was constructed, and the sediments can be differentiated according to depositional environment. Three types of the Dachstein Limestone are of importance:

a. *Bedded Dachstein Limestones.*—Well-bedded, cyclically formed limestones (sometimes dolomites) consisting of several hundred "Lofer cycles" (Fischer, 1964), which indicate fluctuating deposition in inter- and supratidal environments (algal laminites with fenestral fabrics, "loferites"), and in shallow subtidal environments (wackestones with megalodontid lamellibranchs, gastropods, and echinoderms; corals of the *Thecosmilia* type may also be present), followed by an emergent phase recognizable by red or green residual clays and reworked limestone clasts. These bedded limestones may be regarded, with reference to distance to the Dachstein Reef complexes, as a "far-reef" back-reef zone (Zankl, 1971).

b. *Dachstein Reef Limestone.*—Reef complexes were developed predominantly on the southern edge of the Dachstein Limestone platforms (see maps in Zapfe, 1962, and Tollmann, 1976, p. 220), which show differentiation into fore-reef, central reef (upper reef-crest and reef-flat), and back-reef areas. Biota, facies and construction of these reefs are the main topic of the present paper.

c. *Wand Limestone.*—A massive reef limestone of probable Norian age, which is different by lithological and paleontological characteristics from the Dachstein Reef Limestones. This unit is known from the Hohe Wand near Wiener Neustadt, Lower Austria, and yields lamellibranchs and ammonites which indicate connection with the basinal facies. These interconnections with more basinal facies types, and the microfossil content (especially foraminifers and algae), are comparable to the "Furmanec Limestone" described from the Slovakian Carpathians (Kollarova-Andrusova, 1960; Bystricky, 1973). According to Tollmann (1972 and 1976), similar limestones are known from the Emmerberg region near Wiener Neustadt (Plöchinger, 1967), the Hochschwab Mountains (Lobitzer, 1974), and from the Untersberg region near Salzburg (Bittner, 1884). The Hohe Wand Reef is presently the subject of a detailed microfacies study by Sadati (Erlangen). According to preliminary results, the biolithite facies of the central reef area is characterized by hydrozoans, bryozoans, tabulozoans, inozoan and sphinctozoan calcisponges as well as colonial corals. Many framebuilders are unfortunately strongly recrystallized. The distribution of the foraminifers and *microproblematica* (e.g., *Spiriamphorella*, see Figs. 11C and D) within the reef patches seems to be similar to that in upper Rhaetian and Dachstein Limestone reefs of the Gesäuse.

6. *Kössen Beds.*—In the area of the Hauptdolomite facies west and north of the Dachsteinkalk facies, shallow subtidal basinal sediments were deposited, and are now know as "Kössen beds" (type locality is the Weiss lofer Gorge near Kössen, Tyrol). Several lithofacies and biofacies types can be recognized (Fabricius, 1966). These include Kössen Marls (dark marls rich in pyrite, with a low diversity fauna interpreted as gyttialike deposits, Fabricius, 1961); Kössen Limestones (dark-colored limestones mainly calcilutites and bioclastic or oolitic calcarenites); and Kössen coral limestones (thick-bedded limestones with abundant colonies of phaceloid corals). These coral limestones represent biostromes, which often are truncated by erosional surfaces (Zankl, 1971). The *Thecosmilia* biostromes are characterized by a large areal distribution, by relatively little reef derived detritus in the surrounding beds, and by a general lower diversity of framebuilding organisms than is present in the upper Rhaetian Reefs. Since the biota of the Kössen Reef Limestones has not been studied in detail this statement is speculative.

7. *Upper Rhaetian Reef Limestones ("Oberrhätriffkalk").*—In the upper part and at the top of the Kössen Formation, massive to thick-bedded, light-grey to white reef limestones are well developed in the Salzburg region as well as in Tyrol, Vorarlberg, and in the Bavarian Alps. Fabricius (1966) has published a thorough analysis of the various facies types of these reef limestones, and these have been the subject of intensive studies during the last few years. Similar to the Dachstein Limestone reefs, these rather small reefs also show a zonation indicating clearly separated paleoecological environments within the

central reef area, fore-reef, and back-reef. This zonation is only developed in the "framework stage" of the reefs, whereas in the quiet water mud-mound stage facies differentiation is rather negligible. The age of these reefs is regarded as "Rhaetian" or (uppermost) Norian.

8. *Zlambach Beds.*—In the Salzkammergut area the basinal Hallstatt Limestones interfinger with a predominantly marly sequence of Sevatian and Rhaetian age. These Zlambach Beds also overlay basinal limestones (e.g., Pötschen Limestone), or are found above detrital Dachstein Reef Limestones (Gosaukamm). Therefore the Zlambach Beds can be interpreted as sediments deposited in rather shallow, subtidal basins adjacent to Dachstein Limestone Reefs. Of special interest are coral bearing marls ("Korallen Mergel," e.g., at the locality Fischerwiese near Alt-Aussee, Styria) with calcarenites and some calcilutites. According to Bolz (1974) four microfacies types can be distinguished, which show broad grain-size spectra and a different composition with regard to skeletal grain types. Some bioclastic packstones contain abundant angular to subangular fragments of corals, spongiomorphid hydrozoans, bryozoans, molluscs, echinoderms, and calcareous algae. These fossils correspond with those of the intercalated marls, especially where isolated corals can be found that show excellent preservation. Since the faunal and floral aspect of this coarse biodetritus is completely different from that of Dachstein limestones, it seems reasonable to assume that the bioclasts have been derived from shallow water mounds within a subtidal area in front of the fore-reef zones of the Dachstein Limestone reefs (Flügel, 1962; Pistotnik, 1974).

### REEF DEVELOPMENT IN THE TRIASSIC OF THE NORTHERN CALCAREOUS ALPS

All Triassic reefs known from the Northern Calcareous Alps are of Middle and Upper Triassic age. The absence of Lower Triassic (Scythian) buildups is in accordance with the world-wide scarcity of potential reef building organisms during this time. The reasons for the absence of Scythian corals may be related to evolutionary processes which could have been connected with the westward opening of the Tethys and overall enlargement of the marine area (Cuif, 1979; Oliver, 1979).

The oldest Triassic reefs in the Northern Calcareous Alps were described from the Anisian Steinalm Limestone of the Mieminger Mountains, Tyrol (Miller, 1965). The dimensions of these thick-bedded to massive reef limestones are rather small (10 to 100 meters in thickness, with a lateral extension up to several hundred meters). Only sparse colonial corals associated with echinoderm fragments have been found in these rocks. The start of reef development during upper Anisian time is shown by the existence of reefs in other alpine regions (e.g., small buildups in the Upper Alpine Muschelkalk of the eastern Karawanken Mountains, Bauer, 1970; isolated boulders of reef limestones in the upper Anisian Peres Beds of the Olang Dolomites, South Tyrol, Bechstedt and Bradner, 1970; and small patchreefs in the Lienzer Dolomites, eastern Tyrol, Bradner, 1972).

Starting with the Middle Triassic of the Northern Calcareous Alps, especially the Drauzug Range, and in the Southern Alps, where a wide unstable area was formed. Here, sedimentation patterns were affected by changing subsidence, synsedimentary block-faulting, and sea level changes, due to the proximity of a mobile rifting area in the southeast (Bechstädt *et al.*, 1978). Presumably created by tectonic uplift of the Steinalm Limestone Platform, an area of relief was formed upon which patch-reef sequences developed adjacent to basinal regions. In addition, extended reef complexes like the Hafelekar Reef complex near Innsbruck, Tyrol (Bradner and Resch, this volume), show distinct zonation patterns attributed to different water energies. These Wetterstein Limestone Reefs of Ladinian and Coredevolian age have been studied by Ott (1967 and 1972), who emphasized the role of the calcisponges, and the problematic blue-green alga *Tubiphytes* as reef building organisms. This was further emphasized by Wolff (1973), Colins and Nachtmann (1974), and by Bauer (1970). The facies interpretations in most of these studies has been strongly influenced by the facies model developed by Schneider (1964) for the Middle Triassic platform to basin sediments. As is evident from the study by Bradner and Resch (this volume) and by the comparison of various Wetterstein Limestone reefs (Table 1), this scheme has to be modified with respect to the relationship between the reefs and the basins, and with respect to the internal composition of the reefs. In contrast to the model, many Wetterstein Limestone Reefs seem to be characterized by rather gentle fore-reef slopes. All the Wetterstein Limestone Reefs which have been carefully studied are of a regressive type. This is indicated by the general overgrowth of the underlying basinal sediments (Partnach Beds or basinal carbonates) by the Wetterstein Limestone Platform. As can be shown in the Karwendel Mountains, the lagoonal areas of the Wetterstein Limestone reef complexes were enlarged during Ladinian and Cordevolian time with most of the reefs covered by well-bedded lagoonal Wetterstein Limestone. Microfacies and sedimentary structures of these limestones and dolomites indicate that the sediments were deposited dominantly in intertidal and shallow water subtidal environments. Supratidal facies types, and vadose diagenetic patterns are not as well known (Bechstädt, 1974).

The reef building biota of the Wetterstein Limestone Reefs consists of calcisponges, colonial corals, and *Tubiphytes,* a form which seems to have a sediment-binding capability. Not all of the Wetterstein Limestone Reefs may have been formed primarily by calcisponges and *Tubiphytes,* as suggested after initial detailed studies of the reefs (Ott, 1967 and 1972). Corals may be at least of similar importance (Bradner and Resch, this volume). The fauna and the algal flora of the Wetterstein Limestone Reefs is conspicuously different from the biota of the Norian and Rhaetian reefs.

Stronger similarities can be found between the fauna of the Wetterstein Reef Limestones and those of the Carnian *Cidaris* Beds. But in contrast to the Wetterstein Limestone, segmented calcisponges and various encrusting *microproblematica* (E. Flügel *et al.*, 1978) of the *Cidaris* Beds do not contribute to a true-reef framework, but are associated with brachiopods, echinoderms, and some algae, indicating a typical shallow water subtidal environment. These potential reef builders are presently being studied by Dullo (Erlangen), in order to determine whether these organisms have lost their ability to construct reefs, or whether the lack of reefs in the *Cidaris* Beds is controlled by environmental parameters such as unfavorable water depth, or strong influx of terrigenous material following a marked change in the depositional history during the "Reingraben juncture" (Schlager and Schöllnberger, 1974).

Another Carnian rock-unit, rich in framebuilding fossils such as calcisponges and corals, but also with abundant dasycladacean algae (e.g., *Poikiloporella duplicata* (Pia)), is the Tisovec Limestone, which lithofacially can be compared with Middle Triassic or Upper Triassic reef limestones. This is indicated by the older names "Carnian Wetterstein Limestone," or "Carnian Dachstein Limestone." The Tisovec Limestone is of major importance for the unresolved question of relationships between the Ladinian-Carnian and the Norian-Rhaetian reef biotas.

During the Norian, extensive shallow water carbonate platforms were formed which show a differentiation into marginal reef zones with reef complexes and skeletal sand-shoals (Dachstein Reef Limestones), and lagoons with cyclic sedimentational patterns (bedded Dachstein Limestone). On more restricted parts of the platform dolomites formed in predominantly intertidal environments (Müller-Jungbluth, 1970). The basinal facies adjacent to Dachstein Limestone Reefs is represented by a rather thin sequence of vari-

TABLE 1.—CHARACTERISTICS OF WETTERSTEIN LIMESTONE REEFS IN THE NORTHERN CALCAREOUS ALPS

| | Karwndel Mountains (E. Ott, 1967, 1972) | Hafelkar Reef Complex near Innsbruck (Bradner & Resch, 1980) | Wendelstein, Bavaria (Wolf, 1973) | Dobratsch, Drauzug (Colins & Nachtmann, 1974) | Northern Karawanken Carinthia (F. K. Bauer, 1970) |
|---|---|---|---|---|---|
| Reef type | barrier-reefs forming atoll-like structures | patch-reefs at the base, platform edge in the upper part | — | barrier-reef | atoll-like distribution |
| Thickness of the reef carbonate | about 500 m (reef-core), up to 1700 m (lagoon) | patch-reef 30–40 m, | up to 350 m (reef-core) | reef-core up to 700 m, fore-reef up to 100 m | reef-core about 500 m |
| Area of the reef-flat | about 4 to 8 km$^2$ | — | — | area of reef patches on the reef-flat 30–40 m$^2$; width of the reef-flat 800–1000 m, width of the near reef lagoon 800–1000 m | ? |
| Base of the reefs | Partnach marls and limestones (basinal beds) | thick-bedded basinal sediments (Reifling Ls.) | "Übergangskalk," dark colored bedded limestones or Partnach beds (basin facies) | "Fellbacher Ls.," basinal carbonates | Partnach beds, basinal facies |
| Overlying beds | bedded Cordevolian Wetterstein Limestone | bedded Cordevolian Wetterstein Limestone | bedded Wetterstein Limestone | bedded Wetterstein Limestone (lagoonal) | bedded Wetterstein Limestone (lagoonal) |
| Relationship reef/basin | slope about 20–30°, with reefs over-growing basinal sediments | very gentle slope | not described | the reefs overgrow basinal sediments | not described |

Zonation patterns of the reef complex:

| | Karwndel Mountains (E. Ott, 1967, 1972) | Hafelkar Reef Complex near Innsbruck (Bradner & Resch, 1980) | Wendelstein, Bavaria (Wolf, 1973) | Dobratsch, Drauzug (Colins & Nachtmann, 1974) | Northern Karawanken Carinthia (F. K. Bauer, 1970) |
|---|---|---|---|---|---|
| fore-reef | limestone breccias, composed of fragments, of reef builders and lithified reef rock; sparry carbonate cements; many crinoids and solenoporaceans | rubble and block rubble in a well-sorted grainstone matrix, cavities partly filled with graded sediments | dark colored calcirudites | limestone breccia and fine-grained reef debris, intercalated with basinal limestones | poorly bedded limestones |
| reef-core | autochthonous framebuilders, together with sediment-binding organisms, reef detritus, and chemical internal fillings | autochthonous framebuilders, reef debris, calcarenites | autochthonous framebuilders reef debris | autochthonous framebuilders, reef/debris ratio 1:9 | autochthonous framebuilders in small areas, arenitic reef detritus; spar-filled reef cavities; limestones and dolomites |
| near-reef lagoon | well-bedded calcilutites and calcarenites with intercalations of reef detritus, especially dasyclads (*Teutloporella*) | sand shoals with tepee structures | bindstones with fenestral fabrics, stromatolites, grapestones, filamentous blue-green algae, dasyclads, gastropods | two zones: limestones with large oncoids and gastropods, and stromatolithic limestones | |
| far-reef platform | well-bedded limestones with dasyclads | not described | well-bedded dolomitic limestones, laminated, with dasyclads | limestones with coated-grains, peloids and molluscs; special dasyclads (*Teutloporella*) | bedded limestones with intercalations of dolomites; stromatolite bindstones with oncoids, indications of subaerial exposure (vadose pisoids); limestones with fenestral fabrics |

TABLE 1.—CONTINUED

| | Karwndel Mountains (E. Ott, 1967, 1972) | Hafelkar Reef Complex near Innsbruck (Bradner & Resch, 1980) | Wendelstein, Bavaria (Wolf, 1973) | Dobratsch, Drauzug (Colins & Nachtmann, 1974) | Northern Karawanken Carinthia (F. K. Bauer, 1970) |
|---|---|---|---|---|---|
| Reef building biota | main elements: *Tubiphytes*, segmented calcisponges; in some places phaceloid corals; bryozoans, hydrozoans, solenoporacean and porostromate algae | main elements: corals, *Tubiphytes*, calcisponges, several biocoenoses: Codiaceae b., *Tubiphytes* b., Calcisponges b., with Inozoa and Sphinctozoa; *Thecosmilia-Calcispongia* b., Coal-echinoderm *Tubiphytes* b., encrusting elements: algae, *Ladinella* | main elements: colonial corals, calcisponges; various biocoenoses: "*Thecosmilia*"-b., *Holocoelia* b., calcisponges b. Encrusting secondary framebuilders: porostromate algae, tabular sponges, *Tubiphytes*, *Ladinella* | main elements: corals, calcisponges, *Tubiphytes*, secondary framebuilders: spongiostromate algae | main elements: calcisponges, corals, *Tubiphytes*; in some places algae |
| Reef dwelling biota | lamellibranchs, gastropods, echinoderms foraminifers | molluscs, echinoderms, ostracodes, foraminifers | gastropods, lamellibranchs; rare echinoderms, foraminifers | gastropods, lamellibranchs, foraminifers | not described |

colored, bedded limestones and marls (Hatlstätt Limestone) with pelagic faunas. These sediments seem to have been deposited in two or three narrow channels, and in small basins between the various sized platforms (for a discussion of this rather controversial subject see Tollmann, 1976, and Lein, 1975).

Towards the end of Upper Triassic time the Dachstein and Hauptdolomite Platforms were destroyed, and the Hauptdolomite areas were differentiated into basins with terrigenous influx (Kössen Beds "terrigenous mud facies"; see Schäfer and Senowbari-Daryan, this volume), and into areas of small reefs (upper Rhaetian Reef Limestone), which are discussed in detail in the following sections.

Summing up the information concerning reef development in the Triassic of the Northern Calcareous Alps, we can offer the following conclusions:

1. first reefs are known from the upper Anisian as small buildups probably formed in somewhat deeper water environments;

2. these patch-reefs gave rise to the formation of various large reef complexes (Wetterstein Reefs), which overgrew basinal sediments;

3. the reef builders of the Wetterstein Limestone reefs are also known from non-reef shallow water carbonates intercalated within the predominantly clastic Carnian Series (e.g., *Cidaris* Beds);

4. the Ladinian/Carnian reef fauna, and the biota of the Norian and Rhaetian Reefs, show striking differences with regard to species;

5. the transition between these two reef biotas is not well understood and further study of the Carnian Tisovec Limestone may offer an answer to this problem;

6. Norian and Rhaetian Dachstein Limestone Reefs formed at the edges of extensive carbonate platforms, and upper Rhaetian Reefs formed within the Kössen Basin at the margin of the Kössen Basin and also on the Dachstein Limestone Platform, can be compared with respect to the specific and generic composition of their reef biotas.

### UPPER TRIASSIC REEF BIOTAS

The following sections deal with various organisms found in upper Rhaetian and Dachstein Limestone Reefs and their importance in the construction of these reefs. This summary is based on the paleontological investigation of upper Rhaetian reefs in Salzburg (Schäfer and Senowbari-Daryan, this volume), and Tyrol (Piller, this volume) as well as of the Dachstein Reefs, especially of the Göll Reef (Zankl, 1969), and the Gosaukamm and Gesäuse Reefs. Fossils from these reefs have been described by Büchner (1970, and 1973), Dullo (1979), Fabricius (1966), E. Flügel (1960, 1962a, 1962b, 1964, 1967, 1972b, and 1975), E. Flügel and E. Flügel-Kahler (1963), Hohenegger (1974), Hohenegger and Lobitzer (1971), Hohenegger and Piller (1975a, 1975b, 1975c, and 1977), Lobitzer (1974), Lobitzer and Piller (1979), Ohlen (1959), Piller (1976 and 1978), Schäfer (1979), Schäfer and Senowbari-Daryan (1978a, 1978b), Senowbari-Daryan (1978a, 1978b, and 1978c), Senowbari-Daryan and Schäfer (1978, 1979a, 1979b), Zankl (1965, 1969, and 1971), and Zapfe (1962, 1963, 1964, 1965, 1967a, and 1967b).

The organisms of the reef environments can be differentiated according to relative importance in the construction of organic buildups, and their importance in biological destruction. This information is strongly biased with respect to framebuilding organisms; molluscs, brachiopods, and echinoderms need further study.

### Constructive Organisms

The organisms involved in the construction of Upper Triassic reefs in the Northern Calcareous Alps consist of various sessile foraminifers, calcisponges and perhaps sclerosponges, hydrozoans, corals, bryozoans, tabulozoans, as well as blue-green, green, and red algae. In addition many *microproblematica* contribute to the formation of the reef framework mainly as secondary framebuilders, and partly as sediment-binding agents.

The following discussion of the main groups of reef constructing organisms aims to summarize our present knowledge of the framebuilding biota, and the ecological importance of various groups. Foraminifers are not discussed here because of their greater importance as reef dwelling forms rather than as reef builders (see chapter "Distributional Patterns of Micro-organisms").

As can be seen in Tables 2 to 5 there are no essential differences between species found in the Dachstein Limestone Reefs and in the upper Rhaetian Reefs. Comparing faunas of both units on species level, the Simpson Coefficient for calcisponges is 54.5, for corals 86.2, for "hydrozoans" 88.9, and for calcareous algae 85.7. The difference seen in the calcisponges is clearly reflected by the lack of intensive studies of this group, especially in the Dachstein Limestone Reefs. At generic levels, the similarity coefficients indicate very high coincidences; e.g., between 94 and 100 for calcisponges, corals, and hydrozoans, and 77.7 for the calcareous algae. The similarity coefficients for species of the framebuilding organisms known from the upper Rhaetian Adnet, Rötelwand, Feichtenstein, and Gruberalm Reefs range between 66.6 and 100 for the corals, and between 57.1 and 92.8 for the calcisponges. These high similarities can be explained by the rather close proximity of these

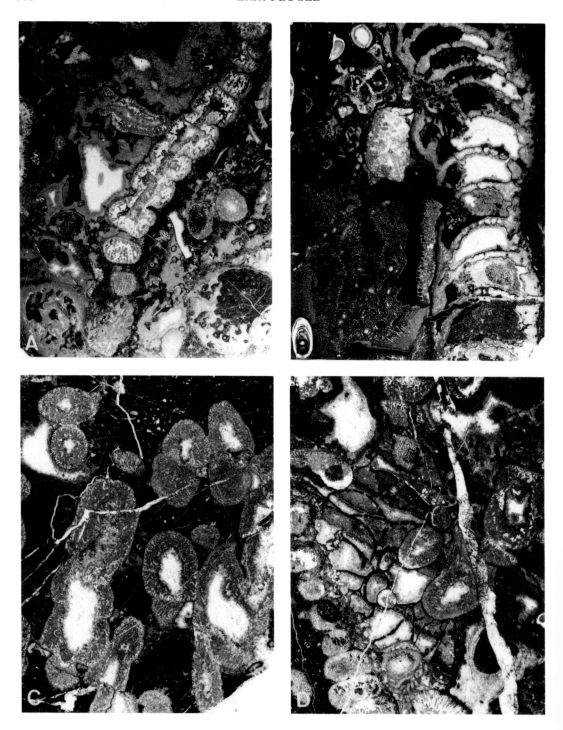

FIG. 4.—Thin-section photomicrographs of various framebuilding calcisponges (Sphinctozoa) from upper Rhaetian Reefs. A. *Paradeningeria weyli* Senowbari-Daryan and Schäfer, with extensive overgrowths of tabulozoans,
sessile foraminifers and various *microproblematica*. Interstitial cavities are filled with fibrous and blocky cements;
Feichteinstein, in the protected part of the central reef area, ×1.5. B. *Colospongia* sp. intergrown with tabulozoans
(at the left of the sponge), spongiostromate algal crusts (at the right), *microproblematica* and serpulids; on the left
is a hydrozoan colony (*Disjectopora* sp.). This community characterizes the inner parts of the reef patches at

reefs and by their corresponding types of reef construction (Schäfer and Senowbari-Daryan, this volume). On the other hand, it is interesting to see that different paleogeographical settings of the reefs (Rötelwand: reef development preceded by a mud-mound stage; and Adnet: reef development initiated on a carbonate platform), seem to have had no influence on the systematic composition of the framebuilding organisms, but only to their diversity.

*Calcisponges (Fig. 4).*—Most calcisponges belong to new species and genera, this is especially so for the Sphinctozoa. These segmented calcisponges are the main framebuilders in the more protected parts of upper Rhaetian Reefs, and probably in many Dachstein Limestone Reefs. In contrast to the Dachstein Limestone Reefs, the Sphinctozoa seem to dominate over the non-segmented Inozoa present in the upper Rhaetian Reefs. The quantitative importance of calcisponges as primary framebuilders is shown in Figure 22.

In the Dachstein Limestone Reefs as well as in the upper Rhaetian Reefs, calcisponges are concentrated in the central reef area. Here, sponges are found in the protected areas behind reef patches ("Riffknospen") formed by high-growing corals. Most calcisponges seemed to have prefered a muddy soft bottom, and only forms of the *Colospongia*-type (see Figs. 4B, and 16) are found growing on calcarenitic substrates adjacent to reef patches. The various growth forms known from upper Rhaetian Reefs (Schäfer, 1979) are correlated with special sediment types: cylindrical forms like *Peronidella fischeri*, or *Paradeningeria alpina* lived on carbonate mud bottoms, whereas flat extended forms (*Colospongia* sp., *Salzburgia variabilis*) are found in biomicrites and biosparites, and have also been found growing on various reef builders. An irregular mesh-like growth form is exhibited by *Follicatena irregularis* (Fig. 4D), which is one of the most important secondary framebuilders. A similar mode of life can be assumed for small pulvinate or dendroid growth types like *Annaecoelia mirabilis,* and *Annaecoelia interiecta.* Dome-shaped massive forms (*Cryptocoelia* sp.) may have lived loosely attached to the sea bottom.

*Corals (Figs. 5 and 6).*—Corals play a dominant role in the construction of reef patches in the central reef areas of upper Rhaetian Reefs, as well as in the Dachstein Limestone Reefs. They are also

the most important group in biostromal intervals of the Kössen coral limestone, and in some parts of the Zlambach Beds. Systematic determinations of corals found in reef limestones is generally difficult because of the often rather poor preservation, and because of the as yet incompleted revision of Triassic corals. As shown by the works of Cuif (1965, 1972, 1974a, 1974b, 1975, 1976, and 1979), Melnikova (1967, 1971, 1972, and 1979), Montanaro-Gallitelli *et al.* (1979), and Roniewicz (1974), there is no general agreement with regards to generic reevaluation of those coral species which have previously been described in classical monographs by Volz (1894; Cordevolian Cassian Beds, Southern Alps), and by Frech (1890; Norian and Rhaetian Zlambach Beds, Northern Alps). This is caused partly by the fact that the new systematics are based on microstructural details which often can not be seen in recrystallized material. Most authors studying Upper Triassic reef corals therefore still use the older nomenclatorial system, taking into account that some "species" may in reality correspond only to growth form types. On the other hand, a recent classification of the corals of the upper Rhaetian Reefs according to their dominant growth forms has resulted in an ecologically significant subdivision (Schäfer, 1979). Similar to Recent hermatypic corals, the distribution of these growth forms seems to have been influenced by fluctuations in water energy, and by an adaption to local biotopes (Table 6).

The distribution of various biocoenoses in upper Rhaetian Reefs (Table 7) is similar in all of the reefs which have been investigated. Differences exist in diversity of biocoenoses; e.g., the reef patches of the Adnet Reef are characterized by *"Thecosmilia"* cyclica (Fig. 13A), and those of the Rötelwand Reef by *"Thecosmilia"* clathrata (Fig. 13B). In the Feichtenstein Reef both species occur together. The framebuilding corals can be classified into several groups according to the type of epibionts, growth form of the corals related to sediment type of epibionts, and growth form of the corals related to sediment type (Fig. 8). The abundance of organic encrustations on the corals, the close associations with dasycladacean algae, and current orientated growth forms of large phaceloid *"Thecosmilia"* colonies suggest that most of the corals were hermatypic shallow water species. This is in contrast to the interpretation given by Stanley (1979) for Upper Triassic buildups in North America, which he believes to

---

Adnet, ×1.3. C. *Paradeningeria alpina* Senowbari-Daryan and Schäfer, one of the most common sponges in the Rötelwand Reef, ×2. D. Typical primary framework composed of *Paradeningeria* sp. stabilized by abundant secondary framebuilders like the encrusting sphinctozoan *Follicatena irregularis* Senowbari-Daryan and Schäfer (black perforated segments), sessile foraminifers, *Microtubus,* and serpulids; Rötelwand Reef, ×1.6.

TABLE 2.—CALCISPONGES IN UPPER TRIASSIC REEF LIMESTONES OF THE NORTHERN CALCAREOUS ALPS

| Calcisponges | Dachsteinkalk Reefs | | | | | Upper Rhaetian Reefs | | | |
|---|---|---|---|---|---|---|---|---|---|
| | Göll | Gosau-kamm | Gesäuse | Sau-wand | Adnet | Rötel-wand | Feich-tenstein | Gruber-alm | Stein-platte |
| **Inozoa** | | | | | | | | | |
| Molengraafia seilacheri FLÜGEL | x | x | | | x | x | | x | x |
| Peronidella communis FLÜGEL | x | x | | x | x | x | x | | x |
| Peronidella fischeri FLÜGEL | x | x | x | | x | x | x | | |
| Peronidella ? sp. 1 ZANKL | x | | | | x | x | | | |
| **Sphinctozoa** | | | | | | | | | |
| Amblysiphonella sp. 1 SENOWBARI-DARYAN | | | x | | | | | x | |
| Amblysiphonella sp. 2 SENOWBARI-DARYAN | | (x) | | | | | | x | |
| Annaecoelia interiecta SENOWBARI-DARYAN & SCHÄFER | | | | | x | | | x | |
| Annaecoelia maxima SENOWBARI-DARYAN & SCHÄFER | | | | | | | x | x | |
| Annaecoelia mirabilis SENOWBARI-DARYAN & SCHÄFER | | | | | x | | x | x | |
| Annaecoelia sp. 1 SENOWBARI-DARYAN | | | | | | | | x | |
| Annaecoelia sp. 2 SENOWBARI-DARYAN | | | | | | | | x | |
| Annaecoelia sp. | | | | | | | | x | |
| Colospongia bimuralis SENOWBARI-DARYAN | | x | | x | x | x | | x | x |
| Colospongia catenulata OTT | | | x | | | x | | | |
| Colospongia dubia (MÜNSTER) | | | | | | | | | |
| Colospongia sp. 1 SCHÄFER | | | | | x | x | x | x | |
| Colospongia sp. 2 SCHÄFER | | | | | x | x | x | x | |
| Colospongia sp. 3 SENOWBARI-DARYAN | | | | | | | x | | |
| Colospongia? sp. 4 SENOWBARI-DARYAN | | | | | | | | x | |
| Colospongia sp. | x | x | | x | | | | | x |
| Cryptocoelia zitteli STEINMANN | x | x | | | x | | (x) | (x) | |
| Cryptocoelia? sp. 1 SCHÄFER | | | | | x | x | | | |
| Cryptocoelia sp. | | | x | | | | | | |
| Dictyocoelia manon (MÜNSTER) | | | x | | | | | | |
| Dictyocoelia manon invesiculosa SENOWBARI-DARYAN | | x | | | | | | | |
| Follicatena irregularis SENOWBARI-DARYAN & SCHÄFER | | x | | x | x | x | x | x | x |
| Paradeningeria alpina SENOWBARI-DARYAN & SCHÄFER | | | | | x | x | x | x | |
| Paradeningeria gruberensis SENOWBARI-DARYAN & SCHÄFER | | | | | x | x | x | x | |
| Paradeningeria weyli SENOWBARI-DARYAN & SCHÄFER | | | | | x | x | x | x | |
| Paradeningeria sp. | x | x | | x | | | | | x |
| Paravesicocaulis concentricus KOVACS | | x | | | | | | | |
| Polytholosia cf. cylindrica SEILACHER | x | | | | | | | | |
| Polytholosia sp. | x | x | | x | x | | | | |
| Salzburgia variabilis SENOWBARI-DARYAN & SCHÄFER | | | | | | x | x | x | x |
| Salzburgia sp. | | x | | | | | | | x |
| Verticillites gruberensis SENOWBARI-DARYAN | x | | x | | | | | x | |
| Verticillites sp. | | | | | | | | | x |
| Welteria repleta STEINMANN | | | | | | | (x) | (x) | x |

TABLE 3.—DISTRIBUTION OF CORALS IN THE UPPER TRIASSIC REEFS OF THE NORTHERN CALCAREOUS ALPS

| Corals | Dachsteinkalk Reefs | | | | Upper Rhaetian Reefs | | | | |
|---|---|---|---|---|---|---|---|---|---|
| | Göll | Gosau-kamm | Gesäuse | Sau-wand | Adnet | Rötel-wand | Feich-tenstein | Gruber-alm | Stein-platte |
| Actinastraea juvavica (FRECH) | x | x | | | | x | | x | |
| Astraeomorpha confusa (WINKLER) | x | x | | x | x | x | x | x | |
| Astraeomorpha confusa minor FRECH | | | | | | x | x | x | |
| Astraeomorpha confusa, form A ZANKL | x | | | | | | | | |
| Astraeomorpha crassisepta REUSS | | x | | | x | x | x | x | |
| Astrocoenia waltheri FRECH | | | | x | x | | x | x | |
| Coccophyllum acanthophroum FRECH | | x | | | | | | | |
| "Coccophyllum" sturi REUSS | x | | | | | x | x | x | |
| Cyathocoenia schafhäutli (FRECH) | | | | | | x | | | |
| Gablonzeria austriaca (FRECH) | | | x | | | | | | |
| Gablonzeria profunda (REUSS) | | x | | | x | x | x | x | |
| Gablonzeria profunda minor (FRECH) | | | | | | x | | | |
| Gablonzeria sp. | x | x | | x | | | | x | |
| Montlivaultia fritschi FRECH | | x | | | | | | (x) | |
| Montlivaultia marmorea FRECH | x | x | x | x | | x | x | x | |
| Montlivaultia norica FRECH | x | x | | x | | x | x | x | x |
| Oedalmia norica (FRECH) | x | | | x | (x) | | x | | x |
| Omphalophyllia ?zitteli VOLTZ | | | x | | | | | | |
| Palaeastraea grandissima (FRECH) | x | | | (x) | | | x | x | |
| Pamiroseris rectilamellosa (WINKLER) | x | x | | | x | x | (x) | (x) | x |
| Pamiroseris rectilamellosa minor (FRECH) | x | | | | | x | | | x |
| Phyllocoenia incrassata FRECH | x | | | | | x | | | |
| Pinacophyllum sp. 1 ZANKL | x | x | | | x | x | x | x | |
| Pinacophyllum sp. 2 ZANKL | x | | | | | x | x | x | |
| Procyclolithes triadicus FRECH | x | x | | | | x | x | x | x |
| Retiophyllia clathrata (EMMERICH) | x | x | | x | x | x | x | x | x |
| Retiophyllia paraclathrata (RONIEWICZ) | x | x | | x | x | x | x | x | x |
| Seriastraea multiphylla SCHÄFER & SENOWBARI-DARYAN | x | | | | x | x | x | x | |
| Stylophyllopsis polyactis FRECH | | x | | x | | x | | x | |
| Stylophyllopsis sp. | | | | x | | | | | |
| Stylophyllum paradoxum FRECH | | x | | | | x | x | x | |
| Stylophyllum polycanthum REUSS | x | | | | | x | x | x | |
| "Thecosmilia" cyclica SCHÄFER & SENOWBARI-DARYAN | | | | | | | | x | |
| "Thecosmilia" defilippi (STOPPANI) | | | x | | | | | | |
| "Thecosmilia" konosensis KANMERA | | | x | | | | | | |
| "Thecosmilia" norica FRECH | | | | | | | x | | |
| "Thecosmilia" subdichotoma VOLTZ | | | | | | | | (x) | |

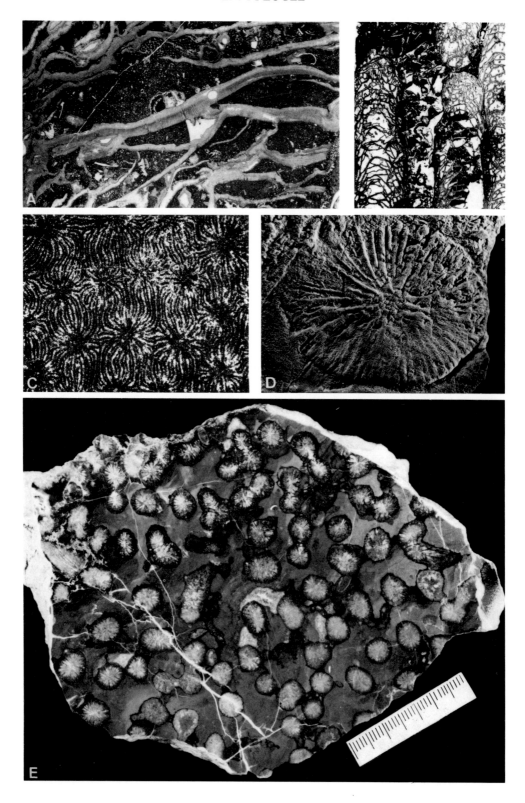

have been formed by ahermatypic corals. In the upper Rhaetian Reefs as well as in the Dachstein Limestone Reefs no corals have been found in primary shadow areas similar to reef cavities, or beneath high-growing framebuilders. These areas are generally occupied by calcisponges and "hydrozoans." A similar situation may have existed in the Middle Triassic Wetterstein Limestone Reefs.

Those areas of the reef-crest and reef-flat, which were occupied by active growing corals and other framebuilding organisms, seem to have been geographically rather small in the Dachstein Limestone Reefs (Göll: 2 to 5 m², maximum 15 m²; Sauwand: about 3 to 5 m²; Hochschwab: smaller than 3 m²; and Gosaukamm: smaller than 3 m²). In the upper Rhaetian Reefs larger geographic areas with dense crops of autochthonous framebuilding organisms are known (Rötelwand: up to several hundred m²; and Adnet: up to 80 m²). The distances between these areas depend on the individual stage of reef development and on the position within the reef complex; e.g., densely-spaced reef patches are found more abundantly near the edge of the central reef area than in the protected reef-flat. Similar differences can be seen in the relationships of reef framework and reef detritus within the central reef areas, which are 1:9 for the Göll, Gosaukamm, and Sauwand Reefs, as well for the Gesäuse Reefs, and 0.5:9.5 for the Hochschwab Reef, but 1:3 to 2:3 for the upper Rhaetian Reefs.

"*Hydrozoans.*"—The systematic affinities of many primary or secondary framebuilders in the Upper Triassic reefs, which formerly had been assigned to the Hydrozoa or to the Stromatoporoidea, have become further obscured because of uncertainty regarding the possible relationships with some sponge groups like sclerosponges or ischyrosponges (Wendt, 1975; G. and H. Termier, 1975). As a result of the varied strong morphological differences noted between the more frequently occurring "hydrozoans" of the Upper Triassic reefs, three groups can be distinguished:

1. *Spongiomorphids (Figs. 7A and 7C).*—According to the observations by Schäfer (1979), Senowbari-Daryan (1978), and Piller (this volume), spongiomorphids are found predominantly in similarly protected environments as most sphinctozoan sponges. Some species such as *Spongiomorpha acyclica* may be classified as a sphinctozoan sponge, others like *Spongiomorpha ramosa* can be compared with sclerosponges; *Spongiomorpha ramosa* seems to have been restricted to low energy environments. This can be assumed from its frequent occurrences within isolated reef patches, which partly represent the somewhat deeper mud-mound stage of upper Rhaetian Reefs (Steinplatte, Rötelwand, Feichtenstein).

2. *Disjectoporids (Fig. 7B).*—Species belonging to this group, which are known from Permian and Triassic rocks, seem to be concentrated in protected parts of the reefs (innermost areas of the reef patches in the Adnet Reef; lower part of the reef-slope, Steinplatte). Some species can occupy an area up to about 20 m² within the individual reef patches. The disjectoporids may be regarded as hydrozoans comparable in their internal morphology with some milleporid forms. All forms known from Upper Triassic reefs represent new species, probably to some degree similar to species described from the Upper Triassic of the Pamir Range by Boiko (1972).

3. *Lamellata wähneri Flügel and Sy (Fig. 7D).*—The tabular colonies of this species are found mainly in the marginal zones of the reef patches (Rötelwand and Feichtenstein), or as important framebuilding elements in small patch-reefs, adjacent to the margin of the reef-crest (Steinplatte).

As shown in Table 8, the "hydrozoans" of the upper Rhaetian Reefs are characterized by various kinds of epibionts and different growth forms, which can be used for recognition of local environments within the reef complexes.

"*Bryozoans*" and "*Tabulozoans*" (Fig. 8).—Very important secondary framebuilders in upper

---

←

FIG. 5.—Photographs of various corals from upper Rhaetian Reefs (A, B, C), and from Dachstein Limestone Reefs (D, E). A. Foliate colony of *Seriastraea multiphyllia* Schäfer and Senowbari-Daryan, strongly bored, with epibionts (small bryozoans and sphinctozoans, foraminifers, serpulids, and *Microtubus*). This coral is restricted to the margins of reef patches and also to the somewhat deeper reef areas; Gruberalm, type material, ×1.3. B. Longitudinal section of the phaceloid coral *Stylophyllum polyacanthum* Reuss with typical abundant dissepiments. Characteristic high-growing framebuilder often found in the high energy marginal zones of the reef patches; Rötelwand, ×1.5 C. *Pamiroseris* cf. *rectilamellosa* (Winkler); tangential section. This coral is found in different environments, and also in the oncoid facies; Gruberalm, ×2. D. *Stylophyllopsis polyacis* Frech, a common solitary coral of the Dachstein Limestone Reefs; Donnerkogel, Gosaukamm Reef, smaller diameter 22 millimeters. E. Polished slab of a colony of "*Thecosmilia*" *clathrata* (Emmrich) form B (=*Retiophyllia paraclathrata* Roniewicz). These dendroid low-growing corals are restricted to protected niches within reef patches, and also in somewhat deeper, muddy environments of the reef-slopes, or in the Kössen biostromal buildups or lagoonal patch-reefs; Gosaukamm (Angerstein), mm-scale.

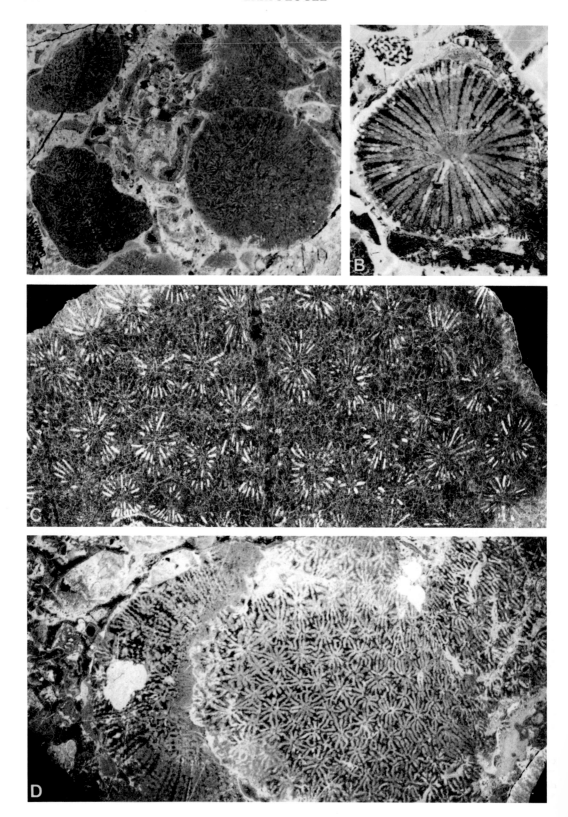

Rhaetian and Dachstein Limestone Reefs encrust colonies of various sizes and are composed of parallel arranged tubes, sometimes with horizontal or arched cross-partitions. Some of these are often strongly recrystallized, but still can be assigned to the Bryozoa. Many other forms are of uncertain systematic position (Flügel, 1963) because of the lack of good diagnostic criteria. Colonial growth forms have been artificially recognized by the diameter and the arrangement of the tubes (Zankl, 1969; Ohlen, 1959; Schäfer, 1979; and Senowbari-Daryan, 1978). Following a proposal by Kühn (1943), tabular or dome-shaped colonies consisting of relatively large, tabulated tubes (mean diameter 0.30 to 0.60 mm) are designated as "Tabulozoa." These organisms may in part represent sclerosponges.

Paleoecological analyses of the bryozoans and tabulozoans of the upper Rhaetian Reefs indicate essential differences between both groups, especially with regard to growth forms (tabulozoans: tabular, with several vertically growing generations; bryozoans: generally small globular or ramose colonies), preference of special types of substrate (tabulozoans on corals, bryozoans on calcisponges and hydrozoans), and various epibiont overgrowths. These differences support the opinion that the distinction of various groups is not too artificial.

Most of these forms are found in biomicrites formed within local quiet water environments of the reef patches, and also in the larger reef cavities. In contrast to most other reef building organisms, the "bryozoans" and "tabulozoans" are not found as detrital grains within the calcarenitic limestones surrounding the reef patches. The limitation of these groups to zones of active reef construction is also known from upper Rhaetian and Dachstein Limestone Reefs (Sauwand, Gosaukamm, and Gesäuse).

*Algae (Figs. 9 and 10).*—Calcareous algae are represented by spongiostromate and porostromate blue-greens, by Solenoporaceae, Dasycladaceae, problematical Codiaceae, and various endolithic algae. Encrusting blue-green algae, possible codiaceans, and solenoporaceans contribute to reef construction. Dasycladaceans are concentrated primarily in the "algal-foraminiferal facies" surrounding or adjacent to the biolithite facies. Algal species found in various upper Rhaetian Reefs are generally similar (Flügel, 1975; Senowbari-Daryan and Schäfer, 1979), and no striking differences can be seen in comparison with the algal flora from the Dachstein Limestone Reefs (Table 5). Differences do exist however, specifically in the distributional patterns of the various algae (Flügel, 1979).

*1. Micritic algal crusts (Figs. 9D and 23).*— Micritic crusts composed of irregularly crumbled layers of micrite or peloidal micrite alternating with thin microsparry or sparry layers are important secondary reef builders in the central reef areas of the Dachstein Limestone Reefs and in the upper Rhaetian Reefs. Various types can be distinguished:

a. Crusts up to 15 milimeters thick can circumscribe or encrust corals, calcisponges, tabulozoans or molluscs. These crusts consist of now parallel micrite layers (150 to 600 microns in thickness), separated by microsparry intervals. No regular dome-shaped structures are developed. Sometimes tiny peloids and a few microfossils (ostracodes, foraminifers) are enclosed within the micritic laminae. These crusts are often parts of biogenous sequences built up by several generations of secondary reef builders (Fig. 16). These crusts often represent the first generation of epibionts, associated with the *microproblematicum Microtubus communis*, but they can also be found as the final generation on the tops of these sequences. In Upper Triassic reefs these crusts seem to have been restricted to relatively low energy environments within and between the reef patches. High-growing phaceloid coral colonies of the more turbulent edge-zones of the reef patches are generally devoid of algal crusts.

b. Crusts, similar in internal structure to type a, and also overgrowing the sediment. Their distributional pattern is similar to that of type-a crusts.

c. Crusts up to several centimeters thick composed of distinct, undulating and densely-packed micritic laminae (50 to 150 microns in thickness). In contrast to the types described above, micritic and pelmicritic layers can be grouped to form relatively flat, dome-shaped structures which are sometimes separated from each other by vertically arranged spar-filled voids (?desiccation cracks). This type of crust is generally represented as angular lithoclasts, especially in the "rud/

←

FIG. 6.—Photographs of common corals from the Dachstein Limestone Reefs; negative prints. A. Low nodular colonies of *Pamiroseris* sp. together with micritized reef-detritus; Sauwand, ×2.5. B. Corallite of the colonial coral *"Thecosmilia" fenestrata* Frech encrusted by algae and by tabulozoans; Gosaukamm, ×5. C. *Palaeastraea grandissima* (Kühn); Steinriese, Gosaukamm, ×1.2. D. *Astraeomorpha crassisepta* Reuss, Donnerkogel, Gosaukamm, ×3.

FIG. 7.—Photographs of various hydrozoans from upper Rhaetian Reefs (A, B, and D), and Dachstein Reef Limestones (C). A. *Spongiomorpha ramosa* Frech generally restricted to low energy environments; Steinplatte, ×3. B. *Disjectopora* sp. 1 of Zankl, characterized by large tubes within an open reticulum; Rötelwand, ×2.2. C. *Stromatomorpha rhaetica* Kühn; Schildmauer, Gesäuse, ×3.1. D. *Lamellata wähneri* Flügel and Sy; Hochiss, Sonnwend Mountains, Tyrol; negative print, ×7.

floatstone facies,'' deposited at the windside flank of Dachstein Limestone Reef complexes (Gosaukamm, and Hochschwab).

All these crusts are generally attributed to the activity of blue-green algae and cyanobacteria. Because these crusts also are on the walls of rather small reef cavities, their growth may not have been light dependent in the manner of calcareous green-algae. The crusts are generally called "Spongiostromata" crusts following the proposal by Pia (1927).

2. *Porostromate Blue-Green Algae (Figs. 9A and 9B).*—This group is dominant in the "algal-foraminiferal facies" and "oncoid facies" of the back-reef zone. Thin, frequently micritized nodular thalli are composed of algal filaments which

show a distinct pattern of branching. On the basis of SEM investigations these algae can be differentiated according to overall size and mode of tube branching (Flügel, 1977). Nevertheless there is no generally accepted classification of these fossils, which have long been regarded as codiaceans (Johnson, 1961; Wray, 1977). The types known from the upper Rhaetian and the Dachstein Limestone Reefs may be assigned to the genera *Garwoodia, Cayeuxia, Ortonella, Zonotrichites,* or *Apophoretella.* These algae are found together with dasycladaceans and with a unique association of foraminifers (involutinids, duostominids, and glomospirids). Their quantitative distribution seems to be different in various upper Rhaetian Reefs: Adnet and Rötelwand Reefs show increase

→

FIG. 8.—Photographs of various bryozoans and tabulozoans from the Dachstein Limestone Reefs (A, C), and the upper Rhaetian Reefs (B). A. Cyclostomid bryozoan *Paralioclema* sp.; Kalkbling, Gesäuse; ×15. B. Various tabular colonies of "bryozoans" and "tabulozoans," these encrusting forms are important secondary framebuilders; Feichtenstein, ×3. C. Tabulozoan colony encrusted by spongiostromate algal crusts (type a) characteristic of the protected parts of the central reef area; Grosser Donnerkogel, Gosaukamm, ×20.

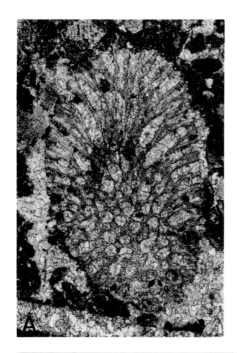

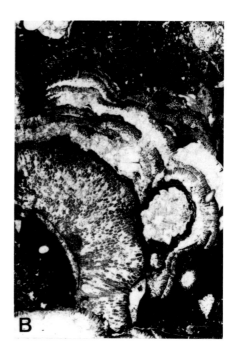

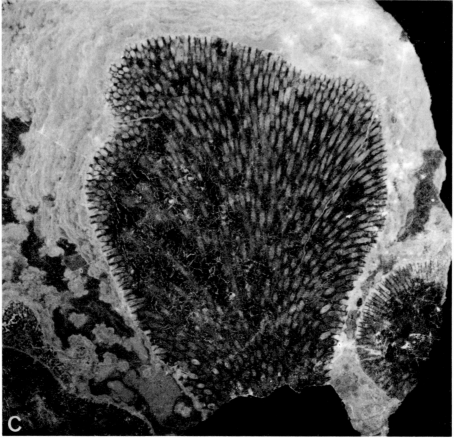

TABLE 4.—DISTRIBUTION OF CORAL GROWTH FORMS AND ASSOCIATIONS IN RECENT REEFS AND IN UPPER RHAETIAN REEFS (RÖTELWAND; ADNET) (AFTER P. SCHÄFER, 1979)

Water energy gradient: decreasing ←   maximal water energy >   decreasing →

base of the slope, reef-slope, leeward reef-crest, reef-flat, leeward reef-margin, lagoonal reefs

| Locality | base of the slope | reef-slope | leeward reef-slope | reef-crest | reef-flat | leeward reef-margin | lagoonal reefs |
|---|---|---|---|---|---|---|---|
| Bermuda, Caribbean (patch-reefs) | dendroid — *Oculina-Madracis* association | | | massive — *Montastraea-Diploria-Porites astreoides* association | | | dendroid — *Oculina-Madracis* association |
| Gulf of Aqaba, Red Sea (fringing-reefs) | massive — zone with increasingly more massive corals (*Platygyra, Porites, Favia, Favites*) | tabular — *Acropora* zone | massive/dendroid — *Millepora* zone | tabular/massive — *Millepora* zone | massive/(dendroid) — reduced fauna | tabular — *Millepora* zone | massive/(dendroid) — lagoonal reefs |
| Jamaica, Caribbean | | massive dendroid — *annularis* zone | massive/(dendroid) — *cervicornis* zone | reef-ridges — *palmata* zone | tabular — *zoanthus* zone | | massive/(dendroid) — back-reef zone |
| upper Rhaetian reefs | phaceloid/massive — *Thecosmilia-Thamnasteria* association | phaceloid/dendroid/massive — *Thecosmilia Astraeomorpha-Thamnasteria* association | | phaceloid — *Thecosmilia Pinacophyllum-Stylophyllum* association | dendroid — *Thecosmilia Astraeomorpha-Seriastraea* association | massive / foliate | solitary — *Montlivaultia/Stylophyllopsis* associations |

TABLE 5.—DISTRIBUTION OF CORAL GROWTH FORM GROUPS IN UPPER RHAETIAN REEFS (ACCORDING TO P. SCHÄFER, 1979; AND B. SENOWBARI-DARYAN, 1978)

| Growth Form Groups of Corals | Species | Distribution on the Rötelwand and Adnet Reef complexes |
|---|---|---|
| High growing, phaceloid and dendroid colonies with relatively thick corallites | *Retiophyllia clathrata* (EMMERICH) "*Thecosmilis*" *cyclica* SCHÄFER & SENOWBARI-DARYAN *Pinacophyllum* sp. *Stylophyllum polycanthum* REUSS | Polymict reef patches of the central reef area; at the high-energy margin of the reef patches, protecting the central areas of the reef patches. Generally poor in epibionts |
| Low growing, dendroid colonies with relatively thin corallites | *Retiophyllia paraclathrata* RONIEWICZ (= "*Thecosmilia*" *clathrata*, form B) | Protected areas within the reef patches, in somewhat deeper areas of the margin of the reef patches, and as a "coral coppice" in Kössen biostromes |
| Solitary corals | *Montlivaultia norica* FRECH *Montlivaultia* cf. *resussi* M.E. & H. *Stylophyllopsis polyacis* FRECH | Protected areas within the reef patches of the central reef-flat; generally together with hydrozoans, calcisponges and tabulozoans. Relatively rich in epibionts, often circumcrusted by algae and foraminifers |
| Massive ceroid colonies | *Astraeomorpha confusa* (WINKLER) *Astraeomorpha crassisepta* REUSS *Isastraea profunda* REUSS *Palaeastraea grandissima* (FRECH) | Reef patches at the periphery of the central reef area; upper reef-slope; leeward detrital bottoms of the reef patches |
| Massive thamnasteroid colonies | *Pamiroseris rectilamellosa* (WINKLER) | Reef patches at the periphery of the central reef area; leeward part of larger reef patches; oncoidal grainstone facies adjacent to the reef area |
| Platy-foliate colonies | *Seriastraea multiphylla* SCHÄFER & SENOWBARI-DARYAN | Peripheral and in deeper parts of the reef patches |

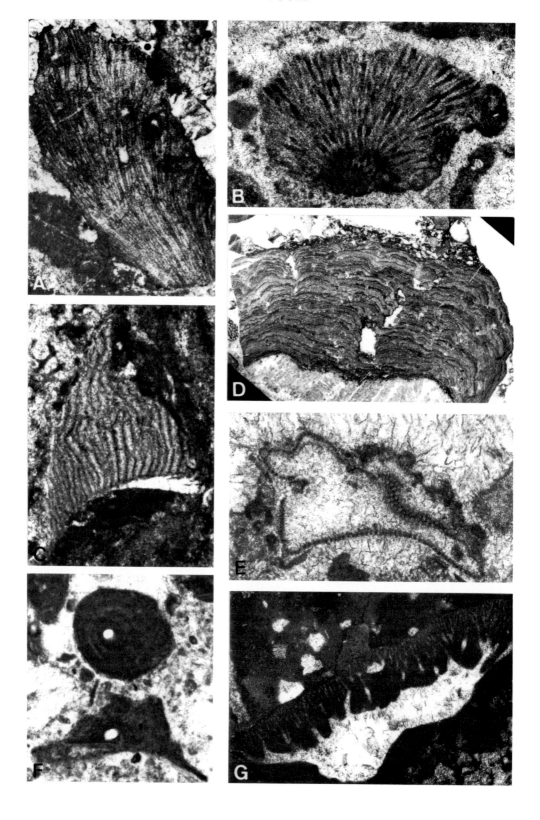

from the oncoid facies to the algal-foraminiferal facies and very rare occurrences in the marginal reef patches; Feichtenstein Reef shows a similar tendency common in the reef patches; at Steinplatte they are common in the lagoonal patch-reef facies. In the Dachstein Limestone Reefs algae of the "*Cayeuxia*"-type are concentrated in back-reef environments, together with dasycladaceans, and with grapestone.

3. *Solenoporaceae (Figs. 10F and G).*—These red algae are represented by the genera *Solenopora* and *Parachaetetes*. They are mainly concentrated in the marginal zones of the central reef area, and in the near-reef oncoid facies. With increasing distance from the reef porostromate blue-green algae seem to take the place of solenoporacean algae. This can be seen in the upper Rhaetian Reefs as well as in the Dachstein Limestone Reefs. Distribution of growth form types, as distinguished in the Rötelwand and Adnet Reefs (Schäfer, 1979), does not seem to be representative for other reefs: e.g., small branched forms like *Solenopora alcicornis* (Fig. 10F) described from the lower and peripheral parts of reef patches have a similar distribution in the Dachstein Limestone Reefs (Göll and Sauwand). However, nodular thalli of *Solenopora* sp. 1, first described by Zankl (1969) from the calcarenitic back-reef environment of the Göll Reef complex, have been found as important framebuilders in reef patches of the central reef area at Adnet, and in the same position in the Schildmauer Reef, Gesäuse (Dullo, 1979). Irregularly nodular and mushroom-shaped forms like *Solenopora endoi* and *Solenopora* cf. *styriaca* are known from the oncoid facies, and from within marginal reef patches of the Rötelwand Reef, and from the central reef and back-reef areas of the Sauwand Reef, whereas in the Gesäuse Reefs these species are concentrated in both the "algal-foraminiferal facies" and "grapestone facies" where both types are characteristic of the lagoonal environment. Dome-shaped thalli like those of *Parachaetetes maslovi* seem to have lived in the marginal areas of the reef patches, or in the transitional areas between the central reef and the fore-reef (Dachstein Limestone Reefs).

4. *Dasycladaceae (Figs. 10A–E).*—In contrast to the Middle Triassic reefs, where dasyclads are concentrated in lagoonal environments (Ott, 1967), in the Upper Triassic reefs this algal group occurs in the more peripheral parts of the reef patches. The oncoid facies is generally rich in *Heteroporella zankli* and *Heteroporella crosi*; and in the algal-foraminiferal facies *Diplopora adnetensis* may occur frequently. In the Dachstein Limestone Reefs endospore species of *Diplopora* similar to *Diplopora phanerospora* characterize the algal-foraminiferal facies of the back-reef environments, together with *Heteroporella* and *Griphoporella curvata*, but it should be noted that all these genera can also be found within the central reef areas (Gosaukamm and Gesäuse).

5. *Codiaceae (Fig. 9G).*—This group is extremely rare in Upper Triassic reefs if we only consider unquestionable codiaceans like *Boueina hochstetteri liasica* (Gosaukamm). But there may be additional codiaceans represented by encrusting fossils like the "*problematicum* A" described by Ohlen (1959), which is now thought to be identical with *Lithocodium* Elliott.

*Microproblematica (Fig. 11).*—Large numbers of micro-organisms known from thin-sections of Upper Triassic reef limestones cannot be assigned with sufficient certainty to a special group of organisms (Flügel, 1972b; Borza, 1975). These fossils have been variously described by specific and generic names, or they have been identified by using informal nomenclature (Senowbari-Daryan, 1978; Schäfer, 1979; Dullo, 1979). About 20 *microproblematica* have been found in upper Rhaetian Reefs of Adnet and the Rötelwand, and 26 from the Feichtenstein and the Gruberalm Reefs. From the Dachstein Limestone Reefs of the Gesäuse, 18 *microproblematica* have been described, and from the Gosaukamm Reef about 15 "species" are known.

Some *microproblematica* are important contributors to the construction of the reef framework because they represent secondary reef builders growing upon corals, calcisponges, and tabulozoans. Of major importance are *Microtubus communis* (Fig. 11I), *Radiomura cautica* (Fig. 11F), *problematicum* A of Ohlen (=*Litho-*

---

←

FIG. 9.—Thin-section photomicrographs of various blue-green algae and problematical algae from the Dachstein Limestone Reefs (A, and C–F) and upper Rhaetian Reefs (B and G). A. *Cayeuxia alpina* Flügel a filamentous blue-green alga; Hohe Wand, ×22. B. "*Cayeuxia*" or "*Garwoodia*," Rötelwand ×40. C. *Pycnoporidium? eomesozoicum* Flügel, probably a red alga, overgrowing a shell fragment; Gosaukamm, ×25. D. Typical reworked algal crust frequently found on the windward side of reefs; Gosaukamm, ×2. E. *Thaumatoporella parvovesiculifera* (Raineri), a form widely distributed in all parts of Upper Triassic reefs; Hohe Wand, ×28. F. *Tubiphytes obscurus* Maslov, probably a cyanophycean alga; Gosaukamm Reef, ×60. G. *Problematicum* A first described by Ohlen (1959) and probably identical to the questionable codiacean alga *Lithocodium* Elliott. This very abundant encrusting form dominates in the high energy marginal zones of the central reef; Adnet, ×30.

TABLE 6.—INTERRELATIONSHIPS BETWEEN GROWTH FORMS OF FRAMEBUILDING CORALS, EPIBIONTIC OVERGROWTHS AND LIMESTONE TYPES IN UPPER RHAETIAN REEFS (AFTER P. SCHÄFER, 1979)

| | Epibionts on Corals | Corals | Sediment Types | Growth Forms of Corals |
|---|---|---|---|---|
| Group A | *problematicum* A, OHLEN *Bacinella irregularis*, partly together with *Nubecularia*, *Alpinophragmium*, and porostromate blue-green algae | *Stylophyllum polyacanthum*, *Stylophyllum paradoxum* "*Thecosmilia*" *cyclica*, *Pinacophyllum* sp. | arenitic detrital limestones | phaceloid |
| Group B | *problematicum* A, OHLEN *Bacinella irregularis* frequent large borings | *Astrocoenia*, *Astraeomorpha*, *Thamnasteria*, *Isastraea* | arenitic detrital limestones and colored biomicritic limestones | massive-ceroid and massive-thamnasteroid |
| Group C | sessile miliolid foraminifers arenaceous foraminifers; sometimes *Thaumatoporella*, small colonies of *Annaecoelia*, and bryozoa, type 1 | "*Thecosmilia*" *clathrata* form A, B and C; *Montlivaultia* cf. *reussi* | arenitic detrital limestones, partly colored biomicritic limestone | dendroid |
| Group D | *Microtubus*, *Follicatena*, tabuluzoans; partly together with serpulids, *Annaecoelia*, *Radiomura*, *Baccanella floriformis* etc. | *Seriastraea multiphylia*, *Montlivaultia norica* | micritic limestones or pelbiosparitic limestone (reef cavities) | tabular and cylindrical |

TABLE 7.—"HYDROZOANS" OF THE UPPER TRIASSIC REEF LIMESTONES IN THE NORTHERN CALCAREOUS ALPS

| "Hydrozoa" and "Ischyrospongia" | Dachsteinkalk Reefs | | | | Upper Rhaetian Reefs | | | | |
|---|---|---|---|---|---|---|---|---|---|
| | Göll | Gosau-kamm | Ge-säuse | Sau-wand | Adnet | Rötel-wand | Feich-ten-stein | Grub-eralm | Stein-platte |
| *Circopora triadica* FLÜGEL & SY | | | | | | X | | | |
| *Disjectopora* sp. 1 ZANKL | X | X | | | X | X | X | X | |
| *Disjectopora* sp. 2 ZANKL | X | X | | | X | X | X | X | |
| *Disjectopora* sp. 3 SCHÄFER | | | | | X | X | X | X | |
| *Disjectopora* sp. | | | | X | | | | | X |
| *Lamellata wähneri* FLÜGEL & SY | X | X | | | X | X | X | X | X |
| *Spongiomorpha acyclica* FRECH | | | | | X | X | X | X | |
| *Spongiomorpha dendroidea* KÜHN | X | | | | | | | | |
| *Spongiomorpha globosa* FRECH | | | | | X | | | | |
| *Spongiomorpha minor* FRECH | X | | | | | | | | |
| *Spongiomorpha ramosa* FRECH | | X | X | X | X | X | | X | X |
| *Stromatomorpha rhaetica* KÜHN | X | X | X | | X | X | X | X | X |
| *Stromatomorpha stylifera* FRECH | X | | X | | | X | X | X | |
| *Stromatomorpha* sp. 1 SCHÄFER | | | | | X | X | | | |

*codium*, Fig. 9G), and *Bacinella irregularis*. Other forms like *Cheilosporites tirolensis* (Fig. 11A), *Muranella sphaerica* (Fig. 11G) and *Baccanella floriformis* seem to have been restricted to protected reef cavities within the framework. Some of the more common *microproblematica* are listed in Table 10.

### Reef Dwelling Organisms

Of the megafossils, megalodontid lamellibranchs have been studied more carefully (Czurda, 1973), although other macrofossils of the reef facies have been studied by Sickenberg (1932; Feichtenstein), Sieber (1937; Rötelwand), Zankl (1969; Göll), and especially Zapfe (1967a, Adnet, and 1967b, Gosaukamm).

*Brachiopoda.*—Most of the brachiopods found in the reef facies are either rhynchonellids or terebratulids. In the Rötelwand and Adnet Reefs, rhynchonellids appear to have been concentrated by water currents within the interstices of the reef framework, whereas terebratulids are found most commonly at the tops of Kössen biostromes. Of special interest is the occurrence of cemented brachiopods as epizoans on calcisponges, hydrozoans and tabulozoans.

*Lamellibranchs.*—Sieber (1937) distinguished a lamellibranch association consisting of *Pecten, Ostrea, Pteria* and *Lima*, that was found in the central part of the Rötelwand Reef, and an association with *Gervilleia* and *Modiola*. According to Schäfer (1979) the first association represents allochthonous elements within the peripheral, higher-energy portions of patch-reefs, but autochthonous in the transition area between the reef facies and the mud facies. The *Gervilleia-Modi-*

*ola*-association is characteristic of lower-energy environments in which the substrate for the initial reef patches had already been formed.

*Gastropods.*—This group is dominant within reef patches (e.g., *Zygopleura* in the biocoenosis with *Paradeningeria*), and also in protected areas between reef patches. Another environment with many gastropods is the near-reef grapestone facies of the Dachstein Limestone Reefs where algal films may have acted as a source of food for these molluscs. Rich occurrences of snails can be interpreted either as autochthonous associations or as a thanatocoenosis (current-swept fauna, or one relocated by the action of predatory organisms such as crabs).

*Cephalopods.*—The few ammonites described from Upper Triassic reefs have generally been found between high-growing corals along the periphery of reef patches in upper Rhaetian Reefs, or at the windward side of some Dachstein Limestone Reefs (Göll, Gosaukamm, Hochschwab, and Gesäuse). All ammonites indicate an upper Norian age, but most forms can not be exactly determined because only juvenile individuals are seen in thin-sections, or because the material is so poorly preserved.

*Serpulids.*—True serpulids are rare with the exception of forms found as epizoans on framebuilders, especially in the inner areas of reef patches.

*Crustaceans.*—Only characteristic fecal pellets of anomuran crustaceans have been found. These occur relatively often in pelbiosparitic sediments of reef cavities located in the central reef area. These "*Favreina*"-like microfossils are probably a new species (Senowbari-Daryan, 1979).

*Ostracodes.*—Smooth-shelled and ornamented

Table 8.—Interrelationships between epibiont overgrowths, and growth forms and occurrences of "Hydrozoan" species in the Rötelwand and Adnet Reefs (after P. Schäfer, 1979)

| | Epibionts on "Hydrozoans" | "Hydrozoan" Species | Growth Forms of "Hydrozoans" | Sediments | Occurrences |
|---|---|---|---|---|---|
| A | Microtubus, Radiomura, Follicatena, ?Girvanella, sessile brachiopods, small sponges (Annaecoelia), boring pelecypods | Disjectopora sp., Spongiomorpha acyclica | nodular, spherical, mushroom-shaped | micrite or pelmicrite (in reef cavities) | Disjectopora: protected innermost parts of the reef patches; S. acyclica: protected cavities |
| B | Alpinophragmium, Thaumatoporella, Nubecularia, Problematicum A. OHLEN, Bacinella | Spongiomorpha ramosa, Spongiomorpha gibbosa, Lamellata wähneri | Spongiomorpha ramosa: dendroid; S. gibbosa: massive dome-shaped; Lamellata wähneri: massive-tabular | colored micrite and arenitic bioclastic limestone | in peripheral parts of reef patches and separated small reef patches at the margin of the central reef area |
| C | epibionts very rare or lacking | Stromatomorpha stylifera, Stromatomorpha rhaetica | Stromatomorpha stylifera: spheroidal; S. rhaetica: tabular | S. stylifera: arenitic bioclastic limestone; S. rhaetica: micrite | S. stylifera: marginal patches; S. rhaetica: central part of the reef and leeward region |

ostracodes are found in small cavities between the skeletons of framebuilders, or in the open spaces within sponges or corals.

*Crinoids.*—The distribution pattern of this group is not clear. In contrast to the upper Rhaetian Reefs, where concentrations of crinoids are found in the "reef-detrital–mud facies," and also serve as a substrate for the initial reef patches, in the Dachstein Limestone Reefs crinoids are known from nearly all facies units.

*Ophiurids and Holothurians.*—Both groups are known primarily from acid residues. They are found chiefly in micritic limestones of the central reef areas. Here, they are concentrated in reef cavities between colonies of *Thecosmilia* (Zankl, 1965). Another occurrence is known from reddish biomicritic limestones of the Adnet Reef. This striking patchy distribution pattern may reflect primary isolated biotopes.

*Vertebrates.*—Remains of fishes or reptiles are extremely rare in Upper Triassic reef limestones (e.g., *Placochelys* in the Adnet Reef; Zapfe 1960).

## Destructive Organisms

Both upper Rhaetian Reefs and Dachstein Limestone Reefs, and to a lesser degree, the Kössen biostromes, were all subjected to intensive biological erosion. This is indicated by borings up to one centimeter in diameter of lithophagous lamellibranchs (concentrated especially in massive coral colonies, and in tabulozoans and hydrozoans), and by abundant grains with micrite envelopes found both in the oncoid facies and in the algal-foraminiferal facies. It is believed that these irregularly-formed tiny borings were probably made by both endolithic algae and marine fungi (Friedman *et al.*, 1971). In some calcarenites of the Dachstein Limestone Reefs, about 80 percent of the angular to subangular intraclasts representing reworked clasts of reef carbonates are micritized, and about 90 percent of all bioclasts appear to be similarly algally bored.

### UPPER TRIASSIC REEFS: PALEOECOLOGY

The Upper Triassic reefs of the Northern Calcareous Alps offer major advantages for paleoecological analyses because the composition of the reef biota is relatively well known. The interrelationships between reef organisms and their environment can be described in terms of reef building organism communities, by the spatial distribution of selected organism groups, by organism diversity gradients, and by the interpretation of unique ecologic niches; e.g., small reef cavities. A future possibility for more detailed classification of various reef environments may be hoped for by analysis of the successions formed by encrusting organisms.

TABLE 9.—CALCAREOUS ALGAE (DASYCLADACEAE AND SOLENOPORACEAE) IN UPPER TRIASSIC REEF LIMESTONES OF THE NORTHERN CALCAREOUS ALPS

| Calcareous Algae | Dachsteinkalk Reefs | | | | | Upper Rhaetian Reefs | | | |
|---|---|---|---|---|---|---|---|---|---|
| | Göll | Gosau-kamm | Gesäuse | Sau-wand | Adnet | Rötel-wand | Feich-tenstein | Gruber-alm | Stein-platte |
| **Dasycladaceae** | | | | | | | | | |
| Acicularia sp. | | x | | | | | | x | |
| Clypeina sp. | | | | | | x | x | x | |
| Diplopora adnetensis FLÜGEL | | | | | x | | x | x | |
| Diplopora phanerospora PIA | | x | | | x | x | x | | |
| Diplopora tubispora OTT | x | x | | | | | x | | |
| Griphoporella curvata (GÜMBEL) | | x | x | x | | x | x | | |
| Gyroporella vesiculifera GÜMBEL | | | x | | | | | x | |
| Heteroporella crosi (OTT) | | x | | | | | x | | |
| Heteroporella zankli (OTT) | x | x | x | | x | x | x | | |
| Macroporella sp. | | x | | | | | | | x |
| Pentaporella rhaetica SENOWBARI-DARYAN | | | | x | | | | x | x |
| **Solenoporaceae** | | | | | | | | | |
| Parachaetetes maslovi FLÜGEL | (x) | x | x | x | | x | x | x | |
| Solenopora alcicornis OTT | | (x) | | | x | x | x | | |
| Solenopora endoi FLÜGEL | | x | x | x | (x) | | | | x |
| Solenopora styriaca FLÜGEL | x | | | | | (x) | | | |
| Solenopora sp. 1 ZANKL | x | | x | | x | x | x | | |
| Solenopora sp. | x | x | | x | | | | | x |

TABLE 10.—FREQUENTLY ENCOUNTERED *MICROPROBLEMATICA* IN UPPER RHAETIAN REEFS AND DACHSTEINKALK REEFS

| Microproblematica | Characteristics | Interpretation | Distribution | Occurrence within Upper Rhaetian Reefs |
|---|---|---|---|---|
| *Aeolisaccus tintinniformis* MISIK, 1971 | small tubes, conical, micritic walls; $\phi$ 40–70 $\mu$m | pteropods or foraminifers | Scythian to Rhaetian | reef cavities within the framework of reef patches |
| *Baccanella floriformis* PANTIC, 1971 | cauliflower-shaped aggregates composed of radially arranged branches, $\phi$ of the form 100–600 $\mu$m | codiacean algae | Ladinian to Rhaetian | protected environments of the central reefs, together with calcisponges |
| *Bacinella irregularis* RADOICIC, 1959 | encrusted sheets, composed of an irregular network of chamberlets | algae | Middle Triassic to Lower Cretaceous | upper reef-slope, high energy environment |
| *Cheilosporites tirolensis* WÄHNER, 1903 | series of cup-shaped chambers, connected by a central channel | calcareous sponge; algae or foraminifers | Norian and Rhaetian | central parts of the reef patches, together with calcisponges |
| *Lamellitubus cauticus* OTT, 1967 | tubes, open at one end, walls composed of oblique lamellae, length of the tubes up to 5 mm | serpulids, hydroid polyps, calcareous sponges | Middle and Upper Triassic | central reef areas |
| *Lithocodium* sp. (= *Problematicum* A, OHLEN, 1959) | crusts on framebuilding organisms; crusts composed of chambers, whose roofs contain numerous pores | algae resp. codiacean algae | Upper Triassic, Upper Jurassic and Lower Cretaceous | marginal reef patches with high-growing corals |
| *Microtubus communis* FLÜGEL, 1964 | small tubes, often curved, tubes segmented with rings 100–200 $\mu$m | cyanophycean algae; foraminifers | Upper Triassic | central areas of the reef patches |
| *Muranella sphaerica* BORZA, 1975 | globular hollow bodies, walls with radial structures, diameter 100–400 $\mu$m | algae | Carnian to Rhaetian | in reef cavities in the central areas of the reef patches |
| *Pycnoporidium?* eomesozoicum FLÜGEL, 1975 | crusts on framebuilders, composed of parallel, curved tubes, thin walls, diameter of the tubes 30–50 $\mu$m | rhodophycean algae | Norian and Rhaetian, Upper Jurassic | reef patches and oncoidal facies |

TABLE 10.—CONTINUED

| Microproblematica | Characteristics | Interpretation | Distribution | Occurrence within Upper Rhaetian Reefs |
|---|---|---|---|---|
| Radiomura cautica SCHÄFER & SENOWBARI-DARYAN, 1979 | rows and aggregates of half-spherical chambers, partly with a central channel, diameter about 500–1200 μm | algae | Norian and Rhaetian | central parts of the reef patches |
| Thaumatoporella parvovesiculifera RAINERI | thin sheets composed of polygonal cells | rhodophycean algae | Middle Triassic to Upper Cretaceous | central parts of the reef patches and oncoid facies; often allochthonous |
| Tubiphytes obscurus MASLOV | nodular bodies, consisting of a very fine, micritic network, sometimes with spar-filled cavities | cyanophycean algae | Carboniferous to Upper Jurassic | central reef patches |

## Reef Communities

As can be seen from Figure 12 the basic framework of the reefs is composed of several biological communities, which are defined in a rather simple way by the quantitative predominance of one or a few reef building species and genera, and by a regular association of these forms with particular successions of organisms found as epibionts upon the primary framebuilders. Because of the difference in purpose of various investigations the biotic associations described by Schafer (1979), Senowbari-Daryan (1978), and Piller (this volume) from upper Rhaetian Reefs, and Flügel and E. Flügel-Kahler (1963), Zankl (1969), Lobitzer (1974), and Dullo (1979) from the Dachstein Limestone Reefs, are difficult to compare. Therefore comparison is made to only those communities listed in Table 11, which seem to have been more widely distributed. Of special interest are 6 communities:

1. *"Thecosmilia" clathrata, Form A (=Retiophyllia clathrata).*—This community shown in Figure 13B is characterized by the predominance of this particular species, and by the overall scarceness or absence of other primary framebuilders. This community is known from Adnet, the Feichtenstein Reef, the Steinplatte, and from the Hochschwab. The community occurs in individually separated reef patches surrounded by areas with oncolithic reef-derived carbonate particles. This community characterizes the high energy marginal parts of the reef-crest.

2. *"Thecosmilia" clathrata, Form B (Retiophyllia paraclathrata).*—This community shown on Figure 20A is known from most upper Rhaetian Reefs, from Kössen coral biostromes, and from some Dachstein Limestone Reefs (Göll, Gosaukamm, Sauwand). It is characterized by relatively low diversity phaceloid and dendroid colonies with thin corallites. As can be seen from the micritic sediment type, from its location within the more protected parts of the central reef (Adnet, Feichtenstein, Göll, and Sauwand), and from the occurrences in the somewhat "deeper" Kössen environments, this community seems to have been adapted to low energy environments. This corresponds with the occurrence of this community in patch-reefs behind the reef-crest zone of the Steinplatte complex (see Piller, this volume).

3. *"Montilauvltia" or "Solitary Coral."*—This community is characterized by generally low-growing but medium- to high-diverse groups of solitary corals associated with calcisponges (e.g., *Paradeningeria*), and hydrozoans. Most of the corals serve as substrates for tabulozoans. This community is known from the lower part of the

TABLE 11.—GENERALIZED COMPARISON OF THE MOST IMPORTANT FRAMEBUILDING BIOCOENOSES. THE BIOCOENOSES ARE NAMED AFTER THE QUANTITATIVELY DOMINANT TAXA. LOCALITIES 1.–9.: 1. ADNET, 2. RÖETELWAND, 3. FEICHTENSTEIN, 4. GRUBERALM, 5. STEINPLATTE, 6. HOHER GÖLL, 7. GESÄUSE, 8. HOCHSCHWAB, 9. SAUWAND; GROWTH FORMS : HF = HIGH-GROWING FORMS, LF = LOW-GROWING FORMS, EF = ENCRUSTING FORMS

| Biocoenosis | Growth Forms | Diversity | Framebuilding Communities | Upper Rhaetian Reefs | | | | | Dachsteinkalk Reefs | | | |
| --- | --- | --- | --- | --- | --- | --- | --- | --- | --- | --- | --- | --- |
| | | | | 1 | 2 | 3 | 4 | 5 | 6 | 7 | 8 | 9 |
| Corals | HF | Low | *Thecosmilia clathrata* A | x | | | | x | | | x | x |
| | | | *Procyclolites* | | | | | | x | | | |
| | | High | *Thecosmilia cyclica* | | x | x | x | | | | | |
| | | | *Thecosmilia clathrata* A | x | | x | x | x | | x | | x |
| | | | *T. clathrata* A + B | x | | x | | | | | | |
| | | | *Thecosmilia* + sponges | | | | | | x | | x | |
| | | | *Astraeomorpha* | | | x | | | x | | | |
| | LF | Low | *Thecosmilia clathrata* B | x | x | x | | x | x | | | x |
| | | High | *Thamnasteria rectilam* | x | | x | | | | | | |
| | | | *Montlivaultia* | x | | x | | x | | x | | |
| | | | *Thecosmilia clathrata* B + *Astraeomorpha* | x | | | | x | | | | |
| | | | *Astraeomorpha* (+*Solenopora*) | | | x | | | x | | | |
| | EF | High | *Seriastraea* | | x | | | x | x | | | |
| Calcisponges | LF | Low | *Peronidella* | x | | | | | | | | x |
| | LF | High | *Colospongia* | x | | x | | x | x | | x | |
| | | | Calcisponges | | x | | | | | | | |
| | | | *Paradeningeria* + *Disjectopora* | x | x | x | | | x | x | | |
| | | | *Paradeningeria* + corals | | | x | | | | | | |
| | | | *Paradeningeria* | | x | x | x | x | | | | |
| | | | *Salzburgia* | | | | | x | | | | |
| S. | HF | High | Solenoporacean algae | | | x | | | x | x | x | |
| Hydrozoa Bryozoa | LF | Medium to Low | Hydrozoa + bryozoa | | | | | | | x | | |

TABLE 12.—DISTRIBUTION OF VARIOUS FORAMINIFERAL ASSOCIATIONS IN SOME UPPER TRIASSIC CARBONATES (AFTER SCHÄFER AND SENOWBARI-DARYAN, 1978; DULLO, 1979; HOHENEGGER AND LOBITZER, 1971; AND HOHENEGGER AND PILLER, 1975B)

| Foraminiferal Associations | Reef Carbonates | | | | Platform Carbonates Totes Gebirge |
|---|---|---|---|---|---|
| | Rötelwand | Feichtenstein | Gesäuse | Hochschwab | |
| sessile arenaceous forms and *Alpinophragmium* | biolithite facies within the high-growing coral community | biolithite facies, in the *Pinacophyllum* community | biolithite facies dominantly on high-growing framebuilders | reef patches in the central reef area | — |
| sessile miliolids | reef detritus between reef patches and Kössen Limestones | protected parts within the biolithite facies; pellet-mud facies | calcarenites with reef derived bioclasts between the reef patches | central reef | — |
| *Galeanella*, "*Sigmoilina*," *Ophthalmidium* | low-growing communities with calcisponges, hydrozoans, and *Montlivaultia*; reef cavities | reef cavities in the *Montlivaultia*-community and in the *Colospongia*-community | reef cavities and in low-growing framebuilders | reef patches and reef detritus | — |
| *Ophthalmidium leischneri, O. martanum* | in reef cavities; low-growing framebuilders | reef cavities | only in reef cavities | Aflenz patch-reefs | |
| *Aulotortus communis, A. tumidus* | grapestone facies | algal-foraminiferal facies; oncoid facies | grapestone facies | reef patches and back-reef areas | algal-foraminiferal-detritus facies, mud facies |
| "*Duostomina*," "*Tetrataxis*" | oncoid facies | widely distributed, common in the pellet-mud facies; in the reef in the *Astraeomorpha* community | grapestone facies | reef patches and Aflenz patch-reef | oolite facies |
| *Aulotortus sinuosus, A. communis, A. tumidus* | algal-foraminiferal facies | algal-foraminiferal facies, oncoid facies | algal-foraminiferal facies | reef patches and back-reef areas | algal-foraminiferal-detritus facies |
| *Aulotortus friedli, A. tenuis*, lagenids | mud facies | in all the facies units, maximum in the basinal mud facies | mud facies | nodosariids; mud facies | mud facies |
| *Trochammina, Endothyra* | near-reef detritus facies | widely distributed, specially in the oncoid facies and the basinal mud facies | pellet-mud facies | mud facies | pellet-mud facies |

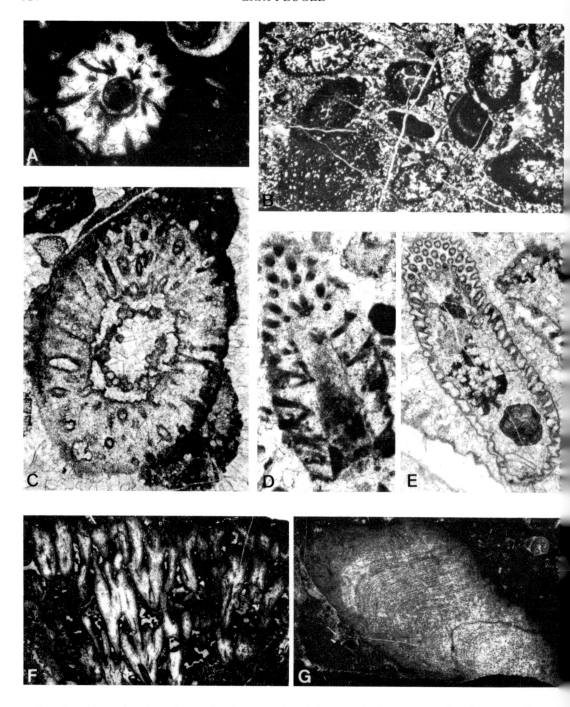

FIG. 10.—Thin-section photomicrographs of common dasycladacean and solenoporacean algae from upper Rhaetian Reefs (A–C and F), and Dachstein Limestone Reefs (C, E, G). A. *Diplopora adnetensis* Flügel, Adnet, ×20. B. Several coated fragments of *Heteroporella zankli* (Ott) one of the most frequent dasyclads from the oncoid facies; Feichtenstein, ×5. C. *Diplopora* cf. *phanerospora* Pia, a characteristic endospore dasyclad found in the algal-foraminiferal facies of the back-reef; Hinterer Gosausee, Gosaukamm, ×30. D. *Pentoporella rhaetica* Senowbari-Daryan, an oblique longitudinal section; Gruberalm, ×25. E. *Griphoporella curvata* Gümbel; Haindlmauergöss, Gesäuse, ×25. F. *Solenopora* cf. *alcicornis* Ott, a characteristic solenoporacean alga from the marginal areas of reef patches; Rötelwand, ×6. G. *Parachaetetes maslovi* Flügel, Gosaukamm, ×6.

TABLE 13.—INTERRELATIONSHIPS BETWEEN FORAMINIFERAL ASSOCIATIONS AND BIOTOPE PREFERENCE, FEICHTENSTEIN REEF (AFTER SENOWBARI-DARYAN, 1978)

| Foraminiferal Associations | Associated Genera | Preferred Communities | Facies Unit |
|---|---|---|---|
| Alpinophragmium association | relatively rare; sessile arenaceous foraminifers and Tetrataxis | high-growing corals ("Thecosmilia" clathrata, forms A–C, Pinacophyllum-communities) | biolithite facies |
| Association with sessile miliolids | Ophthalmidium, Galeanella, "Sigmoilina," Quinqueloculina | low-growing communities (Montlivaultia-C, hydrozoans-C, calcisponges-C) | biolithite facies |
| Ophthalmidium-Galeanella association | "Sigmoilina," Quinqueloculina, "Lituosepta" | reef cavities between framebuilding organisms | central part of the biolithite facies |
| "Involutina" association | Glomospira, Glomospirella, Trocholina, "Duostomina" | various algal associations | oncoid facies and algal-foraminiferal facies |
| Association with lagenids | Trochammina, sessile miliolids, Duostomina, Agathammina | together with ostracodes and molluscs | pellet-mud facies, mud facies (Kössen basinal facies) |

reef-slope (Steinplatte); from the innermost protected reef patches in the central reef area (Feichtenstein), and from protected areas between the corals of the community "Thecosmilia" clathrata, form A (Adnet).

4. "Colospongia"—Community (Figs. 4B and 16).—This community is known from micritic as well as calcarenitic reef limestones. Colospongia and other sphinctozoans are the dominant elements (up to 50 percent of the biota). Organic encrustations are common to frequent, especially by tabulozoans, spongiostromate algal crusts, Microtubus, Radiomura, and serpulids. In contrast to the "Thecosmilia"-communities, the percentage of partly spar-filled open spaces in the organic framework is rather high. This community seems to have been lived in well protected parts of the central reef (Adnet and Feichtenstein), and also in the relative high energy environments of the upper reef-slope (Steinplatte). Some of the communities in the Dachstein Limestone Reefs, identified as "calcisponge-communities," may actually represent the "Colospongia"-community. But in these reefs, especially in the Gosaukamm, Gesäuse and Hochschwab Reefs, non-segmented calcisponges (type Peronidella) are also important framebuilders.

5. "Paradeningeria"-Community (Figs. 4A, C and D).—Described first from the upper Rhaetian Reefs, but known also from the Dachstein Limestone Reefs (Gosaukamm), this community seems to have been concentrated in the protected parts of the reef-flat (Feichtenstein and Rötelwand), the upper part of the reef-crest (Steinplatte), and in protected areas on the leeward side of reefs (Feichteinstein, and Gosaukamm). This community is characterized by the predominance of low-growth forms of sphinctozoan sponges associated with a high diversity of corals, hydrozoans, and bryozoans/tabulozoans. The occurrence of high diversity of corals, hydrozoans, and of the hydrozoan Disjectopora has been used by Schäfer to distinguish a community association of Paradeningeria and corals, and another community association with Paradeningeria and Disjectopora.

6. "Solenoporacean Algal"-Community (Fig. 10F).—The bush-like thalli of solenoporacean red alga together with inozoan calcisponges and hydrozoans, are quantitatively important elements in some Dachstein Limestone Reefs. In contrast to distribution patterns at Göll and Hochschwab (transition area between central reef and fore-reef areas), in the Schildmauer Reef, Gesäuse, this community is also found in the lagoonal area of the central reef (Dullo, 1979). In the upper Rhaetian Feichtenstein Reef a solenoporacean community (with Solenopora alcicornis Ott instead of Solenopora sp. 1, Zankl, 1969) is restricted to the

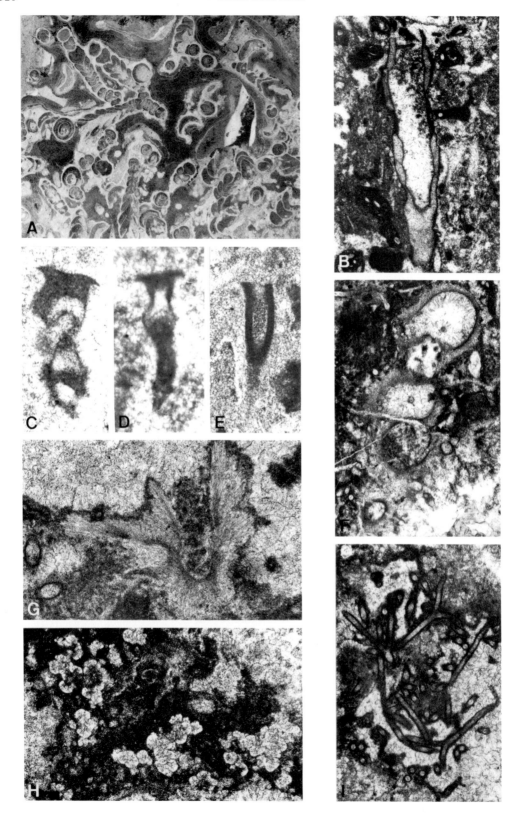

central reef area; this community generally contains high-growing colonies of the coral *Astraeomorpha confusa*. This coral may dominate other species within this community and thus characterizes an individual association that is typical for the more protected low energy areas of some reefs.

A critical comparison of the various communities described from the Upper Triassic reefs is needed in order to reliably recognize regular distributional patterns of these biotic associations. At this time various facies and distribution patterns within the reefs are fairly well known (Göll, Adnet, Rötelwand, Feichtenstein, Steinplatte), but rather incomplete distribution patterns are known from other reefs (Hochschwab and Gesäuse). Schäfer (1979) has shown that in the upper Rhaetian Adnet and Rötelwand Reefs a total of 9 organism communities can be distinguished based on the dominant organic framebuilders and their associated epibionts. The greatest diversity in the epibiont biota is found in the more protected parts of the central reef areas, or on the somewhat deeper reef-slope environments where there is a predominance of calcalcisponges, hydrozoans, and solitary corals.

*Distributional Patterns of Micro-organisms*

Foraminifers, calcareous algae, and various *microproblematica* are concentrated in distinct parts of the reef complexes and in their adjacent facies. This is due to the adaption of various organisms to special micro-environments which can be strongly influenced by different types of substrate and varying levels of water energy. These distributional patterns seem to be rather similar in all the reefs.

*Foraminifera (Figs. 14 and 15).*—Apparent facies control of Triassic foraminifers has been documented by the work of Resch (1972 and 1979), Hohenegger and Lobitzer (1971), Hohenegger (1974), Hohenegger and Piller (1975b), Piller (1978), Schäfer and Senowbari-Daryan (1978), and Dullo (1979). Even considering the often poor preservation of some foraminifers, there is now sufficient evidence for recognition of various facies units on platforms and on basins by distinctive foraminiferal association. It is possible to

differentiate various units of the platform environment (Senowbari-Daryan and Schäfer, this volume), as well as special foraminiferal ecological niches of the central reef area. Within the central reef various communities of reef building organisms correspond to special biotopes containing well defined foraminiferal associations (see Table 12). Strong differences can be seen in the composition of the foraminiferal fauna found within high-growing and low-growing framebuilder communities. Another very specialized foraminiferal association characterizes the biotopes of small reef cavities. The foraminiferal distributional patterns in the various upper Rhaetian Reefs are very similar; the same is true for distributional patterns found within corresponding facies units of the Dachstein Limestone Platform (Hohenegger and Piller, 1975b; and Dullo, 1979). All studies indicate a strong dependence of foraminiferal distribution on the setting within the depositional environment (differentiated into reef patches, reef-slope, shallow water flats, and basins), the type of substrate (biogenic hardgrounds, sandy or muddy bottoms), and the water depth and salinity. The most obvious relations appear to have existed between sediment substrate and foraminiferal distribution.

All foraminiferal studies are based on a quantitative survey of the organisms in which all individuals are counted within a defined thin-section area (9 cm²), and are taken as a reference standard. These counts indicate the occurrence of various foraminifers in different areas of the reefs, and in adjacent units. By comparing the values obtained from some upper Rhaetian and Dachstein Limestone Reefs, differences in abundance can be readily seen. In the upper Rhaetian Reefs, and in the platform environment of the Dachstein Limestone, most foraminifers occur in the "algal-foraminiferal facies." In the Dachstein Limestone Reef complexes (Gesäuse, Gosaukamm, and Sauwand) the greatest frequency of foraminifers occurs within reef cavities, where numerous miliolid individuals (*Ophthalmidium*, *Galeanella*, and *Sigmoilina*) can be recognized. Table 13 shows some of the most common foraminiferal associations.

*Calcareous Algae (Figs. 9 and 10).*—The dis-

←

FIG. 11.—Thin-section photomicrographs of common *microproblematica* from Upper Triassic reef limestones. A. *Cheilosporites tirolensis* Wähner; Feichtenstein, ×1.5. B. *Lamellitubus cauticus* Ott; Gosaukamm, ×15. C.–D. *Spiriamphorella rectilineata districta* Borza and Samuel, common vase-shaped forms frequent in the Dachstein Reef Limestones of the Gesäuse, and in the "Wand Limestone," but also present in Upper Triassic reefs of northwestern Sicily; Lupo di Cuzzo near Palermo, ×80. E. *Aeolisaccus tintinniformis* Misik; Lupo di Cuzzo, Sicily, ×80. F. *Radiomura cautica* Senowbari-Daryan and Schäfer; Feichtenstein, ×20. G. *Muranella sphaerica* Borza; Adnet, ×50. H. *Baccanella floriformis* Pantić; Rötelwand, ×15. I. *Microtubus communis* Flügel; Feichtenstein, ×12.

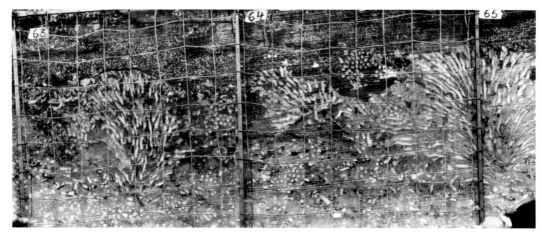

FIG. 12.—Outcrop photograph showing the vertical ecological zonation in the marginal area of the Adnet Reef in an exposed quarry wall (Tropfbruch, Adnet). At the base of the quarry wall oncoidal detrital carbonates are exposed, and these have served as the substrate for some isolated bushy colonies of *"Thecosmilia" clathrata* form A. Between these colonies biogenous detritus derived from other nearby reef patches has been deposited, sometimes occurring with finely laminated carbonate mud. Colonies of *Thecosmilia clathrata* form A are truncated by a solution surface, indicating an interruption of reef development. The upper part of the quarry wall shows a very dense association of coppice-like interwoven colonies of the *"Thecosmilia" clathrata* form B characterized by rather small corallites. Size of the grid is 15 centimeters.

tributional patterns of calcareous algae have been reported for upper Rhaetian Reefs by Senowbari-Daryan and Schäfer (1979), and for the Dachstein Limestone Reefs by Flügel (1979). The algae species may be grouped according to occurrences in different water depths, different water energy, and various substrate types. Differences in algal distributional patterns appear to exist between upper Rhaetian and Dachstein Limestone Reefs, especially with regards to the dasycladacean algae.

*Microproblematica (Fig. 11).*—As shown in Table 10 this large group of organisms of uncertain systematic position can be used to characterize individual facies types. Only a few *microproblematica* such as *Thaumatoporella vesiculifera* (Raineri), probably a red alga, and various types of *Aeolisaccus* Elliott, and similar tiny tubes with micritic walls, occur in nearly all facies units. Within the central reef area, the more protected environments are characterized by biomicrites with *Microtubus communis, Radiomura cautica, Cheilosporites tirolensis*, and *Baccanella floriformis*. The high energy marginal areas of the reef patches are rich in *Lithocodium* and *Bacinella irregularis*. These species also characterize the oncoid facies near the central reef, and in areas between reef patches. The interiors of reef cavities contain small spheroidal bodies with radially structured "walls" (*Muranella sphaerica*), which perhaps represent cyanophycean algal colonies. Similar distributional pat-

terns are known from the Feichtenstein, Rötelwand, Adnet and Steinplatte Reef complexes. The occurrences in the Dachstein Limestone Reefs are not as well known, and many new *microproblematica* still await description (Dullo, 1979).

### Ecological Successions

Since the construction of reefs by organisms consists entirely of framebuilding organisms created by growth of carbonate skeletons, and by framebinding organisms through encrustation and accretion to the framework by sessile organisms, it is also important to study the ecological succession, formed by secondary framebuilders which inhabited dead or partly living primary reef framework (Fig. 16). For the Upper Triassic reefs, E. Flügel and E. Flügel-Kahler (1963), and especially Zankl (1969), and Schroeder and Zankl (1974), have emphasized the necessity to describe successions of various epizoans and epiphytes found as secondary framebuilders or framebinders. According to the detailed investigations by Schäfer (1979) and Senowbari-Daryan (1978), regular but rather complicated biotic successions can be recognized, and these rarely follow the sequence of generations described by Zankl (1969, p. 66). Zankl's complete succession of secondary framebuilders may consist of four biotic generations (sessile foraminifers; encrusting bryozoans, calcisponges and solenoporacean algae; algal crusts and sessile foraminifers; and spongiostro-

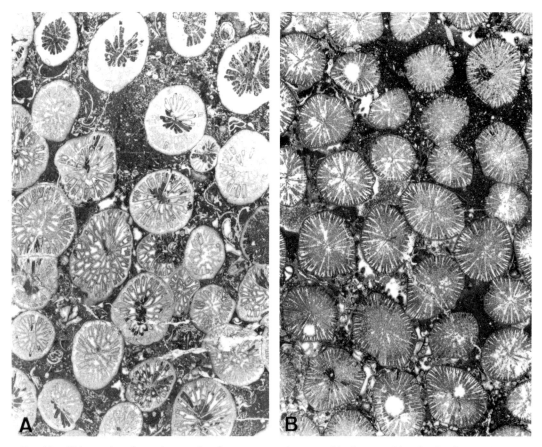

F IG. 13.—Thin-section photomicrographs of common coral communities in upper Rhaetian Reefs. A. *Thecosmilia cyclica*-community, characterized by the predominance of *T. cyclica* Schäfer and Senowbari-Daryan along with abundant encrustations of the foraminifer *Alpinophragmium perforatum* Flügel; Feichtenstein, ×1.5. B. Community with *"Thecosmilia" clathrata* form A; Feichtenstein Reef, ×1.5.

mate algal crusts). The differences or absence of a certain generation may be due to intensified sedimentation, or too rapid cementation. From the statistical data gathered by Schäfer, it seems that the preference of primary framebuilders by distinct secondary reef builders is governed by biological factors such as larval fixation, size of the substrate area, and also the special requirements of coexistence.

Generally, the hydrozoan-, calcisponge- and tabulozoan-dominated associations show a conspicuously higher diverse spectrum of epibionts than coral communities. In upper Rhaetian Reefs up to 10 encrusting organisms may be found growing upon sphinctozoan calcisponges like *Colospongia*, *Polytholosia*, or *Paradeningeria*. But it is evident from investigations of the Rötelwand and the Adnet material, that calcisponges which are similar to other framebuilders may have been selected as to type of epizoans and epiphytes. For

example, the genera cited above are always encrusted by *Microtubus*, *Radiomura*, *Follicatena*, and small *Annaecoelia*; other epibionts may join the community. In contrast a calcisponge group consisting of large *Peronidella*, and the sphinctozoans *Salzburgia* and *Colospongia* sp. may be devoid of biogenous overgrowth. Disjectoporid hydrozoan colonies often are encrusted by *Microtubus*, *Radiomura*, *Follicatena*, *Girvanella*, small brachiopods, and *Annaecoelia*; whereas *Spongiomorpha ramosa* is overgrown by *Alpinophragmium alpinum*, *Thaumatoporella*, *Nubecularia*, *Lithocodium* and *Bacinella*.

Tabulozoans and bryozoans exhibit different intensities and types of biogenous overgrowth, probably indicating various biotopes. Four groups of corals can be recognized as having different types of epibionts. *Lithocodium*, *Bacinella* and porostromate blue-green algae are found growing upon *Stylophyllum*, *"Thecosmilia" cyclica*, and

*Pinacophyllum. Lithocodium* and *Bacinella,* but no other epibionts encrust colonies of *Astraeomorpha, Thamnasteria, Astrocoenia,* and *"Isastraea."* Phaceloid colonies of *"Thecosmilia"* are only overgrown by sessile miliolid and agglutinating foraminifers as well as by small bryozoans, *Annaecoelia,* and *Thaumatoporella.* Extraordinarily common overgrowths are reported for *Montlivaultia norica* and tabular colonies of *Seriastraea multiphyllia.*

High diversities of epibionts as well as primary framebuilding organisms are found in the more central parts of the reef patches, but there are rather low diversities in the marginal parts. This pattern seems to be valid especially for upper Rhaetian Reefs. In the Dachstein Limestone Reefs the low-growing coral communities and calcisponge communities contain other epibiont associations.

### MICROFACIES AND FACIES TYPES

In 1972 the author attempted to describe the most abundant types of Upper Triassic reef and platform carbonates within broadly defined "microfacies types," based primarily on their paleontological characteristics and depositional textures as seen in thin-sections. This classification approach was adopted and modified by Wilson (1975) who erected a "Standard Microfacies Types" classification for limestones, through which the interpretation of carbonate environments has now become somewhat more refined. Some of his "SMF" types are certainly too broad ranging and must be modified in terms of paleontologically defined microfacies types. In the Upper Traissic reefs this is particularly true, especially for the biolithite facies of the reef-crest and reef-slope, where primary and secondary framebuilders together with reef detritus of varying composition may contribute to the formation of microfacies of differing limestone types (frame- and bafflestones; wackestones; packstones; grainstones). By considering the predominance of certain microfossils, such as foraminifers and *microproblematica* in more local ecologically restricted areas (reef cavities, with their special associations of epibionts) these limestone types can give clues to the diversity of reef biotopes (e.g., Schäfer,

1979). On the other hand, it does seem useful to define broader facies units, which can also be recognized in the field, in order to simplify large-scaled facies analysis. If these facies units are described by the same or similar kinds of characteristics, as in Recent depositional environments of comparable sedimentation areas (e.g., ancient platform carbonates and the Great Bahama Bank), the genetic aspect would allow an initial interpretation of the facies. This method of approach has been introduced for study of Upper Triassic platform and reef carbonates by Piller (1975b), who, with analogy to the sedimentation patterns of the Bahama Bank, differentiated 6 facies units (3 to 8, Table 14, Fig. 15) in the platform carbonates of the bedded Dachstein Limestones of the Totes Gebirge, Upper Austria. These facies units are defined by the most frequent occurrence or the most conspicuous carbonate particle type (e.g., grapestone facies), by a mud-to-particle ratio, and by the biota, especially the foraminifers. As shown by investigations of the "Erlangen Reef Research Group," facies units established by Piller (Hohenegger and Piller, 1975) can be used without major modifications to describe sediments adjacent to the framework of upper Rhaetian and Dachstein Limestone Reefs.

The description of reef complexes requires the separation of facies units which characterize the zone of active reef growth ("biolithite facies," Fig. 25), and the high-energy or lower-energy fore-reef zones (sparry rudstone facies or micritic floatstone facies). Furthermore it seems valuable to differentiate a near-reef shallow water zone in which most of the reef derived bioclasts are circumscribed by algae, foraminifers, *microproblematica,* and other sessile organisms. This "oncoid facies" (Schäfer and Senowbari-Daryan, 1978; Schäfer, 1979) is in some ways similar to the "grapestone facies" (Fig. 25).

Other facies types are necessary for platform carbonates in order to include the far-reef algal bindstones, formed in inter- and partly supratidal environments. The terms "algal mat loferite facies" and "pellet-loferite facies" seem appropriate for these carbonates (Fischer, 1964, p. 127; Dullo, 1979). The term "mud facies" introduced by Piller should be changed to "shallow water mud facies" in order to avoid confusion with the

$\rightarrow$

FIG. 14.—Thin-section photomicrograph of typical foraminifers, which may be used as facies indicators in upper Rhaetian Reefs (A and E), and Dachstein Reefs (B–D, F–I). A. *Opthalmidium triadicum* (Kristan); Adnet, ×70. B. *"Sigmoilina"* sp.; Gosaukamm, ×70. C. *Aganthammina austroalpina* Kristan-Tollmann and Tollmann; Gosaukamm, ×150. D. *Alpinophragmium perforatum* Flügel; Kalbling, Gesäuse, ×60. G. *Triasina hantkeni* Majzon; Grimming, Styria, algal-foraminiferal facies, ×50. H. *Glomospirella kuthani* (Salaj) and *Frondicularia* sp.; Gosaukamm, ×20. I. *Galeanella tollmanni* Kristan-Tollmann; Gosaukamm, characteristic element of small reef cavities, ×90.

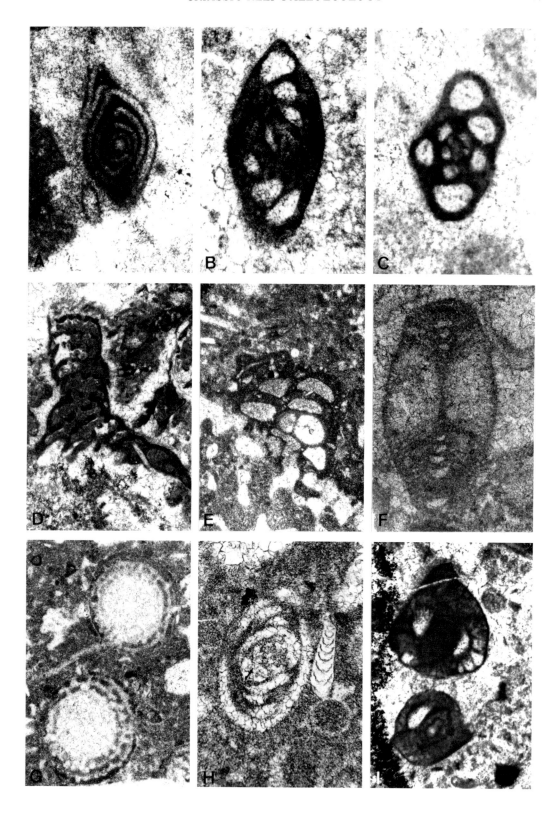

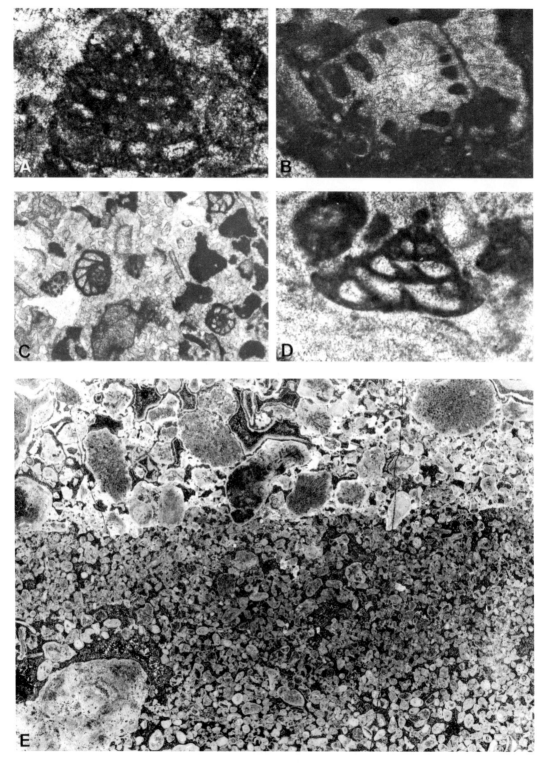

Fig. 15.—Thin-section photomicrographs of various foraminifers and facies types. A. *"Lituosepta"* sp., Schildmauer Reef; Gesäuse, ×70. B. *Trocholina* sp., Gruberalm Reef, ×50. C. Grainstone with several rotaliid foraminifers (Duostominidae); Sauwand Reef, ×15. D. *"Tetrataxis" inflata* Kristan; Adnet, ×40. E. Grapestone facies

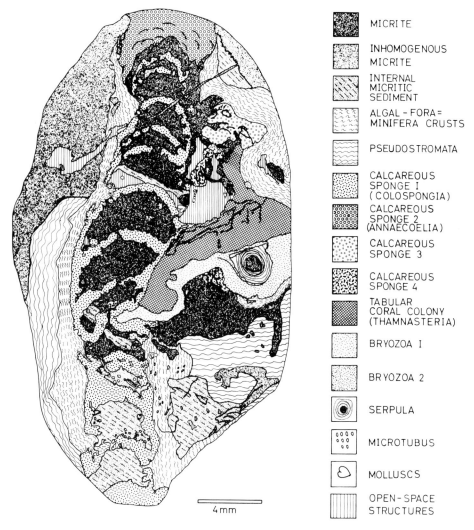

| | MICRITE |
| --- | --- |
| | INHOMOGENOUS MICRITE |
| | INTERNAL MICRITIC SEDIMENT |
| | ALGAL – FORA= MINIFERA CRUSTS |
| | PSEUDOSTROMATA |
| | CALCAREOUS SPONGE I (COLOSPONGIA) |
| | CALCAREOUS SPONGE 2 (ANNAECOELIA) |
| | CALCAREOUS SPONGE 3 |
| | CALCAREOUS SPONGE 4 |
| | TABULAR CORAL COLONY (THAMNASTERIA) |
| | BRYOZOA I |
| | BRYOZOA 2 |
| | SERPULA |
| | MICROTUBUS |
| | MOLLUSCS |
| | OPEN–SPACE STRUCTURES |

4mm

FIG. 16.—Diagram (×4) showing the ecological succession within a *Colospongia*-community, Adnet Reef. This diagram shows various different generations of secondary framebuilders not only growing upon a sponge, but even inside the sponge.

basinal mud facies represented by the Kössen environments ("terrigenous mud facies," Schäfer and Senowbari-Daryan, this volume).

### Facies Units of the Reef and Platform Environments

In Table 14 the basic characteristics of 12 facies units are summarized: these data show which fa-cies can develop in Upper Triassic reef complexes and platform carbonates. These units by no means represent the total facies spectrum, because the marly and the basinal sediments of the Kössen Zlambach Beds are not included and differentia-tion is not made between the mud-mound stage and the "ecological reef"-stage of the biolithite facies.

←

(lower part), and algal-foraminiferal facies (upper part). The grainstone of the grapestone facies is characterized by moderate to well-sorted aggregate grains and micritized bioclasts. At lower left corner blue-green algal nodule (*Zonotrichites*). In the poorly-sorted packstone of the algal-foraminiferal facies many solenoporacean bioclasts, algal lumps, and a few foraminifers can be seen. The interparticle voids are occluded by several generations of fibrous cements; negative print, Steinplatte, ×5.

TABLE 14.—CHARACTERISTICS OF FACIES-UNITS DIFFERENTIATED WITHIN UPPER TRIASSIC REEF- AND CARBONATE PLATFORM-FACIES ON THE NORTHERN CALCAREOUS ALPS

| | Restricted Shelf with Tidal Flats | | | | Open Shelf-Lagoon | |
| | 1 Algal Loferite Facies | 2 Pellet Loferite Facies | 3 Mud Facies | 4 Pellet Mud Facies | 5 Algal-Foraminiferal Calcarenite | 6 Grapestone Facies |
|---|---|---|---|---|---|---|
| **Carbonate particles** | | | | | | |
| peloids | rare | abundant | in patches or layers | common | common | abundant |
| fecal pellets | — | common | common | abundant (size 50–70 $\mu$m) | — | — |
| mud aggregates | — | rare | more frequent than peloids | common | — | — |
| grapestones | — | — | — | — | common to abundant | up to 50% |
| coated-grains | — | — | — | common | rare | abundant |
| superficial ooids | — | — | — | — | — | rare |
| normal ooids | — | — | — | — | — | — |
| intraclasts | — | — | common, up to 5 cm, algal intraclasts | rare | — | rare |
| **Biota** | | | | | | |
| foraminiferal frequency | very rare | very rare | up to 100% of the total biota | common | abundant | rare |
| typical genera | | | thin-shelled Nodosariids, "I." communis, "I." friedli | Agathammina, Trochammina, Frondicularia, Palaeospiro-plectammina | "I." communis, "I." sinuosa, Frondicularia | Quinqueloculina, Miliolipora |
| algal mats | abundant | not well differentiated | | | very rare | rare |
| fenestral fabrics | abundant, laminoid, LF-A | birdseyes irregularly distributed | | | | — |
| porostromate blue-green algae | scarce | scarce | — | — | nodular or encrusted, common | common |

| Biota (continued) | Restricted Shelf with Tidal Flats | | | | Open Shelf-Lagoon | |
|---|---|---|---|---|---|---|
| | 1 — Algal Loferite Facies | 2 — Pellet Loferite Facies | 3 — Mud Facies | 4 — Pellet Mud Facies | 5 — Algal-Foraminiferal Calcarenite | 6 — Grapestone Facies |
| dasycladacean algae | — | — | — | — | abundant in situ Diplopora Griphoporella | common (Heteroporella, Diplopora) |
| solenoporacean algae | — | — | rare | — | rare to common | rare |
| algal oncoids, algal-foraminiferal oncoids | — | — | — | — | — | — |
| calcisponges | — | — | — | — | very rare (reef derived) | rare (reef derived) |
| corals | — | — | — | — | very rare (reef derived) | rare (reef derived) |
| hydrozoans | — | — | — | — | very rare (reef derived) | rare (reef derived) |
| bryozoans, tabulozoans | — | — | — | — | — | — |
| brachiopods | — | — | — | — | — | — |
| gastropods | — | — | smaller gastropods | — | — | common |
| pelecypods | — | — | rare | — | megalodontid forms | rare to common |
| ostracodes | rare | rare | in patches | common | — | rare |
| echinoderms | — | — | holothurian sclerites | rare | — | rare |
| Matrix | micritic matrix and internal sediment | micrite | micrite | micrite | micrite | micrite rare |
| Cement | — | cements A and B | scarce | — | common | common |
| Size of the particles | — | fine arenite | fine to medium arenite | fine to medium arenite | fine to medium arenite | coarse arenite to fine rudite |
| Roundness of the particles | — | good | moderate to good | good | moderate | moderate to good |
| Depositional texture | bindstone | bindstone | mudstone, wackestone | wackestone, packstone | grainstone | grainstone, packstone |

TABLE 14.—CONTINUED

| | Oolite Dunes | | Organic Buildup | | | |
| | 7 | 8 | 9 | 10 | 11 | 12 |
| | Oolitic Facies | Oolite Facies | Reef Detrital Mud Facies | Oncoid Facies | Biolithite Facies | Sparry Rudstone Facies |
|---|---|---|---|---|---|---|
| **Carbonate particles** | | | | | | |
| peloids | abundant | common | common to abundant | common | common | — |
| fecal pellets | rare | rare | common | rare | common | — |
| mud aggregates | — | — | common | — | — | — |
| grapestones | abundant | common | — | common | — | — |
| coated-grains | common | — | common | common | rare | — |
| superficial ooids | common | rare | — | — | — | — |
| normal ooids | rare | abundant | — | — | — | — |
| intraclasts | rare | — | — | common | rare | abundant |
| **Biota** | | | | | | |
| foraminiferal frequency | rare | very rare | common | common | abundant | very rare |
| typical genera | *Frondicularia, Glomospirella, Agathammina, Trochammina* | *Tetrataxis, Glomospirella* | *Trochammina, Duostominidae, Miliolidae* | "*Involutina,*" *Glomospira, Tetrataxis* | *Alpinophragmium, Ophthalmidium, Galeanella, Lituosepta* | — |
| algal mats | rare | rare | — | rare | — | — |
| fenestral fabrics | — | — | — | — | — | — |
| porostromate blue-green algae | — | — | — | rare | rare to common | — |
| dasycladacean algae | very rare | rare, reworked | — | rare | rare | — |
| solenoporacean algae | — | — | sometimes common | common | common | — |
| algal oncoids, algal-foraminiferal oncoids | — | — | — | abundant | common | — |

TABLE 14.—CONTINUED

| | Oolite Dunes | | Organic Buildup | | | |
| | 7 Oolitic Facies | 8 Oolite Facies | 9 Reef Detrital Mud Facies | 10 Oncoid Facies | 11 Biolithite Facies | 12 Sparry Rudstone Facies |
|---|---|---|---|---|---|---|
| calcisponges | — | — | rare (reef derived) | — | abundant | common (reef derived) |
| corals | — | — | very rare | — | abundant | abundant (reef derived) |
| hydrozoans | — | — | rare (reef derived) | reef derived common | abundant | common (reef derived) |
| bryozoans, tabulozoans | — | — | rare | — | abundant | common (reef derived) |
| brachiopods | — | — | — | — | rare | rare to common |
| gastropods | rare | — | — | — | rare to common | rare |
| pelecypods | rare to common | rare | sometimes lumachelles | rare | common | rare |
| ostracodes | — | — | — | — | rare | very rare |
| echinoderms | rare | rare | common to abundant (crinoids) | rare | rare to common | rare to common |
| Matrix | — | — | micrite | rare | micrite | — |
| Cement | common | common | — | common | various cement types | various cement types |
| Size of the particles | fine to coarse arenite | fine to medium arenite | siltite to arenite | arenite and rudite | fine arenite to coarse rudite | rudite |
| Roundness of the particles | good | good | poor to moderate | moderate to good | poor to good | poor |
| Depositional texture | grainstone, packstone | grainstone | wackestone | bindstone, grainstone, wackestone | framestone | rudstone |

Biota (continued)

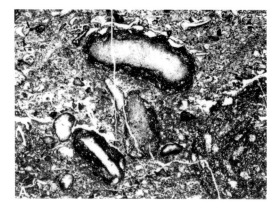

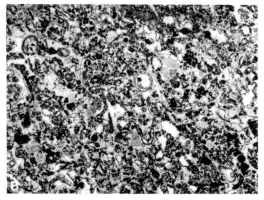

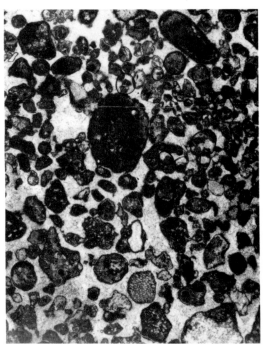

FIG. 18.—Thin-section photomicrograph (×10) of a typical grapestone facies showing poorly-sorted grainstone with many partly micritized grains, and a few algal fragments (solenoporaceans). Some aggregate grains exhibit thin oolitic coatings; Dachstein Reef Limestone; Gosaukamm.

FIG. 17.—Thin-section photomicrographs of various facies types. A. Oncoid facies; large oncoids formed by *Lithocodium,* sessile foraminifers and algal crusts encrusting reef derived bioclasts, Rötelwand, ×4. B. Algal-foraminiferal facies; grainstone with many dasycladacean algae (*Griphoporella*), involutinid foraminifers, small gastropods, echinoderms, micritized bioclasts, peloids, and some aggregate grains; Dachstein Reef Limestone of the Hahnstein, Gesäuse, ×4.

1. *Algal-Mat Loferite Facies.*—These consist of peloidal bindstones with densely-packed tiny peloids and mud aggregates. The laminoid fabric is due to spar-filled pores which tend to be alligned along bedding planes. The thin, wrinkled or flat laminations are separated by irregular sparry blisters, which may correspond to voids formed under the tough algal mats, or in part, perhaps to shrinkage cracks, animal burrows, or gas blisters. Thin laminae of micrite may be present in which the vague outlines of algal threads can be seen.

2. *Pellet-Loferite Facies.*—Consists of peloidal packstones and bindstones (often dolomites) with densely-packed micritic peloids of various sizes, which are generally merged together. This inhomogenous structure may grade into a homogenous

calcilutite. Due to changes in peloid size no conspicuous layers of millimeter laminations are developed. Irregularly distributed spar-filled pores, without internal sediment, are characteristic of these limestones.

Both facies types have been regarded by Fischer (1964) to have had an algal mat origin within an intertidal environment. The same is true for the "homogenous loferites," and for the "loferite conglomerates." The loferite biota is highly restricted; except for sparse algal remains only a few foraminifers and some thin-shelled ostracodes occur.

The algal mat loferite facies and the pellet-loferite facies are the most common sediments of the bedded Dachstein limestones (Sander, 1936) and of the Hauptdolomite (Müller-Jungbluth, 1970). These facies types represent member B of the cyclic sedimentational pattern described by Fischer (for a critical discussion see Zankl, 1971; Piller, 1976; and Lobitzer and Piller, 1979).

3. *Shallow Water Mud Facies.*—Consists of mudstones, bioclastic and peloidal wackestones with abundant skeletal grains, randomly distributed peloids, and many mud aggregates, along

with algal mat derived intraclasts. The biota consists of specialized involutinid foraminifers, thin-shelled nodosariids and ostracodes, as well as small gastropods and holothurian sclerites. Millimeter laminations may occur, and animal burrows are common.

According to Piller (1976) the mud facies environment can be characterized as shallow subtidal and intertidal, with perhaps somewhat higher turbulence than in the pellet-mud facies, and with a slightly lowered salinity. Within the bedded Dachstein limestones this facies is generally found intercalated with carbonates of the pellet-loferite facies, algal-mat loferite facies, and the pellet-mud facies.

4. *Pellet-Mud Facies.*—Consists of peloidal wackestones and packstones with abundant pellets and mud aggregates embedded in micrite. Fossils are not very common, generally only ostracodes and foraminifers (*Agathammina, Trochammina,* and *Frondicularia*) can be found, as well as some thin-shelled molluscs, sometimes forming lumachelles.

This environment is believed to have been a very shallow subtidal one, partly grading to intertidal areas with increased salinity and low turbulence. In the Dachstein Limestones this facies is intercalated within the facies types described above.

5. *Algal-Foraminiferal Facies (Fig. 17B).*— Consists of bioclastic grainstones rich in dasycladacean, solenoporacean, and filamentous blue-green algae, together with large and often thick-shelled involutinid foraminifers (*Involutina, Triasina,* and *Trocholina*). The dasycladacean algae are often found in life position or only slightly reworked. In areas adjacent to Dachstein Limestone Reefs, *Griphoporella* and *Gyroporella* are common, whereas in upper Rhaetian Reefs *Diplopora adnetensis* and *Heteroporella* may occur frequently. Many of the nodular solenoporacean and blue-green algal thalli are strongly micritized. Peloids and aggregate grains are relatively common. Some limestones belonging to this facies type are poorly-washed biomicrites, especially those with abundant solenoporacean and dasycladacean fragments. Schäfer (1979, p. 35) describes three algal-foraminiferal microfacies types which are characterized by dominance of only one or two taxa. Biogenous overgrowths are relatively rare, as are reef derived bioclasts.

This environment can be interpreted as a shallow subtidal one which may be differentiated according to varying water energy and depth, and perhaps to increased salinity. Piller (1976) postulates a sedimentation area in shallow depressions within the grapestone facies, or adjacent to that facies.

This facies is also known from the back-reef

area of the Dachstein Limestone Reefs. Zankl (1969) described calcarenites consisting of up to 50 percent dasycladacean bioclasts, together with solenoporacean and filamentous blue-green algae. A portion of these calcarenites may represent the "oncoid facies" because the average composition of near-reef lagoonal sediments, about 10 to 20 percent oncoids, are also included. Both facies units are known from back-reef areas of other Dachstein Limestone Reefs (Hochschwab, Sauwand, and Gosaukamm), as well as in upper Rhaetian Reefs. Similar microfacies types can be found in the subtidal member of the cyclical sedimentation pattern present in bedded Dachstein Limestone (Fischer, 1964).

6. *Grapestone Facies (Fig. 18).*—Consists of grainstones and packstones with up to 50 percent poorly-sorted aggregate grains (grapestones, lumps) and peloids, together with strongly micritized bioclasts, and sometimes with oncoids and superficial ooids. The biota consists of common porostromate algae, dasycladaceans (*Diplopora,* and *Heteroporella*), abundant gastropods, as well as relatively rare foraminifers (*Involutina, Quinqueloculina,* and *Miliopora*). Reef derived bioclasts may also be found.

This depositional environment is interpreted as shallow subtidal to intertidal, with varying water turbulence. This facies is known from the Dachstein Limestone Platform (Piller, 1976), and from the back-reef areas of some Dachstein Limestone Reefs (Kalbling, Gesäuse and Gosaukamm). Frequent grapestones can also be found in the limestones of the "oncoid facies" of upper Rhaetian Reefs.

7. *Oolitic Facies.*—Consists of peloidal and oolitic grainstones and packstones containing relatively many superficial ooids, peloids and grapestones, together with variously-sized and sorted micritized bioclasts (molluscs and relatively rare echinoderms). Foraminifers and dasycladaceans are scarce.

This facies unit has been described from the Dachstein Platform of the Totes Gebirge (Piller, 1976) and was compared with the "mixed oolite facies" described by Purdy (1963) from the Bahamas.

8. *Oolite Facies.*—Consists of oolitic grainstones with many superficial and "normal" ooids, together with aggregate grains and peloids. Other particles like worn algal fragments and very few foraminifers, are of minor importance. These ooids may have been redeposited in areas adjacent to local oolite shoals.

9. *Reef-Detritus Mud Facies (Fig. 21).*—Schäfer (1979) and Senowbari-Daryan (1978) have described thick-bedded to massive limestones whose matrix consists of silt-sized peloids and small bioclasts as well as micrites. Large mollusc

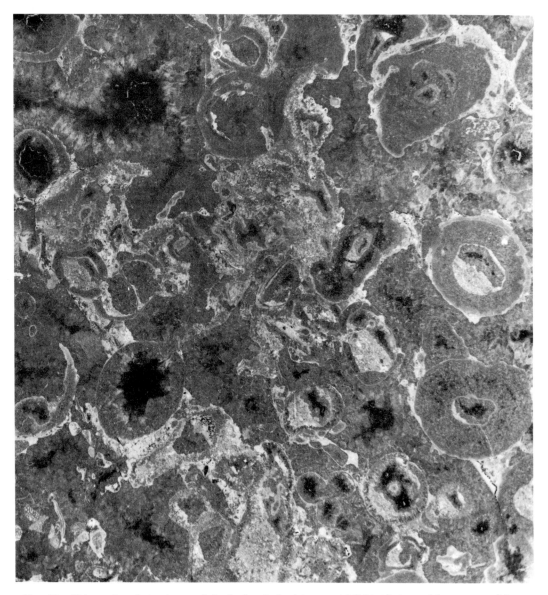

FIG. 19.—Thin-section photomicrograph (×4) of typical calcisponge biolithite facies and framestone of the protected parts of the central reef area. Note abundant sections of the non-segmented calcisponge *Peronidella* sp., together with a few sphinctozoans. All sponges are strongly encrusted by sessile foraminifers (*Alpinophragmium*, and miliolid forms), and spongiostromate algae. The sponges and a few molluscs are partly filled with internal micritic sediment, and interspaces between the sponges are occupied by strongly recrystallized carbonate cements; Donnerkogel, Gosaukamm, Dachstein Reef Limestone.

shells (partly as lumachelles) and crinoids are common. Reef derived detritus may be conspicuous, especially strongly micritized solenoporacean thalli and fragments of calcisponges. The foraminiferal association is characterized by lagenids together with *Trochammina, Planiinvolutina* and *Agathammina*.

This facies interfingers with the terrigenous mud facies of the Kössen Basin, and with the algal-foraminiferal facies and biolithite facies of the Rötelwand Reef. In the Feichtenstein Reef the reef-detritus mud facies also interfingers with the oncoid facies. Perhaps this facies characterizes the protected reef-slope.

10. *Oncoid Facies (Fig. 17B)*.—Consists of oncoidal grainstones, and sometimes wackestones

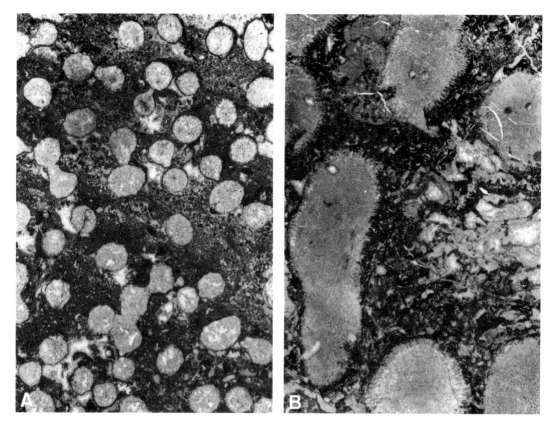

FIG. 20.—Thin-section photomicrographs (×15) of typical *Thecosmilia* biolithite facies. A. Framestone with recrystallized corallites of *"Thecosmilia" clathrata* (Emmrich) form B. B. Framestone with bryozoans and solenoporaceans completely encrusted by foraminifers (*Alpinophragmium*). Both from the Dachstein Reef Limestone, Gosaukamm.

with reef derived poorly-sorted bioclasts, intraclasts, aggregate-grains, and peloids. Most larger bioclasts are fragments of phaceloid or massive colonial framebuilders (mostly corals). These rudite-sized fragments are generally circumscribed by *Lithocodium* and *Bacinella,* forming conspicuous oncoids. Fragments of calcisponges, hydrozoans or tabulozoans are generally absent. Remains of solenoporacean algae as well as dasycladaceans and porostromate algae may be common. The foraminiferal associations are similar to that of the "reef detritus" developed between and within reef patches.

This facies is found on the windward side of the Rötelwand Reef, in the immediate neighborhood of the biolithite facies. Here the oncoid facies is developed between the reef patches in the central reef area. In the Adnet Reef, the reef patches are completely surrounded by limestones of the oncoid facies, which is believed to have been formed in a high energy shallow water environment. In the Feichtenstein Reef this facies interfingers with

the biolithite facies and with the algal-foraminiferal facies.

11. *Biolithite Facies (Fig. 19).*—Consists of framestones with strong or weak biogenic overgrowth, and coral framestones generally with less overgrowth than calcisponge framestones. The sediments within the reef patches can be differentiated according to type of matrix and particles, and according to the depositional fabrics in four microfacies types (Schäfer, 1979), which describe the "reef-detritus." These are: 1. biopelsparite with many well-sorted arenitic bioclasts (chiefly fragments of corals and sessile foraminifers, solenoporacean and dasycladacean algae as well as intraclasts); this microfacies is similar to that of the oncoid facies except for the relative scarcity of biogenic encrustations; 2. micrite with rare fossils generally only spongiostromate crusts with *Microtubus,* and some thin-shelled ostracodes and foraminifers; this microfacies type is found in protected parts of low-growing communities; 3. pelbiosparite composed of abundant, often grad-

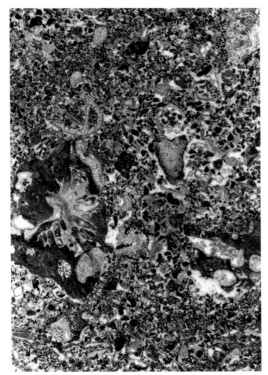

FIG. 21.—Thin-section photomicrograph (×4) of typical poorly-sorted "reef-detritus," found between the coral colonies of the reef patches. The bioclasts consist of up to 70 percent of reef building organisms (corals and calcisponges), and subangular cemented lithoclasts; Steinriese, Gosaukamm.

ed, peloids, foraminifers (mainly *Ophthalmidium, Galeanella, Sigmoilina,* and *Lituosepta,* and biserial agglutinated forms), together with *Muranella* and *Spiriamphorellina*; this sediment type is characteristic for small reef cavities within the inner parts of reef patches, but it is also found in other areas of calcisponge-hydrozoan-tabulozoan associations; 4. colored biomicrites, known only from the Adnet and the Steinplatte Reefs; the micrite contains nodular solenoporaceans often strongly micritized, as well as echinoderms, thick-shelled lagenid and miliolid foraminifers and reef derived lithoclasts; the corals within this sediment show effects of solution and often are truncated by dissolution planes (Fig. 12); this may indicate a temporary phase of subaerial exposure.

These microfacies types are also represented in other upper Rhaetian Reefs (Feichtenstein, Gruberalm, and Steinplatte), as well as in Dachstein Limestone Reefs (Sauwand, Gesäuse, Hochschwab, Hohe Wand, and Gosaukamm). The most widely distributed microfacies type seems to be pelbiosparitic grainstones typical of reef

cavities. As shown by Zankl (1969 and 1971), and Senowbari-Daryan (1978), two types of open-space structures can be recognized in the framework of the biolithite facies:

1. Reef cavities in or beneath framebuilding organisms, generally filled with micritic and pelsparitic sediments. These cavities may also be filled by internal sediment, or by a repeated alternation of internal sedimentation and internal cementation, whereby remaining open spaces are filled by growth of cements A and B. The internal cement microfacies corresponds to type C described above, or it may be characterized by colored biomicrites with many crinoids, foraminifers, ostracodes, and algal filaments; 2. the filling of cavities begins with formation of fibrous cement type A, growing as several generations perpendicular to cavity walls, followed by the formation of the blocky cement type B; the remaining open space may be filled with internal sediment. According to Mirsal (1978) various cement generations may reflect alternating salinity gradients.

12. *Sparry Rudstone Facies/Micritic Floatstone Facies (Fig. 23).*—Consists of reef derived rudstone with centimeter size poorly-sorted, angular litho- and bioclasts cemented by several generations of fibrous and blocky cements type B. All lithoclasts and bioclasts appear to have been eroded in the central reef area, or from more lagoonal areas of the reef complexes; some of these clasts are encrusted by spongiostromate algal crusts.

This facies type characterizes the high energy areas of the fore-reef slope. In the Dachstein Limestone Reefs, the fore-reef zone is generally rather limited in size due to tectonic truncation (see Table 15). In all Dachstein Limestone Reefs which have been studied, the fore-reef is characterized by a lack of reef building organisms that formed *in situ* by the predominance of poorly-sorted reef derived litho- and bioclasts, and by fossils which often indicate a connection to the open marine basinal environment (cephalopods, filaments, and conodonts). Differences can also be seen in the matrix of the fore-reef breccias. In contrast to the Gesäuse, Sauwand, and the Gosaukamm Reefs, which may be regarded as examples of the sparry rudstone facies, the fore-reef zone of the Göll and the Hochschwab Reefs is characterized by breccias with a micritic or finely calcarenitic matrix. This micritic floatstone facies may indicate the lower part of the fore-reef slope.

Decidedly different situations are assumed for upper Rhaetian Reefs. Here no rudstones or floatstones are known from zones regarded as being in a fore-reef position, if we agree with the reinterpretation of the so-called "fore-reef breccia" of the Steinplatte Reef by Piller (this volume). In

TABLE 15.—CHARACTERISTICS OF THE FORE-REEF ENVIRONMENTS OF UPPER TRIASSIC REEFS IN THE NORTHERN CALCAREOUS ALPS

| | | Hoher Göll (ZANKL, 1969) | Gesäuse (DULLO, 1979) | Hochschwab (LOBITZER, 1974) | Sauwand (FLÜGEL & FLÜGEL-KAHLER, 1963) |
|---|---|---|---|---|---|
| width of the preserved fore-reef zone | | 500 m | 50 m | 350 m | 10 m |
| Microfacies — carbonate particles | types | dominantly lithoclasts and reef derived bioclasts | litho- and bioclasts (reef derived material) | dominantly lithoclasts (cemented reef rocks) | dominantly lithoclasts, broken framebuilders |
| | sizes | coarse arenite and rudite, up to 1 m | coarse arenite to rudite, up to 3 cm | coarse arenite to rudite | rudite and coarse arenite |
| | sorting | very poor to poor | very poor | very poor | poor |
| | roundness | angular to subangular | angular to poorly rounded | angular to subangular | angular to subangular |
| | texture | poor bedding | no bedding | no distinct bedding | no distinct bedding |
| | matrix, cement | grey or red micrite, partly laminated, or fine calcarenite | sparite with cement A and cement B | micrite and fine calcarenite | micrite and sparite (open-void filling cements) |
| Biota | | foraminifers, gastropods, "filaments," rare ammonites, holothurian sclerites | foraminifers (derived from the back-reef) brachiopods rare ammonites algal crusts (with spongiostromata and porostromate algae) | foraminifers (lagenids) brachiopods "filaments" echinoderms ostracodes fish teeth | brachiopods, echinoderms |
| limestone type | | floatstone | rudstone | floatstone | rudstone and floatstone |

## DACHSTEINKALK REEFS

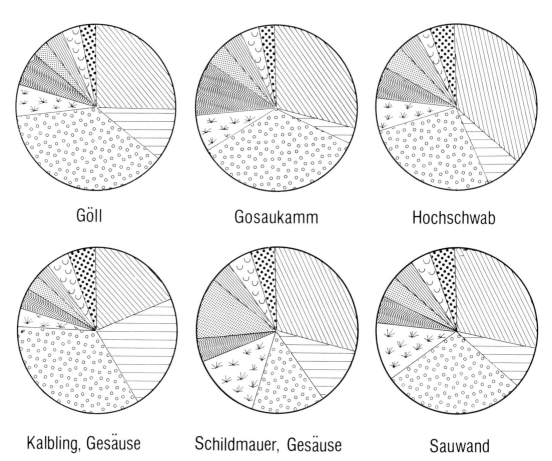

FIG. 22.—Diagram showing the relative frequency of primary and secondary framebuilding groups in the central reef areas of some Upper Triassic reefs of the Northern Calcareous Alps. Most often figures are based on general estimates rather than on a quantitative survey of the reef limestones (like in the Göll and Adnet Reefs). Sources:

→

the Rötelwand and the Feichtenstein Reef complexes an asymmetrical zonation of the facies units is developed. Adjacent to the marginal biolithite facies, characterized by high-growing coral communities, there follows a high energy oncoid facies which seems to indicate the wave exposed part of the reef complexes. The oncoid facies may replace the rudstone/floatstone facies of the Dachstein Limestone Reefs. Since the central reef area of the Adnet Reef is surrounded by the oncoid facies the development of this facies seems to have been governed by the existence of high energy conditions in association with an only slightly inclined slope.

Another situation is described from the Steinplatte Reef (Piller, this volume). Here the reef-slope shows an inclination of about 35 degrees, similar to that of some Dachstein Limestone

Reefs. The facies of this zone is characterized by bioclastic wackestones rich in crinoids; and in the upper part of the fore-reef slope by packstones with frequent reef derived bioclasts (corals, especially *Montlivaultia*, hydrozoans, solenoporaceans, *Thaumatoporella*, and *Alpinophragmium*), together with some lithoclasts. The autochthonous macrofauna consists of crinoids, brachiopods, lamellibranchs, and rare gastropods. The lack of true fore-reef breccias is explained by low energy conditions in the deeper part of the fore-reef zone.

### CONSTRUCTION OF UPPER TRIASSIC REEFS

Facies types, and distribution patterns of various organisms can be used to establish a general model for the construction of Upper Triassic reefs. Since the development of some upper

# UPPER RHAETIAN REEFS

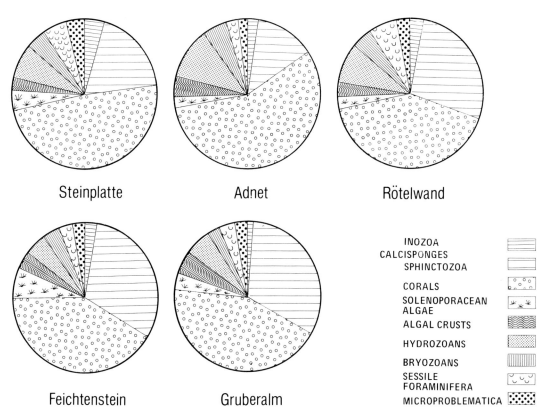

Steinplatte          Adnet          Rötelwand

Feichtenstein          Gruberalm

INOZOA
CALCISPONGES
SPHINCTOZOA

CORALS
SOLENOPORACEAN
ALGAE
ALGAL CRUSTS
HYDROZOANS
BRYOZOANS
SESSILE
FORAMINIFERA
MICROPROBLEMATICA

Göll, Zankl (1969); Gosaukamm, Flügel (1962), and Wurm (pers. comm.); Hochschwab, Lobitzer (1974); Kalbling and Schildmauer, Gesäuse, Dullo (pers. comm.); Sauwand, Flügel and E. Flügel-Kahler (1963); Steinplatte, Piller (pers. comm.); Adnet and Rötelwand, Schäfer (1979); Feichtenstein and Gruberalm, Senowbari-Daryan (1978).

Rhaetian buildups has already been described in detail through the works of Schäfer and Senow-bari-Daryan, as well as of Piller (this volume) only a few summarizing remarks are necessary.

The upper Rhaetian Reefs, especially the Rö-telwand and Feichtenstein Reefs, and probably the Steinplatte Reef, show two stages in their development. These stages are preceded by an initial stage, in which mechanical accumulations of sediments, and accumulations of molluscs and echinoderms formed shoals and local hardgrounds within basins. These stages may be compared with the ideal sequence of carbonate mud-mounds as described by Wilson (1975, p. 366).

*Initial Stage (Basal Bioclastic Wackestone Pile).*—Occurs with the development of shoals within the Kössen Basin by mechanical accumulation of both fine micritic and coarser bioclastic sediments consisting of brachiopod, mollusc, and echinoderm fragments as well as some ooids. Baffling or binding organisms are scarce.

*Stage 1 (Micritic Bafflestone Core).*—Occurs when autochthonous and allochthonous piles of shell debris, which formed during the initial stage, and which continue to develop during stage 1 create the hard substrate necessary for colonization by pioneer communities (mainly coral associations). During this time micritic carbonate is trapped between rather small local reef patches. In the Rötelwand Reef, the mud-mound that formed during stage 1 was covered up by thick piles of lamellibranch accumulations (up to 30 meters) and these correspond to an organic veneer.

*Stage 2 (Boundstone).*—After reaching wave base the lumachelles are colonized by frame producing organisms, and as such, form a wave resistant organic reef. During this stage the intermediate community (exhibited in the base of the Feichtenstein Reef and consisting predominantly of calcisponges), is followed by various high diverse climax communities which show strong zo-

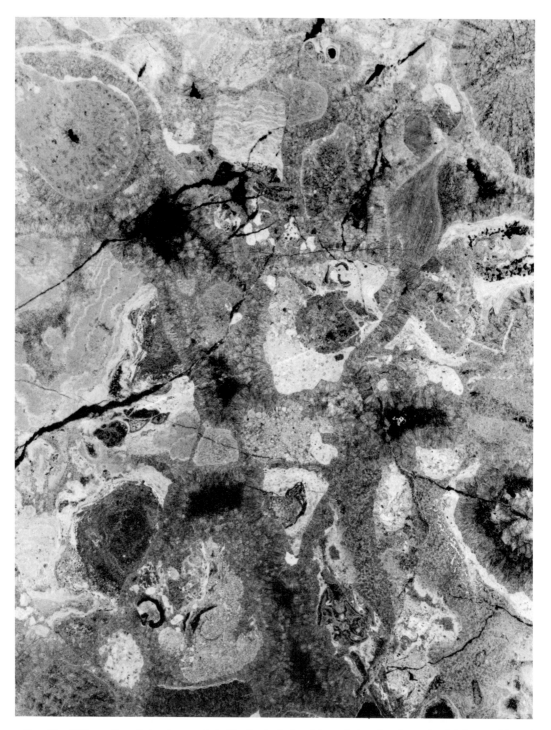

FIG. 23.—Thin-section photomicrograph (×5) of a typical rudstone facies showing polymict fore-reef breccia composed of angular lithoclasts which have been derived from various reef environments: e.g., algal bindstones (upper part of figure), packstones with peloid and rounded bioclasts (lower part), wackestones with dasycladaceans (center), boundstones with encrusting organisms (lower right), broken sponges (center), and relatively well preserved calcisponges (upper left). The bio- and lithoclasts are cemented by fibrous and radiaxial cements; Dachstein Limestone, Donnerkogel, Gosaukamm.

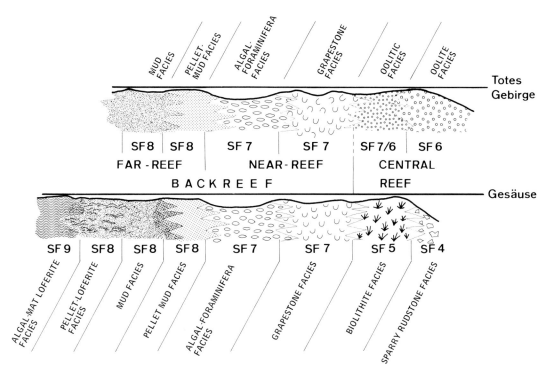

FIG. 24.—Diagram showing position of various facies units of the Dachstein Platform environment. The platform-margins can be characterized by oolitic shoals (type: Totes Gebirge; compare with Piller, 1976), or by organic buildups (type: Gesäuse; compare with Dullo, 1979). SF = Standard Facies Type according to Wilson (1975).

nation patterns, dependent upon bathymetry and position of the reef-crests.

*Final Stage (Flanking Beds).*—This occurred when the Rötelwand and Steinplatte Reefs were covered by calcarenites of reworked reef bioclasts and lithoclasts, in addition to pelecypod luma-chelles. This suggests a drowning of the reefs due to a general subsidence of the platform-edge, which until then had been active in protecting the upper Rhaetian Reefs (Piller, this volume).

Due to the striking paleonotological and facies differences, the "mud-mound stage" (stage 1), and the higher energy "organic reef stage" (stage 2) have been recognized as the standard sequences in the development of upper Rhaetian Reefs. The lack of a high energy stage in the Gruberalm Reef, and of a low energy stage in the Adnet Reef, can be well explained by the paleogeographical and bathymetrical position of these reefs (Gruberalm Reef formed in protected parts and the Adnet Reef formed on mud-mounds upon a Dachstein shallow water platform).

The base of the Dachstein Limestone Reefs has not yet been studied. However, from the interfingering of Aflenz Limestones and the Dachstein Reef Limestones in the Hochschwab region (Lob-

itzer, 1974), and from the superposition of Dachstein Reef Limestones of the Sauwand upon the subtidal Mürztal Beds (E. Flügel and E. Flügel-Kahler, 1963; Lein, 1975), it seems possible that mud-mound stages may have existed. In the Gosaukamm Reef the base of the framestone facies is developed as bedded limestones with cherts, and at Göll Mountain as secondary dolomitized Dachstein Limestone. More detailed investigations of the framebuilding organisms and facies types are necessary for more meaningful comparisons between the Dachstein Limestone Reefs and the upper Rhaetian Reefs.

## UPPER TRIASSIC REEFS IN THE TETHYAN REGION

Norian and Rhaetian Reefs are known from Europe, Asia, and North America (Fig. 26). Not all references which list occurrences of "corals" or "calcisponges" indicate the existence of "true reefs." However, it does seem necessary to include a general summary of the distribution of all known Upper Triassic reefs, and to point out some preliminary results of recent on-going investigations on Upper Triassic reefs in Sicily and Greece.

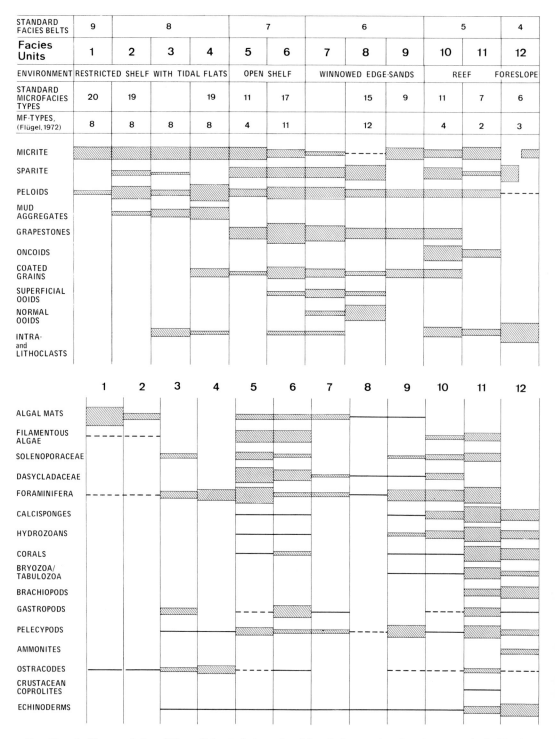

FIG. 25.—A. Characteristics of Upper Triassic facies units of the platform and reef environments in the Northern Calcareous Alps. Facies units include: 1. Algal mat loferite facies, 2. Pellet-loferite facies, 3. Shallow water mud facies, 4. Pellet-mud facies, 5. Algal-foraminiferal facies, 6. Grapestone facies, 7. Oolitic facies, 8. Oolite facies, 9. Reef-detritus mud facies, 10. Oncoid facies, 11. Biolithite facies (not differentiated into calcisponge-reef or coral-

*European Reefs*

*Northern Alps.*—The Dachstein Limestone Reefs and the upper Rhaetian Reefs have been treated in this paper, and in papers by Schäfer and Senowbari-Daryan, as by Piller (this volume).

*Southern Alps.*—Overlying the platform facies of the Hauptdolomite during upper Norian and Rhaetian time several southwest-northeast trending basins and troughs were developed, some of which are characterized by bioclastic and oolithic limestones with small coral bioherms (Tessin; compare Bernouilli, 1964; Wiedenmayer, 1963). In the Lombardian region of northern Italy, limestones with *"Thecosmilia"* are known but have not as yet been studied. In the Tridentinian Trough, and in the Dolomites no reefs of Norian or Rhaetian age are known. On the Dachstein Platform, adjacent to the Carnian Basin in the Julian Alps, and in the Karawanken Mountains reefs are known (Buser, 1974). Only a preliminary study has been published (Flügel and A. Ramovs, 1961), and this indicates a similar reef biotic composition comparable to that of the Dachstein Reef Limestones of the Northern Calcareous Alps.

*Western Carpathian Mountains, CSSR.*—Facies equivalents of the Dachstein Reef Limestone are known as the Furmanec Limestone from the Muran Plateau. These limestones represent patch-reef environments composed predominantly of corals and calcisponges, as well as algae, hydrozoans, sessile foraminifers, and various *microproblematica*. Bioclastic calcarenites surrounding the patch-reefs consist of reef detritus and fragments of reef dwelling organisms. The transition between the central reef area and the back-reef appears to be characterized by an "oncoid facies" and an "algal-foraminiferal facies," whereas the fore-reef facies is characterized by micritic limestones with crinoids (compare Mello, 1974). Algae, foraminifers and *microproblematica* found in bioclastic calcarenites have been described in several papers by Bystricky and by Borza. Data on the primary framebuilders is not available. The same is true for the reef communities of the Fatra Formation, which has been studied by Michalik (1977), and Michalik and Jendrejakova (1978); this formation corresponds to the Kössen Formation of the Northern Calcareous Alps.

*Tatra Mountains, Poland.*—Rhaetian sediments found in the Polish and Slovakian Tatra Mountains, within the Subtatric Krizna Nappe,

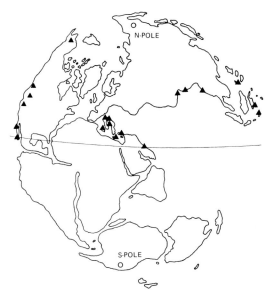

FIG. 26.—Diagram showing the world-wide distribution of Upper Triassic reefs, modified from Seyfert and Sirkin (1973). Only the better known localities are noted on this projection, therefore not all reported localities with Upper Triassic corals have been considered for this figure.

have been studied by Gazdzicki (1974), and Roniewicz (1974). In a predominantly clastic sequence ("Suebian facies"), bedded limestones are intercalated ("Carpathian facies"), along with several other facies types including a "coral facies" characterized by small patch-reefs with high-growing dendroid corals. These generally low diversity patch-reefs may have been formed in lagoons, which were bordered by the Furmanec Reefs of the Muran Plateau.

*Apennines, Italy.*—In the northern and central Apennines, especially in the Tuscany region and in Sicily, well-bedded "calcare massiccio" (upper Rhaetian to Lower Liassic) is widely distributed and generally overlays black limestones with *Rhaetavicula contorta* of Rhaetian age. The "calcare massiccio" represents platform carbonates with alternating subtidal, inter- and supratidal environments. According to microfacies analysis (e.g., Boccaletti *et al.*, 1969), no reefs developed.

*Sicily.*—Very interesting Upper Triassic reefs are known from the tectonically separated Panormide Carbonate Platform (cf. Abate *et al.*,

←

reef facies), and 12. Sparry rudstone facies/micritic floatstone facies. B. The Standard Microfacies Types (Wilson, 1975) are compared with the microfacies types described by Flügel (1972a). Sources: Fischer (1964), Piller (1976), Hohenegger and Piller (1975b), and results of the "Erlangen Reef Research Group."

1977) in northwestern Sicily, and from the Madonie Mountains. Several facies units can be differentiated and these consist of an intertidal zone, a back-reef area, and a central reef zone and upper fore-reef slope. The Cozzo di Lupo Reef near Palermo has recently been studied by members of the "Erlangen Reef Research Group" together with Italian colleagues. Preliminary results can be summarized as follows:

Main framebuilding organisms are calcisponges (especially highly diversified sphinctozoans), bryozoans/tabulozoans, and solenoporacean and cyanophycean algae. Corals and hydrozoans are scarce. Foraminifers and dasycladacean algae show similar distributional patterns as seen in the Upper Triassic reef complexes of the Northern Calcareous Alps. Despite a corresponding age, many species of framebuilders are new (cf. Senowbari-Daryan, 1979).

The lagoonal environments are characterized by an abundance of filamentous cyanophycean algae, dasycladaceans (mainly *Heteroporella*), and specialized foraminifers (involutinids, glomospirellids, endothyranellids, duostominids, and some coarsely agglutinated forms).

*Dinarides, Yugoslavia.*—Upper Triassic reefs are known from the Outer Dinarides and Inner Dinarides region of Yugoslavia (Pantić-Prodanovic, 1975). In the Outer Dinarides, platform carbonates up to 700 meters thick contain reefs. The best are known from the area southeast of Titograd, and in the Tara region of northwestern Montenegro. In the Inner Dinarides, Upper Triassic reefs are known from regions east of Sarajevo and near Valjevska. According to Pantić and Rampnoux (1972), in eastern Montenegro and in southern Serbia, platform and reef carbonates have developed upon structural swells which are separated by chert-bearing basinal carbonates. In contrast to the microfauna of the platform carbonates (compare Pantić-Prodanovic, 1975), no information is available concerning the paleontology and ecology of the framebuilding biota.

*Hungary.*—According to a survey by Balogh (1974), in the western Hungarian Balaton region, Norian Hauptdolomite is overlain by the Kössen Formation and by Dachstein Limestones believed to be of upper Rhaetian age. These Dachstein Limestones contain emergent horizons near the Triassic/Liassic boundary. In the Vertes-, Gerecse-, Pilis- and Buda Mountains, the sections are characterized by Hauptdolomite, bedded Dachstein Limestones, and Liassic Hirlatz Limestones. The microfacies of the bedded Dachstein Limestones of the Gerecse Mountains, and the Balaton region, correspond to the "algal-foraminiferal facies" and "grapestone facies" described above. Reefs seem to be rare, although suggestions of reefs are given by Vegh (1972) for the upper Norian limestones near Csövar, and for the uppermost Dachstein Limestones of the Gerecse Mountains.

*Rumania.*—Massive limestones with corals, overlain by limestones with megalodontid lamellibranchs and by dark-colored limestones with *"Thecosmilia,"* have been described from the southern Rumanian Apuseni Mountains (Mutihac and Preda, 1974). From the Eastern Carpathians, limestones with calcisponges and corals of Norian age are also known.

*Bulgaria.*—In the eastern Balcanian Mountains giant blocks of Rhaetian coral limestones have been found embedded in flysch-like sediments. Dachstein Limestones without reefs are known from the region of Luda-Kamcia.

*Greece.*—The name "Pantokrator Limestone" (Partsch, 1887) is generally used for Upper Triassic to Liassic bedded and massive limestones whose facies are similar to the Dachstein Limestones of the Northern Calcareous Alps. According to Renz (1955), the Pantokrator Limestones are especially well developed in the Adriatic-Ionic region, but they are also known from other regions such as the southeastern Aegean Sea. Investigations by members of the "Erlangen Reef Research Group" have shown that true reefs are developed only on the Island of Hydra, and in the region of Epidauros, Argolis. The Pantokrator Limestones of the type locality on Korfu, and in the Southern Didymi Mountains (Süsskoch, 1967; Bachmann and Jacobshagen, 1974), correspond to typical platform carbonates (algal mat loferite facies, algal-foraminiferal facies, limestones with large oncoids and those with megalodontid lamellibranchs). Reef facies appear to be developed in Chios (Besenecker *et al.*, 1968; Lüdtke, 1969).

The reefs of Hydra Island show striking differences as compared to Upper Triassic reefs in the Northern Calcareous Alps. These differences seem to be due primarily to difference in age. The reef biota indicates an upper Carnian age. Many of the corals and calcisponges show close affinities to corals described from Cordevolian patch-reefs of the Southern Alps, to calcisponges known from the Carnian *Cidaris* Beds of Austria, and to a Carnian patch-reef in Northern Slovenia (Senowbari-Daryan, 1980). The highest diversity is in the calcisponges which correspond in part to the fauna present in the Middle Triassic Wetterstein Limestone Reefs of the Northern Calcareous Alps. About 40 species have been differentiated. The Sphinctozoa are represented predominantly by the families Celyphiidae, Cystothalamiidae, and Cryptooeliidae; the inozoan sponges by the genera *Eudea*, *Peronidella* and *Sestrostomella*. The coral fauna comprises about 30 species most of which are identical or closely related to Cassian corals of the Southern Alps. Bryozoa and Tabulozoa are very common, and bryozoans are represented by new cyclostome and trepostome species; new species and genera of hydrozoans have also been found. Differences in comparison with the reefs of the Northern Calcareous Alps can also be seen in the composition of the algal flora, in which solenoporacean red algae, porostromate cyanophyceans, and the problematical alga *Tubiphytes* are of great importance within reef patches. The foraminifers are strongly facies controlled.

### Asian Reefs

*Anatolia.*—A Triassic to Liassic carbonate platform facies, comparable to the Italian "calcare massiccio," and the Ionia "Pantokrator Limestone," is known from the Lycian Taurus Mountains and from the southeastern Aegean region "Gereme Limestone" (Bernoulli *et al.*, 1974). These carbonate platforms show a distinctive microfacies differentiation with several emergent phases; reefs seem to be very scarce (Brinkmann, 1976).

*Caucasus Mountains.*—Only sparse information is available for the Upper Triassic reef facies described by Moiseev (1944) from the Caucasus Mountains. These reefs appear to be formed predominantly by corals and calcisponges together with bryozoans and tabulozoans, as well as red algae acting as secondary framebuilders.

*Central Iran.*—Within the "Naiband Formation" near Naiband, north of Kerman in central Iran, limestones are intercalated in a 2000 meters thick predominantly clastic terrigenous sequence (Douglas, 1929; Brönnimann *et al.*, 1972). These limestones yield many corals, calcisponges and hydrozoans, and appear to form biostromal structures. Similar occurrences are known from the Shotori Range in the Tabas region (Stocklin *et al.*, 1965), and between Kerman and Sagand (Huckriede *et al.*, 1962). Facies and fauna are similar to that of the Kössen Formation in the Northern Calcareous Alps.

*Pamir and Karakorum Ranges, Central Asia.*—Upper Triassic reefs described by Kushlin (1973), and by Melnikova (1967 and 1971), from the northern Pamir Ranges seem to be very similar in biotic composition to reefs in the Northern Calcareous Alps, especially in considering the coral communities. Hydrozoans and bryozoans belong to unique species, not known from the Alps. In the Karakorum Range, Upper Triassic corals, hydrozoans, and calcisponges have been described (Vinassa de Regny, 1932), but more detailed studies are lacking.

*Himalayas.*—In 1906 Arthaber pointed out the existence of Upper Triassic reefs comparable to European reefs in the Himalayan region. As shown by Yang and Wang (1976) this is also true for the Mt. Everest Range.

*Yunnan, Southwestern China.*—Some Upper Triassic corals have been described from Yunnan, as well as from northern Burma and northern Thailand, but detailed information concerning facies of the coral-bearing beds is lacking.

*Szechuan, China.*—Chi and Peng (1940) have described abundant occurrences of "*Thecosmilia*" in thin-bedded upper Triassic limestones, but there is no mention of reefs.

*Seran, Moluccas.*—Thick limestones up to 150 meters, with many sphinctozoan sponges, corals, and hydrozoans are known from the central and eastern parts of Seran Island (Wilckens, 1937). These limestones can be compared with the Fatu Limestones of Timor; the reefs are believed to be upper Norian in age.

*Timor.*—Various localities have yielded Norian reef building organisms; Vinassa de Regny (1915) described 57 species of predominantly sphinctozoans, high-growing colonies of "*Thecosmilia*," other colonial corals and solitary corals, and some hydrozoans.

*Japan.*—An Upper Triassic (probably Carnian) reef fauna is known from the Konose Group of Kyushu, southern Japan (Kanmera, 1964). The corals, calcisponges, hydrozoans, bryozoans, and calcareous algae are embedded in pyroclastic sediments and seem to have been derived from reefs which grew on submarine volcanic ridges.

### North American Reefs

Coral buildups are widely distributed in Middle and Upper Triassic rocks of Nevada, California, Oregon, Idaho, British Columbia, and Alaska (Smith, 1912, 1927; Squires, 1956). These buildups have recently been studied in detail by Stanley (1978), who interprets some of them as organic deeper water structures formed by ahermatypic corals. Associated with the corals, which are related to the Tethyan coral faunas, are spongiomorphids, calcisponges, brachiopods, molluscs, and echinoderms, which characterize a distinct zoogeographic province within North America. Perhaps because of the paleogeographical setting of Triassic North American buildups (on relatively stable carbonate platforms or on active volcanic arcs), only small patch-reefs without ecological zonation have formed.

As shown by the above review a good deal more in the way of paleontologically and facies orientated investigations have to be accomplished in the future, in order to understand development and ecological evolution of Upper Triassic reefs on a worldwide scale.

## CONCLUSIONS

1. Reefs developed in the Triassic of the Northern Calcareous Alps as small buildups of upper Anisian age; as yet, these are unstudied.

2. The reef building fauna and flora of large Middle Triassic and Cordevolian Wetterstein Limestone Reefs seem to have continued in shallow water but non-reef environments are represented by limestones of the Carnian *Cidaris* Beds.

3. Ladinian and Carnian reef biotas, and framebuilders of Norian and Rhaetian Reefs, show striking differences with regard to specific biotic composition. This is true in the Northern Calcareous Alps, and also in differences seen in the biotas of Carnian Reefs on Hydra Island, and Norian to Rhaetian Reefs of the Pantokrator facies in Greece.

4. The evolution of the Norian/Rhaetian Reef biota is still not well understood. Investigation of the Carnian Tisovec Limestone of the Western Carpathians and the Northern Calcareous Alps may help to clarify this problem in the future.

5. Norian and Rhaetian Reefs in the Northern Calcareous Alps are represented by large extended Dachstein Limestone Reefs of platform-edge setting, and by rather small "upper Rhaetian Reefs" (formed upon shoals within the intra-platform Kössen Basin, or at the margins of the Kössen Basin upon the Dachstein Platform), and by the somewhat "deeper-water" biostromes of the Kössen Formation.

6. The fauna and flora contributing to the framework of the Dachstein Limestone and upper Rhaetian Reefs is very similar with respect to specific and generic biotic composition.

7. Primary framebuilders are corals, calcisponges, hydrozoans, and solenoporacean algae. Secondary framebuilders are represented by various algal crusts, sessile foraminifers, bryozoa, and "tabulozoa," as well as many organisms of uncertain systematic position.

8. Some reef dwelling organisms need further study (e.g., brachiopods, lamellibranchs, gastropods, ammonites, crustaceans, ostracodes, echinoderms, and vertebrates).

9. Upper Rhaetian and Dachstein Limestone Reefs yield several communities which are characterized by the predominance of one or more species, or by regularly recurring associations of framebuilding organisms. Some reef communities seem to have flourished in similar niches within different reefs.

10. Reef communities can also be defined by special associations of epizoans and epiphytes. The greatest diversity of epibiontic organisms is found in protected parts of the reef-crests, and in the low energy reef-slope environments.

11. Due to a strong adaptation to micro-environments within reefs and in adjacent lagoonal and shallow water areas, foraminifers, calcareous algae, and *microproblematica* show distinct distributional patterns which can be used to identify various reef zones and smaller ecological niches (like reef cavities).

12. Similar zonation patterns have been found to exist in upper Rhaetian Reefs and in the Dachstein Limestone Reefs in Gesäuse, Styria.

13. Ecologic successions, formed by several generations of secondary framebuilders growing upon dead or living primary reef building organisms, need additional study with respect to recognition of generally valid distributional patterns.

14. Microfacies types, various grain types, and the qualitative and quantitative composition of the foraminiferal fauna can be used to distinguish 12 facies units within reef and platform carbonates. These facies units (Figs. 24 and 25) have been developed in various parts of the inner restricted or open-shelf area, within smaller or larger reefs, and on fore-reef slopes.

15. In upper Rhaetian Reefs a distinct zonation of facies units can be recognized, but the arrangement of zones does not seem to be similar in all reefs (see the contributions by Schäfer and Senowbari-Daryan, and by Piller, this volume).

16. The facies model proposed for the Dachstein Limestone Reefs (Göll, Sauwand, and Hochschwab) needs critical re-evaluation with regard to a more detailed study of facies relationships within the central reef area, and a thorough investigation of the transition to basinal facies.

17. Little new information is available concerning the initial phases of reef development, and the construction of patch-reefs and biostromal buildups formed in the "deeper" environments (Kössen coral limestone; Aflenz Limestone).

18. Comparison of Upper Triassic reefs of the Northern Calcareous Alps with Norian and Rhaetian Reef structures described from various regions in Europe, Asia, and North America, is difficult due to the present state of investigations, since only the Northern Alpine Reefs, some reefs in northwestern Sicily and in Greece, as well the somewhat aberrant Upper Triassic buildups in western North America, have been studied with comparable detail.

## ACKNOWLEDGMENTS

Financial assistance that was provided by the Deutsche Forschungsgemeinschaft (projects Fl 42/25–27 and Fl 42/31) to members of the "Erlangen Reef Research Group" is gratefully acknowledged. I wish to thank Dr. W. Piller and Dr. H. Lobitzer (Vienna) for critical discussions. Valuable suggestions were also offered by Wolf-Christian Dullo, Massoud Sadati, Priska Schäfer, Baba Senowbari-Daryan, and Detlef Wurm (Erlangen). Thanks are also due to Mrs. Chr. Sporn for technical support.

# REFERENCES

ABATE, B., CATALANO, R., D'ARGENIO, B., DI STEFANO, P., AND RICCOBONO, R., 1977, Relationships of algae with depositional environments and faunal assemblages of the Panormide Carbonate Platform, Upper Triassic, Northwestern Sicily, *in* Flügel, E., Ed., Fossil Algae, Recent Results and Developments: New York, Springer-Verlag, p. 301–313.

ARTHABER, G., 1906, Die alpine Trias des Mediterrangebietes, *in* Frech, F., Ed., Lethaea geognostica, part 2, v. 1: Schweizerbart, p. 223–472.

BACHMANN, G. H., AND JACOBSHAGEN, V., 1974, Zur Fazies und Entstehung der Hallstätter Kalke von Epidauros (Anis bis Karn; Argolis, Griechenland): Zeitschrift Deutsch. Geol. Ges., v. 125, 195–223.

BALOGH, K., 1974, Kurzfassung der triassischen Stratigraphie in Ungarn, *in* Zapfe, H., Ed., Die Stratigraphie der alpin-mediterranen Trias: Österr. Akad. Wiss., Schriftenreihe Erdwiss. Komm., v. 2, 41–43.

BAUER, F. K., 1970, Zur Fazies und Tektonik des Nordstammes der Ostkarawanken von der Petzen bis zum Obir: Jahrbuch Geol. Bundesanstalt, v. 113, p. 189–246.

BECHSTÄDT, T., 1978, Faziesanalyse permischer und triadischer Sedimente des Drauzuges als Hinweis auf eine gross räumige Lateralverschiebung innerhalb der Ostalpins: Jahrbuch Geol. Bundesanstalt, v. 121, p. 1–121.

———, AND BRANDNER, R., 1970, Das Anis zwischen St. Vigil und dem Höhlensteintal (Pragser- und Olanger Dolomiten, Südtirol): Festband Geol. Inst. 300-Jahr-Feier Univ. Innsbruck, p. 9–103.

———, ———, AND MOSTLER, H., 1976, Das Frühstadium der alpinen Geosynklinalentwicklung im westlichen Drauzug: Geol. Rundschau, v. 65, p. 616–648.

———, ———, ———, AND SCHMIDT, K., 1978, Aborted rifting in the Triassic of the Eastern and Southern Alps: Neues Jahrbuch Geol. Paläont. Abhandlungen, v. 156, p. 157–178.

BERNOULLI, D., 1964, Zur Geologie des Monte Generoso (Lombardische Alpen): Beiträge Geol. Karte Schweiz, v. 118, p. 1–134.

———, AND JENKYNS, H. C., 1974, Alpine, Mediterranean, and Central Atlantic Mesozoic facies in relation to the early evolution of the Tethys: Soc. Econ. Paleontologists and Mineralogists Spec. Pub., No. 19, p. 129–160.

BITTNER, A., 1883, Der Untersberg und die nächste Umgebung von Golling: Jahrgang, Verhandlungen Geol. Reichsanstalt, p. 200–204.

BOIKO, E. V., 1972, Pozdnetriasovye spongiomorfidy (Hydrozoa) yugo-vostochnogo Pamira: Paleont. Zhurnal, v. 1972, no. 2, p. 20–25.

BORZA, K., 1975, Mikroproblematika aus der oberen Trias der Westkarpaten: Geol. Zbornik Slov. Akad. Vied, Bratislawa, v. 26, p. 199–236.

———, 1977, New genera and species (*incertae sedis*) from the Upper Triassic in the West Carpathians: Geol. Zbornik, Geol. Carpathica, v. 28, p. 95–119.

BOSELLINI, A., AND ROSSI, D., 1974, Triassic carbonate buildups of the Dolomites, Northern Italy: Soc. Econ. Paleontologists and Mineralogists Spec. Pub. No. 18, p. 209–233.

BRADNER, R., 1972, "Südalpines" Anis in den Lienzer Dolomiten (Drauzug) (ein Beitrag zur alpin-dinarischen Grenze): Mitteilungen Gesellschaft Geologie-und Bergbaustudenten Österreich, v. 21, p. 143–152.

———, AND RESCH, W., 1980, Reef development in the Middle Triassic (Ladinian and Cordevolian) of the Northern Limestone Alps near Innsbruck, Austria: this vol.

BRINKMANN, R., 1976, Geology of Turkey: Stuttgart, Enke, 158 p.

BRÖNNIMANN, P., ZANINETTI, L., BOZORGNIA, F., DASHTI, G. R., AND MOSHTAGHIAN, A., 1972, Lithostratigraphy and Foraminifera of the Upper Triassic Naiband Formation, Iran: Revue Micropaleont., v. 5, 7–16.

BYSTRICKY, J., 1973, Fazieverteilung der mittleren und oberen Trias in den Westkarpaten: Mitt. Geologie- und Bergbaustudenten Österreich, v. 21, p. 289–310.

———, 1979, Dasycladaceae of the Upper Triassic of the Stratenska Hornatina Mountains (the West Carpathians): Geol. Zbornik, Geol. Carpathica, v. 30, p. 321–340.

BUECHNER, K.-H., 1970, Geologie der nördlichen und südwestlichen Gesäuse-Berge (Ober-Steiermark, Österreich) [Ph.D. Dissertation]: Universität Marburg/Lahn, 118 p.

———, 1973, Ergebnisse einer geologischen Neuaufnahme der nördlichen und südwestlichen Gesäuseberge (Ober-Steiermark, Österreich): Mitteilungen Gesellschaft Geologie- und Bergbaustudenten Österreich, v. 22, p. 71–94.

BUSER, ST., 1974, Die Entwicklung der Triasschichten in den westlichen Karawanken, in Zapfe, H., Ed., Die Stratigraphie der alpin-mediterranen Trias: Österr. Akad. Wiss., Schriftenreihe Erdwiss. Komm., v. 2, 63–68.

COLINS, E., AND NACHTMANN, W., 1974, Die permotriadische Schichtfolge der Villacher Alpe (Dobratsch), Kärnten: Geol. Paläont. Mitteilungen, Innsbruck, v. 4, p. 1–43.

CUIF, J.-P., 1965, Sur les rapports du genres *Montlivaultia* Lam. et *Thecosmilia* M.-Edw. & Haime et leur présence au Trias: Bull. Soc. Géol. France, v. 7, p. 530–536.

———, 1972, Recherches sur les Madréporaires du Trias. 1. Famille des Stylophyllidae: Bull. Mus. Nat. Hist. Nat., v. 97, Science Terre, v. 17, p. 211–291.

———, 1974a, Recherches sur les Madréporaires du Trias. 2. Genres *Montlivaultia* et *Thecosmilia*: Bull. Mus. Nat. Hist. Nat., Sciences Terre, v. 40, p. 293–400.

———, 1974b, Indices de la nature aragonitique des fibres chez les Madreporaires paléozoiques: Comptes rendus sommaire. Soc. Géol. France, p. 162–164.

———, 1975a, Caractères morphologiques, microstructuraux et systématiques des Pachythecalidae, nouvelle famille de Madrêporaires triasiques: Geobios, v. 8, p. 157–180.

————, 1975b, Recherches sur les Madréporaires triasiques. 3. Étude des structures pennulaires: Bull. Mus. Nat. Hist. Nat., no. 310, Science Terre, v. 44, p. 45–127.

————, 1976, Recherches sur les Madréporaires du Trias. 4. Formes cerioméandroides et thamnastérioides du Trias des Alpes et du Taurus su-anatolien: Bull. Mus. Nat. Hist. Nat., no. 381, Science Terre, v. 53, p. 65–196.

————, 1979, Microstructure versus morphology in the skeleton of early Scleractinian Corals: Internat. Symposium Fossil Cnidarians, Third, Polska Akad. Nauk, Warszawa, p. 107–121.

CZURDA, K., 1973, Fazies und Stratigraphie obertriadischer Megalodontenvorkommen der westlichen Nördlichen Kalkalpen: Verhandlungen Geol. Bundesanstalt, Wien, v. 1973, p. 397–409.

————, AND NICKLAS, L., 1970, Zur Mikrofazies und Mikrostratigraphie des Hauptdolomites und Plattenkalk-Niveaus der Klostertaler Alpen und des Rhätikon (Nördliche Kalkalpen, Vorarlberg): Festband Geol. Institut. 300-Jahr-Feier Univ. Innsbruck, p. 165–253.

DOUGLAS, J. A., 1929, A marine Triassic fauna from eastern Persia: Quart. Jour. Geol. Soc. London, v. 85.

DULLO, W.-CHR., 1979, Fazies und geologischer Rahmen der Dachsteinkalke (Obertrias) in den südwestlichen Gesäuse-Bergen (Steiermark) [Diplom thesis]: Erlangen, West Germany, Paleontological Institute, Erlangen Univ., 197 p.

————, 1980a, Paläontologie, Fazies und Geochemie der Dachsteinkalke (Ober-Trias) im südwestlichen Gesäuse, Steiermark: Facies, v. 2, in press.

————, 1980b, Über ein neues Vorkommen von Tisovec-Kalk in den südwestlichen Gesäuse-Bergen (Admont, Steiermark): Mitt. Ges. Geol. Bergbaustud. Österreich, v. 26, p. 155–165.

FABRICIUS, F. H., 1959, Vorschlag zur Umbenennung von "Oberrhätkalk" in "Rhätolias-Riffkalk" (Nördliche Kalkalpen): Neues Jahrbuch Geologie Paläontologie, Monatshefte, v. 1959, p. 546–549.

————, 1960, Sedimentation und Fazies des Rät und der Lias-Überdeckung in den Bayerisch-Tirolischen Kalkalpen [Ph.D. Dissertation]: Munich, Technische Hochschule München.

————, 1961, Die Strukturen des "Rogenpyrits" (Kössener Schichten, Rät) als Beitrag zum Problem der "vererzten Bakterien": Geologische Rundschau, v. 51, p. 647–657.

————, 1966, Beckensedimentation und Riffbildung an der Wende Trias/Jura in den Bayerisch-Tiroler Kalkalpen: Internat. Sedim. Petrograph. Series, v. 9, 143 p.

————, 1974, Die stratigraphische Stellung der Rät-Fazies: Schriftenreihe Erdwissenschaftlichen Kommissiion Österreichische Akademie der Wissenschaften, v. 2, p. 87–92.

FISCHER, A. G., 1964, The Lofer cyclothems of the alpine Triassic: Bull. Geological Survey Kansas, v. 169, p. 107–149.

FLÜGEL, E., 1960, Untersuchungen im obertriadischen Riff des Gosaukammes (Dachsteingebiet, Oberösterreich). 2. Untersuchungen über die Fauna und Flora des Dachsteinriffkalkes der Donnerkogel-Gruppe: Verhandlungen Geol. Bundesanstalt, Wien, v. 1960, p. 241–252.

————, 1962a, Beiträge zur Paläontologie der nordalpinen Riffe. Neue Spongien und Algen aus den Zlambach-Schichten (Rät) des westlichen Gosaukammes, Oberösterreich: Annalen Naturhistor. Museum Wein, v. 65, p. 51–56.

————, 1962b, Untersuchungen im obertriadischen Riff des Gosaukammes (Dachsteingebiet, Oberösterreich). 3. Zur Mikrofazies der Zlambach-Schichten am W-Ende des Gosaukammes: Verhandlungen Geol. Bundesanstalt Wien, v. 1962, p. 138–144.

————, 1962c, Untersuchungen über den Fossilinhalt und die Mikrofazies der obertriadischen Riffkalke in den Nordalpen: Habilitationsschrift. Univ. Wien, 279 p.

————, 1963, Zur Mikrofazies der alpinen trias: Jahrbuch Geol. Bundesanstalt, Wien, v. 106, p. 205–228.

————, 1964, Mikroproblematika aus rhätischen Riffkalken der Nordalpen: Paläontologische Zeitschrift, v. 38, p. 74–87.

————, 1967, Eine neue Foraminifere aus den Riff-Kalken der nordalpinen Ober-Trias: *Alpinophragmium perforatum* n.g., n.sp.: Senckenbergiana Lethaea, v. 48, p. 381–402.

————, 1972a, Microfazielle Untersuchungen in der alpinen Trias: Mitteilungen Gesellschaft Geologie- und Bergbaustudenten Österreich, v. 21, p. 9–64.

————, 1972b, Mikroproblematika in Dünnschliffen von Trias-Kalken: Mitteilungen Gesellschaft Geologie- und Bergbaustudenten Österreich, v. 21, p. 957–988.

————, 1975, Kalkalgen aus Riffkomplexen der alpin-mediterranen Obertrias: Verhandlungen Geol. Bundesanstalt, Wien, v. 1974, p. 297–346.

————, 1977a, Untersuchungen über die Beziehungen zwischen mikrofaziellen und technologischen Merkmalen steirischer Dachsteinkalke (Obertrias; Grimmingstock, Gesäuse): Mitteilungen Abteilung Geol. Paläont. Bergbau Landesmuseum Joanneum, Graz, v. 38, 193–204.

————, 1977b, Verkalkungsmuster porostromater Algen aus dem Malm der Südlichen Frankenalb: Geologische Blätter Nordost-Bayern, v. 27, p. 131–140.

————, 1978, Mikrofazielle Untersuchungsmethoden von Kalken: Berlin - Heidelberg - New York, Springer-Verlag, 454 p.

————, 1979, Paleoecology and microfacies of Permian, Triassic and Jurassic algal communities of platform and reef carbonates from the Alps: Bull. Centre Rech. Explor.-Prod. Elf-Aquitaine, v. 3, p. 569–587.

————, AND FLÜGEL-KAHLER, E., 1963, Mikrofazielle und geochemische Gliederung eines obertriadischen Riffes der Nördlichen Kalkalpen (Sauwand bei Gusswerk, Steiermark, Osterreich): Mitteilungen Museum Geologie Paläont. Bergbau, Landesmuseum Joanneum, v. 24, p. 1–129.

———, LEIN, R., AND SENOWBARI-DARYAN, B., 1978, Kalkschwämme, Hydrozoen, Algen und Mikroproblematika aus den Cidarisschichten (Karn, Ober-Trias) der Mürztaler Alpen (Steiermark) und des Gosaukammes (Oberösterreich): Mitteilungen Gesellschaft Geol. Bergbaustudenten Österreich, v. 25, p. 153–195.

———, AND RAMOVŠ, A., 1961, Fossilinhalt und Mikrofazies des Dachsteinkalkes (Ober-Trias) im Begunjsica-Gebirge, S-Karawanken (NW-Slowenien, Jugoslawien): Neues Jahrbuch Geol. Paläont., Monatschefte, v. 1961, p. 287–294.

———, AND SY, E., 1959, Die Hydrozoen der Trias: Neues Jahrbuch Geologie Paläontologie, Abhandlungen, v. 109, p. 1–108.

FRECH, F., 1890, Die Korallen der Trias. Die Korallen der juvavischen Tirasprovinz: Palaeontographica, v. 37, p. 1–116.

FRIEDMAN, G. M, GEBELEIN, C. D., AND SANDERS, J. E., 1971, Micritic envelopes of carbonate grains are not exclusively of photosynthetic algal origin: Sedimentology, v. 16, p. 89–96.

FÜRSICH, F. T., AND WENDT, J., 1977, Biostratinomy and Palaeoecology of the Cassian Formation (Triassic) of the Southern Alps: Palaeogeography, Palaeoclimatology, Palaeoecology, v. 22, p. 257–323.

GAZDZICKI, A., 1974, Rhaetian microfacies, stratigraphy and facial development in the Tatra Mts.: Acta Geol. Polonica, v. 24, p. 17–96.

———, KOZUR, H., AND MOCK, R., 1979, The Norian-Rhaetian boundary in the light of micropaleontological data: Geologija, Razprave in Porocila, Ljubljana, v. 22, p. 71–112.

GÖKDAG, H., 1974, Sedimentpetrographische und isotopenchemische (O¹⁸, C¹³) Untersuchungen im Dachsteinkalk (Obernor-Rät) der Nördlichen Kalkalpen [Ph.D. Dissertation]: Marburg Univ., 156 p.

HOHENEGGER, J., 1974, Über einfache Gruppierungsmethoden von Fossil-Vergesellschaftungen am Beispiel obertriadischer Foraminiferen: Neues Jahrbuch Geol. Paläont. Abhandlungen, v. 146, p. 263–297.

———, AND LOBITZER, H., 1971, Die Foraminiferen-Verteilung in einm obertriadischen Karbonatplattform-Becken-Komplex der östlichen Nördlichen Kalkalpen: Verhandlungen Geol. Bundesanstalt, Wien, v. 1971, p. 458–485.

———, AND PILLER, W., 1975a, Diagenetische Veränderungen bei obertriadischen Involutinidae (Foraminifera): Neues Jahrbuch Geol. Paläont. Monatschefte, v. 175, 26–39.

———, AND ———, 1975b, Wandstrukturen und Grossgliederung der Foraminiferen: Sitzungsberichte Österr. Akademie Wissenschaften, mathematisch-naturwissenschaftliche Klasse, Abteilung 1, v. 184, p. 67–96.

———, AND ———, 1975c, Ökologische und systematische Stellung der Foraminiferen im gebankten Dachsteinkalk (Obertrias) des Nordöstlichen Toten Gebirges (Oberösterreich): Palaeogeography, Palaeoclimatology, Palaeoecology, v. 18, p. 241–276.

———, AND ———, 1977, Die Stellung der Involutinidae Bütschli und Spirillinidae Reuss im System der Foraminiferen: Neues Jahrbuch Geol. Paläont., Montschefte, v. 1977, p. 407–418.

JOHNSON, J. H., 1961, Limestone-Building Algae and Algal Limestones: Pub. Colorado School Mines, 297 p.

KANMERA, K., 1964, Triassic Coral Faunas from the Konose Group in Kyushu: Mem. Fac. Sci. Kyushu Univ., Ser. D, Geol., v. 15, p. 117–147.

KOLLAROVA-ANDRUSOVOVA, V., 1960, Récentes trouvailles d'Ammonoides dans le Trias des Karpates occidentales: Geol. Sbornik Slov. Akad. Vied, v. 11, p. 105–110.

———, 1967, Cephalopodenfaunen und Stratigraphie der Trias der Westkarpaten: Geol. Sbornik Slov. Akad. Vied, v. 19, p. 267–275.

KÜHN, O., 1942, Zur Kenntnis des Rhät von Vorarlberg: Mitt. Alpenländ. Geol. Ges. (Mitt. Geol. Gesellschaft Wien), v. 33, p. 111–157.

———, 1962, Autriche: Lexique Stratigraphique International, v. 1, fascicle 8, 646 p.

LEIN, R., 1975, Neue Ergebnisse über die Stellung und die Stratigraphie der Hallstätter Zone südlich der Dachsteindecke: Sitzungsberichte Österr. Akad. Wissenschaften, mathematisch-naturwissenschafliche Klasse, Abteilung 1, v. 183, 197–235.

———, AND ZAPFE, H., 1971, Ein karnischer "Dachsteinkalk" mit Pachyodonten in den Mürztaler Alpen, Steiermark: Anzeiger Österr. Akad. Wissenschaften, mathematisch-naturwissenschaftliche Klasse, v. 108, p. 133–139.

LOBITZER, H., 1972, Fazielle Untersuchungen an triadischen Karbonatplattform/Becken-Gesteinen des südöstlichen Hochschwabgebietes (Wetterstein-und Reiflingerkalk, Dachstein- und Aflenzer Kalk): Anzeiger Österr. Akad. Wiss., mathematisch-naturwissenschaftliche Klasse, v. 109, p. 201–203.

———, 1975, Fazielle Untersuchungen an norischen Karbonatplattform—Beckengesteinen (Dachsteinkalk—Aflenzer Kalk) im südöstlichen Hochschwabgebiet, Nördliche Kalkalpen, Steiermark: Mitteilungen Geol. Gesellschaft Wien, v. 66/67, p. 75–91.

MELNIKOVA, G. K., 1967, New species of Triassic Scleractinia from the Pamirs (Russian): Paleont. Zhurnal, v. 1, p. 18–26.

———, 1971, New data on the morphology, microstructure and systematics of the Late Triassic Thamnasteroidea: Paleont. Zhurnal, v. 5, p. 156–169.

———, 1972, The genus Cyathocoenia (Hexacoralla): Paleont. Zhurnal, v. 6, p. 9–15.

————, 1979, The Pamirian Late Triassic Facies Reconstruction: Internat. Symposium Fossil Cnidarians, Third, Warszawa, Polska Akad. Nauk, p. 49–50.

MICHALIK, J., 1977, Paläogeographische Untersuchungen der Fatra-Schichten (Kössen Formation) des nördlichen Teiles des Fatrikums in den Westkarpaten: Geol. Zbornik Slov. Akad. Vied, Bratislawa, v. 28, p. 71–94.

————, AND JENDREJAKOVA, O., 1978, Organism communities and biofacies of the Fatra Formation (Uppermost Triassic, Fatric) in the West Carpathians: Geol. Zbornik, Geol. Carpathica, Bratislawa, v. 29, p. 113–137.

MILLER, H., 1965, Die Mitteltrias der Mieminger Berge mit Vergleichen zum westlichen Wettersteingebirge: Verh. Geol. Bundesanstalt, Wien, v. 1965, p. 187–212.

MIRSAL, I. A., 1978, Zementmineralisation in fossilen Korallenriffen. Eine petrographische und geochemische Faktorenanalyse [Ph.D. Dissertation]: Marburg University, 120 p.

MOISSEEV, A. S., 1939, New data on the Upper Triassic of North Caucasus and the Crimea: Comptes Rendus (Doklady) Acad. Sciences URSS, v. 23, p. 816–818.

MOJSISOVICS, E., 1879, Die Dolomit-Riffe von Südtirol und Venetien: Wien, Hölder, 552 p.

MONTANARO-GALLITELLI, E., 1979, Upper Triassic Coelenterates of Western North America: Boll. Soc. Paleont. Italiana, v. 18, p. 133–156.

MOSTLER, H., SCHEURING, B., AND URLICHS, M., 1978, Zur Mega-, Mikrofauna und Mikroflora der Kössener Schichten (alpine Obertrias) vom Weiss loferbach in Tirol under besonderer Berücksichtigung der in der suessi- und marshi-Zone auftretenden Conodonten: Österr, Akad. Wiss. Schriftenreihe Erdwiss. Komm., v. 4, p. 141–174.

MÜLLER-JUNGBLUTH, W.-U., 1970, Sedimentologische Untersuchungen des Hauptdolomites der östlichen Lechtaler Alpen, Tirol: Festband Geol. Institut, 300-Jahr-Feier Univ. Innsbruck, p. 255–308.

OHLEN, R. H., 1959, The Steinplatte Reef complex of the Alpine Triassic (Rhaetian) of Austria [Ph.D. Thesis]: Princeton, NJ, Princeton Univ., 122 p.

OLIVER, W. A., 1979, The relationship between Rugosa and Scleractinia: Internat. Symposium on Fossil Cnidarians, Third, Warszwawa, Polska Akad. Nauk, p. 132–174.

OTT, E., 1967, Segmentierte Kalkschwämme (Spinctozoa) aus der alpinen Mitteltrias und ihre Bedeutung als Riffbildner im Wettersteinkalk: Abhandlungen Bayer. Akad. Wissenschaften, mathematisch-naturwissenschaftliche Klasse, v. 131, p. 1–96.

————, 1972, Mitteltriadische Riffe der Nördlichen Kalkalpen und altersgleiche Bildungen auf Karaburun und Chios (Ägäis): Mitteilungen Gesellschaft Geologie- und Bergbaustudenten Österreich, v. 21, p. 251–276.

PANTIĆ-PRODANOVIC, S., 1975, Les microfacies trasiques des Dinarides: Soc. Sciences Arts Montenegro, Monographies, v. 4, 257 p.

————, AND RAMPNOUX, P., 1972, Concerning the Triassic in the Yugoslavian Inner Dinarids (Southern Serbia, Eastern Montenegro): Microfacies, Microfaunas, an attempt to give a paleogeographic reconstruction: Mitteilungen Gesellschaft Geologie- und Bergbaustudenten Österreich, v. 21, p. 311–326.

PARTSCH, J., 1887, Die Insel Korfu: Petermanns Mitteilungen, Ergänzungsheft, v. 88, p. 12–16.

PIA, J., 1927, Thallophyta, in Hirmer, M., Ed., Handbuch der Paläobotanik, v. 1, p. 31–136.

PILLER, W., 1976, Fazies und Lithostratigraphie des gebankten Dachsteinkalkes (Obertrias) am Nordrand des Toten Gebirges (S Grünau/Almtal, Oberösterreich): Mitteilungen Gesellschaft Geologie- und Bergbaustudenten Österreich, v. 23, p. 113–152.

————, 1978, Involutinacea (Foraminifera) der Trias und des Lias: Beiträge Paläontologie Österreich, v. 5, p. 1–164.

————, 1980, The Steinplatte Reef Complex part of an Upper Triassic Carbonate Platform near Salzburg, Austria: this volume.

————, AND LOBITZER, H., 1979, Die obertriassische Karbonatplattform zwischen Steinplatte (Tirol) und Hochkönig (Salzburg): Verhandlungen Geol. Bundesanstalt, Wien, v. 1979, Verh. Geol. Bundesanst. Wien, Jg. 1979, p. 171–180.

————, AND SENOWBARI-DARYAN, B., 1980, Foliotortus spinosus n.g., n.sp.—ein neues Mikrofossil (Foraminifera?) aus obertriadischen Riff-Kalken von Sizilien: Facies, v. 2, in press.

PISTOTNIK, U., 1972, Zur Mikrofazies und Palaögeographie der Zlambachschichten (O.Nor-?U.Lias) im Raum Bad Goisern—Bad Aussee (Nördliche Kalkalpen): Mitteilungen Gesellschaft Geologie- und Bergbaustudenten Österreich, v. 21, p. 279–288.

PLÖCHINGER, B., 1967, Erläuterungen zur Geologischen Karte des Hohe Wand—Gebietes (Niederösterreich): Geol. Bundesanstalt Wien, 142 p.

RENZ, C., 1955, Die vorneogene Stratigraphie der normal sedimentären Formationen Griechenlands: Athen, Inst. Geol. Subsurface Res., 637 p.

RESCH, W., 1979, Zur Fazies-Abhängigkeit alpiner Trias-Foraminiferen: Jb. Geol. Bundesanst. Wien, v. 122, p. 181–249.

RICHTHOFEN, F., 1860, Geognostische Beschreibung der Umgegend von Predzaao, St. Cassian und der Seisser Alpe in Südtirol: Gotha, Perthes, 327 p.

RONIEWICZ, E. W. A., 1974, Rhaetian corals of the Tatra Mountains: Acta Paleont. Polonica, v. 24, p. 97–116.

SANDER, B., 1936, Beiträge zur Kenntnis der Anlagerungsgefüge (Rhythmische Kalke und Dolomite aus der Trias): Min. Petrograph. Mitteilungen, v. 48, p. 27–139, 141–209.

SCHÄFER, P., 1979, Fazielle Entwicklung und palökologische Zonierung zweier obertriadischer Riffstrukturen in den Nördlichen Kalkalpen ("Oberrhät"-Riff-Kalke, Salzburg): Facies, v. 1, p. 3–245.

———— AND SENOWBARI-DARYAN, B., 1978a, Neue Korallen (Scleractinia) aus Oberrhät-Riffkalken südlich von Salzburg (nördliche Kalkalpen, Österreich): Senckenbergiana Lethaea, v. 59, p. 117–135.

———, AND ———, 1978b, Die Häufigkeitsverteilung der Foraminiferen in drei oberrhätischen Riff-Komplexen der Nördlichen Kalkalpen (Salzburg, Österreich): Verhandlungen Geol. Bundesanstalt, Wien, v. 1978, p. 73–96.

———, AND ———, 1980a, Globochaeten-Zoosporen aus obertriadischen Riffkalken südlich von Salzburg (nördliche Kalkalpen): Verh. geol. Bundesanst. Wien, Jg. 1980, in press.

———, AND ———, 1980b, *Abatea culleiformis* n. g., n. sp., eine neue Rotalge (Gymnocodiaceae) aus den "oberrhätischen" Riff-Kalken südlich von Salzburg (Nördliche Kalkalpen, Österreich): Verh. Geol. Bundesanstalt Wien, Jg. 1979, p. 393–399.

———, AND ———, 1980c, Facies Development and Paleoecological Zonation of four Upper Triassic patch-reefs, Northern Calcareous Alps near Salzburg, Austria: this volume.

SCHLAGER, W., AND SCHOLLNBERGER, W., 1974, Das Prinzip stratigraphischer Wenden in der Schichtfolge der Nördlichen Kalkalpen: Mitteilungen Geol. Gesellschaft Wien, v. 66/67, p. 165–193.

SCHNEIDER, N.-J., 1964, Facies differentiation and controlling factors for the depositional lead-zinc concentration in the Ladinian Geosyncline of the Eastern Alps: Developments in Sedimentology, v. 2, p. 29–45.

SCHÖLLNBERGER, W., 1973, Zur Verzahnung von Dachsteinkalk-Fazies und Hallstätter Fazies am Südrand des Toten Gebirges (Nördliche Kalkalpen, Österreich): Mitteilungen Gesellschaft Geologie- und Bergbaustudenten Österreich, v. 22, p. 95–153.

SCHROEDER, J., AND ZANKL, H., 1974, Dynamic reef formation: a sedimentological concept based on studies of Recent Bermuda and Bahama Reefs: Internat. Coral Reef Symposium, Second, Proc., v. 2, p. 413–428.

SENOWBARI-DARYAN, B., 1978a, *Pentaporella rhaetica* n.g. n. sp., eine neue Kalkalge (Dasycladaceae) aus dem oberrhätischen Gruber-Riff (Hintersee/Salzburg): Paläont. Zeitschrift, v. 52, p. 6–12.

———, 1978b, Neue Spinctozoen (segmentierte Kakschwämme) aus den "oberrhätischen" Riffkalken der nördlichen Kalkalpen (Hintersee/Salzburg): Senckenbergiana Lethaea, v. 59, p. 205–227.

———, 1978c, Fazielle und paläontologische Untersuchungen an "oberrätischen" Riffen—Feichtenstein- und Gruber-Riff - bei Hintersee/Salzburg (Nördliche Kalkalpen) [Ph.D. Dissertation]: Erlangen, West Germany, Erlangen Univ., 352 p.

———, 1978d, Ein neuer Fundpunkt von *Placklesia multipora* Bilgütay aus den Kössener Schichten des Feichtensteins bei Hintersee (Salzburg, Österreich): Mitt. Ges. Geol. Bergbaustud. Österreich, v. 25, p. 197–203.

———, 1979, Anomuren-Koprolithen aus der Obertrias der Osterhorngruppe (Hintersee/Salzburg, Österreich): Annalen Naturhist. Museum Wien, v. 82, p. 99–107.

———, 1980a, Neue Kalkschwämme (Sphinctozoen) aus obertriadischen Riffkalken von Sizilien: Mitt. Ges. Geol. Bergbaustud. Österreich, v. 26, p. 179–203.

———, 1980b, *Barbafera carnica* n. g., n. sp., ein Problematikum aus den *Cidaris*-Schichten (Gosaukamm, Oberösterreich) und Amphyclinen-Schichten (Slowenien, Jugoslawien)—Karn (Ober-Trias): Verh. Geol. Bundesanst. Wien, Jg. 1980, in press.

———, 1980c, Fazielle und paläontologische Untersuchungen in oberrhätischen Riffen (Feichtenstein- und Gruber-Riff bei Hinteresse, Salzburg, Nördliche Kalkalpen): Facies, v. 3, in press.

———, Kurze Mitteilung über die Sphinctozoen-Fauna eines kleinen Riffes innerhalb der Amphyclinen-Schichten (Lokalität: Južna Huda, Slowenien): Geologija, Ljubljana, in press.

———, AND DULLO, W.-CHR., 1980, *Cryptocoelia wurmi* n. sp., ein Kalkschwamm (Sphinctozoa) aus der Obertrias (Nor) der Gesäuseberge (Obersteiermark/Österreich): Mitt. Ges. Geol. Bergbaustud. Österreich, v. 26, p. 205–211.

———, AND SCHÄFER, P., 1978, *Follicatena irregularis* n. sp., ein segmentierter Kalkschwamm aus den "Oberrhät"-Riffkalkem der alpinen Trias: Neues Jahrbuch Geologie Paläontologie, Monatshefte, v. 1978, p. 314–320.

———, AND ———, 1979a, Distributional patterns of Calcareous algae within Upper Triassic patch-reef structures of the Northern Calcareous Alps (Salzburg): Bull. Centre Rech. Explor.-Prod. Elf-Aquitaine, v. 3, p. 811–820.

———, AND ———, 1979b, Neue Kalkschwämme und ein Problematikum (*Radiomura cautica* n. g., n. sp.) aus Oberrhät-Riffen südlich von Salzburg (Nördliche Kalkalpen): Mitt. Österr. Geol. Ges., v. 70, p. 17–42.

———, AND ———, 1980a, *Abatea culleiformis* n. g. n. sp., eine neue Rotalge (Gymnocodiaceae) aus den "oberrhätischen" Riffkalken südlich von Salzburg (Nördliche Kalkalpen, Österreich): Verh. Geol. Bundesanst. Wien, Jg. 1980, p. 393–399.

———, AND ———, 1980b, *Paraeolisaccus endococcus* n. g. n. sp., eine Alge (?) aus den obertriadischen Riffkalken von Sizilien/Italien: Verh. Geol. Bundesanst. Wien, Jg. 1980, in press.

———, AND ———, 1980, *Helicerina siciliana* n. sp., a new anomuran coprolite from Upper Triassic reef limestones near Palermo (Sicily): Boll. Soc. Paleont. Ital., in press.

SICKENBERG, O., 1932, Ein rhätisches Korallenriff aus der Osterhorngruppe: Verhandlungen Zool.-Botan. Gesellschaft Wien, v. 82, p. 35–40.

SIEBER, R., 1937, Neue Untersuchungen über die Stratigraphie und Ökologie der alpinen Triasfaunen. 1. Die Fauna der nordalpinen Rhätriffkalke: Neues Jahrbuch Min. Geol. Paläont., Beilageband, v. 78, p. 123–188.

SMITH, J. P., 1912, The occurrence of coral reefs in the Triassic of North America: Am. Jour. Sci., v. 33, p. 92–96.

———, 1927, Upper Triassic marine invertebrate faunas of North America: U.S. Geological Survey Professional Paper, v. 141, 262 p.

SQUIRES, D. F., 1956, A new Triassic coral fauna from Idaho: Amer. Mus. Novitates, no. 1797, p. 1–21.

STANLEY, G. D., 1979, Paleoecology, structure, and distribution of Triassic coral buildups in Western North America: Univ. Kansas Paleont. Contrib., v. 65, p. 1–58.

STÖCKLIN, J., EFTEKHAR-NEZHAD, J., AND HUSHMAND-ZADEH, A., 1965, Geology of the Shotori Range (Tabas Area, East Iran): Geological Survey Iran, Rept., v. 3, p. 1–68.

STUR, D., 1871, Geologie der Steiermark: Graz, Geognostisch-montanistischer Verein, 654 p.

SÜSSKOCH, H., 1967, Die Geologie der südöstlichen Argolis (Peleponnes, Griechenland) [Ph.D. thesis]: Marburg Univ., 114 p.

TERMIER, H., AND TERMIER, G., 1973, Stromatopores, Sclérosponges et Pharétrones: les Ischyrospongia: Ann. Mines Géol., v. 26, p. 285–297.

TOLLMANN, A., 1967, Tektonische Karte der Nördlichen Kalkalpen. 1. Teil: Der Ostabschnitt: Mitteilungen Geol. Gesellschaft Wien, v. 59, p. 231–253.

———, 1972, Der karpatische Einfluss am Ostrand der Alpen: Mitteilungen Geol. Gesellschaft Wien, v. 64, p. 173–208.

———, 1976, Analyse des klassischen nordalpinen Mesozoikums: Wien, Deuticke, 580 p.

———, 1978, Bemerkungen zur Frage der Berechtigung der rhätischen Stufe: Schriftenreihe Erdwissenschaftl. Kommsission Österr. Akad. Wissenschaften, v. 4, 175–177.

VEGH-NEUBRANDT, E., 1960, Petrologische Untersuchungen der Obertrias-Bildungen des Gerecsgebirges in Ungarn: Geologica Hungarica, Ser. Geol., v. 14, p. 1–110.

VINASSA DE REGNY, P., 1915, Triadische Algen, Spongien, Anthozoen und Bryozoen aus Timor: Paläont. Timor, v. 4, p. 75–118.

———, 1932, Hydrozoen und Korallen aus der oberen Trias des Karakorums, in Trinkler, E., and Terra, H. de, Eds., Wissenschaftliche Ergebnisse der Dr. Trinkler-schen Zentralasien-Expedition, v. 2: Berlin, Reimer-Vohsen, p. 192–196.

VOLZ, W., 1896, Die Korallen der Schichten von St. Cassian in Süd-Tirol: Palaeontographica, v. 43, p. 1–123.

VORTISCH, W., Oberrhätischer Riffkalk und Lias in den nordöstichen Alpen. Teil 1: Jahrbuch Geol. Bundesanst., v. 76, p. 1–64.

WAEHNER, F., 1903, Das Sonnwendgebirge im Unterinntal, ein Typus eines alpinen Gebirgsbaues, v. 1: Leipzig—Wien, Deuticke, 356 p.

WENDT, J., 1975, Aragonitische Stromatoporen der alpinen Obertrias: Neues Jahrbuch Geologie Paläontologie, Abhandlungen, v. 150, p. 111–125.

WIEDENMAYER, F., 1963, Obere Trias bis mittlerer Lias zwischen Saltrio und Tremona (Lombardische Alpen). Die Wechselbeziehungen zwischen Stratigraphie, Sedimentologie und syngenetischer Tektonik: Eclogae Geol. Helvetiae, v. 56.

WIEDMANN, J., FABRICIUS, F., KRYSTYN, L., REITNER, J., AND URLICHS, M., 1979, Über Umfang und Stellung des Rhaet: Newletters Strat., v. 8, p. 133–152.

WILCKENS, O., 1937, Korallen und Kalkschwämme aus dem obertriadischen Phaeretronenkalk von Seran (Molukken): Neues Jahrbuch Miner. Geol. Paläont., Beilageband, v. 77, p. 171–211.

WILSON, J. L., 1975, Carbonate Facies in Geologic History: Berlin-Heidelberg-New York, Springer-Verlag, 471 p.

WOLFF, H., 1973, Fazies-Gliederung und Paläogeographie des Ladins in den bayerischen Kalkalpen zwischen Wendelstein und Kampenwand: Neues Jahrbuch Geologie Paläont., Abhandlungen, v. 143, p. 246–274.

WRAY, J. L., 1977, Calcareous Algae: Amsterdam, Elsevier, 185 p.

YANG, JING-ZHI, AND WANG, CHENG-YUAN, 1979, The Stromatoporoidea and Hydrozoa from the Mount Jolmo Lungma: Rep. Scientific Exped. Mount Jolmo Lungma Region, Palaeont., fasc. 1, Nanking Inst. Geol. Paleont., p. 71–82.

ZANKL, H., 1965, Zur mikrofaunistischen Charakteristik des Dachsteinkalkes (Nor/Rhät) mit Hilfe einer Lösungstechnik: Zeitschrift Deutsch. Geol. Gesellschaft, v. 116, p. 549–567.

———, 1967, Die Karbonatsedimente der Obertrias in den nördlichen Kalkalpen: Geol. Rundschau, v. 56, p. 128–139.

———, 1969, Der Hohe Göll: Aufbau und Lebensbild eines Dachsteinkalk-Riffes in der Obertrias der nördlichen Kalkalpen: Abhandlungen Senckenbergischen Naturforsch. Gesellschaft, v. 519, p. 1–123.

———, 1971, Upper Triassic carbonate facies in the Northern Limestone Alps, in Müller, G., Ed., Sedimentology of Parts of Central Europe, Guide Book: Frankfurt, Kramer, p. 147–185.

ZAPFE, H., 1959, Faziesfragen des nordalpinen Mesozoikums: Verhandlungen Geol. Bundesanstalt, Wien, v. 1959, p. 122–128.

———, 1960, Untersuchungen im overtriadischen Riff des Gosaukammes (Dachsteingebiet, Oberösterreich). 1. Beobachtungen über das Verhältnis der Zlambach-Schichten zu den Riffkalken im Bereich des Grossen Donnerkogels: Verhandlungen Geol. Bundesanstalt, Wien, v. 1960, p. 236–241.

———, 1962, Untersuchungen im obertriadischen Riff des Gosaukammes (Dachsteingebiet, Oberösterreich). 4. Bisher im Riffkalk des Gosaukammes aufgesammelte Makrofossilien (excl. Riffbildner) und deren stratigraphische Auswertung: Verhandlungen Geol. Bundesanstalt Wien, v. 1962, p. 346–361.

———, 1963, Beiträge zur Paläontologie der nordalpinen Riffe. Zur Kenntnis der Fauna des oberrhätischen Riffkalkes von Adnet, Salzburg (excl. Riffbildner): Annalen Naturhist. Museum Wien, v. 66, p. 207–259.

———, 1964a, Untersuchungen im obertriadischen Riff des Gosaukammes (Dachsteingebiet, Oberösterreich). 6. Das Alter der Nornsteinkalke im Liegenden des Riffes: Verhandlungen Geol. Bundesanstalt, Wien, v. 1964, p. 177–181.

————, 1964b, Beiträge zur Paläontologie der nordalpinen Riffe. Zur Kenntnis der Megalodontiden des Dachstein-kalkes im Dachsteingebiet und Tennengebirge: Annalen Naturhist. Mus. Wien, v. 67, p. 253–286.

————, 1965, Beiträge zur Paläontologie der nordalpinen Riffe. Die Fauna der "erratischen Blöcke" auf der Falmbergalm bei Gosau, Oberösterreich: Annalen Naturhist. Mus. Wien, v. 69, p. 279–308.

————, 1967a, Untersuchungen im obertriadischen Riff des Gosaukammes (Dachsteingebiet, Oberösterreich). 8. Fragen und Befunde von allgemeiner Bedeutung für die Biostratigraphie der alpinen Obertrias: Verhandlungen Geol. Bundesanst., Wien, v. 1967, p. 13–27.

————, 1967b, Beiträge zur Paläontologie der nordalpinen Riffe. Die Fauna der Zlambach-Mergel der Fischerwiese bei Aussee, Steiermark: Annalen Naturhist. Mus. Wien, v. 71, p. 413–480.

————, 1974, Trias in Österreich: Schriftenreihe Erdwissenschaftl. Kommission Österr. Akad. Wissenschaften, v. 2, p. 245–251.

SEPM SPECIAL PUBLICATION NO. 30, P. 361–369, MAY 1981

# AN UPPER JURASSIC REEF COMPLEX FROM SLOVENIA, YUGOSLAVIA

DRAGICA TURNŠEK
Institut za Paleontologijo SAZU
Ljubljana, Yugoslavia

STANKO BUSER AND BOJAN OGORELEC
Geoloski zavod
Ljubljana, Yugoslavia

## ABSTRACT

An Upper Jurassic (Oxfordian and lower Kimmeridgian) reef complex is well exposed in the area of Slovenia, northwestern Yugoslavia. This reef complex is thought to be a barrier-reef that developed along the shelf-margin of an ancient carbonate platform. From basin to lagoon the following subdivisions have been delineated: a fore-reef area characterized by carbonate breccias and blocks of reef debris; a central reef area with abundant hydrozoans and corals that can be further subdivided into actinostromariid and parastromatoporid zones; and, a back-reef area with locally developed lagoons and patch-reefs defined as the *Cladocoropsis* zone.

### EXTENT OF REEF COMFLEX

In the province of Slovenia, coral-hydrozoan reefs of Middle Devonian, Upper Carboniferous, Upper Triassic, Upper Jurassic, Lower and Upper Cretaceous, and Oligocene ages are well exposed. Of these, the Upper Jurassic barrier-reef complex has the largest geographical extent. This barrier-reef complex can be traced, in a more or less continuous series of outcrops, from the Soča Valley across the central portion of Slovenia to Metlika into neighboring Croatia, and then along the entire Yugoslav Dinaric region into Albania. The reef complex also extends westward from the Seča Valley into Italy. In Slovenia, the reef complex extends for approximately 140 kilometers in length, is about 20 kilometers in width, and ranges in thickness from 200 to 600 meters (Fig. 1). Postulated restored length of the reef trend possibly amounted to at least several hundred kilometers, a distance comparable to that of the present Australian Barrier Reef.

### STRATIGRAPHIC SETTING

It is thought that at the beginning of Upper Jurassic time the environment of this region changed considerably due to a pronounced warming trend. In Slovenia, the marginal portion of the Dinaric carbonate shelf passed laterally to the north and northeast into deeper water environments. Along the shelf-edge margin, between the carbonate platform and the deeper water areas to the north and northeast, a large barrier-reef complex was initiated. In places, reefal formations overlie lower Oxfordian platy, cherty limestones (Dogger micrite) with oolitic and crinoidal limestones. Occasionally, reefal limestones may lie unconformably on Liassic limestones. The basal portion of the reef complex consists of light-colored Oxfor-

dian limestones composed of crinoids and echinoid spines. The limestones range up to 10 meters in thickness. These pass upward into massive, non-bedded reef units composed chiefly of corals and hydrozoans, which can attain a thickness of up to 600 meters. Faunal comparison of this reef interval with similar intervals in Europe and Asia indicates that these sediments range in age from Oxfordian to lower Kimmeridgian time. The reef interval is conformably overlain by limestones containing the alga *Clypeina jurassica* Favre, a form whose known stratigraphic range is from upper Kimmeridgian to Portlandian time (Buser, 1965 and 1978). It should be noted however, that some Croatian workers consider the algal interval as a reef equivalent, and regard it Tithonian in age (Nikler, 1978).

### SUBDIVISIONS OF REEF COMPLEX

The reef complex is situated on the transitional portion of an ancient carbonate platform that faced towards the deeper open ocean of Upper Jurassic time. In cross-section (Fig. 2) lithological and faunal zonation can be followed from the deeper water basinal sediments of the north and northeast to lagoonal settings of the south and southeast. In the reef itself, a distinctive fore-reef zone can be distinguished, followed by a central reef area subdivided into two hydrozoan zones, this in turn giving way to a wide back-reef area with local patch-reefs.

#### Fore-Reef Area

This area consists of a reef-slope on which fore-reef sediments were deposited and which grades into basinal sediments. Fore-reef sediments are represented by calcareous breccias consisting of allochthonous blocks and fragments of reef lime-

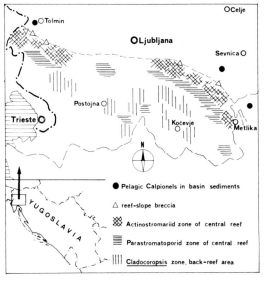

FIG. 1.—Index map showing location and extent of the Upper Jurassic barrier-reef complex exposed in Slovenia, northwestern Yugoslavia.

stone, usually cemented with reef detritus, crinoidal fragments, and echinoid spines. In some fore-slope beds graded breccias may be seen. Basinward of the breccias there is a conspicuous two mile wide belt of limestone consisting almost entirely of crinoidal fragments and echinoid spines. Some areas within this zone contain sporadic corals and hydrozoans. Deeper water basinal sediments consist mainly of marly limestones in which pelagic tintinnids and cephalopods are found.

## Central Reef Area

This is a massive barrier-reef complex approximately 15 kilometers in width, which extends, with several interruptions, across the entire area of central Slovenia. The principal barrier-reef framebuilders are corals and hydrozoans. Based mainly upon hydrozoan distribution, two faunal subzones have been delineated. These are: actinostromariid zone on the platform-edge, and the parastromatoporid zone directly behind the above zone (Turnšek, 1966 and 1969).

The actinostromariid zone is 6 to 10 kilometers in width and is principally organogenic, consisting primarily of corals and hydrozoans. Typical forms of actinostromariids and sphaeractinids are shown on Figure 3. Identified genera include: *Astrostylopsis, Actinostromina, Coenostella, Sporadoporidium, Tubuliella, Sphaeractinella, Ellipsactinia,* and *Cylicopsis.* Of the above hydrozoan genera 26 species have been determined, all char-

acterized by the orthogonal microstructure of the skeletal elements (Turnšek, 1966). Among the corals, representatives of all groups can be found with a total of 27 genera representing 38 species. Typical coral forms from the actinostromariid zone are illustrated on Figure 4. Most frequently occurring corals are species of the genera: *Pseudocoenia, Heliocoenia, Stylosmilia, Complexastraea, Clausastraea, Amphiastraea, Schizosmilia, Mitrodendron, Donacosmilia, Microsolena, Microphyllia, Dermosmilia, Calamophylliopsis,* and others (Turnšek, 1972). Massive and phaceloid coral colonies are dominant with solitary and ramose corals and chaetetids less abundant. Total hydrozoan biomass is much greater than total coral biomass, even though more coral species are represented. Other organisms occur only rarely within the reef limestones, among these are: nonskeletal algae, foraminifers, bryozoans, crinoid debris, and echinoid spines. Occasionally, brachiopods (*Terebratula formosa* Suess) can also be found. In the highest part of the reef interval numerous pelecypods (*Diceras*) and gastropods (*Nerinea*) occur associated with such problematical forms as *Tubiphytes morronensis* Crescenti, *Marierella dacica* Dragastan, and *Baccanella* cf. *B. parvissima* (Dragastan), according to Buser, 1978.

Lithological composition of the limestones in the actinostromariid zone is variable. Large fossil organisms are dominant, while the interstices contain debris of breccia and calcarenite. Rock cement is mainly sparite suggesting a high to very high water energy index. Well-washed sediments occur in the main reef areas marginal to the slope into the basin. Locally, areas of low energy may also occur; these are characterized by intervals of pelmicrite and biomicrite.

The parastromatoporid zone occurs in an approximately 5 kilometer wide belt directly behind the actinostromariid zone (Fig. 2). This zone differs from the actinostromariid zone primarily in the occurrence of different hydrozoans, and to some extent, differences in the coral assemblages. The limestones of this zone are composed of hydrozoans, corals, and chaetetids. Among the hydrozoans, parastromatoporid types are dominant. Some typical forms are illustrated on Figure 5. The following genera, represented by 9 species, are most common: *Parastromatopora, Dehornella, Hudsonella, Reticulina,* and *Shuqraia.* These hydrozoan forms are also characterized by orthogonal microstructure of the skeletal elements. Numerous chaetetids are also present, represented by species of the genera: *Bauneia* (Fig. 5F), *Chaetetopsis* (Fig. 5G), and *Ptychochaetetes* (Turnšek, 1966). Within this zone coral biomass is dominant, as well as numbers of species. In all,

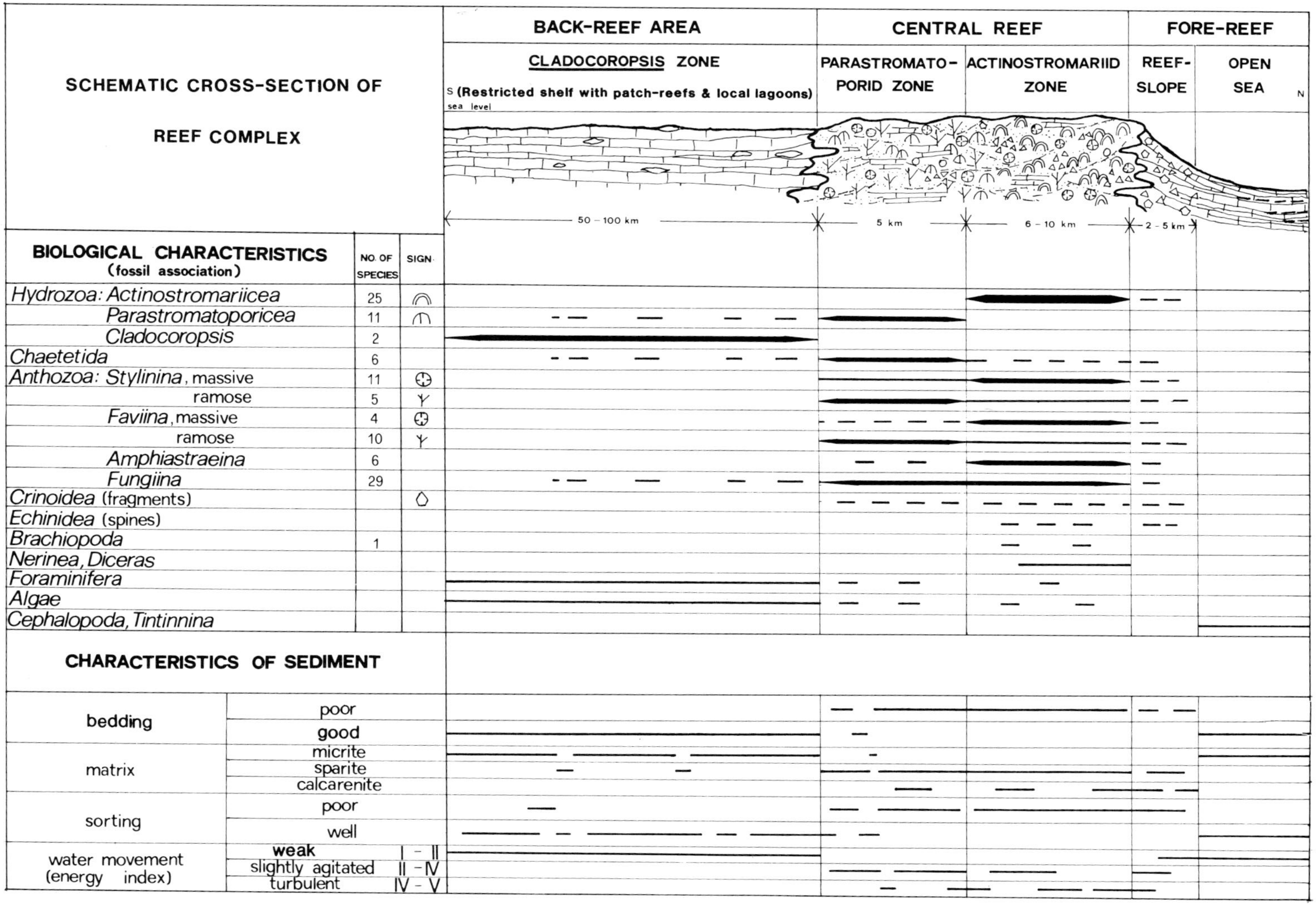

**SCHEMATIC CROSS-SECTION OF REEF COMPLEX**

**FORE-REEF:** OPEN SEA, REEF-SLOPE (N)

**CENTRAL REEF:** PARASTROMATO-PORID ZONE, ACTINOSTROMARIID ZONE

**BACK-REEF AREA:** CLADOCOROPSIS ZONE (Restricted shelf with patch-reefs & local lagoons) (S, sea level)

scale markers: 50 – 100 km, 5 km, 6 – 10 km, 2 – 5 km

**BIOLOGICAL CHARACTERISTICS (fossil association)**

| | NO OF SPECIES | SIGN. |
|---|---|---|
| Hydrozoa: Actinostromariicea | 25 | |
| Parastromatoporicea | 11 | |
| Cladocoropsis | 2 | |
| Chaetetida | 6 | |
| Anthozoa: Stylinina, massive | 11 | |
| ramose | 5 | |
| Faviina, massive | 4 | |
| ramose | 10 | |
| Amphiastraeina | 6 | |
| Fungiina | 29 | |
| Crinoidea (fragments) | | |
| Echinidea (spines) | | |
| Brachiopoda | 1 | |
| Nerinea, Diceras | | |
| Foraminifera | | |
| Algae | | |
| Cephalopoda, Tintinnina | | |

**CHARACTERISTICS OF SEDIMENT**

| | |
|---|---|
| bedding | poor / good |
| matrix | micrite / sparite / calcarenite |
| sorting | poor / well |
| water movement (energy index) | weak I–II / slightly agitated II–IV / turbulent IV–V |

FIG. 2.—Schematic cross-section of the Upper Jurassic barrier-reef complex in Slovenia, northwestern Yugoslavia, showing its facies zonation, faunal associations, and characteristic sediments.

this zone contains 36 identified coral species, some of the more typical forms are illustrated on Figure 6. Most frequently occurring genera are: *Pseudocoenia, Stylina, Heliocoenia, Goniecora, Stylosmilia, Aplophyllia, Montlivaltia, Thecosmilia, Ceratothecia, Dermosmilia, Calamophylliopsis, Microsolena, Comoseris, Meandrophyllia, Thamnasteria, Fungiastraea,* and others (Turnšek, 1972). Compared to the corals of the actinostromariid zone, corals in this zone are dominantly ramose stilinids and faviids, and crust-like fungids. Less common are massive colonial growth forms, while amphiastraeids are very rare. Of a total of 65 coral species identified from the central reef area, only 6 coral species are common to both zones. In addition to the above organisms, only rare occurrences of red algae, foraminifers, brachiopods, gastropods, pelecypods, and echinodermal debris have been identified. These fossil remains are often coated with non-skeletal algae.

The lithological characteristics of the limestones in the parastromatoporid zone are similar to those noted in the actinostromariid zone. Here too, biolithites predominate with the skeletons of reef builders broken and deposited in place. Some calcarenitic debris, from poor- to well-sorted, is also present. The water energy index appears to have been variable reflecting varying depositional settings. For areas of moderately agitated water environment biomicrites are common. The occurrences of geopetal cements in the interstices of some biolithic material suggests that the reef complex was periodically emerged above sea level,

and that cementation occurred under vadose conditions. At the Otlica locality oolites have been found; however, these are believed to have been transported from a different source area.

Faunal differentiation within the central reef area is the result of different ecological conditions. The reef grew along the shelf-edge or margin of a carbonate shelf-platform. Reef subsidence seems to have been continuous with the top of the barrier remaining within a constant depth range interval. On the ocean-facing side water movement was strongest, hence leading to an abundant supply of nutrients, accompanied by thorough water aeration. These conditions appear to have been ideally suited to the actinostromatiid hydrozoans, especially those forms with massive coenostea. Here too, massive colonial coral forms are abundant. Conversely, in areas farther back from the reef-front, as in the parastromatoporid zone, ramose and crust-like forms of reefs organisms thrived. During growth, the barrier-reef was constantly subjected to wave and current erosion and destruction, and broken reef blocks and fragments were redeposited and incorporated with skeletal debris forming calcarenites. It is to be noted that the present writers, on the basis of this study, regard hydrozoans as more environmentally sensitive organisms than corals, and thus more meaningful indicators of paleoecological environments.

### Back-Reef Area (Cladocoropsis Zone)

The parastromatoporid zone of the central reef passes landward into the lagoonal shelf-area (Fig.

→

FIG. 3.—Thin-section photomicrographs of some forms of hydrozoans from the central reef actinostromariid zone; all ×4. A. *Actinostromina germovsheki* Turnšek, transverse section showing massive coenosteum within sparry matrix; Mačkevec locality (P-30). B. *Ellipsactinia polypora* Canavari, massive colony with strong lamellar elements; Radovica locality (P-125). C. *Astrostylopsis circoporea* (Gemovšek); longitudinal section of coenosteum; Ojstrovca locality (P-83). D. *Ellipsactinia ellipsoidea* Steinmann, massive coenosteum of hydrozoan overgrown by chaetetid; Slamna vas locality (P-132). E. *Cylicopsis lata* Turnšek, vertical and partly oblique section of coenosteum; Ojstrovca locality (P-203).

FIG. 4.—Thin-section photomicrographs of corals from the actinostromariid zone of the central reef; all ×4. A. *Complexastraeopsis lobata* (Geyer), transverse section of massive colony; Lokovec locality (1902/8B). B. *Complexastraea seriata* Turnšek, transverse section of massive colony; Mrzovec locality (P-245). C. *Microphyllia bachmayeri* Geyer and *Mitrodendron ogilvie* Geyer, transverse sections of both colonies showing both massive and ramose forms; Ivanja vas locality (P-311). D. *Amphiastraea piriformis* Gregory, transverse section of cerioid colony; Ivanja vas locality (P-309). E. *Stylosmilia corallina* Keby, transverse section of a phaceloid-dendroid colony; Lokovec locality (1902/12).

FIG. 5.—Thin-section photomicrographs of some forms of hydrozoans and chaetetids from both the parastromatoporid and *Cladocoropsis* zones of the central reef; all ×4. A.–B. *Cladocoropsis mirabilis* Felix within a micritic matrix; Racna gora locality (Ri/63-1). C.–D. *Parastromatopora compacta* Turnšek, longitudinal and transverse sections of coenosteum; Otlica locality (P-96). E. *Hudsonella otlicensis* Turnšek, transverse section of coenosteum; Otlica locality (P-114b). F. *Bauneia multitabulata* (Deninger), longitudinal and partly oblique section of a chaetetid; Luče locality (P-143). G. *Chaetetopsis krimholzi* Yaworsky, transverse section of colony; Čušperk locality (P-41).

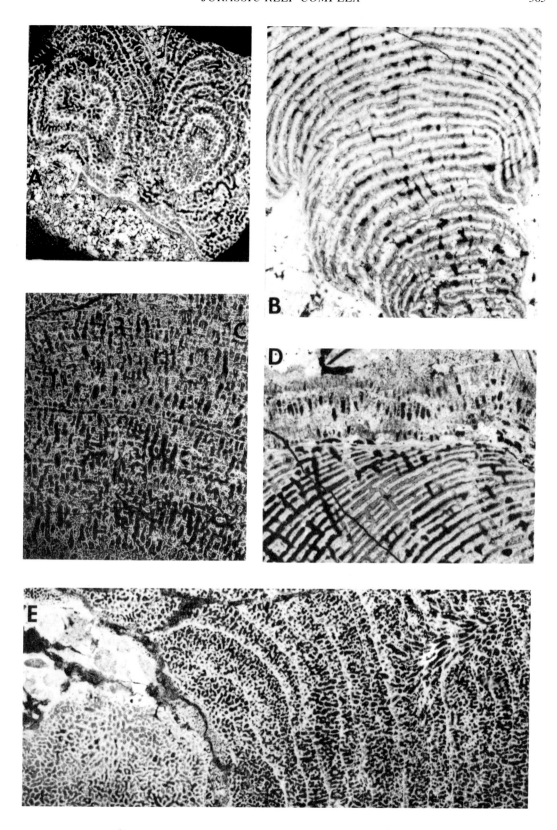

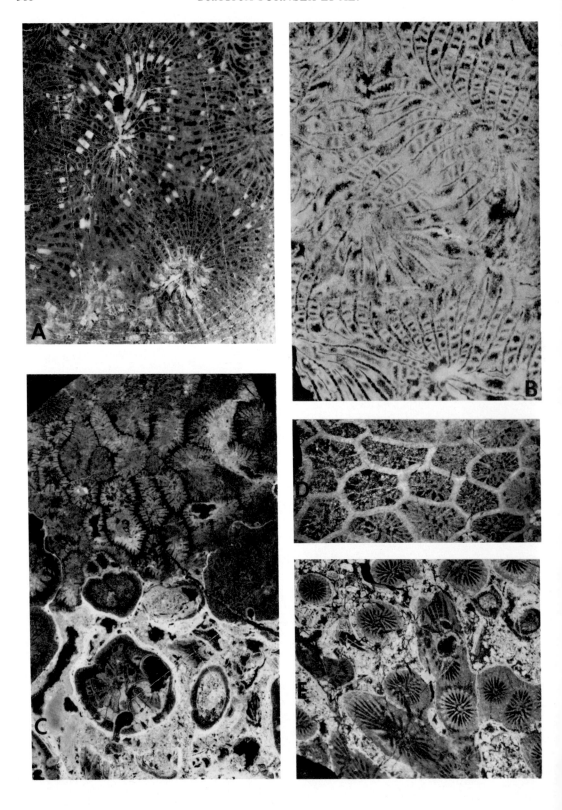

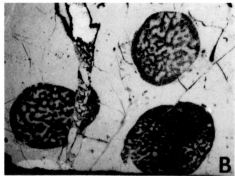

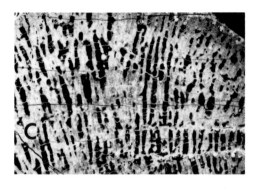

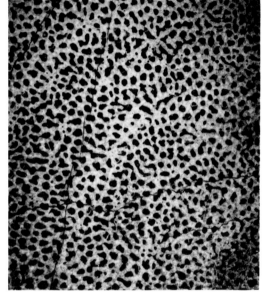

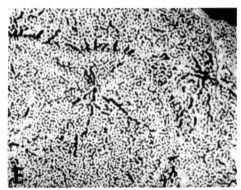

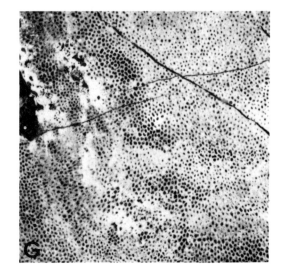

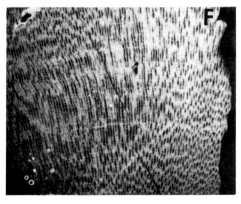

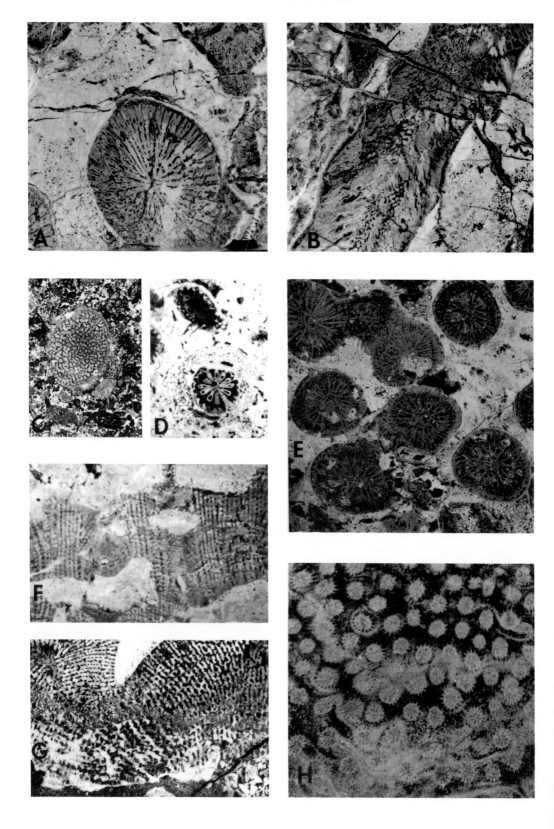

2). This area is characterized by quiet, low energy sedimentary environments. In this zone dark-grey massive, micritic limestones formed. Locally these may be bituminous or marly, and may even be peloidal. The dominant organism is the hydrozoan *Cladocoropsis mirabilis* Felix (Fig. 5A–B). Associated with this form are various foraminifers and algae. The foraminifers include: *Kurnubia palastiniensis* Henson, *K. wellingsi* Henson, *Pfenderina trochoidea* Smouth and Sugden, and *P. salernitana* Sartoni and Crescenti. Identified algae include the dasyclads *Macroporella sellii* Crescenti, *Thaumatoporella parvovesiculifera* (Raineri), and others. During short time intervals both the lagoons and shallow shelf areas were covered by oolitic deposits. In addition, some small patch-reefs, composed of a biota similar to that seen in the parastromatoporid zone, developed in this zone. These small patch-reefs are often only several meters thick, and are overlain by typical lagoonal deposits.

The authors anticipate that several more years of detailed paleontological and sedimentological study will be needed to finally analyze all of the parameters of the Upper Jurassic barrier-reef complex of Slovenia.

### ACKNOWLEDGMENTS

The authors are grateful to Dr. Simon Pirc for help with the English translation and to Milojka Huzjan for the drafting and Carmen Narobe for the photography.

## REFERENCES

BUSER, S., 1965, Neue Forschungsergebnise über die Juraschichten in Südslowenien: Wien, Anzeiger Mat. Nath. Cl. Österr. Akad. Wiss., Jg. 1965, p. 161–165.

———, 1978, Razvoj jurskih plasti Trnovskega gozda, Hrušice in Logaške planote (The Jurassic strata of Trnovski gozd, Hrušica and Logaška planota): Ljubljana, Rudarsko-metalurški zbornik, 1978, 4, p. 385–406.

NIKLER, L., 1978, Stratigrafski položaj grebenskog facijesa malma u sjeverozapadnim djelovima Dinarida (Stratigraphic position of the Malmian reef facies in northwestern Dinarids): Zagreb, Geološki vjesnik, 30, 1, p. 137–150.

TURNŠEK, D., 1966, Zgornjejurska hidrozojska favna iz južne Slovenije (Upper Jurassic Hydrozoan Fauna from Southern Slovenia): Ljubljana, Razprave IV. Razr. SAZU, 9, p. 335–428.

———, 1969, Prispevek k paleoekologiji jurskih hidrozojev v Sloveniji (A Contribution to the Palaeoecology of Jurassic Hydrozoa from Slovenia): Ljubljana, Razprave IV. Razr. SAZU, 12, p. 209–237. Tab. 1.

———, 1972, Zgornjejurske korale iz južne Slovenije (Upper Jurassic Corals of Southern Slovenia): Ljubljana, Razprave IV. Razr. SAZU, 15, p. 145–261.

←

FIG. 6.—Thin-section photomicrographs of some forms of corals from the parastromatoporid zone of the central reef area; all ×4 except C. which is ×3. A.–B. *Dermosmilia laxata* (Étallon), phaceloid-dendroid colony showing transverse and longitudinal sections of a single corallite; Col locality (P-299). C. Biocalcarenite with hydrozoan fragment derived from breccia within parastromatoporid zone; Otlica locality (6a/630). D. *Goniocora pumila* (Quenstedt), dendroid colony showing transverse sections of two corallites; Frata locality (P-406). E. *Calamophylliopsis flabellum* (Michelin), phaceloid colony showing transverse sections of corallites; Hrušica locality (F-297). F.–G. *Microsolena thurmanni* Koby, encrusting colony showing longitudinal and transverse sections of septa; Frata locality (P-399 and P-354). H. *Heliocoenia variabilis* Étallon, massive small colony showing transverse section of corallites and peritheca; Predole locality (P-381).

SEPM Special Publication No. 30, p. 371–397, May 1981

# AN UPPER JURASSIC SPONGE-ALGAL BUILDUP FROM THE NORTHERN FRANKENALB, WEST GERMANY

ERIK FLÜGEL AND TORSTEN STEIGER
Institut für Paläontologie
Universität Erlangen-Nürnberg, Erlangen, West Germany

## ABSTRACT

During Upper Jurassic time (middle-upper Oxfordian) a number of relatively small organic buildups, principally composed of siliceous sponges and algae, developed in the area of southern Germany. This study concentrates on one of these organic structures, the Müllersfelsen buildup (northern Franconian Alb near Streitberg). It is demonstrated that the dominant buildup constructional organisms are siliceous sponges, principally of two morphologies (cup-shaped and dish-shaped forms), and cyanophycean algae, which aided in forming the mound configuration. The Müllersfelsen buildup developed during several cyclic stages which occurred in relatively deeper water subtidal depositional environments, without strong current and wave action. Three facies types have been recognized, these are: 1. sponge-crust boundstone facies characterized by micritic boundstone rich in calcified siliceous sponges, tuberoids, and crusts; 2. lithoclastic packstone facies in which spheroidal sedimentary particles (lithoclasts, tuberoids, and bioclasts) are the dominant grain-supported allochems; and 3. tuberolitic wackestone facies in which mud-supported sedimentary particles are dominant. The spatial distribution of these microfacies indicate that the sponge-crust boundstone facies is the mound constructional facies, whereas both the tuberolitic wackestone and packstone facies are only developed in those areas marginal to the main organic buildup. The uppermost portion of the Müllersfelsen buildup is dolomitized.

## INTRODUCTION

During Upper Jurassic time two principal rock facies developed in the epicontinental seas of central Europe. These are: 1. a dominant, so-called "normal facies" composed of well-bedded limestones and marls, and containing a representative ammonite succession, and 2. a "reef facies" composed of buildups formed by sponges and corals, and rarely, algae.

Previously studied Jurassic sponge buildups from Europe include the following: from Poland, Oxfordian age (Wisniewska-Zelichowska, 1971; Laptas, 1974); from Franconia, West Germany, Callovian/Oxfordian to lower Tithonian age (Roll, 1932; Dorn, 1934; Meyer, 1975); from Swabia, West Germany, upper Oxfordian to middle Kimmeridgian age (Fisher, 1910; Fritz, 1958; Hiller, 1964; Gwinner, 1977); from the southern Jura Mountains of Switzerland and France, Oxfordian age (Oppliger, 1910; Schrammen, 1936; Burri and Bolliger, 1970; Gaillard, 1972); from Spain, Oxfordian age (Behmel, 1970; El Khoudary, 1974); from Roumania (central Dobrogea), lower Oxfordian to Kimmeridgian age (Draganescu, 1976).

This present contribution deals with a microfacies analysis and genetic interpretation of the Upper Jurassic Müllersfelsen sponge-algal buildup located near Streitberg, in northern Franconia, West Germany. This particular small sponge-algal buildup is one of the "classic" Upper Jurassic "sponge reef" exposures, and has been the subject of previous works by v. Gümbel (1862), Dorn (1932), and Roll (1934).

## STRATIGRAPHY AND PALEOGEOGRAPHICAL SETTING

The Upper Jurassic of Franconia, West Germany, the so-called "Malm," is similar to the "White Jurassic" (Weisser Jura *sensu* Quenstedt) of Swabia, including two principal facies types: the "reef" facies, and the "normal" stratified facies.

The lower part of the stratified facies that outcrops in the Northern Frankenalb is made up of bedded marly limestones alternating with calcareous marls (local term: Untere Mergelkalke, of lower and middle Oxfordian age). Upwards marls diminish and give way to deposition of well-bedded limestones (local term: "Werkkalk," upper Oxfordian). Marls appear again and form an interbedded sequence similar to that at the base (local term: Obere Mergelkalke, lower Kimmeridgian). The upper part of the northern Franconian Malm (middle-upper Kimmeridgian) is generally dominated by the reef facies except where thin lenses of partly dolomitized units of the stratified facies occur (local term: Engelhardtsberger Schichten).

Stratigraphic and chronological subdivision has been accomplished with ammonites and foraminifers (Zeiss, 1977; and Seibold, 1955). Time-equivalent reef facies commences with the initiation of small sponge-algal buildups in the lower Oxfordian. The buildups contain interbedded marls and increase in size and frequency in time, forming massive dome-shaped structures in the middle and upper kimmeridgian. Here, they are generally huge dolomitized blocks of the so-called "Fran-

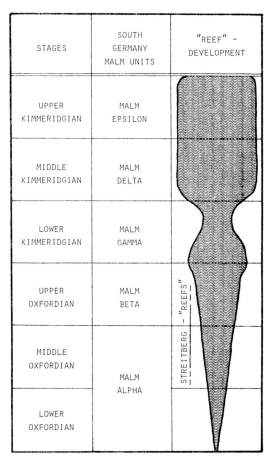

| STAGES | SOUTH GERMANY MALM UNITS | "REEF" – DEVELOPMENT |
|---|---|---|
| UPPER KIMMERIDGIAN | MALM EPSILON | |
| MIDDLE KIMMERIDGIAN | MALM DELTA | |
| LOWER KIMMERIDGIAN | MALM GAMMA | |
| UPPER OXFORDIAN | MALM BETA | |
| MIDDLE OXFORDIAN | MALM ALPHA | STREITBERG – "REEFS" |
| LOWER OXFORDIAN | | |

FIG. 1.—General stratigraphy of the Upper Jurassic sequence in Franconia, West Germany.

FIG. 2.—General view of the Müllersfelsen buildup east of Streitberg, Northern Frankenalb. The lower part of the dome-shaped structure consists of limestones and marls with abundant siliceous sponges. The upper steep slope is strongly dolomitized. At the left, the bedded limestones of the "normal facies" can be seen; some small buildups are embedded in this facies (see Fig. 11A and B).

kendolomit'' (''Franconian Dolomite,'' Meyer, 1972).

North of the river Wiesent the stratified facies with thicknesses of 20 to 50 meters is called the ''Feuerstein-Sequence'' (from Feuerstein-Quarry near Ebermannstadt). It is separated from the southern ''Hartmannshof-Sequence'' (from Hartmannshof-Quarry near Hersbruck) by the Wiesent barrier-reef (v. Freyberg, 1966). The difference between these two sequences is in the higher percentage of marls in the north.

The western part of the Wiesent barrier-reef belongs to the lower portion of the stratigraphic series embracing Oxfordian to lower Kimmeridgian time. The Müllersfelsen buildup developed from middle Oxfordian to upper Oxfordian time (Fig. 1). In this area the so-called normal facies is made up of a well-bedded sequence of limestones and marls grading laterally into the reef facies dominated by rocks of the sponge-algal boundstone facies.

Thick layers of marls which continue through both the stratified facies and the boundstone facies demonstrate the equivalency of these two facies types. The Müllersfelsen buildup can be regarded as the example of an ''early'' buildup, considering both its stratigraphic age and structural development.

At this location the thickness of the reef facies exceeds that of the stratified facies by about 12 meters. This thickness difference increases gradually to a maximum, at which a flat mound-like structure is formed (Fig. 2). The base of the Upper Jurassic is several meters below the rocks outcropping above the clays (Ornatenton of the Callovian), and can be recognized by a line of springs where limestone tufa is precipitated.

The location of the Müllersfelsen is shown in Figure 3 (inset). Well exposed rock of this buildup is situated 1 kilometer east of Streitberg and north of the river Wiesent, and can be reached on the Erlangen-Forchheim-Pegnitz Road. The distance between Erlangen and Streitberg is about 30 kilometers.

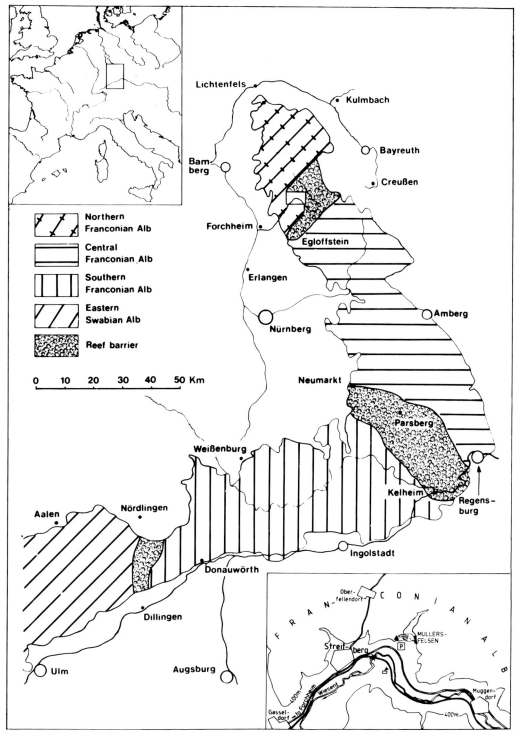

FIG. 3.—Location of the Müllersfelsen buildup (modified after Zeiss, 1977). Different facies areas are separated by barrier-reefs.

W

E

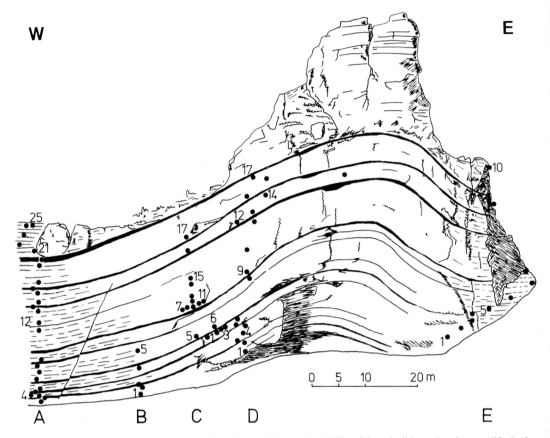

A          B          C          D                              E

FIG. 4.—Location of samples and stratigraphic profiles at the Müllersfelsen buildup; drawing modified after Gottwald, 1958.

## METHODS

Sampling was accomplished along the well exposed outcrops of the southern wall of the Müllersfelsen buildup. Five profiles, A–E, from west to east were studied, starting at the foot path which runs along the top of the rubble slope. Profiles A, B, and E are situated within tuberolitic sediment of the stratified beds. Profile C was taken at the western margin of the buildup, and profile D is situated in the center of the boundstone facies. The Location of profiles and sample sequences can be seen in Figure 4. Large thin-sections (50 to 150 cm²) were prepared and studied. The microfacies analyses and paleontological determinations are based on these thin-sections, peels, and scanning electron microscope (SEM) observations.

### MICROFACIES AND PALEONTOLOGY

#### Biota

*Sponges.*—Siliceous sponges are the dominant biotic constructional elements (hexactinellids and lithistids) of the buildup. However, due to pervasive calcification of the original siliceous skeleton, most characteristics necessary for generic or specific determinations are obscured. Therefore it is only possible to distinguish between Hexactinosa and Lychniscosa (groups of Hexactinellidea) and the Rhizocladina (Lithistida), primarily on structures visible in thin-section. The Lychniscosa are characterized by a skeleton meshwork of triaxon spicules with an open cross-structure (Fig. 12B). The Hexactinosa, conversely, have filled cross-points. On the other hand Rhizocladina form a skeleton meshwork of desma spicules. In thin-section this group can be recognized by the irregular construction of the meshwork. During calcification, the siliceous skeleton may be replaced totally or only partly by calcite. Therefore the growth form of the sponge is usually preserved, whereas the taxonomically important criteria (wall structures as apo- and epirhyses, ostia and postica, Müller, 1972) are diagenetically altered. Calcification of the sponges

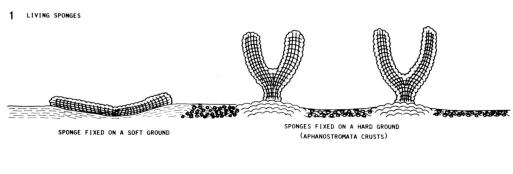

**1** LIVING SPONGES

SPONGE FIXED ON A SOFT GROUND

SPONGES FIXED ON A HARD GROUND
(APHANOSTROMATA CRUSTS)

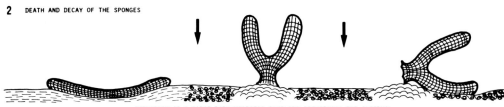

**2** DEATH AND DECAY OF THE SPONGES

FORMATION OF A FIRST DARK MICRITIC ENVELOPE PRESERVING THE INTERNAL
SKELETON AND THE SHAPE OF THE SPONGES - THE ORGANIC MATTER IS PROTECTED

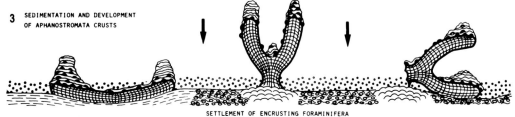

**3** SEDIMENTATION AND DEVELOPMENT
OF APHANOSTROMATA CRUSTS

SETTLEMENT OF ENCRUSTING FORAMINIFERA

**4** CALCIFICATION OF THE SPONGES

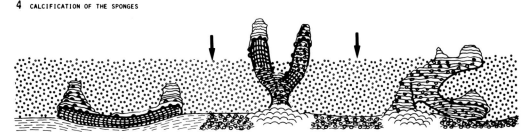

COMPLETE CALCIFICATION OF
THE SILICEOUS SKELETON,
BEING TOTALY PRESERVED

OPENING OF THE CLOSED
SYSTEM - DECAY OF THE
ORGANIC MATTER AND PARTIAL
SOLUTION AND DESTRUCTION
OF THE SILICEOUS SKELETON
- PENETRATION OF SEDIMENT
- MIXTURE OF SPICULA,
TUBEROIDS AND FORAMINIFERA,
BEING ENCRUSTED BY CYANO-
PHYCEANS - FORMATION OF
APHANOSTROMATA CRUSTS -
CALCIFICATION OF THE
SILICEOUS SPICULA

DESTRUCTION OF THE DARK
CRUST - OPENING OF THE
SYSTEM - TOTAL SOLUTION OF
THE SKELETON - PENETRATION
OF SEDIMENT, BEING BOUND
BY ENCRUSTING MICROORGANISMS
WHICH FORM APHANOSTROMATA
CRUSTS

FIG. 5.—Postmortem alterations of siliceous sponges.

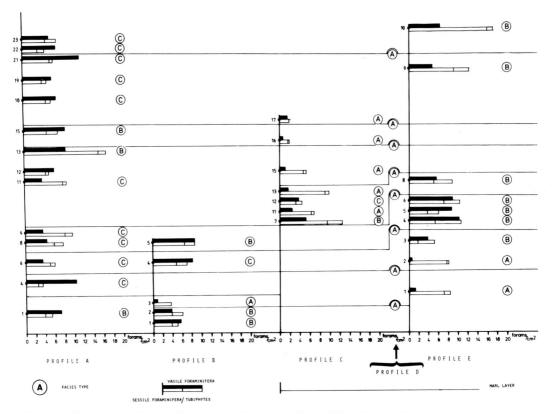

FIG. 6.—Distribution of foraminifers in several sections of the Müllersfelsen buildup; frequency has been calculated from thin-sections (not to scale).

can be explained by early diagenetic processes during sedimentation. According to the criteria seen in thin-section, and the hypothesis developed by Fritz (1958), concerning the alterations of the chemical environment during decay of the sponges, a clear-cut sequence can be outlined (Fig. 5). This hypothesis is dependent on the composition of the substrate, where encrusting nubeculinellid foraminifers modified the substrate to a hard-ground. It is thought that the original sponge ectoderm is then able to stabilize the sponge shape during internal decay of organic matter and diagenetic solution of the siliceous skeleton.

*Foraminifers.*—Vagile-benthonic, sessile and planktic foraminifers can be distinguished in the tuberolitic sediment of the Müllersfelsen buildup. They also occur with some frequency in the stratified facies and in the boundstone facies as well.

The dominant vagile-benthonic group consists of miliolids, rotallids, nodosariids and textulariids. The miliolids are represented mainly by two genera: *Ophthalmidium* (species *O. strumosum* (Gümbel)) and *Quinqueloculina*. Concerning the

rotaliids, diverse species of the genus *Spirillina* (*S. polygyrata* (Gümbel)) and *Trocholina* have been identified. The nodosariid group with *Lenticulina* and other uniserial genera (*Pseudonodosaria,* and *Dentalina*), as well as a group of agglutinating forms like *Reophax, Ammobaculites* and *Textularia* (species (*T. jurassica* (Gümbel)) are rather rare.

Sessile foraminifers belonging to the miliolids encrust sedimentary particles such as tuberoids, lithoclasts, bioclasts, and sponges. They are as numerous as the benthonic vagile miliolids. Planktic globigerinids occur only rarely in the micritic sediment of the Müllersfelsen buildup.

It is impossible to subdivide the boundstone facies and the stratified facies utilizing foraminiferal assemblages. Both facies contain the same foraminiferal microfauna. Sponges are only encrusted by forms of *Thurammina* and *Tolypammina*.

Figure 6 shows the frequency of foraminifers as seen in thin-section. In the sponge crust boundstone facies (type A), sessile foraminifers prevail. Vagile forms are very rare because the thin-sectioned area is usually occupied by sponges (with-

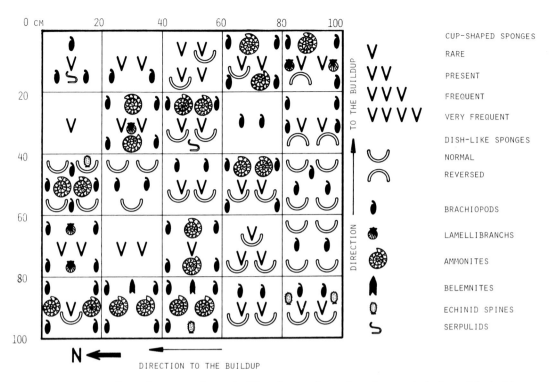

FIG. 7.—Frequency of organisms occurring on bedding-plane surfaces along the marginal edges of the buildup.

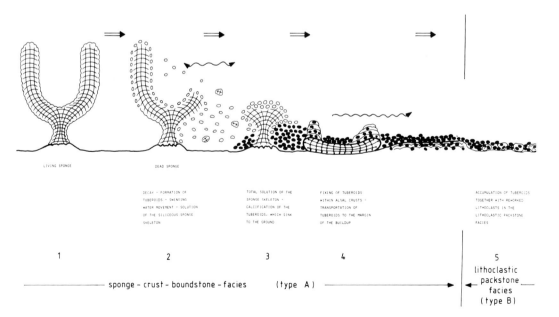

FIG. 8.—Schematic diagram showing development of tuberoids resulting from the decay and synsedimentary calcification of sponge fragments.

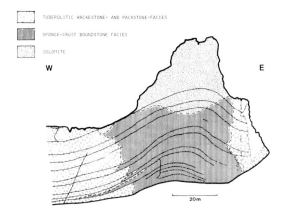

FIG. 9.—Distribution of facies types within the Müllersfelsen buildup.

out vagile foraminifers), and crusts (with very few vagile foraminifers). It is difficult to clearly separate the lithoclastic packstone facies (type B) from the tuberolitic wackestone facies (type C), with regard to the frequency of foraminifers. There seems to be a slight tendency for sessile foraminifers, together with sessile *Tubiphytes,* to be found in the packstones of the buildup margins. Their frequent occurrence is the result of the accumulation of encrusted lithoclasts, bioclasts and tuberoids. With the transition from packstone facies to wackestone facies the quantity of the vagile foraminifers increases, but in type C the ratio vagile/sessile forms is variable.

*Miscellaneous Organisms.*—Various miscellaneous organisms can be differentiated according to their mode of life: vagile organisms are not limited to special biotopes or facies, whereas sessile organisms are affixed to certain substrates.

1. *Vagile Organisms.*—Echinoderms, as echinoids and ophiuroids, are dominant (Fig. 18D and E). The echinoids are represented by spines and by complete tests (Schauertal near Streitberg, Gümbel, 1862); ophiuroids occur as isolated vertebral elements (*Lombardia*). Ostracodes are irregularly distributed in all facies. In some places ostracodes accumulated within internal sediment within cavities formed by boring organisms (Fig. 19D and E). Cephalopods are represented by belemnites and ammonites, the latter occur within the stratified facies, but also in the marginal areas of the buildup. Figure 7 is a frequency plot showing relationships of ammonites and other organisms encountered on the margin of the buildup.

2. *Sessile Organisms.*—These are mainly found attached to sponge surfaces. Two groups can be distinguished: those encrustations on the upper and the bottom sides of the sponges (and on the bottom side of all sessile groups, such as serpu-

lids, bryozoans, pelecypods, brachiopods and calcareous sponges can be recognized), and those on the upper sides of sponges, usually encrusted by some of the above biota in addition to foraminifers. Bryozoans are found as epizoans and fragments in the tuberolitic wackestone facies (Fig. 18A), whereas brachiopods occur on sponges and on aphanostromata crusts as well. The brachiopods are concentrated in patches 2 m² in area within the buildup. Most of the brachiopods belong to the terebratulids with only a few rhynchonellids present. Serpulids are restricted only to sponge surfaces (Fig. 21F). Of the pelecypods, only two of the four ecological groups recognized by Wagenplast (1972) could be found in the buildup. Sessile forms of pelecypods are attached to sponge surfaces (*Liostrea, Plicatula*), and to vagile pelecypods like *Pholadomya* and *Nucula* (Fig. 19F).

*Destructive Organisms.*—In the tuberolitic wackestone facies as well as in the boundstone facies organism burrowing (Fig. 18F–I and 19A) is common. Borings occur only in the boundstone facies (Fig. 19D and E). Burrows are represented as black tubes with distinctive agglutinated walls, these tubes range from 0.25 milimeters in diameter and 4 milimeters in length (within sponges and crusts), to 1 milimeter in diameter and several centimeters in length (within tuberolitic sediment). The origin of these tubes can be attributed to the activity of worms binding the surrounding soft sediment. A soft bottom is also indicated by nuculid lamellibranches in life-position within the sediment (Fig. 19F). Borings can be recognized as circular cavities with distinct boundaries. They are usually filled with layered internal sediments as crusts, or mechanically deposited bioclasts and pellets. Some borings may be caused by lamellibranches which can dissolve lithified sediment.

*Crusts*

Several types of crusts can be recognized within the sponge facies. These crusts are defined by the differences in sediment texture and predominance of various encrusting organisms.

1. *Laminoid, Peloidal Crusts* (*Aphanostromata* Nitzopoulos, 1964, figs. 17A, 19, 20A and B, 21A–D, 23A and B).—These crusts are planar to dome-shaped structures composed of loosely packed micritic peloids, growing upon sponge surfaces. The thickness of single layers is about 100 to 150 microns. The base of most of these crusts is composed of relatively large peloids; in the upper part the peloids are smaller. The top of this type of crust is characterized by a thin micritic layer corresponding to a crypto-hardground upon which foraminifers and other encrusting organisms settle. In some sponge beds there is a predominance of laminoid crusts that formed dur-

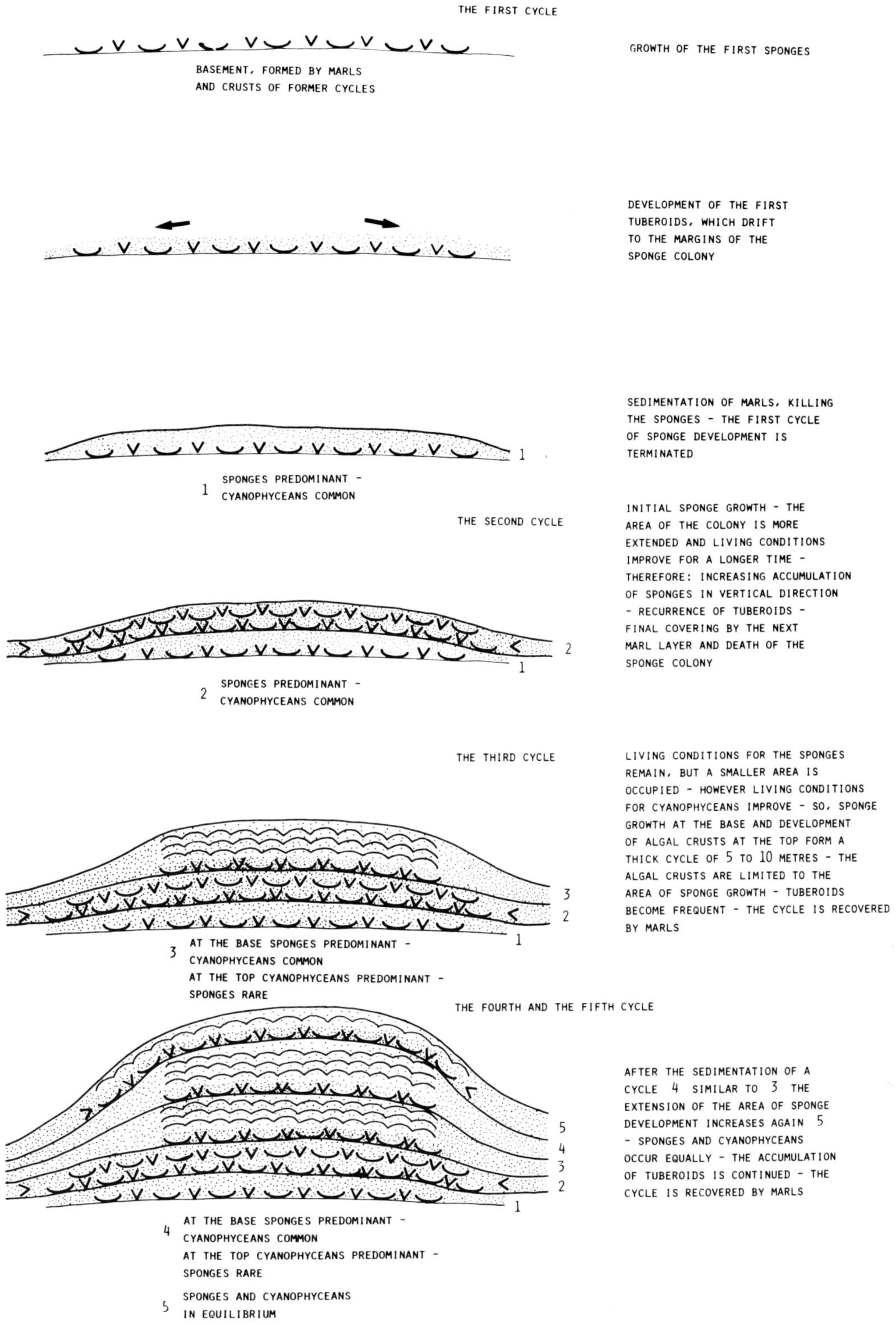

THE FIRST CYCLE

BASEMENT, FORMED BY MARLS
AND CRUSTS OF FORMER CYCLES

GROWTH OF THE FIRST SPONGES

DEVELOPMENT OF THE FIRST
TUBEROIDS, WHICH DRIFT
TO THE MARGINS OF THE
SPONGE COLONY

SPONGES PREDOMINANT –
1  CYANOPHYCEANS COMMON

SEDIMENTATION OF MARLS, KILLING
THE SPONGES – THE FIRST CYCLE
OF SPONGE DEVELOPMENT IS
TERMINATED

THE SECOND CYCLE

INITIAL SPONGE GROWTH – THE
AREA OF THE COLONY IS MORE
EXTENDED AND LIVING CONDITIONS
IMPROVE FOR A LONGER TIME –
THEREFORE: INCREASING ACCUMULATION
OF SPONGES IN VERTICAL DIRECTION
– RECURRENCE OF TUBEROIDS –
FINAL COVERING BY THE NEXT
MARL LAYER AND DEATH OF THE
SPONGE COLONY

SPONGES PREDOMINANT –
2  CYANOPHYCEANS COMMON

THE THIRD CYCLE

LIVING CONDITIONS FOR THE SPONGES
REMAIN, BUT A SMALLER AREA IS
OCCUPIED – HOWEVER LIVING CONDITIONS
FOR CYANOPHYCEANS IMPROVE – SO, SPONGE
GROWTH AT THE BASE AND DEVELOPMENT
OF ALGAL CRUSTS AT THE TOP FORM A
THICK CYCLE OF 5 TO 10 METRES – THE
ALGAL CRUSTS ARE LIMITED TO THE
AREA OF SPONGE GROWTH – TUBEROIDS
BECOME FREQUENT – THE CYCLE IS RECOVERED
BY MARLS

AT THE BASE SPONGES PREDOMINANT –
3  CYANOPHYCEANS COMMON
AT THE TOP CYANOPHYCEANS PREDOMINANT –
SPONGES RARE

THE FOURTH AND THE FIFTH CYCLE

AFTER THE SEDIMENTATION OF A
CYCLE 4 SIMILAR TO 3 THE
EXTENSION OF THE AREA OF SPONGE
DEVELOPMENT INCREASES AGAIN 5
– SPONGES AND CYANOPHYCEANS
OCCUR EQUALLY – THE ACCUMULATION
OF TUBEROIDS IS CONTINUED – THE
CYCLE IS RECOVERED BY MARLS

AT THE BASE SPONGES PREDOMINANT –
4  CYANOPHYCEANS COMMON
AT THE TOP CYANOPHYCEANS PREDOMINANT –
SPONGES RARE

SPONGES AND CYANOPHYCEANS
5  IN EQUILIBRIUM

FIG. 10.—Facies model showing the generalized growth and development of the sponge-algal Müllersfelsen buildup.

Fig. 11.—Smaller organic buildups within the "normal facies" and central part of the Müllersfelsen buildup. A. Small organic buildup ("Kleinstotzen") embedded in micritic wackestones and marls of the "normal facies," about 30 meters west of the Müllersfelsen buildup; diameter of the globular limestone body about 50 centimeters. B. Slightly domed buildup within sponge-bearing marly sediments overlain by limestones and marls of the "normal facies." Located on path between the Müllersfelsen and Muschelquelle near Streitberg; length of the stick 1 meter. C. Central part of the dome-shaped Müllersfelsen buildup, consisting of marly wacke- and packstones and marls with dense accumulation of sponges; scale 1 meter, the arrow points to the location of Figure 11D. D. Detail of the sponge bafflestone, with many saucer-shaped and some cup-shaped siliceous sponges which have been calcified ("Kalk-Mumien"). Most of the saucer-shaped sponges are in life position, the cup-shaped sponges are overturned.

FIG. 12.—Different sponge growth forms from the sponge-crust boundstone facies. A. Siliceous sponge *Sporadopyle favrei* Etallon, a portion of a saucer-shaped form. B. Calcified siliceous sponge with superficial annulations. C. Siliceous cup-shaped sponge referred to *Platychonia* Zittel. D. Siliceous sponge treated with HCL showing pores and spicular meshwork; scale 3 centimeters.

ing more active times of the development (Fig. 10).

2. *Filamentous Algal Crusts.*—These crusts are restricted to sponge surfaces, are milimeter-thick, and composed of vertically arranged filaments of porostromate blue-green algae (Fig. 21F–H). It should be noted that the organisms responsible for the origin of laminoid crusts are also believed to be algae (Wagenplast, 1972). This is demonstrated by the striking vertical growth habit of these structures (Fig. 21B). On the other hand the widely distributed planar shape of the crusts is attributed to the combined sedimentary activities of bacteria and cyanophycean algae (cf. Krumbein, 1979).

3. *Foraminiferal Crusts.*—Crusts composed of encrusting foraminifers (nubeculinellid type) are well represented in the sponge and tuberolitic facies; they are perhaps more common in the sponge facies (Fig. 17D). Here accumulations upon laminoid crusts and sponge surfaces are conspicuous features.

*Sedimentary Particles*

Various sedimentary particles can be differentiated according to their shape and size. Angular particles are represented by lithoclasts and bio-

clasts, whereas with rounded grains there is a continuum in size ranging from peloids to tuberoids to oncoids.

1. *Peloids.*—These are small micritic grains (Fig. 20E and F) composed of equal-sized calcite crystals (<1 micron) that are found exclusively in the laminoid crusts of the sponge facies. The diameter of the peloids ranges from 20 to 200 microns. The largest peloids are of comparable size to small tuberoids.

2. *Tuberoids.*—This category of sedimentary particle has been defined by Fritz (1958) for irregularly-shaped, dark calcareous lumps of varied size, which show diverse internal microstructures (micritic, Fig. 22B, and peloidal). Some tuberoids are composed of fragments of sponge skeletons (Type A); their size ranging from 100 microns to several milimeters in diameter. The origin and development of tuberoids is diagrammatically shown in Figure 8.

3. *Lithoclasts.*—Angular lithoclasts (Fig. 22E) are generally very rare, whereas rounded particles composed of tuberoids, foraminifers, spicules and peloids embedded in a micritic matrix, occur frequently in the lithoclastic packstone facies (Type B). These lithoclasts are interpreted as reworked tuberoids although the sediment of the lithoclasts

---

$\rightarrow$

FIG. 13.—Various thin-section photomicrographs of siliceous sponge fragments showing different preservational features. A. Meshwork of a hexactinellid siliceous sponge characterized by cross-shaped spicules; (1) calcified sponge ("Kalkmumie") encrusted by serpulids, and (2) organism borings into sponge wall. B. Meshwork of a hexactinellid siliceous sponge of the group Lychniscosa, characterized by open cross-shaped spicules; dark components are tuberoids (see Fig. 22). C. Cross-section of a lithistid sponge of the group Rhizocladina with a strongly interwoven spicular meshwork, a large osculum and distinct ostia. D. Altered siliceous sponge of the group Rhizocladina with encrusting foraminifer and a biogenous crusts (2); tuberolitic wackestone facies. E. Sponge fragment covered by a calcareous crust; original siliceous sponge structure has been partly leached and replaced by calcite during decay of organic material ("Verwesungsfällungskalk"). F. Synsedimentary cavity filled with calcified sponge spicules; sponge-crust boundstone facies with tuberoids. G. Sponge spicule (1), bryozoa (2) and foraminifers, forming a bioclastic packstone unit near the sponge buildup. H. Authigenic dolomitization is a common diagenetic phenomenon in the sponge-crust boundstone facies as well as in the near-by marly lithoclastic packstone facies, as shown in this photomicrograph. Length of the scales: A, B, F, G and H = 1 milimeter; C and E = 10 milimeters; D = 3 milimeters.

FIG. 14.—SEM-micrographs of etched limestone samples from the Müllersfelsen buildup. A. Portion of a siliceous sponge spicule embedded in carbonate micrite; scale 100 microns. B. Broken triaxon spicules; scale 10 microns. C. Triaxon sponge spicule, partly calcified; scale 100 microns. D. Siliceous sponge spicule, partly replaced with calcite; center of spicule still possesses its original siliceous composition; scale 100 microns. E. "Kalkmumie" (calcified sponge) with calcified spicule and inhomogenous micrite; scale 100 microns. F. Borings with micritic walls and infilled inhomogenous micrite; scale 100 microns. G. Micritic tuberoids enveloped by clay minerals; scale 100 microns. H. Sessile foraminifer on the surface of a crust; scale 100 microns.

FIG. 15.—Common foraminifers of the near-buildup tuberolitic wackestone facies (facies type C). A.–D. Miliolids: A. *Ophthalmidium strumosum* Gümbel; axial section, together with tuberoids. B. *Ophthalmidium* sp., oblique tangential section. C. *Ophthalmidium* sp., sagittal section. D. Miliolid foraminifer, quinqueloculinid-type (upper part), together with agglutinated forms. E.–I. Rotaliids: E. *Spirillina* cf. *helvetica* Kübler & Zwingli; axial section, together with *Ophthalmidium* sp. F. *Spirillina polygyrata* Gümbel; sagittal section. G. *Patellina* sp., sagittal section. H. *Paalzowella* sp., oblique section. I. *Paalzowella?* sp., oblique section. J. Lagenid, *Lenticulina* sp., sagittal section. K.–L. Agglutinated form cf. "*Reophax*," longitudinal sections. Length of scales: A–I and L = 0.5 milimeter; J and K = 0.25 milimeter.

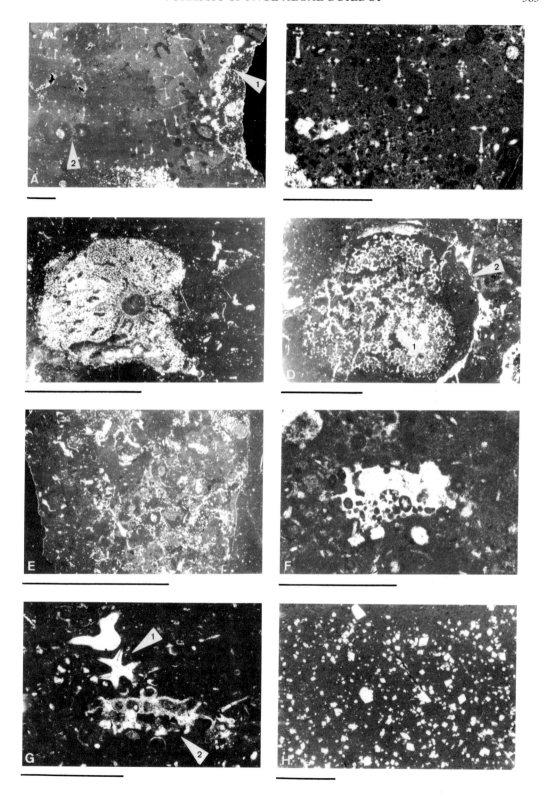

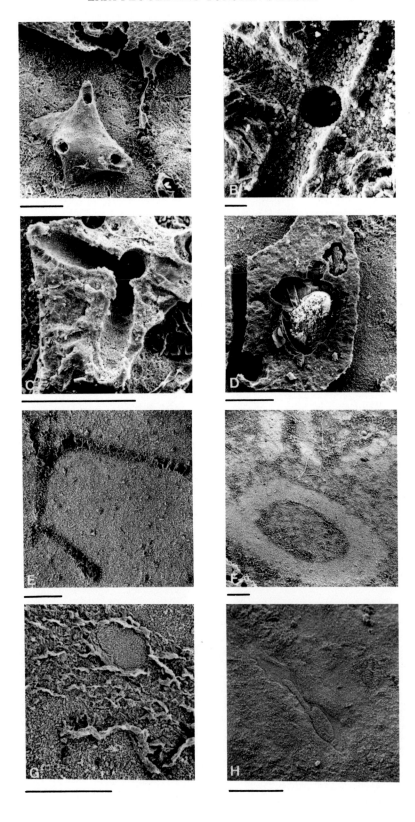

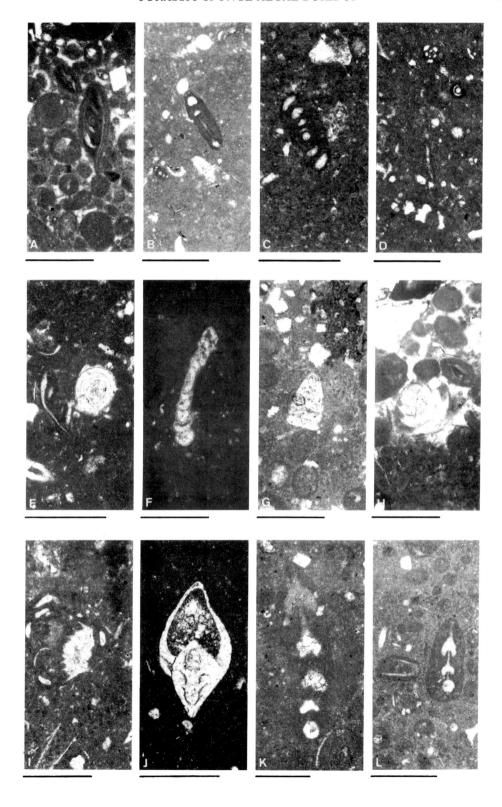

may be different in texture and color. Hence, transport and allochthonous sedimentation are indicated for these sedimentary particles.

4. *Oncoids.*—In the tuberolitic facies irregularly formed oncoids are found to be composed of encrusting foraminifers enmeshed within irregularly wrinkled micrite layers. These concentric structures are similar to those constructed by cyanophycean algae generally attributed to *Tubiphytes* Maslov. Frequently the outer surface of these oncoids are encrusted by nubeculinellid foraminifers.

## FACIES TYPES

Based on both microfacies and paleontological data, three distinctive facies-types can be recognized, these are distinguished by their characteristic fabrics (Fig. 9).

1. *Sponge-Crust Boundstone Facies.*—This facies is characterized by micritic boundstone rich in calcified siliceous sponges, tuberoids and crusts (Fig. 23A and B, Facies-type A). Two sponge morphologies occur. These are: 1. relatively flat dish-shaped forms, and 2. cup-shaped forms. Most of the dish-shaped sponges are found in life position, whereas the cup-shaped forms are often found displaced. The sediment between the sponges is similar to that found in the tuberolitic wackestone facies. The crusts appear in the more mature stage of boundstone formation. Sponges are the dominant constructive elements of the buildup, therefore the boundstone facies consists of a lower portion with bafflestones and of an upper portion dominated by boundstones (Fig. 10).

2. *Lithoclastic Packstone Facies.*—Diagnostic criteria for this facies are densely-packed spheroidal particles, most commonly, lithoclasts, tuberoids, foraminifers and other bioclasts, as well as oncoids (Fig. 22F and G, Facies-type B). Of special interest is the scarcity of rock matrix. This facies is restricted to areas adjacent to the boundstone facies, forming the transitional deposits between the buildup and the normal stratified facies. It can also occur in depressions between buildups.

3. *Tuberolitic Wackestone Facies.*—In contrast to the lithoclastic packstone facies, the particles of this facies are mud-supported (Figs. 17F, 18E, and 22D, Facies-type C). Tuberoids are most abundant, and other components such as lithoclasts, bioclasts and oncoids can also occur. These wackestones are typical for the normal stratified facies occurring in a near reef position.

The distribution of the three major facies types within the Müllersfelsen buildup is shown in Figure 9.

## DIAGENESIS

Early diagenetic processes are documented by the fossilization patterns shown by the sponges, and by hard-ground development (Fig. 19G). Late diagenetic events are indicated by subsidence, pressure-solution and dolomitization. Synchronous with rapid cementation the buildup appears to have acted as a rigid mass sinking into the underlying sediment. However, due to differences in texture, and the higher content of clay, cementation of the tuberolitic wackestones appears to have lagged behind.

Effects of pressure-solution are conspicuous within the boundstone facies as well as the tuberolitic facies. This is demonstrated by the fact that within the buildup portions of sponges are

→

FIG. 16.—Relatively rare foraminifers of the near-buildup tuberolitic wackestone facies (A.–J.), and sessile foraminifers found in siliceous sponges (K.–L.). A. *Ammobaculites* sp. B. *Textularia* sp. C.–D. *Glomospira* sp. E. *Ammodiscus* sp. F. *Lagena* sp. G.–H. *Nodosaria* sp. I. *Dentalina* sp. J. *"Globigerina"* sp. K.–L. *"Thurammina"* sp., encrusted within the canals of siliceous sponges. Length of scales 0.5 milimeter.

FIG. 17.—Encrusting foraminifers, growing on the surface of laminoid crusts or on oncoids. A. Laminoid crusts, consisting of peloids and tuberoids, encrusted by foraminifers; encrustation indicates the existence of sporadic hardgrounds. B. Sessile foraminifer encrusting surface of a tuberoid. C. Encrusting foraminifer on the surface of a lithoclast. D. Sequence showing siliceous sponge fragment, encrusted by foraminifer, serpulids, and an encrusting foraminifer again (1–4). E. Sponge encrusted by foraminifer of the nubeculariid-type; micritic matrix is strongly burrowed, and cavities are filled with fine-grained peloidal micrite. F. Oncoid with nodophthalmidiid foraminifer with porcellaneous wall (center), and irregularly wrinkled micrite layers around the foraminifer, and nubeculinellid encrusting foraminifer on the surface of the oncoid (1–3). G. Oncoid with nodophthalmidiid foraminifer, longitudinal section. Length of scales 1 milimeter.

FIG. 18.—Miscellaneous micro-organisms including boring organisms, found in the limestones of the tuberolitic wackestone facies. A. Holothurian sclerite. B. Ostracode. C. *Microproblematica* (?Dasycladacean). D. Echinoid spine, cidarid-type, surrounded by authigenic dolomite. E. Vertebrae of ophiuroids, together with tuberoids. F.–I. Borings with distinct micritic, often agglutinated walls, filled with sedimentary particles. These structures are widely distributed in the tuberolitic wackestone facies as well as in the carbonate crusts or in the "Kalkmumien." Length of the scales: A., C.–I. = 1 milimeter; B. = 0.5 milimeter.

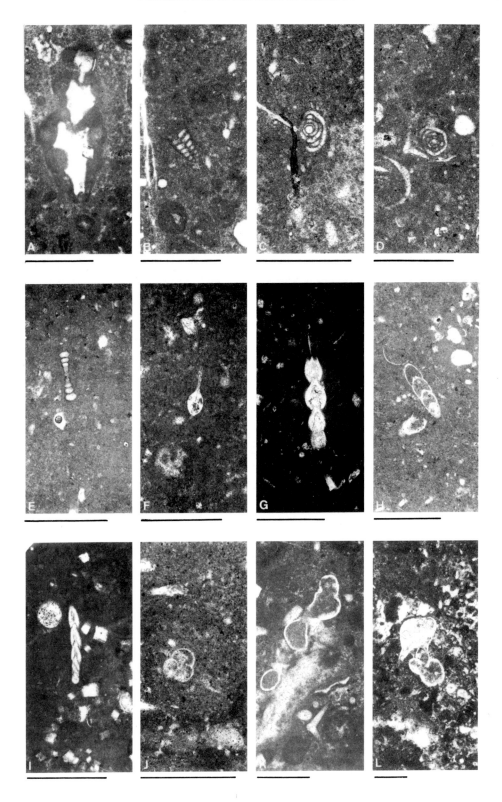

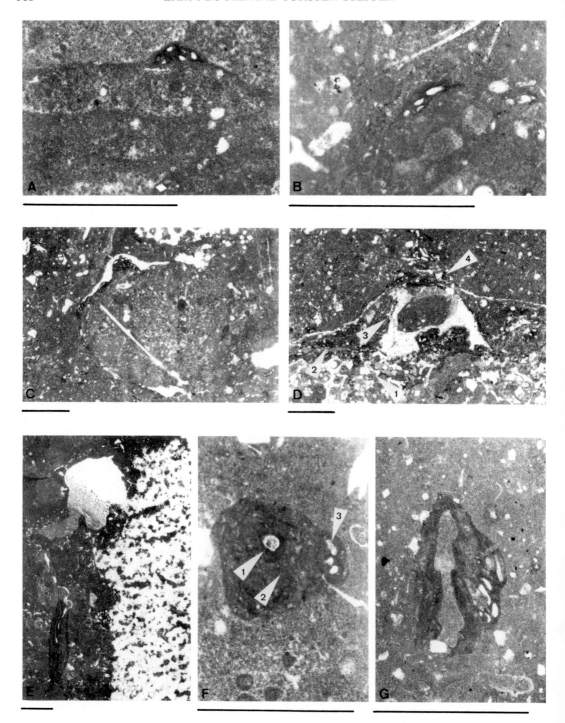

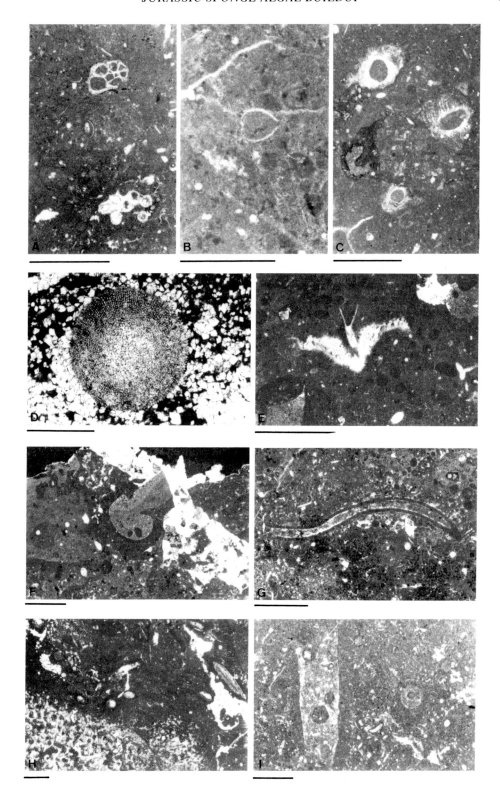

often dissolved (Fig. 22H). Argillaceous laminae that developed around calcareous components, as seen in SEM-micrographs (Fig. 14G), may have been caused by differential pressure-solution.

The most recent diagenetic process concerns dolomitization, and this can only be seen in the boundstone facies. Frequent scattered authigenic dolomite crystals occur in the lower part of the buildup, but overlying this is a completely dolomitized interval (Fig. 9).

### FACIES MODEL

The ecological parameters of Jurassic sponge-algal reefs have been discussed by (Roll, 1934; Fritz, 1958; Hiller, 1964; Nitzopoulos, 1974; and Gwinner, 1976). All of these authors have taken into consideration the following paleoecological considerations:

1. Comparison with Recent biotopes containing abundant siliceous sponges. Unfortunately, the bathymetry of most Recent siliceous sponges (tetraxons 150–300 meters, and triaxons 500–1000 meters) cannot be used to estimate the water depth of Upper Jurassic sponge buildups, because similar siliceous sponges are also known from Recent shallow water environments (Wiedenmayer, 1978). According to the paleogeographical setting it is probable that during Upper Jurassic time a gradual shallowing of the sea took place and this appears to be reflected both in the biota and sediments (Meyer, 1977).

2. The use of algae as depth indicators, since micritic and peloidal crusts generally are attributed to the sediment binding activity of cyanophycean algae. These, in contrast to the calcareous green and red algae, are not all photosynthesizing forms. Some of them have bacterial affinities and are independent of light penetration (Riding, 1976). Somewhat better criteria for possible bathymetry are afforded by the filamentous algal crusts and by some dubious dasycladaceans (Fig. 18C).

3. The relative water energy appears to control the presence and shape of angular sedimentary particles and well-rounded lithoclasts, and thus serves as a good index. The clasts and the special conditions necessary for the formation of tuberoids (see Fig. 14B), show that temporary periods of agitation of the bottom waters alternated with phases of both higher and lower water energy throughout.

4. An alternation of both soft and hard substrates is indicated by crypto-hardgrounds with an encrusting biota (Fig. 19B–G), and the presence of sediment burrowing organisms (Fig. 19F).

The Müllersfelsen sponge-algal buildup seems to have formed in water depths generally not affected by wave energy. The buildup was initiated and grew upward from the bottom of the basin (at Müllersfelsen locality, about 12 meters). Therefore the water depth must have been below the wave base for at least this depth range.

### SUMMARY

A summary of the growth and development of the Jurassic Müllersfelsen sponge-algal buildup is schematically shown in Figure 10. This diagram shows a postulated cyclicity in the overall evolution of mound growth.

The pioneer stage of development of the Müllersfelsen buildup was initiated by growth of siliceous sponges on a substrate of marl and algal crusts formed during previously sedimentary cycles. As sponges lived and died, in association

→

FIG. 19.—Cavities due to burrowing and boring organisms mainly found on the crusts and crypto-hardgrounds. A. Transverse section of a boring with distinct agglutinated walls. B. Bioturbate structure with peloidal internal sediment. C. Different types of internal fillings within sedimentary cavities. D.–E. Borings filled with layered sediment and ostracodes. F. Nuculid lamellibranch fossilized in life position indicating a soft bottom. G. Cryptohardground, characterized by encrustation of serpulids and sessile foraminifers. Length of scales 2 milimeters.

FIG. 20.—SEM micrographs of micrites of the tuberolitic wackestone facies; broken and etched samples. A.–B. Boundary between two layers of a micritic peloidal crust, characterized by the encrustation of nubeculinellid foraminifers. For the radial patterns in micrite see E. and F. C. Small tuberoids, consisting of inhomogenous micrite. D. Larger tuberoids. E.–F. Radially arranged calcite crystals around globular micritic peloids. This pattern corresponds to the structures interpreted as calcified Rivulariacean blue-green algae (see Behr and Behr, 1976). G. Miliolid foraminifer with spherulitic filling. H. Authigenic quartz formed in micritic wackestone. Length of scales: A.–E. = 100 microns; F.–H. = 10 microns.

FIG. 21.—Various types of carbonate crusts. A. Laminoid crust, partly dense, partly peloidal with irregularly distributed open-space structures; sponge can be seen at the upper right. B.–C. Dome-shaped crusts. D. Sharply defined micrite crusts with some filament structures. E., F., H. Crusts with distinct algal filaments, partly bifurcated; these crusts can be found on the surface of micritic crusts (E), or on calcified sponges (F and H). G. Siliceous sponge with a crust consisting of spar replaced algal filaments. Length of scales 2 milimeters.

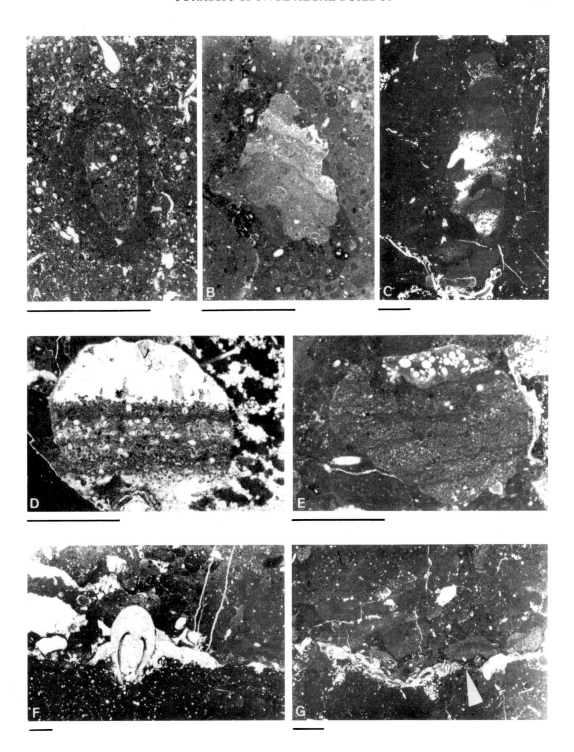

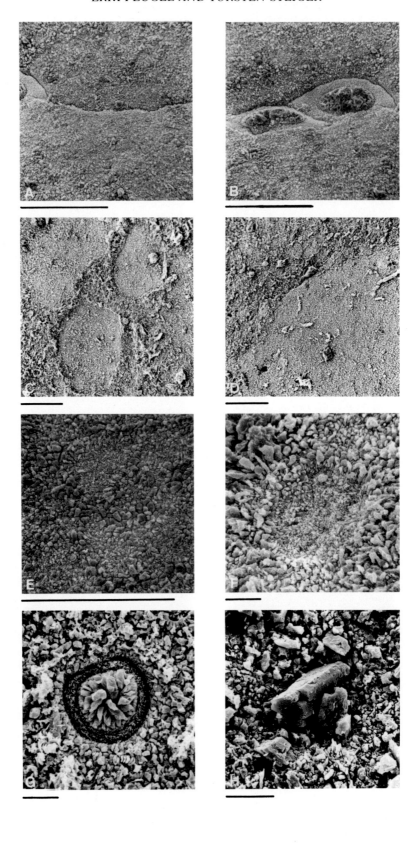

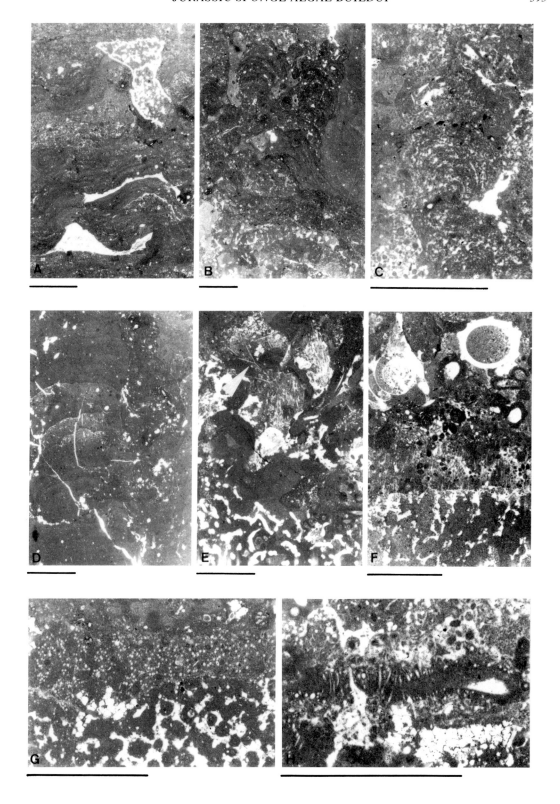

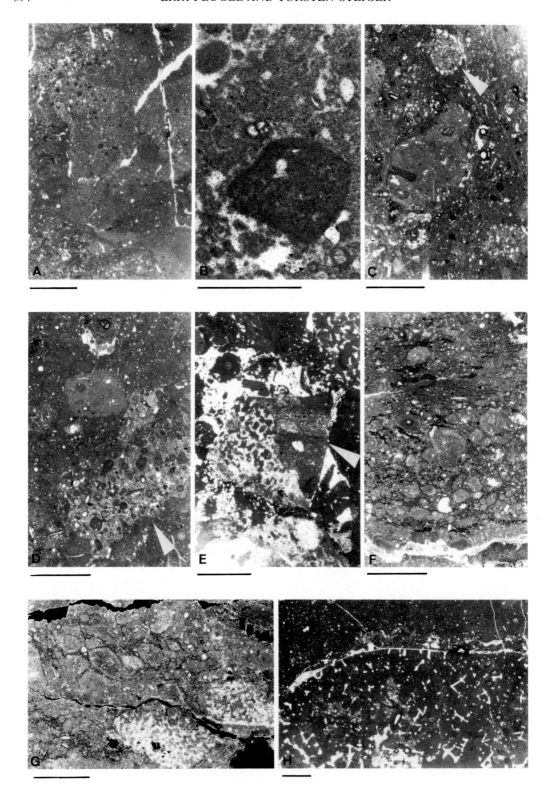

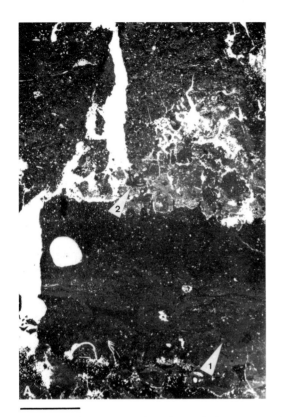

FIG. 23.—Sponge-crust boundstone facies. A. Irregularly distributed crusts (1), strongly burrowed and bored (2), and alternating with siliceous sponge fragments. B. Dome-shaped algal crusts (1), together with sponges (3), and borings (2). Length of scales 1 centimeter.

with cyanophycean algae, tuberoids developed and appear to have been selectively sorted so that they come to rest along the margins of dominant sponge growth. This first cycle of sponge growth appears to have been killed-off by sediment influx. On the resulting sedimented surface a second cycle of sponge-algal growth occurred and appears to have flourished and grown upward for an extended time period, only to be killed-off by another influx of marly sediment. At some later time a third cycle of upward growth of sponge-algal development commenced. However, the substrate area of growth appears to have been substantially reduced, and there was an increased abundance of sedimentary particles settling along the periphery of sponge-algal growth. As the mound grew upward into more relatively shallower waters, algae became a more common constitutent. Sponge-algal growth finally gave way to the development of a series of thick (up to 10 meters) algal crusts eventually forcing out most of the sponges. At some point, sponges again became abundant and were able to assert their dominance over a wider area of the sea bottom then in the previous cycle, and algae also became more abundant at this stage of development. In turn, a third layer of thin algal crusts was laid down extinguishing this last sponge-rich interval. Growth

←

FIG. 22.—Sedimentary textures present in the wacke- and packstones surrounding the algal-sponge buildup. A. Large lithoclasts consisting of smaller tuberoids and peloids. B. Angular tuberoid. C. Lithoclasts, irregularly-shaped and with encrustations of sessile foraminifers. D. Lithoclasts consisting of poorly-sorted tuberoids. E. Lithoclast made up of sponge and crust debris. F. Packstone with densely-packed lithoclasts; this facies type is found near the buildup generally alternating with the tuberolitic wackestone facies. G. Lithoclastic packstone facies. H. Calcified siliceous sponge in micritic matrix; effect of pressure-solution is indicated by a sharp upper boundary of the sponge. Length of scales: A., C.–H. = 2 milimeters; B. = 0.5 milimeter.

of the buildup was terminated with another influx of marly sediment.

It is thought that the overall growth pattern of the Müllersfelsen buildup reflects repeated cyclic sedimentation generally grading upward from a relatively deeper water base into a somewhat shallower depositional environment, although never quite being exposed to wave base, and then subsiding again. This typical cycle appears to have been repeated a number of times. This type of cyclic sequence was originally recognized by Roll (1932). We firmly believe that the sediment binding activity and capability of the algae was more important in buildup growth than the trapping and baffling of suspended sediment by sponges. Shape of the buildup appears to have been governed by early diagenetic cementation, especially in those areas of prolific sponge devel-

opment. Comparison of the biotic evolution of the Müllersfelsen buildup with other ancient mud-mounds described by Wilson (1975) shows that many mud-mounds have a relatively simple organic composition and do not appear to go through both a developmental stage and a climax community stage. These organically diverse and more complex stages appear to be characteristic of organic buildups that grow upward into extremely shallow water, culminating in subaerial exposure.

### ACKNOWLEDGMENTS

The study was supported by the DFG (Deutsche Forschungsgemeinschaft, Project FL 42/20 a). Special thanks are given to Dr. H. Keupp, Erlangen, for preparing the SEM micrographs.

## REFERENCES

BEHMEL, H., 1970, Beiträge zur Stratigraphie und Paläontologie des Juras von Ostspanien. 5. Stratigraphie und Fazies im präbetischen Jura von Albacete und Nord-Murcia: Neues Jahrb. Geol. Paläont. Abh. 137, p. 1–102.

BEHR, K., AND BEHR, H.-J., 1976, Cyanophyten aus oberjurassischen Algen-Schwammriffen: Lethaia, v. 9, p. 283–292.

BOLLINGER, W., AND BURRI, P., 1970, Sedimentologie von Schelf-Carbonaten und Beckenablagerungen im Oxfordien des zentralen Schweizer Jura. Mit Beiträgen zu Stratigraphie und Ökologie: Beitr. zur Geol. Karte der Schweiz, N.F. 140.

DORN, P., 1932, Untersuchungen über fränkische Schwammriffe: Abh. d. Geol. Landesunters. am Bayer, Oberbergamt, 6, p. 13–44.

DRAGANESCU, A., 1976, Constructional to corpuscular sponge-algal, algal and coralalgal facies in the Upper Jurassic Carbonate formation of central Dobrogea (the Casimcea Formation), in Patrulius, D., et al., Carbonate Rocks and Evaporites: International Coll. Carb. Rocks/Evapor. Roumania, Guidebook, ser. 15, p. 13–43.

EL KHOUDARY, R. H., 1974, Beiträge zur Stratigraphie und Paläontologie des Juras von Ostspanien. 4. Untersuchungen im Oberjura der südwestlichen Iberischen Kordillere unter besonderer Berücksichtigung der Mikrofauna (Provinz Teruel und Rincón de Ademuz): Neues Jahrb. Geol. Paläont. Abh. 144, p. 196–241.

FISCHER, E., 1913, Geologische Untersuchung des Lochengebietes bei Balingen: Geol. u. Paläont. Abh., N.F. 11, p. 267–336.

FREYBERG, B. VON., 1966, Der Faziesverband im Unteren Malm Frankens. Ergebnisse der Stromatometrie: Erlanger Geol. Abh., 62.

FRITZ, G. K., 1958, Schwammstotzen, Tuberolithe und Schuttbreccien im Weiss en Jura der Schwäbischen Alb: Arb. Geol. Paläont. Inst. TH Stuttgart, 13, p. 1–119.

GAILLARD, C., 1972, Les éponges siliceuses des Calcaires lités du Jura méridional (Oxfordien supérieur): Docum. Lab. Géol. Fac. Sci. Lyon, no. 50, 1972, p. 103–141.

GOTTWALD, H., 1958, Stratigraphische und tektonische Spezialaufnahmen im Jura nördl. von Muggendorf (Fränkische Alb): Erlanger Geol. Abh., v. 25.

GÜMBEL, C. W., 1862, Die Streitberger Schwammlager und ihre Foraminifereneinschlüsse: Jh. Ver. Vaterl. Naturkde. Württ., 18, p. 192–238.

GWINNER, M. P., 1976, Origin of the Upper Jurassic Limestones of the Swabian Alb (Southwest Germany), in Füchtbauer, H., et al., Contributions to Sedimentology, 5, p. 1–75.

HILLER, K., 1964, Über die Bank-und Schwammfazies des Weiss en Jura der Schwäbischen Alb (Württemberg): Arb. Geol. Paläont. Inst. TH Stuttgart, 40, p. 1–190.

KRUMBEIN, W. E., 1978, Algal Mats and their Lithification, in Krumbein, W. E., Ed., Environmental Biochemistry and Geomicrobiology, The Aquatic Environment: Ann Arbor, Ann Arbor Science Pub., v. 1, p. 209–225.

LAPTAS, A., 1974, O dolomitach w wapieniach skalistiych okolic Krakowa. The dolomites in the Upper Jurassic limestones in the Area of Cracow (Southern Poland): Rocznik Polk. Towaszystwa Geol.-Ann. Soc. Géol. Pologne, 44, p. 247–273.

MEYER, R. K. F., 1972, Stratigraphie und Fazies des Frankendolomits 1. Teil: Nördliche Frankenalb, mit 4 geologischen Karten 1:50 000: Erlanger Geol. Abh., 96, 34 p.

———, 1975, Mikrofazielle Untersuchungen in Schwamm-Biohermen und -Biostromen des Malm Epsilon (Ober-Kimmeridge) und obersten Malm Delta der Frankenalb: Geol. Bl. NO-Bayern, 25, p. 149–177.

———, 1977, Mikrofacies im Übergangsbereich von der Schwammfazies zur Korallen-Spongiomorphiden-Fazies im Malm (Kimmeridge-Tithon) von Regensburg bis Kelheim: Geol. Jb., A 37, p. 37–69.

MÜLLER, W., 1972, Beobachtungen an der hexactinelliden Juraspongie *Pachyteichisma lamellosum* (Goldf.): Stuttgarter Beitr. Naturk., B, 2, 13 p.

NITZOPOULOS, G., 1974, Faunistisch-ökologische, stratigraphische und sedimentologische Untersuchungen am Schwammstotzenkomplex bie Spielberg am Hahnenkamm (Ob. Oxfordien, Südliche Frankenalb): Stuttgarter Beitr. Naturkde., B, 16, p. 1–143.

OPPLIGER, F., 1915, Die Spongien der Birmensdorferschichten des schweizerischen Jura: Abh. d. Schweiz. Paläont. Ges., 40, p. 1–84.

RIDING, R., 1977, Systematics of *Wetheredella*: Lethaia, v. 10, p. 94.

ROLL, A., 1934, Form Bau und Entstehung der Schwammstotzen im süddeutschen Malm: Paläont. Z., 16, p. 197–246.

SCHRAMMEN, A., 1936, Die Kieselspongien des Oberen Jura von Süddeutschland: Paläontographica, Abt. A., v. 84, p. 149–194; v. 85, p. 1–114.

SEIBOLD, E., AND SEIBOLD, J., 1955, Revision der Foraminiferenbearbeitung C. W. Gümbels (1962) aus den Streitberger Schwamm-Mergeln (Oberfranken, Unterer Malm): Neues Jahrb. Geol. Paläont., Abh. 101, p. 91–134.

WAGENPLAST, P., 1972, Ökologische Untersuchung der Fauna aus Bank- und Schwammfazies des Weiss en Jura der Schwäbischen Alb: Arb. Geol. Paläont. Inst. Univ. Stuttgart, N.F., 67, p. 1–99.

WIEDENMAYER, F., 1978, Modern sponge bioherms of the Great Bahama Bank: Eclogae Geol. Helv., v. 7⅓, p. 699–744.

WILSON, J. L., 1975, Carbonate Facies in Geologic History, New York, Springer-Verlag, 471 p.

WIŚNIEWSKA-ZELICHOWSKA, M., 1971, Fauna bioherm jurajskich w Rudnikach pod Częstochowa (Fauna of the Jurassic Bioherms at Rudniki, near Częstochowa (Central Poland)): Biul. Inst. Geol., 243, Z badań geologicznych regionu Śląsko-Krakowskiego (Geological Research in the Silesian-Cracovian Region), pt. 11, p. 5–77.

ZEISS, A., 1977, Jurassic stratigraphy of Franconia: Stuttgarter Beitr. Naturk., B, 31, p. 1–32.

SEPM Special Publication No. 30, p. 399–426, May 1981

# CRETACEOUS CORAL-RUDIST BUILDUPS OF FRANCE

JEAN-PIERRE MASSE and JEAN PHILIP
University of Marseille, France

## ABSTRACT

Coral-rudist buildups are known from three outcrop areas of Cretaceous age (Berriasian to Maestrichtian) in France. These are: 1. Paris Basin, 2. Sud-Est Basin, and 3. Aquitaine-Pyrénées Basin. During Berriasian time coral-rudist buildups were limited to the Sud-Est Basin, and here they only occur locally in relation to "Purbeck-type" sediments. The Valanginian Stage corresponds to an extensive phase of marine shallow water carbonate sedimentation containing coral-rudist buildups. These deposits are recorded from both the Jura Subalpine and Provence-Pyrénées Platforms, while isolated coral beds have been reported from the Paris Basin. During the Hauterivian, regional tilting of former carbonate platforms caused a reduction of coral-rudist buildups, nevertheless isolated coral beds have been reported from the Paris Basin. Barremian and lower Aptian times were marked by a considerable extension of platform-type carbonates in the Sud-Est Basin and in the Pyrénées. In these areas "Urgonian" limestones are better represented than coral beds. During the upper Aptian deepening of the Sud-Est Basin led to the disappearance of coral-rudist buildups in most areas, except in the Aquitaine-Pyrénées region. Albian tectonism caused the disappearance of most shallow water carbonates. The Cenomanian transgression followed, leading to reestablishment of coral-rudist buildups, especially in the Aquitaine-Pyrénées and Provence regions. In lower Turonian time another deepening phase occurred and caused the disappearance of shallow water carbonates. During upper Turonian time a new surge of rudist buildups occurred coincident with an increasingly more important influx of terrigenous sediments. Only relatively small carbonate platforms containing lenses of rudists have been recorded from the lower Senonian of the Pyrénées-Provence region. The upper Senonian is mainly a regressive sequence characterized by an extension of deltaic deposits and restriction of coral-rudist lenses primarily to the Aquitaine-Pyrénées Basin.

Two groups of coral-rudist formations have been distinguished: 1. those associated with off-shore "highs," and 2. those associated with carbonate platforms. Three types of off-shore "highs" have been recognized. These are: 1. coral "highs," known only from the Lower Cretaceous, 2. oobioclastic/coral "highs" from the lower Barremian of the subalpine area, and 3. rudist banks present only in the Upper Cretaceous. We regard "highs" as topographic units of limited lateral extent that can be divided into a small number of ecological and sedimentological zones, and are surrounded by deeper water sediments. Platforms are regarded as morphological units of regional extent composed of several adjacent biofacies and lithofacies related to the hydrodynamic properties of their aqueous environment, and which may laterally grade into a continental facies. A platform may be subdivided into two zones: 1. an outer zone with high energy deposits and organic buildups (mainly corals in the Lower Cretaceous and corals and rudists in the Upper Cretaceous), and 2. an inner zone of quiet to moderate energy deposits characterized by abundant rudists.

Development of coral-rudist formations appears to be governed by six important factors. These are: 1. shallow water conditions, mainly infralittoral, 2. relative basement stability (although in some instances tectonism may create "highs" on which organisms may thrive), 3. eustatic stability or transgression, 4. low terrigenous influx, 5. absence of organism restricting oceanographic conditions, and 6. a warm climate of tropical to subtropical nature.

## INTRODUCTION

Although Cretaceous rudists and corals from France have long been studied from a paleontological point of view, paleoecological and sedimentologic analyses have been undertaken only recently. But even as far as the paleontology of these groups is concerned, rudists are better known than corals. Nevertheless, in spite of some insufficient data, our knowledge of both paleontology and paleoenvironments of the French coral-rudist formations allows us to present a summarized account of this subject. In this paper we will discuss a number of pertinent facts. These include: 1. a general paleogeographic setting of the French Cretaceous sedimentary basins, 2. a general stratigraphic and paleogeographic history of coral-rudist formations of the Cretaceous rocks of France, 3. a description and an analysis of depositional models, and 4. a discussion of the paleoenvironmental factors controlling the development of these formations.

Forthcoming additional investigations will necessitate modifications or revisions of the results given in this study. However, we hope that in spite of this inherent difficulty this review will be useful to all geologists who may be interested in comparing similar formations with those present in the Cretaceous rocks of France.

### FRENCH CRETACEOUS SEDIMENTARY BASINS

During the Cretaceous, previous basins that evolved during Jurassic times (Paris Basin, Aquitaine-Pyrénées Basin, and Sud-Est Basin) remained and continued to evolve (Fig. 1).

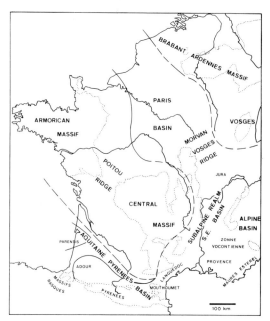

FIG. 1.—Geological setting of Cretaceous basins in France.

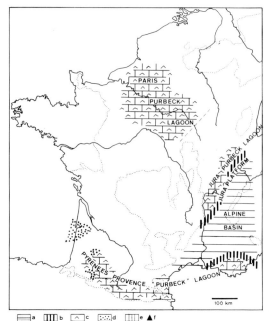

FIG. 2.—Paleogeographic map of the Berriasian Stage. Standard legend for Figures 2 to 12. a. Basinal facies, neritic to bathyal, fine terrigenous or calcareous deposits. b. Carbonate platform with coral-rudist build-ups or offshore- "highs." c. Marginolittoral to paralic facies, mainly carbonates with associated marls or lignites (low clastic content). d. Detrital littoral facies (high clastic content) in association with soil profiles. e. Emergent areas. f. Bauxitic deposits.

## The Paris Basin

Lower Cretaceous sedimentation is character-ized by marine deposits in the southeastern part with marly to calcareous sedimentation during the Neocomian, and sandy to argilaceous deposits during the Barremian. In the northwestern part of the basin continental argillaceous to sandy depos-its (of "Purbeck-Weald" type) developed. Terrig-enous clastic influx came from the emerged Her-cynian massifs (Brabant-Adrennes, Central and Armorican Massifs). During the Upper Creta-ceous, coral-rudist formations are unknown. Rel-atively deep marine sedimentation occurs in the center of the basin (chalky facies dominant), while clastics were deposited along the border. During the entire Cretaceous, the Paris Basin was con-nected with the Sud-Est Basin by the Vosges-Morvan Strait (or Vosges-Morvan Ridge). Com-munication with the Aquitaine-Pyrénées by the "Poitou Ridge" was only effective during Upper Cretaceous time.

## The Sud-Est Basin

This basin extends from Jura to Provence and is characterized during the Lower Cretaceous by a belt of shallow water sediments, mainly carbon-ates (i.e., "Urgonian" limestones) which devel-oped on the western and southern part of the Al-pine Geosyncline. The "zone voncontienne," a pelagic area dependant on the Alpine Basin, sep-arates the Jura-Subalpine shallow water area from

the corresponding Languedoc-Provence area. Middle Cretaceous tectonic movements (Albian-lower Cenomanian), in connection with the "Aus-trian phase," gave rise to the "Bombement dur-ancien," an emerged area covering the main part of the Provence-Languedoc region, on which bauxitic deposits formed. During the Upper Cre-taceous, this land remained as a barrier separating the Alpine Basin from the Provence Basin, which may be considered as the eastern part of the Aquitaine-Pyrénées Basin. Important terrigenous influx came to the basins from the Hercynian land masses which were intensively eroded. Clastic sedimentation impeded the development of coral-rudist buildups which grew only on offshore "highs," far away from deltaic fans. At the end of the Cretaceous the Sud-Est Basin emerged (Laramian phase).

## Aquitaine-Pyrénée Basin

During the main interval of the Lower Creta-ceous, the Aquitaine-Pyrénées Basin was com-posed of two subsiding zones: 1. the Parentis Ba-sin, and 2. the Adour-Mirande Basin extending toward the east in the northern Pyrénées area. The two basins are separated by the "Landes Ridge." The Neocomian Stage is characterized by very shallow water sedimentation, but in some places carbonate deposits are absent. The Barremian Stage corresponds to the main shallow carbonate phase, whereas during Aptian-Albian time the deepening of the basin is quite general, notably at the end of the Albian with the opening of the "Pyrénées Rift." Shallow water carbonates remain locally, especially during the upper Aptian. The Upper Cretaceous is marked by marine transgressions, so that the Aquitaine-Pyrénées Basin was enlarged and joined with the Paris Basin by the Poitou Strait. Following the Albian rifting phase, and in connection with the opening of the Gulf of Gascogne, there developed a narrow northwest trending deep furrow, parallel to the Hercynian axis of the Pyrénées ("Fosse du Flysch"). During this period the northern and southern edges of the basin were relatively stable so that the development of carbonate platforms with rudist buildups was enhanced. At the end of the Cretaceous (Senonian), the eastern part of the basin was filled with clastic sediments and became emergent (as a consequence of the Laramian tectonic phase), while the western part underwent continuous marine sedimentation.

### GENERAL STRATIGRAPHIC AND PALEOGEOGRAPHIC HISTORY OF CORAL-RUDIST FORMATIONS

#### Berriasian

Following extensive marine carbonate sedimentation at the end of the Jurassic, the initial Cretaceous deposits were generally deposited in low salinity waters known as Purbeck-Wealdian environments (Rat, 1965), Fig. 2.

Terrigenous fluvial deltaic or lacustrine sedimentation ("Wealdian" type) are confined to the western part of the Aquitano-Pyrénées Basin (Barreyre and Delfaud, 1956; Seronie-Vivien et al., 1965; Bouroullec and Deloffre, 1970). Marginolittoral carbonates ("Purbeckian" type) are widespread in the Paris Basin, as in the Pyrénées-Provence and western Jura areas (Donze, 1958; Peybernes, 1976). Deep water marine sediments occur in the Alpine Basin, whereas shallow marine carbonates developed along the margins. In the northern part, the "Jura Platform" contains coral-rudist formations. Diceratid beds are mentioned in the vicinity of Salève Mountain (Joukowski and Favre, 1913) while coral beds are known to form a band from Geneva to Grenoble (Steinhauser and Charollais, 1971). To the south, the "Provence Platform" plays a similar role, with an outer zone occupied by coral beds and an inner zone with a diceratid and/or requieniid rudist populations (Fournier, 1890; Mongin and Trouve, 1953; Cotillon, 1973).

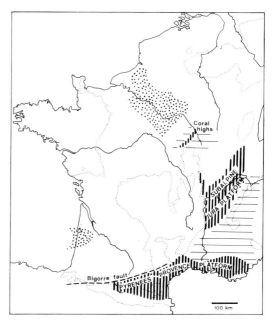

Fig. 3.—Paleogeographic map of the Valanginian Stage.

#### Valanginian

Nearly everywhere marine conditions were restored during this time (Fig. 3). However, the main part of the Paris Basin remained under clastic influence of a continental environment (Corroy, 1925; Mathieu, 1965). Similar conditions prevailed in the Parentis Basin (Seronie-Vivien et al., 1965), while no deposits are known from the western part of the Pyrénées area (Bouroullec and Deloffre, 1970). It is believed that this paleogeographic boundary may correspond to the "Bigorre Fault." Isolated corallian beds surrounded by marly sediments occur in the southeastern part of the Paris Basin (Leymerie, 1842; Lambert in Corroy, 1925). On the other hand, a wide carbonate platform with well developed coral and rudist beds extends from the Jura to the Subalpine area, with a northwest to southeast direction of progradation. Indeed, requieniid/monopleurid rudist limestones are recorded in the southern Jura (Guillaume, 1966), as well as in the Subalpine area (Munier-Chalmas, 1882; Douville, 1887). Isolated rudistid limestone outcrops of the Charollais Mounts (Munier-Chalmas, 1882) may be interpreted as a part of the inner zone of the Jura Platform. Scarce rudists are also recorded in marly sediments, as the "marnes d'Arzier" of the Jura (de Loriol, 1868). Similar to the Berriasian Stage, coral beds are located on the southeastern part of the carbonate complex bordering the Alpine Basin (Guillaume, 1966; Steinhauser and Charollais,

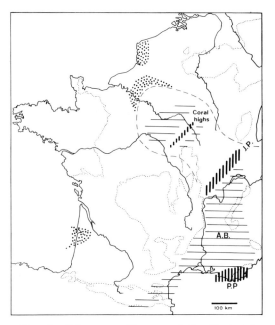

FIG. 4.—Paleogeographic map of the Hauterivian
Stage.

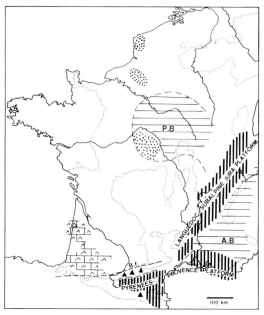

FIG. 5.—Paleogeographic map of the Barremian
Stage.

1971; Steinhauser, 1970). Scarce hexacorals and
stromatoporoids may also occur in marly sediments (de Loriol, 1868; Schnorf-Steiner, 1960–
1963). In the southern part of the Sud-Est Basin,
i.e., the Provence-Pyrénées Platform, rudistid
limestones (requieniids/monopleurids) are well
developed while stromatoporoid/hexacoral beds
are rare (Turnšek and Masse, 1973; Masse, 1976;
Peybernes, 1976).

The Valanginian Stage corresponds to an extensive phase of shallow carbonate sedimentation
in connection with a marine transgression (which
does not relate to the western Aquitaine area) and
tectonic stability. These conditions furnished a
proper environment for the development of coral-
rudist buildup.

### Hauterivian

Deepening of the basins and transgressions on
their borders are the principal features of this time
period (Fig. 4). In relation to the paleogeographic
setting marine sediments of the southeastern part
of the Paris Basin were enlarged, so the continental "Wealdian" facies was confined to the northwestern part of the basin. "Purbeck-Wealdian"
deposits also remained in some parts of western
Aquitaine. As for the Valanginian, enlarged isolated coral beds surrounded by marly sediments
occur in the southeastern part of the Paris Basin.
However, carbonate platforms were reduced in
size and are limited specifically to the Jura and
Provence areas. In these two regions corallian as

well as rudistid limestones are known. The so
called "Urgonian" limestones (previously considered to be of Barremian age) belong to the Hauterivian in part. Such is the case of the "*Pachytraga* beds" (Astre, 1961; Guillaume, 1966;
Masse, 1976). Hexacoral/stromatoporoid/chaetetid
associations are particularly well developed in
limestones, whereas occurrences in marls are
only local (Marcou, 1846 in Guillaume, 1966).

The Hauterivian transgression, concurrent with
distal tilting of previous platforms, allowed enlargement of coral development within the Paris
Basin, but also led to the reduction in size of the
perialpine platforms. Hence, coral-rudist buildups
are less well developed than during Valanginian
time.

### Barremian

Coral beds disappeared from the Paris Basin
during this time, and were replaced by marly sediments which spread over the marine part of this
area simultaneously with continental deposition
along the borders (Fig. 5). In the Sud-Est Basin
and in the Pyrénées carbonate platforms were of
considerable extent. "Urgonian" limestones
formed a subcontinuous belt on the western side
of the Alpine Basin (Jura Subalpine region, Languedoc and Provence), (Conrad, 1969; Guillaume,
1966; Arnaud-Vanneau and Arnaud, 1976; Masse,
1976). The western part of the basin is known as
the "vocontian zone." Rudistid limestones (requienids, monopleurids and caprotinids) are ex-

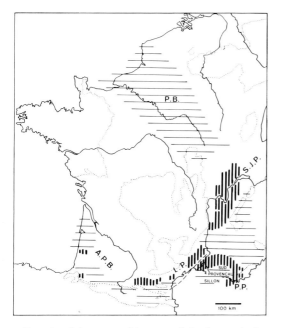

FIG. 6.—Paleogeographic map of the lower Aptian Stage.

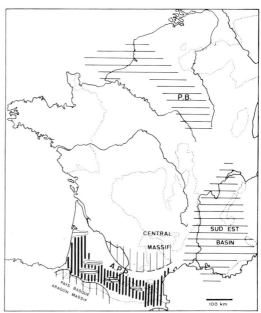

FIG. 7.—Paleogeographic map of the upper Aptian Stage.

tensively developed, whereas coral beds play a significant but minor role. The "Urgonian" limestones extend to the Pyrénées, where the "carbonate basin" is rimmed by bauxitic deposits (Combes, 1969; Peybernes, 1976). Non-typical "Urgonian" rocks are found in Aquitaine, where evaporitic and terrigenous horizons are interbedded with carbonates (Schroeder and Poignant, 1968; Bouroullec and Deloffre, 1970; Senonie-Vivien et al., 1965). Note, that for the Valanginian Stage, the "Bigorre Fault" may be considered as a paleogeographic frontier between the true "Urgonian" facies of the Pyrénées and those of Aquitaine.

*Lower Aptian (Bedoulian)*

Nearly everywhere deepening of the basin is recorded accompanied by overall enlargement of previous basinal areas (Fig. 6). The Paris Basin is overspread by marly sediments containing cephalopods. This is also the case in the Aquitaine-Pyrénées region and the southern part of Provence. However "Urgonian" platforms remain as a perivocontian belt from the Geneva region to Provence (Paquier, 1900; Conrad, 1969; Arnaud-Vanneau and Arnaud, 1976; Masse, 1976). Shallow water carbonates are also found in the eastern part of the Pyrénées (Corbières, Pays de Sault, La Clape) (Peybernes, 1976) and locally in western Aquitaine (Arbailles, Parentis), (Esquevin et al., 1971; Seronie-Vivien et al., 1965). Coral-ru-

dist buildups are chiefly developed on the "basin du Sud-Est" platforms and in the Pyrénées. In a general way, requienids/monopleurids and caprinids/caprotinids compose distinct faunal associations, the latter are often adjacent to coral beds. Note that at the end of this stage, deepening of the basin is almost complete and led to the disappearance of all shallow water carbonates.

*Upper Aptian (Gargasian-Clansayesian)*

Terrigenous sediments overspread the Paris and Sud-Est Basins (mainly littoral sediments with cephalopods, Fig. 7). In the Sud-Est Basin overspreading of Vocontian sediments was followed by a brief shallow water carbonate phase ("Clansayesian Limestones"), which are devoid of coral-rudist buildups. Urgonian-type carbonates with well developed corals and rudists (requienids/monopleurids/caprotinids and radiolitids) are limited to the Aquitaine-Pyrénées realm where deep water sediments remained as isolated patches. The borders of the previous basin were transgressed, which led to the incorporation and fossilization of pre-existing bauxites.

*Upper Albian*

Relatively deep water terrigenous sediments with cephalopods are recorded in the Paris and Sud-Est Basins during the main part of this stage (Fig. 8). However, upper Albian carbonates with corals and rudists (caprinids and radiolitids) are

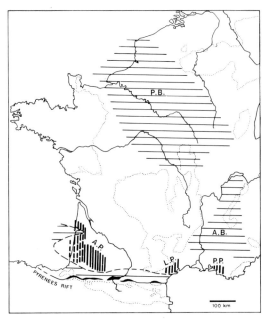

FIG. 8.—Paleogeographic map of the upper Albian-Vraconian Stages.

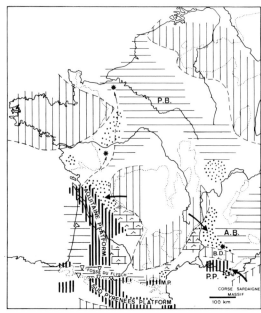

FIG. 9.—Paleogeographic map of the Cenomanian Stage. M.P. Mouthoumet Platform, P.P. Provence Platform, B.D. "Bombement durancien," A.B. Alpine Basin, P.B. Paris Basin. In Figures 9 to 12 star-symbol indicates isolated rudists, triangle symbol indicates breccia facies.

known locally in the southern part of Provence (Philip, 1970; Masse and Philip, 1976). Following the upper Clansayesian/lower Albian deep water phase, shallow water limestones developed locally in the Aquitaine-Pyrénées region, and the borders of previous emerged areas were transgressed (Poignant, 1964; Peybernes, 1976; Seronie-Vivien et al., 1965).

In the whole of the south of France, the upper Albian-Vraconian Stage (Fig. 8) is characterized by important paleogeographical modifications. In Provence at the end of the Albian time appeared the "Bombement durancien," an emergent area covered with bauxite deposits (Philip, 1970; Masse and Philip, 1976), while siliciclastic deposits originating from the Corso-Sarde Massif, were deposited in the southern part of this region. In the Pyrénées, the east-west "sillon orogenique nord-pyrénéen" formed, and here different types of flysch deposits accumulated and metamorphism occurred (limited to the central rift), and submarine volcanic activity and synsedimentary brecciation are associated with flysch sedimentation (Peybernes, 1976; Souquet et al., 1977). The deepening of the central zone of the basin was followed by a marine regression along the edges, which were intensively eroded. These paleogeographical features led to the disappearance of many coral-rudist buildups which remained only locally in the Pyrénées, but were well developed in Aquitaine (Adour and Parentis Basins).

## Cenomanian

Following the unstable period corresponding to the end of the Lower Cretaceous, the Cenomanian Stage is marked by a transgressive tendency over previously emerged and eroded areas (such as the "Bombement durancien"), in relatively quiet tectonic conditions (Fig. 9). Shallow water carbonates developed on the edges of the basins and coral-rudist buildups grew on the Aquitaine, Sud-Pyrénées and Provence Platforms.

The greater extent of the Aquitaine Platform of middle and upper Cenomanian age, is a consequence of increased transgression with decreasing terrigenous influx. The transition zone between the carbonate platform and the land (Central Massif) was occupied by paralic (e.g., organic clays) sediments. Typical inner-platform-type deposits are composed of caprinid and radiolitid beds and bioclastic limestones with large foraminifers (orbitolinids and praealveaolinids); no coral development has been reported. No data is available on the external platform deposits and their relationship to the deeper part of the basin. Local occurrences of rudists on the southern edge of the Paris basin are interpreted as a consequence of the opening of the Poitou Strait, allowing migration of the rudist fauna from the Mesogea to the "Chalky-basin."

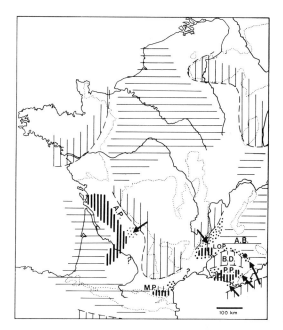

FIG. 10.—Paleogeographic map of the upper Turonian Stage; L.O.P. = Languedoc oriental Platform, other symbols as Figure 9.

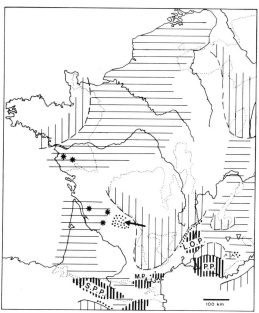

FIG. 11.—Paleogeographic map of the lower Senonian Stage.

In the Pyrénées, carbonate platforms rest on the edges of Paleozoic massifs; Ebre Massif for the southern border of the Sud-Pyrénées Platform, or on the margin of Hercynian massifs and axis of the Pyrénées. Their ages range between Vraconian to Cenomanian time (Basques massifs; Feuillee, 1971), and middle to upper Cenomanian time (Corbières and southern part of the Pyrénées Mountains; Souquet, 1967; Bilotte, 1973). In the Sud-Pyrénées Platform, the dominant facies are bioclastic limestones with caprinids and alveolinids. True organic buildups seem to be absent (Souquet, 1967; Feuillee, 1971). Nevertheless, a typical zonation may be observed between the Ebre Massif and the "Fosse du flysch," where inner-platform deposits are almost entirely dolomitized, while in the external-platform deposits, rudists and alveolinids are dominant (Feuillee, 1971). Allodapic carbonates (rudist fragments, limestone breccias) lie along the platform-basin transition zone (Feuillee, 1971; Souquet, 1967; Bouroullec and Deloffre, 1976). These sediments originated from the synsedimentary folding of early lithified carbonate shelf-edges.

In the eastern Pyrénées, coral-rudist buildups developed on a topographical anomaly corresponding to the present Mouthoumet Massif, within a high clastic content environment. Towards the south this carbonate complex grades to basinal facies (Bilotte, 1973–1977).

On the Provence Platform, situated in the southern part of the "Bombement durancien," well developed middle and upper Cenomanian coral-rudist buildups are known (Philip, 1970–1972), while siliciclastic deposits with scarce isolated rudists occur in the lower Cenomanian. During the middle Cenomanian, important coral-caprinid bioherms mark the transition zone between the south margin of the "Bombement durancien" and the "sud provençal" basin. During the upper Cenomanian interval, the carbonate platform enlarged as a consequence of an extensive transgression on the "bombement durancien" area associated with decreasing terrigenous influences from the southeast. Rudist (radiolitids-requienids) associations are well developed whereas corals play a minor role in this area.

### Turonian

Proper conditions for the development of shallow water carbonates, which seemed to be prevalent at the beginning of Turonian time (Philip, 1978), disappear as marly pelagic sediments became widespread (Fig. 10). These general paleogeographic modifications originated from basin deformation which affected paleolands, such as the "Bombement durancien" which had been elevated during the lower Turonian (Philip, 1978). Formation of carbonate platforms began again during the upper Turonian (Angoumian) along some parts of the basin-margin, in relation to de-

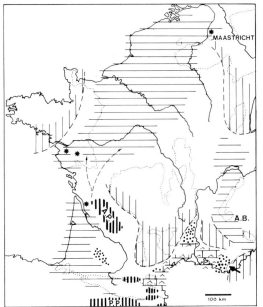

FIG. 12.—Paleogeographic map of the upper Senonian Stage.

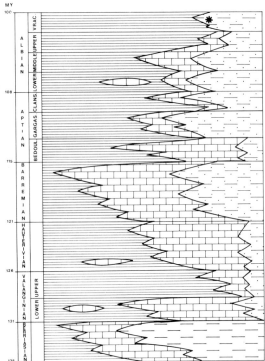

FIG. 13.—Schematic representation of the stratigraphic distribution of coral-rudist occurrences in the Lower Cretaceous of France. a., b., c. marginolittoral to continental sediments: a. marls or marly limestones, sometimes associated with lignite (Purbeck beds of the Lower Cretaceous). b. marls and sandstones (Weald beds of the Lower Cretaceous). c. sandstones dominant. d. coral-rudist carbonate occurrences. e. basinal facies (marls or marly limestones), f. isolated corals or rudists; geochronologic data from Van Hinte (1976).

creasing water depth and tectonic stability. Nevertheless, their lateral extension is of less importance than for Cenomanian time, so that in the major part of the Pyrénées region pelagic sediments remain well represented ("*Fissurina* Limestones" in Souquet, 1967). Shallow water carbonates developed in four areas. These are: 1. Aquitaine Platform (only in the northern part of this region, according to Cassoudebat and Platel, 1976), 2. Mouthoumet Platform (Bilotte, 1974–1977), 3. Eastern Languedoc Platform or "plateforme gardoise" (Monleau and Philip, 1972) and 4. the Provence Platform (Philip, 1978). In these facies, rudists (hippuritids/radiolitids) are dominant, whereas coral buildups are rare or absent. Important terrigenous influx coming from eroded adjacent lands impeded development of true coral communities.

### Lower Senonian (Coniacian-Santonian)

The North Aquitaine carbonate Platform (Fig. 11) disappeared during this time and shallow water limestones with rudists were replaced by a deeper water facies (cherty glauconitic limestones of bryozoa facies for the Coniacian chalks, with ammonites and inoceramids for the Santonian (Seronie-Vivien, M., 1972). Rudists occur only sparsely in Perigord and Vendée. The southern Pyrénées Platform was restored (hippuritids and Vidalina limestones for the Coniacian, and Lacazina and coral-rudist limestones for the San-

tonian, Souquet 1967). On the Mouthoumet Platform, increasing terrigenous influx caused spatial reduction of rudist limestones, reduced only to local lenses (Freytet, 1973; Bilotte, 1977).

In the Sud-Est Basin carbonate platforms with rudists are well developed during the Coniacian, but are reduced to lenses during the Santonian as a result of deltaic sedimentation. At the end of the Santonian, Provence and Languedoc were abandoned by marine waters and at first replaced by paralic lagoons, and then by lakes. Allodapic limestones with sparse rudists are known in bathypelagic sediments of the Alpine Basin, notably in the Argentera zone (Sturani, 1962) and in Ubaye (El Kholy, 1972). They originated from the destruction of previous platforms, the fragments of which were carried into deep waters by submarine currents.

### Upper Senonian (Campanian-Maestrichtian)

Incipient regression which began during the lower Senonian continued (Fig. 12). Shallow water carbonates remained only in the Aquitaine-Pyrénées region. On the North Aquitaine Platform (Seronie-Vivien, 1972), the Campanian is represented by relatively deep water facies (chalks with cherts), while in the Maestrichtian more shallow water coral-rudist buildups (hippuritids, radiolitids) once again appeared. In the Pyrénées, such types of buildups occur both in the Campanian and Maestrichtian, between the "Fosse du Flysch" and the south Pyrénées Platform which was covered by deltaic deposits.

Schematic representations of the stratigraphic distribution of coral-rudist occurrences from the Cretaceous of France are summarized in Figures 13 and 14.

#### GENERAL ORGANIZATION OF CORAL-RUDIST FORMATIONS

Coral-rudist formations are defined as lithostratigraphic units in which these organisms play a significant role, hence the "coral-rudist formation" concept is essentially descriptive. These coral-rudist assemblages developed in shallow water environments. In a general way, corals and rudists formed distinctive and relatively independent biosedimentologic units, often adjacent to each other, but sometimes isolated. Beds made up mostly of corals are common, while beds entirely of rudists are rare, especially during the Lower Cretaceous. The zonation of shallow water carbonates is based on two principal concepts: firstly, vertical benthic successions ("étagement benthique" *sensu* Peres and Picard, 1957), who defined the two main parameters: exposure and light. The four benthic stages of the phytal zone are: supralittoral, mediolittoral, infralittoral and circalittoral. When the tidal effect is significant, supralittoral and mediolittoral are quite similar to supratidal and intertidal zones, while infra and circalittoral may be regarded as subtidal areas. This classification is essentially bathymetric. In this sense, carbonate highs and platforms belong to supra, medio and infralittoral stages, whereas the circalittoral stage may be considered as the deeper part of these units, or as a part of the "basinal zone" (corresponding to circalittoral and bathyal stages). Secondly, horizontal zones characterized by hydrodynamic (e.g., energy) and hydrologic parameters (physicochemical properties of the water such as salinity, oxygenation, even terrigenous suspensions, *etc.*) independent of the bathymetry.

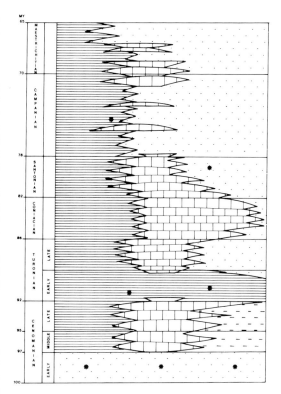

FIG. 14.—Schematic representation of the stratigraphic distribution of coral-rudist occurrences from the Upper Cretaceous of France (legend is the same as for Fig. 13).

On shelves with regular slopes and limited lateral continuity, vertical zonation is parallel to horizontal zonation, so the reference to relative bathymetry is explicit. Whereas on shelves with irregular morphology and considerable lateral extent, bathymetric zonation is not significant by itself and may be associated with horizontal zonation, especially in the shallow water realm (e.g., infralittoral regime). In platform areas shallow water zones of considerable lateral continuity may be distinguished, according to water energy and salinity anomalies, e.g., "inner platform" (more or less restricted and with low energy), and an "outer platform" (with normal sea water, and high energy conditions). Lateral transition to the continental realm is possible through the medium of a marginolittoral or paralic area, according to the amount of terrigenous influx passing through a deltaic environment.

In this two dimensional zonation, coral-rudist formations are principally situated in the infralittoral zone. Differences between platforms and "highs" are based on relative dimensions and polarity which determine sedimentological and ecological differences. These "highs" are of limited

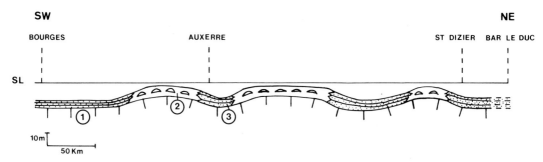

FIG. 15.—Schematic cross-section through southeastern part of the Paris Basin during Hauterivian time (modified after G. Corroy, 1925). 1. deformed pre-Cretaceous basement (Portlandian Limestones), 2. offshore coral "highs," 3. marly limestones; SL = sea level.

lateral extent and contain only a small number of ecological parameters (for example only coral populations), while platforms are larger and show several adjacent ecological units (e.g., both coral and rudist populations).

### Offshore "Highs"

According to faunal/sedimentological properties and paleogeographic setting, several types of offshore "highs" may be distinguished.

*Coral Highs.*—These types of offshore "highs" are restricted to Lower Cretaceous time. The best examples are recorded in the Hauterivian Stage from the southeastern margin of the Paris Basin, and in the Clansayesian/Albian interval from Aquitaine (Bearn). On coral "highs" of the Hauterivian (Fig. 15), corallian beds developed from Yonne méridionale to Haute-Marne (Corroy, 1925) and are surrounded by ferruginous oolitic limestones (Sancerrois), glauconitic limestones (Auxerre) or marly limestones (Troyes area). Coral formations are made up of isolated hexacoral beds, the thicknesses of which never exceed a few meters, and with limited lateral continuity. The different isolated coral patches are situated along a northeast-southwest band (about 100 kilometers long and 10 kilometers wide). These organic buildups lie on Portlandian Limestones separated from the base by thin marly layers. Coral colonies are also interbedded in marls or marly bioclastic limestones. Evidence of a very slow rate of sedimentation associated with coral growth and low energy is indicated by the high percentage of terrigenous mud in surrounding sediment, low coral debris content (indeed the bioclastic fraction is made up of numerous relict grains bored and ferruginized and probably corresponding to the lower Hauterivian marine transgression on a previous Eo-Cretaceous hardground), and the high degree of bioencrustation (serpulid worms, bryozoa, etc.), and boring of the coral colonies.

A rich coral fauna has been recorded (more than 100 species, according to Corroy, 1925), associated with an abundant epifauna, bryozoans, sponges, bivalves, and echinoids. The infauna is well represented by various echinoids, bivalves and gastropods. Importance of sciaphil organisms (bryozoa—sponge group) and absence of typical photophil organisms (except hexacorals) such as calcareous green algae, in association with the well known circalittoral fauna (foraminiferal genera *Lenticulina, Dorothia, Toxaster etc.*) show that the coral assemblage may be referred more accurately to the circalittoral zone than to the infralittoral one. No rudists have been recorded. Either a vertical or horizontal zonation has not been observed for sediments or fauna. As for the Oxfordian reefs of the nearby region (Lorraine, Humbert, 1975), it may be inferred that Hauterivian coral beds grew on topographic "highs" corresponding to Hercynian faults ("direction varisque"), so that this relief anomaly is not the result of "reef growth."

A coral high of the Clansayesian Albian from southern Aquitaine, specifically the Arudy Com-

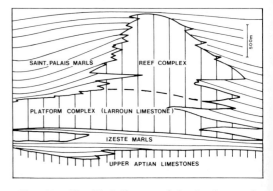

FIG. 16.—Simplified diagram of the Arudy complex and associated formations (after C. Boltenhagen, 1967).

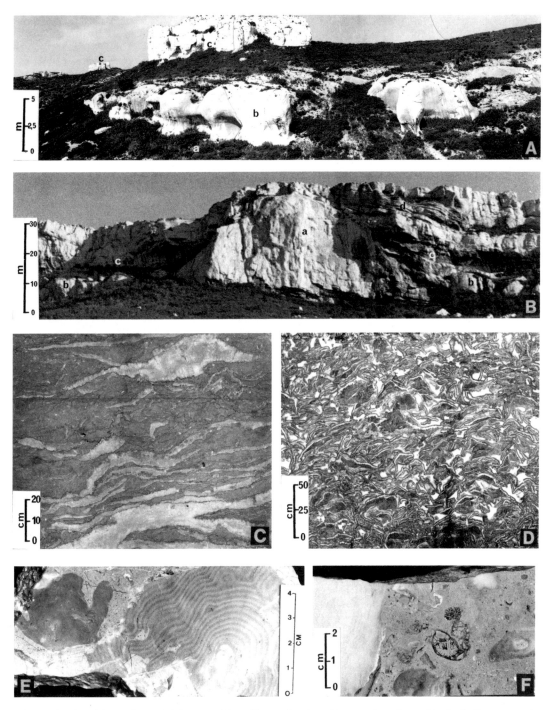

FIG. 17.—Coral formations: A. Biostromal lens (Barremian from Provence): a. lower bioclastic limestones, b. hexacoral biostrome lens, c. upper bioclastic limestone. B. Coral bioherm (Cenomanian from Provence). Typical formation of the outer-platform zone: a. coral bioherm, b. laterally equivalent breccias, c. terrigenous beds, d. bioclastic limestones. C. Hexacoral bafflestone to bindstone in muddy matrix (Clansayesian-Algian from Arudy Limestone Complex, Aquitaine). D. Hexacoral bindstone with open space white cement and muddy matrix (Clansayesian-Albian from Arudy Limestone Complex, Aquitaine). E. Bulbous stromatoporoids in muddy limestone, Hauterivian, Marseille area. F. Massive dendroid hexacorals in fine-grained matrix, Cenomanian from Provence.

plex (Fig. 16), is known as a large limestone build-up (more than 1000 meters thick) surrounded by marls. Visible outcrops laterally extend east-west for about 10 kilometers. The north-south lateral extension is very difficult to assess but may be of the same length. Detailed examination of the internal structure has been described (Boltenhagen, 1967), who regards it as the result of the interconnecting of more or less distinct coral lenses. The basement corresponds to the "Izeste Marls," which grade laterally (westward) into limestones so that the Arudy Complex is not isolated.

The starting phase of the "reef complex" may be related to distal progradation of the Larroun Platform. The individualization phase is correlated with lateral tilting of the earlier platform so shallow water limestones occur only in the Arudy area. Subsequently, the platform complex was reduced to a narrow offshore "high." These topographic modifications were followed by ecological changes (e.g., rudist communities are replaced by coral/algal communities). Interconnected coral buildups show a clear vertical faunal zonation in which the lower part is composed of relatively large, scattered domal, colonies (bafflestone type) (Fig. 17C), whereas the upper part shows lamellar close-packed coral colonies (bindstone type) (Fig. 17D). This upward evolution may be interpreted as a progressive deepening of depositional environment, leading to the progressive narrowing of the buildups. Domal morphology of individual buildups may be explained in the following way. The sedimentary matrix is micritic, coral debris and coralline algae (free or encrusting forms) compose the sand fraction and no textural changes follow the ecological vertical succession. Nevertheless the tops of limestone lenses are sometimes devoid of corals while coralline algae remain well represented. As for the Hauterivian buildups of the Paris Basin, sedimentary and biological properties show a quiet water environment of possibly circalittoral aspect. Indeed, coralline algae are not typically photophil. Contact with superimposed marls is sharp. Evidence of early lithification is provided by penecontemporaneous open-space cements (Fig. 17D) and reworked (e.g., allodapic) "reef blocks" in the surrounding marls. Early leaching features are well developed in the upper part of the sequence (Figs.

18 and 19), which show occurrences of coral synsedimentary emersion. Synoptic morphology (*sensu* Hoffmann, 1969) is difficult to observe, whereas early emersions indicate that the coral buildups may have had some relief among adjacent sediments. Geophysical data suggests that the location of the Arudy Reef complex has been determined by the positive basement anomaly related to a north-south Hercynian fault. In this case reef growth is very important and structural control is limited to the creation of stable area which initiated proper conditions for coral development, whereas adjacent areas are tilted and buried under pelagic marls.

Oobioclastic/coral offshore "highs" (Fig. 20) are recorded in the lower Barremian of the Vercors area (Subalpine region) and have been described by Arnaud-Vanneau and Arnaud (1976). These correspond to a semi-elliptical body about 30 kilometers wide, surrounded by marly limestones. The central part is occupied by oolitic grainstones, trending west-northwest to east-southeast, while bioclastic/oolitic limestone, followed by bioclastic limestones, form successive asymmetric rims. This irregular spatial distribution is associated with similar sedimentary modifications, so that coral lenses (a few meters high and some hectometers long) are only developed on the narrow southwestern edge, and oomicritic sediments only on the enlarged northeastern part. Occurrence of hexacoral beds with calcareous green algae (dasyclads) and large foraminifers (Orbitolinids) suggests that this type of offshore "high" had an infralittoral aspect. Sedimentological properties (sedimentary structures, facies types and their paleogeographic patterns) indicate a typical hydrodynamic control, e.g., southwest-northeast trending currents that originated from a dominant swell from the southwest vocontian zone, which induced offshore progradation towards the northeast. The location of the oobioclastic/coral zone corresponds to a stable area surrounded (at least on its southeastern edge) by unstable areas under the influence of active paleofaults. Upward growth is the result of sedimentary embankment rather than coral bioconstruction, which only played a minor role.

Rudist banks (Fig. 21) are a type of offshore "high" known only during Upper Cretaceous

$\rightarrow$

FIG. 18.—Coral formations: A. Coarse grainstone with coral debris, Barremian from Provence. B. Mudstone with lamellar microsolenid corals (bindstone), Cenomanian from Provence. C. Fine bioclastic packstone with chaetetid colony, Barremian from Provence. D. Coarse bioclastic wackestone with bored coral colony. Cavity shows geopetal fill of lime mud, Cenomanian from Provence. E. Cavity with early gray druse cement. A later generation of mosaic druse fills the inner part of the cavity; Cenomanian from Provence. F. Inclined lamellar hexacoral colonies resting on finely-bedded muddy to pelletoidal sediment with coralline algal fragments, Clansayesian to Albian-Arudy complex, Aquitaine.

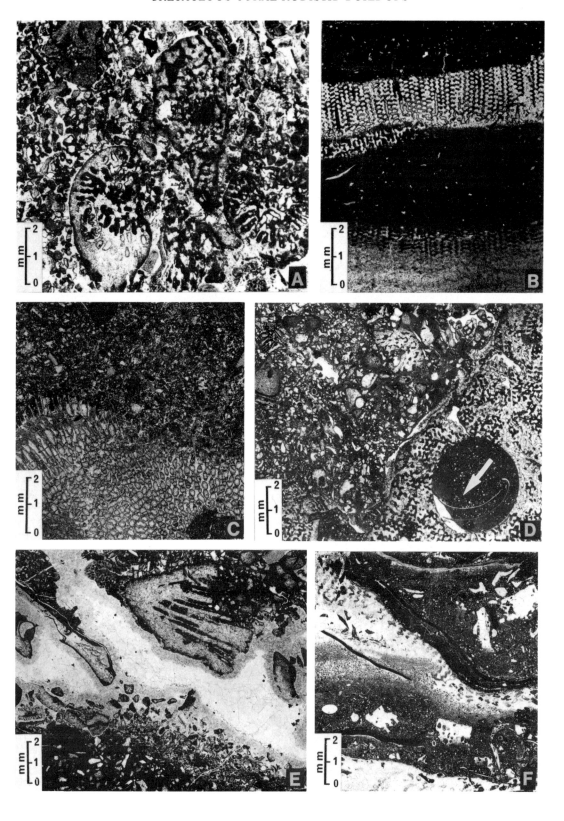

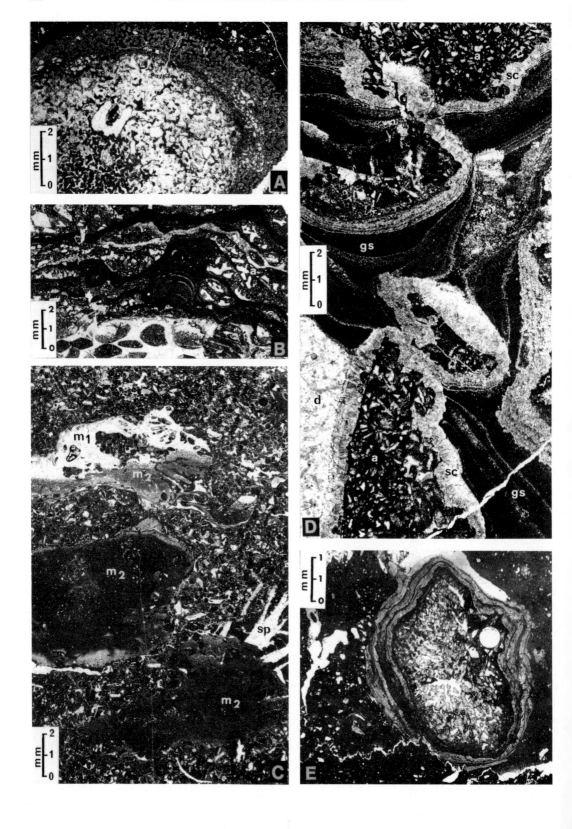

time, and have been reported during the Santonian from the Languedoc and Provence areas, and during Campanian and Maestrichtian times in the Pyrénées and Aquitaine regions. Good examples have been described from Provence (Philip, 1970), where "rudist banks" correspond to bioconstructed lenses some several meters high (exceptionally several tens of meters), and some hundreds of meters wide, surrounded by marls or fine sands to marls. Rudist faunas (hippuritids) are the dominant bioconstructed beds, they are generally classified as bafflestone types. Hippuritids (Skelton, 1976) are the pioneer organisms of the buildups (Philip, Amico and Allemann, 1978), in which other groups may develop, e.g., radiolitids, chaetetids and coralline algae. In some cases (Cenomanian from Provence, and the lower Senonian from the Pyrénées-Provence region), hexacoral beds may be associated with rudists. The sedimentary matrix of the organic framework is micritic (mudstone to wackestone) with variable amounts of coarse skeletal grains, mainly rudist fragments. Terrigenous horizons may have disturbed the upward carbonate growth. There is no horizontal biotic zonation, but vertical ecological and sedimentological zonation is apparent with partial replacement of hippuritids by other rudists, associated with upward decreasing terrigenous influx. Vertical stacking of individual carbonate lenses is frequent.

The rudist banks must have been of low relief over the surrounding area for they show no evidence of wave resistance, nor peripheral paleoslopes. Consequently, they may not be interpreted as "reefs," as has been done for Santonian rudists lenses from Languedoc (in Freytet, 1972). The ecologic significance of these rudist banks is that they are composed of a relict reef community. Dominance of rudists among other groups (especially hexacorals) may result from the influx of fine terrigenous clastics. The rudist banks are located in quiet water environments, e.g., the sheltered part of bays or unexposed basin-margins in the lower infralittoral to upper circalittoral zones.

*The Platform Model.*—Cretaceous platforms are well developed in southern France. General organization through the entire Cretaceous is al-

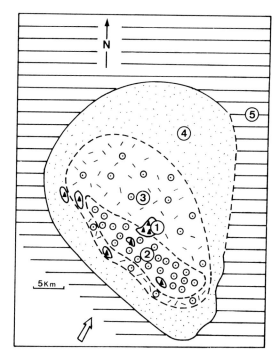

Fig. 20.—Oobioclastic/coral offshore "high." Paleogeographic map of lower Barremian to Barremian from Vercors (Subalpine Realm); after A. Arnaud Vanneau and Arnaud (1976). 1. Coral patches, 2. Oolitic sand, 3. Oobioclastic sand belt, 4. Pure bioclastic to micritic sands, 5. Fine muddy terrigenous limestones. Arrow indicates main current direction.

most always the same (Fig. 22). Typically, a platform is composed of an outer zone of variable lateral extent (larger in Lower Cretaceous models than in Upper Cretaceous areas) with high energy deposits (bioclastic to oolitic sands), or bioconstructed bodies (mainly corals). The inner zone is relatively wide and characterized by rudist populations. Several biosedimentary environments are in close relationship with these rudist communities. Associated sediments are generally of quiet to moderate energy. Some variations may be observed in the platform/basin or platform/marginolittoral area transition zones. Thus the

← 

Fig. 19.—Coral formations: A. Encrusting spongiomorph hydrozoan from the Cenomanian of Provence. B. Encrusting foraminifers (a) interlaminated with a coralline crust developed on caprinid debris, Cenomanian from Provence. C. Bioclastic to peloidal packstone with leached coral colonies, open cavities filled with muddy matrix, m 1, sparite coral colony, m 2, coral cavity filled by sediment, sp., sparry septal fragment from partially preserved coral colony; Barremian from Provence. D. Leached and filled-up cavities from a corallian limestone (partly dolomitized), Clansayesian-Albian, from Arudy complex, Aquitaine: a, bioclastic packstone-wackestone with fine coral debris—sc, vadose stalactitic early cement, gs, geopetal sediment, d, dolomite (late diagenesis). E. Encrusting algae (*Ethelia*) surrounding coral debris; muddy matrix, Cenomanian from Provence.

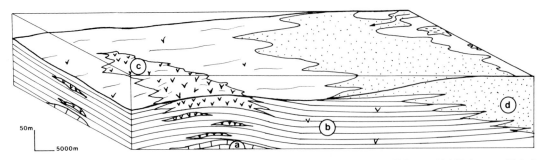

FIG. 21.—Schematic reconstruction of an Upper Cretaceous "rudist bank" (offshore "high") in a perideltaic environment. a. gently folded substratum. b. shallow water marls with isolated corals and rudists, *Inoceramus*, sponges and sparse ammonites. c. Rudist bank. d. Perideltaic sandstones.

platform unit *sensu stricto* is connected to the basinal area (mainly marls or marly limestones with cephalopods and planktic foraminifers) by coarse to medium-size biocalcarenites with echinoderms, bryozoa and various detrital grains derived from the platform, or micritic to peloidal limestones with chert nodules (well developed during the Lower Cretaceous), and sponge spicules. Locally, especially during the Middle and Upper Cretaceous (Cenomanian/Turonian in the Pyrénées-Provence region) allodapic limestones rim the platform/basin transitional zone.

The inner zone of the platform may be linked to the continental area by means of paralic, marly, sediments with a brackish water fauna, or a perideltaic zone of sandy deposits. During the Lower Cretaceous, platforms were far from the continental edge and the inner zone was rimmed by a marginolittoral unit which contains beach deposits of sandy sediments with vesicular structures (Fig. 26G), and stalactitic to meniscus cement

(Fig. 26H). In addition, there are marsh sediments, muddy carbonates, sometimes dolomitized miliolid muds, intraformational breccias, fenestrate structures, e.g., loferites, algal mats, and microstromatolites. The distribution of the main biosedimentary zones demonstrates that coral development was limited to the outer part of the platform and therefore well separated from rudist development, which lies in the inner zone. The mixing of the two types of organisms is locally possible in a "transition zone" containing stromatoporoid-chaetetid assemblages, especially in Berriasian to lower Aptian time. During Cenomanian time caprinid rudists and corals generally occur together on the outer part of the platform. Coral formations may occupy the whole outer zone or only a part, or may be absent and replaced by oobioclastic sands.

*Coral Formations.*—The best examples are found in Provence, in the "Urgonian" limestones, Barremian to lower Aptian interval

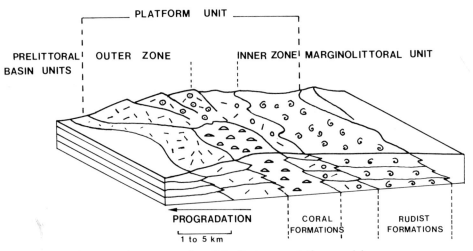

FIG. 22.—Idealized diagrammatic representation of the Cretaceous platform model.

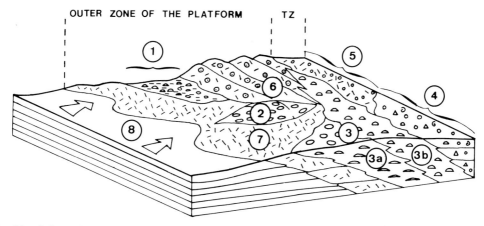

FIG. 23.—Schematic reconstruction of a Lower Cretaceous platform, specifically the outer zone with coral growth. 1. Hexacoral bank under terrigenous influence. 2. Isolated radial hexacoral biostrome lens. 3. Marginal outer hexacoral growth (complex biostrome) 3a. lower zone, 3b. upper zone. 4. Hydrozoa, chaetetid development. 5. Coarse bioclastic sands and gravels with broken corals. 6. Oolitic/bioclastic sands dunes. 7. Fine to medium bioclastic sands. 8. Muddy limestones. Current direction shown by arrows. TZ is the transition zone between outer and inner parts of the platform.

(Masse, 1977) (Fig. 23), and also in the Cenomanian-Turonian beds (Philip, 1965, 1970, 1972, and 1975) (Fig. 24). On the basis of their dimensions, morphology, internal structure, associated sediment, dominant fauna, sequential position, and mode of association with surrounding deposits a number of types of coral growth have been distinguished. These include: thickets and banks, isolated biostromal lenses (Fig. 17A) and complex biostromes of greater lateral extent, and varied bioherms (Fig. 17B). Unquestionable and well developed bioherms are limited to Upper Cretaceous time, while other types of buildups are recorded throughout the entire Cretaceous. The major characteristics of coral formations are summarized in Table 1.

The rock matrix is mainly micritic (Figs. 18B–

D) wackestone/packstone types, although in some buildups it may be terrigenous (e.g., marly or coarsely bioclastic, Fig. 18A). The origin of the micritic matrix is unknown. Locally especially in the Cenomanian bioherms from Provence, "stromatolitic" structures may act as traps or sites for micrite production. In most of the Lower Cretaceous coral formations, especially in the "Urgonian" limestones from Provence, hexacoral fragments are dominant averaging 40 percent of the sand-size fraction. Bivalves, foraminifers and echinoderms are less important and only average from 2 to 15 percent. The same results are recorded from the Upper Cretaceous, but here rudist fragments are more frequent in Cenomanian bioherms. Nevertheless in some cases, e.g., lamellar hexacoral biostromes from the Cenoma-

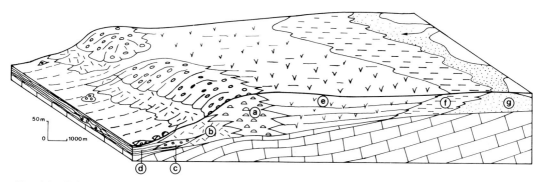

FIG. 24.—Schematic reconstruction of an Upper Cretaceous carbonate platform with coral-rudist development. a. Coral/rudist/algal buildup. b. Bioclastic talus. c. Reef breccias. d. Basinal marls. e. Rudists-foraminiferal carbonate muds. f. "Lagoonal" marls. g. Fluvio-deltaic sandstones.

TABLE 1.—SUMMARY OF VARIOUS CRETACEOUS CORAL BUILDUPS IN FRANCE

| | Lateral Extent | Thickness | Internal Structure | Contact With Surrounding Sediments | Matrix | Sequential Position | Physiographic Position | Stratigraphic Horizon |
|---|---|---|---|---|---|---|---|---|
| thickets | several hundred meters | 1 to 2 m | bafflestones with large colonies | transitional | biocalcarenite | variable | shallower part of the outer zone of the platform-top of complex biostromes | Barremian Bedoulian |
| anthozoa/hydrozoa banks | hecto to kilometric | 1 to 3 m | bafflestone nodular to rounded colonies | sharp (at the base) | biomicrite with clay fraction | interbedded in bioclastic sediments | outer zone of the platform under terrigenous influence | Barremian |
| hydrozoa banks | hecto to kilometric | 1 to 5 m | bafflestone branching to nodular colonies | transitional | biomicrite | possible transition between anthozoa and rudist beds | between outer and inner zone of the platform top of complex biostromes | Hauterivian Barremian and Bedoulian |
| lamellar-hexacorals biostromes or banks | several hundred meters | some meters | bafflestone/bindstone:lamellar colonies | transitional | clay biomicrite | muddy bottom | external part of the platform (pioneer position) | Cenomanian Turonian |
| isolated coral biostromes | several hundred meters | some meters | bafflestone/with low and small colonies | sharp | biomicrite | muddy bottom/sandy top | isolated buildups in outer zone—depth intermediate | Barremian |
| complex biostromes | several kilometers | 15 to 30 m | bafflestones/bindstone vertical ecologic zonation | transitional (with breaks) | biomicrite to biosparite | transition between bioclastics and rudist beds | between outer and inner zone of the platform depth decreasing from bottom to top of the buildup | Barremian Bedoulian |

TABLE 1.—CONTINUED

| | Lateral Extent | Thickness | Internal Structure | Contact With Surrounding Sediments | Matrix | Sequential Position | Physiographic Position | Stratigraphic Horizon |
|---|---|---|---|---|---|---|---|---|
| anthozoa/hydrozoa bioherms | hectometric | up to 20 m | bindstone/framestone | sharp with high angle paleoslopes | biomicrite | may derive from lamellar hexacorals biostromes | external part of the platform | Cenomanina |
| coral/rudists bioherms | hectometric | up to 20 m | large coral colonies and rudists | sharp with high angle paleoslopes | biomicrite | may be a variety of anthozoa/hydrozoa bioherms | external part of the platform | Cenomanian Turonian |

nian-Turonian interval, bioclastic production is very low. Evidence from penecontemporaneous cements (Fig. 26E), as well as preservation of high angle paleoslopes, shows that early lithification was very common in many Cretaceous buildups (Fig. 18E). In some cases (Cenomanian bioherms) brecciation of previous lithified organic limestones may occur. In addition, early leaching features are often recorded on the tops of relatively large bioherms or biostromes, which show evidence of penecontemporaneous emersion (Fig. 19D).

The different coral formations are dominated by hermatypic hexacorals (Figs. 17C, D, and F), which suggest that they lived in an infralittoral environment. The main coral groups are of the families Microsolenidae, Stylinidae, Calamophyllidae, Thamnasteriidae and Montlivaltiidae. Associated fauna is composed of: foraminifers (encrusting forms dominant, Fig. 19B), but also orbitolinids, especially during Cenomanian time, sponges (mainly calcareous types), stromatoporoids and chaetetids (Figs. 17E, 19A, and 18C), few bryozoa, brachiopods, and bivalves (especially ostreids and lithodoms). Rudists associated with Upper Cretaceous buildups may be caprinids, radiolitids, or hippuritids.

During the Lower Cretaceous, algae play only a minor role and are represented by scarce corallines (Archaeolithothamnium, Polystrata) and Udoteacea (Boueina), whereas dasycalds are very rare, as are schizophytes. But during upper Aptian and Albian times corallines become more abundant (Fig. 18F), and especially play an increasing role in the sediments surrounding coral colonies during the Upper Cretaceous (Figs. 19B and E). In biostromal or biohermal bodies, clear vertical faunal zonation is recorded. Basal strata show scattered or closely-packed lamellar coral colonies, medium strata are composed of branching or domal colonies in loose assemblages, while the top strata are composed of various colony shapes sometimes associated with scarce rudists, and chaetetids/stromatoporoids. Sedimentary changes follow vertical ecological zonation with a micritic basal matrix grading upward into micritic/bioclastic sediments, then into almost pure carbonate sands, essentially a coarsening upward sequence. These textural and faunal modifications may be interpreted as a shallowing environment coincident with varied bioconstruction and sediment types.

Topographic relief of the Lower Cretaceous biostromal bodies was always low. This may be explained by the following reasons: a high degree of stylolitizations of bioconstructed formations compared to the surrounding sediment (diagenetic reduction); the low relief of the coral colonies upon the substrate (which may be interpreted as a low rate of growth of the organism in relation to adjacent sedimentation); and the loose organic frame, especially for massive head-shaped corals with radial growth. This structural organization implies a weak wave resistance capability of the organic body. Nevertheless evidence of early emersions show that the top of some buildups may have been very near the water/air interface.

In connection with the low biostromal relief, it should be noted that wave talus is generally absent. However, horizontal dispersion of bioclastic material (e.g., coral fragments) derived from the coral buildups, occurs quite frequently in the adjacent areas. Therefore there is some degree of control of the surrounding environment by coral formation. Considering synoptic morphology and evidence of some degree of control over the adjacent environment (according to Hoffman, 1969 and Heckel, 1974), biohermal bodies as well as biostromal ones can be considered as reefs (ecological reefs *sensu* Dunham, 1970). This is not the case of banks and thickets which represent flat bodies, and which sometimes have a very loose structure.

The external zone of the platform is not always occupied by coral formations, and organic buildups may be associated with or replaced by bioclastic and/or oolitic sands. Usually coral buildups are absent in areas of high sediment mobility, which may be represented by megaripples or submarine bioclastic fans. In the same way they can be absent, especially during the Upper Cretaceous, on flat platform-margins with low energy where they may be replaced by rudist biostromes which grade distally to fine bioclastic micrites with sponge assemblages.

*Rudist Formations.*—In the Lower Cretaceous rudist development may be classified according to paleontological parameters, e.g., shell types (related to functional morphology *sensu* Kauffman, 1969) corresponding to taxonomic position. Some paleontologic parameters appear to be closely related to sedimentological ones. During the Lower Cretaceous, two main groups of rudist communities are distinguished (see Table 2). Rudists of Type 1 are thick-shelled, large-form communities with variable internal structures (often loose), as-

sociated with medium-sized pack/grainstone substrates (Figs. 25A and D). These types of rudist assemblages are known throughout the entire Lower Cretaceous. Biologic renewal of rudists occurs at specific, generic, and family levels with all families involved.

Rudists of type 2 are thin-shelled, small-form communities, with variable internal structures, often tightly packed, associated with muddy limestones (wacke/packstones), or even marly substrates (Fig. 25B). This type of assemblage is known throughout the entire Cretaceous. Biologic renewal is less important than for the former community. Requieniids and monopleurids are the principal rudist groups during the Upper Cretaceous, and three main groups of communities are recorded (see Table 3, and Figs. 25E and F). These are: 1. isolated clusters ("constructions en bouquets" according to Philip, 1970), 2. loose frame biostromes, and 3. tabular buildups.

These communities are characterized by the dominance and abundance of rudists. Only some ostreiform bivalves (*Liostrea, Chondrodonta*) are associated with the pachyodonts. In some cases scattered rudists occur, isolated or mixed with other organisms, such as various bivalves, gastropods (nerineids), chaetetids and rare hexacorals. Calcareous algae are relatively uncommon: for instance, corallines are completely absent in the Lower Cretaceous, although they are sometimes present in the Upper Cretaceous.

The textural and granulometric parameters of the substrate changes with the type of rudist community. Pure sandy or pure muddy substrates are very uncommon, except in Cenomanian caprinid biostromes, where the matrix is grainstone (Fig. 26D). More frequently substrates are muddy sands with the medium size of the sandy fraction ranging between 0.10 to 0.60 millimeters (Figs. 26A–C). In the Lower Cretaceous beds sand content has almost the same composition, a high percentage of peloids (average 50 percent), and foraminifers, mainly miliolids (Fig. 26F) (average 25 percent) also occur. Rudist fragments are infrequent (15 percent), and echinoids, calcareous algae and other molluscs are rare.

All the rudists encountered are associated with soft substrates. These sessile bivalves possessed

---

→

FIG. 25.—Rudist growths: A. Community type 1 from the Lower Cretaceous monopleurid and requienid rudist beds, Barremian of the Marseille area, length of the pencil is 13 centimeters. B. Community type 2 from the Lower Cretaceous requienid facies—loosely-packed whole and broken shells in peloidal to muddy matrix, Barremian from the Marseille area. C. Coarse grainstone with rudist fragments (storm deposit), Barremian from the Marseille area. D. Rudist Community type 1 from the Lower Cretaceous radiolitid bed (clusters) with isolated requienids and ostreiform bivalves; Clansayesian to Albian from Arudy, Aquitaine. E. Loose caprinid assemblage; Cenomanian from Provence. F. Tabular hippuritid buildup with closely-packed shells; Senonian, inner-platform from Provence.

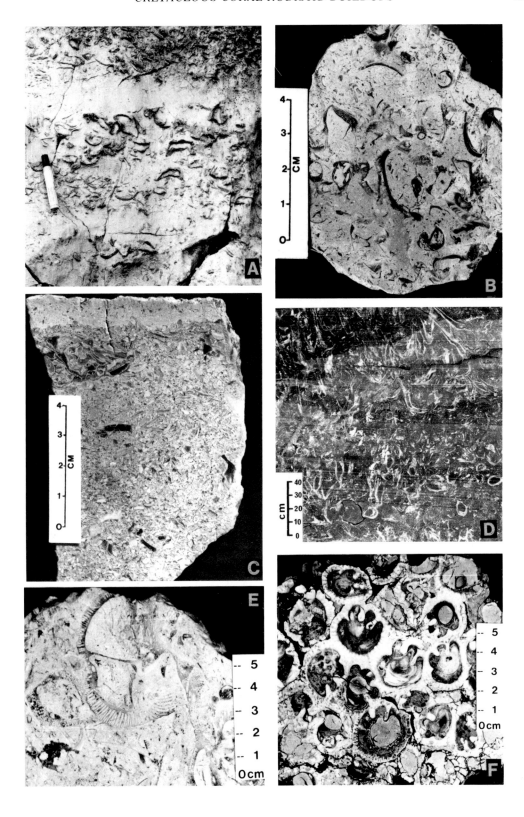

TABLE 2.—GENERIC COMPOSITION OF TYPES 1 AND 2 RUDIST ASSEMBLAGES DURING THE LOWER CRETACEOUS

| Stage | Communities type 1 | Communities type 2 |
|---|---|---|
| Albian | Valletia, Monopleura, Requienia, Agriopleura, Toucasia, Matheronia, Pachytraga, Praecaprina, Offneria, Pseudotoucasia, Eoradiolites, Horiopleura, Polyconites, Caprina | Pachytraga, Matheronia, Monopleura, Toucasia, Requienia, Agriopleura |
| Gargasian-Clansayesian | | |
| Bedoulian | | |
| Barremian | | |
| Hauterivian | | |
| Valanginian | | |
| Berriasian | | |

an attached valve as an initial small support. This is aided by resting on the substrate (sediment support), or by fixing onto other individuals (shell support). Gregariousness is effective in some cases thus leading to bioconstructed lenses, especially during Upper Cretaceous time.

During the Lower Cretaceous, sediment support was dominant and no true bioconstructions occur. Flat bioconstructed bodies of limited lateral extent were made by tubular rudists, such as some monopleurids (*Monopleura* and *Agriopleura*) and caprotinids (*Pachytraga*), or by cylindrical or conical rudists such as the hippuritids and elevated radiolitids, especially in the Albian and Upper Cretaceous (Fig. 25D). Thickness of these banks is of a few meters, and lenses are sometimes less than 1 meter thick. No typical reefs have been recorded, nevertheless rudist communities may have been able to control the adjacent environment by mean of shell fragmentation by which shell debris may be incorporated into adjacent biotopes (Fig. 25C). Possibly by peloid production since this type of grain is very abundant in the sediment substrate (Fig. 26F), and by a reduction of a truly marine fauna, such as hexacorals. The abundance of miliolids, and the presence of adjacent typical marginolittoral sediments with beach deposits (Figs. 26G–H), and paleogeo-

TABLE 3.—CLASSIFICATION AND MAIN CHARACTERISTICS OF UPPER CRETACEOUS RUDISTS COMMUNITIES

| | Lateral Extent | Thickness | Internal Structure | Matrix | Main Rudists | Stratigraphic Occurrence | Ecologic Significance |
|---|---|---|---|---|---|---|---|
| Clusters | Some meters | 1 to 2 m | Bafflestone | Mudstone Wackestone | Cylindrical or long conical hippuritids and (or) radiolitids | Turonian Senonian | Low energy Moderate and irregular micritic sedimentation |
| Loose network | Some ten to hundred meters | Some meters | Bafflestone | Wackestone Mudstone | Cylindrical and conical hippuritids and (or) radiolitids | Upper Cenomanian/ Turonian/ Senonian | Low energy Moderate micritic sedimentation |
| Tabular buildups | Some ten to hundred meters | Some meters | Framestone | Packstone with filling sediment or grainstone | Cylindrical and conical hippuritids and (or) radiolitids recumbent rudists (caprinids) | Middle and upper Cenomanian Caprinids Turonian Senonian Hippuritids and Radiolitids | Moderate or high energy/Lack or low micritic sedimentation |

BIOSTROMES

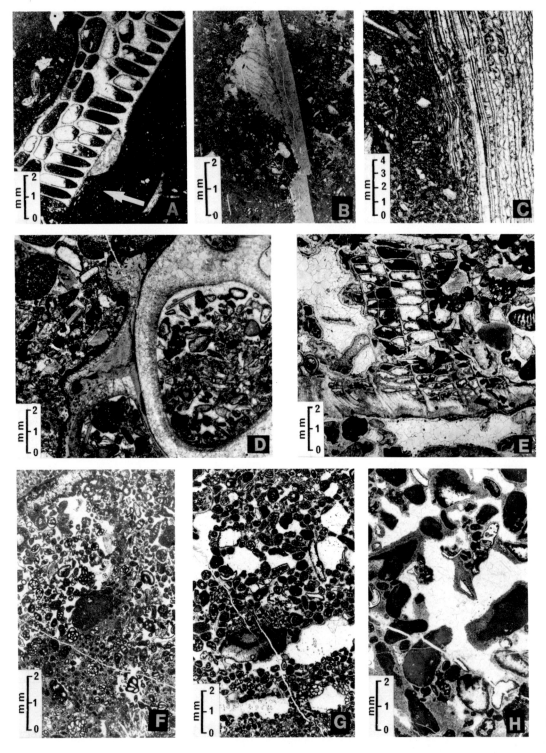

FIG. 26.—Rudist growths: A. Pallial-canal shell layer of caprinid in a muddy matrix. Canals filled with geopetal peloidal sediment; arrow indicates top of sample; Cenomanian reef from Provence. B. Broken monopleurid shell in muddy matrix (miliolid micrite), Barremian from the Marseille area. C. Broken radiolitid cellular shell layer in

graphic setting, suggests that rudist communities may be interpreted as inner-platform populations (low or moderate energy and sometimes restricted environments).

## PALEOENVIRONMENTAL FACTORS OF CORAL-RUDIST FORMATIONS

Shallow water conditions are needed for coral-rudist development and growth (mainly infralittoral). Thus rapid subsidence (lowering the bottom) or tectonic uplift may be inhibiting factors. Consequently, relative basement stability is one of the main factors necessary for coral-rudist development. This condition is imperative for platform initiation, since the wide lateral extent is related to sedimentary embankment and progradation. However, tectonic activity may be a favorable parameter for the development of offshore "highs" when a positive geophysical anomaly creates local stable conditions in an area of general bottom lowering (mole effect), e.g., the Albian of southern Aquitaine. Tectonic activity may also lead to the reactivation of paleofaults which induce topographic modifications, e.g., the Hauterivian of the southeastern part of the Paris Basin.

In the same way as sea level stability is needed for platform development, eustatic changes may in some cases offer favorable conditions for offshore "highs" to develop. Hence, Hauterivian transgression on an irregular surface of the western part of the Paris Basin induced establishment of coral patches located on topographic "highs." In the compressive Upper Cretaceous tectonic environment, elevated relief induced development of rudist lenses on small platforms in areas which deltaic sedimentation dominated (e.g., Provence-Pyrénées). During the time when eustatic transgressions (especially those that occurred in Cenomanian and Coniacian time) were favorable to coral-rudist development, they inhibited terrigenous influx and induced biogeographical migration between platforms.

No coral-rudist development is known from sandy siliciclastic deposits, which suggests that active terrigenous sedimentation inhibits the development of these communities. However, coral beds, like rudist beds, are recorded in fine clay sediments (marls or marly limestones). Upper Cretaceous terrigenous influences related to regressions or tectonic movements are important, and colonization of substrates composed of detrital material can be accomplished by hippuritids. Isolated rudists are present in perideltaic environments.

No coral-rudist formations are recorded in non-marine sediments. However for some rudist communities, slightly restricted oceanographic conditions may be assumed, especially during Turonian and Senonian time when stenobiontic communities such as hexacoral/caprinid communities were absent in France.

General ecologic characters, mineralogical data (clay minerals of marine sediments as well as those of paleosoils), and sedimentological data, suggest a warm climate with contrasted seasons (tropical to subtropical type). No major climatic changes seem to have occurred during the Lower and Upper Cretaceous time interval in France.

### CONCLUSIONS

From this review of Cretaceous coral-rudist formations in France the following conclusions may be offered:

1. Coral-rudist formations are recorded during most of Cretaceous time, but some stratigraphic gaps occur especially during the Vraconian and lower Turonian. These formations may occur isolated on offshore "highs," which represent small local paleogeographic units surrounded by marls or marly limestones. They may also be found together on carbonate platforms, which represent large regional paleogeographic units with typical horizontal zonation.

2. From Berriasian to lower Aptian time, carbonate platforms developed on the western and southern parts of the Alpine Basin as well as in the central and eastern parts of the Pyrénées region. During upper Aptian to Albian time, carbonate platforms were considerably reduced in the southeastern basin while they continued to develop in the Aquitaine-Pyrénées region. During the Upper Cretaceous, carbonate platforms developed on the northern and southern edges of the "Sillon pyrénéen" and in the Provence-Languedoc region.

3. Lower Cretaceous offshore "highs" are known from the Valanginian to Hauterivian interval of the southeastern edge of the Paris Basin, and in the Albian of south Aquitaine. Rudist

---

←

wackestone matrix; upper Turonian, inner-platform from Provence. D. Shell assemblage of caprinids (right valves) in grainstone matrix, Cenomanian reefs from Provence. E. Early cemented bioclastic radiolitid-coral grainstone; reef talus, upper Turonian from Provence. F. Miliolid-peloidal packstone, typical facies of requienid monopleurid beds from the Barremian ("Urgonian") limestone of Provence. G.–H. Facies adjacent to rudist beds in the Barremain of Provence. G. Foraminifer-peloid grainstone with keystone-vugs (swash-zone). H. Bioclastic mud lump grainstone with stalactitic vadose early cement.

banks are limited to Senonian time in the Pyré-nées-Provence region.

4. Three types of offshore "highs" have been distinguished: coral "highs," oobioclastic/coral and a special type, e.g., rudist banks. These formations created morphological anomalies on the sea bottom which may have originated from previous topographic irregularities, or from original bioconstruction. When organic construction was predominant a clear horizontal zonation is usually lacking, while vertical zonation is generally present.

5. Platform systems are characterized by an outer-zone occupied by coral or coral/rudist communities and bioclastic or oolitic sediments, which grade to basinal deposits by various means; and an inner-zone in which rudist communities developed on muddy bottoms, associated with miliolid foraminifers; this inner-zone may grade to marginolittoral or continental sediments.

6. Shallow water conditions, relative tectonic and eustatic stability, and low terrigenous content are needed for coral-rudist formation and development.

# REFERENCES

ALVINERIE, J., GAYET, J., PRATVIEL, L., AND VIGNEAUX, M., 1972, Esquisse structurale et faciologique du Crétacé supérieur nord-aquitain: Bull. Inst. Géol. Bass. d'Aquitaine, No. 12, p. 79–100.

ANONYMOUS, 1975, Géologie du Bassin d'Aquitaine: Bur. Rech. Geol. Min.

ARNAUD-VANNEAU, A., AND ARNAUD, H., 1976, L'évolution paléogéographique du Vercors au Barremien et à l'Aptien inférieur (Chaines subalpines septentionales, France): Géologie Alpine, 52, p. 5–30.

ASTRE, G., 1954, Radiolitidès nord-pyrénéens: Mem. Soc. Géol. France, n.s., 33, Mém. 71, 148 p.

——, 1961, *Pachytraga* tubuleux du Barremien du Doubs. Bull. Soc. Hist. Nat. Toulouse, 96, 3–4, p. 1–18.

BARREYRE, M., AND DELFAUD, J., 1965, Etude stratigraphique du Néocomien recontré par les sondages de la Société Nationale des Pétroles d'Aquitaine (S.N.P.A.) en Aquitaine occidentale. Colloq. sur le Crétacé inf., Lyon 1963: Mem. Bur. Rech. Géol. Min., 34, p. 625–635.

BAUDRIMONT, A. F., AND DUBOIS, P., 1977, Un bassin mésogéen du domaine péri-alpin: le Sud-Est de la France. Bull. Centre Rech. Explor. Prod. Elf-Aquitaine I, 1, p. 261–308.

BILOTTE, M., 1973, Le Cénomanien des Corbières méridionales (Pyrénées): Bull. Soc. Hist. Nat. Toulouse 109, fasc. 1–2, p. 7–22.

——, 1974, Exemple d'environnement périrécifal dans le Crétacé supérieur des Corbières, le récif-barrière du Turonien de Rennes-les-Bains (Pyrénées audoises). C. Rend. 90° Congr. Nat. Soc. Sav. Toulouse, II, p. 239–244.

——, 1977, Evolution paléorécifale sur la plate-forme de Mouthoumet pendant le Crétacé supérieur. Bull. Soc. Géol. France, (7), 19, no. 2, p. 277–280.

BOLTENHAGEN, C., 1967, Etude des formations calcaires de l'Albo-Aptien entre le Gave d'Aspe et la vallée de l'Ouzom: Rapp. Int. S.N.P.A., inedit 86 p.

BOUROULLEC, J., AND DELOFFRE, R., 1970, Interprétation sédimentologique et paléogéographique, par microfaciès, du Crétacé inférieur basal d'Aquitaine sud-ouest: Bull. Centre Rech. Pau S.N.P.A., 4, 2, p. 381–429.

——, AND ——, 1976, Relations faciès-environnement au Crétacé moyen en Aquitaine occidentale: Bull. Centre Rech. Pau S.N.P.A., 10, 2, p. 535–583.

CARBONE, F., PRATURLON, A., AND SIRNA, G., 1971, The Cenomanian shelf-edge facies of Rocca di Cave (Prenestini Mts., Latium): Geol. Romana, vol. 10, p. 131–198.

CASSOUDEBAT, M., AND PLATEL, J. P., 1976, Sédimentologie et paléogéographie du Turonien de la bordure septentrionale du bassin aquitain: Bull. Bur. Rech. Géol. Min., 2e sér. Sect. 1, no. 2, p. 85–102.

COOPER, M. R., 1976, Eustacy during the Cretaceous: its implication and importance: Palaeogeography, Palaeoclimatology, Palaeoecology, v. 22, p. 1–60.

COMBES, P. J., 1969, Recherches sur la génèse des bauxites dans le Nord-Est de l'Espagne, le Languedoc et l'Agriège (France): Montpellier, Mém. Centr. Etud. Rech. Géol. et Hydrogéol., 3–4, 335 p.

CONRAD, M. A., 1969, Les calcaires urgoniens dans la région entourant Genève Eclog: Géol. Helv., 62/1, 79 p.

CORROY, G., 1925, Le néocomien de la bordure orientale du Bassin de Paris: Nancy, Coubé Pils, 334 p.

COTILLON, P., 1973, Le passage du Jurassique au Crétacé dans les faciès néritiques de la haute Provence orientale (Var, Alpes de haute-Provence, Alpes-Maritimes): Coloq. sur la limite Jurassique-Crétacé, Lyon, Sept. 1973, Mém. Bur. Rech. Géol. Min, no. 86, p. 315–322.

CUMINGS, E. R., 1932, Reefs or bioherms?: Geol. Soc. America. Bull., v 3, p. 25.

DEBOURLE, A., AND DELOFFRE, R., 1975, Etude du complexe récifal d'Arudy in livre guide Excursion D. Sud-Aquitaine: Internat. sur les Coraux et Récifs Coralliens fossiles, Paris, Symp., p. 1–5.

DECHASEAUX, C., AND SORNAY, J., 1959, "Récifs" à Rudistes: Bull. Soc. Géol. France, (7) 1, p. 399–402.

DENIZOT, G., 1934, Description des massifs de Marseilleveyre et de Puget: Ann. Mus. Hist. Nat. Marseille, 26, 5, 229 p.

DONZE, P., 1958, Les couches de passage du Jurassique au Crétacé dans le Jura français et sur les pourtours de la "fosse vocontienne": Thèse Sci. Nat., Lyon, 221 p.

DOUVILLE, M., 1887, Sur quelques formes peu connues de la famille des Chamidès: Bull. Soc. Géol. France, 15, (3), p. 756–802.

——, 1889, Rudistes du Crétacé inférieur des Pyrénées: Bull. Soc. Géol. France, (3), 17, p. 627.

DUNHAM, R. J., 1970, Stratigraphic reefs versus ecologic reefs: Am. Assoc. Petroleum Geologists Bull., v. 54, p. 1931–1932.

EL KHOLY, Y., 1971, Données nouvelles sur l'âge et les faciès du Crétacé supérieur entre Col-Bas et Dormillouse (Alpes de Hte Provence): C. Rend. Somm. Soc. Géol. France, p. 119–120.

ESQUEVIN, J., FOURNIE, D., AND DE LESTANG, J., 1971, les séries de l'Aptien et de l'Albien des régions Nord-Pyrénées et du Sud-Aquitain: Bull. Centre Rech. Pau S.N.P.A., 5, 1, p. 87–151.

FEUILLEE, P., 1971, Les Calcaires biogéniques de l'Albien et du Cénomanien pyrénéo-cantabrique: problèmes d'environnement sédimentaire: Palaeogeography, Palaeoclimatology, Palaeoecology, v. 9, no. 4, p. 277–312.

FOURNIER, E., 1890, Esquisse géologique des environs de Marseille: Marseille 8°, 106 p.

FREYTET, P., 1973, Edifices récifaux développés dans un environnement détritique: exemple des biostromes à *Hippurites* (Rudistes) du Sénomien inférieur du sillon languedocien (région de Narbonne, sud de la France): Palaeogéography, Palaeoclimatology, Palaeoecology, v. 13, p. 65–76.

GIDON, P., 1948, Sur l'extension des faciès coralligènes dans le Berriasien des environs de Chambery: C. Rend. Som. Soc. Géol. France, p. 284–285.

GUILLAUME, S., 1966, Le Crétacé du Jura français: Bull. Bur. Rech. Géol. Min., nos. 1, 2, 3, p. 1–43, p. 7–69, p. 11–79.

HART, M. B., AND TARLING, D. H., 1975, Cenomanian palaeogeography of the North Atlantic and possible mid-ceno-manian eustatic movements and their implications: Palaeogeography, Palaeoclimatology, Palaeoecology, v. 15, p. 95–108.

HECKEL, P. H., 1974, Carbonate buildups in the geologic record: a review, *in* Laporte, L. F., Ed., Reefs in time and space: Soc. Econ. Paleontologists Mineralogists, Spec. pub. No. 18, p. 90–154.

HOFFMANN, H. J., 1969, Attributes of stromatolites: Geol. Surv. Canada Paper 69-39, 43 p.

HUMBERT, L., 1975, Dynamique biosédimentaire de la formation et de l'évolution d'une plateforme carbonatée: Congr. Internat. Sediment, 9, Nice, Thème 5, p. 225–229.

JOUKOWSKY, E., AND FAVRE, J., 1913, Monographie géologique et paléontologique du Salève (Haute-Savoie, France): Mém. Soc. Phys. Hist. Nat. Genève, v. 37, fasc. 4, p. 295–523.

KAUFFMAN, E. G., 1969, Form, function and evolution, *in* Moore, R. C., Ed., "Bivalvia," Mollusca 6, Part N., v. 1 (of 3). Treatise on Invertebrate Paleontology, p. N129–N205.

———, AND SOHL, N. F., 1974, Structure and evolution of Antillean Cretaceous Rudist frameworks: Verhandl. Naturf. Ges. Basel, v. 84, no. 1, p. 399–467.

LORIOL, P., DE, 1868, Monographie des couches de l'étage Valangien des carrières d'Arzier (Vaud) *in* Pictet: Mat. Paleont. Suisse, Liv. 10–11, 110 p.

MARIE, P., AND MONGIN, D., 1957, Le Valanginien du Mont-Rose de la Madrague (Massif de Marseilleveyre, B.du-Rh.): Bull. Soc. Géol. France, 6, 7, p. 401–425.

MASSE, J. P., 1976, Les calcaires urgoniens de Provence (Valanginien—Aptien inférieur): Stratigraphie, Paléon-tologie, les eents et leur évolution: Thèse paléoenvironnements et leur évolution: Thèse Sci. Nat., Marseille, 445 p.

———, 1977, Les constructions à Madrépores des calcaires urgoniens (Barrimien-Bédoulien) de Provence (S.E. de la France): Internat. coraux et récifs fossiles, 2nd., Paris, Symp., 1975: Mém. Bur. Rech. Géol. Min., 89, p. 322–335.

———, AND PHILIP, J., 1969, Sur la présence de brèches et de klippes sédimentaires dans l'Albien de la région de Sainte-Anne d'Evénos (Var). Conséquences paléogéographiques: Bull. Soc. Géol. France, 7, 11, p. 666–669.

———, AND ———, 1973, Mise en évidence de l'Albien au Mont-Combe (Nord de Toulon, Var). Implications paléontologiques, paléogéographiques et tectoniques: Bull. Bur. Rech. Géol. Min., 2 (10), p. 207–214.

———, AND ———, 1976, Paléogéographie et tectonique du Crétacé moyen en Provence: révision du concept d'isthme durancien: Rev. de Géogr. Physiq. et Géol. Dynam. (2), 18, 1, p. 49–66.

MONGIN, D., AND TROUVE, PH., 1953, Le Valanginien inférieur calcaire du Grand Canon du Verdon (Basses-Alpes): Bull. Soc. Géol. France, 3, p. 223–239.

MONLEAU, C., AND PHILIP, J., 1972, Reconstitution paléogéographique des formations calcaires à Rudistes du Turonien supérieur de la Basse Vallée du Rhône à partir d'une étude de microfaciès: Rev. Micropal., no. 1, p. 45–56.

MOULLADE, M., AND PORTHAULT, B., 1970, Sur l'âge précis et la signification des grès et conglomérats crétacés de la vallée du Toulourenc (Vaucluse). Répercussions de la phase orogénique "autrichienne" dans le Sud-Est de la France: Géol. Alpine, 46, p. 141–150.

MUNIER-CHALMAS, M., 1882, Etudes critiques sur les Rudistes: Bull. Soc. Geol. France, 3, 10, p. 472–494.

PEYBERNES, B., 1976, Le Jurassique et le Crétacé inférieur des Pyrénées franco-espagnoles, entre la Garonne et la Méditerranée: Thèse Sc. Nat., Univ. Paul Sabatier, Toulouse, 459 p.

PHILIP, J., 1965, Présence de biohermes à Madréporaires dans le Crétacé supérieur des Martigues (B.du-Rh.): C. Rend. Acad. Sci., France, 260, p. 5841–5843.

———, 1970, Les formations calcaires à Rudistes du Crétacé supérieur provençal et rhodanien: Thèse Sci. Nat. Marseille, 438 p.

———, 1972, Paléoécologie des formations à Rudistes du Crétacé supérieur l'exemple du Sud-Est de la France: Palaeogeography, Palaeoclimatology, Palaeoecology, v. 12, 3, p. 205–222.

———, 1974, Les formations calcaires à Rudistes du Crétacé superieur provençal et rhodanien: stratigraphie et paléogéographie: Bull. Bur. Rech. Géol. Min., 2e sér., no. 3, p. 107–151.

———, 1975, Indicateurs sédimentologiques et paléoécologiques dans les milieux de plates-formes carbonatées du

Crétacé supérieur. L'exemple du Cénomanien de la Basse-Provence occidentale: Congr. Intern. Sédiment. 9, Nice, Thème 1, p. 143–147.

———, 1978, Stratigraphie et paléoécologie des formations à Rudistes du Cénomanien: l'exemple de la Provence: Géol. Méditerranéenne, t. 5, no. 1, p. 155–168.

———, AMICO, S., AND ALLEMANN, J., 1978, Importance des Rudistes dans la sédimentation calcaire au Crétacé supérieur: Lyon, Livre Jubilaire J. Flandrin, p. 343–360.

PLATEL, M. P., 1974, Un Modèle d'organisation des biotopes à Rudistes: l'Angoumien de l'Aquitaine septentrionale: Bull. Soc. Linn. Bordeaux, t. 4, no. 1, p. 3–13.

POIGNANT, A. P., 1964, Révision du Crétacé inférieur en Aquitaine occidentale et méridionale: Thèse Sci. Nat. Paris, 317 p.

PORTHAULT, B., 1974, Le Crétacé supérieur de la ''fosse vocontienne'' et des régions limitrophes (France, Sud-Est): Thèse Sci. Nat., Lyon, 342 p.

RAT, P., 1965, Rapport sur les formations non marines du Crétacé inférieur français. Colloq. sur le Crétacé inf., Lyon 1963: Mém. Bur. Rech. Géol. Min. 34, p. 333–343.

———, 1978, La France au Cénomanien: schémas paléogéographiques: Géol. Mediterranéenne, t. 5, no. 1, p. 207–213.

SERONIE-VIVIEN, M., 1972, Contribution à l'étude du Sénonien en Aquitaine Septentrionale. Edit. Centre Nat. Rech. Sci.: Les Stratotypes Français, v. 2, 195 p.

SERONIE-VIVIEN, R., SENS, J., AND MALMOUSTIER, G., 1965, Contribution à l'étude des formations du Crétacé inférieur dans le Bassin de Parentis (Aquitaine): Colloq. sur le Crétacé inf. Lyon, 1963, Mém. Bur. Rech. Géol. Min., 34, p. 669–692.

SCHNORF-STEINER, A., 1960, *Disparistromaria,* un Actinostromariidae nouveau du Valanginien d'Arzier (Jura Vaudois): Eclogae Géol. Helv., 53, no. 1, p. 439–442.

———, 1960, Les Milleporidiidae des marnes valanginiennes d'Arzier: Eclogae Géol. Helv., 53, no. 2, p. 716–727.

———, 1960, Les *Actinostromaria* des marnes valanginiennes d'Arzier: Eclogae Géol. Helv., 53, no. 2, p. 733–746.

———, 1963, Les *Steinerella* des marnes Valanginiennes d'Arzier: Eclogae Géol. Helv., 56, no. 2, p. 1131–1139.

SCHROEDER, R., AND POIGNANT, A. F., 1968, Sur la position stratigraphique des ''calcaires à *Dictyoconus*'' rapportés au Valanginien en Aquitaine: C. Rend. Acad. Sci. Paris, t. 266, D, p. 992–993.

SKELTON, P. W., 1976, Functional morphology of the Hippuritidae: Lethaia, v. 9. p. 83–100.

SORNAY, J., 1950, Etude stratigraphique sur le Crétacé supérieur de la vallée du Rhône: Thèse Sci. Nat., Grenoble, Imp. Allier, 254 p.

SOUQUET, P., 1967, Le Crétacé supérieur sud-pyrénéen en Catalogue, Aragon et Navarre: Thèse Sci. Nat., Toulouse, 529 p.

———, PEYBERNES, B., BILOTTE, M., DEBROAS, J. E., AND (REY, J., CANEROT, J., COLL.), 1977, Nouvelle esquisse structurale des Pyrénées: Trav. Labo. Géol., Univ. Paul Sabatier, Toulouse 17 p.

STEINHAUSER, N., 1970, Recherches stratigraphiques dans le Crétacé inférieur de la Savoie occidentale (France): Thèse no. 1506, Univ. Genève, Edit. ''Médecine et Hygiène,'' p. 1–18.

———, AND CHAROLLAIS, J., 1971, Observations nouvelles et réflexions sur la stratigraphie du ''Valanginien'' de la région neuchateloise et ses rapports avec le Jura méridional: Géobios., v. 4, fasc. 1, p. 7–59.

STURANI, C., 1962, Il complesso sedimentario autoctono all'a estremo hord-occidentale del Massiccio del Argentera: Mém. Inst. Géol. Min., Univ. Padova, 22.

TURNŠEK, D., AND MASSE, J. P., 1973, The Lower Cretaceous Hydrozoa and Chaetetidae from Provence (Southeastern France). Razprave, Acad. Sci. et Art: Slovenica, Ljubljana, 4, 16-6, p. 219–244.

VAN HINTE, J. E., 1976, A cretaceous time scale: Am. Assoc. Petroleum Geologists Bull., v. 60, no. 4, p. 269–287.

SEPM Special Publication No. 30, p. 427–445, May 1981

# UPPER CRETACEOUS REEF MODELS FROM ROCCA DI CAVE AND ADJACENT AREAS IN LATIUM, CENTRAL ITALY

FEDERICO CARBONE AND GIUSEPPE SIRNA
Centro di Studio Italia Centrale CNR and
Instituto di Geologia e Paleontologia Università di Roma

## ABSTRACT

Deposition of Upper Cretaceous carbonate sequences in central Italy (westcentral Latium) appears to be directly related to the tectonic evolution of an epioceanic Central Apennine Platform. From early Cenomanian to late Senonian time organic buildups, along with skeletal shelf-edge deposits, were laid down along the present northwestern margin of the Lepini Mountains, and the southern portion of the Prenestini Mountains. These sediments overlie rocks of the Cretaceous (Aptian-Albian) restricted platform facies. Late Cretaceous shelf-edge deposits are characterized by sequences of rudistid and coral communities laid down within varied depositional settings.

The rapid development of these organic communities along a linear tectonically induced shelf-margin caused pronounced increases in the production of bioclastic debris, and the distribution of these sediments both on the marginal slope and in the back-reef areas. This bioclastic debris accumulated in offshore shoals and in tidal banks connecting patch-reefs and organic banks.

The continued presence during Cenomanian time of a linear shelf-edge organic buildup complex appears to have been the dominant controlling factor in regards to inner-platform sedimentation. Here, nerineids (gastropods), ostreids (pelecypods), and radiolitid rudistid pelecypod communities, occurring in muddy skeletal wackestones, suggests deposition within both open marine shelf-lagoons and back-reef environments. In addition, the presence of mudstones (some laminated), and peloidal grainstones containing benthonic microfossils, suggests somewhat sheltered shelf-lagoonal to tidal flat depositional settings.

Differential tectonism during Turonian time, causing regional sea level fluctuations, produced varied facies patterns due to changes in water circulation and sedimentation over the shelf area. An open marine shelf-lagoon facies with abundant microfossils was widespread during this time. Conditions of sea level fluctuations and influx of clean marine waters onto the platform led to the development of rudistid banks and winnowed skeletal sands along the platform-margin. The tectonism of early Turonian time caused subaerial exposure of the western portion of the Rocca di Cave area, followed by sedimentary progradation.

Various changes in overall regional bathymetry, with resultant restriction in water circulation, increased during Senonian time. Isolated platform areas became outer "highs" surrounded by deeper water, although these "highs" were partially covered by shoaling transgressive sediments (Rocca di Cave). In some instances the shelf area was drowned bringing shelf-margin facies into former platform areas, and forming organic shoals and barrier islands during repeated progradation (Lepini Mountains).

## INTRODUCTION

In westcentral Latium, east of Rome, rise the Prenestini and Ernici Mountains, while to the southeast, in the direction of the Apennines, stretch the Lepini Mountains. All of these mountain ranges are composed chiefly of shallow water carbonates of Cretaceous to Miocene age (Fig. 1). The bordering depressions (Latina Valley and Pontina Plain) are covered by Cenozoic terrigenous flysch deposits. These deposits are the result of a number of factors including Alpine orogenesis, Quaternary volcanic deposits from the Alban Hills, and deposition of marine to aeolian sediments of Pliocene to Pleistocene age.

This area has previously been studied by several authors including: Alberti (1952); Negretti (1954); Boni (1962); Carbone, Praturlon, and Sirna (1971); Praturlon and Sirna (1976); and Carbone and Catenacci (1978). However, recent road cuts, particularly near Rocca di Cave (Prenestini Mountains), along the southern slope of Mt. Scalambra (Ernici Mountains), and in the Lepini Mountains, have exposed fossiliferous carbonate sequences previously not seen, and whose study has contributed new data on stratigraphy and paleoenvironments. Sedimentological studies have also been carried out, and from all of this new information it is now possible to describe a more meaningful paleogeographical model for this area.

### GEOLOGICAL SETTING

The present structural configuration of this area is the result of a post-Miocene compressive tectonic phase that has altered the original paleogeographic order of the carbonate units. In fact, eastward displacements of different magnitudes took place during tectonic phases, with Mesozoic units overriding Cenozoic units. Final tectonic movements determined the actual regional geomorphology and initiated Quaternary volcanism. Mesozoic successions exposed in central-southern Latium exhibit characteristics of carbonate sedimentation within a shallow sea. This type of sedimentational setting is related to the existence of

427

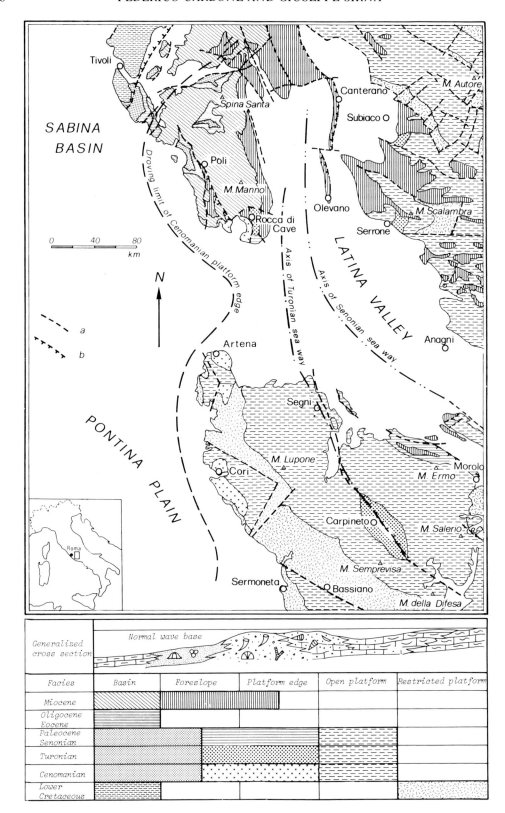

| Facies | Basin | Foreslope | Platform edge | Open platform | Restricted platform |
|---|---|---|---|---|---|
| Miocene | | | | | |
| Oligocene Eocene | | | | | |
| Paleocene Senonian | | | | | |
| Turonian | | | | | |
| Cenomanian | | | | | |
| Lower Cretaceous | | | | | |

FIG. 2.—Outcrop view showing distinctive sedimentary features within a restricted platform depositional sequence in the Lepini Mountains. The photograph shows an interformational conglomerate filling a channel (position of hammer) within a vertical sequence of laminated dolomitic limestone deposited in an intertidal environment.

an epioceanic carbonate platform during the entire Mesozoic; a platform subjected to considerable variation in areal development and facies distribution.

Hence, it is impossible to describe Mesozoic sedimentation of Latium as representing a single sedimentary model, although several of the stratigraphic sequences reflect outlines of Wilson's (1975) general scheme of carbonate facies distribution.

Development of carbonate sedimentation during Jurassic-Cretaceous time suggests general and progressive reduction in size of an epioceanic carbonate platform, especially during the Lower Cretaceous. This in turn led to the formation of a Cenomanian organogenic ridge along the northwestern edge of the structure (Artena, Colle Illirio, Cori, Monte Arrestino, Rocca di Cave), as noted by Carbone et al. (1971); Praturlon and Sir-

na (Fig. 1, 1976). Hence, open platform facies (usually sands and muddy sands) generally overlie facies characteristic of restricted-lagoonal environments; in a few cases these are even evaporitic (Carbone et al., 1978). Lime or dolomitic muds characteristic of the subtidal, intertidal, and supratidal zones (stromatolites and loferites) may also be present (Fig. 2). Further disintegration of the Cenomanian platform, by now strongly divided and reduced compared to Jurassic-Lower Cretaceous time, led during Turonian-Senonian time to its complete disintegration.

The pelagic facies that developed as a result of this phenomenon, reflects detrital debris derived from the platform, which was partly emerged during the process of destruction. In the entire Sabina area (around Tivoli), pelagic sequences ("scaglia") consist of layers of detrital-organogenic limestone intercalated with channels of Paleocene

←

FIG. 1.—An Upper Cretaceous facies map of westcentral Latium showing the Prenestini Mountains to the north, Lepini Mountains to the south, and the Simbruini-Ernici Ridge to the east; a. the trace of normal faults, b. the presence of overthrust and reverse faults.

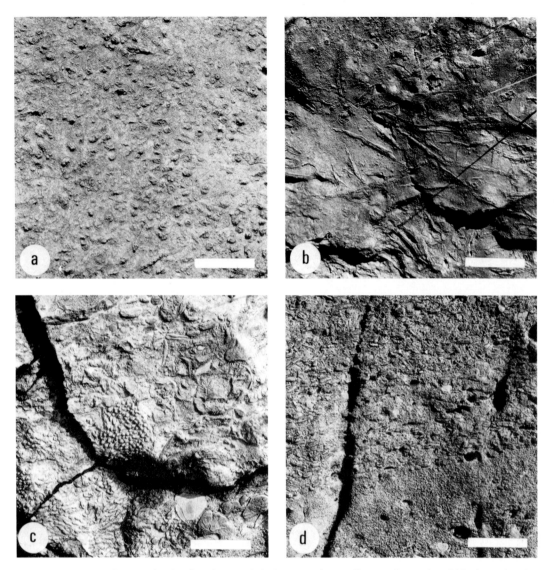

FIG. 3.—Outcrop photographs showing characteristic features of some Cenomanian rocks within the region; bar scale 5 centimeters in length. A. Gastropod mudstone with a limited fauna characterizing quiet, restrictive environments of the shelf-edge. B. Ostreid (oyster) mudstone characterized by the form *Perna*? sp. and indicating a shelf-lagoon facies. C. Reef rudstone consisting of massive bioclastic deposits composed of caprinid rudistids and corals derived from stable bottom growing communities. D. Winnowed bioclastic grainstone containing well-rounded skeletal deposits with reverse graded-bedding indicating a shifting substrate within a shoal beach environment.

sediment. Within the influence of the relict platform we find restricted lagoon-tidal marsh deposits, also referable to the Paleocene.

Periods of emersion, mainly in connection with tectonic uplift are recorded over the entire area. At Rocca di Cave, for instance, there existed emerged areas at the end of the Cenomanian, with karsted holes filled-in with pelagic mudstones of Maestrichtian and Paleocene age. In the zone of

Cori (Lepini Mountains) the same phenomenon occurs, but here the cavities are only filled with Maestrichtian pelagic muds. Paleocene emergence, occasionally indicated by karst phenomena containing *Microcodium* associated with continental breccias, is a more general phenomenon.

Deposition of carbonate sediments during the Eocene and Oligocene appears to have been limited to the basinal areas and to the more periph-

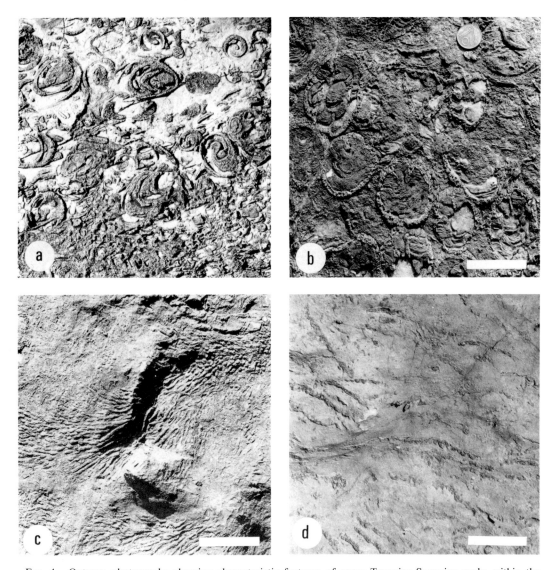

Fig. 4.—Outcrop photographs showing characteristic features of some Turonian-Senonian rocks within the region; bar scale 5 centimeters in length. A. An acteonid (molluscan) wackestone showing matrix-supported whole fossils suggesting deposition in a sheltered shelf-lagoonal environment; actual size. B. Hippuritid rudistid bank built up by clumps of *Vaccinites boehmi* in growth position suggesting a platform shelf-edge environment sheltered from wave and current action. C. Fine bioclastic packstone containing ramose corals in fine sediment indicating a more sheltered sea bottom below normal wave base. D. Microbioclastic wackestone with stromotoporoids associated with echinodermal remains suggesting a deeper high slope environment.

eral zones of the emerged carbonate units. Facies present generally belong to the platform-edge and are characterized by organic calcarenites with macroforaminifers, interfingering with pelagic sediments (calcareous-marly) rich in planktic foraminifers (''scaglia cinerea''). Miocene sediments transgressively overlie carbonate-platform units, and correspond to a phase of general transgression.

FACIES ANALYSIS OF THE STRATIGRAPHIC SEQUENCE

*Aptian-Albian Time*

*Restricted Platform Environments.*—This interval is generally characterized by the prevalence of restricted platform deposits, referable to lagoonal and tidal flat environments. The main facies are represented by mudstones and wackestones with peloids and intraclasts. Locally the

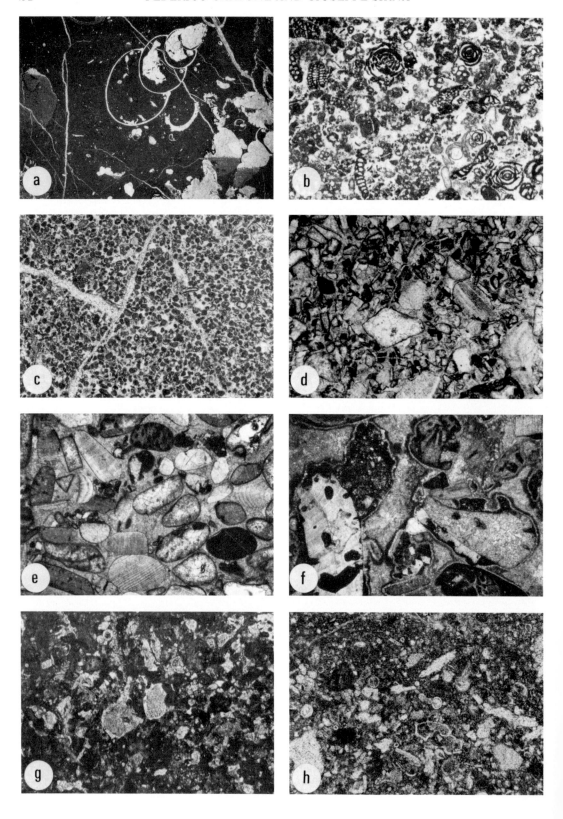

transition to the sheltered shoal facies can be identified by packstones made up of peloids and benthic microfossils. Fenestral structures and algal loferites occur in some horizons, often associated with dolomitization and reflecting tidal flat environments. Clayey-marly episodes occur in the lower part of the interval, and these are rich in charophytes and orbitolinids suggesting the existence of both fresh and brackish water environments during Aptian time.

*Major Facies.*—There are three dominant facies, these consist of:

1. Miliolid and ostracode mudstones containing tests of charophytes (*Atopochara trivolvis* Peck, and *Flabellochara harrisi* Peck), suggesting existence of restricted ponds and lagoons; 2. Requienid mudstones characterized by whole rudistids scattered in micrite, and indicative of open shelf-lagoons; and 3. Nerineid grainstones containing unbroken gastropod shells and marking the inner shoal environment of shelf-lagoons.

### Cenomanian Time

*Restricted Platform Environments* (Fig. 5).— Characterized by calcareous-dolomitic mudstones containing miliolids, ostracodes and benthic foraminifers. These rock types are associated with various sedimentary structures such as geopetal fillings in desiccation planar cavities and loferitic laminations (Fischer, 1964). Floatstones can also be recognized, characterized by dark intraclasts filling-up small channels.

*Major Facies.*—Three major facies characterize this interval, they are:

1. Laminated and bioturbated mudstones and wackestones locally grading into pelsparites. These are thought to reflect tidal flat deposition.

2. Lithoclastic rudstones and floatstones composed of micritic and dolomitic intraclasts formed as tidal channel fillings (Fig. 2).

3. Homogeneous and scarcely fossiliferous mudstones containing rare ostracodes and miliolids, and thought to be indicative of restricted marine shelf-lagoons.

*Open Platform Environments.*—Environments chiefly represented by muddy lime sediments of varying texture dependent on the depositional agent. The faunal association is dominated by indigenous communities of benthic foraminifers and green algae; local accumulations of molluscs can also be present. The greater part of the grain fraction is represented by peloids, intraclasts and fecal pellets.

*Major Facies.*—Two major facies characterize this interval, these are:

1. Foraminiferal wackestones-packstones (Fig. 5B–C) characterized by *Pseudedomia vialii* (Colalongo) and *Cisalveolina fallax* Reichel, and various miliolids. These are thought to have been deposited on shoals within open marine shelf-lagoons;

2. Ostreid mudstones-wackestones (Fig. 3B) characterized by molluscs (*Perna*?), occasionally found in association with nerineids or with radiolitid rudistids. Facies of this type reflect interaction between typical lagoonal and back-reef environments.

*Platform-Edge Environments.*—These environments are characterized by extreme variability of sedimentary textures and faunal composition. This is a result of numerous sub-environments created by the interaction of the inner-platform realm of low energy, and the platform-edge which may be susceptible to strong energy variations due to its complicated morphology. This is shown by the occurrence of muddy episodes within lagoonal faunal associations, repeatedly intercalated with marginal deposits. The scarcity of planktic forms among the faunal associations confirms the general trend towards the back-reef position.

---

←

FIG. 5.—Thin-section photomicrographs of various textural aspects of significant Upper Cretaceous facies from the studied area. A. Gastropod mudstone with scattered ostracodes (×6); gastropod shows geopetal filling. Rocks of this type are deposited in restricted bays within the marginal platform areas. B. Foraminiferal packstone with concentrations of foraminiferal tests and peloids (×10), suggesting deposition within slightly agitated waters of sheltered shelf-lagoon shoals. C. Peloidal grainstone consisting of hardened fecal pellets with small intraclasts and scattered small foraminifers (×10). The winnowed aspect is due to water movement in a shallow restricted shelf-lagoonal environment. D. Bioclastic grainstone consisting of skeletal hash of various platform-edge organisms (×10); grains circumscribed by micrite envelopes. This sediment was formed in a shoal environment approaching wave base. E. Winnowed bioclastic grainstone-rudstone (×6) containing well-rounded bioclastic grains often coated with micrite envelopes, and replaced by sparry calcite. Sediments of this type commonly occur within areas of shifting substrates on the platform-margin. F. Reef rudstone (×6) consisting of coarse gravel composed of the skeletal remains of frame building organisms. Sediments show well developed drusy cement rims and cavities filled with fine matrix typical of beach rock. G. Bioclastic packstone (×10); an off-bank sediment composed of micritized bioclastic debris partly indigenous to the area and partly derived from outside the immediate depositional area. Photomicrograph shows echinodermal debris surrounded by syntaxial rim cement. H. Microbioclastic wackestone with abundant micritic clasts containing pelagic foraminifers (*Globotruncana tricarinata* and *Stomiosphaera*), indicating deposition within the fore-slope environment (×10).

| SPECIES | APTIAN ALBIAN | CENOM. | TURONIAN | SENONIAN |
|---|---|---|---|---|
| **PELECYPODS and GASTROPODS** | | | | |
| Chondrodonta joannae (CHOFFAT) | | ─ | | |
| Chondrodonta munsoni (HILL) | | ── | | |
| Neithea aequicostata (LAMARCK) | | ─ | | |
| Neithea incostans (SHARPE) | | ───── | | |
| Caprina carinata (BOEHM) | | ─ | | |
| Caprina schiosensis BOEHM | | ─ | | |
| Neocaprina gigantea PLENICAR | | ─ | | |
| Neocaprina nanosi PLENICAR | | ─ | | |
| Sphaerucaprina forojuliensis BOEHM | | ─ | | |
| Schiosia carinatoformis POLSAK | | ─ | | |
| Schiosia schiosensis BOEHM | | ─ | | |
| Orthoptychus striatus FUTTERER | | ─ | | |
| Ichthyosarcolites bicarinatus (GEMMELLARO) | | ─ | | |
| Ichthyosarcolites monocarinatus SLISKOVIC | | ── | | |
| Ichthyosarcolites poljaki POLSAK | | ─ | | |
| Ichtyosarcolites rotundus POLSAK | | ─ | | |
| Ichtyosarcolites tricarinatus PARONA | | ─ | | |
| Radiolites lusitanicus (BAYLE) | | | ── | |
| Radiolites praesauvagesi TOUCAS | | | ──── | |
| Radiolites trigeri (COQUAND) | | ─── | ── | |
| Sauvagesia contorta CATULLO | | | ── | |
| Sauvagesia nicaisei (COQUAND) | | | ── | |
| Sauvagesia raricostata POLSAK | | | | ── |
| Sauvagesia sharpei (BAYLE) | | | ── | |
| Vaccinites boehmi DOUVILLE' | | | | ── |
| Vaccinites narentanus SLISKOVIC | | | | ── |
| Vaccinites sulcatus DEFRANCE | | | | ── |
| Hippurites colliciatus WOODWARD | | | | ──── |
| Cossmannea annulata (GEMMELLARO) | | ── | | |
| Cossmannea edoardi (PARONA) | | ── | | |
| Plesioplocus grandis PCHELINCEV | | ───── | | |
| Plesioptygmatis nobilis (MUNSTER) | | | ── | |
| Poliptyxis requieni (D'ORBIGNY) | | ── | | |
| Poliptyxis schiosensis (PIRONA) | | ── | | |
| Plesioptyxis olisiponensis (SHARPE) | | ── | | |
| Trochactaeon ellipsoides (FITTIPALDI) | | | ── | |
| Trochactaeon obtusus (ZEKELI) | | | ── | |
| **ALGAE and FORAMINIFERS** | | | | |
| Atopochara trivolvis PECK | ── | | | |
| Flabellochara harrisi (PECK) | ── | | | |
| Ethelia alba PFENDER | ──── | | | |
| Pseudedomia vialii (COLALONGO) | | ── | | |
| Cisalveolina fallax REICHEL | | ── | | |
| Cuneolina pavonia parva HENSON | | ───── | | |
| Dicyclina schlumbergeri MUNIER-CHALMAS | | ─────────── | | |
| Accordiella conica FARINACCI | | | ──── | |
| Globotruncana stephani GANDOLFI | | ───── | | |
| Globotruncana alpina BOLLI | | ───── | | |
| Globotruncana tricarinata QUEREAU | | | | ── |
| Globotruncana lapparenti PROTZEN | | | | ── |

Major Facies.—Five major facies dominate these environments, they are:

1. Foraminiferal mudstones-wackestones containing associations of *Cuneolina pavonia parva* Henson, *Thaumatoporella parvovesiculifera* (Raineri), miliolids, ostracodes and green algae. These microfossils are typical of sheltered lagoonal occurrences, where they occur locally in great abundance. Protected environments are occasionally established in sheltered zones of the ridge, with consequent development of specialized monotypical faunas (Carbone *et al.*, 1971; Fig. 3A).

2. Bioclastic wackestones-packstones where the fossil community is derived from the interaction between indigenous lagoonal populations and bioclastic material transported from a sea floor colonized by typical rudistid reef populations (radiolitids and caprinids). Occasionally nerineids settle on this mobile substrate in a location lagoonward from the reef.

3. Caprinid (rudistid) banks are mainly sandy bodies containing specimens of *Caprina schiosensis* Boehm and *Chondrodonta munsoni* (Hill). This type of facies reflects a high energy environment with the existence of a shifting substrate occasionally colonized by these mollusc species.

4. Caprinid and radiolitid rudistid rudstones (Fig. 3C) consist of bioclastic accumulations of coarse skeletal material neither worn nor sorted. The sparse sandy-muddy matrix observed seems to be secondary cavity filling in many cases (Fig. 5B). Besides abundant fragmentary biogenic material, one can find in these accumulations a considerable number of unbroken or slightly fragmented molluscan specimens belonging to the following species (Fig. 6): *Chondrodonta joannae* (Choffat), *C. munsoni* (Hill), *Neithea aequicostata* (Lamarck), *N. incostans* (Sharpe), *Caprina carinata* (Boehm), *C. schiosensis* Boehm, *Neocaprina gigantea* Pleničar, *N. nanosi* Pleničar, *Sphaerucaprina forojuliensis* Boehm, *Schiosia carinatoformis* Polšak, *S. schiosensis* Boehm, *Orthoptychus striatus* Futterer, *Ichthyosarcolites bicarinatus* (Gemmellaro), *I. poljaki* Polšak, *I. rotundus* Polšak, *I. tricarinatus* Parona, *Sauvagesia nicaisei* (Coquand), *S. sharpei* (Bayle), *Cossmannea annulata* (Gemmellaro), *C. edoardi* (Parona), *Plesioplocus grandis* Pchelincev, *Plesioptygmatus nobilis* (Munster), *Polyptyxis requieni* (D'Orbigny), *P. schiosensis* (Pirona), *Plesioptyxis olisiponensis* (Sharpe), *Trochactaeon obtusus* (Zekeli). In association with these molluscs are species of branching corals, red algae

(*Ethelia alba* Pfender, *Pycnoporidium* sp., and *Marinella* sp.), and orbitolinids of the IV–V group of Hofker. This facies accumulated below wave base in zones continguous to bottoms colonized in clear water by sessile framebuilding organisms which may or may not be dominant.

5. Winnowed bioclastic grainstones-rudstones (Fig. 3D), which are well-washed and fairly-sorted skeletal sands, formed in an environment of constant wave action, above wave base, and forming worn and abraded coquinas of benthic reef organisms. Reverse graded structures and beach rock cementation associated with frequently emerged sandy shoals testify to a beach environment.

### Turonian Time

*Open Platform Environments.*—The beginning of the Turonian sequence is characterized in the entire region by a predominance of lagoonal open platform facies, composed of mudstones and wackestones with benthic foraminifers, and in some instances, a local increase of nerineids. High energy facies in this environment are represented by abraded and well-rounded bioclastic sands, and organogenic banks consisting of practically monotypical populations of radiolitid rudistids. These sediments take the form of high energy inner-shoals, and beach and tidal bars. Indicative of changed environmental conditions, especially in the platform areas, is the presence of wackestones with much echinodermal debris and scarce globotruncanid remains. Increased water energy is demonstrated by well-washed skeletal sands, and the appearance in the platform faunal community of pelagic species suggesting lagoonal environments with more direct contact to the open sea.

*Major Facies.*—Three major facies characterize these environments, they are:

1. Foraminiferal mudstones and wackestones occurring in sheltered lagoonal communities and characterized by miliolids, cuneolinids, *Dicyclina schlumbergeri* Munier-Chalmas, and green algae. Some wackestones contain whole nerineid shells (*Plesioptygmatis bassani* (Fittipaldi), or *Acteonella* shells (Fig. 4A).

2. Radiolitid rudistid banks where these organisms are preserved in living position, and in some cases even retain the opercular valve. This is an open marine shelf-lagoon environment containing the following species: *Radiolites lusitanicus* (Bayle), *R. praesauvagesi*, *R. trigeri* (Coquand), and *Sauvagesia contorta* (Catullo).

3. Echinodermal wackestones where echino-

---

←

FIG. 6.—Stratigraphic distribution of various distinctive organisms found in the Upper Cretaceous rocks of westcentral Latium, central Italy.

derms are dominant and globotruncanids rare (*Globotruncana stefani* Gandolfi, and *G. alpina* Bolli). This assemblage characterizes shelf-lagoons floored below wave base and in direct contact with the open sea.

*Platform Edge Environments.*—These are widespread bioclastic accumulations and sandy beach facies. They interfinger in several ways, testifying to the existence of both mobile and stable shoals colonized by rudistids, corals, and red algae exposed to high wave energy.

*Major Facies.*—Two major facies characterize these environments, they are:

1. Winnowed bioclastic grainstones-rudstones where the sediments consist of abraded and rounded coquinal material and are strongly sparized and circumscribed by micrite rims. They may be formed *in situ* along barriers and sandbanks or they may be shifted by wave and current motion (Fig. 5E). Sorting and grain-size distribution of these sediments varies as a function of water energy and depth (Fig. 5D). These sediments may constitute the substrate where specialized communities such as acteonids (*Trochactaeon ellipsoides* Fittipaldi) develop.

2. Hippuritid and radiolitid rudistid rudstones containing poorly sorted bioclastic deposits with specimens of *Vaccinites narentanus* Slišković, *Distefanella bassani* Parona, and *D. salmojraghi* Parona, either unbroken or as coarse fragments. For the first time variable percentages of sand and mud appear in the rock-texture. This depositional environment is related to *in situ* organic growth, subject to destruction.

### Senonian Time

*Restricted Platform Environments.*—These environments occur at several levels within the Upper Cretaceous sequence, and are represented by calcareous or dolomitic muddy facies containing a very limited fauna. Important episodes are located near the end of the stratigraphic sequence, and these indicate the persistence of restricted lagoon-tidal marsh environments in areas protected by sandy barrier-islands.

*Major Facies.*—Two major facies characterize these environments, they are:

1. Bioturbated dolomitic and lime mudstones lacking a significative fauna in which only charophyte oogonia occur in abundance locally. This facies indicates fine sedimentation in very shallow lagoons and ponds which have restricted water circulation and hypo-hypersaline conditions.

2. Open platform environments.—These environments are most widespread in the area and are represented by wackestones with benthic foraminifers and green algae, intercalated with sandy intervals rich in peloids and lithoclasts. These are related to increase in water energy near sand-

banks and lagoonal beaches, in areas where such conditions favor the settling of hippuritid rudistid communities.

*Major Facies.*—Two major facies have been delineated, these are:

1. Peloidal wackestones and packstones in which the fossil association is represented by *Accordiella conica* Farinacci, *Dicyclina schlumbergeri* Munier-Chalmas, and *Cuneolina pavonia parva* Henson suggesting an open marine shelf-lagoonal environment.

2. Hippuritid rudistid banks where these deposits have packstone texture and contain peloids, intraclasts, and benthic foraminifers, and do not differ appreciably from the preceding facies. The hippuritid rudistid population is chiefly represented by *Vaccinites boehmi* Douville. The resultant rudistids colonized mobile sea floors and is indicative of shelf-lagoonal environments with clear and well-oxygenated water conditions.

*Platform-Edge Environment.*—These environments are represented by sandy shoal facies exposed to wave action, and by facies of rapid bioclastic accumulation, usually situated below wave base. The fossil community appears to be particularly rich in corals (massive as well as branching), and rudistids (radiolitids, hippuritids, and caprinids). The last two groups make up the main body of small bioherms, and it is interesting to note that with both these forms living position is horizontal not vertical. It is probable that this growth position favored colonization on a sandy sea floor. The organic buildups indicate relatively quiet water environments sheltered from wave action. In certain cases, wave and current action caused the sediments to be distributed in the form of barrier-islands, that controlled facies patterns along and behind the platform-edge, as is the case in the Lepini Mountains. Here we find sandy Maestrichtian sediments containing a fauna of macroforaminifers, red algae, and echinoderm remains.

*Major Facies.*—Three major facies have been delineated, these are:

1. Winnowed bioclastic grainstones and rudstones that are well-rounded and well-sorted sands and gravels; graded-bedding and cross-bedding structures can also be recognized. Few indigenous organisms occur in these sediments because of the shifting substrate.

2. Organic buildups of corals, radiolitids, hippuritids and caprinid rudistids constructing bioherms of different textural characteristics. In some cases this association can colonize more or less stabilized sandy bottoms (Fig. 4B–C), in other instances they produce patch-reefs in association with corals and red algae. The following species have been identified from this facies: *Vaccinites boehmi* Douvillé (Santonian-lower Cam-

panian), *V. sulcatus* Defrance (Santonian-lower Campanian), *Hippurites colliciatus* Woodward (Campanian), *Sauvagesia raricostata* Polšak (Santonian-lower Campanian), and *Dictyoptychus* sp.

3. Orbitoid foraminiferal packstones-grainstones composed of muddy sands with a rich benthic microfauna of macroforaminifers (*Orbitoides media* D'Arc., *O. apiculatus* Schul, *Fallotia* sp., and *Loftusia* sp.), in association with echinoderm remains, red algae (*Solenomeris*), and *Stomiosphaera*, along with rare globotruncanids. These skeletal sands display remarkable variations in composition, texture and fauna, and are generally referrable to platform-edge environments, and offshore shoals and back-shoal sub-environments.

*Slope Environments.*—These environments are represented by microbioclastic lime material mixed with mud, and intercalated with coarser debris from up-slope. The fossil community consists of indigenous organisms such as echinoderms, stromatoporoids, and planktic foraminifers.

*Major Facies.*—Two major facies have been delineated, they are:

1. Very fine grainstones-packstones with echinoderms and stromatoporoids (Fig. 5G). These are microbioclastic sediments rich in echinodermal remains, sponge spicules, *Stomiosphaera,* and locally, stromatoporoids (*Actinostromaria* sp.) (Fig. 4D). With increase of the muddy fraction the sediments become pelagic wackestones in which globotruncanids appear (*Globotruncana tricarinata* Quereau (Fig. 5H), and *G. lapparenti* Brotzen of the Campanian-Maestrichtian). This facies, located seaward, below wave base, marks the transition from the outer shelf-edge to the toe of the slope.

2. Floatstone sediments consisting of organic debris of gravel size often lithified, but loose in a fine matrix similar to the preceding facies. This material was deposited seaward on the slope below wave base.

### PALEOECOLOGICAL FEATURES

As described previously the complexity of facies distribution increases gradually with destruction of the carbonate platform, and with differentiation in tectonic behaviour of the several disjointed areas.

During the Cretaceous, the first indications of change from the typical Jurassic-Cretaceous sedimentary model arose during Aptian-Albian time with a general supplanting of the restricted platform facies by a more open platform facies.

In the studied area, Aptian outcrops are characterized by predominantly muddy sheltered to restricted shelf-lagoon facies. The sediments are marked by faunal associations of small benthic

FIG. 7.—Outcrop photograph showing a typical Turonian facies sequence representing deposition on an open marine platform (Carpineto). Base of section contains clusters of abundant radiolitid rudistids attributed to a rudistid bank, overlain by coarse, winnowed grainstone containing large whole nerineids (*Plesioptygmatis nobilis*). The upper portion of the section is a shelf-lagoonal wackestone containing *Dicyclina schlumbergeri.*

foraminifers (miliolids and cuneolinids), ostracodes and requienid rudistids. On account of ecological conditions reflecting increase or decrease of salinity, there are local organism transitions into highly specialized associations. These generally consist of ostracodes, sometimes accompanied by charophyte oogonia, Which suggest restricted shelf-lagoonal environments. During Cenomanian time lateral facies variations occur showing an increase to open-platform environments over the entire area. Sands and muddy sands, rich in orbitolinids, miliolids, dasyclad algae, and alveolinids can be recognized. Rudistid buildups appear in a few outcrops, indicating the existence of a platform-edge to the west of the Lepini and Prenestini Mountains. These are in-

FIG. 8.—Outcrop photograph showing unsorted coarse bioclastic debris rich in branched and massive corals, rudistids (hippuritids, radiolitids, and caprinids), and red algae, all of which characterize the Upper Cretaceous reef complex at Rocco di Cave.

variably associated with rounded and well-sorted sands and gravels typical of winnowed substrates.

The lower Turonian (with nerineids and rudistids, Fig. 7) records abrupt changes in the paleogeography of the region. Cessation of subsidence and differential tectonic uplift led to emergence of vast sectors of the platform, with the shelf-edge facies prograding towards the bordering basin. From upper Turonian to Maestrichtian time there is a progressive development of platform-edge to fore-slope facies overriding platform deposits. These Upper Cretaceous marginal facies are distributed over the entire area as relatively small

outcrops containing a more or less complete series (from Santonian to Campanian), as at Rocca di Cave. These deposits sometimes contain thin high water energy episodes interfingered with platform deposits, e.g., Serrone, and at Scalambra Mountain. The Upper Cretaceous succession at Rocca di Cave is in fact clearly transgressive and is represented from Santonian to Campanian time by assemblages of hippuritids and radiolitid rudistids, and corals (Fig. 8). The evolution of the sedimentary environment is shown by the superposition of platform-edge to fore-slope facies prograding over open platform-edge facies. There is

FIG. 9.—Outcrop photograph showing a Senonian hippuritid rudistid bank (*Vaccinites boehmi*) overlain by well-washed and rounded sandy beach deposits, near Rocca di Cave.

a faunal transition from associations characterized by *Vaccinites boehmi* Douvillé, *Hippurites colliciatus* Woodward, *Clausastrea* sp., *Actinastrea* sp. and red algae (*in situ* deposits as well as bioclastic accumulations (Figs. 9–10)) into associations of echinoderms and stromatoporoids of the slope, and planktic microfossils such as globotruncanids (*Globotruncana tricarinata*, and *G.*

FIG. 10.—Close-up photograph of Figure 9 showing hippuritid rudistid bank deposit with rudistids in growth position (parallel to bedding), indicating horizontal growth position.

FIG. 11.—Outcrop photograph of nerineid-ostreid-radiolitid limestone association indicating a typical Cenomanian transitional environment connecting shelf-lagoonal deposits to shelf-edge facies.

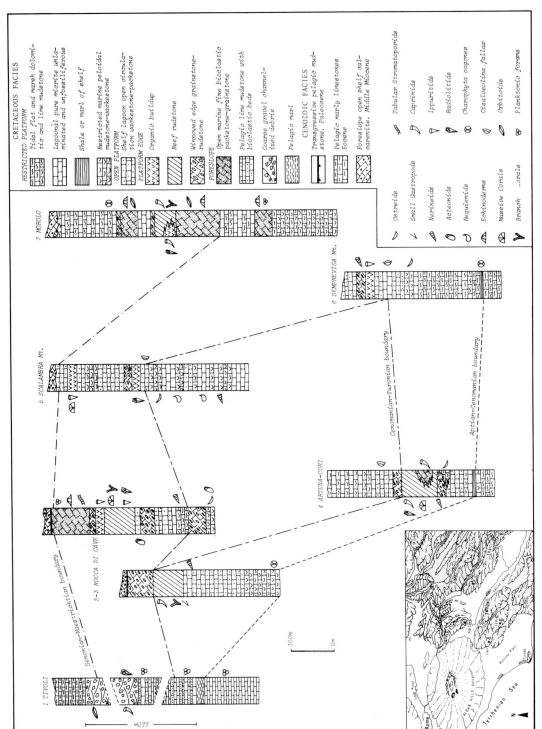

Fig. 12.—Diagram showing stratigraphic relationships of various key Upper Cretaceous sections in westcentral Latium, central Italy. Correlations are based on facies interfingering and distinctive stratigraphic markers.

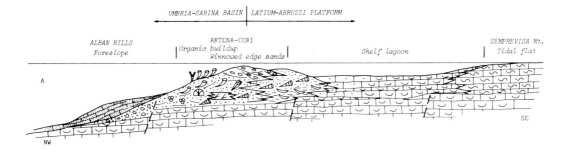

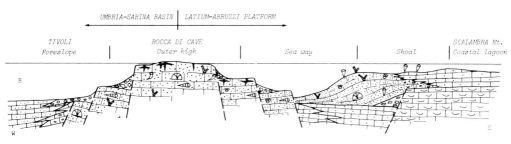

FIG. 13.—Schematic cross-sections showing the regional evolution of shelf-to-basinal areas. A. Distribution of facies in the Lepini Mountains during Cenomanian time. B. Distribution of facies and tectonics at Rocca di Cave, Prenestini Mountains, during Senonian time.

*lapparenti*), and *Stomiosphaera*. More to the north there is a change towards pure pelagic environments, yet still retaining indications of proximity to a productive biogenic edge (Tivoli). The eastward marginal facies are subordinate to the development of muddy or sandy open-platform facies, with faunal associations typical of shelf-lagoons. Marginal facies of uppermost Cretaceous age (Campanian-Maestrichtian) are represented by caprinid rudistids, orbitoids and red algae; a community assemblage belonging to a barrier-island complex that formed during this age along the northeastern border of the Lepini Mountains. Upper Cretaceous shelf-lagoon facies of open circulation can be found widespread, especially in the Simbruini Mountains, whereas in the Lepini Mountains Maestrichtian deposits are exclusively represented by restricted lagoon-tidal marsh facies, closely related to the existence of barrier-islands.

The Cretaceous platform-edge sequence of Rocca di Cave is transgressively overlain by sporadic outcrops of Paleocene mudstone with planktic foraminifers (*Globigerina eugubina*, and *Glo-*

*borotalia pseudobulloides*); locally, very thin veins of Maestrichtian mudstone with *Globotruncana tricarinata* can be found.

Through comparative analyses of texture, fossils, and the content of the sediments, it was possible to establish the existence of numerous sub-environments, gradually connecting the platform-edge facies to the isochronous inner-platform facies. This situation is emphasized by the presence of common faunal elements in different facies. Generally, species characteristic of a certain facies occur at the same time as accessories in other facies. Thus the transition from high water energy environments into quiet platform environments is accomplished by way of gradual substitution. Significantly, in the Cenomanian, there are connecting subenvironments between the platform-edge facies characterized by radiolitid-caprinid communities, and shelf-lagoonal facies with nerineid sands and ostreid (*Perna*?) lime muds. In fact, associations of nerineids and ostreids, nerineids and radiolitids, and ostreids and radiolitids, were observed (Fig. 11), in a few cases only. Finding different species in association does

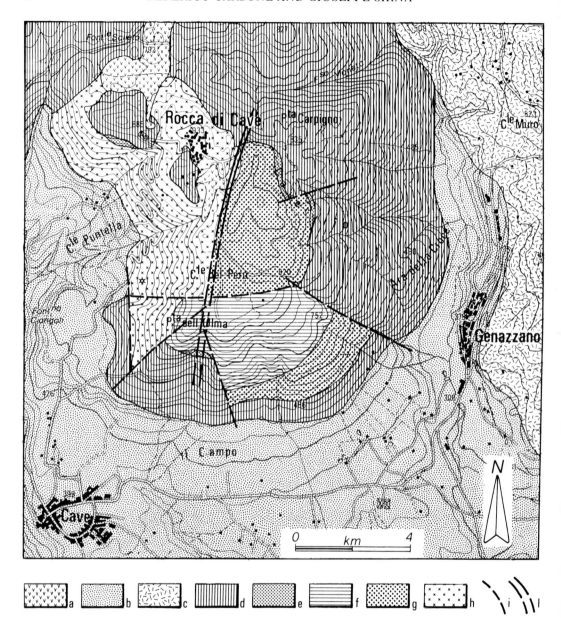

FIG. 14.—Geologic map of the Upper Cretaceous reef complex of Rocca di Cave. A. Unconsolidated red soils (Quaternary). B. Pyroclastic deposits (Quaternary). C. Terrigenous clastic flysch deposits (Tortonian to Messinian). D. Transgressive carbonate-ramp calcarenite (Aquitanian to Helvetian). E. Pelagic lime mudstone (Paleocene). F. Fore-slope microbioclastic wackestone-packstone and lithoclastic floatstone rich in echinodermal debris, stromatoporoids, and planktic forams (Campanian to Maestrichtian). G. Platform-edge with coral and rudistid buildups and winnowed shoal grainstones interfingering with back-reef and off-shore muddy deposits (Turonian to Santonian). H. Open platform skeletal packstone-grainstone, and reef rudstone with caprinids, radiolitids, corals, and red algae, interfingering with restricted mudstone-wackestone (Aptian to Cenomanian). I. Probable paleotectonic line of differential movement during early Turonian time. J. Trace of normal fault.

not necessarily indicate a transitional facies between two environments, but may well mark the invasion of one species into the ecological realm of another species. This is the case in the Cenomanian where small radiolitid rudistids seem to gradually take over the ecological niche of requienid rudistids, and in the end completely supplanting them.

A tentative reconstruction of several evolutionary phases of Cretaceous paleogeography along the northwestern margin of the Latium-Abruzzi Carbonate Platform is based upon analyses of sedimentary environments and fossil communities (Fig. 12). The reconstruction of facies development has been complicated by compressive tectonics, modifying spatial relations between several carbonate units and concealing much of the sedimentary package that constitutes transitional belts between platform and basinal facies. The obliteration of relevant segments of the carbonate platform does seriously influence the paleogeographical reconstruction of the Lower Cretaceous, since the facies identifying the paleogeography during this time span are uniform and widespread. However a more meaningful reconstruction of Upper Cretaceous paleogeography is even more complicated, since compressive tectonics disarranged the continuity of the environments and altered facies relationships.

Several evolutionary stages will be discussed in order to attain a more accurate description of the paleogeographical model developed herein.

### Aptian-Albian Time

Representative facies of this interval belong to a widespread and monotonous paleogeography typical of epioceanic platforms of southcentral Italy. In the studied area, these represent the restricted platform facies which make up the lower interval of the carbonate structure. In the more southerly areas (Lepini Mountains) the dominant sedimentary sequences are dolomitic-loferitic lithotypes of the tidal-flat, grading northward into more markedly lagoonal environments. This situation might indicate a slow transition (towards the north and east) to environments with more open water circulation.

### Cenomanian Time

This time span marks the appearance of high energy facies connected to the open sea and offering considerable paleogeographical change (Fig. 13A). Organic buildups and sandy shoals, located in the southern section of the Prenestini Mountains, delimit the western boundary of the pelagic realm against the Cenomanian platform. The presence of this continuous marginal-platform determined the development of several platform facies to the east of it. Here, the transition to lagoonal environments (Mt. Scalambra), and tidal-flat environments (Mt. Semprevisa), is represented by a series of facies typical of the open platform as either sandy shoals (nerineid facies), or sheltered lagoons (facies with ostreids). Pelagic limestones, in outcrop or in the substratum of the volcanic complex of the Alban Hills (Funiciello

and Parotto, 1978), confirm the trend of the platform-edge.

### Turonian Time

The Cenomanian transgression, indicated by progradation of the edge facies containing rudistids over the lagoonal facies, is followed in Turonian time by: 1. a shifting westwards of the old Cenomanian margin, with the lagoonal facies overlying the shelf-edge facies; and 2. contemporaneous rupture of the old shelf-margin, with invasion of the platform realm by clear water and deposition of higher energy facies. The rupture of the old Cenomanian shelf-margin is linked to a tectonic phase that led to uplift of parts of the platform (Rocca di Cave) (Fig. 14) and to the formation of sea-ways that penetrated onto the carbonate platform (Carpineto Valley to the south, Olevano Valley to the north).

As noted previously, the Turonian is often characterized by high energy facies on large areas of the carbonate platform, and occasionally with the formation of organic buildups. This is the case at Rocca di Cave, where a hydrozoan and rudistid association indicates a shelf-edge environment, with transgression from east to west. This suggests the presence of another ridge to the east of the emergent structure of Rocca di Cave. This event is confirmed by the presence of facies with pelagic foraminifers and echinoderms at Mt. Scalambra. No traces exist of the old western margin, which probably moved more westwards and was later buried under a Pliocene-Quaternary cover.

### Senonian Time

To a great extent this time span is represented by micritic facies of open or restricted shelf-lagoons. Locally, along the periphery of the structures a typical shelf-edge succession appears (Fig. 13b). At Rocca di Cave this interval is characterized by a marked increase in the coral fauna. Sedimentation is similar to that of the beginning of upper Turonian time, and tends in this area towards a slope facies characterized by mudstones and floatstones rich in pelagic and slope organisms. This reveals the transgressive character of the sedimentary succession. Towards the end of the Senonian important bathymetric variations become typical, with a branch of the sea penetrating into the area of the Latina Valley permitting formation of a sandy barrier-island complex behind an emerged platform area (Morolo-Segni). At this time, zones like Rocca di Cave were cut-off from the platform area and surrounded by pelagic environments, and represented zones of reduced sedimentation, similar to Recent sea mounts (Fig. 14). This is confirmed by thin intervals of Maestrichtian and Paleocene muds randomly occurring on these structures.

ACKNOWLEDGMENTS

This work has been carried out with a grant of C.N.R. (Paleobenthos), and the Faculty of Science at the University of Rome (Interdisciplinary Project GEOLAZIO). We are particularly indebted to Professor C. Caputo for help with the manuscript, to the technicians of the Cartographic Section (Dr. G. Bigi, Mr. M. Albano, and Mr. M. Salvati), and to the technicians of the rock preparation laboratory (A. d'Artino, C. d'Artino, and A. Mancini).

## REFERENCES

ALBERTI, A., 1952, Osservazioni sulla zona di transizione dalla facies umbromarchigiana alla facies abruzzese nei Monti Tiburtini, Prenestini e Lepini (Lazio): Boll. Serv. Geol. It., v. 74, p. 181–190.

BALDONERO CARRASCO, V., 1977, Albian sedimentation of submarine autochthonous and allochthous carbonates, east edge of the Valles San Luis Potosi Platform, Mexico: Soc. Econ. Paleontologist Mineralogists Spec. Pub., No. 25, p. 263–272.

BATHURST, R. G. C., 1971, Carbonate sediments and their diagenesis: Developments in Sedimentology, v. 12, 620 p.

BEIN, A., 1976, Rudistid fringing-reefs of Cretaceous shallow carbonate platform of Israel: Am. Assoc. Petroleum Geologists Bull., v. 60, no. 2, p. 258–272.

BONI, C. F., 1967, La geologia dei Monti Tiburtini (Lazio): Geologica Romana, v. 6, p. 165–188.

CARBONE, F., AND CATENACCI, V., 1978, Facies analysis and relationships in Upper Cretaceous carbonate beach sequences (Lepini Mts., Latium): Geologica Romana, v. 17, p. 191–231.

———, ———, AND GIARDINI, G., 1978, I carbonati di mare sottile con quarzo delle sequenze cretaciche dei Monti Lepini (Lazio): considerazioni genetiche: Soc. It. Miner. e Petrog., p. 115–127.

———, PRATURLON, A., AND SIRNA, G., 1971, The Cenomanian shelf-edge facies of Rocca di Cave (Prenestini Mts., Latium): Geologica Romana, v. 9, p. 131–198.

CONRAD, M. A., 1969, Les calcaires urgoniens dans la région entourant Genève: Eclogae Geol. Helvetiae, v. 62, no. 1, 79 p.

COOGAN, A. H., 1972, Recent and ancient carbonate cyclic sequence, in Elam, J. C., and Chuber, S., Eds., Cyclic sedimentation in the Permian Basin (2nd ed.): West Texas Geol. Soc. Pub. 69-56, p. 5–16.

———, BEBOUT, D. G., AND MAGGIO, CARLOS, 1972, Depositional environments and geologic history of Golden Lane and Poza Rica Trend, Mexico, an alternative view: Am. Assoc. Petroleum Geologists Bull., v. 56, no. 8, p. 1419–1447.

COOPER, M. R., 1977, Eustacy during the Cretaceous: Its implications and importance: Palaeogeography, Palaeoclimatology, Palaeoecology, v. 22, 60 p.

DICKINSON, K. A., BERRYHILL, H. L., AND HOLMES, C. W., 1972, Criteria for recognizing ancient barrier coastlines, in Rigby, J. K., and Hamblin, W. K., Eds., Recognition of ancient sedimentary environments: Soc. Econ. Paleontologists and Mineralogists Spec. Pub., No. 16, p. 192–214.

DUNHAM, R. J., 1962, Classification of carbonate rocks according to depositional texture, in Ham, W. E., Ed., Classification of carbonate rocks: Am. Assoc. Petroleum Geologists Mem. v. 1, p. 108–121.

ENOS, P., 1977, Tamabra Limestone of the Pozo Rica Trend, Cretaceous, Mexico: Soc. Econ. Paleontologists and Mineralogists Spec. Pub. No. 25, p. 273–314.

FUNICIELLO, R., AND PAROTTO, M., 1978, Il substrato sedimentario nell'area dei Colli Albani: Considerazioni geodinamiche e paleogeografiche sul margine tirrenico dell'Appennino centrale: Geologica Romana, v. 17, p. 233–287.

HART, M. B., AND TARLING, D. H., 1974, Cenomanian palaeogeography of the North Atlantic and possible mid-Cenomanian eustatic movements and their implications: Palaeogeography, Palaeoclimatology, Palaeoecology, v. 15, p. 95–108.

HECKEL, P. H., 1972, Recognition of ancient shallow marine environments, in Rigby, J. K., and Hamblin, W. K., Eds., Recognition of ancient sedimentary environments: Soc. Econ. Paleontologists and Mineralogists Spec. Pub. No. 16, p. 226–286.

HOYT, J. H., 1967, Barrier-island formation: Geol. Soc. America Bull., v. 78, p. 1125–1136.

HUGHES CLARKE, M. V., AND KEIJ, A. J., 1973, Organism as producers of carbonate sediment and indicators of environment in the Southern Persian Gulf, in Purser, B. H., Ed., The Persian Gulf, Holocene carbonate sedimentation and diagenesis in a shallow epicontinental sea: New York, Springer-Verlag, p. 35–56.

KAUFFMAN, E. G., AND SOHL, N. F., 1974, Structure and evolution of Antillean Cretaceous Rudist frameworks: Verhandl. Naturf. Ges. Basel, v. 84, no. 1, p. 399–467.

MONLEAU, CL., AND PHILIP, J., 1972, Reconstitution paléogéographique des formations calcaires à Rudistes du Turonien supérieur de la basse vallée du Rhône à partir d'une étude de microfaciès: Revue de Micropaléontologie, No. 1, p. 45–54.

NEGRETTI, G. C., 1954, Nuovi dati stratigrafici e paleontologici sul versante meridionale del Monte Scalambra (tra Serrone e Piglio, Lazio): Publ. Ist. Geol. Univ. Roma, No. 15, 10 p.

———, 1956, Appunti sulla evoluzione paleogeografica della Valle Latina settentrionale dal Cretaceo superiore al Miocene superiore, con particolare riguardo alla trasgressione miocenica: Pub. Ist. Geol. Univ. Roma, No. 27, 27 p.

Peres, J. M., and Picard, P., 1964, Nouveau manuel de bionomie benthique de la Mer Méditerranée: Recueil des travaux de la Station Marine d'Endoume, v. 31, 137 p.

Philip, J., 1974, Les formations calcaires à Rudistes du Crétacé supérieur provençal et rhodanien: stratigraphie et paléogéographie: Bur. Rec. Geol. of Min., Bull., 2nd ser., no. 3, p. 107–151.

Polšak, A., 1967, Macrofaune crétacée de l'Istrie méridionale (Yougoslavie): Palaeontologia Jugoslavica, v. 8, 219 p.

Praturlon, A., and Sirna, G., 1976, Ulteriori dati sul margine cenomaniano della plattaforma carbonatica laziale-abruzzese: Geologica Romana, v. 15, p. 83–111.

Purser, B. H., 1973, Sedimentation around bathymetric highs in the Southern Persian Gulf, *in* Purser, B. H., Ed., The Persian Gulf, Holocene carbonate sedimentation and diagenesis in a shallow epicontinental sea: New York, Springer-Verlag, p. 157–177.

Reineck, H. E., and Singh, I. B., 1973, Depositional sedimentary environments: New York, Springer-Verlag, 439 p.

Schwartz, M. L., 1973, Barrier-islands: Benchmark Papers in Geology, 451 p.

Servizio Geologico d'Italia, 1966, Carta Geologica d'Italia alla scala 1/100.000, Foglio 159, Frosinone, 2 ed.

———, 1975, Carta Geoloica d'Italia alla scala 1/50.000, Foglio 389, Anagni.

Sirna, G., 1968, The Lower Cretaceous Charophyta and the paleogeography of Mediterranean Basin: Acc. Naz. Lincei, ser. 8, v. 44, no. 4, 7 p.

Slišković, T., 1971, Les nouveaux Rudistes de l'Herzegovine: Wissen. Mitteil. Bosnisch-Herzeg. Landesmuseums, v. 1, 103 p.

Tebbut, G. E., Conley, C. D., and Boyd, D. W., 1965, Lithogenesis of a distinctive carbonate rock fabric, *in* Parker, R. B., Ed., Contributions to Geology: Laramie, Univ. Wyoming, p. 1–13.

Wagner, C. W., and Van der Togt, C., 1973, Holocene sediment types and their distribution in the Southern Persian Gulf, Holocene carbonate sedimentation and diagenesis in a shallow epicontinental sea: New York, Springer-Verlag, p. 123–155.

Wilson, J. L., 1974, Characteristics of carbonate-platform margins: Am. Assoc. Petroleum Geologists Bull., v. 58, no. 5, p. 810–824.

———, 1975, Carbonate facies in geologic history: New York, Springer-Verlag, 471 p.

SEPM SPECIAL PUBLICATION NO. 30, P. 447–472, MAY 1981

# UPPER CRETACEOUS BIOLITHIC COMPLEXES IN A SUBDUCTION ZONE: EXAMPLES FROM THE INNER DINARIDES, YUGOSLAVIA

ANTE POLŠAK
University of Zagreb, Yugoslavia

## ABSTRACT

An Upper Cretaceous biolithic complex exposed at Donje Orešje on Mt. Medvednica, Yugoslavia (northern Croatia), is composed of a central portion designated as a barrier-reef made up of rudistid and coral bioherms. Associated with the barrier-reef are fore-reef and peri-reefal breccias (reef talus) that originated from destruction of the reef-front. This breccia is unsorted, and contains the rubble of reef building organisms and biocalcarenite fragments. The breccia grades into detrital limestones, and more distally, the sediments are basinal hemipelagic and pelagic limestones with conspicuous turbidity features ("*Scaglia*"-beds). The backreef area contains nerineid (gastropod) biostromes and detrital limestone. The lagoon is mainly represented by clastic terrigenous deposits, through some rudist patch-reefs ("*Gosau*"-beds) occur rarely. Sporadically, the lagoon also exhibits fresh water characteristics. The reef-flat was successively contaminated by terrigenous clastic material transported from the lagoon, e.g., material derived from erosion on an island arc. The porous reef-frame acted as a barrier, preventing transport of clastic material toward the open sea, but trapping it in the lagoon. Areas with no barrier-reefs, or the areas where the reef was breached or destroyed, turbidity currents could and did deposit flysch-type sediments into the basin. Numerous fossils (rudists, corals, benthic and pelagic foraminifers, and nannoplankton) indicate that the Donje Orešje biolithic complex is Upper Cretaceous in age (Santonian to lower Campanian).

Barrier-reefs of the Donje Orešje type developed in the Upper Cretaceous and lower Paleogene of the Inner Dinarides on the slopes of island arcs. These island arcs and correlative trenches, as well as the inter-arc basins, were formed in a subduction zone of Tethyan oceanic crust beneath the Panonian, e.g., Rhodope Plate. The subduction zone is characterized by the occurrence of ophiolites and mélanges. The Adriatic Plate (the Outer Dinarides) is marked by uniform shallow water carbonate sedimentation ("Carbonate platform"). During its eastward motion this plate was uplifted at the end of Cretaceous time ("Laramian Orogeny"), and accordingly sedimentation reflects regression and bauxite deposition. However, a continuous depositional sequence between Cretaceous and Paleogene time has been determined for the Inner Dinarides. Apparently, there was not one simple slope between the Outer and Inner Dinarides, with transition from shallow water to basin sediments in the sense of a "classical geosynclinal concept," but shallow water and basinal facies were successively reestablished due to the influence of island arcs.

Distribution of the biolithic complexes and other Upper Cretaceous facies in regular bands demonstrates that since Cretaceous time there has not been any regionally extended nappes. Such overthrustings would certainly have dislocated the original paleogeographical distribution of Upper Cretaceous facies belts. The main Dinarides "orogenic phase" occurred by the end of the Eocene, when intensive northward and northeastward movement of the African Craton caused a maximal compression and junction of the Adriatic and Panonian Plates. As a consequence, tectonic movements on the plates, as well as in the area of the "oceanic suture" of the Inner Dinarides, brought about the uplift of the Dinarides.

## INTRODUCTION

Upper Cretaceous sediments of the Inner Dinarides in Yugoslavia exhibit a great variety of facies. The main facies are: 1. lagoonal clastic deposits (so-called *Gosau*-deposits), 2. biolithic (reef) complexes, 3. pelagic carbonate sediments (so-called "*Scaglia*"-deposits), and flysch deposits.

Recent investigations have shown that these deposits could occur repeatedly in both space and time. The biolithic complexes, although small in absolute dimensions, play a significant role in the sedimentary model. Apparently the geometry of a basin that was created by reef development, could provide relief which controlled facies genesis and differentiation.

## MODEL OF BIOLITHIC COMPLEX IN A SUBDUCTION ZONE

The term biolithic complex is meant to include the biolithic reef ("ecological reef" in the sense of Dunham (1970), and the deposits of the peri-reefal zone where sedimentation is closely controlled by the reef environment. These deposits include peri-reefal breccia (reef talus), and various types of detrital limestone and reef bioclasts. The *biolithic reef* itself is a biolithite (Folk, 1959, 1962), i.e., autochthonous limestone composed of skeletal remains in growth position.

Biolithic complexes of the subduction zone exhibit some features which differ essentially from biolithic complexes that developed along the margins of a continental shelf, e.g., in Texas,

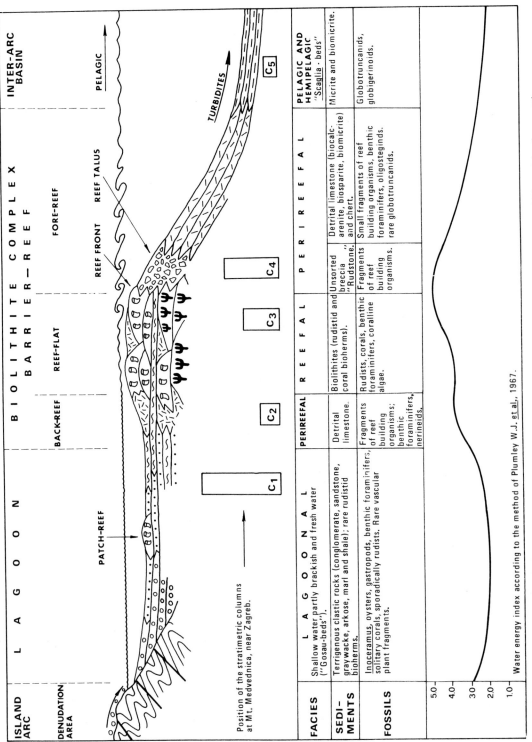

Fig. 1.—Sedimentational model of a Cretaceous biolithitic complex in a subduction zone within an island arc setting.

Perkins (1974), or on "carbonate platform," e.g., in the Outer Dinarides (Polšak, 1965b).

The model of biolithitic complex developed within the subduction zone is shown in Figure 1, and has been established on the basis of detailed investigation of the biolithitic complex at Donje Orešje (Fig. 2) in the area of Mt. Medvednica, northern Croatia (Polšak *et al.*, 1978; Polšak, 1979).

### The Biolithitic Reef

The reef in the subduction zone is a barrier type reef characterized by an elongated shape and separated from land by a lagoon. The reef complex is composed of a series of coral and rudistid bioherms. The coral bioherms in the reef complex at Donje Orešje are composed of colonies of massive and branching hermatypic corals. At some places the coral colonies are so densely spaced so as to form a compact reef structure. However, there are numerous cavities among the coral colonies so that the reef frame is very porous. Epibionts are very scarce and consist of poorly developed coralline algae and bryozoans, and occasionally, single rudistids of the family Hippuritidae occur (stratigraphic unit 7). From stratigraphic units 7, 11, 14, and 16; Fig. 5) the following corals have been identified: *Plocladocora tenuis* (Reuss); *P. simonyi* (Reuss); *Actinastraea ramosa* (Michelin), Fig. 3; *A. octolamellosa* (Michelin), Fig. 4B; *Astraraea media* (Sowerby), Fig. 4A; *Astinacis martiniana* d'Orbigny; *Mycetophylliopsis antiqua* (Reuss); *Columnastraea striata* (Goldfuss), Fig. 4C; *C. formosa* (Goldfuss); *Phyllocoeniopsis pediculata* (Deshayes); *"Dendrosmilia" crassa* (Reuss); *Thamnoseris hoernesi* (Reuss); *Synastraea procera* (Reuss); *Dermosmiliopsis tenuicostata* (Reuss); and *Columactinastraea pygmaea* (Felix).

Many of the above species are also found in Santonian and Campanian deposits of the Gosau area in Austria (Beauvais and Beauvais, 1974).

The porous reef framework of the coral bioherms in the barrier-reef of Donje Orešje, lacking abundant epibionts, appears to have been most susceptible to wave destructive action. This inherent lack of strength of the reef frame is the primary reason that the coral bioherms did not attain larger dimensions. Accordingly, coral colonies migrated from place to place and frequently developed on a substrate of dense, compact reef sand, which had a higher strength load capacity. As a result there is an interfingering of bioherms and bioclastic limestones. Because of coral colony massiveness, the lower part of a bioherm is frequently impacted into the underlying bioclastic substrate. Bioclastic limestone is frequently deposited in and amongst coral bioherms and rep-

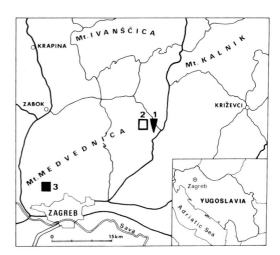

FIG. 2.—Index map showing investigated localities on Mt. Medvednica, near Zagreb. 1. Biolithite complex at Donje Orešje. 2. Basin sediments at Donje Orešje. 3. Lagoonal sediments at Vrapče.

resents the lithified reef sand. This limestone substrate consists predominantly of densely-packed, poorly-sorted bioclasts. Among these bioclasts fragments of corals, bivalves and gastropods are common, whereas fragments of hydrozoans, echinoid spines, coralline algae, crinoids, and benthic foraminifers are rare. Size of the bioclasts ranges from 1 to 3 millimeters. The bioclasts are frequently bored, recrystallized, and are circumscribed by micrite envelopes (Fig. 6A). The cement is frequently calcarenitic, more rarely sparitic (e.g., stratigraphic units 10 and 13). Such limestone appears to have been deposited on the reef-flat with fairly intensive wave action.

Occasionally, wave and current energy was low and resulted in deposition of fossiliferous micrite. The dense micritic matrix contains large fragments of reef building organisms (e.g., stratigraphic units 12 and 15). Similar micrites can also be deposited amongst coral colonies, particularly in the inner, protected parts of the reef-flat, where water energy is usually low. Wave and current energy coming from the open sea was probably neutralized at the barrier-reef front, and concurrently systematically eroded. Hence, only a small fraction of directed wave and current energy from the fore-reef reached the inner area of the reef-flat.

The reef-flat at Donje Orešje was contaminated by clayey and silty material derived from the back-reef. As a rule, only the finest clayey silt ever reached the reef-flat. Coarser clastic fractions were deposited in the lagoon. Here, where water was frequently contaminated by clastic in-

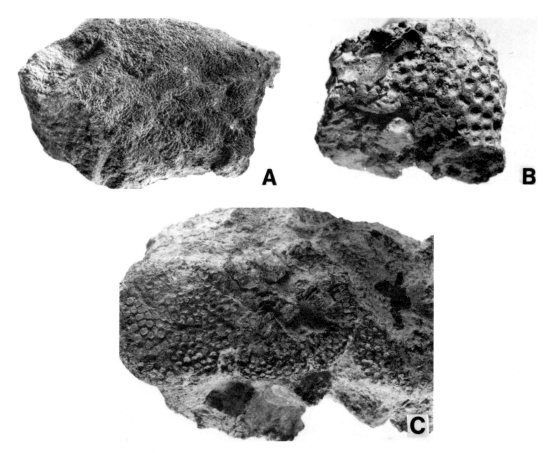

FIG. 4.—Reef-flat hermatypic corals, from Donje Orešje. A. *Astraraea media* (Sowerby), massive colony, ×2, stratigraphic unit O-7. B. *Actinastraea octolamellosa* (Michelin), bulbous, cerioid colony, ×2, stratigraphic unit O-7. C. *Columnastraea straiata* (Goldfuss), massive, cerioid colony, natural size, stratigraphic unit O-7.

flux, was not a particularly suitable environment for the development of the reef biocoenosis. Small-sized and slow-growing organisms were particularly hindered in such a restricted environment, and such organisms were soon overwhelmed by influx of clastic sediments. Those organisms primarily affected were algae, bryozoans, hydrozoans, small bivalves, and various epibionts. Consequently, cavities within the reef structure are most frequently filled with calcarenitic or clayey material.

The primary reef building organisms were fast-growing, and were thus able to escape the relative instability of a muddy substrate by growing up-ward into comparatively clearer water free of mud and its clogging influence on organism life processes. Hence, upward directed growth of organism colonies was strongly emphasized in such "contaminated" reefs, surpassing by far morphological growth by lateral expansion over the muddy bottom. Organism expansion towards the back-reef was hindered by the existence of a lagoon undergoing clastic sedimentation, and alternatively, towards the open sea (fore-reef) where the reef-front was constantly subjected to the destructive activity of waves and currents. The predominance of upward organism growth, and reduced lateral organism expansion, resulted in

←

FIG. 3.—Reef-flat hermatypic corals from Donje Orešje, *Actinastraea ramosa* (Michelin), stratigraphic unit O-7. A. Massive branching colony with each "branch" composed of massive cerioid corallites; natural size. B. Thin-section photomicrograph of a tangential section of a "branch," ×4. C. Thin-section photomicrograph of a transverse section of the same "branch," ×4.

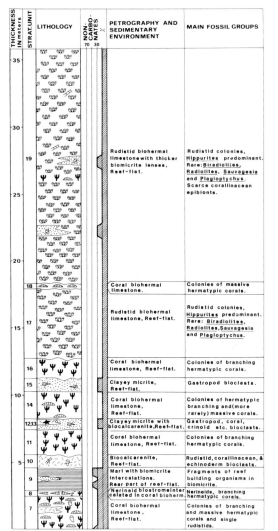

FIG. 5.—Stratigraphic section C₃ of the Donje Orešje biolithitic complex, reef-flat deposits.

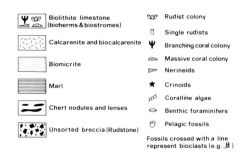

**EXPLANATION**

FOR STRATIGRAPHIC COLUMNS C₂(FIG. 9), C₃(FIG. 5) AND C₄(FIG. 10)
OF THE DONJE OREŠJE BIOLITHITE COMPLEX

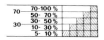

reefs that were narrow and high and with extremely steep slopes. Ergo, they represent classic examples of barrier-reefs.

Only when a particularly strong clastics influx occurred from the lagoon onto the reef-flat was there a periodical cessation in the growth of coral colonies and deposition of silty marl. This is exemplified in stratigraphic unit 9, which is composed of marl consisting of non-diluted clay particles, grading to quartz, muscovite and feldspar, and rare green tourmaline, chromspinel, epidote and zircon. Extensive bioturbation has also been observed in this marl.

The upper part of the reef-core at Donje Orešje is composed of rudistid bioherms (stratigraphic units 17 and 19). Their total thickness is about 25 meters. Rudistid specimens in these bioherms consist mostly of members of the family Hippu-

→

FIG. 6.—Thin-section photomicrographs (×15) of reef-flat deposits. Donje Orešje: A. Biosparite (biocalcarenite) of unsorted bioclasts (fragments of hermatypic corals and rudists predominate) intermingled with entire fossil tests (a *Nerinea* near the bottom and *Idalina* sp. in the upper left corner). Bioclasts and fossils are recrystallized into coarse mosaic calcite, which also fills most of the interstices (cement B). Fossils and fossil fragments are circumscribed with micrite envelopes, fibrous calcite (cement A) is also partly developed. Sample from reef-flat deposited in high water energy; sample O-10-1. B. Biolithite, showing fragment of a colonial coral encrusted by coralline alga *Pseudolithothamnium*. Cavities in coral polyp are filled with micrite. Thin-section made from a limestone fragment of the unsorted peri-reefal breccia ("rudstone"), sample O-25-2. C. Biomicritic rock composed of rather densely-packed bioclasts, in which fragments of rudistids and corals are predominant. Bioclasts contain numerous boring attributed to the sponge *Cliona*. A portion of a recrystallized dasyclad algal thallus is also shown. Thin-section made from a limestone fragment in the unsorted peri-reefal breccia ("rudstone"). This limestone is derived from the reef-flat and represents lithified "reef sand" deposited among the rudistid and coral colonies; sample O-25-2.

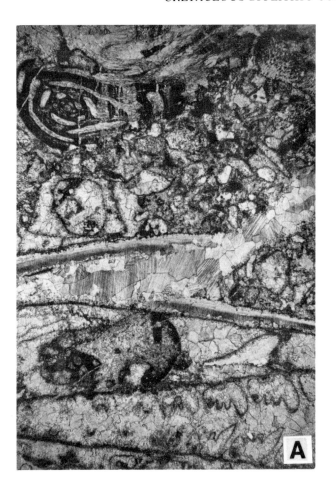

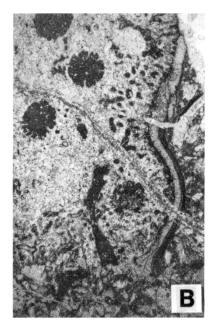

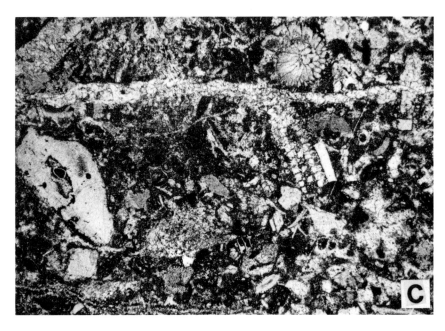

ritidae. The subgenus *Vaccinites* is by far the most predominant in both size and total numbers of specimens.

The following rudistid species have been identified: *Hippurites* (*Vaccinites*) *inaequicostatus* Münster (Fig. 8B and C), *H.* (*V.*) *giganteus* d'Hombres-Firmes (Fig. 8D), *H.* (*V.*) *vredenburgi* Kühn (Fig. 8A), *H.* (*V.*) *oppeli santoniensis* Kühn, *H.* (*V.*) *atheniensis* Ktenas, *H.* (*V.*) *salcatus* Defrance, *H.* (*V.*) *cornuvaccinum* Bronn (Fig. 7A), *H.* (*V.*) *archiaci* Munier-Chalmas, *H.* (*V.*) *carinthiacus* Redlich (Fig. 7B).

Among the above mentioned species *H.* (*V.*) *inaequicostatus* is the most abundant, followed by *H.* (*V.*) *giganteus.*

Representatives of the subgenus *Orbignya* may often develop dense colonies, but these rudistids are nearly always smaller than representatives of the subgenus *Vaccinites.* The following rudistid species were identified: *Hippurites* (*Orbignya*) *nabresinensis* Futterer (Fig. 7G), *H.* (*O.*) *matheroni* Douvillé, *H.* (*O.*) *crassicostatus* Douvillé, *H.* (*O.*) *toucasianus* d'Orbigny, *H.* (*O.*) *carezi* Douvillé, *H.* (*O.*) *praesulcatissimus* Toucas (Fig. 7F), *H.* (*O.*) *sulcatissimus* Douvillé, *H.* (*O.*) *variabilis* Munier-Chalmas (Fig. 7C), *H.* (*O.*) *sulcatoides* Douvillé (Fig. 7E), *H.* (*O.*) *socialis irregularis* Douvillé, *H.* (*O.*) *striatus* Defrancé (Fig. 7I), *H.* (*O.*) *heberti* Munier-Chalmas (Fig. 7H).

Rudistids belonging to the families Radiolitidae and Caprinidae occur only rarely. Only *Radiolites mammillaris* Matheron, *Biradiolites martellii* (Parona), *Sauvagesia meneghiniana* (Pirona) and *Plagioptychus paradoxus* Matheron have been identified. The reason that rudistids of the two above mentioned families occur only rarely is the contamination of sea water on the reef-flat by clayey and silty clastic material, which hampered development of those rudistids possessing a compact smooth upper valve. On the contrary, the perforated upper valve in rudistids of the family Hippuritidae (Fig. 8B) probably served to filter the muddy water.

Of the rudistid species listed above, the greatest

number are known to occur during Santonian-lower Campanian time, while some are restricted to Santonian time and others only to Campanian time (Kühn, 1932; Kaumanns, 1962; Polšak, 1967). This rudistid community belongs to the fifth assemblage zone (Santonian-lower Campanian) as referred to the biostratigraphic subdivisions for the Outer Dinarides given by Polšak (1965b), Polšak and Mamužić (1969); and Slišković (1971).

Occasionally, hermatypic corals occur within the rudistid bioherms and coralline algae may also be present (Fig. 6B). Smaller colonies of massive or branching corals sometimes grew attached to lower rudistid valves. The greatest development of corals took place between the 17th and 19th rudistid bioherm assemblage zones, where a lens-shaped coral bioherms developed. Because of scanty epibionts on the reef-frame, and because the framework was built up mostly of rudistid colonies, it remained empty of co-inhabitants, and accordingly is highly porous.

Cavities among rudistids are less frequently filled with reef sand (calcarenite), instead they are more frequently filled with micrite and clayey material brought in from the back-reef area. Only when growth of the rudistids slowed down did more intense erosion of the bioherms take place. This was accompanied by the deposition of large quantities of reef sand, now preserved as calcarenite lenses. This calcarenite is composed of rather densely-packed bioclasts of reef building organisms. Fragments of gastropods, dasyclad algae (Fig. 6C), and benthic foraminifers are less frequent, whereas spines of irregular echinoids are extremely rare. The sparse matrix is dominantly micrite, which suggests comparatively low water energy in the protected parts of the reef-flat.

### Back-Reef Sediments

In the biolithitic complex of Donje Orešje, back-reef sediments are preserved to a lesser extent. These sediments formed in the distal part of the lagoon, in areas of low water energy, where there was interaction of terrigenous clastic sedi-

$\rightarrow$

FIG. 7.—Photographs of reef-flat rudistids, Donje Orešje. A. *Hippurites* (*Vaccinites*) *cornuvaccinum* Bronn, transverse section of right valve, natural size; stratigraphic unit O-17-19. B. *Hippurites* (*Vaccinites*) *carinthiacus* Redlich, transverse section of right valve viewed from below, natural size; stratigraphic unit O-17-19. C. *Hippurites* (*Orbignya*) *variabilis* Munier-Chalmas, transverse section of right valve, natural size; stratigraphic unit O-17-19. D. *Hippurites* (*Vaccinites*) *oppeli oppeli* Douvillé, transverse section of right valve ×⅔; stratigraphic unit O-26. E. *Hippurites* (*Orbignya*) *sulcatoides* Douvillé, transverse section of right valve, natural size; stratigraphic unit O-17-19. F. *Hippurites* (*Orbignya*) *praesulcatissimus* Toucas, natural size; stratigraphic unit O-17-19: a. transverse section of right valve, b. outside view of right valve. G. *Hippurites* (*Orbignya*) *nabresinensis* Fütterer, transverse section of right valves, natural size; stratigraphic unit O-26. H. *Hippurites* (*Orbignya*) *heberti* Munier-Chalmas, transverse section of right valve, natural size; stratigraphic unit O-19-19. I. *Hippurites* (*Orbignya*) *striatus* Defrance, transverse section of right valves, natural size; stratigraphic unit O-17-19.

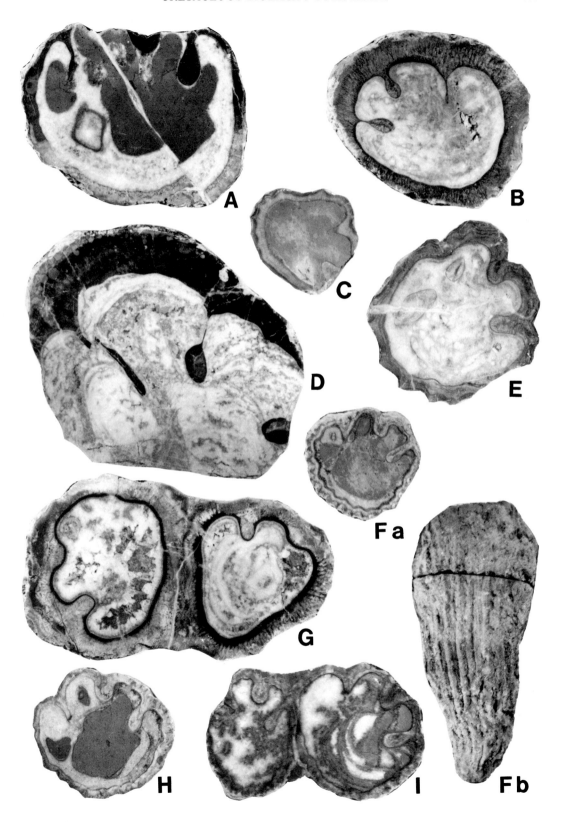

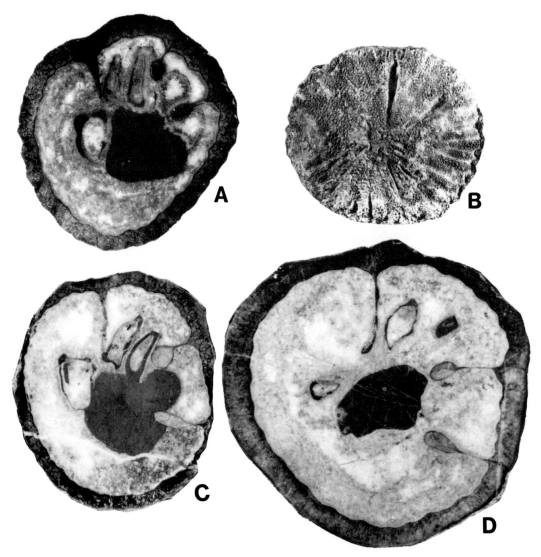

FIG. 8.—Photographs of reef-flat rudistids, Donje Orešje. A. *Hippurites (Vaccinites) vredenburgi* Kühn; transverse section of right valve, natural size, stratigraphic unit O-17-19. B. *Hippurites (Vaccinites) inaequicostatus* Münster; view of left valve, ×⅔; stratigraphic unit O-17-19. C. *Hippurites (Vaccinites) inaequicostatus* Münster; transverse section of right valve, natural size, stratigraphic unit O-17-19. D. *Hippurites (Vaccinites) giganteus* d'Hombre-Firmas; transverse section of right valve, natural size, stratigraphic unit O-17-19.

ments and bioclastic material derived from the reefs (Fig. 9). The oldest sediments are calcarenites with poorly preserved benthic foraminifers. A part of the sediment is bituminous, which suggests a protected and partly reducing environment. Overlying these sediments are 2–3 meters thick biostromes composed of closely-spaced nerineid gastropod shells, bound by sparse calcarenitic cement. The species identified are: *Nerinea (Simploptyxis) nobilis* (Münster) and *N. (S.) buchi* (Keferstein), both forms known to occur

during the Coniacian-Santonian time span, and *N. (S.) ampla* (Münster) characteristic of Santonian time (Tiedt, 1958). It can be concluded that this biostrome is of Santonian age. The shells are complete and well preserved which indicates an environment with comparatively low water energy. Only rarely are small colonies of massive corals present in these sediments.

The nerineid biostromal interval is overlain by 3–4 meters thick deposits of fine-grained marl in which coral bioherms occur as intercalated small

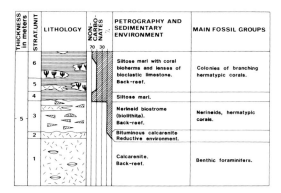

FIG. 9.—Stratigraphic column C$_2$ of the Donje Orešje biolithitic complex, showing development of back-reef deposits.

lenses. The following coral species have been identified: *Neocoeniopsis lepida* (Reuss), *Elasmophyllia deformis* (Reuss) and *Pelurocora haueri* Milne-Edwards and Haine. Heavy mineral assemblages derived from the marl points to the transport of clastic material from an island source region composed of basic and ultrabasic rocks and serpentines.

### Fore-Reef Sediments

The fore-reef area includes a relatively broad marine zone, which begins immediately at the reef-front and extends towards the open sea. This zone is characterized by sediments which show a strong reefal influence, e.g., which contain a great amount of bioclastic reef derived detritus.

*Peri-Reefal Breccia (Reef Talus).*—Deposited immediately along the outer reef-slope are unsorted, coarse peri-reefal breccia (reef talus). In Recent reef-slopes, especially in the upper portion, inclined dips of 30–60 degrees occur, and at depths of more than 50 meters the slopes become almost vertical (Maxwell, 1968; Goreau and Land, 1974; Purdy, 1974). Peri-reefal breccia was formed by intensive destruction of the reef-front by the continuous action of open sea waves and currents. The breccia consists of the rubble of reef building organisms (corals and rudistids), along with coarse fragments of calcarenite that were originally deposited on the reef-flat. The coarse fragments are in excess of 10 centimeters in diameter. These large fragments are mixed with smaller ones forming a chaotic mass. The sparse matrix consists of calcarenitic material mixed with a large percentage of black clayey-silty material. Such coarse-grained and unsorted clastic sediment has been called "rudstone" by Embry and Klovan (1971), in their addition to Dunham's (1962) carbonate classification. This breccia is included in stratigraphic units 20 to 25 (Fig. 10).

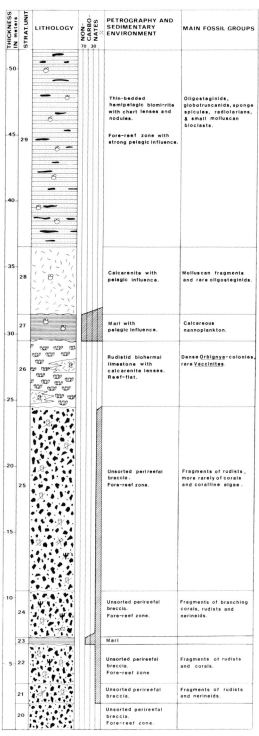

FIG. 10.—Stratigraphic column C$_4$ of the Donje Orešje biolithitic complex; fore-reef deposits.

FIG. 11.—Thin-section photomicrographs of fore-reef deposits; typical detrital limestones from Donje Orešje. A. Biomicrite (biocalcarenite) with abundant poorly-sorted intraclasts. Fragments with prismatic structure are fragments of rudistids of the family Radiolitidae, and those with "compact" prismatic structure belong to the family Hippuritidae. Interstices are filled with micritic matrix containing rare oligosteginids; sample O-30-12, ×15. B. Poorly-washed biosparite (biocalcarenite); this rock was deposited in a high energy environment and consists of densely-packed and partly-sorted bioclasts derived from the reef. In the middle of the photomicrograph there is a fine-grained calcarenite portion with oligosteginids. This portion of the sediment corresponds to a phase of quiet sedimentation and lesser influx of bioclastic material from the reef; sample O-33-1, ×20.

Cretaceous rudistid reefs, due to negligible occurrence of epibionts coupled with a highly porous loose framework, were subjected to rapid and intense wave and current destruction giving rise to thick deposits of reef talus. Usually, abrupt disintegration of the reef-core occurred soon after interruption of reef "growth," hence biologic growth potential also ceased at that time. Thus failing to protect the reef from the destructive activity of wave and current action. At this point waves and currents were able to erode the morphological organic body of the reef. This is thought to be the explanation of why a great portion of the Cretaceous reefs in the Inner Dinaric region have been destroyed and "transformed" into reef talus deposits and detrital limestone.

$\rightarrow$

FIG. 12.—Thin-section photomicrographs of fore-reef deposits; detrital limestones from Donje Orešje. A. The lower part of the thin-section is a hemipelagic biomicrite with oligosteginids, rare bioclasts, globotruncanids, and sponge spicules. In the upper part, fine-grained biocalcarenite deposits reflect a turbidity current deposit with bioclastic material derived from the reef; sample O-30-2, ×20. B. Hemipelagic biomicrite with *Globotruncana globigerinoides* Brotzen, and recrystallized radiolarians; sample O-30-⅔, ×70. C. Hemipelagic biomicrite with *Globotruncana lapparenti tricarinata* (Quereau), oligosteginids, radiolarians, and very small bioclasts derived from the reef; sample O-30-8, ×70. D. Mud balls formed by the inclusion of mud during transportation of fine-grained bioclastic material from the reef. Note concentrically arranged bioclastic material accumulated during rolling; sample O-30-2, ×20.

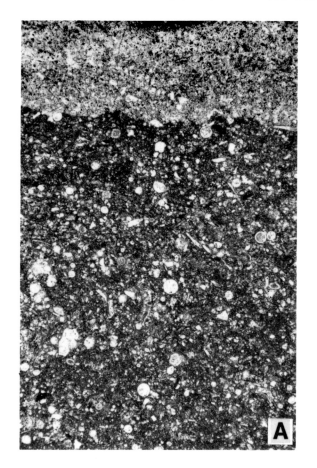

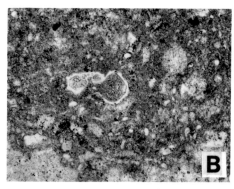

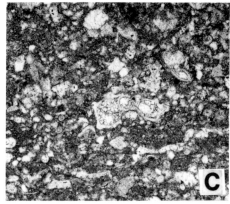

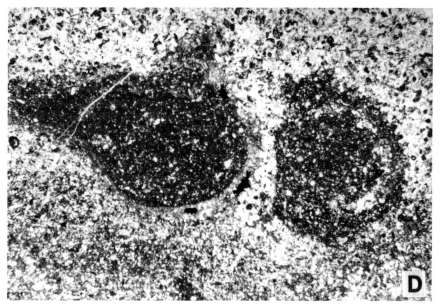

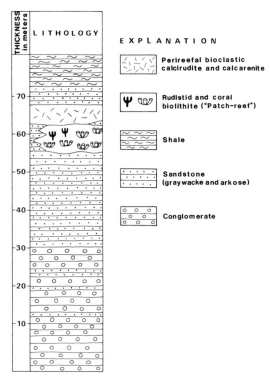

FIG. 13.—Schematic stratigraphic column through Santonian-Campanian lagoonal sediments ("*Gosau-beds*") at Vrapče, on Mt. Medvednica, near Zagreb (modified after Crnjaković, 1977).

*Peri-Reefal Detrital Limestone.*—In the biolithitic complex of Donje Orešje the peri-reefal detrital limestone represents a lateral continuation of the above reef talus deposits. These sediments can be described as alternating calcarenitic and calcilutitic limestones (Figs. 10, 11, and 12). Calcarenites are mostly fine-grained with well-sorted and densely-packed allochems, mostly grain-supported ("packstone"). In such a densely-packed calcarenite the cement is sparitic. Among the allochems subangular bioclasts derived from the reef predominate, e.g., fragments of rudistids, corals and other organisms. Intraclasts are not abundant but when they occur they are mostly particles of biomicritic sediment. Benthic foraminifers are rare and represented by specimens of the family Anomalinidae. Specimens of *Gavelinella lorneiana* (d'Orbigny) and *Goupillaudina daugini* Marie, which are characteristic of the Senonian, have been identified in association with *Minouxia* sp. and various Rotaliidae.

Less frequently coarse-grained calcarenites occur, and these contain particles of ruditic size (Fig. 11A). The latter are bioclasts of rudistids and to a lesser extent fragments of corals and cor-

alline algae. In these coarse-grained calcarenites the allochems are unsorted, angular, and dispersed within a micritic matrix which contains numerous oligosteginids.

The above mentioned calcarenitic rocks alternate with calcilutites (biomicritic limestones) which contain rare very small reef derived bioclasts (Fig. 12). Smaller bioclastic lenses and intercalations of chert also occur. Pelagic microfossils are especially abundant in the micritic matrix, with oligosteginids most numerous. The following species have been identified: *Calcisphaerula innominata* Bonet, *C. innominata lata* Adams, Khalili, Khosrovi, and Said, *Stomiosphaera sphaerica* (Kaufmann), *Pithonella ovalis* (Kaufmann), *P. perlonga* Andri, *Cadosina* cf. *C. fusca* (Wanner), and *C.* cf. *C. semiradiata* Wanner.

In addition to the oligosteginid assemblage there are also globigerinaceans, rare and poorly preserved globotruncanids, sponge spicules, and recrystallized radiolarians, which all suggest a typical pelagic biota. The following globotruncanid species have been identified: *Globotruncana lapparenti tricarinata* (Quereau), *G. lapparenti bulloides* Vogler, *G. lapparenti inflata* Bolli, *G. globigerinoides* Brotzen, and *G. fornicata* Plummer.

All of the listed *Globotruncana* species are characteristic of the Senonian, but not a single species is characteristic of the upper Senonian. Therefore, according to the known stratigraphic ranges of these species in the Tethyan Realm, it can reasonably be concluded that the limestones are of lower Senonian age.

The above described detrital limestones were deposited not far out from the reef-front where the sea bottom was relatively steep, and bioclastic detritus from the reef was continually moving down the slope. At greater depth the detritus mixed with "autochthonous" mud. It should be emphasized that no large accumulations of detrital material occur, and consequently no turbidity current deposition is believed to have been involved. Only occasional strong influxes of detritus from the reef occur, and this as a result of continual active erosion of the reef-front. This detritus produced densely-packed calcarenites with sparite cement, erosional channels, and mud balls (Fig. 12D). During the periods of weak influx of bioclastic material from the reef, deposition of biomicrite predominated (Fig. 12A–C).

### Marl with Calcareous Nannoplankton

A 2 meter thick bed of marl (stratigraphic unit 27, Fig. 10) overlies the reef and subreef deposits, indicating an abrupt change of sedimentational conditions and suggesting cessation of the reef environment on the biolithitic complex of Donje Orešje. Large quantities of clayey and silty material from the lagoon appears to have engulfed the entire reef-flat area, even reaching into the fore-reef. The composition of the heavy mineral association in the marl points to a source area of basic and ultrabasic rocks.

A fairly rich association of calcareous nanno-plankton has been found in this marl signifying a strong pelagic influence. The following nannofos-sil species have been identified from this marl: *Cretarhabdus ingens* (Gorka), *Cribrosphaerella ehrenbergi* Arhangelsky, *Micula surophora* (Gar-det), *Eiffelithus turriseiffeli* (Deflandre), *E. exi-mius* (Stover), *Broinsonia parca* (Stradner), *Pre-discosphaera spinosa* (Bramlette and Martini), and *Watznaueria barneae* (Black).

The presence of *Broinsonia parca*, which oc-curs in both Campanian and Maestrichtian depos-its, and *Eiffelithus eximius* which does not occur as high as the Maestrichtian (Thierstein, 1976), suggests that the marl is of Campanian age.

The marl containing calcareous nannoplankton is overlain by thin-bedded biomicrites with abun-dant hemipelagic oligosteginids, rare globotrun-canids, globigerinids, calcitized radiolarians, and sponge spicules (Fig. 10, stratigraphic units 28 and 29). Bioclasts in the marls contain bivalves which are very small and rare. This type of sediment suggests that the reef was of some distance away and that the reef environment in the region of Donje Orešje had come to an end.

### LAGOON

Upper Cretaceous barrier-reefs were located in a subduction zone along the outer-slopes of island arcs, in which a lagoon of variable width existed between the shore and the reef. Various types of terrigenous clastic sediments were derived from the island arcs and transported into the lagoon (Fig. 13). Based on sediment composition it can be concluded that these islands, e.g., in the broad-er Mt. Medvednica area, were composed of sed-imentary rocks (sandstones and shales) and low-grade metamorphic rocks (phyllites) with some contribution from basic and ultrabasic rocks. Is-land arc denudation coupled with tectonic activity and erosion was intensive, and led to rapid sedi-ment deposition in the lagoon. The most frequent lagoonal sediments consist of various sandstones (graywackes and arkoses), marls and shales, and in the proximal parts of the lagoon, conglomerates as well (Crnjaković, 1977). Molluscan fossils rep-resented in these sediments include various pe-lecypods: *Inoceramus, Pycnodonta, Neithea, Cardium,* and *Tapes,* and gastropods: *Actae-onella, Natica,* and *Turritella* (Herak and Nedéla, 1963). Solitary corals, particularly the genera *Cunnolites, Rennensismilia* and *Diploctenium* (Turnšek, 1978) are also found in some areas. It is important to point out that brackish and fresh-

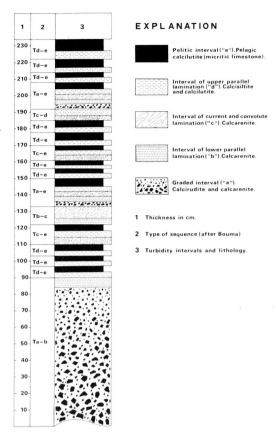

FIG. 14.—Schematic stratigraphic column through Senonian "carbonate turbidites" ("*Scaglia*-beds"), ex-posed at Donje Orešje, on Mt. Medvednica near Zagreb, Yugoslavia.

water pelecypods and gastropods, also occur in lagoonal sediments, e.g., *Cyrena, Glauconia,* and *Pyrgulifera.* Terrestrial (vascular) plants have also been collected from these sediments. In some places rudists and hermatypic corals built small patch-reefs. The included sandstones frequently exhibit cross-bedding structures. In the literature these sediments have been called "*Gosau*-depos-its" since they show paleontological and litho-logical similarities with Upper Cretaceous beds from Gosau in the Northern Alps of Austria. Hence lagoonal sediment demonstrate features of both a shallow marine sedimentary environment as well as fresh water and brackish water envi-ronments.

### BASIN SEDIMENTS AND THEIR RELATION TO THE BIOLITHITIC COMPLEXES

Laterally, the subreef detrital limestones grad-ually pass into thin-bedded and platy grey lime-stones with small chert nodules. These deposits

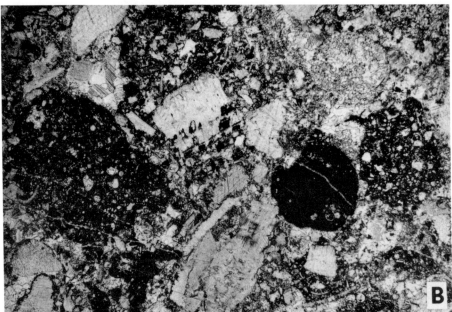

FIG. 15.—Thin-section photomicrographs (×20) of carbonate turbidites at Donje Orešje. A. Poorly-sorted biocal-carenite with common large angular to subangular fragments of rudistids intermingled with densely-packed small bioclasts of reef building organisms. Along the right margin of the photomicrograph is a fragment of the coralline alga *Archaeolithothamnium*; sample O-31-1 (middle part of the graded interval "a"). B. Poorly-sorted biocalca-renite with several spherical clasts of oligosteginid biomicrite; same sample as in Figure 15A.

belong to the so-called "*Scaglia*"-deposits and originated through the action of turbidity currents and, to a lesser extent, through the deposition of autochthonous pelagic sediments (Fig. 14). The depositional sequence consists of numerous abbreviated and truncated turbiditic sequences, 5 to 10 centimeters thick (Fig. 17). The most frequently occurring cycles are Td–e and Tc–e sequences,

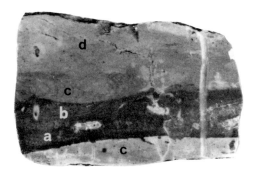

FIG. 17.—Slab of a carbonate turbidite from Donje Orešje. The lower part of the sample belongs to the "c" interval of the sequence O-31-6. The dark calcarenitic part belongs to the sequence O-31-7, and to the graded interval ("a") and to the interval of lower parallel lamination ("b"). The lighter part of the O-31-7 sequence shows current convoluted laminations ("c") and, at the top, indistinct upper parallel laminations ("d"); polished slab is natural size; sample O-31-6,7.

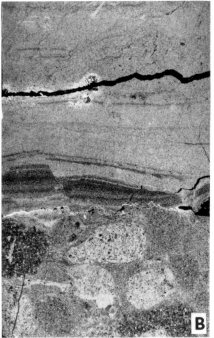

FIG. 16.—Samples of carbonate turbidites from Donje Orešje. A. Transition between the graded interval ("a") and the interval of parallel lamination ("b"); polished slab, sample O-31-1, natural size. B. Thin-section photomicrograph of the above sample, showing transition between the "a" and "b" intervals. The graded interval is composed of calcarenitic matrix with rounded inclusions of biomicrite and fine-grained calcarenite. The "b" interval is composed of an alternating sequence of laminae of densely-packed fine-grained calcarenite, and fine laminae of biomicrite containing small bioclasts; negative print, ×4.

with a few Ta–e, Tb–c and Tc–d truncated cyclic sequences occurring as well (Bouma, 1961). Several thick beds (1–1.5 meters) occur with a clearly graded interval designated "a" (Figs. 15 and 16). This interval contains fragments of reef building organisms of ruditic size, up to several centimeters in diameter. These large bioclasts are dispersed within fine-grained calcarenitic detritus mixed with micrite. Among the bioclasts, fragments of rudistids of the family Hippuritidae predominate. Fragments of corals occur less frequently, and corallinacean algal fragments (*Archaeolithothamnium*) are very rare. Some bioclasts have micritic envelopes, and some have a rim of well developed fibrous calcite cement (sparitic cement A). In the upper portions of the graded beds there are frequent inclusions of biomicritic oligosteginid mud. A comparatively large amount of the bioclastic material in graded intervals indicates a strong periodical influx of detritus from the reef, and transportation toward the center of the basin by the action of turbidity currents.

In the higher, "b," "c," and "d" intervals, reef detritus becomes more and more defined. In the "b" interval, quite frequently there is well pronounced parallel lamination (Figs. 16 and 18C). In the final "e" interval, reef detritus is completely absent. This interval consists of fossiliferous micrite and biomicrite containing only pelagic fossils. Here we can distinguish between the so-called "eupelagic limestone" (*fide* Kuenen, 1960), and limestone deposited under the influence of turbidity currents. The eupelagic limestones ($e^p$) have a micritic matrix in which rare pelagic microfossils are irregularly dispersed

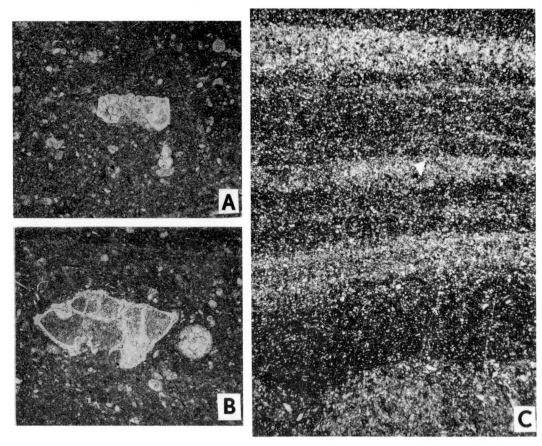

FIG. 18.—Thin-section photomicrograph of carbonate turbidites at Donje Orešje. A. and B. Pelagic fossiliferous micrite of the ''e'' interval; outcrop sample O-31-3, ×60. A. *Globotruncana lapparenti bulloides* Vogler, and Heterohelicidae. B. *Globotruncana rosetta* Carsey with calcitized radiolarians. C. Interval ''b'' showing parallel laminations, an alternating sequence of laminae of densely-packed, fine-grained calcarenite and biomicrite laminae with small bioclasts; sample O-31-1, ×20.

(''pelagic rain''). Most numerous are the globotruncanid foraminifers, next are heterohelicids with other globigerinoid foraminifers, and calcitized radiolarians (Fig. 18A–B). The turbidite portion of this interval ($e^t$) is in turn represented by biomicrites containing densely-packed, uniformly oriented, and partly-sorted pelagic microfossils. This is believed to be a result of limited transportation by the action of turbidity currents.

The following species of globotruncanid foraminifers have been identified: *Globotruncana fornicata* Plummer, *G. lapparenti tricarinata* (Quereau), *G. lapparenti lapparenti* Brotzen, *G. lapparenti bulloids* Vogler, *G. arca* (Cushman), *G. conica* White, *G. elevata* Brotzen, *G. rosetta* Carsey, *G. angusticarinata* Gandolfi, *G. lapparenti coronata* Bolli, and *G. caliciformis* (de Lapparent).

## THE ROLE OF BIOLITHITIC COMPLEXES IN FACIES DIFFERENTIATION WITHIN THE UPPER CRETACEOUS OF THE INNER DINARIDES

Upper Cretaceous biolithitic complexes of the Inner Dinarides are dominated by barrier-reefs. Owing to their biological growth potential these reefs can attain relatively large dimensions, and grow up from the sea bottom as morphological barriers. Hence these reefs can control the differentiation of sedimentary environments (Fig. 19). This differentiation provides a basis for the recognition of several facies, each characterized by particular attributes.

In the lagoon, situated between the shore and barrier-reef, clastic terrigenous ''*Gosau*-deposits'' predominate. This formation has all the features of typical lagoonal facies with periodic brackish water interludes.

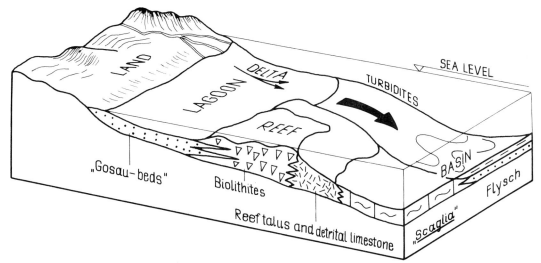

FIG. 19.—Schematic sedimentational model of Upper Cretaceous facies of the Inner Dinarides, Yugoslavia.

The barrier-reefs acted as shields, or "filters" somewhat restricting the transport of terrigenous sediments from the lagoon towards the open sea. In this environment of relatively clear water, "Scaglia"-deposits of carbonate turbidites and autochthonous pelagic limestones were deposited. Based on their composition these deposits can be compared favorably to similar deposits designated by Italian geologists as "Scaglia bianca" (Desio, 1973).

Carbonate turbidites were deposited in the basin, which was bordered by a barrier-reef that filtered the clastic material deposited in the backreef area, e.g., in the lagoon. Thus, the fore-reef received only insignificant quantities of terrigenous clastic material. Hence, close to the reeffront subreef breccia and detrital limestone are in evidence, whereas farther away from the reef, and at greater depths, basinal deposits of the "Scaglia" type formed.

Large accumulations of reef detritus occur only at the transition from the relatively steep fore-reef slope to the gently dipping slope of the proximal portion of the basin. The sea bottom was probably inclined at about 5–10 degrees. When balance was disturbed, freshly accumulated detrital sediment formed turbidity currents. Due to the comparatively small amount of material moved, these turbidity currents were mostly of short duration and of small magnitude.

A schematic sedimentary model of the reef, fore-reef, and adjacent basin (Figs. 1 and 19), shows that the "Scaglia" deposits were formed only in those parts of Upper Cretaceous basins which had an actively growing barrier-reef in the background. These barrier-reefs acted as shields

or "filters," protecting basinal pelagic carbonate sedimentation from the influx of terrigenous clastic material. In those parts of the basins which lacked a barrier-reef, terrigenous clastic deposits were able to form abundant turbidites, and eventually flysch deposits. It can be recognized that Upper Cretaceous flysch deposits may pass, both laterally and vertically, into "Scaglia"-deposits, and *vice versa*.

The described model helps to explain more clearly those processes which might cause essential changes in sedimentational patterns. Thus, flysch deposits overlying pelagic "Scaglia"-deposits do not necessarily imply a period of stronger uplift in the denudation area. It could equally imply the cessation of barrier-reef growth and eventual disintegration and destruction. The area of deposition of terrigenous clastic material, most frequently flysch, is thus expanded into the area where formerly pelagic "Scaglia"-sediments had been deposited. These changes occurred repeatedly in both space and time.

Such an explanation of facies differentation in "geosynclinal" areas is essentially different from existing interpretations. For example, according to Auboin (1965 and 1973), the cherty pelagic limestones (defined in this paper as the "Scaglia"-deposits), are deposited in the so-called "pre-flysch period," e.g., prior to Upper Cretaceous flysch sedimentation. Yet, it is obvious that these deposits can frequently be contemporaneous with the flysch, and even overlying the flysch as well.

Finally, it should be noted that the described model also contributes to a better understanding of the bathymetry of the basinal "Scaglia"-deposits. Generally, a relatively pure carbonate with only negligible admixtures of terrigenous components present within the basin indicates, as a rule, a more distant pelagic environment at comparatively great depth. However, according to the model described above, the "*Scaglia*"-deposits were deposited not far distant from the fore-reef, which also means near the reef itself. At such a distance, that area probably represented only the upper bathyal regime. The lack of terrigenous deposits does not necessarily mean a great distance from the land, but may be the result of the already mentioned filtering role of the barrier-reef.

### BIOLITHIC COMPLEXES AND PLATE TECTONICS

An analysis of spatial position of the main biolithic complexes of the Upper Cretaceous and lower Paleogene in the area of the Dinarides, demonstrates clearly their marked linear distribution. As a rule these complexes exhibit a slight arc shape. In the Inner Dinaric region there are several such lines which represent the remnants of previous island arcs formed in the subduction zone. In some places these arcs are broken, removed, and dislocated to some extent by transform faults.

Island arcs and correlative deep sea trenches are characteristics of many Recent subduction zones. Sedimentation in Recent arc-trench systems has been a subject of many papers (Karig, 1971; Dickinson, 1974; School and Marlov, 1974; and others). Island arcs are characterized by andesitic and basaltic volcanism (Miyashiro, 1972). Interarc basins generally are formed between island arcs. Terrigenous turbidites, as well as pelagic and hemipelagic deposits, are conspicuous elements of interarc sedimentation. Numerous biolithic reef complexes are developed on Recent island arcs of tropical and subtropical regions. The association of island arcs and biolithitic reefs has occurred during evolution of many past orogenic systems, for example, they have been recognized in the Paleozoic of the western United States (Churkin, 1974).

Upper Cretaceous and Paleogene island arcs of the Inner Dinarides show an analogy with Recent island arcs, and with those from the geological past as well.

Island arcs within the Upper Cretaceous of the Inner Dinarides were formed as a consequence of northeastward movement of the Adriatic Plate (McKenzie, 1970 and 1972). This plate has also been called the "Apulian Plate" by Dercourt (1970), and the "Apulian Microcontinent" by Dewey (1973). The Adriatic Plate appears to have been an integral part of the large Greco-Italian "Microcontinent." Hsü (1971) claims on the basis of paleomagnetic anomalies and correlation of lithofacial complexes, that this "microcontinent"

together with the African Craton, in relation to the European continent, has been steadily moving eastwards. This movement caused subduction of Tethyan oceanic crust which was situated on the east in relationship to the Greco-Italian "Microcontinent." The subduction process caused occurrences of ophiolite and mélanges in the Inner Dinarides as reported by: Herak (1960), Crnković (1963), Dimitrijević, M. D. (1974), Dimitrijević, M. D., and Dimitrijević, M. N. (1973 and 1975), Pamić and Jelaska (1975), Majer (1978), and others. This eastward movement lasted until the end of Upper Cretaceous time. Argyriadis (1974) thinks that both the "Euroasian and African blocks" were very close to each other in Cretaceous time, thus causing the orogenic uplifts in the eastern Mediterranean. Aubouin (1975 and 1977) considers that the subduction and contraction process formed island arcs of the Pacific type. The Adriatic Plate was temporarily uplifted along its eastward movement at the end of Cretaceous time ("Laramian Orogeny"), thus providing areas susceptible to bauxite deposition (Polšak, 1965b and 1971). According to Tapponnier (1977), in the Paleogene Epoch, the African Craton changed direction of movement towards the northeast and north. This movement of the Greco-Italian "Microcontinent" approached the Panonian Plate (or according to Dercourt (1970) "Rhodope Craton") at the nearest position in the Eocene. This plate juncture caused intensive orogenic processes including uplift of the Dinarides. The suture produced between these "microcontinents" or plates includes the Inner Dinarides which are characterized by great facies differentiation, and the occurrences of ophiolites and mélanges.

### BIOLITHIC COMPLEXES ON ISLAND ARCS

Island arcs offer suitable environmental and ecological conditions for development of reef growth, and barrier-reefs are commonly formed on such arcs, especially in warm climate regions. Barrier-reefs are separated from the mainland by lagoons of various widths. The Upper Cretaceous sediment of the Dinarides reflect such environments. Here the annual mean temperature was 23 degrees C, with a maximum of 25 degrees C (Polšak and Leskovšek, 1975), thus indicating a tropical sea. Favorable conditions for reef growth on island arcs could and did occur during particular geological stages in numerous places over a wide area, thus the development of a series of reefs of similar ages. However, favorable conditions could have existed earlier in one place and somewhat later elsewhere. Hence, within a particular sedimentary interval reefs of different age can appear within the series, respectively within the same island arc. The main ecological variable in barrier-reefs on island arcs appears to be water

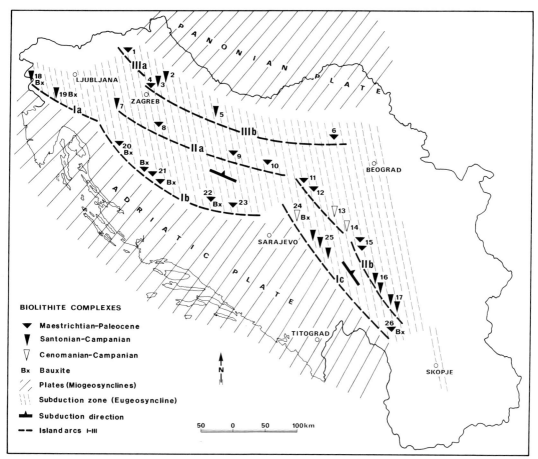

FIG. 20.—Island arcs of the Upper Cretaceous biolithite complexes in the subduction zone of the Inner Dinarides, Yugoslavia. Locations are: 1. Slovenske Konjice; 2. Kalnik; 3. Donje Orešje (Medvednica); 4. Vidovec (Medvednica); 5. Jovanovica (Papuk); 6. Fruška gora; 7. Brašljevica (Žumberak); 8. Medurače (Banija); 9. Brusnica (Motajica); 10. Tinja (Majevica); 11. Kozluk; 12. Gučevo; 13. Poćuta; 14. Kosjerić; 15. Dragačevo-Ivanjica; 16. Golija; 17. Kosovska Mitrovica; 18. Sabotin; 19. Hrušica; 20. Slunj; 21. Grmeč; 22. Bešpelj (Jajce); 23. Vlašić; 24. Vlasenica; 25. Višegrad-Priboj; 26. Grebnik (Metohija).

clarity, because intensive influx of terrigenous clastic material could prevent growth and development of the primary reef building organisms.

The Upper Cretaceous island arc system which extended along the margin of the Adriatic Plate from Sabotin to Vlašić (Fig. 20, I-No. 18 to No. 23) had two biolithitic complexes of Santonian-Campanian age, and these occur at Sabotin (No. 18) and Hrušica (No. 19) in Slovenia (Buser et al., 1976). Further southeastward, there followed Maestrichtian biolithitic complexes at Slunj (No. 20), Grmeč (No. 21), Bešpelj near Jajce (No. 22) and Vlašić (No. 23). As a rule these biolithitic complexes, together with Maestrichtian clastic and pelagic sediments, are transgressive on Lower Cretaceous or on Cenomanian and Turonian basement rocks (Devidé-Neděla and Polšak, 1961;

Grandić and Vuksanović, 1973; Sakač, 1969; Sliškovic, 1975; Vujnović and Vrhovčić, 1975). It is important to note that the basement is composed of carbonate sediments identical to those in the Adriatic Plate ("carbonate platform") of the Outer Dinarides. Therefore, the island arc system and correlative basins were formed during the Upper Cretaceous by "faulting" of the Adriatic Plate under the dominant processes of subduction. These basins have also been included in the interarc basins by Dickinson (1974). The southeastern extension of this arc system has been displaced by the Sarajevo transform fault in the area of Kladanj-Vlasenica. Following this, in the direction of Vlasenica (No. 24) to Grebnik (No. 26) in Metohia, the arc changes direction towards south-southeast (Fig. 20, IC). With this directional change other features of the arc also changed. As a rule, the biolithic complexes are older (Cenomanian-Turonian and Santonian-Campanian), except at Grebnik (No. 26), where the reef environment was established in Maestrichtian time. Cretaceous sediments are transgressive on the Triassic basement in the Vlasenica area (No. 24), while at Višegrad and Priboj (No. 25), they are transgressive on the foothills of Zlatibor where they overlay a serpentinite massif (Miladinović and Zivaljević, 1976). At Grebnik, Maestrichtian biolithic complexes overlay limestone deposits of Cenomanian and Turonian age, and these in turn are transgressive on the serpentinite (Pejovic and Radoičić, 1973).

An important feature of the Sabotin-Grebnik island arc is the frequent occurrence of beds of Upper Cretaceous bauxite. Deposits of bauxite are found at all localities where the basement of transgressive Upper Cretaceous deposits are carbonates of Lower Cretaceous, or Cenomanian and Turonian, and even Triassic in age. Genesis of bauxite in this region appears to be related to an Upper Cretaceous emergence phase associated with deep weathering. Transgressive Upper Cretaceous deposits generally overlay the bauxite beds. Bauxite outcrops are found at Sabotin and Hrušica in Slovenia, from Dubrovčani to Dobra-Slunj, on Mt. Grmeč, at Bosanska Krupa, on Bešpelj near Jajce, at Vlasenica, and at Grebnik in Metohia as well (Grubić, 1971; Sakač, 1973; Zupanič and Papeš, 1973). Upper Cretaceous sediments associated with biolithic complexes near Višegrad and Prijob do not have bauxite deposits, since these deposits are directly transgressive on the Zlatibor serpentinites.

A second island arc (Fig. 20,II) extended from Brašljevica in Žumberak (No. 7) to Kosovska Mitrovica (No. 17). At Zvornik, this arc line is displaced by the Sarajevo transform fault, and changes direction towards the southeast (IIb). This arc appears to have been in a zone of maximum subduction, therefore its evolution had close ties with ophiolite and mélange occurrences. During the Upper Cretaceous this arc was composed of a line of barrier-reefs. Their age varies considerably. Sometimes the reef environment lasted from Cenomanian to Campanian time, e.g., at Pócuta (No. 13) and Kosjerić (No. 14) in western Serbia (Pasić, 1957; Pejović, 1957). At Brašljevica the time span ranged from the Santonian to lower Campanian (No. 7) and Golija (No. 16), whereas at Kosovska Mitrovica (No. 17) the time spans the Santonian to Campanian (Milovanović, 1953 and 1954). Elsewhere, favorable conditions for reef formations occurred much later, e.g., at Kozluk (No. 11), Gučevo (No. 12), and at Dragačevo and Ivanjica (No. 15) as well. Biolithic

formations occurred at these localities during Maestrichtian time (Milovanović, 1960; Milovanović and Grubić, 1969; Sladić-Trifunović, 1972). Exceptionally, reef environments persisted into the lower Paleogene as occurred at Tinja (No. 10) on Mt. Majevica (Jelaska and Bulić, 1975), or initiated in the Paleogene as shown at Medurače (No. 8) in Banija (Babić et al., 1976). The basement of transgressive Upper Cretaceous deposits at these localities generally consists of Paleozoic clastic sediments and ultramafic rocks. These rock types have been identified on Motajica, Majevica, Kosjerić, Golija and in the vicinity of Kosovska Mitrovica. In this island arc area no bauxite occurrence have been found in association with Upper Cretaceous sediments.

The third island arc (Fig. 20) extended from Slovenske Konjice on the foothills of Pohorje (No. 1) to Mt. Fruška Gora (No. 6). In the area of Mt. Medvednica the arc has been dislocated by the Zagreb-Balaton transform fault. In this island arc (III a and b), situated along the margin of Panonian Plate, the biolithic complexes were developed in Santonian and Campanian time, as in the case of Donje Orešje in Mt. Medvednica (No. 3), Kalnik (No. 2), and Jovanovica (No. 5) on Mt. Papuk (Šikić et al., 1975; Polšak, 1979). Maestrichtian biolithic complexes were identified at Mt. Fruška Gora (No. 6) and at Slovenske Konjice (No. 1), here reef growth continued from the Maestrichtian into the lowermost Paleocene (Polšak, 1965a; Petković et al., 1976; Pleničar, 1971). Upper Cretaceous sediments are transgressive on various rocks, and these can be Paleozoic clastic and metamorphic rocks, or Triassic dolomites and serpentinites.

Due to intensive subduction, the formation of this island arc was closely associated with ophiolite volcanism and penetration of ultramafic rocks. Therefore, the islands that resulted are composed of serpentinite masses, e.g., in the area of Lepavina-Kalnik (northwestern Croatia), and Mt. Fruška Gora (No. 6). This is supported by evidence of heavy mineral associations derived from fragments of terrigenous clastics in Santonian and Campanian biolithic complexes at Donje Orešje (No. 3) on Mt. Medvednica (Polšak, 1979). The heavy mineral associations are typical for serpentinite rocks, and Upper Cretaceous clastics appear to have been derived from these serpentinites.

In some portions of this island arc system, active andesite volcanism took place during the Cretaceous. This is indicated by occurrences of andesite-trachyte pyroclasts within Upper Cretaceous deposits taken from several boreholes in middle Banat and southern Bačka in Vojvodina (Kemenci and Čanović, 1975).

## ISLAND ARCS AND FACIES REPETITION

In a direct progression of island arc to barrier-reef to basin there is a series of described facies, which, since the Inner Dinarides exhibits several island arcs, are successively repeated. As an example, we might correlate facies extending from Bešpelj, near Jajce (the biolithitic complex No. 22), over to Motajica (No. 9), to Mt. Papuk (No. 5). This line of sections includes all three island arcs. In the area between Jajce (No. 22) and Motajica (No. 9), so-called "Scaglia" sediments and flysch are deposited. Considering that the entire area extended into the zone of maximum subduction, ophiolite rocks and mélanges also occur. Between the biolithitic complexes of Mt. Motajica and Papuk, basinal sediments predominate, particularly those corresponding to the "Scaglia"-deposits and flysch-like sediments of Požeška Gora. Similar regular repetition of the mentioned facies can be found in other parts of the island arcs and their correlative basins. Therefore, transition between the so-called Outer Dinarides (Adriatic Plate) and the Inner Dinarides of the Upper Cretaceous was not represented by simple slope relationship, as is commonly believed, but instead this region developed a rather complex system of island arcs and basins, thus providing locations for multiple repetition of facies deposited in both shallow and deep water environments.

## ROLE OF BIOLITHITIC COMPLEXES IN OIL GEOLOGY

It is estimated that from 35–50 percent of world oil production is derived from carbonate rocks (Harbaugh, 1967; Chilingar, Mannon and Rieke, 1972). Therefore these rocks are important as petroleum reservoirs. Biolithitic complexes and associated bioclastic deposits are particularly favorable potential reservoirs since these rocks are frequently characterized by marked primary porosity and permeability. This is due primarily to various reef building organisms, in this case corals and rudists, which contain cavities both within and surrounding their skeletons. Cavities between colonial organisms which build the reef framework, can be partially empty, e.g., not filled with reef sands, limestone cements or clastics, whereas cavities within reef building skeletons are particularly large and may have a very porous honeycombed development.

Significant primary porosity may also be found in bioclastic limestones within the peri-reefal area. These limestones are mostly composed of the fragments of reef building organisms with their initial inherent porosity. In the unconsolidated state these are the porous reef sands. In addition to primary porosity, biolithitic complexes can also exhibit various secondary porosity features as well, i.e., subaerial erosional surfaces and evidence of subsurface leaching.

Biolithitic complexes associated with a subduction zone and island arcs, as in the Inner Dinarides, can laterally grade into various rock facies, some of which may be good cap rocks or seals for oil and gas deposits (e.g., "Scaglia" and "Gosau" beds, and flysch). These relationships increase the possibility that biolithitic complexes can be an important asset in the development of stratigraphic oil traps.

## ACKNOWLEDGMENTS

For identification of various fossils presented in this paper I am indebted to Professor Ivan Gušić, Professor Donata Nedela-Devidé, and Josip Benić, M.Sc., of the Department of Geology and Paleontology, University of Zagreb, as well as to Dr. Dragica Turnšek of the Paleontological Institute of the Slovenian Academy of Sciences and Arts in Ljubljana, Yugoslavia.

I am also grateful to Dr. Biserka Šćavničar of the Institute of Geology in Zagreb for her analyses of clastic rocks.

Finally, I acknowledge the help of Mrs. Andjela Truhan for her fine drawings and to Mrs. Roza Pavlešić who prepared the photographs.

This research was partially funded by INA-Naftaplin in Zagreb, Yugoslavia.

## REFERENCES

ANDRI, E., 1972, Mise au point et donnees nouvelles sur la famille des Calcisphaerulidae Bonet 1956: les genres *Bonetocardiella, Pithonella, Calcisphaerula* et "*Stomiosphera*": Rev. Micropaleont., v. 15/1, p. 12–34.

ARGYRIADIS, I., 1974, Sur l'orogenèse mésogénne des temps Crétacés: Rev. Géogr. Physique et Géol. Dynamique, 2, v. 16/1, p. 23–60.

AUBOUIN, J., 1965, Geosynclines: Amsterdam, Elsevier, 335 p.

———, 1973, Des tectoniques superposées et de leur signification par rapport aux modèle géophysiques: l'example des Dinarides: paléotectonique, tectonique, tarditectonique: Bull. Soc. Géol. France, 7, v. 15, p. 426–460.

———, 1975, De la position structurale des zones de subduction frontale et subduction radicale: Comptus Rendu, Acad. Sci., 281-D, p. 99–102.

———, 1977, Alpine tectonics and plate tectonics: thoughts about the Eastern Mediterranean, *in* Ager, D. V. and Brooks, M., Eds., Europe from crust to core: London, Wiley, p. 143–158.

BABIĆ, LJ., GUŠIĆ, I., AND ZUPANIČ, J., 1976, Grebenski paleocen u Baniji, središnja Hrvatska, [Paleocene reef
  limestone in Banija, central Croatia]: Geol. Vjesnik, v. 29, p. 11–47.
BEAUVAIS, L., AND BEAUVAIS, M., 1974, Studies on the world distribution of Upper Cretaceous corals: Interna-
  tional Coral Reef Symp., 2nd, Great Barrier Reef Com., Proc., p. 475–493.
BISSELL, H. J., AND CHILINGAR, G. V., 1967, Classification of sedimentary carbonate rocks, Developments in
  Sedimentology 9A (Carbonate rocks): Amsterdam, Elsevier, p. 87–167.
BOUMA, A. H., 1959, Some data on turbidites from the Alpes Maritimes, France: Geologie en Mijnbow, v. 21, p.
  223–227.
———, 1962, Sedimentology of some Flysch Deposits: Amsterdam, Elsevier, 168 p.
———, AND NOTA, D. J. G., 1961, Detailed graphic logs of sedimentary formations, in Sorgenfrei, T., Ed.,
  International Geol. Congress, 21 Sess., Repts., p. 52–74.
BUSER, S., PEJOVIĆ, D., AND RADOIČIĆ, R., 1976, Biostratigrafija boksitonosnih zgornjekrednih plasti v jugozahodni
  Sloveniji, [Biostratigraphie der Bauxit Führenden Oberkreide-Schichten in Südwest-Slowenien]: 4 Jugosl.
  Simp. Istraž. Eksploat. Boksita, p. 67–70.
CELET, P., 1977, The Dinaric and Aegean arcs: the geology of the Adriatic, in Nairn, A. E. M., Kaines, W. H.,
  and Stehli, F. G., Eds., The Ocean basins and margins: v. 4A, chapter 5A, p. 215–259, Plenum Pub. Corp.
  New York.
CHILINGAR, G. V., MANNON, R. W., AND RIEKE, H. H., 1972, Oil and Gas Production from Carbonate Rocks:
  Amsterdam, Elsevier, 408 p.
CHURKIN, M., JR., 1974, Paleozoic marginal ocean basin-volcanic arc systems in the Cordilleran foldbelt, in Dott,
  R. H., and Shaver, R. H., Eds., Modern and ancient geosynclinal sedimentation: Soc. Econ. Paleontologists
  Mineralogists Spec. Pub. No. 19, p. 174–192.
CRNKOVIĆ, B., 1963, Petrografija i petrogeneza magmatita sjeverne strane Medvednice, [Petrography and petro-
  genesis of the magmatites of the northern part of Medvednica Mountain]: Geol. Vjesnik, v. 16, p. 63–160.
CRNJAKOVIĆ, M., 1977, Klastiti gornje krede Medvednice zapadni dio, [Upper Cretaceous clastic beds of Mt.
  Medvednica Western part]: Ph.D. Dissertation, Sveučilište u Zagrebu, 185 p.
CUMINGS, R. J., 1932, Reefs or bioherms?: Geol. Soc. America Bull. v. 43, p. 331–352.
DERCOURT, J., 1970, L'expansion océanique actuelle et fossile: Bull. Soc. Géol. France, 7, v. 12, p. 261–309.
DESIO, A., 1973, Geologia dell'Italia: Torino, Unione Tipogr. Editr. Torinese, 1081 p.
DEVIDÉ-NEDÉLA, D., AND POLŠAK, A., 1961, Mastriht kod Bešpelja sjeverno od Jajca, [Sur la présence du Mae-
  strichtien dans les environs de Bešpelj au nord de Jajce en Bosnie]: Geol. Vjesnik, v. 14, p. 355–376.
DICKINSON, W. R., 1974, Plate tectonics and sedimentation, in Dickinson, W. R., Ed.; Soc. Econ. Paleontologists
  Mineralogists Spec. Pub. No. 22, p. 1–27.
———, 1974, Sedimentation within and beside ancient and modern magmatic arcs, in Dott, R. H., and Shaver, R.
  H., Eds.: Soc. Econ. Paleontologists Mineralogists, Spec. Pub. No. 19, p. 230–239.
DIMITRIJEVIĆ, M. D., 1974, The Dinarides: a model based on the new global tectonics, in Jankovic, S., Ed., Metall.
  and concepts in the geotectonics development of Yugoslavia: Belgrade, Faculty Mining and Geology, p. 141–178.
———, AND DIMITRIJEVIĆ, M. N., 1973, Olistostrome mélange in the Yugoslavian Dinarides and Late Mesozoic
  plate tectonics: Jour. Geology, v. 81, p. 328–340.
———, AND ———, 1975, Ofiolitski melanž Dinarida i Vardarske zone: geneza i geotektonsko značenje, [Ophiolite
  mélange of the Dinarides and the Vardar zone: origin and geotectonic significance]: 2 God. Znan. Skup Sekcije
  za Primjenu Geol. Geofiz. Geokem. Znan. Savjeta za Naftu JAZU, A, v. 5, p. 39–46.
DEWEY, J. F., PITMAN, W. C., RYAN, W. B. F., AND BONNIN, J., 1973, Plate tectonics and the evolution of the
  alpine system: Geol. Soc. America Bull., v. 84, p. 3137–3180.
DUNHAM, R. J., 1962, Classification of carbonate rocks according to depositional texture in Classification of
  carbonate rocks, a symposium, W. E. Ham, Ed., Am. Assoc. Petroleum Geologists Mem. 1, p. 108–121.
———, 1970, Stratigraphic reefs versus ecologic reefs: Am. Assoc. Petroleum Geologists Bull., v. 54, p. 1931–
  1932.
EMBRY, A. F., AND KLOVAN, J. E., 1971, A Late Devonian reef tract on northeastern Banks Island, Northwest
  Territories: Canadian Petroleum Geol. Bull., v. 19, p. 730–781.
FOLK, R. L., 1959, Practical petrographic classification of limestones: Am. Assoc. Petroleum Geologists Bull., v.
  43, p. 1–38.
———, 1962, Spectral subdivision of limestone types: in Classification of carbonate rocks, a symposium, W. E.
  Ham, ed., Am. Assoc. Petroleum Geologists Mem. 1, p. 62–84.
GOREAU, F. G., AND LAND, L. S., 1974, Fore-reef morphology and depositional processes, north Jamaica, in
  Laporte, L. E., ed., Reefs in time and space; selected examples from the Recent and ancient: Soc. Econ.
  Paleontologists Mineralogists Spec. Pub. No. 18, p. 77–89.
GRANDIĆ, S., AND VUKSANOVIĆ, B., 1973, Kordun kao novo boksitonosno područje na teritoriju SR Hrvatske,
  [Kordun als ein neues Bauxitgebiet auf dem Terrain der Sozialistischen Republik Croatien]: 2 Jugosl. Simp.
  Eksploat. Boksita, A-13, p. 1–8.
GRUBIĆ, A., 1971, Kratak pregled stratigrafskog položaja jugoslovenskih boksita, [Revue de la position stratigraph-
  ique des bauxites dans la Yougoslavie]: Zbornik Radova Rud.-Geol. Fak., Beograd, v. 13, p. 131–139.
HADŽI, E., PANTIĆ, N., ALEKSIĆ, V., AND KALENIĆ, M., 1976, Un modèle préliminaire de l'evolution tectonique
  de la Péninsule balkanique dans le cadre du développement de Méditerranée entière au cours cycle alpin: Bull.
  Soc. Géol. France, 7, v. 18/2, p. 199–203.

HARBAUGH, J. W., 1967, Carbonate oil reservoir rocks: Developments in Sedimentology 9A (Carbonate rocks): Amsterdam, Elsevier, p. 349–398.

HECKEL, P. H., 1974, Carbonate buildups in the geologic record: A review, in Laporte, L. F., Ed., Reefs in Time and Space, Selected Examples from the Recent and Ancient: Soc. Econ. Paleontologists Mineralogists Spec. Pub. No. 18, p. 90–154.

HERAK, M., 1960, Kreda s ofiolitima u Ivanščici, [Kreide mit Ophiolithen in der Ivanščica, NW Kroatien]: Acta Geol. v. 2, Prir. istraž. JAZU, 29, p. 111–120.

————, AND NEDĚLA-DEVIDÉ, D., 1963, Geologija Zagrebačke regije: Geogr. Inst. Prir.-matem. Fakulteta, Zagreb, p. 130.

HSÜ, K. J., 1971, Origin of the Alps and Western Mediterranean: Nature, v. 233, p. 44–48.

JELASKA, V., AND BULIĆ, J., 1975, Paleogeografska razmatranja gornjokrednih i paleogenskih klastita sjeverne Bosne i njihovo moguće naftnogeološko značenje, [Paleogeographic considerations of the Upper Cretaceous and the Paleogene clastic rocks in northern Bosnia and their petroleum-geological importance]: Nafta, v. 26/7–8, p. 371–385.

KARIG, D. A., 1971, Origin and development of marginal basins in the western Pacific: Jour. Geophys. Res., v. 76, p. 2542–2561.

KAUFFMAN, E. G., AND SOHL, N. F., 1974, Structure and evolution of Antillean Cretaceous Rudist Frameworks: Verh. Naturf. Ges., v. 84/1, p. 399–467.

KAUMANNS, M., 1962, Zur Stratigraphie und Tektonik der Gosauschichten. 2 Die Gosauschichten des Kainachbeckens: Sitzungsber. Österr. Akad. Wiss., Abt. 1, Bd. 171, H. 8–10, p. 199–314.

KEMENCI, R., AND ČANOVIĆ, M., 1975, Preneogena podloga vojvodanskog dela Panonskog bazena, prema podacima iz bušotina, [Pre-neogene basement in the Panonian basin of Vojvodina]: 2 God. Znan. Skup Znan. Savjeta za Naftu, Sekc. za Primjenu Geol. Geofiz. Geokem., p. 248–256.

KUENEN, PH. H., 1950, Marine Geology: New York, Wiley, 568 p.

KÜHN, O., 1932, Fossilium Catalogus I Animalia. Rudistae: Berlin, W. Junk, 199 p.

MAJER, V., 1978, Stijene "dijabaz-spilit-keratofirske asocijacije" u području Abez-Lasinja u Pokuplju i Baniji, Hrvatska, Jugoslavija, [Rocks of the "Diabase-spilite-keratophyre association" in the Abez-Lasinja area in Pokuplje and Banija, Croatia, Yugoslavia]: Acta Geol. 9/4, Prir. istraž. JAZU, 42, p. 137–158.

MAXWELL, W. G. H., 1968, Atlas of the Great Barrier Reef; Amsterdam, Elsevier, 258 p.

McKENZIE, D. P., 1970, Plate tectonics of the Mediterranean region: Nature, v. 226, p. 239–243.

————, 1972, Active tectonics of the Mediterranean region: Geophys. Jour. Astron. Soc., v. 30, p. 109–185.

MILADINOVIĆ, M., AND ŽIVALJEVIĆ, T., 1976, Prilog poznavanju krednih slojeva istočne Bosne, [Contribution à la connaissance des sédiments du Crétacé dans la Bosnie orientale]: Geol. Glasnik, v. 8, p. 161–174.

MILLIMAN, J. D., 1974, Marine Carbonates: Berlin, Springer-Verlag, 375 p.

MILOVANOVIĆ, B., 1953–54, Evolucija i stratigrafija rudista, [Evolution und Stratigraphie der Rudisten]: Zbornik Radova Geol.-rud. Fak. Beograd, p. 163–187.

————, 1960, Stratigraphie du Sénonien dans les Dinarides Yougoslaves d'après les Rudistes: Bull. Soc. Géol. France, 7, p. 366–375.

————, AND GRUBIĆ, A., 1969, Gornji senon s rudistima u Dinaridima, Vrbovački slojevi, [Sénonien supérieur à Rudistes dans les Dinarides, Couches de Vrbovac]: 3 simp. Dinarske asoc., p. 103–114.

MIYASHIRO, A., 1972, Metamorphism and related magmatism in plate tectonics: Am. Jour. Sci., v. 272, p. 629–656.

NELSON, H. F., BROWN, C. W. M., AND BRINEMAN, J. H., 1962, Skeletal limestone in Classification of carbonate rocks, a symposium, W. E. Ham, Ed., Am. Assoc. Petroleum Geologists Mem. 1, p. 224–252.

PAMIĆ, J., AND JELASKA, V., 1975, Pojave vulkanogeno-sedimentnih tvorevina gornje krede i ofiolitskog melanža u sjevernoj Bosni i njihov značaj u geološkoj gradi unutrašnjih Dinarida, [Upper Cretaceous volcanic-sedimentary formation and ophiolite mélange in northern Bosnia and their significance in geology of the Inner Dinarides]: 2 God. Znan. Skup Sekcije za Primjenu Geol., Geofiz. Geokem. Znan. Savjeta za Naftu JAZU, p. 109–117.

PAŠIĆ, M., 1957, Biostratigrafski odnosi i tektonika gornje krede šire okoline Kosjerića, zapadna Srbija, [Biostratigraphische Verhältnisse und Tektonik der Oberkreide in der veiteren Umgebung von Kosjerić, Westserbien]: Posebna Izd. Geol. Inst. "Jovan Žujović," v. 7, p. 1–208.

PEJOVIĆ, D., 1957, Geološki i tektonski odnosi terena šire okoline Poćute, zapadna Srbija, s naročitim obzirom na biostratigrafiju gornjokrednih tvorevina, [Geologie und Tektonik der veiteren Umgebung von Poćuta, Westserbien, mit besonderer Berücksichtigung der Biostratigraphie der Oberkretazischen Bildungen]: Posebna Izd. Geol. Inst. "Jovan Žujović," v. 8, p. 1–147.

————, AND RADOIČIĆ, R., 1973, Stratigrafske i paleogeografske metode u proučavanju boksitnog terena Grebnika, [Stratigraphic and paleogeographic methods in investigation of the Grebnik bauxite area, Metohija]: 2 Jugosl. Simp. Istraž. Eksploat. Boksita, A-6, p. 1–11.

PERKINS, B. F., 1974, Paleoecology of a Rudist reef complex in the Comanche Cretaceous Glen Rose Limestone of Central Texas: Geosci. and Man: Louisiana State Univ., Baton Rouge, v. 8, p. 131–173.

PETKOVIĆ, K., ČIČULIĆ-TRIFUNOVIĆ, M., PAŠIĆ, M., AND RAKIĆ, M., 1976, Fruška gora. Monografski prikaz grade i tektonskog sklopa, La Fruška gora, [Exposé monographique de sa structure géologique et de son systéme tectonique]: Novi Sad, Matica Srpska, 267 p.

PHILIP, J., 1972, Paleoecologie des formations à Rudistes du Crétacé supérieur—l'exemple du Sud-Est de France: Palaeogeography, Palaeoclimatology, Palaeoecology, v. 12, p. 205–222.

PLENIČAR, M., 1971, Hipuritna favna iz Stranic pri Konjicah, [The Hippurites fauna of Stranice near Konjice]: Razprave Slov. Akad. Znan. Umetn., 4 v. 14/8, p. 241–263.

PLUMLEY, W. J., RISLEY, G. A., GRAVES, W. W., JR., AND KALEY, M. E., 1962, Energy index for limestone interpretation and classification *in* Classification of carbonate rocks, a symposium, W. E. Ham, Ed., Am. Assoc. Petroleum Geologists Mem. 1, p. 85–107.

POLŠAK, A., 1965a, Rudisti mastrihta iz sjeveroistočnog dijela Zagrebačke gore, [Les rudistes maestrichtiens dans la partie NE de la Zagrebačka gora en Croatie, Yougoslavie]: Geol. Vjesnik, v. 18/2, p. 301–308.

———, 1965b, Geologija južne Istre s osobitim obzirom na biostratigrafiju gornjokrednih naslaga, [Géologie de l'Istrie méridionale spécialement par rapport à la biostratigraphie des couches Crétacées]: Geol. Vjesnik, v. 18/2, p. 415–510.

———, 1971, Über laramische Strukturen von Istrien und Lička Plješevica: I Simp. o Orogen. Fazama u Prostoru Alpske Evrope, Beograd-Bor, p. 41–44.

———, 1979, Stratigrafija biolititnog kompleksa senona kod Donjeg Orešja, SI Medvednica, [Stratigraphy and paleogeography of the Senonian biolithitic complex at Donje Orešje, Mt. Medvednica, northern Croatia]: Acta Geol., v. 9/6.

———, DEVIDÉ-NEDĚLA, D., TURNŠEK, D., GUŠIĆ, I., AND BENIĆ, J., 1978, Biostratigrafski odnosi grebenskih, prigrebenskih i bazenskih naslaga gornje krede u području Donjeg Orešja, SI Medvednica, [Biostratigraphy of Upper Cretaceous reef, subreef and basin deposits at Donje Orešje, Mt. Medvednica, northern Croatia]: Geol. Vjesnik, v. 30/1, p. 189–197.

———, AND LESKOVŠEK, H., 1975, Relations paléothermométriques dans le Crétacé de la Yougoslavie à la base des compositions isotopiques de l'oxygène; reflet sur l'évolution sédimentologique et paléoécologique: International Sediment. Theme 1, 9th. Congress Indicateurs Sédimentologiques, p. 161–167.

———, AND MAMUŽIĆ, P., 1969, Nova nalazišta rudista u gornjoj kredi Vanjskih Dinarida, [Les nouveaux gisements de Rudistes dans le Crétacé Supérieur des Dinarides Externes]: Geol. Vjesnik, v. 22, p. 229–245.

PRUDY, E. G., 1974, Reef configuration: Cause and effect, *in* Laporte, L. F., Ed., Reefs in time and space, Selected examples form the Recent and ancient: Soc. Econ. Paleontologists Mineralogists Spec. Publ. No. 18, p. 9–76.

SAKAČ, K., 1969, O stratigrafiji, tektonici i boksitima planine Grmeč u zapadnoj Bosni [On stratigraphy, tectonics and bauxites of the Grmeč Mountains in western Bosnia]: Geol. Vjesnik, v. 22, p. 269–301.

———, 1973, Stratigrafski položaj i opće karakteristike boksitnih ležišta Dinarida, [Stratigraphic position and general characteristics of the Dinaric bauxite deposits]: 2 Jugosl. Simp. Istraž. Eksploat. Boksita, A/15, p. 1–20.

SCHOLL, D. W., AND MARLOW, M. S., 1974, Sedimentary sequence in modern pacific trenches and the deformed circum-Pacific eugeosynclinal sedimentation, *in* Dott, R. H., and Shaver, R. H., Eds., Modern and ancient geosynclinal sedimentation: Soc. Econ. Paleontologists Mineralogists Spec. Pub. No. 19, p. 193–211.

ŠIKIĆ, K., BRKIĆ, M., ŠIMUNIĆ, AL., AND GRIMANI, M., 1975, Mezozojske naslage Papu˘ckog gorja, [Mesozoic deposits of Papuk highlands]: 2 God. Znan. Skup Sekc. za Primjenu Geol., Geofiz. Geokem. Znan. Savjeta za Naftu JAZU, p. 87–96.

SLADIĆ-TRIFUNOVIĆ, M., 1972, Senonski krečnjaci sa orbitoidima i rudistima Kozluka, severoistočna Bosna, [Senonian limestones with Orbitoides and Rudists from Kozluk, Northeastern Bosnia]: Geol. Anali Balk. Poluostr., v. 37/2, p. 111–150.

SLIŠKOVIĆ, T., 1971, Biostratigraphie du Crétacé supérieur de l'Herzégovine méridionale: Wiss. Mitt. Bosn.-Herz. Landesmus., 1, H.C, p. 13–72.

———, 1976, Deux espèces nouvelles de Rudistes du genre *Petkovicia* Kühn et Pejovic provenante de couches du Sénonien supérieur de la montagne de Vlašic en Bosnie centrale: Wiss. Mitt. Bosn.-Herz. Landesmus., 6, H.C, p. 59–65.

———, AND GUŠIĆ, I., 1973, O stratigrafskom položaju boksita i razvoju krede u istočnoj Bosni, [Stratigraphic position of bauxites and Cretaceous development in Eastern Bosnia]: 2 Jugosl. Simp. Istraž. Eksploat. Boksita, A–X, p. 1–10.

TAPPONNIER, P., 1977, Evolution tectonique du système alpin en Méditerranée: poinconnement et écrasement rigide-plastique: Bull. Soc. géol. France, 7, v. 19/3, p. 437–460.

THIERSTEIN, H. R., 1976, Mesozoic calcareous nannoplankton biostratigraphy of marine sediments: Marine Micropaleontology, v. 1, p. 325–362.

TIEDT, L., 1958, Die Nerineen der österreichischen Gosauschichten: Sitzungsber. Österr. Akad. Wiss., Abt. 1, Bd. 167, H. 1–10, p. 484–517.

VUJNOVIĆ, L., AND VRHOVČIĆ, J., 1975, Neke tektonske karakteristike u prostoru izmedu planina Vlašić i Grmeč, [Some tectonic characteristics in the area between the Vlašić and Grmeč Mountains]: 2 God. Znan. Skup Sekcije za Primjenu Geol., Geofiz. Geokem. Znan. Savjeta za Naftu JAZU, p. 218–222.

WILSON, J. L., 1975, Carbonate Facies in Geologic history: Berlin, Springer-Verlag, 471 p.

ŽIVALJEVIĆ, T., AND PAPEŠ, J., 1973, Geološki pregled ležišta boksita Bosne i Hercegovine, [A geological survey of the Bauxite deposits in Bosnia and Herzegovina]: 2 Jugosl. Simp. Eksploat. Boksita, A-14, p. 1–17.

ZUPANIČ, J., 1976, Senonske naslage tipa Scaglia u zapadnim Dinaridima: postanak sedimenata, lateralne promjene i varijabilnost paleogeografskog zhavenja [Genesis and lateral variability of Senonian "*Scaglia* type" sediments in western part of Dinarids]: Jugosl. Geol. Kongr., 8, v. 2, p. 335–342.

SEPM SPECIAL PUBLICATION NO. 30, P. 473–482, MAY 1981

# VARIOUS PORE TYPES IN A PALEOCENE REEF, BANIJA, YUGOSLAVIA

LJUBO BABIĆ AND JOŽICA ZUPANIČ
University of Zagreb, Yugoslavia

## ABSTRACT

Various pore types and their fillings are described and related to the processes in the developmental history of a Paleocene reef exposed near Banija, Yugoslavia. Several pore types are recognizable and include: 1. large spaces between biolithic constructions, 2. small growth cavities, 3. intraskeletal pores, 4. cavities produced by organic borings, 5. skeletal moldic pores, 6. solution cavities, and 7. fissure cavities. Organic growth and encrustation, internal deposition and cementation are regarded as constructional processes in pore formation whereas abrasion, organism borings, dissolution and fracturing represent destructive processes that produced and modified various types of pore spaces. Included in this discussion is an outline of the paleogeography and depositional history of Paleocene marine carbonates in the Banija region of Yugoslavia.

## INTRODUCTION

The purpose of this study is to describe how various pore types were generated and their later consequences of development. The study area is a Paleocene reef complex exposed near Banija, southeast of the city of Zagreb, Yugoslavia (Fig. 1). This study has taken into account a wide range of processes which characterize the development of various reef complex facies, e.g., organic growth, mechanical disintegration, sediment transport, organism boring, solution, vadose circulation, and internal sedimentation.

## PALEOGEOGRAPHY AND FACIES

The paleogeographic setting of the Paleocene reef complex exposed near Banija is shown in Figure 2. Here a rather long "inner" Paleogene marine zone stretched from Mt. Medvednica (close to Zagreb) across the Banija region to the lower Drina River Valley (Fig. 1). During the Paleocene this marine zone consisted of basinal sediments (with flysch), and sediments laid down in shallow marginal areas. Along the northern marginal areas siliceous clastic sediments were dominant in both coastal and shallow marine settings. These were located along the "Pannonian" land mass, a structural high area composed primarily of metamorphic and acidic intrusive rocks. Marginal to the land mass Mesozoic eugeosynclinal deposits were laid down along with minor amounts of other sediment types. Oceanward from this depositional realm sporadically developed carbonate depositional areas existed in the form of reefs and shoals. These latter carbonates were only developed in those areas not influenced by a river system carrying terrigenous material down from the land mass. This paleogeographic setting is similar to some of the so-called "classic" reef settings presented by Henson (1950) from the Middle East.

## Outcrops and Facies Units

Widespread sedimentary cover of most Neogene and Quaternary deposits has isolated Paleogene outcrops, but field relationships suggest that the reef complex sediments are intercalated within outcropping terrigenous sediments in close proximity. These sediments are represented by conglomerates and sandstones that probably originated in coastal and shallow marine environments. Some of these sediments contain mollusc remains and biolithite fragments.

The reef complex outcrops are well exposed along quarry walls (Figs. 3–5), and consist principally of two distinctive facies units: 1. massive reef limestones, and 2. bedded limestones representing near-reef shoal environments. These units are separated by a fault most clearly seen in the northern part of the quarry, but whose contact is not clearly discernible at the southern end. As a result, facies transitions are not recognizable. Detailed analysis of these facies may be found in Babić, Gušić and Zupanič (1976), but a short summary is presented below.

The massive reef limestone, with an exposed thickness of 20 meters, has been altered by strong tectonics and its reef characteristics and fabric is seen in only a few places. The reef framework is composed of massive growths of hermatypic corals and associated red algae. Accessory reef dwellers consist of encrusting algae (both red and blue-green), foraminifers, and bryozoans. Red (corallinacean) algal crusts have frequently been observed as the final boundstone phase of small biolithitic structures (Figs. 6–7). Another type of sediment appears to have been deposited as "fillings" between and within biolithic constructions. These sediments usually have either a micritic or sparry calcite matrix but are generally grain-supported; although some mud-supported fillings have been noted. All "fillings" have poor to mod-

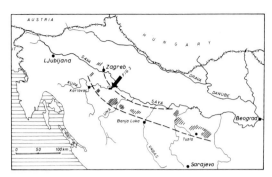

FIG. 1.—Index map showing the distribution of Paleocene-Eocene marine deposits (oblique lines) within the "inner" Paleogene marine zone (dashed lines); compiled from various sources.

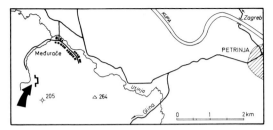

FIG. 3.—Index map of rock quarry showing position of outcrops along quarry walls (squares).

erately poor sorting and contain various particles, some attaining a diameter of 4 centimeters. The various particles include skeletal fragments (red algae, sessile foraminifers, corals, and bryozoans), biolithite fragments, and both fragments and whole specimens of pelecypods and gastropods, loose benthic foraminifers, rare rhodoliths, dasycladaceans, and echinodermal fragments. Siliceous detritus is present only sporadically, but in some cavities it may attain an abundance of 20 percent. Quartz silt is also agglutinated onto some foraminiferal tests.

The bedded limestone facies is at least 12 meters in thickness, but like the massive reef facies, is seen only sporadically due to apparent tectonics. The dominant rock type is a grain-supported skeletal packstone (Fig. 8). Grain sorting is generally poor-to-good and the matrix may be sparry calcite, both sparry calcite and micrite (including microspar), and rarely, only micrite. In most instances micrite recrystallizes to microspar. Grain size averages from 2 to 4 milimeters, but large particles, usually biolithite fragments, may attain a diameter of 100 milimeters. Generally the grains are fragmented skeletal debris of red and blue-green algae, encrusting foraminifers, corals, bryozoans, pelecypods and gastropods. Pelecypod fragments are usually angular, most others are

subangular or rounded. Whole dasycladacean algae, as well as free benthic foraminifers, whole ostracodes, rhodoliths, and blue-green algal nodules are usually undamaged. Composition of the bedded limestones facies, calculated to percents on the basis of thin-section point counts, is shown on Table 1. It should be noted that many loose carbonate grains have been locally stabilized by thin corallinacean algal crusts (crusts up to 30 centimeters long have been observed), and small (up to 10 centimeters in height) biolithic constructions can develop sporadically.

It is thought that this environment reflects deposition within very shallow water marine environments characterized by frequent periods of agitated waters and less frequent spans of quiet water sedimentation. Skeletal grain production

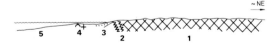

FIG. 2.—Generalized section (not to scale) showing the paleogeographic setting of the Banija Reef complex: 1. Pannonian igneous and metamorphic rocks, 2. location of Mesozoic eugeosynclinal sediments, 3. shallow marine coastal terrigenous sediments, 4. reef (left) and near-reef shoal facies (+), and 5. basinal area in which flysch and marls contain a few turbidite beds.

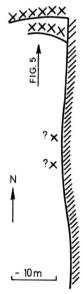

FIG. 4.—Detailed map of the quarry wall showing distribution of massive reef rock (crosses) and bedded sediments (oblique lines); continuous lines delineate quarry walls.

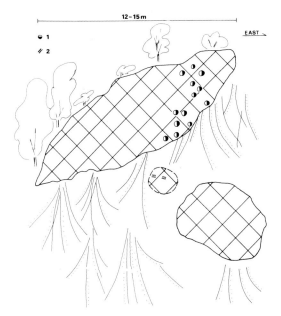

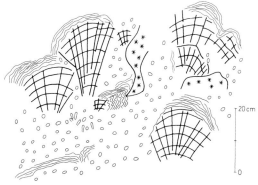

FIG. 6.—Sketch map showing details of the reef framework and location of various components: massive reef corals (heavy diagonal lines), smaller corals (asterisks), algal crusts (thin wavy lines), and various "filled" deposits (circles and ellipses).

FIG. 5.—Sketch map showing outcrop of main reef body (heavy crossed lines): 1. shows the distribution and approximate orientation of geopetal filled solution cavities (not to scale); 2. location of Neptunian dikes.

within this depositional setting appears to have been high, and also appears to have been supplemented by reef detritus brought into the depositional environment. Accordingly, this environment can be thought of as a near-reef, probably back-reef to shoal environment.

TABLE 1.—COMPOSITION OF BEDDED LIMESTONES (FACIES 2) ON THE BASIS OF THIN-SECTION POINT COUNTS

| Samples: | A All constituents | | | B Particles only | | |
|---|---|---|---|---|---|---|
| | 1 | 2 | 3 | 1 | 2 | 3 |
| Cyanophyceae | 3.3 | 1.6 | 1.6 | 5.8 | 2.6 | 2.1 |
| Dasycladaceae | 0.3 | 2.7 | 3.7 | 0.5 | 4.4 | 5.0 |
| Rhodophyta | 21.8 | 14.3 | 17.9 | 37.2 | 23.0 | 24.0 |
| Foraminifera | | | | | | |
| —encrusting | 1.8 | 6.2 | 1.5 | 3.1 | 10.0 | 2.0 |
| —free | 0.4 | 1.2 | 1.2 | 0.6 | 1.9 | 1.6 |
| Anthozoa (Corallia) | — | 2.6 | 11.3 | — | 4.2 | 15.2 |
| Pelecypoda + Gastropoda | 0.9 | 5.1 | 6.5 | 1.5 | 8.3 | 8.7 |
| Ostracoda | 0.2 | 0.1 | — | 0.3 | 0.2 | — |
| Bryozoa | 1.7 | 2.3 | 3.1 | 2.9 | 3.7 | 4.2 |
| Brachipoda | — | — | 0.4 | — | — | 0.5 |
| Echinoidea + Crinoidea | 10.5 | 4.9 | 2.8 | 17.8 | 7.9 | 3.8 |
| Unidentified skeletal particles | 16.6 | 19.0 | 19.7 | 28.1 | 30.7 | 26.3 |
| *Total skeletal particles* | *57.5* | *60.0* | *69.7* | *97.8* | *96.9* | *93.4* |
| Biolithite fragments | 1.0 | 0.9 | 4.4 | 1.6 | 1.4 | 5.9 |
| Intraclasts (?) | 0.2 | 1.0 | 0.3 | 0.4 | 1.6 | 0.4 |
| Siliceous particles | 0.1 | 0.1 | 0.2 | 0.2 | 0.1 | 0.3 |
| *Total non-skeletal particles* | *1.3* | *2.0* | *4.9* | *2.2* | *3.1* | *6.6* |
| Micrite | 3.9 | 0.4 | 1.4 | | | |
| Sparite (incl. microsparite) | | | | | | |
| —among packed particles | 16.7 | 17.7 | 24.0 | | | |
| —masses | 20.6 | 19.9 | — | | | |

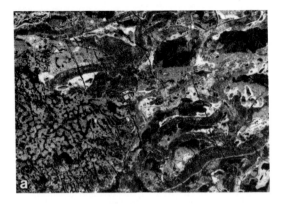

FIG. 7.—Deatiled structures of the reef boundstone framework with emphasis on various small growth cavities. A. Negative print thin-section photomicrograph illustrating typical reef fabric. B. Line drawing of the above field of view with identification of various components: COR, coral; S, crusts of red algae; P, *Pseudolithothamnium album* crust; C, corallinacean alga; black areas represent encrusting porcellaneous foraminifers, possibly including blue-green algae; F, free benthic foraminifers; G, gastropod; B, bored cavity filled with sparry calcite; R, recrystallized sparry vein calcite; small dots, fine-grained calcite cement; and crossed diagonal lines representing sparry calcite cement.

### VARIOUS PORES: FEATURES AND PROCESSES

Seven types of pores and their fillings are described below, and related to processes recognized in the developmental history of the Paleocene Banija Reef.

### Large Spaces between Biolithite Constructions

By necessity, normal growth of a coral reef framework provided numerous open spaces between the individual coral colonies. Figure 6 shows the developments of spaces between coral colonies on the Paleocene Banija Reef. Variations in the size of the interspace areas can vary markedly, and depends to a great deal on coral growth rates and overall morphology. Coral colony inter-

spaces can be as great as 2 meters, and probably represent areas of nonskeletal growth. These interspace areas may even represent former channel ways through the reef mass. Sediments, largely exceeding the biomass of the biolithite clusters, may be filled with sediments that have been derived through the erosion and destruction of neighboring organic buildups. Thus these sediments have a distinctive texture composed of biolithite fragments (Zankl, 1969). Common organism borings into coral heads and reef detritus, both in the massive reef rock and in the bedded near-reef shoal sediments, suggests that a combination of biological (organism borings) and mechanical (wave and current abrasion) processes were continuously acting on and destroying a great portion of the organic framework, and transforming this material into detritus thus creating and adding material to fill-in any available interspace areas.

In addition to reef framework detritus, interspace filling may be composed of the remains of nonattached reef dwellers, detritus derived from neighboring near-reef or back-reef environments, together with some locally derived terrigenous admixtures. It is thought that many of these large primary spaces were filled-in rather rapidly, and probably cemented early in the diagenetic history of the reef sediments.

### Small Growth Cavities

These types of cavities are genetically comparable to the above, but are much smaller in size. In essence they are small pores, up to 10 millimeters in depth, between algal crusts or on the reef surface. Filling of this type of cavity probably began very early and consists mainly of fine-grained carbonate sediment, which may include rare silt-size quartz particles. On the Banija Reef some cavities have been only partially filled, and a few not at all. Apparently, the youngest cavity-fill stage occurred rather soon after initial sediment fill. This final cavity-fill stage generally consists of chemically precipitated calcium carbonate in the form of rim and mosaic sparry calcite cement.

### Intraskeletal Pores

This is not an important type of pore filling process on the Banija Reef and will not be extensively treated. Suffice it to say that these pores disappeared very early, usually by filling-up with fine-grained carbonate and/or sparry calcite cement. In this study the primary organism group affected by filled intraskeletal pores were massive hermatypic corals. Generally these coral pores were filled-in with fine-grained internal cement.

FIG. 8.—Thin-section photomicrograph of a typical bedded limestone facies. This rock can be classified as a grain-supported skeletal packstone; bar scale 1 millimeter: A. fragments of red algae; B. bryozoan; C. gastropod with a coated crust of skeletal algae; D. corallinacean algal fragment coated by another corrallinacean; and E. crinoid particle circumscribed by algal-foraminiferal encrustation. Other skeletal grains, not clearly seen in this photomicrograph, include benthic foraminifers, other sessile foraminifers, and various bryozoans and echinoderms.

FIG. 9.—Negative print thin-section photomicrograph (nonoriented) showing varied organism bored cavities and their fillings; bar scale 2 millimeters. In the upper left there are two narrow tubular borings filled with fine-grained sediment; in the upper center is a dark, bored tubular cavity filled with sparry calcite cement; in the lower right is a large cavity probably formed by the boring action of several pelecypods (photomicrograph shows three probable boring pelecypods within the cavity, and with probable worm encrustations on the upper valve of the largest pelecypod and on the left wall of the cavity).

## Cavities Created by Organism Borings

A variety of organisms (algae, sponges, worms?, and pelecypods) bored and ultimately destroyed skeletal material (Figs. 9–11). Most frequently bored organisms appear to be corals, and pelecypods seem to be the active boring agents. These pelecypod borings clearly postdate internal sedimentation within coral intraskeletal pores. The boring cavities attain a diameter of 12 millimeters, but even larger cavities attributed to pelecypods have been observed, and these appear to have a rather complex boring and filling history (Fig. 9). The activity of borers, especially pelecypods, not only create a distinctive morphology to the cavity, but obviously play an important part in creating rather abundant reef detritus that is eventually deposited in both the reef and near-reef environments. The frequency of borings suggests a conspicuous intensity of boring organisms. The resultant cavities may be completely or partially filled with fine-grained sediment, which may or may not include coarser skeletal sands. Only rarely are cavities free of any sediment. Some "empty" cavities, and the upper portions of partially filled cavities, are characterized by sparry calcite cement fillings which sometimes show increase of calcite crystal size outward from the

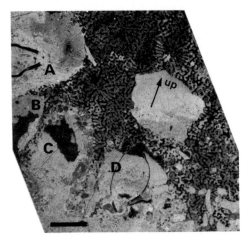

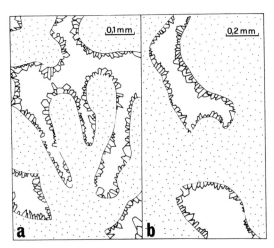

Fig. 10.—Negative print thin-section photomicrograph of four organism-bored cavities; bar scale 1 millimeter. Cavities A–D are separated by thin walls of a bored coral. Cavity C has an organic encrustation on the cavity "ceiling" and along the right wall. The origin of sparry calcite cement in cavities B and C is not clear. Note concentration of small scale borings in lower right corner.

Fig. 12.—Drawings prepared from thin-sections showing calcite fillings in skeletal-moldic pores; dots represent fine-grained sediment, while the blank space represents mosaics of sparry calcite. A. detail from a coral colony, and B. a boring pelecypod (detail from Fig. 9).

cavity walls. Both these filling processes are considered as very early diagenetic features.

Similar deposition and cementation processes have been observed in the interiors of boring pelecypod shells (Figs. 9 and 10).

### Skeletal Moldic Porosity

This type of porosity development commonly occurs when aragonite skeletons have been replaced by calcite, during which time a cavity stage precedes final calcite replacement. This calcite replacement process is observed in corals and in some boring pelecypods. Generally, small tooth-

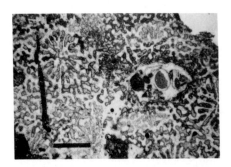

Fig. 11.—Thin-section photomicrograph of internal sediment present in two organism-bored cavities; bar scale 1 millimeter. Narrow tube on left is filled with fine-grained sediment; cavity on the right has skeletal sand cemented by sparry calcite.

shape calcite crystals line the internal cavity wall and then equidimensional sparry calcite crystal mosaics occupy the remaining cavity void (Fig. 12). In these instances, no relict fabrics have been observed, and earlier micrite filling of intraskeletal pores does not appear to have been altered. According to Bathurst (1964) the fabric of replacement sparry calcite indicates cementation of moldic pores, this is in contrast to replacement *in situ*. Other features noted by Bathurst (1964) and Dodd (1966), known as indicators of the cementation process, have not been found. Such features and "enfacial contacts" are rare, but the small dimensions of the pores, which approach the dimensions of individual calcite crystals, may be the cause. No fillings have been observed which display detrital or geopetal arrangements. On the other hand, these features have only been observed sporadically because later recrystallization, especially in corals, produced large crystals which obliterated previous fabrics to varying degrees. In some instances this process produced uniform sparry calcite masses (Figs. 9–11). It seems probable that examples of uniform equidimensional fabrics of crystals, which occupied the space of former skeletal aragonite, have also been generated by recrystalization of primary cement. This replacement, if present, is clearly younger than intraskeletal sedimentation in coral skeletons and internal sedimentation in boring cavities, and the lithification of these cavity fillings.

Aragonite dissolution is generally considered to be caused by fresh water (Bathurst, 1971), but

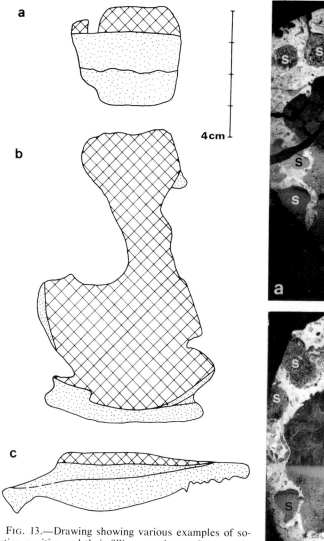

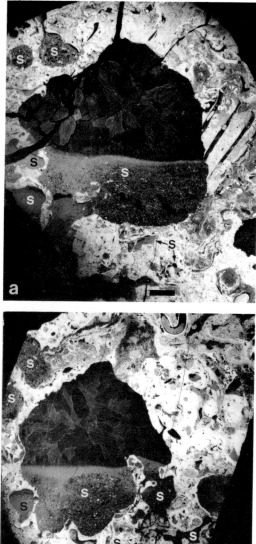

Fig. 13.—Drawing showing various examples of solution cavities and their fillings as observed on weathered reef surfaces; dots represent predominantly fine-grained sediment whereas crossed diagonals represent sparry calcite cement.

other features which would be caused by fresh water circulation have not been identified. For example, there is no enlargement of pores nor any genetic relationship to solution pores cutting included lithified sediment together with calcite replacement cement. Consequently, cementation must have been a rather early phenomenon. Following the conclusion by Schroeder (1973, p. 197) that a "cement in a rock of marine bio- and lithofacies should be expected to be submarine unless evidence to the contrary is found," we may assume submarine conditions for at least the rim-phase cement.

Fig. 14.—Negative print thin-section photomicrographs of typical geopetal-filled solution cavities (S) present in reef rock; bar scale 2 millimeters. Cavities A and B are only spaced about 5 millimeters apart on the reef surface. Important differences in cavity morphology and overall size suggests the presence of a rather complex and irregular channel and cavity system. The light areas of the photomicrographs represent skeletal wackestone. Note presence of small cavities in B and the inclined internal sediment within.

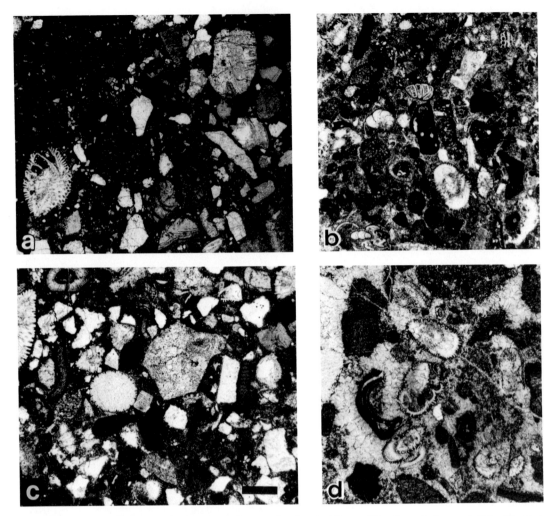

FIG. 15.—Representative thin-section photomicrographs of typical fissure-fill sediments; bar scale 0.5 millimeter. Approximately one-third of the grains shown in photomicrographs A and C are of terrigenous origin. Note the presence of clasts and varied skeletal debris in photomicrographs B and D.

## Solution Cavities

This is a nonselective fabric porosity (*sensu* Choquette and Pray, 1970), that cuts both reef framework, cavity sediment, and replacement calcite of former aragonite skeletons. Solution cavities are younger than the "lithification" process and also younger than "mineral stabilization." Cavity diameter may range from several milimeters to 30 centimeters, and cavity shape is mostly irregular. Quite narrow interconnections between cavities may also be observed (Figs. 13 and 14). The photomicrographs in Figure 14 suggest the existence of a rather complex cavity and channel system, which in some cases may be densely developed. This characteristic suggests a karstification process with consequent emersion

and vadose water circulation. It should be emphasized that no enlargement of, nor genetic connection with probable skeletal-moldic pores (noted above) has been found, and we have seen no indications of contemporaneous dissolution of skeletal aragonite.

Structural features observed in the quarry, i.e., joints, faults, and tectonic breccias, are not related to solution cavities and are obviously younger features. Similarly, the occurrence of oriented geopetal fillings of solution cavities (Fig. 5) indicating vertical reef orientation, suggest that the main tectonic phase postdates cavity fillings. In addition, the identical orientation of geopetal fillings, in both organism borings and in solution cavities, indicates that the reef maintained the same orientation before emergence and after cavity filling. Since solution cavities have only been identified in the eastern portion of the exposed reef facies, we may assume that this period of emergence represented a temporary break in reef history.

Usually, the filling of solution cavities begins with sediment detritus, which may be laminated. This initial cavity-fill is usually composed of a mixture of sand-to-silt-sized skeletal and terrigenous particles, and may even range to lime mud. The base or "floors" of some solution cavities may be level or slightly inclined. Some of the cavity filling may produce wedge-shaped "beds," (Fig. 13B and C). This possibly suggests sediment supply from one side of the cavity, and the trace of several "bedding planes" may possibly indicate some microerosion (see bottom sediments in Figure 13B). These features indicate intermittent vadose water circulation and sediment transport. The cavity shown in Figure 13B shows fine-grained sediment present on the side walls of the cavity, which is not connected with sediment in the lower portion of the cavity. This may represent a relict or remnant of an earlier, more complete infilling, which has been subsequently eroded. Conversely, this could possibly represent only a small "niche" that was infilled at the same time that deposition took place in the bottom portion of the cavity. A younger void stage was eliminated by sparry calcite cementation which produced an even rim and equidimensional fabric, probably under submarine conditions.

### Fissure Cavities

Fissures and their fillings, similar to Neptunian dikes, may range in width from several millimeters up to 20 centimeters, and may be traced laterally up to several meters. These structures can best be seen on large reef blocks that have fallen from the quarry walls, although one rather poorly preserved structure was seen in place (Fig. 5). The Neptunian dikes may run straight, they may curve, or they may change orientation abruptly. The openings of these fissures postdate lithification of indigenous reef rock and precede the main tectonic disturbances, because youngest tectonic features do not appear to be at all connected with these dikes. It is possible that these fissures were initiated during time of reef emersion, but no definitive evidence has been found.

The presence of Neptunian dikes is easily seen due to their dark color (red, grey, green, or brown) in relationship and contrast to the light-colored reef rock. Representative thin-section photomicrographs of typical fissure-fill sediments

are shown on Figure 15. These sediments are usually skeletal wackestones and packstones, but imbricated grainstones are also present. All rock types do contain terrigenous admixtures, and may even grade to calcarenitic sandstones. Still, some dikes contain a few "beds" slightly differing in composition and texture. The different textural properties and the sharp contact between these "beds" suggest that at least a few of these fissures were probably filled during several phases, which may have been separated by periods of nondeposition. The fissure filling processes were probably accomplished within a submarine environment, and likely to have been coeval with youngest reef growth. This is thought to be the case because of the presence of whole skeletons and similar bioclastic particles occurring in the fissures, in both the reef and near-reef shoal facies. It would also reinforce the idea that the fissures were opened during this phase of reef history.

We have given some thought to the question as to the derivation and source of the contained detritus found within the fissure fillings. As we have noted, there are textural differences in fissure sediments, and there is a pronounced increase in the amount of terrigenous particles. In addition, there is a slightly higher frequency of rotallid foraminifers in these sediments. We believe that these differences are a result of sediment particles derived from a different environment and source area, than those of the reef and back-reef shoal environments. It is our opinion, that a possible fore-reef environment, not seen in outcrop or even suspected until now, seems to be the depositional environment and source area needed to produce this type of fissure filling.

### CONCLUSIONS

Several pore types and cavities were generated and modified by various processes that characterize a small Paleocene coral reef in northeastern Yugoslavia. These are: 1. large spaces between biolithite constructions, 2. small growth cavities, 3. intraskeletal pores, 4. cavities created by boring organisms, 5. skeletal-moldic pores, 6. solution cavities, and 7. fissure cavities. Although outcrop exposures of the Banija Reef are not ideal they did furnish abundant data relating to pore genesis and modification, which suggest a rather complex diagenetic history of reef environment and facies. We have noted that diagenetic processes in reefs, especially those concerning various pore types, aid in the recognition of events that are the basis of understanding reefs as geological features.

## REFERENCES

BABIĆ, LJUBO, GUŠIĆ, IVAN, AND ZUPANIĆ, JOŽICA, 1976, Grebenski paleocen u Baniji (Središnja Hrvatska)—Paleocene reef-limestone in the region of Banija, Central Croatia: Geol. Vjesnik, v. 29, p. 11–47.

BATHURST, R. G. C., 1964, The replacement of aragonite by calcite in the molluscan shell wall, *in* Imbrie, J., and Newell, N., Eds., Approaches to paleoecology: New York-London-Sydney, Wiley, p. 357–376.

———, 1971, Carbonate sediments and their diagenesis, Developments in Sedimentology 12: Amsterdam-London-New York. Elsevier, 620 p.

CHOQUETTE, P. W., AND PRAY, L. C., 1970, Geologic nomenclature and classification of porosity in sedimentary carbonates: Am. Assoc. Petroleum Geologists Bull., v. 54, p. 207–250.

DODD, J. R., 1966, Processes of conversion of aragonite to calcite with examples from the Cretaceous of Texas: Jour. Sed. Petrology, v. 36, p. 733–741.

HENSON, F. R. S., 1950, Cretaceous and Tertiary reef formations and associated sediments in Middle East: Am. Assoc. Petroleum Geologists Bull., v. 34, p. 215–238.

SCHROEDER, J. H., 1973, Submarine and vadose cements in Pleistocene Bermuda reef rock: Sedimentary Geology, v. 10, p. 179–204.

ZANKL, HEINRICH, 1969, Der Hohe Göll. Aufbau und Lebensbild eines Dachsteinkalk-Riffes in der Obertrias der nördlichen Kalkalpen: Abh. Senckenberg. Naturforsch. Ges., No. 519, 123 p.

SEPM SPECIAL PUBLICATION NO. 30, P. 483–539, MAY 1981

# OLIGOCENE REEF CORAL BIOFACIES OF THE VICENTIN, NORTHEAST ITALY

STANLEY H. FROST
Gulf Research and Development Company
Houston, Texas U.S.A.

## ABSTRACT

Outcrops of Oligocene reef facies strata in the Vicentin area of northeastern Italy are well known because of the extensive collections of fossil reef corals that have been made from them. This paper presents an overview of the paleoecologic distribution of these reef corals and identifies coral communities indicative of different environments, or of different stages in ecologic successions.

One of the largest and best exposed of outcropping Tertiary reefs is the Berici barrier-reef/lagoonal complex of the Castelgomberto Limestone in the Colli Berici and eastern Monte Lessini. The barrier-reef, which now forms much of the southeast face of the Colli Berici, has a core facies which is 150–200 meters thick × 800–900 meters wide and about 8 kilometers long. The lagoonal facies represents most of the Castelgomberto Limestone and extends northwestward back of the barrier-reef for about 30 kilometers. Other buildups of this area include a. patch-reefs and coppices in the shelf-lagoon, b. a basal Oligocene sequence alternating between thickets of branching corals in terrigenous beds and coral head biostromes in carbonates, c. coral carpets and biostromes high in the Castelgomberto Limestone which postdate the Berici Barrier-reef and d. biostromal assemblages adapted to soft mud substrates in marine tuffaceous mudstones which cap the Castelgomberto Limestone.

In the Marósticano area, coral thickets, coppices and thin fringing-reefs formed along the northeastern margin of a large Oligocene volcanic complex. These structures developed in a region dominated by terrigenous clastic sedimentation in some cases adjacent to centers of submarine volcanic eruption.

Variations of three basic coral assemblages are responsible for the wide spectrum of buildups: a. low diversity "pioneer" or thicket assemblage; b. high diversity reef-core and flank assemblage; and c. soft mud substrate assemblage. The environmental factors which appear to have been important in determining the areal mix of species, the community diversity and the abundance of the reef corals, are adaptation to water flow, potential to colonize unstable substrates and to survive sedimentation, and species-specific dominance relationships. Interaction of these factors can explain the seral succession of pioneer-to-intermediate-to-climax communities observed in many Tertiary reef buildups.

Most of the biomass that existed at any given time when these ancient reefs were living left no trace in the fossil record of these buildups. Particularly important in the ecology of these reefs (by analogy with modern reefs) were the benthic algae, sponges and octocorals. For the most part, each of these groups (except coralline red algae) is unrepresented or greatly underrepresented in the Vicentin reef fossil record. This is because they lacked mineralized skeletal parts or because these skeletons were destroyed before they became part of the sediments during diagenesis. Most of the reef dwelling biota which *did* leave an extensive fossil record in the buildups actually aggregated a relatively minor part of the reef biovolume, either as standing crop or over a period of time.

## INTRODUCTION

The extensive outcrops of coral-bearing Oligocene limestone in the Vicentin area of Italy have long been known to geologists and paleontologists. Indeed, the abundant and well preserved corals were studied as early as 1842 by Michelin, and virtually all of the known species were described before the 20th century. These assemblages, as well as corals from the Piemonte area of northwestern Italy, from southern Bavaria and from northwestern Yugoslavia, include more than 520 named species and have formed the standard of comparison for Oligocene coral assemblages on a worldwide basis. Thus, Vaughan (1919, p. 199 and 202) in considering the abundant and diverse Oligocene coral assemblages from Antigua, used the Vicentin faunas as a standard of reference for the middle Oligocene.

However, little has been written about the paleoecological relationships of these species or their distribution in the thick carbonate sequence. In addition, until recently relatively little effort has been made to apply modern taxonomic principles to this vast and bewildering array of species, most of which were defined in accord with 19th century typological concepts.

This paper involves the following objectives:

1. to present an overview of the paleoecologic distribution of Oligocene corals in northeastern Italy and to identify coral communities indicative of different environments or of different stages in ecologic successions;

2. to reconstruct the ecological and sedimentological development of several types of Oligocene reef buildups in the Vicentin; coral biostromes, coppices, patch-reefs, and a well preserved barrier-reef shelf-lagoon complex.

In order to adequately analyze the regional and local habitats of Oligocene coral communities in the Vicentin it is first necessary to examine the relationships of Oligocene sedimentary facies there.

## Monte Lessini-Colli Berici

Major outcrops of Oligocene coral-bearing strata in northeast Italy are confined to the eastern part of the Monte Lessini, to the Colli Berici, and to a triangular outcrop belt northeast and northwest of Maróstica (Fig. 1). The most important single Oligocene sedimentary unit in this area is a thick, resistant light-gray to cream limestone known to pioneering workers as the "strati di Castelgomberto" (Maraschini, 1824), and to modern workers (Bosellini et al., 1967) as the "calcareniti di Castelgomberto" or "calcare de Castelgomberto" (Coletti et al., 1973). This formation crops out over some 200 square kilometers in the hills of the eastern Monte Lessini between Montécchio Maggiore and Malo and in the Colli Berici between San Donato and Vicenza (Fig. 1). Over most of its outcrop area in the Monte Lessini and Colli Berici, the Castelgomberto Limestone consists of highly fossiliferous biomicrite and biosparite with local lignite beds and lenses of volcanic tuff or sedimentary breccia. It is estimated by Bosellini et al. (1967, p. 25) to be about 200 meters thick in the eastern Monte Lessini, while Coletti and others (1973, fig. 8) show it to be 300 meters thick in the same area. A composite stratigraphic section prepared by the author from field work in the eastern Monte Lessini (Fig. 2) indicates that the estimate of Bosellini et al. (1967) is the most accurate. In the southeastern face of the Colli Berici, the Castelgomberto Limestone includes an Oligocene barrier-reef and/or carbonate shelf-edge buildup which is at least 200 meters thick in maximum development (Fig. 1, and Rossi and Semenza, 1958).

The base of the barrier-reef in the Castelgomberto Limestone of the southeast margin of the Colli Berici appears to more or less coincide with the Eocene-Oligocene boundary (Francavilla et al., 1970; Ungaro, 1972). In the rest of the Colli Berici, the basal Oligocene appears to be represented by skeletal limestones especially rich in nodular coralline algae. To the northwest in the eastern Monte Lessini, earliest Oligocene sediments grade upwards from upper Eocene (Priabonian) shales and marls, and consist of beds of dense skeletal limestone rich in miliolid foraminifers and coralline algae alternating with marls and calcareous shales or sandstones characterized by thickets of branching poritid scleractinian corals (Figs. 2, 3, 4 and 5A).

The uppermost part of the Castelgomberto Limestone is preserved in the summits of the higher hills in the southern part of the Monte Lessini outcrop belt, as at San Urbano, and in the central Colli Berici. At several localities in the southern Monte Lessini the uppermost beds of limestone are overlain by gray to pinkish sedimentary tuffs and mudstones. At Monti Grumi a heavily weathered calcareous tuff about 25 meters thick contains a diverse and beautifully preserved coral assemblage which is clearly still Oligocene in age. The marine tuffs appear to be the youngest Oligocene marine sediments in the region. At several localities such as Covolo and Sovizzo in the southern Monte Lessini, a clean quartzose sandstone ("saldame"), in part with nummulites and echinoids, interfingers with or overlies the tuffs (Coletti et al., 1973, figs. 8 and 9, p. 26) and at others blankets a karst solution surface developed on the Castelgomberto Limestone (Bosellini et al., 1967, p. 25).

Suprajacent sandstone and carbonates at San Urbano in the southern Monte Lessini and at Valmarana and Altavilla in the Colli Berici contain an abundant lower Miocene (Aquitanian) marine fauna including Lepidocyclina (Bosellini et al., 1967, p. 26; Fabiani, 1915, p. 57).

Oligocene rocks are also exposed in the Colli Euganei, about 10 kilometers southeast of the Colli Berici. However, instead of reef-lagoonal carbonates these rocks are latitic-trachyandesite submarine flows, agglomerates, sedimentary tuffs and thin, fossiliferous marls (Proto Decima and Sedea, 1970; Dieni and Proto Decima, 1964) which contain deep water early Oligocene assemblages of planktic and benthic foraminifers and molluscs. Thus, the Colli Euganei appears to preserve basinal facies volcanogenic sediments equivalent to at least the lower part of the Colli Berici Barrier-reef.

## Marósticano

About 15 kilometers northeast of the eastern margin of the Monte Lessini, Oligocene sediments of volcanogenic and shallow water biogenic origin are exposed in a triangular outcrop pattern between Bassano del Grappa at the eastern end and Thiene at the western end (Fig. 1). These rocks, totaling some several hundred meters in thickness, are exposed in an east-northeast plunging monoclinal structure and are the product of two major Oligocene events affecting this area of the Vicentin: 1. regional marine transgression, and 2. regional volcanism, resulting here in basaltic submarine flows and tuffaceous sediments (Piccoli, 1958 and 1965) which also form much of the southwestern Marósticano.

The Oligocene sedimentary sequence in the Marósticano (Oppenheim, 1909; Piccoli, 1967;

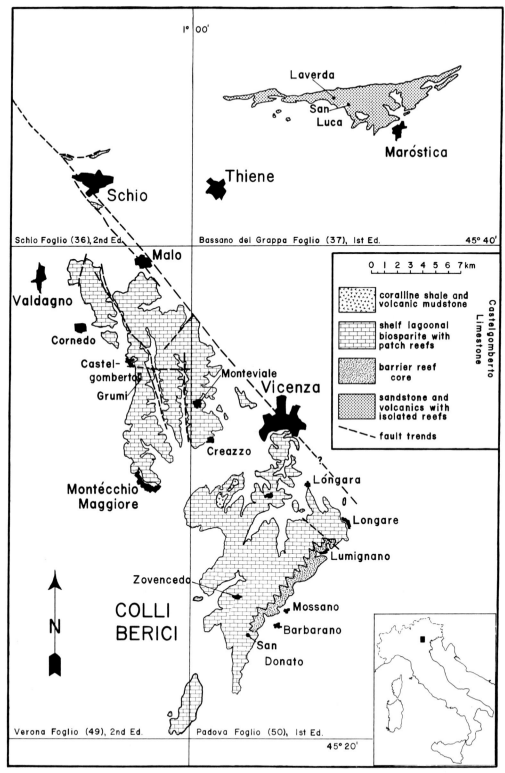

Laverda

San
Luca

Maróstica

Thiene

Schio

Schio Foglio (36), 2nd Ed.    Bassano del Grappa Foglio (37), 1st Ed.    45° 40'

Malo

Valdagno

Cornedo

Castel-
gomberta
Grumi

Monteviale

Vicenza

0  1  2  3  4  5  6  7km

coralline shale and
volcanic mudstone

shelf lagoonal
biosparite with
patch reefs

Castelgomberto
Limestone

barrier reef
core

sandstone and
volcanics with
isolated reefs

fault trends

Creazzo

Lóngara

Montécchio
Maggiore

Longare

Lumignano

N

Zovenceda

COLLI
BERICI

Mossano

Barbarano

San
Donato

Verona Foglio (49), 2nd Ed.    Padova Foglio (50), 1st Ed.

45° 20'

FIG. 1.—Outcrops of Oligocene strata in the Vicentin, after Bosellini *et al.*, 1967, Picolli, 1967, and Rossi and Semenza, 1958.

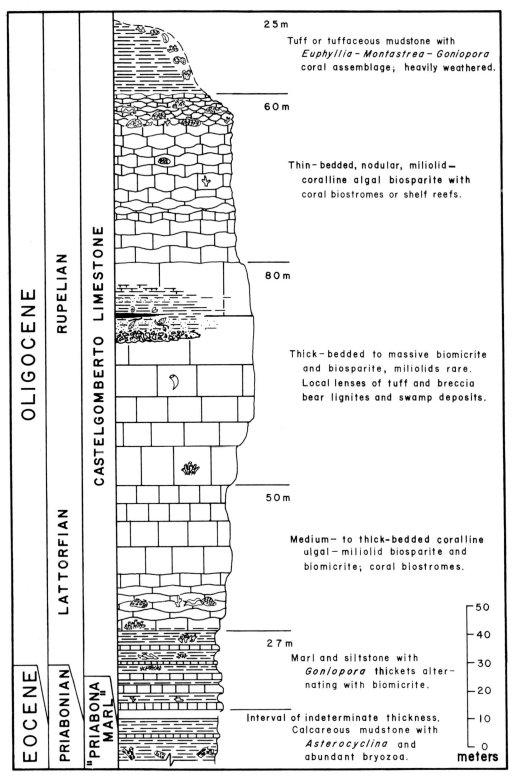

FIG. 2.—Composite Oligocene stratigraphic sequence for the eastern Monte Lessini.

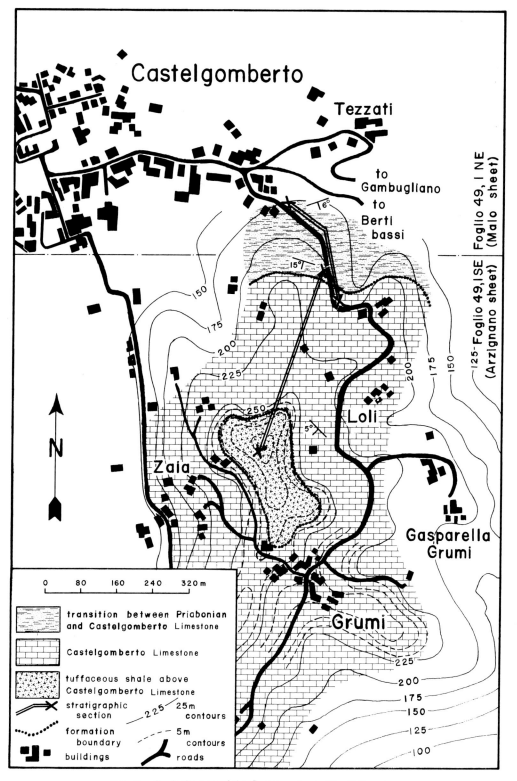

FIG. 3.—Geologic map of the Castelgomberto-Monti Grumi area.

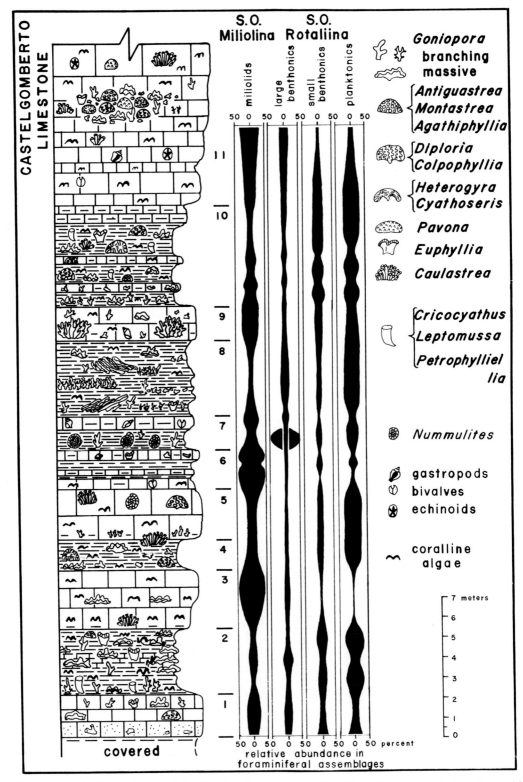

Fig. 4.—Basal Oligocene strata of the Castelgomberto area; stratigraphic section measured along the road between Castelgomberto and Loli.

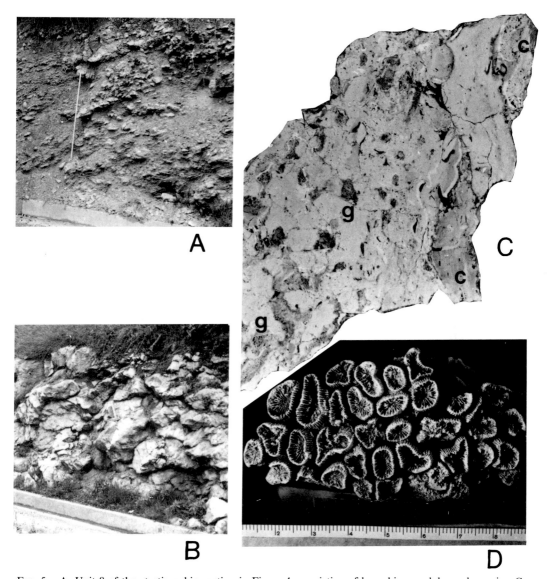

FIG. 5.—A. Unit 8 of the stratigraphic section in Figure 4, consisting of branching, nodular and massive *Goniopora* and *Actinacis* in a marly mudstone and siltstone matrix; the stick is one meter long. B. Lower beds of the Castelgomberto Limestone, Unit 11 of Figure 4. These are thick-bedded to massive biosparites, packed biosparites and biomicrites with small oysters, echinoids and masses of the "organ-pipe coral" *Caulastrea*. The hammer in the center of the photograph is 35 centimeters long. C. Sawed slab surface, natural size, and parallel to bedding, of lagoonal biomicrite formed in a *Goniopora-Caulastrea* thicket. Micritic mass in upper right corner is a lime mud fill of the cavity formed by the decay of a soft-bodied organism (probably a sponge) growing among the *Caulastrea* corallites. Part of outer surface was encrusted by coralline algae. *C. Caulastrea*, transverse and longitudinal sections of corallites. *G.* = *Goniopora*, transverse section of thin branches. Colli Berici, exposure along road between Zovencedo and Pederiva. Specimen collected by Dr. R. Herb. D. Upper surface of a mass of corallites of *Caulastrea pseudoflabellum*. Specimen from Catullo Collection, University of Pisa. Photograph by Dr. T. Pfister.

Ritondale Spano, 1967 and 1969) consists of a basal series (Mortisca Sandstone) of nearshore marine or in part terrestrial terrigenous sediments; sandstone, siltstone and conglomerate with thin lenses of skeletal biosparite. The middle part (Calvene Formation) contains marine calcareous sandstone and marl, beds of skeletal biosparite and interbedded basaltic pillow lavas and tuffs (Dal Pra, 1965; Dal Pra and Medizza, 1965; Piccoli, 1967). The upper unit (Salcedo Formation) marks the culmination of the Oligocene transgressive cycle, with the development of relatively thick calcareous units, some of which include coral banks and reef structures, but basaltic lavas and tuffs or tuffaceous mudstones account for a substantially greater volume of the rock. The volcanic component of the Marósticano Oligocene, especially flows or pillow lavas, appears to decrease substantially or to disappear in the eastern quarter of the outcrop belt (Piccoli, 1967). Surprisingly, all the coral-bearing localities reported in early studies (Michelin, 1842; Catullo, 1856; d'Achiardi, 1867 and 1868A and B; Reuss, 1869; de Angelis, 1894) and in modern investigations (Barta-Calmas, 1973; Pfister, 1974, 1977A and in press; Frost, this study) are concentrated in the northcentral and northwestern parts of the outcrop belt and are close to the major Oligocene complex of the southwestern Marósticano. Moreover, Oligocene sediments exposed 15 kilometers to the northeast in the Possagno area (Venzo, 1938; Masperoni Bastianutti, 1964) are only locally present, and where present are near-shore glauconitic sandstone with thin marly interbeds. These strata contain abundant bryozoa (Braga, 1965) in addition to benthic foraminifers (Masperoni Bastianutti, 1964) and molluscs (Venzo, 1938) but very few, if any, corals.

## OLIGOCENE BIOSTRATIGRAPHY IN THE VICENTIN

As of this writing, few data exist to enable a detailed biostratigraphic zonation of Oligocene sediments in the Vicentin. The lowermost part of the Castelgomberto Limestone in the eastern Monte Lessini and in the Colli Berici can accurately be dated as early Oligocene (Oppenheim, 1900; Kranz, 1910; Fabiani, 1915), on the basis of sedimentological continuity with upper Eocene (Priabonian) strata which are now reasonably well dated in the Monte Lessini and Colli Berici areas (Hardenbol, 1968; Cita and Piccoli, 1964; Francavilla et al., 1970; Ungaro, 1972; Herb and Hekel, 1973). However, placing the exact Eocene-Oligocene stratigraphic boundary, especially in the eastern Monte Lessini, is difficult. Coletti et al. (1973, figs. 8 and 9) include all of the marly sediments at the base of the Castelgomberto Limestone in the upper Eocene. Field investigation by the writer along the western edge of the outcrop belt of the Castelgomberto Limestone between Castelgomberto and Montécchio Maggiore confirms that Oligocene strata are present beneath the base of the Castelgomberto Limestone (Fig. 3). A new road cut up the eastern face of Monti Grumi south of Castelgomberto exposes a sequence more than 25 meters thick of alternating coralline marl, claystone, and limestone below the Castelgomberto Limestone (Figs. 4 and 5A). These strata contain an abundant and diverse scleractinian coral fauna clearly indicative of the Oligocene in the Alpine Tethyan area. Moreover,

the marly beds bear moderately well preserved planktic foraminifers which on preliminary investigation appear to be indicative of the lower Oligocene Cassigerinella chipolensis/Hastigerina micra planktic foraminiferal zone of Bolli (1966) or lower Globigerina ampliapertura zone of Blow (1969). Upper Eocene hantkeninid planktic foraminifers, with their distinctive tubulospinose chambers, are conspicuous by their absence (Berggren, 1972, figs. 5 and 6), although they do occur in the upper Eocene at Ghenderle, 6 kilometers north of Castelgomberto. Finally, nummulitids occur throughout this section, being especially abundant in Unit 7 of Figure 4, and are of the Oligocene reticulate Nummulites intermedia-fichteli and striate Nummulites vascus-boucheri groups. The contact of these earliest Oligocene strata with the Priabonian is obscured in the eastern end of Castelgomberto. In outcrops at Priabona and Ghenderle to the North (Hardenbol, 1968, fig. 2) supposed uppermost Priabonian sediments are capped by a bed of algal-bryozoan limestone 2–4 meters thick which directly underlies the main mass of the Castelgomberto Limestone. This unit, as well as the subjacent bryozoan claystones, may also be earliest Oligocene.

The thick-bedded to massive lower part of the Castelgomberto Limestone, 120–140 meters thick, has long been considered to be lower Oligocene (Lattorfian) on the basis of similarity of mollusc, bryozoan, coral and other faunas with those of the "Sagonini beds" of the northwestern Marósticano (e.g., Oppenheim, 1900, p. 250; Kranz, 1910, p. 184–85; Fabiani, 1915, p. 39–41; Coletti et al., 1973, Table 1). The upper thin-bedded part bearing abundant corals, is about 50 meters thick and is regarded as the middle Oligocene (Rupelian) stratigraphic standard for the western Tethys. Kranz (1910, p. 186) demonstrated that prolific faunas of the suprajacent volcanic tuffs, as at Monti Grumi, were virtually identical to those of the Castelgomberto Limestone and should be considered to be Rupelian also.

It is important to note, however, that modern detailed investigations of taxonomic relationships and stratigraphic distribution of the major fossil groups used by early workers to subdivide the Veneto Oligocene indicate that the stratigraphic range of these species is much greater than had been appreciated. Thus, a recent analysis of the stratigraphic distribution of extensive middle Oligocene molluscan assemblages in the eastern Monte Lessini by Coletti et al. (1973) shows that of 32 species of bivalves, only 2 have not been recorded from the early Oligocene or Eocene, and 5 range into the late Oligocene or early Miocene. Of 75 species of gastropods, only 16 are not also known from the early Oligocene or Eocene, and 13 range into the late Oligocene or early Miocene.

It appears, on the basis of this information, that the stratigraphic ranges of most of these mollusc species are too long to satisfactorily subdivide Oligocene strata of the Vicentin into stages. The same relationship appears to be true for Oligocene bryozoa described by Braga (1965) from the Possagno area. Of 48 species (one new), only 7 appear to range through less than the entire Oligocene. Biostratigraphic zonation of the Veneto Oligocene by means of the stratigraphic distribution of scleractinian corals is no more precise, although there have been more than 520 species described from the Oligocene of the Alpine Tethyan region. Of these supposed 520 species, less than 120 appear to be valid at this writing, and almost all of these range through the entire marine Oligocene section of the Alpine Tethys. Closely comparable Oligocene coral assemblages occur in the Caribbean, and most key species have recently been demonstrated (Frost, 1972; Frost, 1974; Frost and Langenheim, 1974) to range from the uppermost Eocene to the uppermost Oligocene, being replaced in the lower Miocene by a coral fauna (Frost, 1977a) of substantially different aspect. Comparison of Alpine Tethyan Oligocene coral faunas with Miocene faunas of the same region (Chevalier, 1977) indicates that a similar sweeping faunal turnover occurred in the latest Oligocene or earliest Miocene. On this basis, it appears that Oligocene strata can be clearly differentiated from lower Miocene strata by their associated coral faunas.

Oligocene coral-bearing strata in the northcentral Marósticano ("Sagonini strata") were considered by most early workers (e.g., Reuss, 1869; Bittner, 1883; Munier-Chalmas, 1891; Oppenheim, 1900; Canestrelli, 1908; Kranz, 1910) as lower Oligocene (Lattorfian), while Fabiani (1915) placed them in the middle Oligocene. But Piccoli (1967), in a detailed investigation of these strata in the Laverda Valley area, concluded that the lower and middle Oligocene could not be differentiated, even though these sediments are richly fossiliferous.

In total it thus appears that stage-level subdivision of the shallow marine Oligocene sequence in the Vicentin is impractical on the basis of the stratigraphic distribution of the fossils that have been used. The most precise zonation of these strata would ideally be based on the stratigraphic distribution of planktic organisms such as calcareous nannoplankton and planktic foraminifers. Thus Berggren (1972, fig. 6) regards both the Lattorfian and Rupelian Stages as early Oligocene and the Chattian as late Oligocene. He defines the Rupelian-Chattian datum on the basis of the youngest occurrence of reticulate *Nummulites* and the earliest occurence of *Miogypsina* (among other taxa), and correlates this boundary with the extinction of the planktic foraminifer *Pseudohastigerina*. Striate and reticulate *Nummulites* range through the entire marine Oligocene sequence in the Vicentin, including the fossiliferous marine tuffs capping the Castelgomberto Limestone in the southeastern Monte Lessini, and on this basis the barrier-reef-shelf lagoon sediments of this region and perhaps those of the Marósticano accumulated during the early Oligocene, as defined by Berggren.

## DEVELOPMENT OF OLIGOCENE REEFS IN THE VICENTIN

### *Colli Berici-Monte Lessini*

The Berici Barrier-Reef–Castelgomberto lagoon complex represents the culmination of a major mid-Cenozoic tectonic and sedimentary cycle in northeastern Italy that began in the late Eocene and ended in the late Oligocene. As such it comprises a cycle of a duration of at least 8–10 million years, including a protracted period of subaerial erosion during its latter part. Upper Priabonian (upper Eocene) strata in the Vicentin grade southeastward from blue claystone and argillaceous limestone characterized by the great abundance of bryozoa in the eastern Monte Lessini (Seuss, 1868; Hardenbol, 1968; Bosellini *et al.*, 1967) to bryozoan-crustose coralline algal limestone in the Colli Berici (Fabiani, 1908; Schweighauser, 1953; Francavilla *et al.*, 1970; Ungaro, 1969; 1972). These sediments were deposited on a broad submarine platform that sloped gently southeastward, with the source of fine-grained terrigenous sediment apparently in the Alpine highlands to the north. The change in lithofacies, which is characterized by decrease in volume of argillaceous sediment and increase in volume of biogenic debris, is gradual from northwest to southeast. Uppermost Eocene sediments in this region can be reasonably well differentiated from the suprajacent Oligocene by the widespread presence of larger foraminifers such as *Discocyclina* and *Asterocyclina* which became extinct at the close of the Eocene, as well as by characteristic species of *Operculina*, *Heterostegina*, and *Nummulites* (Roveda, 1961; Francavilla *et al.*, 1970; Ungaro, 1969 and 1972). In the Colli Berici there is clear evidence of a progressive shallowing of the depositional environment in the latest Eocene. This is demonstrated by the stratigraphically upward expansion in abundance of crustose coralline algae, particularly *Archaeolithothamnium*, *Lithothamnium*, and *Mesophyllum*, and the upward decrease and final disappearance of *Discocyclina* and *Asterocyclina* in the upper 30 meters of strata beneath the reefoid limestone at Barbarano (Francavilla *et al.*, 1970, Table 1) and at Lumignano (Ungaro, 1972; Geister and Ungaro, 1977).

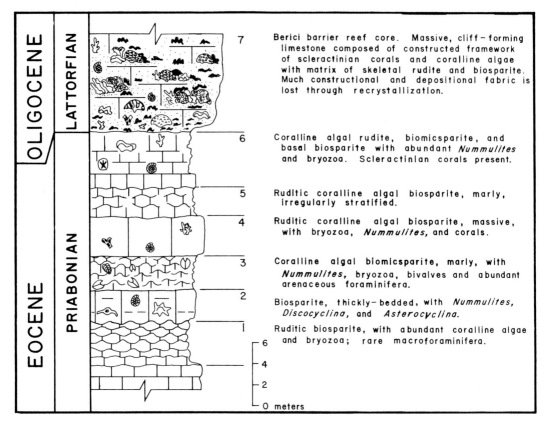

FIG. 6.—Stratigraphic section through the base of the Berici Barrier-Reef at Lumignano.

Conspicuous by their rarity in Priabonian strata are hermatypic (reef building) scleractinian corals. This is somewhat surprising in the light of their abundance and diversity in Oligocene strata deposited under essentially similar conditions. Priabonian corals do occur in the Possagno area about 45 kilometers to the northeast and are summarized by Oppenheim (1901, p. 49–77). These assemblages are dominated by large solitary hermatypic (?) genera such as *Cycloliltes* and *Placosmilia,* and ahermatypic corals such as *Parasmilia* and *Flabellum.* Potential reef framework constructors are rare, being represented by a few specimens of *Astreopora, Actinacis* (or *Haimesastrea*), and *Goniopora.* The general aspect of the assemblage is that of a community adapted to life on unstable terrigenous mud substrates. Similar Eocene communities occur in the Caribbean-Gulf of Mexico region (Frost, 1972; 1977a). It appears that at least part of the ecologic niche occupied by thickets of ramose scleractinian corals on terrigenous substrates in the early Oligocene was occupied by abundant and diverse cheilostome and cyclostome bryozoan assemblages during the latest Eocene.

The Berici Barrier-Reef developed near the southeast margin of a northwest-southeast trending submarine platform (Fig. 1) in earliest Oligocene time (Fig. 6). Although the southeast margin of this platform sloped gently basinward during the Eocene, there is evidence to indicate that a substantial break in submarine topography developed seaward of the barrier-reef during the latest Eocene, thus accentuating the southeastern slope of the platform and deepening the basin to the southeast. This appears to have been due to vertical movement in a fault zone which parallels the trend of the barrier-reef (Fig. 1) and is shown by Coletti *et al.* (1973, fig. 1) to lie 3.5–4.0 kilometers southeast of the Colli Berici. Thus the block northwest of the fault(s) either did not move or was uplifted only a few meters while the southeastern block subsided several tens of meters or more.

The upper 20 meters of strata beneath the base of the reef along its northeastern extent (Fig. 6) display clear evidence of the shallowing trend that led to the establishment of communities of framework building scleractinians. These strata correspond to units H and I of Francavilla *et al.* (1970,

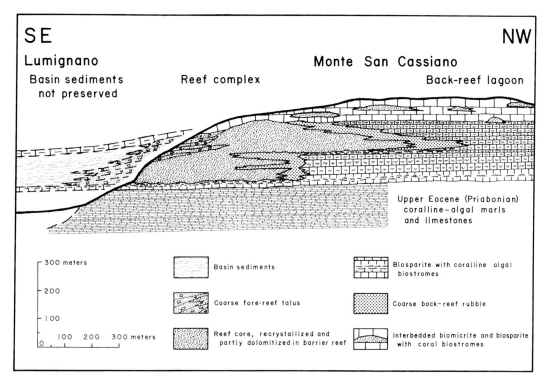

SE                                                                    NW

Lumignano                              Monte San Cassiano

Basin sediments          Reef complex                    Back-reef lagoon
not preserved

Upper Eocene (Priabonian)
coralline-algal marls
and limestones

300 meters

[ ] Basin sediments

[ ] Biosparite with coralline algal biostromes

200

100

[ ] Coarse fore-reef talus

[ ] Coarse back-reef rubble

0   100   200   300 meters

[ ] Reef core, recrystallized and partly dolomitized in barrier reef

[ ] Interbedded biomicrite and biosparite with coral biostromes

FIG. 7.—Cross-section through the Berici Barrier-Reef at Lumignano.

Table 1) and to unit C of Ungaro (1972). The lower 10 meters of the sequence (units 1–3 of Fig. 6) are rubbly and marly coralline algal rudite and biosparite in indistinct beds about 1 meter thick, deposited under shallow marine conditions of strong wave and current action. Larger foraminifers include characteristic upper Eocene species of *Nummulites* as well as rare *Discocyclina* and *Asterocyclina* (Francavilla *et al.*, 1970) and abundant bryozoa, including the cheilostomes *Vibracella*, *Steginoporella*, *Porina* and *Tubucella* (Ungaro, 1972, p. 78). Unit 3, marly rudite of coralline algae, also contains bivalves and abundant arenaceous smaller foraminifers. The upper half of the sequence (units 4–6 of Fig. 6) is dominantly coralline algal rudite, biosparite and argillaceous biomicrite in beds about 0.5 meter thick. *Discocyclina* and *Asterocyclina* appear to be absent in these beds, while actinacid and poritid scleractinian corals appear in the beds of cleanly winnowed biosparite. Bryozoans, such as the cyclostomes *Crisia* and *Hornera*, and cheilostomes *Steginoporella*, *Biflustra*, and *Porina*, are abundant throughout this interval (Ungaro, 1972, p. 79). Francavilla *et al.* (1970) consider this upper interval to be transitional between the Eocene and Oligocene, while Ungaro (1972, p. 79) assigns an

uppermost Eocene age, based on foraminiferal and bryozoan assemblages.

Investigations of the north end of the barrier-reef indicate that initial construction of the shelf-edge buildup began in the earliest Oligocene with the colonization of frame building scleractinian corals of a substrate of coarse coralline algal rubble. The basal beds of the reef at Monte San Cassiano and Monte Broion in the Lumignano area contain a relatively small volume of coral skeletons per volume of sediment. These corals include massive species of such genera as *Montastrea*, *Antiguastrea*, *Agathiphyllia*, *Goniopora*, *Colpophyllia*, *Pavona*, and *Diploria*. Much of the volume of the initial reef framework is skeletal rubble bound into place by encrusting coralline algae and by submarine cements. Bryozoa are poorly preserved, but appear to have been abundant in the early stages of the reef, and were progressively squeezed out of surface sites of attachment by the competitively superior reef biota.

The reef complex consists of 1. a northeast-southwest trending reef-core which is 800–900 meters in maximum width (Figs. 1, 7, and 8) and about 8 kilometers long (Rossi and Semeza, 1958, p. 59; 1962); and 2. a sheet of coarse skeletal rubble swept northwestward behind the reef by waves

FIG. 8.—Oblique view, looking northwest, of erosional section through the Berici Barrier-Reef front, Monte San Cassiano at Lumignano. Lower slopes are underlain by uppermost Eocene coralline algal-bryozoan limestone and marl. Two stages of reef-core underlie the middle and upper slopes. Re-entrants in cliff faces are formed by differential weathering of rubbly beds formed predominantly by coralline algae; note heavy vegetation cover.

and currents (Fig. 7). The reef-core facies is a maximum of 150–200 meters thick, and at Lumignano actually comprises two superposed cores, a basal core about 500 meters wide and 100 meters thick and a second, plano-convex core as much as 700 meters wide and 110–115 meters thick, which was established 150–200 meters northwest (shelfward) of the subjacent lower core. Field examination of both reef-cores exposed in the east flank of Monte San Cassiano and west flank of Monte Broion indicates that much of the original depositional and constructional fabric of the reef-core has been obscured by a complex series of events. These include: a. penecontemporaneous destruction of coral and other skeletal structures by bioeroders; b. diagenetic solution of originally aragonitic skeletal microstructures and imperfect replacement by calcite pseudomorphs; c. formation of several phases of internal sediment and aragonite and/or magnesian calcite cavity-fill (now represented by calcite pseudomorphs); d. dolomitization of irregular masses of the original fabric; e. at least two phases of pervasive flushing by meteoric water and formation of karstic terrain; and f. extensive deposition of travertine by meteoric water on cave surfaces and along joints. The combination of factors b, c, d, and e tend to give the Berici Barrier-Reef complex considerable remnant primary, biomoldic, intercrystalline, and fracture porosity and permeability in the core facies. The pervasive alteration, which makes the reef-core facies an excellent example of carbonate reservoir rock, also makes it very difficult to determine exactly how much reef coral boundstone framework volume is present in the core of the reef. By analogy with Oligocene reef masses in the Marósticano as well as those from Mexico,

Jamaica, Antigua, Puerto Rico, and the U.S. Gulf Coast investigated by the author, we might estimate a maximum of 50–70 percent coral skeletal volume for typical Oligocene reef-crest and buttress zones with substantially less coral framework in the skeletal sand sheets back of the reef.

A census was made of the reef-core by the author on both vertical and bedding-plane exposures on Monte San Cassiano and Monte Broion. In most parts of the reef-core, recognizable massive corals account for less than 15 percent of the rock volume present. This figure might be increased to 20 percent if large sawed and polished surfaces could be examined. However, lenses at a number of levels within the reef-core are composed of as much as 85 percent by volume of columniform, nodular, or ramose species of *Goniopora, Astreopora, Actinacis, Dendracis, Stylocoenia*, and possibly *Alveopora*. These fossil coral thickets and coppices weather as prominent reentrants on exposure faces (Fig. 8). Other reentrants are formed by differential weathering of coarse rubble masses composed of nodular coralline algae.

Geister and Ungaro (1977) contend, after an extensive field study, they could not find strong evidence to confirm the presence of a barrier-reef in the southeast face of the Colli Berici. They base their conclusions primarily on the fact that they were not able to discern laterally continuous interlocking coral boundstone framework in many of the exposures along the reef-front. In addition, they note that where they found masses of interlocking reef framework there is evidence to indicate that these did not rise appreciably above the surface of the surrounding sediments. They suggest that if the barrier-reef did occur at the margin of the platform, its location must have been to the southeast and that the face of the Colli Berici has been eroded back from the shelf-edge reef site. Another alternative which they pose is that there was no barrier-reef and that the southeastern margin of the carbonate platform was a carbonate ramp which sloped gently into the basin.

It appears that Geister and Ungaro have an unrealistic view of what they would expect to see as fossil evidence for a platform-margin, fringing or barrier-reef. Their comparative model of a platform-margin reef must be that of a linear wall reef which is growing on the upper slope and crest of a steep bathymetric scarp. Here the topography would promote efficient downslope removal of skeletal sediments and where coral settlement and growth is concentrated along a narrow strip, forming reef-core rock in which a very high proportion is composed of coral skeletons. The field evidence for the front of the Colli Berici suggests to this writer that the slope across the reef-crest, fore-reef and upper basin slope was less steep, especially during the final phase of barrier-reef

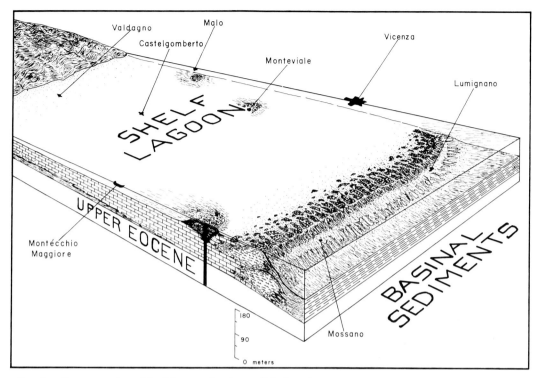

FIG. 9.—Reconstruction of the Oligocene barrier-reef/platform-lagoon complex, final stage; patch-reefs in lagoon have been omitted.

growth (Fig. 7). Secondly, with regard to the lack of depositional relief between the coral framework masses and the surrounding sediments, it appears that Geister and Ungaro have in mind large patch-reefs or possibly the reef-front buttresses typical of Pleistocene and Holocene reefs of the Caribbean. From their map (1977, fig. 1) of the massive coral buildups they investigated in the Colli Berici, it is apparent that virtually all of these grew in the shallow central and rear part (lagoonal side) of the reef complex where the slope northwestwards into the lagoon was gentle and where the high rate of sedimentation skeletal debris swept across the reef and the shallowness of the water limited vigorous coral growth. Finally, as Geister and Ungaro admit, good exposures of Oligocene carbonates in the Colli Berici are limited by the dense growth of bush that covers most of the area and thin caliche-like crusts on bare rock surfaces. In summary, their tentative disclaimer of the existence of a platform-margin reef tract in the southeastern Colli Berici is unconvincing.

Erosional cross-sections of the Berici Barrier-Reef occur in the flanks of deep valleys incised in the southeast face of the Colli Berici. The most extensive of these exposures are in the valleys

north and south of Lumignano where the reef-core can be seen to grade northwestward through wedges of coarse back-reef rubble (Fig. 7) to coralline algal-miliolid foraminiferal skeletal biosparites and biomicrites similar to that exposed to the northwest in the Monte Lessini. Analysis of the relationships between the rear margin of the reef-core and the plume of skeletal sediments shed behind it indicates that the rate of deposition of the sediments immediately back of the reef more or less kept pace with the constructional upward growth of the reef-core. Thus no recognizable reef-flank beds developed back of the reef and the difference in water depth between the rear margin of the reef-flat (which was essentially at sea level) and the margin of the back-reef lagoon was only a very few meters (Fig. 9).

The upper surface of the reef-core is irregular, with isolated mounds of core that project upward into horizontally-bedded coralline algal rudites and biosparites. The final phase of the reef, as represented by the upper reef-core, fore-reef talus and back-reef rubble, apparently consisted of a deeply embayed reef trend in which the upper fore-reef slope was much gentler than in the first phase of the reef (Figs. 7 and 9). This was because the fore-reef slope of the upper part of the reef was developed on the upper surface of the subjacent first phase reef-core. Thus the final phase of the Berici Barrier-Reef must have been similar in plan view to the present day Swains Reef complex of the southern Great Barrier Reef or to the deeply embayed southern part of the Belize (British Honduras) Barrier Reef between Queen Cay and Sapodilla Cay (Fig. 10). In these examples, extensive linear "wall" reef trends characteristic of steep shelf-breaks are absent. Instead the reef trend is discontinuous, consisting of isolated crescentic or elliptical reef mounds which usually lack lagoons. Reefs of this type have been termed shallow ring-reefs or composite apron-reefs, depending on the slope back of them. The isolated reef mounds of the upper Berici Reef appear to be ancient examples of apron-reefs.

The northwestward displacement of upper reef-core away from the steep shelf-margin appears to have been in response to general subsidence which resulted in deepening conditions across the entire carbonate platform. Back-reef lagoonal strata equivalent to this phase of the reef also show evidence of this deepening trend. This lateral migration of reef building communities in response to change in sea level is an excellent example of a transgressive reef.

*Fore-Reef.*—In contrast with the massive exposures of the barrier-reef cores and back-reef sediments of the Castelgomberto Limestone, relatively little remains of the fore-reef sediments shed on the slope of the subsiding basin to the southeast. Erosion has stripped off the softer basinal sediments to expose the surfaces of the barrier-reef cores more or less as they must have been near the culmination of the reef complex (Figs. 7 and 8). This is analogous to the differential weathering of basinal sediments to expose the well known Permian El Capitan Reef complex of West Texas. Upper fore-reef sediments shed from the upper phase of the reef-core are preserved as a cap over the lower reef-core in the southeast face of Monte San Cassiano. These are mixtures of skeletal and reef framework rubble and marly or argillaceous basinal sediments.

*Basinal Facies.*—Oligocene basinal sediments deposited southeast of the reef trend appear to have been completely eroded away, forming the lowlands between the Colli Berici and the Colli Euganei. Most of the upper Priabonian sediments have also been removed (Fig. 7).

In the low hills of Montegalda about 5 kilome-

ters east-southeast of the extreme eastern edge of the Colli Berici, Oligocene coral-bearing limestones overlie basaltic breccias and tuffs. These carbonates are apparently equivalent to the Berici Barrier-Reef, and may have formed a fringing-reef or bank constructed on the top of a submarine volcanic buildup in the basin.

Early Oligocene sediments apparently equivalent to the lower part of the barrier-reef occur in the volcanogenic terrain of the Colli Berici. These include marls and tuffaceous mudstones with abundant planktic and benthic foraminifers at Cava Cementifico Zillo (Proto Decima and Sedea, 1970), and thin tuffs and marls with molluscs and planktic and benthic foraminifers interbedded with latitic-trachyandesitic submarine flows and breccias (Dieni and Proto Decima, 1970).

*Back-Reef Lagoonal Facies.*—A broad shelf-lagoon extended northwestward back of the Berici Barrier-Reef for at least 30 kilometers during the early Oligocene and early middle Oligocene (Fig. 9). This occupied a carbonate platform of at least 470 square kilometers area in a regional terrigenous and volcanogenic depositional terrain. The exact margins of the platform are difficult to determine because the western margin of the Castelgomberto Limestone outcrop has been eroded back from what must have been its original extent, while the eastern margin is defined by a major northwest-southeast trending fault zone (Fig. 1). From the evidence at hand, it appears that the barrier-reef developed along the crest of the platform, being hindered in lateral expansion to the northeast and southwest by increasing depth of water on the flanks of the platform. This relationship clearly occurs on the southwestern end of the reef trend at San Donato, where scleractinians disappear and the reef-core abruptly disappears into coralline algal rudites. The shoreward margin of the lagoon is not preserved. Indeed, stratigraphic sections figured by Coletti and others (1973, fig. 8) from Monte Saglio, Faedo and Campi Piani in the extreme northwestern end of the outcrop belt, indicate almost no change in lithofacies between there and localities to the southeast.

Oligocene lagoonal sediments deposited in the Colli Berici back of the barrier-reef appear to be essentially similar throughout the existence of the reef. These consist of thick-bedded to massive coralline algal-miliolid foraminiferal biosparite and rudite, interbedded with biomicrite containing local thickets or coppices of branching and mas-

$\rightarrow$

FIG. 10.—Holocene equivalent of final stage of Berici reef/lagoon complex: deeply embayed Belize Barrier Reef at Queen Cay. Reef has migrated westward away from shelf-edge because of eastward tilting of southern one-fourth of Belize Shelf. O.N.I. unclassified photographs.

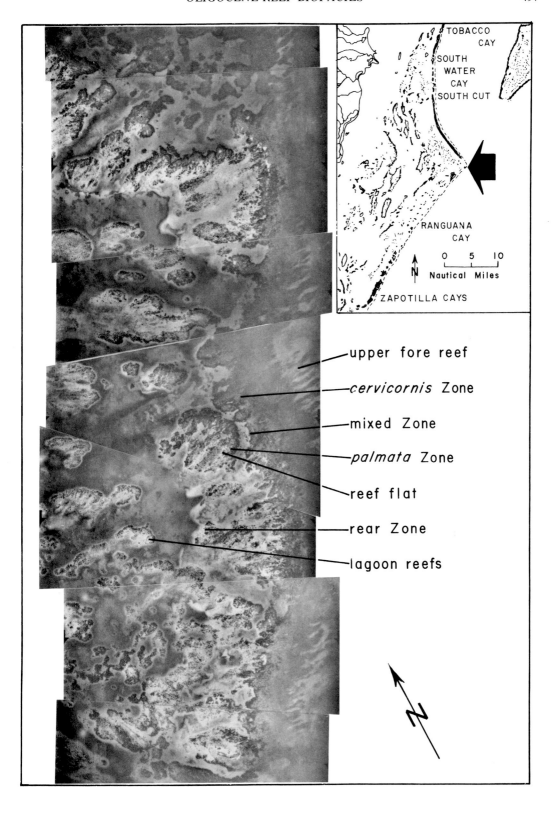

upper fore reef

*cervicornis* Zone

mixed Zone

*palmata* Zone

reef flat

rear Zone

lagoon reefs

sive corals such as *Goniopora, Astreopora, Alveopora, Actinacis* and *Euphyllia* (Fig. 5C and D). Earliest Oligocene lagoonal sediments of the Lessini outcrop area however, consist of coralline silts, mudstones and marls alternating with resistant beds of miliolid foraminiferal biosparite and biomicrite (Figs. 4, 5, and 11B and C). These sediments were deposited in the initial stages of the northwestward expansion of the carbonate shelf-lagoon. The local depositional environment alternated cyclically between shallow marine terrigenous substrates with thickets of branching corals (10–30 meters depth) and carbonate lagoonal substrates with biostromes (and, perhaps patch-reefs) containing massive corals. Foraminiferal assemblages, as well as the corals, effectively underscore these differences in depositional environment. Figure 4 depicts relative abundance in foraminiferal assemblages between strata in the lower slopes of Monti Grumi, determined from counts of specimens both matrix-free and in thin-section. Foraminifers of the suborders Miliolina and Rotaliina were most useful, with the inverse relationship between miliolids and planktic rotaliid foraminifers particularly sensitive to the alternation in lithology.

The earliest Oligocene silts and clays represent incursions of terrigenous sediments from the north onto the platform at a time when the Berici Barrier-Reef was just being established at its southeastern margin. Their characteristic foraminiferal assemblage, dominated by small benthic bolivinids and uvigerinids and by planktic foraminifers, has been widely reported from terrigenous middle- to outer-shelf environments, both modern and ancient (e.g., Bandy, 1964). Various mixtures of these foraminifers are indicative of a relatively wide range in depth from about 15 meters to as much as 200 meters. The presence of abundant hermatypic scleractinian corals, however, indicates that the maximum depth must have been at or near the shallowest part of this range because of the light requirements of symbiotic plants (zooxanthelle) in the tissues of the corals. The relatively turbid conditions that must have been present in the water column during the deposition of these terrigenous sediments probably limited the maximum depth of vigorous scleractinian coral growth here to less than 15–20 meters.

The interbedded limestones bear a lagoonal facies miliolid assemblage characterized by *Austrotrillina paucialveolata* Grimsdale, *Praerhapydionina delicata* Henson, and *Heterellina hensoni* Grimsdale. Abundant and diverse miliolid foraminiferal assemblages are particularly characteristic of tropical shelf-lagoonal environments. These early Oligocene assemblages are closely comparable to modern miliolid assemblages described by Wantland (1975) from the British Honduras (Belize) back-reef/shelf-lagoon and by Collins (1958) from the northern region of the Great Barrier Reef. Also important in the lagoonal-facies foraminiferal assemblage are conical rotaliids with complex internal structure such as *Chapmania* and *Halkyardia* (or *Crespinina*). These typically occur in reef associated carbonate of the ancient Tethys and modern Indo-Pacific provinces. Other rotaliids include abundant *Amphistegina, Cymbalopretta*, encrusting *Falsocibicides* and *Dyocibicides* and *Nummulites*.

The lagoonal facies foraminiferal assemblages occur throughout the Castelgomberto Limestone (Fig. 11D). While planktic foraminifers do occur in the limestones of the basal Oligocene mixed terrigenous-carbonate sequence, they diminish in abundance in the lower beds of the Castelgomberto Limestone and are rare to absent in characteristic samples of this formation from both the Colli Berici and Monte Lessini. This suggests that the establishment of the barrier-reef effectively blocked calcareous plankton in the upper water column of the open basin to the southeast from being swept into the lagoon by surface currents and wave action, and also that their development in the lagoon must have been relatively minor.

Other fossils also occur in the interbedded terrigenous-carbonate sequence beneath the Castelgomberto Limestone. Molluscs, including the bivalves *Ostrea, Spondylus, Pecten,* and *Chama* and the gastropod *Ampullina* occur sparsely throughout the entire sequence (written comm.,

$\rightarrow$

FIG. 11.—Photomicrographs of reef-lagoon facies carbonates: A. Skeletal rubble apron behind Berici Barrier-Reef. Packed biosparite with coralline algae (dense grains), cyclostome bryozoa, and the foram *Amphistegina*, ×25. Road exposures between Zovencedo and Pederiva, Colli Berici, collected by Dr. R. Herb. B. Marl and argillaceous biomicrite, ×10, of Unit 2 of Figure 4, lowermost Oligocene. Encrusting foraminifera *Eorupertia* (with porous chamber walls) intergrown with masses of coralline algae. C. Carbonate beds of lowermost Oligocene sequence, Unit 5 of Figure 4. Algal-foraminiferal biomicrite, ×30, with the dasycladacean *Neomeris* (?). D. Basal beds of the Castelgomberto Limestone, Unit 11 of Figure 4. Lagoonal biomicrite, ×30, with small and large porcelaneous forams, including *Praerhapydionina* (left margin) and *Peneroplis*. E. Uppermost Castelgomberto Limestone at Monti Grumi, coral biostrome facies postdating the Berici Barrier-Reef. Packed biomicrite, ×30, with coralline algae and the codiacean green alga *Ovulites* (?). F. Calcareous tuffs or tuffaceous mudstone, ×30, capping the Castelgomberto Limestone at Monte Grumi.

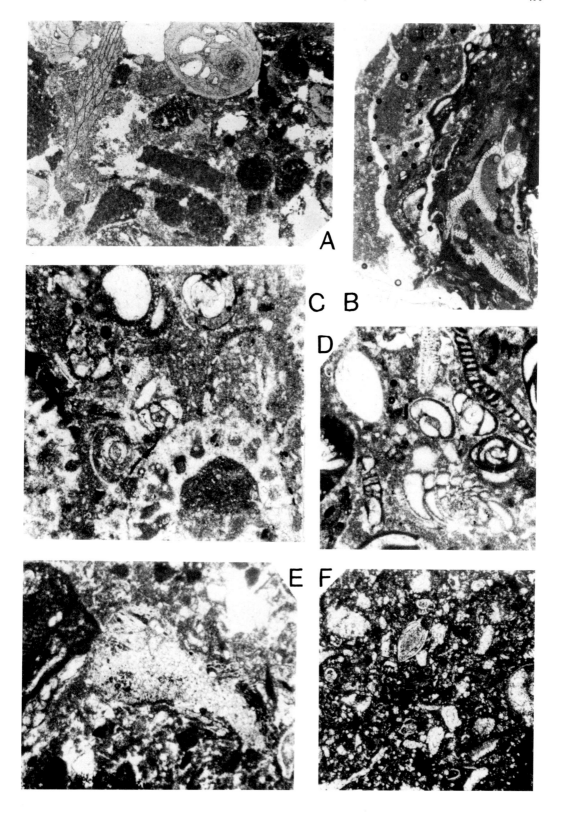

Peter Jung). Molluscs are abundant only in the upper part of Unit 7, Fig. 4, where many large specimens of *Spondylus* and a few *Ampullina* occur in the limestone. Bryozoa (Fig. 16C) range through the section, but appear to be most abundant in the terrigenous beds.

The major volume of the Castelgomberto Limestone in the eastern Monte Lessini and the Colli Berici, corresponding to the lower two-thirds of the formation, was deposited back of the Berici Barrier-Reef. This corresponds to the lower 200 meters of the formation as depicted by Coletti *et al.* (1973), but measurement of the Castelgomberto-Monti Grumi section by the author indicates that this interval is about 130 meters thick (Fig. 2). The same thickness, 130 meters, was determined by Kranz (1910, p. 198) for these strata. The thin-bedded upper third of the formation, depicted by Coletti and others as about 100 meters thick, and measured by the author to be 60 meters thick, apparently postdates the barrier-reef and correlates with strata suprajacent to the reef in the southeastern Colli Berici (Fig. 7). Thus the back-reef lagoonal equivalents of the barrier-reef in the Castelgomberto Limestone apparently thin from about 200 meters thick at the rear margin of the reef to about 130 meters thick in the northwestern part of the outcrop belt. This is a reasonable relationship based upon the demonstrably higher skeletal sediment productivity of modern reef tracts in comparison with their associated lagoons, e.g., Matthews (1965) for British Honduras (Belize), and Maxwell (1968, p. 13–232) for the Great Barrier Reef.

This lower two-thirds of the limestone is surprisingly uniform over the Lessini and Colli Berici outcrop areas, indicating that the large scale sedimentological regime remained more or less constant on the lagoonal platform during the existence of the barrier-reef. The lower 40–50 meters of the formation is miliolid-coralline algal biomicrite and biosparite, with local argillaceous or marly partings, in irregular beds 50–200 centimeters thick. Coletti *et al.*, (1973, p. 24 and fig. 9) show that the abundance of coralline algae and *Nummulites* in the lower beds decreases northwestward in the Lessini outcrop area, disappearing altogether in the outcrops north of Cornedo. At these localities (Campi Piani and Faedo) miliolid foraminifers and the boring bivalve *Toredo* dominate the biota. In the central and southern Lessini outcrop area the basal units bear mollus-

can assemblages dominated by gastropods such as *Ampullina, Cypraea, Conus, Cryptoconus,* and *Terebellum* which Coletti and others estimate to be indicative of a maximum depth of 30 meters. In general, these lower beds contain only isolated colonial corals, and especially characteristic are phaceloid colonies of *Caulastrea pseudoflabellum* (Catullo) which reach a meter or more in diameter. However Coletti *et al.* (1973, fig. 9 and p. 24) report a small patch-reef at Monte Monedo, near Montécchio Maggiore, and a small biostrome of massive frame building corals occurs about 10 meters above the base of the formation in the Monte Grumi Road section (Fig. 2).

There appears to have been a gradual deepening of the lagoon, represented by limestones about 80 meters thick which are more or less similar in texture to the lower strata but in much thicker beds. Coletti *et al.* (1973, p. 26) suggest that these strata were deposited at a maximum depth of about 70 meters, based on molluscan assemblages dominated by carnivorous gastropods such as *Turritella* and *Xenophora* and the substantial decrease in abundance at some localities of miliolid foraminifers. Scleractinian corals are sparse in these strata, assemblages being mixtures of hermatypic and solitary ahermatypic species.

Details of the physiography of the shelf-lagoon were apparently quite complex during the existence of the Berici Barrier-Reef. Strong submarine eruptions of basaltic tuff and breccia occurred sporadically on the platform over most of the time it existed (Fabiani, 1908; 1915; Kranz, 1910; Piccoli, 1966; Coletti *et al.,* 1973, fig. 1). Kranz (1910, p. 189) reports unfossiliferous tuff masses low in the Castelgomberto Limestone at Monte Cugola (northwest of Creazzo), and at Monte Crocetta, both in the southeastern part of the Lessini outcrop belt. Fabiani (1908, p. 29; 1915, p. 51) reports an extensive mass of yellowish tuff interbedded with the barrier-reef core about 60 meters above its base near Mossano while Coletti *et al.* (1973, figs. 8 and 9) depict masses of tuff associated with a volcanic neck in the Castelgomberto Limestone near Priabona. Apparently most of the masses of tuff and breccia poured out on the lagoon floor were entirely submarine, or, if they built up above the water surface, existed for only short periods of time. However, a few of the eruptions built extensive subaerial cays or islands in the lagoon. These existed for fairly substantial periods of time, long enough to accumulate beds

→

FIG. 12.—A. Southeastern part of reef-core and flank beds of small fringing or platform-reef, San Luca, Maróstica. Distance covered in view is about 20 meters. B. Reconstruction of reef in A. Data from author's field notes and Pfister, 1974, Figure 3.

**A**

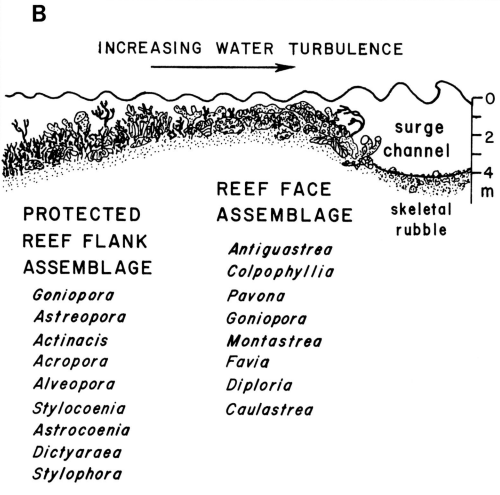

**B**

INCREASING WATER TURBULENCE

surge channel

0
2
4
m

skeletal rubble

**REEF FACE ASSEMBLAGE**

*Antiguastrea*
*Colpophyllia*
*Pavona*
*Goniopora*
*Montastrea*
*Favia*
*Diploria*
*Caulastrea*

**PROTECTED REEF FLANK ASSEMBLAGE**

*Goniopora*
*Astreopora*
*Actinacis*
*Acropora*
*Alveopora*
*Stylocoenia*
*Astrocoenia*
*Dictyaraea*
*Stylophora*

of lignite as much as 30 centimeters thick, as a Monteviale (Kranz, 1914, p. 289). Three lignites have been reported from the Castelgomberto Limestone. These occur at Zovencedo in the Colli Berici and at Monteviale and Monte Piano in the Lessini (Fabiani, 1915, p. 56–68; Kranz, 1910, p. 191, and 1914, p. 289; Coletti et al., 1973, figs. 8 and 9) and apparently occur at more or less the same stratigraphic horizon in the upper middle part of the formation. The lignite-bearing sequences occur on substrates of basaltic breccias, some of which contain marine fossils such as the gastropods *Natica* and *Strombus*. These grade upward through thin gray-green clay, sand and marl (derived from the erosion of the volcanic rocks) with fresh or brackish water molluscs, fish remains, wood, leaves and stems, to dark brown lignites. The fossil leaves provide abundant evidence of the type of vegetation that accumulated to form the lignite. Surprisingly, these indicate a floral assemblage which includes bog myrtle, poplar, cinnamon, dogwood, honeysuckle, eucalyptus, and mountain ash (Mossalongo, cited in Fabiani, 1915, p. 57), all forms which are restricted to fresh water or only slightly brackish conditions (P. D. Sorenson, pers. comm., 1974). Moreover, the lignites contain the fresh water turtle *Trionyx*, as well as *Anthracotheriium*, a swine-like mammal, and *Trigoniaas*, a rhinoceros (Beggiato, 1865; Dal Piaz, 1930). Thus, it is reasonable to infer that these islands were high enough above the sea so that swamps developed in low-lying areas over fresh water lenses in the porous volcanic rocks. Transition from the swamp environment back to marine lagoonal conditions usually involved muddy sands and clays with abundant fresh or brackish water molluscs (*Potamides, Corbula,* etc.).

Figure 9 summarizes the major environmental and sedimentological units that occurred on the platform during the existence of the Berici Barrier-Reef.

*Biostromal Facies.*—Minor warping of the carbonate platform produced profound changes in the distribution of reef building and reef associated communities. Downwarp or tilting of the southeastern margin of the platform made the benthic environment too deep to support vigorous framework construction by hermatypic scleractinian corals (perhaps 50–60 meters). These were displaced to shallow water environments on the platform to the northwest, and the upper part of the shelf-margin reef core grades into suprajacent rubbly limestones with great volumes of nodular coralline algae. Coarse rubble coralline algal pavements such as these have been reported by various authors (e.g., Thomassin, 1971, p. 386–388, fig. 4; Adey and Macintyre, 1973) to be typical of modern fore-reef environments beneath the

zone of active coral growth at 30–60 meters, extending in cases of clear water to as much as 200 meters depth.

It appears that there was concomitant upwarp of much of the platform northwest of the barrier-reef rather than simple subsidence of the entire platform. While the barrier-reef on the southeastern margin of the platform was killed by progressive drowning, the shelf-lagoon to the northwest became appreciably shallower than it had been during the existence of the final phase of the barrier-reef.

The sedimentological products of this phase of the development of the platform comprise the upper part of the Castelgomberto Limestone. These are thin-bedded, nodular miliolid-coralline algal biosparite and biomicrite totalling 40–50 meters in thickness (shown by Coletti et al., 1973, figs. 8 and 9, to be 90 meters thick). In the central and northern part of the Lessini outcrop area, extensive flat shelf-reefs or biostromes, with a diverse frame building scleractinian assemblage, occur in the upper part of this unit. These reefs seldom are more than 1–1.5 meters in thickness and essentially consist of "pavements" of massive and encrusting colonial coral heads. The sediments become finer to the southeast in the Colli Berici, reflecting somewhat deeper conditions, and thickets or coppices constructed by branching or phaceloid corals replace the coralline "pavement". The coral coppices in the upper Castelgomberto Limestone are comparable to similar *Goniopora-Actinacis* biostromes or thickets of the earliest Oligocene, but are taxonomically more diverse. This difference appears not to be due to evolutionary change in diversity but rather to the absence of appreciable terrigenous sediment in the upper Castelgomberto Limestone.

Coletti and others conclude, on the basis of analysis of the petrology and molluscan assemblages of this upper unit, that it was deposited under substantially shallower conditions than the subjacent middle part of the formation. Miliolid foraminifers and coralline algae constitute most of the volume of skeletal sediment, while mollusc assemblages are dominated by gastropods such as *Xenophora, Strombus, Cryptoconus,* and *Turritella*. Although they report that *Nummulites* is absent from these beds, the author has found numerous specimens of both *N. vascus* and *N. intermedius* in the uppermost limestone beds at Monti Grumi.

*Volcanogenic Sediments.*—The final phase of Oligocene marine sedimentation on the platform involved the deposition of water laid basaltic tuffs, now represented by scattered erosional remnants of grayish-blue or pinkish bentonitic mudstone and claystone (Bosellini et al., 1967, p. 25; Piccoli, 1966). These remnants are preserved in the north end of the Colli Berici and in the south end of the Lessini Plateau, and it is difficult to determine from them if the entire platform was smothered with volcanic products. While the shallow marine carbonate shoal community characteristic of the upper Castelgomberto Limestone was buried by the rain of volcanic ash, a few marine communities survived. These were adapted to soft mud substrates and were capable of evading or withstanding burial by fine sediment or were types which would be unaffected by these conditions. One such community occurs in the volcanic mudstone underlying the rounded crest of Monti Grumi, south of Castelgomberto (Fig. 11F). At this classic locality, great numbers of well preserved fossils weather out of a grayish volcanic mudstone. These have been the subject of numerous monographs starting in the early 19th century. Kranz, in a series of papers, summarized the stratigraphic distribution of molluscs (1910), bryozoa, brachiopods, echinoids and larger foraminifers (1911), and corals (1914) in the Castelgomberto Limestone and the suprajacent tuffaceous sediment at Monti Grumi. Coletti et al. (1973, Tables 2 and 3) demonstrate that the molluscan fauna of the Monti Grumi tuff beds is essentially the same as that of the upper Castelgomberto Limestone, substantiating Kranz's (1910, p. 186) contention for the entire fauna and flora. However, detailed analysis by the author of the scleractinian coral assemblage from Monti Grumi indicates that it represents a somewhat different community than those occurring during the later stage of the carbonate platform. These relationships are discussed below.

*Emergence of the Platform.*—Uplift and emergence of the platform was apparently linked to the culminating phase of Oligocene volcanic activity. Most likely a period of widespread subaerial erosion ensued, with the erosion of most of the volcanic sediment and development of a karst solution surface on the Castelgomberto Limestone (Bosellini et al., 1967, p. 25). At least part of this surface was blanketed by a pure quartzose sandstone ("saldame") of probable aeolian origin.

If the upper Castelgomberto Limestone and the fossiliferous tuffs at Monti Grumi are considered to be Rupelian in age, then a substantial period of time must have elapsed between the initial emergence of the platform and its subsequent flooding in the early Miocene. In employing an absolute time scale such as the one proposed by Berggren (1972) to this problem, it becomes apparent that a time period of at least 7–8 million years was involved in this subaerial erosion and terrestrial deposition, a period perhaps longer than the existence of the reef-lagoon platform itself.

The reef and lagoon facies rocks of the Castelgomberto Limestone make an excellent standard of comparison for Tertiary reef facies reservoir rocks of the subsurface (e.g., the Kirkuk reef-platform complex of Iraq, and Miocene fringing-reefs of Indonesia). Examples of virtually the complete range of biofacies and lithofacies, as well as diagenetic and postdiagenetic phenomona are present. Examples of Oligocene reefs constructed under conditions of terrigenous clastic sedimentation include: a. those of Liguria in northwestern Italy, outcropping about 35 kilometers west of Genoa and referred to in the literature as Carcare, Dego, Sassello, and Cassinelle; b. in extreme southeastern Bavaria near the border with Austria at a locality referred to in the older literature as Reit im Winkel; c. those of northwesternmost Yugoslavia north of Ljubljana at Poljsica and at Gornji Grad (referred to in the old literature as Oberburg), and d. in the Marósticano region of northeast Italy, discussed below.

### Marósticano

Although fossil scleractinian corals from the western end of the Marósticano outcrop belt were the subject of a monograph by Reuss (1869) and were incorporated in studies of broader nature by d'Achiardi (1866, 1868A and B), Felix (1885), de Angelis (1894), Oppenheim (1900), Osasco (1902), and Barta-Calmus (1973), until recently little research has been conducted on the paleoecology of the assemblages or sediments from which the specimens were collected. Pfister (1974, 1977a, 1977b, and 1980, in press) has recently investigated both the paleoecology and taxonomy of Oligocene reef corals of the area.

Coral-bearing strata crop out at a number of localities west and northwest of Maróstica (Fig. 1). These include scattered exposures at Sagonini (Santa Maria de Lugo), Lonedo (Lugo di Vicenza), Salcedo and Calvene, but the largest and most fossiliferous exposures occur in the Laverda Valley and along the crest of the ridge to the east-southeast of San Luca. These latter two areas of exposure were included by Reuss (1869) under the locality name Crosara.

Recent research by Professor R. Herb and Dr. T. Pfister (Pfister, 1974; 1977a, unpub., and 1980, in press) and reconnaissance field work by the author in the San Luca-Laverda Valley area established that two major types of Oligocene reef-oid communities occur: 1. thickets or coppices dominated by great numbers of only a few species of branching or nodular corals, and 2. thin patch-reefs or perhaps shelf-reefs which consist of core facies constructed by a diverse scleractinian assemblage of massive, hemispherical or encrusting species and flank beds dominated by the rubble of the thicket scleractinian assemblage.

Most of the outcrops of coralliferous strata exposed at present and those apparently available to early workers, are of coral thickets and coppices in the Calvene and Salcedo Formations of Ritondale Spano (1967). Good exposures occur in the valley flanks and incised bank of the Laverda River west of the town of Crosara, and as detailed by Pfister (1974, 1977a) at several localities along the crest and slopes of the San Luca Ridge. The thicket communities are usually established on beds of tuff or tuffaceous sandstone, and the base of the typical thicket is composed of a prolific growth of two or three species of branching and nodular *Goniopora* and *Actinacis,* with lesser amounts of *Stylocoenia, Alveopora,* and *Astreopora.* Some thickets were buried by fresh incursions of volcanogenic sediment while others evolved a more diverse community structure evidenced by the upward increase in abundance and diversity of massive scleractinians and the concomitant increase in carbonate content of the matrix material. A few of these structures developed into small coppices in which the main coral skeletal volume is composed of massive and encrusting species. Matrix material in these buildups is usually biomicrite. These structures and associated biota are closely comparable to the earliest Oligocene *Goniopora* thickets of the Castelgomberto-Grumi area discussed above and illustrated in Figures 4 and 5. They are not as similar to the *Euphyllia-Montastrea-Goniopora* inner-shelf community (Fig. 19) occurring in the tuff above the Castelgomberto Limestone at Monti Grumi, even though both communities established on and existed during the deposition of volcanic sediments.

An extensive recent road cut along the crest of the San Luca Ridge exposes elements of a table-reef or patch-reef carbonate bank complex which can be traced for about 2 kilometers in a north-south direction. Pfister (1974, p. 22–28, p. 37–51 and 1977a, p. 73–121) details results of an intensive paleoecologic and petrologic investigation of these exposures. The crest of ridge southeast of San Luca coincides with the strike of relatively resistant carbonate units interbedded in the Salcedo Formation sequence of volcanogenic sediments and basalt, and the Crosara-Maróstica Highway follows this crest for about a kilometer. Thus, highway and subsidiary road cuts expose a sequence about 25 meters thick of reefoid strata along their depositional strike and allow the complex lateral relationships between different contemporaneous communities to be determined. Four local reef buildups of the carbonate bank reef occur:

1. At the extreme southeastern end of the first road cut about 400 meters southeast of San Luca (Pfister's 1974, 1977a, exposure No. 7). Here, a bed of massive biomicrite about 2 meters thick with a depositional eastward dip component contains massive, hemispherical, platy and encrusting frame bulding scleractinians in growth position in what appears to be the core of a small patch-reef. Discordant dips in the outcrop to the northwest of this structure define low mounds of marl or biomicrite a meter or two high and as much as 20 meters in diameter which may have been a series of coalescing carbonate mudmounds. Some of the mounds were capped by thickets of branching corals such as *Goniopora* and *Acropora,* while others are rich in octocoral spicules and were probably alcyonarian coppices. Still others contain a profusion of *Nummulites* and may have been foram/seagrass mud-banks.

2. At the northwestern and the southeastern ends of the next highway cut, about 600 m southeast of San Luca (Pfister's 1977a, exposure No.4). The northwesternmost structure contains a reef-core about 25 meters long and 1.5 meters thick composed of a diverse assemblage of framework scleractinians in growth position. While covered at its northwestern end, the southeast margin of the core grades into rubbly biomicrite with only scattered corals. This structure may be the lateral equivalent of the patch-reef discussed in 1. located about 60 meters to the northwest.

The southeastern end of the highway cut exposes a more complex structure (Figs. 12A and B) which displays striking lateral differences in coral communities. This appears to be a cross-section through a large patch-reef or platform-reef which was elongated in a dominantly north-south direction and which was flanked laterally by channels of similar orientation, perhaps similar in plan view to one of the elongate individual reef masses of the Holocene example in Figure 10. The reef-core structure, which is about 15 meters across × 1.5 meters thick, is constructed by framework scleractinians of an assemblage similar to those in adjacent reef-cores. The main framework builders include massive species of *Pavona, Colpophyllia, Diploria, Cyathoseris, Montastrea, Antiguastrea,* and *Goniopora* as well as the encrusting octocoral *Parapolytremacis.* These appear to be concentrated in greatest density in packed biomicrite and biosparite matrix along and in the steep northwestern margin of the reef-core, where the latter abruptly passes laterally into rubbly argillaceous marl (Fig. 12A). From the outcrop relationships, it appears that the marl was deposited in a reef surge-channel roughly 50 meters wide and of unknown length, one of many which served as conduits for the transportation of skeletal, terrigenous, and volcanoclastic sediment through the reef tract, presumably to the basin to the south. The luxuriance of coral growth on the side of the reef adjacent to the channel was apparently due to more vigorous current and wave generated circulation in the slightly deeper water of the channel (Fig. 12B). The eastern side of the reef-core contains fewer corals in a relatively pure biomicrite matrix, and grades laterally into rubbly flank beds composed of coralline algae and prolific branching and nodular scleractinians. These include *Goniopora, Acropora, Actinacis, Alveopora,* and *Astreopora,* with subsidiary *Stylocoenia* and *Dictyaraea* (Fig. 12B), all efficient sediment movers or evaders and rapid skeleton producers, but with skeletons more subject to breakage by strong wave action. The flank beds are inclined away from the core structure at attitudes of 4–6 degrees and, while there may have been some increase in dip of the sediments due to compaction, it appears that most of the inclination represents the original depositional attitude. The southeastern side of the channel and part of the reef-core is capped by an indurated biomicrite with abundant coralline algae and scattered heads of massive corals. This may have represented some sort of seagrass bank which became established after the surge-channel was filled with sediments.

3. Two other reefoid masses representing an earlier stage of the reef-bank complex discussed above, are exposed in the cut of the first switchback below highway level of a secondary road which descends the northeast face of the San Luca Ridge into the Inverno Valley. This is near the southeast end of the exposure discussed above under 2. (Pfister, 1974, fig. 2, p. 22, p. 36–40, detail drawings of Exposure 5; 1977a, figs. 7–12). These structures occur at two stratigraphic levels in an exposed sequence about 6 meters thick, the top of which is essentially contiguous with the base of the sequence discussed in 2.

The internal structure and scleractinian assemblages of the lower reefoid mass suggests that this might be most accurately termed a coral coppice (Fig. 13A). The total mass is roughly 2 meters thick and is exposed for a horizontal distance of about 15 meters, and displays the control by substrate of the abundance and species composition of coral assemblages. The reef structure was constructed on a substrate of muddy, tuffaceous sand (exposed at the extreme northwest end of the outcrop) by a "pioneer" stabilizing assemblage composed of coralline algal rubble and massive or nodular species of *Goniopora, Colpophyllia, Actinacis, Astreopora,* and *Antiguastrea* as well as broken branches of *Acropora.* The stabilizing assemblage contains relatively few scleractinian species, all able to resist inundation by coarse terrigenous sediment. These constituted what might be termed a sparse "coral pavement" (Fig. 13B). Wave generated turbulence must have been fairly intense during this phase because the primary thicket constructor on unstable substrates, branching *Goniopora ramosa,* is conspicuous by its rarity or absence. Scleractinian diversity increases above this basal bed, with the appearance in abundance of large heads of meandroid species of *Colpophyllia* and *Diploria,* cerioid *Favia* and plocoid *Montastrea* as well as other massive corals. The maximum coral diversity occurs about 1.2 meters above the base, with the establishment of a coppice "climax community" (Fig. 13B) constructed by massive, columniform or encrusting scleractinians of meandroid, cerioid, and plocoid corallum structure. Scattered between the heads of massive corals (maximum diameter 40–50 centimeters) are clumps of branching *Goniopora, Stylocoenia, Dendracis* and *Dictyaraea,* flabellate *Euphyllia,* phaceloid *Caulastrea,* and solitary corals such as *Petrophylliella.* Terrigenous sediment volume increases substantially in the upper 80–90 centimeters of the structure, with concomitant decrease in scleractinian diversity and the appearance of a coral thicket. Large clumps of thickly branched *Actinacis* in growth position and as much as 50 centimeters high and 1.4 meters in diameter dominate this part of the structure, although sediment resistant meandroid corals such as *Colpophyllia* persist (Fig. 13B) and small, branching colonies of *Acropora* and *Goniopora* are densely packed to form a coral "understory" in the *Actinacis* thicket. This community, although similar to the basal "pioneer" community in lithology of sediment matrix, obviously existed under substantially less turbulent conditions. Acceleration in the rate of terrigenous sediment deposition finally overwhelmed the final vestiges of the thicket, and the reef structure was buried under a bed of marly tuffaceous sandstone about 1 meter thick, which contains scattered meandroid corals.

A second reefoid structure about 2.5 meters thick × 15 meters long is relatively poorly exposed at the southeastern end of the road cut (Pfister, 1977a, figs. 10–12). It was established on a muddy or sandy substrate by a "pioneer" community similar to that of the lower reef. Scleractinian diversity also similarly increases upward in the structure. However this reef-core contains a substantially greater volume of massive frame building scleractinians than the subjacent structure, although roughly the same species are present in smaller growth forms. Only a portion of the core facies of this reef is exposed, so that flank beds (if present) or peripheral structures cannot be investigated. From the volume of frame building corals and the thickness of the core, it appears that this structure may have been a large patch-reef or a portion of a more extensive fringing-reef complex.

The local buildups described above occur in a 15 meters thick sequence in the lower Salcedo Formation and can be traced in exposures over an area of about 2.5 square kilometers (Pfister, 1977a, figs. 2 and 3) in the San Luca Ridge south of Crosara. As such, the San Luca thickets, biostromes, mounds and patch- or shelf-reefs must have comprised part of a local carbonate platform that persisted for some time despite the close proximity of extensive volcanic activity. Indeed, Pfister (1977a, fig. 5) establishes that tuffs and submarine basalt flows contemporaneous with the buildups occur in the southern outcrop area on the San Luca Ridge less than one-half kilometer from the exposures discussed above. The platform developed on the northeastern margin of the huge Oligocene volcanic complex that forms the southwestern Marósticano (Piccoli, 1967), perhaps fringing the volcanic mass that is exposed south and west of San Luca.

The San Luca buildups and communities are closely comparable to those of the earliest Oligocene terrigenous-carbonate sequence of the Castelgomberto area, and perhaps to the flat shelf-reefs and biostromes of the upper Castelgomberto Limestone which represent the final phase of the Berici Barrier-Reef lagoonal complex. In these cases the absence of major depositional relief precluded the development of local fringing-reefs into barrier-reef tracts.

The Marósticano coral occurences may be generally correlated with those of the eastern Monte Lessini and Colli Berici. The early Oligocene "Crosara fauna" thickets and biostromes of the Laverda Valley (lower Calvene Formation) of Pfister (1977) are contemporaneous with the terrigenous-carbonate sequence below the Castelgomberto Limestone in the Monte Lessini (Figs. 2 and 4) and with the base of the barrier-reef in the Colli Berici (Fig. 6). The San Luca reef-platform carbonates of the lower Salcedo Formation are probably equivalent to the upper part of the Castelgomberto Limestone on the Monte Lessini and Colli Berici (Figs. 2 and 7), and are thought to postdate the Berici Barrier-Reef.

### OLIGOCENE REEF CORAL COMMUNITIES
*Biogeographic Relationships of Oligocene Reef Corals*

In terms of the range of reef communities developed, luxuriance of their growth, and number of buildups reported, the Oligocene was clearly the height of development of Tertiary reefs (Frost, 1972; 1977a; Schafersman and Frost, 1979). Oligocene reef coral communities have been widely reported from the Caribbean biogeographic province (e.g., Vaughan, 1919; Frost and Langenheim, 1974) and from the length and breadth of the Western Tethyan belt (Frost, 1977b): France, Italy, southern Germany (Bavaria), Yugoslavia, Greece, southern Russia, Iraq, Iran, Israel, Syria, Somalia, Tanzania, Libya, Tunisia, and probably Spain. Many more doubtless remain to be found, especially in the subsurface.

The Western Tethyan corals which are so abundant in the Monte Lessini, Colli Berici and Marósticano are part of a cosmopolitan mid-Tertiary fauna of virtually pantropical distribution (Frost, 1972). The present author (Frost, 1977b; Schafersman and Frost, 1979) postulates that during the Paleogene, when the northwestern Indian Ocean was connected with the eastern Mediterranean by a shallow seaway across what is now the Middle East, and with an open Isthmus of Panamá, there was a tropical/subtropical westward flowing surface current through the Western Tethys across the central Atlantic and through the Caribbean into the eastern Pacific Ocean.

Frost (1977c) therefore contends that dispersal patterns of Western Hemisphere Paleogene Tertiary hermatypic corals were from east to west, with their planula larval stages drifted from the western Tethys into the Caribbean and eastern Pacific; that assemblages collected from various localities throughout the regions should be essentially similar for the Eocene and Oligocene. In the initial comparisons of Oligocene hermatypic coral species of the Mediterranean area with those of the Caribbean/Gulf of Mexico, Frost (1977c) found that similarity between various taxonomic groups varied. For the Suborder Astrocoeniina, of 23 species in both biogeographic provinces, 9 pairs (18 species) or 78 percent of the species are closely related enough to be considered potentially conspecific. Most of the species of Astrocoeniina that occur in the two regions are of branching growth habit and occur in high population densities, so that it is likely that a fairly complete fossil record has been preserved. The

$\rightarrow$

FIG. 13.—A. Portion of a small patch-reef about 1.7 meters thick and 4 meters in diameter which is exposed below road level about 400 meters southeast of San Luca, Marósticano. The basal 0.5 meter is composed of massive heads of *Goniopora* in a muddy and sandy limestone matrix. Flank beds are absent, perhaps due to the small size of the buildup and to its protected shelf position. Sediments surrounding the patch-reef are poorly sorted volcanic sandstone and siltstone. Dip is to left of photograph. B. Reconstruction of Development: 1. Massive corals with strong sediment rejection ability colonize volcanic sand substrate under turbulent conditions. Branching corals appear to have been excluded. 2. Diverse patch-reef develops on stabilized substrate of former biostrome. 3. Influx of terrigenous sediment under moderate turbulence smothers patch-reef and is replaced by tiered thicket of rapidly growing branching corals. Subsequently, this is also buried by sediment.

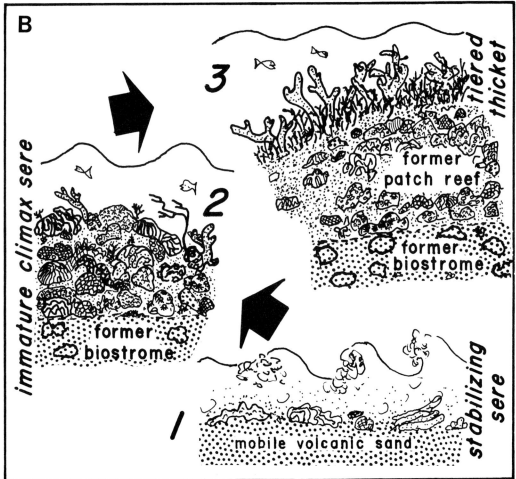

Suborder Fungiina shows different paleobiogeography. Of 35 species in the Suborder both provide only 7 pairs (14 species) or 40 percent which are closely related enough to be considered potentially conspecific, and most of these species belong to *Actinacis, Goniopora* and *Alveopora*. In the Family Agariciidae, corals with foliose and frondose growth form suited for quiet water environments (especially the fore-reef ecozone) show substantial differences between the two provinces. Species of *Pavona* (massive growth form), *Heterogyra,* and *Cyathoseris* dominate Mediterranean agariciid assemblages while abundant and diverse species of foliose *Leptoseris* and less diverse foliose *Pavona* characterize the Caribbean. In the Family Siderastreidae, *S.* (*Siderastrea*) with cerioid corallites occurs in the Caribbean while semimeandroid *S.* (*Siderofungia*) is present in the Mediterranean.

Considerable overlap appears to occur at the species level in corals of the Suborder Faviina. Of 62 species listed by Frost for both biogeographic provinces, at least 13 pairs (26 species) are very closely related. This is more significant than the 40 percent figure would at first indicate because, of 42 species of potential framework corals of the Family Faviidae, at least two-thirds are closely related enough to be potentially considered as conspecific. Moreover, the total overlap is probably greater than indicated by the preliminary comparison. Some groups, especially the large, solitary hermatypic corals of the Families Faviidae and Mussidae which occur in low population densities, are probably more extensive than shown by the presently known fossil record because Caribbean communities have not been exhaustively collected. The Suborders Caryophylliina and Dendrophylliina are predominantly ahermatypic and are not extensively represented in Oligocene reef dwelling coral assemblages. An exception to this is *Euphyllia contorta,* a phaceloid hermatypic coral of the Caryophylliina which occurred prolifically in shallow lagoonal environments of the Mediterranean area (Fig. 18U and V). An unnamed species of *Euphyllia* does occur in the Caribbean area, but it occurs sparsely and is not closely comparable to *E. contorta.* Also, a rare species of *Turbinaria* (S.O. Dendrophylliina) occurs in the Caribbean (Frost and Langenheim, 1974), but the genus has not been found in the Mediterranean Oligocene even though it is fairly common there in the Miocene.

Sweeping changes occurred in the composition and diversity of Tertiary reef coral assemblages at the end of the Oligocene (Chevalier, 1977). In the Caribbean, there is evidence that this changeover occurred within the *Globorotalia kugleri* planktic foraminiferal zone (Frost, 1977c). Evidence for the Mediterranean area is not yet as precise, but can be considered for all practical purposes to have taken place at the Oligocene-Miocene boundary.

*Problems in Identification and Classification*

The great range of skeletal morphologic variation in reef building scleractinian corals has long been known to researchers of both modern and ancient species. Intraspecific variation occurs at all levels—between corallites on the same colony (especially in those near the base of the colony which are shaded or crowded), between colonies in the same ecozone, and across the ecological range of the species. In some cases, substantial overlap in morphological characteristics between two similar species occurs because they overlap in the same ecozone. In addition, ontogenetic differences may make the identification of juveniles difficult, especially in the solitary corals.

This actual or suspected variation causes great problems in scleractinian coral taxonomy. For example Bernard (1901), in the course of his research on the bewildering array of modern Indo-Pacific corals with light, fast growing skeletons, concluded that distinct genetic species groups in reef scleractinians are not discernable. He further (1902) contended that the concept of species, as applied to higher animals, was not applicable to reef corals. Gardinier (1903) suggested that it might require the examination of thousands of living specimens in order to properly identify species. Although much progress has been made in defining taxonomic categories and applying modern concepts of population biology, many subsequent papers (e.g., Vaughan, 1907 and 1919; Hoffmeister, 1925; Wells, 1954; Chevalier, 1971 and 1975) underscore the significant problems that still remain in identification and classification of hermatypic scleractinia.

Some of the best evidence for the range of morphic plasticity in reef corals is in the recent work of Wijsman-Best (e.g., 1972, 1974a, 1974b, 1976, and 1977). She demonstrates that in modern species and genera of southwestern Pacific Faviidae, the morphology of skeletal features having major taxonomic importance (such as number of septa, depth of calices, septal dentitions, corallite density, diameter and spacing, length and breadth of meanders) changes as a function of environmental factors, especially increasing water depth. A few conservative (i.e., genetically programmed) morphologic characters, constant regardless of environmental effect, remain to permit delimitation of species (1974a, p. 225). Because of the great range of Faviid variability, substantial problems occur in defining genera: plocoid *Favia* and cerioid *Favites* are gradational because (1972, p. 67) species of *Favites* assume a plocoid growth form in certain habitats while species of *Favia* occasionally become cerioid. A similar gradation exists between species of *Favites* and *Goniastrea* (cerioid) and between *Goniastrea* and *Platygyra* (meandroid).

Similar patterns of morphologic change are reported by Wallace (1978) for species of *Acropora* in the central and southern Great Barrier Reef Province. Untangling the true species relationships among the multiplicity of names (more than 250, *fide* Wells, 1969) applied to modern Indo-Pacific corals of this genus is one of the major taxonomic problems regarding living hermatypic scleractinia. Wallace (1978, p. 275) notes high levels of variability within *Acropora* species; detailed measurements for analysis of variation often show similar variability within the colony equal to that in the entire population. However, the application of detailed ecologic data to the taxonomic studies shows that in reef-slope *Acropora,* growth form and other morphologic features alter gradually with depth, and that characteristic growth form is developed for certain reef ecozones in many co-occurring species.

On the other hand, in some groups of corals variation appears to be only slightly related or unrelated to environmental influences. Brakel (1977) conducted a quantitative study of 20 corallite characters on modern Jamaican *Porites* from a range of reef habitats. Morphological variation was found to be nearly continuous. Corals from very different environments were found in some cases to have very similar corallite structure, while in other instances specimens from the same or very similar environments had radically different corallite morphologies. Brakel (1977, p. 460) concluded that the pattern of variation observed in *Porites* had an overwhelming genetic component. If this pattern of variation was due to truly conservative species characters in the sense of Wijsman-Best (1974a, p. 225), then *Porites* variation may be "inherently so complex and subtle that it precludes any simple taxonomic resolution at the subgeneric level" as Brakel (1977, p. 458) has postulated.

Adding to problems of interpretation is the ever present possibility that ecomorphic end members of a supposed species population may actually be separate sibling species. Thus, Lang (1971) discovered that the larger of the two supposed ecovariants (Wells, 1973a) of Caribbean *Scolymia* was able to kill the tissues of the other variant by extracoelenteric feeding response. Both had long been lumped together as *Scolymia lacera.* Similar results have been reported by Wells (1973b) for Caribbean colonial reef corals. A meandroid coral with short corallite series was previously assumed to be an ecological variant of *Colpophyllia natans,* but has been shown to be a different species (*C. breviserialis*), on the basis of further taxo-

nomic and ethological research. Another widely recognized meandroid species, *Mycetopophyllia lamarckiana,* is now considered (Wells, 1973a, p. 35–36) to have previously included in it *M. danaana, M. ferox, M. aliciae,* and *M. reesi,* while the monocentric variety of *Dichocoenia stokesi* has been shown to be a separate species, *D. stellaris.*

In applying the general concepts of the taxonomy of living reef corals to their fossil counterparts we are therefore faced with two conflicting relationships. On the one hand, we have extensive proof that over the ecological range of many species we have gradual morphologic change with changing environment, especially that of increasing water depth and that certain morphic features are specific to certain environments. On the other hand there is much ecological and ethological evidence to show that some widely recognized solitary and colonial species are actually two or more morphologically similar sibling species. In other species groups morphologic variation appears to be independent of discernable environmental factors.

In fossil reef scleractinia the problems of imperfect preservation further complicate matters. Not only is it difficult to sample a fossil species across its full range of habitat because of the lack of suitable exposures, there is also loss of information in the way fossil reef corals are preserved. Usually the fossils are calcite replacements of the original aragonite skeleton with the attendant loss of much of the skeletal ultrastructure and microstructure. The range of preservation of the Italian Oligocene corals considered herein follows the experience of the author with Tertiary fossil corals from the Caribbean and Indo-Pacific. The best preservation of corallum and corallite surface details occurs in terrigenous mudstones or calcareous shales with fidelity of preservation generally decreasing with decreasing argillaceous content of the rock matrix. Pure carbonate reefs, such as the Colli Berici shelf-edge buildup, the Oligocene Browns Town Reef complex of Jamaica or the Amerada-Hess Pliocene fringing-reef of St. Croix are almost always characterized by poorly preserved reef corals—as coarse calcite pseudomorphs or as leached casts and molds. Few Tertiary scleractinian specimens still retain their original aragonite skeletons. Almost all of those that do are ahermatypic corals which are encased in impermeable terrigenous claystone or mudstone matrix.

### Vicentin Oligocene Reef Corals

Collections of well preserved Oligocene fossil reef corals from the eastern Monte Lessini, Colli Berici, and Marósticano can be found in a great many museums. Taxonomic descriptions of these scleractinia began early in the 19th century with work by Defrance (1826) and Michelin (1840–1847). For about two decades Oligocene corals were described in great numbers from the Lessini and Piemonte regions by Catullo (1852 and 1856), Michelotti (1847 and 1861), Gümbel (1861), d'Achiardi (1867, 1868A, and 1868b), Reuss (1868, 1869, and 1872) and by Sismonda and Michelotti (1871). In addition, Oligocene corals from what is now Yugoslavia were described by Reuss (1864) and from southern Bavaria by Reis (1889). At the time little was known about the morphologic ranges of modern or fossil reef scleractinia. Species names proliferated and certain abundant and widely distributed corals were described over and over under different names. Moreover, other new species were erected which were based on inadequate fossil material and many named species were inadequately described and illustrated.

Revisions of or additions to these pioneer studies were made by Felix (1885), de Angelis (1894), Osasco (1898 and 1902), Bernard (1906), Canestrelli (1908), Kranz (1914) and Felix (1916), for the most part employing typological approaches to the problems of coral taxonomy. Kranz (1914), however, applied the awakening realization of the substantial range of intraspecific morphologic variation of modern reef scleractinia to the re-evaluation of massive Faviid corals from the classic Monte Lessini localities of the Castelgomberto Limestone. Also, Felix (1925) in his comprehensive catalog of Eocene and Oligocene corals, compiled most of the described species and revised their taxonomy.

By the time Prever (1921 and 1922) completed his research on massive and foliaceous corals from Liguria, more than 520 species had been erected for Oligocene reef corals of the northwestern Mediterranean area, with an additional 40+ species of ahermatypic corals identified or described. Virtually no additional taxonomic research was done until Chevalier (1955) described Oligocene hermatypic corals from southern France and (in Brunn et al., 1955) from northwestern Greece. Sestini (1960) included corals in her description of an Oligocene assemblage from the Genoa region of northwestern Italy. Barta-Calmus (1973) laid the groundwork for modern taxonomic revision of the whole fauna with her extensive dissertation on the type collections of Eocene and Oligocene corals of the Nummulitique. Pfister (1977b) has demonstrated the complex range of morphology in Antiguastrea from the Marósticano, and the impact this range has had on the taxonomy of the coral.

This paper is meant to be an overview of the range of Vicentin Oligocene reef buildups and their associated coral communities. In order to accurately portray which Oligocene reef coral species are found in the various communities, it must also in effect be a progress report on the current state of our knowledge about the taxonomic affinities of the immense array of named species.

Table 1 presents the author's current views on the species composition of the western Tethyan Oligocene reef coral fauna. Although it is based on collections from a wide range of localities, virtually all of the species listed do occur in the Vicentin. The list has been modified from the array utilized in an initial comparison of Oligocene reef building coral faunas from the Caribbean and western Tethyan provinces (Frost, 1977b). This is done in accordance with new research data and the conclusions of Pfister (1977a, 1977b, and 1980, in press). The species presented in Table 1 represent the author's interpretation of results derived from the ongoing research project on the systematic paleobiology of the Oligocene western Tethyan reef coral fauna being conducted by Drs. Thérèse Pfister and Jörn Geister of Bern, Switzerland, and the author. It should be noted that not all of the taxonomic conclusions of Table 1 are necessarily shared by Drs. Pfister and Geister.

### Range of Communities

Oligocene reef corals in the Vicentin occur in a complete spectrum of buildups: biostromes or coral pavements, thickets, coppices, patch-reefs, and fringing or barrier-reefs. Figure 14 depicts the estimated relative abundance of the most common coral species in these various structures. Although various mixtures of these species produce a variety of communities, three basic assemblages are present:

1. low diversity *Goniopora/Actinacis* "pioneer" or thicket assemblage;
2. high diversity reef-core and flank assemblage;
3. soft mud substrate assemblage.

*"Pioneer" or Thicket Assemblage.*—Many Tertiary reef communities developed on and stabilized terrigenous sand and mud substrates, even those which were subject to vigorous wave surge and current action (e.g., Frost and Weiss, 1979). Stabilizing or "pioneer" communities compose low diversity (paucispecific) biostromes or coral carpets composed of species which can successfully settle as planula larva to the mobile substrate, grow and develop in spite of the constant rain of sediments (Frost, 1977d). In the Vicentin, stabilizing communities are dominated by species of *Goniopora* (Fig. 17K–N, and T–U; Fig. 18T) of thickly branching, massive and encrusting growth form (Fig. 16D–E) and by *Actinacis rollei* which grows as superposed plates (Fig. 16F) or

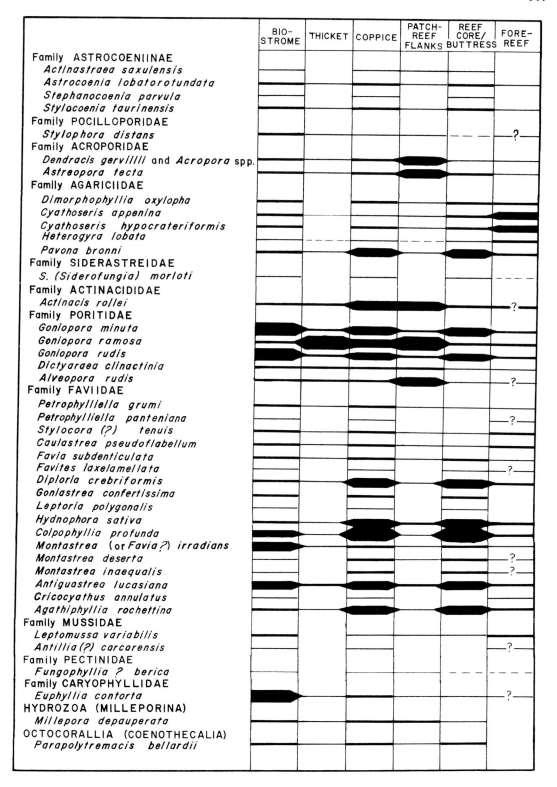

FIG. 14.—Estimated ecologic distribution of selected Oligocene coral species in the Vicentin.

thick palmate branches similar to the modern Caribbean *Acropora palmata* under conditions of agitated water flow, and as finer branches (Fig. 18P–Q) under calmer conditions. Less abundant species include those of *Astreopora* (Fig. 17O–P), *Antiguastrea* (Fig. 18V–W) and, rarely, *Colpophyllia* (Fig. 19O).

Thickets are usually preserved as masses of closely appressed branches which range in thickness from a few tens of centimeters to more than a meter in thickness. The coral communities are usually paucispecific under conditions which tend to exclude the more aggressive massive corals, and in the Vicentin (Frost, 1977d) are most often dominated by branching *Goniopora ramosa* (Fig. 16E and Fig. 17T–U) with lesser abundance of branching and massive corals. The branching colonies are often broken due to sediment dewatering and compaction when they occur in fine-grained terrigenous rocks. Morphological details of the calices and other coral surface features, however, are usually well preserved in such matrix. Examples of paucispecific thicket communities occur low in the Calvene Formation of the Laverda Valley, Marósticano, and in the terrigenous beds (Fig. 16E) of the mixed terrigenous-carbonate sequence below the Castelgomberto Limestone in the eastern Monte Lessini.

*High Diversity Reef-Core and Flank Assemblage.*—Generally speaking, Oligocene reef coral diversity tends to increase as a function of the increasing carbonate content of the associated sediments. This is the result of the fact that many of these species of corals were physiologically stressed in environments with high rates of terrigenous sand and silt sedimentation.

In carbonate substrate habitats which were beneath or sheltered from direct wave surge and/or strong currents, species of rapidly growing corals with light, porous skeletons such as branching *Acropora* (or *Dendracis*) (Fig. 18N–O), *Actinacis* (Fig. 18P–Q), *Goniopora* (Fig. 17T–U), *Dictyaraea* (Fig. 17A–B) and *Stylocoenia* (Fig. 17D–E) as well as nodular *Alveopora* and *Astreopora* (Fig. 17O–P) occur with great growth luxuriance. These typically are found in fringes on the lee side of patch-reefs (Fig. 12), lining the deeper parts of passes through fringing-reefs, and in deeper environments of the reef-front and fore-reef. Under these conditions many ramose corals that were excluded from the *Goniopora/Actinacis* thickets because they lacked strong sediment rejection ability, or were crowded out of the main reef framework by massive corals, occur in great profusion. The protected reef flank or small coppice represent optimum habitats for corals such as these (Fig. 14). Other subsidiary species of this community include the "organ pipe" phaceloid corals *Caulastrea* (Fig. 17S), *Euphyllia* (Fig.

18U–V) and *Cladocora* (?) (Fig. 17H–I) and the large solitary hermatypic corals *Leptomussa* (Fig. 17Q–R), *Cricocyathus* (Fig. 17C–D) and *Petrophylliella* (Fig. 18C).

Lagoonal patch-reefs are scattered throughout the middle and lower two-thirds of the Castelgomberto Limestone that was deposited back of the Berici Barrier-Reef. The smaller buildups are patches and thickets dominated by branching *Goniopora* and *Caulastrea* with *Acropora* (or *Dendracis*) and other elements of the flank assemblage generally being less important. Others are larger masses that appear to have been large coppices or lagoon-reefs which stood as much as a meter or two above the surrounding sediments. Examples of these are exposed in the northern Colli Berici about 4 kilometers south of Vicenza near Arcugnano (Geister and Ungaro, 1977, p. 819) and in the eastern Monte Lessini near Monte Mondeo, about 2 kilometers northwest of Montécchio Maggiore. Such reefs were large enough to have core facies composed of massive corals in a bound framework and branching species in fringing flank facies.

Western Tethyan reef-core or framework coral assemblages are more diverse and apparently grew more luxuriantly than their Eocene or Miocene counterparts. Figure 14 is an estimate of the relative abundance of species in the reef-core and buttress zones of Vicentin Oligocene Reefs. It is derived from observations made on the shelf-reefs of the basal Oligocene in the Monte Lessini, on the San Luca platform-reefs of the Marósticano, and on the poorly preserved reef-front of the Colli Berici. Although Figure 14 shows that a wide range of species might be expected to occur in the core facies of any given reef in the Vicentin, it also shows that a relatively few species aggregate a disproportionately large part of the constructed and bound framework. Thus a minority of the hermatypic coral species associated with these reefs are actually major *reef builders*. The majority of species are *reef dwellers* and subordinate in their contribution because they produce skeletons that are too small or too fragile to be incorporated *in situ* into the primary reef framework structure. Instead they are deposited as secondary fill into the framework or may be transported to other zones of the reef or, out of it altogether. In an example from modern reefs, Goreau and Goreau (1973) found that for the well known fringing-reefs of north Jamaica, of about 42 species of hermatypic corals that occur in the buttress zone of the reef-front, 7 species together constitute more than 75 percent of the reef framework. We thus might consider the remaining 35 or so species to be *accessory reef dwellers* rather than *reef builders*.

The primary Oligocene coral framework constructors of the region are species of *Antiguastrea*

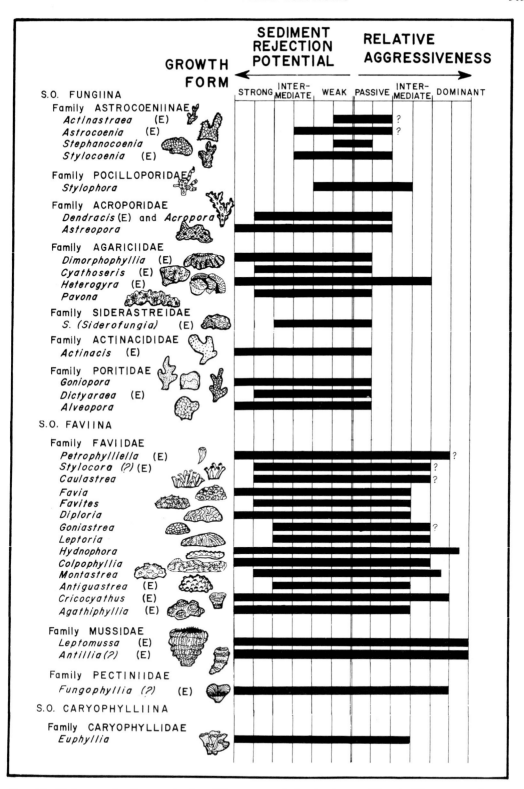

Fig. 15.—Estimates of sediment rejection ability *versus* relative dominance, Vicentin Oligocene coral genera.

(Fig. 17V–W; Fig. 19B–C; I), *Colpophyllia* (Fig. 19F–G, J–K, and O), *Pavona* (Fig. 19N), *Montastrea* (Fig. 18AA), *Diploria* (Fig. 19A, M), massive species of *Goniopora* shown on Figure 17, and *Favia* (Fig. 18R–S). Secondary framework builders include branching species of *Goniopora* and *Actinacis,* the massive corals *Astreopora* (Fig. 17O–P), *Agathiphyllia* (Fig. 18Y and Fig. 19H) and *Heterogyra* (Fig. 18W–X) and explanate *Cyathoseris* (Fig. 18Z; Fig. 19D–E) and *Dimorphophyllia* (Fig. 18L–M). The encrusting octocoral *Parapolytremacis* (Fig. 18J–K) is also important in framework construction.

Relatively little data exists on the species composition or relative abundance of fore-reef coral assemblages in the region. The only reef structure found to date which had relief enough to have developed a true fore-reef zone is the Berici Barrier-Reef. Unfortunately the front of the Colli Berici is weathered back (Fig. 7) so that fore-reef deposits have been extensively removed. Fore-reef talus of the upper reef-core *is* poorly exposed on the south slopes of Monte San Cassiano above Lumignano, but the outcrops are heavily weathered and few of the corals found in the rock matrix were actually in place.

By analogy with comparable Oligocene communities of the Caribbean/Western Atlantic province (Frost, 1977D, p. 372) we would expect fungiids and faviids of foliose or explanate growth form such as *Pavona, Cyathoseris, Fungophyllia,* and possibly *Leptoseris* to characterize the fore-reef ecozone. However, it should be noted that while foliose *Leptoseris* is abundant enough to form a well defined ecozone in several Caribbean Oligocene reefs, it has not been found in the Vicentin buildups. Also typical of the Oligocene fore-reef are colonies of delicately branched *Stylocoenia, Stylophora, Dictyaraea,* and *Goniopora* and solitary hermatypic corals such as *Leptomussa, Petrophylliella,* and *Cricocyathus.* Also expected is the characteristic flattening response of massive species in the deepest zones of coral growth. The base of the fore-reef zone is commonly denoted by the prolific presence of coralline algae in nodular growth form as in the Berici Reef exposures at Zovencedo.

*Communities of Soft Mud Substrates.*—One of the most famous localities for Oligocene reef corals is on the rounded crest of Monte Grumi, 1.5 kilometers south of Castelgomberto (Figs. 2 and 3). At this site, great numbers of well preserved

$\rightarrow$

FIG. 16.—Reef dwelling biota: A. Skew-section, ×30, of a skeletal element of the codiacean green alga, *Halimeda,* now represented by a calcite pseudomorph after its original aragonite composition; Unit 9 of Figure 4, basal Oligocene. B. Cross-section, ×30, through a small branch of a scleraxonoid gorgonian octocoral. The dark rind across the top is encrustation by coralline algae. Note the sediment filled center which was filled with horny axis cortex material in life; Unit 10 of Figure 4, basal Oligocene. C. Skew-section, ×30, of a cyclostome bryozoan (*Smittoidea*?) branch, and miliolid foraminifers in a biomicrite matrix; Unit 5 of Figure 4, basal Oligocene. D. Transverse section, ×30, of rhythmically intergrown coralline algae (thin dark layers) and an encrusting foraminifer (lighter layers with vertical partitions), which is probably *Gypsina.* Mutualistic intergrowths such as these form large rhodolith-like nodules; Unit 2 of Figure 4, basal Oligocene. E. Outcrop of a thicket of finely branched *Goniopora ramosa* in growth position. Length of scale is approximately 4 centimeters; Unit 8 of Figure 4, basal Oligocene. F. Plate-like masses of *Actinacis* sp. cf. *A. rollei* and nodular *Goniopora* in outcrop of calcareous siltstone. Scale is 15 centimeters long; outcrop same as in E.

FIG. 17.—*Goniopora* thicket and biostrome community: A and B (×0.6) *Leptomussa*(?) *cingulata* (Catullo). C and D (×0.6) *Cricocyathus annulatus* (Reuss). E (×0.6) and F (detail, ×5) *Stylophora conferta* Reuss. G and H (×0.6) *Ceratotrochus* (*Conotrochus*) *aequicostatus* (Schauroth), an ahermatypic coral. I (×0.6) and J (detail, ×5) *Reussangia* (?) *brevissima* (Catullo), may also have been ahermatypic. K (×0.6) and L (detail, ×5) *Goniopora minuta* (Reuss). M (detail, ×5) and N (×0.6) *Goniopora rudis* (Reuss). O (detail, ×5) and P (×0.6) *Astreopora tecta* (Catullo). Q and R (×0.6) *Leptomussa variabilis* d'Achiardi. S (×0.6) *Caulastrea pseudoflabellum* (Catullo). T (×0.6) and U (detail, ×5) *Goniopora ramosa* (Catullo). V (detail, ×5) and W (×0.6) *Antiguastrea lucasiana* (Defrance). Illustrations are copied from Reuss (1864, 1868, 1869).

FIG. 18.—*Euphyllia*/mixed soft mud substrate community: A (×0.6) and B (detail, ×5) *Dictyaraea clinactinia* (Michelotti). C (×0.6) *Petrophylliella subcurvata* (Reuss). D (×0.6) and E (detail, ×5) *Stylocoenia taurinensis* (Michelin). F and G (×0.6) *Cricocyathus annulatus* (Reuss). H and I (×0.6) *Cladocora* (?) *biformis* (Catullo). J (detail, ×5) and K (×0.6) *Parapolytremacis bellardii* (Haime), an encrusting octocoral. L and M (×0.6) *Fungophyllia* (?) *berica* (Catullo). N (×0.6) and O (detail, ×5) *Dendracis* (?) *gervillii* (Defrance). P (detail, ×5) and Q (×0.6) *Actinacis rollei* Reuss, branches of this species grow to great size. R (detail, ×5) and S (×0.6) *Favia subdenticulata* (Catullo). T (×0.6) *Goniopora ramosa* (Catullo). U (×0.6) and V (×0.6) *Euphyllia contorta* (Catullo). W and X (×0.6) *Heterogyra lobata* Reuss. Y (×0.6) *Agathiphyllia rochettina* (Michelin), growth form with large corallites. Z (×0.6) *Cyathoseris appenina* (Michelin). AA (×0.6) *Montastrea* (or *Favia*?) *irradians* (Milne Edwards and Haime). Illustrations are copied from Reuss (1864, 1868, 1869).

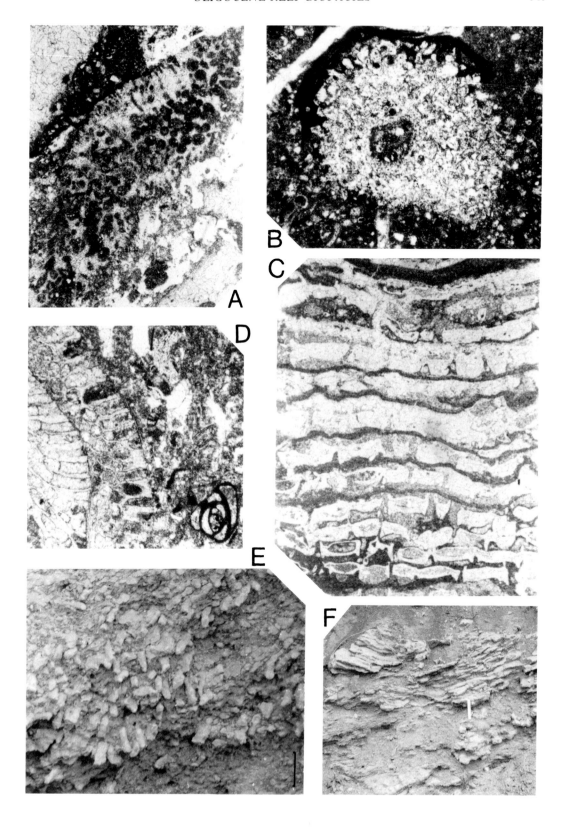

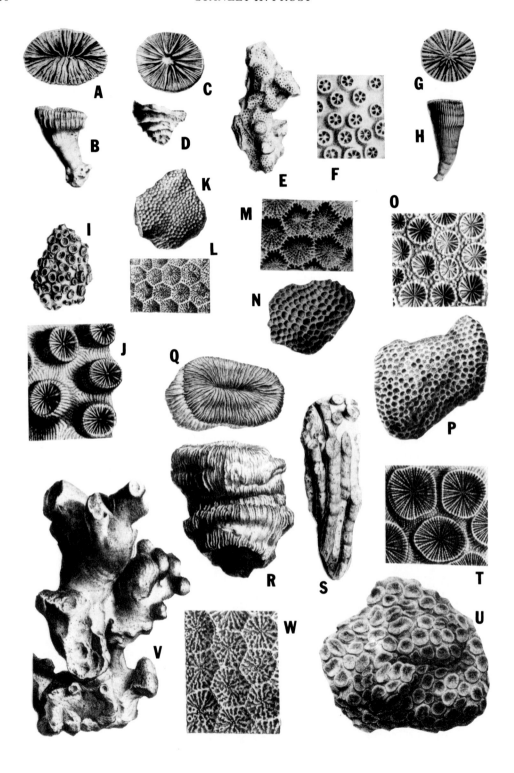

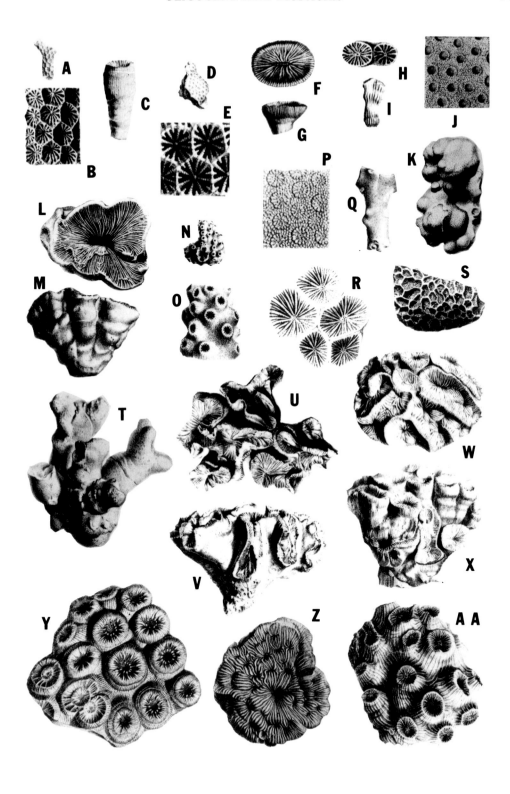

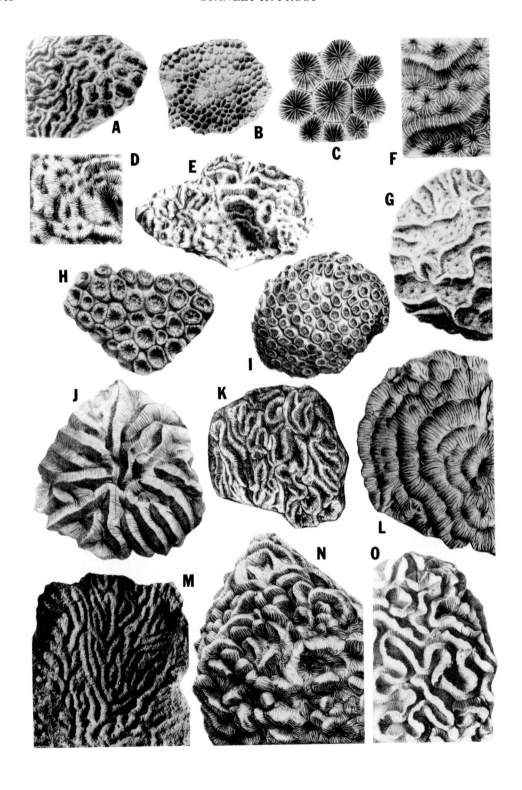

fossils weather out of a volcanic mudstone. These have been the subject of numerous monographs starting in the early 19th century, and the type descriptions for a number of species of widespread and important Oligocene reef dwelling corals were based on specimens collected from these deposits.

The bentonitic mudstones are diagenetically altered water laid basaltic tuffs which contain a high percentage of calcareous skeletal debris, and were deposited during the final phase of Oligocene marine sedimentation on the former Colli Berici-Lessini carbonate platform. Although the carbonate shoal coral biostromal community characteristic of the upper Castelgomberto Limestone was buried by the rain of volcanic ash, a few marine communities survived. These were adapted to soft mud substrates and were capable of evading or withstanding burial by fine sedment or were types which would be unaffected by such conditions (Fig. 15). The most characteristic corals of this assemblage are illustrated in Figure 18. By far the most numerous specimens are the phaceloid coral *Euphyllia* (Fig. 18U–V) and branching *Goniopora* (Fig. 18T), which apparently occur in thicket-like masses. Relatively small-size specimens typify the assemblage, whether they be branching, massive colonial, or solitary, and its general aspect is of a coral and coralline algal community growing in a sea grass bank. The binding effect of sea grass roots and rhizomes would stabilize the fine mud sediment and provide firm places of attachment for many smaller solitary and branching corals such as *Dictyaraea* (Fig. 18A–B), *Stylocoenia* (Fig. 18D–E), and *Cladocora* (?) (Fig. 18H–I). Some massive corals also occur. These in general had large polyps growing in shallow calices and were of plocoid growth [e.g., *Agathiphyllia* (Fig. 18Y) or *Favia* (?) (Fig. 18AA)] or were semimeandroid such as *Heterogyra* (Fig. 18W–X). Modern corals with colony morphologies such as these are known to be efficient sediment evaders. There is no evidence of any bound framework associated with these communities and their heaviest concentrations appear to have been as low thickets or thin biostromes or coral carpets.

## Ecologic Controls of the Coral Communities

*First Order Factors.*—The net accumulation rates of reef buildups and associated sediments as well as the rate of change in framework communities are affected by the interplay of three first-order controls (Frost, 1977d):

1. rates of net skeletal growth by the frame building and frame filling biota (taking into account the destructive process bioerosion and the constructive process submarine cementation); 2. subsidence rates and shape of the foundation the reef is growing on; 3. rates of eustatic sea level rise. For Pleistocene, fossil Holocene and living reefs, rates of net skeletal growth have been demonstrated by a number of papers (e.g., Adey, 1978) to be directly determined by a combination of subsidence rates and eustatic sea level rise or by sea level rise alone. There is strong evidence to indicate that within certain threshold limits, net rates of late Cenozoic reef tract growth and ecological succession are mostly attributable to rates of eustatic sea level change. Mesolella *et al.* (1970, p. 1914) estimated that for Barbados rates of eustatic sea level change during the Pleistocene were 20 times that of tectonic uplift. In applying these concepts to the problem of predicting where Tertiary reefs might be expected to occur in the subsurface, estimates of ancient eustatic sea level changes such as those of the Exxon group (Vail *et al.*, 1977) become extremely important, especially for stable shelf-margin areas.

Drilling of various late Cenozoic fossil reefs has demonstrated that different reef ecozones display different net accumulation rates. In general, the upper or reef-crest ecozones of these reefs accrete most rapidly. This is because the late Cenozoic scleractinian faunas of these zones are dominated on a worldwide basis by at least some rapidly growing species of the families Acroporidae, Poritidae and Seriatoporidae. In the Caribbean, rigid Holocene reef framework of the *Acropora palmata* ecozone accreted at rates as high as 15 m/1000 years in St. Croix (Adey, 1975) and 10.8 m/1000 years in Panamá (Macintyre and Glynn, 1976), compared to rates of about 1 m/1000 years

←

for the reef-flat. Accretion rates of reef-front eco-zones of pre-Pleistocene Tertiary reefs are esti-mated by Frost (1972, 1977D) to have been sub-stantially lower (perhaps 1–3 m/1000 years) because they appear to have lacked the great numbers of competitively successful rapidly growing branching corals such as Caribbean *A. palmata* or the great species diversity of Indo-Pa-cific *Acropora, Montipora* and *Astreopora.*

*Second-Order Factors.*—These factors which directly define or limit the coral reef biotope are reasonably well known and include sea water temperature and salinity, light penetration (or depth), sedimentation, emergence as well as wave and current intensity. The present state of knowl-edge about these ecological forcing functions is summarized by Wells (1957), Stoddart (1969), and Yonge (1963). Second-order controls such as these are broadly selective, affecting the whole reef biota. Many of the controls listed above can now be approximated with reasonably satisfac-tory precision for ancient reefs, based on evi-dence remaining in the rock record. On the other hand, vital factors such as material pools of dis-solved nitrate, nitrite and phosphate, oxygen and carbon dioxide concentration, as well as suspend-ed plankton and organic detritus (pseudoplank-ton) leave no direct trace of their existence for the fossil record. Further, the vital relationship of symbiosis which directly facilitates coral reef buildups, leaves no direct trace in the fossil record and must be inferred by comparison with similar modern communities. Finally, as Frost (1977a) has detailed, a very major proportion of the reef biota has (or had) no mineralized skeletal parts capable of being fossilized, and the presence of whole trophic/functional categories of the reef ecosystem may be inferred only by analogy with the living counterpart, as we will consider below.

Some ecological factors operate very specifi-cally at the community level. These affect the de-tails of reef ecozone development and mainte-nance by determining the actual frame building and frame filling community composition. This in-volves the areal mix of species, species diversity, and species abundance. Many of these factors strongly affect skeletal morphology in modern reef dwelling corals and hence are of importance not only to the taxonomy of modern and ancient species, but also in the reconstruction of detailed paleoecology of ancient reef communities:

*1. Illumination Intensity.*—The characteristic flattening response of massive reef corals to di-minished levels of daily illumination (a function of increasing depth or shading) has been noted by many authors (e.g., Goreau, 1959; Goreau and Goreau, 1973) for modern reefs, and is a factor in estimating the paleobathymetry of reef-front and fore-reef zones of fossil Tertiary reefs. The flat-

tened, plate-like or shingle-like growth form is a mechanism by which the coral increases the sur-face area/skeletal weight ratio of the colony, com-pensating for the decrease in calcification rate as photosynthesis rate decreases in the zooxanthel-lae. On steep slopes of the upper fore-reef zone of Caribbean fringing-reefs, the shingle-like growth form of massive plocoid, cerioid and meandroid corals also appears to aid in draining off the rain of carbonate sediment derived from the shallow reef.

The fore-reef zone of modern and Tertiary reefs is also characterized by the great development of hermatypic corals with frondose or foliaceous growth form ("lettuce corals"), exemplified by species of *Agaricia* and *Helioseris* on modern Caribbean reefs (e.g., Wells, 1973b) and by *Lep-toseris, Oxypora,* and *Echinophyllia* on Indo-Pa-cific reefs (e.g., Dinesin, 1977). The foliaceous growth form is a genetically constant trait in these lineages, not environmentally induced in only portions of the ecologic ranges of the species. This growth strategy is interpreted to be not only an adaptation to growth in environments of lower illumination, but also one alternative to solving the problem of making corals with small polyps such as these more efficient at capturing zoo-plankton in the deeper reef zones (Porter, 1976). The fragileness of the foliaceous skeletons of such corals exclude them from or reduce their abundance in the turbulent shallow reef-crest and reef-flat environments. Their *in situ* presence in Tertiary fossil reefs is one biotic criterion for identifying quiet water coral communities.

*2. Adaptation to Water Flow.*—Wave and cur-rent induced water motions influence the ecology of reef corals in shallow reef tract environments. Many of the shallow reef tract ecozones are de-termined by the hydrodynamic effect of breaking waves (e.g., Geister, 1977). The wave base of nor-mal oceanic swells is the decisive limiting factor in the distribution and abundance of many reef dwellers and defines the boundaries of ecological/morphological zones. The low coral diversity of the *Acropora palmata* zone of Pleistocene and Holocene Caribbean reefs is, for example, a re-sponse to the effects of hydromechanical skeletal breakage and abrasion of polyps and skeletons by coarse sediment churned in the wave surge. The characteristic branching size, shape and orienta-tion of *A. palmata* has been shown by Graus *et al.* (1977) to be an adaptive strategy delicately tuned to reduce hydraulic stress. Although *Ac-ropora* occurs in Eocene-Pliocene reef coral com-munities in the Caribbean and Indo-Pacific and Eocene (?), Oligocene-Miocene communities in the Mediterranean area, it is only rarely abundant (e.g., in some patch-reef assemblages) and has not been found with the thickly branched *palmata*-

type growth form. Of the Oligocene assemblages considered in this report only one species, *Actinacis rollei,* developed growth form or colony size similar to *palmata.* Where *A. rollei* occurs in low diversity *Goniopora* thickets and coppices, the growth form consists of flat, superposed plates inclined at a low angle to what was apparently the direction of wave surge (Fig. 16F and Frost, 1977d). In less agitated environments *Actinacis rollei* grows in great masses of thick branches. The thickly branched growth forms of Oligocene *Actinacis* do not appear to have been as ecologically successful as the Pleistocene and Holocene *Acropora palmata,* at least in terms of abundance.

3. *Effect of Substrate.*—In many coral reef communities the species composition and diversity of sessile plants and invertebrates are controlled by substrate characteristics and rates of sedimentation. This is because the further distribution of such species is dependent upon successful settlement, growth and development of a pelagic larval stage. Kinzie (1973) found that substrate type was the limiting factor in octocoral community distribution and diversity on modern Jamaican reefs. For modern scleractinians of the Great Barrier Reef, Goreau and Yonge (1968) noted that communities on shallow, unstable terrigenous muddy sand substrates were dominated by ahermatypic species, and that of hermatypic corals, only large solitary types such as *Trachyphyllia* (similar to the Vicentin Oligocene *Leptomussa*) were present. On carbonate sand substrates in Madagascar reefs (Pichon, 1974) the assemblage is substantially more diverse, including solitary and colonial fungiids such as *Fungia, Halomitra, Herpolitha,* and *Polyphyllia* in addition to the corals listed above.

The effect of substrate on the composition of Vicentin coral communities can be seen in Figure 14. Both thickets and biostromes are essentially sheet-like deposits whose three dimensional shape is controlled largely by the relatively gentle slope and low relief of their substrate. The striking difference in coral diversity between, for example, the lowermost Oligocene *Goniopora* thickets of the Grumi Road section (Fig. 4) and the head-coral biostromes or coral carpets of the uppermost Castelgomberto Limestone appears to be due largely to substrate mineralogy. The matrix sediments associated with both communities are essentially similar in their grain size and sorting, and apparently grew under similar intensities of water motion. The thicket community, however, is found in sandy and silty calcareous mudstones (Fig. 5A) while the coral carpets occur in variously sorted biosparites and packed biomicrites (Fig. 11E). Similar substrate control over Oligocene scleractinian assemblages is found in the Caribbean/Gulf

of Mexico region, in the Pueblo Nuevo fringing-reef of southern Mexico (Frost and Langenheim, 1974), the Friars Hill patch-reef of Antigua (Frost and Weiss, 1979) and the Damon Mound, Texas, reef-cap of a salt dome (Frost and Schafersman, 1978).

At the individual species or colony level, the firmness and type of the substrate exerts an important modifying effect (often causing great problems for taxonomists) on the basic skeletal plan coded in the genes of the species. The effect of increasing rates of soft sediment accumulation on colonial corals produces striking changes in skeletal morphology. Species which are normally ceriod form plocoid corallites by the separation of adjacent polyps and their consequent upward growth to elevate themselves above the encroaching mud; meandroid species develop hydnophoroid and plocoid morphology while normally plocoid species may develop "organ-pipe" (phaceloid) colonies because of a similar response. In some cases, autoecological differences in morphology between individual colonies and others that live in the less stressed environmental range of the species are great enough that correct identification is difficult.

The Oligocene coral assemblage, which adapted to the soft volcanic mud, substrates of the Monti Grumi tuffs or volcanic mudstones (discussed earlier) is a good example of morphic response to physiological stress. In general, coral species which normally had massive growth form were the most affected by the soft substrate and the apparently high sedimentation rate. This is manifested in plocoid corals such as *Favia* (or *Montastrea*) *irradians* which develop calices highly elevated above the intercorallite coenosteum (Fig. 18AA) or, under conditions of greater stress, become phaceloid (e.g., Reuss, 1868, pl. 11, fig. 1) or even solitary. Similar morphology is developed in *Antiguastrea* (e.g., Reuss, 1868, pl. 12, fig. 4, Pfister, 1977b) as well as *Agathiphyllia* and *Montastrea.* In even the extreme cases of morphic variation, internal morphology of the corallites remained relatively unchanged and it was, therefore, possible to properly identify the species.

In other cases, taxonomic problems are more difficult to solve. For example Reuss (1968) described a new genus of more-or-less meandroid fungiid coral, *Heterogyra,* from the Monte Grumi Tuffs. As can be noted in Reuss' original illustrations of the holotype of the type species, *Heterogyra lobata* (Fig. 18W–X), one of the most distinctive features of this coral are the short meandrine series which are in most cases laterally free from each other and contained in plocoid-elevations. When other topotype specimens were collected by the author and colleagues at Monte

Grumi, the basic plan of organization for this genus turned out to be meandroid with broad, more-or-less radial valleys, which may be two or more corallites wide, and which are separated by acute collines. It appears that the generic concept of *Heterogyra* was based on a specimen (or specimens) which had reacted to physiological stress from the sediments by elevating the corallite series and thus generating nontypical morphology. Anstey and Chase (1974, fig. 5.4) illustrate similar effects on the growth form of the modern brain coral *Isophyllia multiflora* (=*I. sinuosa*) of soft mud substrate and high rates of sediment accumulation in Florida Bay. Polyps of the normally meandrine series have shrunk back from the colony margins to form highly elevated plocoid series or even subphaceloid individual corallites. Taxonomic problems of similar magnitude are involved in determining if *Dimorphophyllia* with its large initial corallite (Fig. 18L–M) and smaller radial meandroid series is actually an end member of variation in *Cyathoseris* (Fig. 18Z; Fig. 19D–E, L). Both corals develop flat or explanate colonies from vasiform initial corallites. It might be argued that the exceptionally large initial corallite (which also overlaps the Caribbean and Pacific Oligocene pectiniid genus *Fungophyllia* in morphology) may have been formed by the necessity for upward growth to maintain the polyps above the mud surface.

In contrast to the pronounced effect of the mud on most massive corals, branching and massive *Goniopora, Actonacies, Alveopora,* and *Astreopora,* as well as phaceloid *Caulastrea* and *Euphyllis,* show little or no discernable morphic adaptation to Monti Grumi Tuff substrates. In addition, solitary hermatypic and ahermatypic corals of Figures 17 and 18, with relatively large corallites and trochoid or cylindrical growth form, appear to be unaffected by these conditions. The giant hermatypic solitary coral *Leptomussa* (Fig. 17Q–R) may be somewhat less abundant at Monte Grumi than in the coarser clastic sediments of the Laverda Valley communities.

*4. Sediment Rejection.*—Although the morphology and hydrodynamics of coral reef tracts provide efficient ways to drain off, stabilize, or lithify the carbonate skeletal sediments they produce (e.g., Goreau and Goreau, 1973), the introduction of large amounts of terrigenous sediments to reef communities exerts ecologic stress on the system, commonly slowing growth and sometimes causing mortality in corals unable to cope with sediment fouling (Dodge *et al.,* 1974). High rates of sedimentation will kill corals, but most species can withstand a low rate of fine sediment deposition by active physical removal of them (Marshall and Orr, 1931; Roy and Smith, 1971).

There is abundant evidence in the fossil record

that Tertiary reefs could grow, even flourish, in terrigenous sedimentary regimes. In addition to the examples from the Marósticano and basal Oligocene of the Monte Lessini detailed herein, western Tethyan Oligocene reefs developed in the province of Liguria (northwestern Italy), southeastern Bavaria, and northwestern Yugoslavia under conditions of terrigenous clastic sedimentation. In the Caribbean/Gulf of Mexico region, Oligocene patch-reefs, coppices, and fringing-reefs grew on the margins of an active volcanic complex in Antigua (Frost and Weiss, 1979), as a salt dome cap in the terrigenous clastic Anahuac Formation of Texas (Frost and Schafersman, 1978), and as a major buildup, the Pueblo Nuevo barrier-reef (Frost and Langenheim, 1974, p. 328–332), which formed on the seaward side of a deltaic-alluvial plain complex in southern Mexico.

Modern examples of luxurious coral growth in muddy, turbid water include the Bay of Batavia, Java (Umbgrove, 1928; and Verwey, 1932), bays of New Caledonia (Wijsman-Best, 1972), the southern Philippine bay of Quezon (Nemenzo, 1955), the Low Isles of the Great Barrier Reef (Marshall and Orr, 1931), and part of Fanning Island Lagoon (Roy and Smith, 1971). Coral lists provided for these communities indicate that the overwhelming majority of coral skeletal mass is composed of branching and foliaceous species, with lesser amounts of phaceloid as well as massive plocoid and meandroid corals (only those with large polyps) and minor amounts of hermatypic and ahermatypic solitary corals. High rates of plankton productivity (Roy and Smith, 1971; Wijsman-Best, 1972) related to increased availability of nitrates and phosphates released from the sediments, are attributed to be the cause of the coral luxuriance.

Experimental evidence on the sediment rejection ability of modern reef corals by Hubbard and Pocock (1972) and Hubbard (1973) provides the basic theory to help us understand the species diversity, abundance, and areal mix of coral communities. Corals dislodge sediment by 1. stomadaeal (gullet-like structure under the mouth) uptake of water and retraction of tentacles to make the polyps distend, 2. manipulation of tentacles by species with small polyps to remove individual particles, or 3. secretion of mucous baffles. As Hubbard and Pocock (1972) have noted, skeletal features readily observable in fossil corals, such as mode of colony growth and the geometry of individual corallite calices, are definitive criteria in determining which corals are efficient sediment evaders or removers. Branching modes of growth are successful in avoiding sediment loading because they offer a minimum of horizontal surface area to intercept settling particles. In massive or encrusting corals, those with meandroid organization are the most efficient at removing sediment, with plocoid species having relatively large polyps being next. Cerioid corals having deep, steep-sided calices with relatively few septa were found to be the least efficient.

Hubbard (1973) found that modern Caribbean *Porites* were unusually efficient at rejecting sediment because the relatively small polyps were able to remove particles by tentacular manipulation or by distending the polyps by stomadaeal uptake of water. The present author has observed that larger polyps of Indo-Pacific *Goniopora* are able to distend outward as much as ten times their diameter so that sediments are simply "shrugged off."

Figure 15 is a plot of estimated sediment rejection potential *versus* estimated relative interactive dominance (discussed below) for western Tethyan Oligocene hermatypic coral genera. Sediment rejection potential is estimated on the basis of Hubbard and Pocock's (1972) and Hubbard's (1973) experimental data on modern Caribbean corals and on the ecological distribution of the modern-day Indo-Pacific descendents of these genera. The prolific occurrence of poritid and *Actinacis* thickets as "pioneer" or stabilizing initial communities (discussed above and by Frost, 1977d) appears to be in large part due to their ability to evade much of the sediment because of their branching growth habit and to remove sediment from the polyp surfaces by a combination of tentacular manipulation and polypal distension. In cases of wave swept habitats where water turbulence is too great for branching poritids to survive, species of the same genera but with encrusting, massive or nodular growth form comprise the stabilizing coral communities, as in the Marósticano ecological succession depicted in Figure 13. Sediment rejection potential was obviously also a major limiting factor in determining the species composition of the soft substrate coral assemblage from the tuffs of Monti Grumi (discussed above). Inability to cope with high rates of sedimentation may be a factor why some Tethyan Oligocene species with massive growths form, and deep, cerioid calices such as *Stephanocoenia parvula, Favites laxelamellata* and *Favites intermedia* are rare, especially so at the classic localities of the Alpine region which occur in terrigenous sediments.

5. *Interspecific Dominance.*—Another important coral interaction at the community or individual colony level is the hierarchy of species-specific digestive dominance relationships among hermatypic scleractinia, discovered by Lang (1970, 1971, and 1973). Digestive interactions are apparently the means by which many slow-growing massive coral species compete in densely packed coral communities. Highly dominant corals, such as those of the Family Mussidae, ex-

trude mesentarial filaments from inside the polyp and dissolve the tissues of less dominant neighbors within reach by means of extracoelenteric forms of digestion. Most of the most highly dominant species construct relatively small encrusting or massive skeletons or are solitary, and use digestive interactions as a defense against competition by means of overgrowth or overtopping by the more rapidly growing foliose and branching corals, and to clear space for their own growth.

The estimated digestive dominance potential of Western Tethyan Oligocene reef coral genera is plotted on Figure 15. It is based on analogy with the modern counterparts of these genera (or to their closest living relatives) and includes data from Lang (1970, 1971, and 1973), Gravier (1911), Catala (1964), and Connell (1973), the author's own field observations on modern Caribbean corals and on field data on the fossil reefs. The evidence for digestive dominance left in the fossil record consists of overgrowth relationships; and of these, only in instances where it can be demonstrated that the coral which was overgrown, was actually living at the time of the interaction. The most satisfactory evidence that the coral overgrown had previously been dead is if its upper skeletal surfaces were bored (fungi, filamentous algae and/or sponges) and encrusted (serpulids and calcareous algae). One of the most spectacular examples of Oligocene dominance and overgrowth relationships occurs in a specimen from Poljisca, Yugoslavia figured by Barta-Calmus (1973, pl. 56, fig. 6). The core of the specimen is formed by a massive species of *Gonopora* cf. *G. minuta* which has been encrusted by *Antiguastrea lucasiana*. The *Antiguastrea* has been overgrown by *Agathiphyllia rochettina*; and *Heterogyra lobata* has partially overgrown both species. The contact surfaces between each pair of interacting species appear to have been fresh at the times of interaction.

Examination of the dominance estimates of Figure 15 shows, that of the total range of western Tethyan Oligocene genera which could be expected to occur at any given locality, the upper half of the diagram (Suborder Fungiina) is composed of relatively passive corals which, with the notable exceptions of *Pavona* and *Goniopora,* are predominately subordinate framework fillers (secondary hermatypes *sensu* Goreau); genera of the Suborder Faviina constitute virtually all of the framework builders. Of the Faviina, the major frame building genera are those which are intermediate in dominance interactions, while the corals which are estimated to have been the most dominant (e.g., *Leptomussa*) aggregate only a minor percentage of coral skeletal volume in any of the communities sampled.

It should be noted that in the Faviina, representatives of the most digestively dominant families, Mussidae and Meandrinidae (Lang, 1873) are, respectively, rare or even absent. Mussidae are important primary hermatypes in the modern western Atlantic/Caribbean region. Meandroid, subcerioid, and phaceloid forms of *Mussismilia* are major framework builders in reefs of Brazil (Laborel, 1969; 1970) and several meandroid *Mycetophyllia* species form much of fore-reef framework in Caribbean reefs (Goreau and Goreau, 1973). Indo-Pacific mussid potential frame builders are *Acanthastraea, Lobophyllia,* and *Symphyllia.* We have found no evidence that any of these genera occur in the western Tethyhan Oligocene. The same is true for the meandrinid genera *Meandrina, Dichocoenia,* and *Dendrogyra,* all important contributors to modern Caribbean reef buildups.

Branching and foliaceous, rapidly growing members of the Fungiina which are relatively low in digestive-dominance hierarchies also engage in interspecific aggression with other corals. Shinn (1972) found that *Acropora palmata* and *Acropora cervicornis* of the Florida reef tracts occasionally caused the gradual death of massive framework corals such as *Montastrea* and *Diploria* by overtopping them, and presumably, reducing the penetration of light and circulation of water to them. Connell (1973, p. 226) reports similar interactions on the Great Barrier Reef between *Acropora* and encrusting species of *Montipora.*

As in sediment rejection potential, concepts of interspecific aggression, especially digestive dominance hierarchies, have important predictive value in understanding the species composition of the communities listed in Figure 14. Species which have relatively low digestive dominance potential, but which are efficient sediment evaders and/or rejectors, dominate communities where more aggressive corals are excluded from the habitat by nature of substrate, sedimentation rate, wave action or by other factors. The specialized paucispecific *Goniopora* thickets are an example. Patch-reef or coppice flank communities are another example. In these sheltered environments, coral diversity is usually fairly high (Figs. 12 and 14) and assemblages are predominantly composed of branching, nodular/columniform, and phaceloid corals with light, fast growing skeletons. Some corals with small and delicate branching skeletons, such as *Stylophora, Stylocoenia, Dictyaraea,* and to some extent *Astrocoenia,* form a minor "understory" in the stands of branches.

Because of the myriad of microhabitats within the growing framework of the reef-core, total coral diversity is highest there. In this habitat the most successful corals (on the basis of abundance

and contribution to reef-core framework) are massive plocoid and meandroid forms of the S. O. Faviina which combine intermediate to strong sediment rejection ability with at least intermediate levels of digestive dominance. In terms of Oligocene paleobiogeographic distribution (discussed above) species of this suborder appear to have been more cosmopolitan than those of the Fungiina or Caryophylliina, being distributed from the Indian Ocean across the western Tethys into the western Atlantic/Gulf of Mexico/Caribbean region (Frost, 1977b).

*6. Other Relationships.*—Many other relationships involving hermatypic scleractinia can be identified in the incredibly complex web of species interactions (Dahl et al., 1974) in modern reef ecosystems. These include important mutualistic relationships with endoskeletal blue-green and green algae (e.g., Odum and Odum, 1955), and with sponge and bryozoan epitheca encrusters (Cuffey, 1973), obligatory relationships with ostracodes, and barnacles (Newman and Ladd, 1974), gall crabs, (Schuhmacher, 1977), some gastropods (Robertson, 1970), predation by errantid polychaete worms (Ebbs, 1966), and parrot fish (many authors). The live factors treated above are of special paleoecological application because their effects may be recorded directly in coral skeletal growth or may be inferred in ancient species because of their close genetic relationship with living descendants.

*Coral Community Succession*

A number of Tertiary reefs studied to date show, on a more-or-less worldwide basis, a sequence of similar ecological and morphological developmental stages or seres (e.g., Frost and Langenheim, 1974; Frost, 1977d). These seral successions, which are particularly well developed in the earlier phases of regional transgressive cycles, typically involve three distinctive types of scleractinian coral assemblages:

1. substrate-stabilizing or "pioneer" communities of low diversity, usually dominated by poritids or other corals with light porous skeletons. Dominant ecological factors are the unstable substrate and ability to efficiently reject sediment;

2. fairly diverse transition assemblages, consisting of branching and massive "pioneer" species as well as species which are somewhat less tolerant of sediments (e.g., *Antiguastrea*) but which are moderately aggressive in digestive dominance hierarchies. Some ecological zonation into reef-core and flank habitats is developed;

3. Climax seral assemblage of high diversity, dominated by massive and encrusting, more-or-less digestively dominant frame building corals of the Family Faviidae; poritid corals are also abundant. A full set of ecologic/morphologic zones

ranging from reef-flat to fore-reef is usually developed.

The ecological evolution of 1. to 3. is accompanied by the potential change in reefoid deposits from coral biostromes and thickets to coppices and patch-reefs to climax lagoon, fringing or barrier-reefs, assuming at each stage the minimum conditions to maintain reef growth and development are met.

In many cases the sequence is interrupted before it can develop to completion (e.g., San Luca, coral coppice of Fig. 13). Continued growth of the climax reef with a static relative sea level leads to seaward progradation of the reef deposits. Relative sea level rise at rates slightly faster than the upward constructional activity of the reef (because of rapid increase in rates of subsidence or eustatic sea level rise) causes shelf-margin reefs to migrate landward (e.g., Fig. 7, barrier-reef phase). If the deepening rates are substantially faster, the shelf-margin reef may be drowned, and the reef biota will migrate landward into the former back-reef lagoon as a series of coral biostromes or patch-reefs (e.g., Fig. 7, upper biostromal phase).

## Biotic Associates

Modern day coral reef ecosystems are considered by many authors to be the most complex of all marine ecosystems in regard to the number of species involved, amount of productivity, and the intricacy of interactions. For example, the compartment-type model developed by the CITRE working group (Dahl *et al.*, 1974) comprises 104 compartments or ecologic units, each of which is defined as plant or animal processors of carbon or carbon-equivalent energy storage units or as material pools through which substances pass as they cycle in the system. These ecologic units involve 10,186 potential interactions (or flows between compartments), of which about 2000 are identified in the model to actually take place. Each biotic compartment, such as epiphytic filamentous algae or reef dwelling scleractinian corals, may have many species in each reef ecozone. In addition, outside variables (forcing functions) which influence the types or rates of flows between the compartments (such as wave surge or light penetration) are also part of the ecosystem.

We have previously examined some of the factors and interactions that affect modern day coral community structure and which are thought to have affected the ancient reef communities of the Vicentin as well. We cannot hope to reconstruct more than a tiny fraction of the ecologic interactions that took place on these and other Tertiary reefs. In order for us to place the Vicentin coral communities in the context of the spectrum of major biotic components with which they are associated in these ancient ecosystems, we will briefly consider the probable reef community.

### Preservation Potential of the Total Reef Community

Frost (1977a) has shown that much of the modern reef community biomass has little chance to be recognized in the future fossil record. The Tertiary fossil record (with the exception of hermatypic scleractinia) is greatly skewed toward some trophic categories of reef dwelling animals which probably aggregated a strikingly small fraction of the standing crop biomass at any given instant (Frost, 1977, fig. 3). Major biovolume in all four modern Caribbean reef ecozones, for which he made estimates, is concentrated in benthic plants; most of these leave no identifiable skeletal fragments in reef framework or sediments *before* diagenesis. The disparity in fossilization potential of the components of the biota standing crop is magnified many times over the duration of a reef buildup. This is because the turnover rate (ratio of annual biomass produced to standing crop) is more than 10 to 1 (e.g., Odum and Odum, 1955, p. 317) for benthic plants of the reef community and less than 1 to 1 for benthic animals.

*Benthic Algae.*—Of the hundreds of benthic plant species found on modern reefs, two groups, calcareous green algae and articulate and crustose coralline red algae account for virtually all of the skeletal production and hence, the potential fossil record of benthic producers. Lightly calcified green algae of the Order Codiaceae such as *Penicillus, Udotea,* and *Rhipocephalus* produce sheaths of aragonite needles associated with the tangle of branched tubular filaments that comprise the plant body. Although calcareous green algae produce prodigious amounts of aragonite skeletal sediment in modern reef associated environments (e.g., Stockman *et al.*, 1967; Neumann and Land, 1975) surprisingly little of these skeletal fragments are recognizable in the sedimentary rock record of their ancient counterparts. This is probably due to the well known instability of aragonite in the diagenesis of carbonate sediments. Because of the evolutionary relationships of this group (e.g., Wray, 1977) we can reasonably expect that they existed on Tertiary reefs, and that at least some part of the micritic matrix component of Oligocene Vicentin reef tracts originated as codiacean aragonite needles.

Another codiacean green alga, the well known *Halimeda,* produces a hard, segmented skeleton of aragonite needles. The cushion-shaped skeletal segments consist of acicular aragonite crystals deposited on the surface and between the tubular filaments of the internal plant structure. Much of the skeletal sand produced in modern reefs, particularly in the back-reef and fore-reef zones, is

composed of *Halimeda* skeletal plates (e.g., Goreau and Goreau, 1973, p. 445, fig. 21). The plates
have fairly good preservation potential (about the
same as that of a similar size piece of *Porites* or
*Goniopora*) and occur very abundantly in fossil
Pleistocene and Holocene reef rock. It has been
the author's observation that fossil codiaceans of
this type are rare in pre-middle Miocene Tertiary
reefs of the western Tethys, Caribbean, and Indo-
Pacific. In the hundreds of thin-sections of reef
sediments examined in the course of this study,
only a few specimens of *Ovulites* (?) (Fig. 11E)
and *Halimeda* (Fig. 16A) were identified. There
is a strong possibility that these codiaceans were
not as abundant in mid-Tertiary reefs as in their
modern counterparts.

Dasycladacean green algae also incorporate
aragonite as a sheath around the stalk and as coatings and partitions in the radially symmetrical,
disk-shaped or cylindrical thallus composed of
whorls or branches (Wray, 1977). In effect, the
rigid aragonite encrustations are calcareous molds
of the stem, branches, and reproductive organs.
Calcified spherical gametangia are called calcispheres. In Oligocene sediments of the Vicentin,
dasyclads are moderately abundant in biomicrites
deposited under quiet water conditions (Fig. 11C)
and are preserved as calcite pseudomorphs after
the orginal aragonite sheaths.

Articulate and crustose red algae (mostly of the
Order Corallinaceae) comprise a substantial component of benthic plant biomass across the entire
range of modern reef ecozones (Frost, 1977a, fig.
2) and have left an extensive fossil record (e.g.,
Adey and Macintyre, 1973). The cellular Mg-calcite internal microstructure of the skeleton makes
it resistant to abrasion as sedimentary particles as
well as ensuring that it can be recognized as a
fossil in sedimentary rocks after diagenesis. In the
Vicentin Reefs, coralline algae are ubiquitous in
every type of reef associated sediment; in lagoonal coral thickets and coppices (Fig. 11A), as secondary framework binders of the reef-core (Fig.
16C) and as rhodolith primary sediment formers
at the base of the Berici Barrier-Reef at Lumignano and in the deep fore-reef at Zovencedo.

*Foraminifers.*—Benthic foraminifers fill the
trophic category of microherbivores and micro-
bacteriovores in modern reef ecosystems. They
have a high preservation potential and, in comparison with benthic plants, are disproportionately well represented in ancient reef sediments.
Their distribution in Vicentin reef associated sediments has been outlined earlier in this paper.
Here we wish to examine the fossil record of encrusting foraminifers as secondary framework
binders in these buildups.

Schuhmacher (1977) concluded that the encrusting foraminifers *Acervulina* and *Planorbuli-*
*noides* (along with encrusting calcareous algae)
were the most important framework binding organisms in modern reef communities developing
on shallow artificial substrates in the Red Sea. In
modern *Porites furcata* thickets of the Caribbean
side of Panamá, Glynn (1973, Table 2) found the
encrusting foraminifer *Homotrema rubrum* in
prolific abundance, recording densities of more
than 13,000 per square meter and dry weight biomass volumes of as much as 135.6 g/m². In coralline algal cup-reefs of Bermuda, Ginsburg, and
Schroeder (1973, p. 589) note that *H. rubrum* is
by far the most numerous internal cavity dweller,
being locally abundant enough to serve as a
framework builder. Goreau and Goreau (1973, p.
457) considered coralline algae and the encrusting
foram *Gypsina* to be the major framework ce-
menters in the fore-reef framework of Jamaican
reefs.

Although we have not made a quantitative census to determine encrusting foraminiferal abundance in the Vicentin buildups, there is some data
to suggest that they may have been as important
in these Oligocene reefs as in their modern counterparts. In the course of sectioning dozens of
coral specimens for taxonomic studies, the author
has observed that encrusting foraminifers are
abundant (along with coralline algae) as epiphytes
on the lower (epithecal) surfaces of massive coral
colonies. These include forms such as *Gypsina,*
*Planorbulina, Crespinina* (?), *Falsocibicides, Dy-*
*ocibicides,* and the large highly porous *Eorupertia*
(Fig. 11B). Encrusting foraminifers appear to be
most abundant in collections from communities
that were living under conditions of fine-grained
terrigenous sedimentation. In the earliest Oligocene mixed carbonate-terrigenous sequence of the
Grumi Road section (Fig. 4) the marls and calcareous mudstones are characterized by the great
abundance of coralline algal rhodoliths. Upon
sectioning, many of these turn out to be formed
by complexly intergrown layers of encrusting foraminifers and coralline algae (Fig. 16F), and
bound sediment. In some cases, these crusts may
be more than 2 centimeters thick and involve dozens of rhythmic or cyclical alternations between
laminae formed by the coralline algae and the foraminifers; the latter of which appear to be species
of *Gypsina* and *Dyocibicides* (?). The crusts appear
to be the result of mutualistic interaction between
the organisms; and the strong regularity in thickness of each type of lamina (present in many of
the specimens) may be the result of their respec-
tive life cycles. Another interpretation of the alternations might be that they resulted from sea-
sonal fluctuations similar to those between
coralline algae and mats of colonial anemones
(zoanthids) on modern reefs (e.g., Kinzie, 1973,
p. 102).

*Sponges.*—By analogy with living reefs (e.g., Frost, 1977a), sponges and octocorals may have aggregated a major part of the biovolume of the animals associated with corals on the Oligocene reefs under study. The overwhelming majority of sponges on modern reefs and most probably Tertiary reefs, are those of the class Demospongia, which have skeletons with opaline silica spicules, collagenous fibers, or both elements combined (Hartman, 1977). Although sponges are distributed throughout all of these reef ecozones, they form a very substantial part of the reef fauna below a depth of about 15 meters, and occur in standing crop biomass concentrations on suitable substrates of perhaps as much as 500 gm/m², dry weight (e.g., Reiswig, 1973; Rützler and Macintyre, 1978, p. 157). We can convert these maximum biomass values into a rough estimate of the maximum standing crop of siliceous spicules at a given time by means of Rützler and Macintyre's (1978, Table 2) spicule content determinations for ten of the most common reef species at Carrie Bow Bay, Belize. If we use a value of 25 percent spicule content of total sponge dry weight (a value slightly less than the average content of the ten species), we derive that the spicule volume which could be potentially liberated into the sediments upon mortality could be as high as 125 grams of spicules/m² of sea bottom. Even allowing for low rates of populations turnover and a fraction of this concentration of spicules, the rates of deposition of reef sediments are slow enough so that silica sponge spicules should be an important component, at least in the deeper reef ecozones.

Generally speaking, though, silica sponge spicules are *much* less abundant in modern reef sediments than the living sponge biovolume would indicate (e.g., Hartman, 1977). Moreover, Land (1976) found in Jamaica that the silica content (mostly sponge spicules) of deeper water sediment such as the fore-reef was less than 20 percent of that in lagoonal carbonate muds; a relationship which is essentially *inverse* to the distribution of sponge biomass. The cause of the apparent shortfall between spicule production and occurrence in the reef sediments is that they are selectively dissolved after liberation from the sponge mass, especially so in sediments that have high interstitial concentration of dissolved calcium carbonate and are undergoing submarine lithification (Land, 1976; Rutzler and Macintyre, 1978).

Selective dissolution during modern day carbonate diagenesis (and presumably, in the past) of the only potentially fossilizable skeletal clues to reef sponge abundance and diversity means that details of their role in Tertiary reef ecosystems (such as those considered herein) will probably remain unknown. Excavating sponges such as *Cliona* and *Siphonodictyon* (e.g., Hartman, 1977; Pang, 1973) have left an extensive though indirect record of their existence (and perhaps relative abundance) in fossil reefs as old as mid-Mesozoic (Warme, 1977).

Very few sponge spicules are present in samples from the Vicentin Oligocene buildups, either in thin-sections or in washed residues, and the few that do occur are mostly triactines and tetractines of calcisponges. Borings made by excavating sponges are present on the undersides of corals that were preserved in growth position, but they are most abundant on corals, bivalves, echinoids, *etc.*, that were part of exposed skeletal rubble on the sea floor. In general, corals and other large skeletal elements of the buildups which grew under conditions of at least some terrigenous clastic sedimentation (Marósticano, basal Oligocene, and Monti Grumi Tuff assemblage) are less bored by sponges and other epibionts than those from solely carbonate units, such as the coral carpets or biostromes of the upper Castelgomberto Limestone.

*Octocorals.*—These coelenterates occur in biomass concentrations on modern reefs which are as great or greater than those of demosponges (e.g., Kinzie, 1973; Opresko, 1973; Preston and Preston, 1975). In contrast to sponges, the greatest octocoral abundance and diversity occurs in shallower reef sites—typically in the coral buttress zones of Caribbean reef-fronts (Kinzie, 1973). Kinzie concluded that the most important factor limiting reef tract octocoral abundance and diversity was the amount and nature of substratum available for settlement and growth.

The fossilization potential of octocorals is a limiting factor in any attempt to reconstruct their paleosynecology or paleoautecology. Most octocoral colonies (except the Order Coenothecalia) bear Mg-calcite spicules or sclerites dispersed throughout the branching or encrusting colonies (Bayer, 1956 and 1961). These sclerites are concentrated at the base of each tentacle of the polyp and in its body wall between septa in a symmetrical octameral arrangement. Denser concentrations of spicules occur in the lower part of each polyp and spicules are distributed in the various layers and zones of the interpolypal coenenchyme. The taxonomy of octocorals is based primarily on the morphology of their sclerites and their arrangement and distribution on the polyps and in the coenenchyme. Although the fossil spicules of some orders of octocorals such as the Stolonifera and the Alcyonacea are difficult to classify or identify accurately (Bayer, 1956), those of the Gorgonacea bear elaborate ornamentation which is of taxonomic value. Moreover, the Gorgonacea and Pennatulacea produce distinctive axial structures, many of which have a high potential for fossilization.

Cary (1918) demonstrated that gorgonian octocorals are prolific producers of skeletal calcium carbonate. He estimated the spicule content of the gorgonian standing crop of the Dry Tortugas Reef (near Key West, Florida) to be as much as 5.38 tons per acre, and that about one ton of skeletal material was released as potential sedimentary particles per acre/per year, assuming a 20 percent annual turnover rate. Although Cary's estimates of gorgonian attrition rates may be too high, the conclusion is inescapable that given the rates of deposition of reef tract sediments, octocoral remains should be expected to be abundant.

Reference to recent studies of modern Caribbean/western Atlantic reef tract sediments (e.g., Steiglitz, 1972 and 1973) suggests that the abundance of sclerites is much less than would be estimated, based on the abundance of octocorals in the areas from which the sediments were taken. Analysis of sediments derived from lagoonal patch-reefs (with abundant octocorals) of Glovers Reef Atoll, Belize, by the author and associates (e.g., Wallace and Schafersman, 1977) confirms that sclerites are indeed rare in these samples; this even though conscious search was made for them using for reference spicules extracted from gorgonian specimens as well as standard works on sclerite morphology. The apparent shortfall between Mg-calcite sclerite production and abundance in substrate sediments most probably can be explained by their solution, either during decomposition of the soft colony tissue or of the disarticulated sclerites at the sediment-water interface. The solution phenomenon does not appear to be as pervasive, however, as in opal demosponge spicules.

On reefs of the Indo-Pacific, however, fleshy alcyonaceans such as *Sarcophyton, Sinularia,* and *Litophyton* are the predominant shallow water octocorals (e.g., Wells, 1957). In these, the interpolypal tissue is even more densely spiculate than that of gorgonians, and Cary (1931) has shown that substantial quantities of alcyonacean spicules are incorporated into reef limestone and fossilized, as in Utelei Reef, Samoa. The greater preservation potential of alcyonaceans appears to be due to the fact that the lower parts of massive colonies may be cemented into the lithified reef rock while the upper parts are still living (Cary, 1931).

Evidence of fossil octocorals does occur in the Vicentin buildups. Most numerous is the massive or encrusting *Parapolytremacis* (Fig. 18J–K) which belongs to the Order Coenothecalia and is closely related to the modern day "blue coral" *Heliopora* of Indo-Pacific reefs. Octocorals such

as these, lack spicules and have a fibrous aragonite skeleton composed of sclerodermites arranged similar to that of a scleractinian coral, and have about the same preservation potential. *Parapolytremacis* occurs as a secondary framework builder or encruster in the reef-core facies. Remains of gorgonian octocorals occur in the terrigenous beds of the lowermost Oligocene sequence in the Grumi Road section. These include fine branches of scleraxonids (Fig. 16B) which have the cortical spicule mass fossilized more-or-less intact, dense masses of elongate gorgonian(?) spicules on bedding planes, each of which apparently records the decay of a single colony and finally, isolated spicules from washed residues of marls which are strikingly similar to those of the modern *Iciligorgia*. Moreover, the calcified axial structures of scleraxonid gorgonians such as *Parisis* occur at San Luca (Pfister, 1977, pl. 13, figs. 5 and 6) and in the Colli Berici (Reuss, 1869, pl. 38, figs. 14–16). These calcified axial structures have been widely found in the Tertiary. Somewhat surprising is the absence in any of the collections of any trace of alcyonacean octocorals with their great masses of ornamented, fat spindle-shaped sclerites and cemented bases.

*Bryozoans.*—On modern reefs bryozoans (which are mostly sheet-like cheilostomes) aggregate relatively small biovolumes and have been relegated to cryptic ecologic roles such as "hidden encrusters" (Cuffey, 1977) and cavity dwellers (Cuffey and Fonda, 1977). This is because of their apparent inability to compete successfully with algae, scleractinians, sponges, octocorals, zoanthids, *etc.,* for exposed surfaces of attachment (Cuffey and Kissling, 1973). Cuffey (1973) has shown that in both Pacific and Caribbean lagoonal reefs, encrusting cheilostomes profusely overgrow the undersides of massive scleractinia, and successfully exclude such competitors as the encrusting foraminifer *Homotrema*. In the reef-flat and shallow buttress zones of Florida Keys Reefs, sheet-like cheilostomes are most abundant in patches of rubble where they encrust the undersides of stable coral boulders.

Bryozoans appear to be similar in their distribution in Vicentin buildups to the encrusting foraminifers. That is, they are most abundant in units which have fairly substantial amounts of fine-grained terrigenous sediment (e.g., Fig. 16C). They are discussed here as associates of the corals because they appear to have served as minor framework binders in the Marósticano and possibly in the basal Oligocene sequences of the Colli Berici and the Grumi Road section. In the course of this investigation, no detailed studies were made on the abundance or distribution of bryozoans in the buildups. An overview of the types of bryozoans associated with the reef-bearing sequence in the Marósticano is afforded by Reuss' (1869) early paper.

*Other Biota.*—A great diversity of other benthic animals and plants (e.g., Frost, 1977a, fig. 2) must have accompanied the corals and their associates considered herein. An extensive study of molluscan faunas from the lagoonal facies of the Castelgomberto Limestone in the Monte Lessini has been made by Coletti *et al.* (1973). Kranz (1910) provides an illustrated survey of similar molluscan faunas and (1911) illustrates and summarizes the bryozoans, brachiopods, echinoids, and large foraminifers of the Castelgomberto Limestone in the central and southern Monte Lessini. Undoubtedly traces of other reef dwellers; perhaps, ostracodes, crinoids, sipunculids, sabellarians, barnacles, polychaetes, scaphopods, large crustacea, ophiuroids, and sea stars have been found in the 150+ years of collecting from the Oligocene strata of the Vicentin.

## SUMMARY

Outcrops of Oligocene reef facies strata in the Vicentin afford the opportunity to document the response of ancient reef coral communities to a broad range of habitats and substrates in a variety of depositional settings and in different stages of ecologic succession. Evidence of the spectrum of morphologic variation in each biological species is vital to an understanding of the true taxonomic relationships of the immense array of named species (more than 520) from the Oligocene of the western Tethys.

The Berici Barrier-reef/lagoon complex of the Castelgomberto Limestone in the Colli Berici and the eastern Monte Lessini, Italy, is one of the largest and best exposed of outcropping Tertiary reef buildups. In addition to the diverse examples of reef tract biota and communities growing under conditions of relatively pure carbonate sedimentation, studies of the exposures provide valuable comparative models of carbonate sedimentary facies and diagenetic sequences which can be applied to Tertiary reefs in the subsurface. Examples of analogous subsurface reefs which are major hydrocarbon reservoirs include the Oligocene Kirkuk Reef complex of Iraq, the Oligo-Miocene Anahuac fringing-reef of southeastern Louisiana, and Miocene fringing and pinnacle-reefs of the Philippines and Indonesia. Outcrops of the Castelgomberto Limestone display abundant examples of carbonate porosity preservation, creation and destruction.

In the Marósticano, reefs formed along the northeastern margin of a huge Oligocene volcanic complex. These structures include examples of thickets, coppices, patch-reefs and thin fringing or platform-reefs which, unlike the Berici Barrier-reef/lagoonal complex, developed in a dominantly terrigenous clastic regime of sedimentation. In some cases coral buildups appear to have grown directly adjacent to centers of submarine and subaerial volcanic eruption. The reefs are preserved as local carbonate rich lenses in the volcanic sandstone-mudstone-tuffite sequence.

Three basic coral assemblages account for the broad spectrum of buildups which occur in the Vicentin: 1. low diversity "pioneer" or thicket assemblage, composed primarily of branching or nodular species of *Goniopora* and *Actinacis*; under conditions of extreme turbulence on wave-swept substrates massive or encrusting species of the same genera may occur; 2. high diversity reef-core and flank assemblage which may include a total of more than thirty species; the flank assemblage is characteristic of protected habitats and includes many branching corals that were excluded from the *Goniopora/Actinacis* thickets because they lacked strong sediment-rejection ability, or were crowded-out of the main reef framework by massive corals; although diversity in the reef-core facies is the highest for any of the reef habitats, a relatively few species (5 or 6) aggregate most of the constructed and bound framework, the rest of the species are reef dwellers because their skeletons are too small or too fragile to be incorporated into the primary reef framework structure; 3. communities of soft mud substrates are dominated by branching and phaceloid ("organ-pipe") corals and those massive corals with elevated corallites (plocoid corallum structure); response of individual colonies to the soft substrates and fine mud sedimentation produces distinctive growth responses in many species; although these assemblages did not produce bound reef framework and their most prolific concentrations appear to have been as biostromes or coral carpets.

In addition to environmental factors which directly define or limit the coral reef biotope, some factors operate very specifically at the community level by determining the actual frame building and frame filling species composition of the assemblages. Three factors, 1. adaptation to water flow, 2. potential to colonize unstable substrates and survive the rain of sediments, and 3. species-specific digestive dominance relationships appear to be major factors in determining the areal mix of species, the diversity and the abundance of Vicentin Oligocene reef corals. Interaction of these factors resulted in the widely observed succession of Tertiary coral communities from pioneer to transitional to diverse climax, with the accompanying change in type of buildups from coral biostromes or thickets to patch-reefs and coppices to climax lagoon, fringing or barrier-reefs.

Most of the biomass of animals and plants growing on modern reefs has little chance to be recognized in the future fossil record. The same was undoubtedly true for their ancient counterparts in Tertiary reefs. Three groups, the benthic algae, sponges and octocorals are postulated by analogy with living reefs to have constituted a major part of the noncoral biovolume. Each of these groups (with the possible exception of the coralline red algae) is unrepresented or greatly underrepresented in the fossil record of Oligocene Vicentin Reefs. This is because they either lacked mineralized skeletal parts or because these parts were chemically unstable and were mostly dissolved before they were incorporated into the sediments or during diagenesis. Encrusting benthic foraminifers and coralline algae are important in these reefs as framework binders and have left an extensive fossil record. Aside from the corals, coralline algae, and foraminifers, most of the Oligocene reef dwelling biota which left an extensive fossil record in these deposits (e.g., molluscs, echinoids, bryozoa, *etc.*) probably comprised a relatively minor part of the living biovolume at any given time.

### ACKNOWLEDGMENTS

The writer gratefully acknowledges the help of many individuals and institutions in the course of this investigation. The field work and much of the initial research on western Tethyan corals was conducted in 1972 while the writer was on academic sabbatical leave from Northern Illinois University. Prof. Dr. Rene Herb, Dr. Therese Pfister, and Dr. Jorn Geister of the Geologisches Institut, University of Bern, Switzerland, guided the writer to key localities in the Marósticano, made available for study large collections of corals from there, and have been the writer's colleagues in research on the taxonomic complexities of the Vicentin Oligocene corals. The Naturhistorisches Museum of Basel, Switzerland provided the writer with research space and technical help, and Drs. Peter Jung and Felix Wedenmayer accompanied him during field work in the eastern Monte Lessini. Most of the subsequent research was conducted during the writer's tenure in the Department of Geology, Northern Illinois University, and this author wishes to warmly acknowledge his colleagues for the many fruitful discussions of modern and ancient reef systems.

TABLE 1.—OLIGOCENE REEF CORALS, WESTERN TETHYS

Suborder ASTROCOENIINA

Family ASTROCOENIINAE

*Actinastrea saxulensis* (d'Achiardi, 1868) = *Stylophora micropora* Sismonda and Michelotti, 1871.

*Astrocoenia lobatorotundata* (Michelin, 1842) = *Syringopora flabellata* Catullo, 1856; *Astrea septemdigitata* Catullo, 1856; *Astrea tuberosa* Catullo, 1856; *Astrocoenia nana* Reuss, 1868; *Stephanocoenia ramea* d'Achiardi, 1866.

*Astrocoenia multigranosa* Reuss, 1868.

*Stephanocoenia parvula* (Sismonda and Michellotti, 1871) = *Prionastrea intermedia* Osasco, 1898; *Stylocoenia minuscola* Osasco, 1902; *Lophoseris anteacta* DeAngelis, 1894; *Mesomorpha gombertina* Kranz, 1914; *Thamnastraea maraschinii*, DeAngelis 1894.

*Stylocoenia taurinensis* (Michelin, 1842) = *Astraea bistellata* Catullo, 1856; *Astraea palmata* Catullo, 1856; *Porites tuberosa* Catullo, 1856; *Astrocoenia laminosa* d'Achiardi, 1866, also probably *Astrocoenia parvistellata* d'Achiardi, 1866.

*Stylocoenia monocycla* (d'Achiardi, 1866). May be a different growth form of *Stylocoenia taurinensis* (Michelin).

Family POCILLOPORIDAE

*Stylophora thirsiformis* (Michellotti, 1847) = *Astraea raristella* Michellotti, 1847 (pars); *Stylophora annulata* Reuss, 1864; *Pocillopora granulosa* Gumbel, 1861; *Stylophora microstyla* d'Achiardi, 1866.

*Stylophora conferta* Reuss, 1868 = *Stylophora tuberosa* Reuss, 1868; *Astraeopora exigua* Reuss, 1869.

Family ACROPORIDAE

*Acropora haidingeri* (Reuss, 1864) = *Dendracis distincta* Osasco, 1902; *Acropora pseudolavandulina* Chevalier, 1955.

*Acropora miocenica* (Sismonda and Michelotti, 1871) = *Acropora pachymorpha* Chevalier, 1955.

*Dendracis gervillii* (Defrance, 1828) = *Seriatopora cribrara* Catullo, 1856; *Astrohelia michelottii* Meneghini in Michelotti, 1861; *Astrohelia coralloidea* Meneghini in Michelotti, 1861; *Dendracis nodosa* Reuss, 1868; *Dendracis mammillosa* Reuss, 1868; *Dendracis granulocostata* d'Achiardi, 1866; *Dendracis seriata* Reuss, 1868.

*Astreopora tecta* (Catullo, 1856) = *Astreopora decaphylla* (Reuss, 1868; *Astrea cylindrica* Defrance. Catullo, 1856; *Astreopora auvertica* (Michelin) d'Achiardi, 1868; *Astreopora auvertica* (Michelin) d'Achiardi, 1868;

*Astreopora deperdita* (Michelotti, 1861); *Astraeopora compressa* Reuss, 1864; *Madrepora discors* Sismonda and Michelotti, 1871.

*Astreopora meneghiniana* (d'Achiardi, 1866) = *Astraea brevissima* Deshayes. Catullo, 1856; *Astrangia minima* d'Achiardi, 1868; *Astraeopora patula* Sismonda and Michelotti, 1871.

*Astreopora* questionable species: *A. paniceoides* Reis, 1889; *Madrepora crispa* Sismonda and Michelotti, 1871; *A. pulchra* d'Achiardi, 1867.

Suborder FUNGIINA

Superfamily AGARICIICAE

Family AGARICIIDAE

*Trochoseris* ? *cornucopia* Michelotti, 1861.

*Dimorphophyllia* (?) *oxylopha* Reuss, 1864 = *Meandrina collinaria* Catullo, 1856; *Lobophyllia formossissima* Catullo, 1856; *Meandrina costata* Catullo, 1856; *Trochoseris distorta* (Michelin). Schauroth, 1865; *Cyathoseris subregularis* Reuss, 1872; *Trochoseris* ? *difformis* Reuss, 1868; *Leptophyllia tuberosa* Reuss, 1868; *Trachosmilia acutimargo* Reuss, 1868; *Trochosmilia profunda* Reuss, 1868.

*Cyathoseris appenina* (Michelin, 1842) = *Cyathoseris catulliana* d'Achiardi, 1867; *Cyathoseris applanata* Reuss, 1872; *Leptoseris antiqua* Reuss, 1869; *Dimorphastraea irradians* Reuss, 1868; *Dimorphastraea depressa* Reuss, 1868; *Dimorphastraea exigua* Reuss, 1869; *Cyathoseris affinis* Reuss, 1869; *Cyathoseris centrifuga* Reuss, 1868; *Podabacia prisca* Reuss, 1864 (pars); *Tham-*

*nastraea pulchella* Reuss, 1869; *Mycetoseris hypocrateriformis* (Michelotti). Prever, 1921 (pars); *Mycetoseris adscita* DeAngelis, 1894; *Mycetoseris conferta* (Reuss). Prever, 1921 (pars); *Lophoseris prisca* DeAngelis, 1894; *Halomitra ambigua* DeAngelis, 1894; *Mycetophyllia stellifera* Michelotti, 1861; *Mycetophyllia radiata* Michelotti, 1861; *Comoseris conferta* Reuss, 1868; *Mycetoseris appenina* (Michelin) (pars). Prever, 1921; *Comoseris alternans* Reuss, 1868; *Comoseris distincta* Osasco, 1902; *Leptoseris raristella* Oppenheim. Prever, 1921; *Dimorphastraea monticularia* Osasco, 1902; *Thamnastraea heterophylla* Reuss, 1868; *Hydnophyllia dalpiazi*, Prever, 1923. *Dimorphastraea minuta* Prever, 1921; *Mycethoseris patula* var. *ornata* Osasco, 1902; *Mycethoseris incerta* Osasco, 1902; *Monticulastraea minima* Prever, 1922; *Comoseris paronai* Prever, 1921; *Comoseris minor* Prever, 1921; *Mesomorpha elegans* Prever, 1921; *Mycetoseris pseudohydnophora* Reis, Prever, 1921 (pars).

*Cyathoseris hypocrateriformis* (Meneghini in Michelotti, 1861) = *Mycedium profundum* Reuss, 1868; *Cyathoseris pseudomaeandra* Reuss, 1869; *Oroseris* (?) *d'Achiardii* Reuss, 1869; *Maeandrina* (?) *subcircularis* Catullo, 1856; *Comoseris ruvida*; Prever, 1921; *Thamnastraea heterophylla* Reuss, 1868; *Thamnastraea eocenica* Reuss. Prever, 1921; *Protoseris miocaenica* Sismonda and Michelotti, 1871; probably *Cyathoseris scripta* Sismonda and Michelotti, 1871; *Mesomorpha appenina* Prever, 1921; *Mycetoseris minuta* Prever, 1921; *Mycetoseris pseudohydnophora* Reis, 1889; *Thamnastraea adscita* DeAngelis, 1894; *Thamnastraea patula* (Michelotti, 1961) = *Plerastraea volubilis* Gumbel, 1861; *Podabacia prisca* Reuss, 1864 (pars); *Thamnastraea pulchella* Quenstedt, 1881; *Oroseris regularis* Osasco, 1902; *Oroseris alpina* Michelotti, 1861; *Leptoseris undata* Prever, 1921.

*Cyathoseris* questionable species: *Cyathoseris parvistella* Sismonda and Michelotti, 1871; *Oroseris deperdtia* Michelotti, 1861; *Comoseria cistaeformis* Sismonda and Michelotti, 1871.

*Heterogyra lobata* Reuss, 1868 = *Mussa ? leptophylla* Reuss, 1868; *Latimaeandra micheloti* Haime in Bellardi, 1852; *Monticularia meandrinoides* Michelin, 1842; *Meandrina sublabyrinthica* Catullo, 1866; *Monticularia granulata* Gümbel, 1861; *Symphyllia microlopha* Reuss, 1868; *Symphyllia cristata* Reuss, 1868; *Meandrina bicarenata* Catullo, 1856; *Latimaeandra morchelloides* Reuss, 1868; *Monticularia guettardi* Michelotti, 1838; *Meandrina cristata* Catullo, 1856.

*Pavona bronni* (Haime, 1850) = *Monticularia venusta* Catullo, 1856; *Latimaeandra limitata* Reuss, 1872; *Mycetophyllia multistellata* Reuss, 1864; *Monticularia meandrinoides* Quenstedt, 1881; *Hydnophora contorta* Osasco, 1902; *Hydnophyllia maeandrinoides* (Michelin). Prever, 1922; *Pavonia dubia* Catullo, 1856; *Agaricia falcifera* Catullo, 1856; *Thamnastraea volvox* Michelotti, 1861; *Tridacophyllia contorta* DeAngelis, 1894; *Mycetoseris involuta* Prever, 1921; *Mycetoseris dachiardii* Reuss var. *cerebriformis* Prever, 1921.

### Family SIDERASTREIDAE

*Siderastrea (Siderofungia) morloti* (Reuss, 1964) = *Columnastraea bella* Reuss, 1869; *Thamnastraea leptopetala* Reuss, 1864; *Pseudastraea columnaris* Reuss, 1864; *Cyathoseris crispa* DeAngelis, 1894.

## Superfamily PORITICAE

### Family ACTINACIDIDAE

*Actinacis rollei* Reuss, 1864 = *Actinacis lobata* DeAngelis, 1894. *Actinacis conferta* Reuss, 1868 = *Actinacis delicata* Reuss, 1869; *Actinacis michelottii* DeAngelis, 1894.

*Actinacis* questionable species: *Actinacis deperditia* Sismonda and Michelotti, 1871.

### Family PORITIDAE

*Goniopora minuta* (Reuss, 1868) = *Porites micrantha* Reuss, 1869; *Rhodaraea dissita* DeAngelis, 1894; *Goniopora vicenza 5* Bernard, 1903; *Litharaea michelottii* Meneghini in DeAngelis, 1894; *Goniopora vicenza 3* Bernard, 1903; *Goniopora genoa 3* Bernard, 1903. *Astrea microsiderea* Catullo, 1856 or *Astrea rotundata* Catullo, 1856 may be the senior synonym of this species.

*Goniopora ramosa* (Catullo, 1856) = *Porites nummulitica* Reuss, 1864; *Goniopora oberburg 1* Bernard, 1902; *Goniopora vicenza*

*4* Bernard, 1903; *Goniopora vicenza 7* Bernard, 1903; *Litharaea lobata* Reuss, 1864; *Porites microtheca* d'Achiardi, 1867; *Goniopora oberburg 2* Bernard, 1903.

*Goniopora rudis* (Reuss, 1869) = *Goniopora vicenza 6* Bernard, 1903; *Litharaea oblita* DeAngelis, 1894; *Rhodaraea ambigua* DeAngelis, 1894; *Goniopora genoa 4* Bernard, 1903; *Prionastraea inaequalis* d'Achiardi, 1867; *? Astrea funesta* Catullo, 1856.

*Goniopora* questionable species: *Litharaea pulvinata* Michelotti, 1861; *Rhodaraea neglecta* DeAngelis, 1894; *Goniopora genoa 1* Bernard, 1903; *Goniopora genoa 2* Bernard, 1903; *Goniopora genoa 5* Bernard, 1903; *Goniopora vicenza 12* Bernard, 1903; *Goniopora vicenza 13* Bernard, 1903.

*Dictyaraea clinactinia* (Michelotti, 1861) = *? Vincularia rhombiphora* Catullo, 1856; *Stephanocoenia elegans* Reuss, 1864; *Dictyaraea excentrica* Quenstedt, 1881; *Astrocoenia irregularis* Osasco, 1902; *Stephanocoenia cellaroides* Michelotti, 1861; *Dictyaraea garauxi* Chevalier, 1955; *Dictyaraea gaasensis* Chevalier, 1955; *Dictyaraea poritoides* Chevalier, 1955; *Dictyaraea superficialis* Osasco, 1902

*Alveopora rudis* Reuss, 1864 = *Cyathophora minor* Osasco, 1902.

*Alveopora* questionable species: *Alveopora septula* Meneghini in d'Achiardi, 1868; *Alveopora papyracea* DeAngelis, 1894.

## Suborder FAVIINA

### Superfamily FAVIICAE

#### Family FAVIIDAE

*Petrophylliella grumi* (Catullo, 1847) = *Caryophyllia pedata* Catullo, 1847; *Caryophyllia dolium* (?) Catullo, 1847; *Montlivaultia brongniartana* Milne-Edwards and Haime, 1849; *Leptophyllia abbreviata* Reuss, 1872; *Turbinolia lingulata* Catullo, 1847; *Dendrophyllia abnormis* Sismonda and Michelotti, 1871; *Dasyphyllia conferta* Michelotti, 1861; *Dasyphyllia erectisulca* Michelotti, 1861; *Trochoseris venusta* Michelotti, 1861; *Trochocyathus elegans* Michelotti, 1861; *Montilivaultia humilis* Sismonda and Michelotti, 1871. *Leptophyllia compressa* Quenstedt, 1881, *Leptophyllia brevis* Quenstedt, 1881; *Trochocyathus exaratus* Michelotti, 1861, *Balanophyllia fallax* DeAngelis, 1894, may be a synonym.

*Petrophylliella panteniana* (Catullo, 1847) = *Turbinolia mitella* Catullo, 1847; *Turbinolia inflata* Catullo, 1847; *Turbinolia mutica* Catullo, 1847; *Turbinolia multilobata* Schauroth, 1895; *Trochosmilia varicosa* Reuss, 1869; *Trochosmilia* sp. Reuss, 1869; *Leptophyllia ambigua* DeAngelis, 1894; *Turbinolia castellinii* Catullo, 1847; *Leptophyllia zitteli* Reis, 1889.

*Indosmilia* (?) *glabrata* (Reuss. 1868) = *Montlivaultia brongniartiana* d'Achiardi, 1868 (pars).; *Leptomussa elliptica* (Reuss), Felix, 1885.

*Grumiphyllia diploctenium* (Oppenheim, 1899).

*Caulastrea pseudoflabellum* (Catullo, 1847) = *Dasyphyllia michelottii* Milne-Edwards, 1857; *Calamophyllia fasciculata* Reuss, 1864; *Cricotheca gemina* Quenstedt, 1881; *Cricotheca trigona* Quenstedt, 1881; *Rhabdophyllia stipata* d'Achiardi, 1867; *Calamophyllia (Rhabdophyllia) trinitensis* Kranz, 1914; *Rhabdophyllia tenuis* Reuss, 1868; *Rhabdophyllia intercostata* Reuss, 1868; *Aplophyllia paucicostata* Reuss, 1868; *Rhabdophyllia crenaticosta* Reuss, 1869; *Dasyphyllia compressa* d'Achiardi, 1868. *Cladocora subintricata* d'Achiardi, 1867 and *Cladocora oligocenica* Quenstedt, 1881, are possible synonyms.

*Caulastrea ?* questionable species: *Dasyphyllia ? taurinensis* Milne-Edwards and Haime, 1849; *Lobophyllia contorta* Michelin 1842; *Dasyphyllia miocenica* Michelotti, 1861; *Dasyphyllia meneghiniana* d'Achiardi, 1861.

*Favites laxelamellata* (Michelotti, 1861) = *Prionastraea fromenteli* Sismonda and Michelotti, 1871; *Metastraea incerta* d'Achiardi, 1868.

*Favites intermedia* (d'Achiardi, 1867) = *Prionastraea (Metastraea) elegans* DeAngelis, 1894; *Septastraea minuslamellata* Osasco, 1898.

*Favia subdenticulata* (Catullo, 1856) = *Favia cylindracea* Michelotti, 1861; *Favia circumscripta* Sismonda and Michelotti, 1871; *Favia meneguzzii* d'Achiardi, 1868 (pars); *Favia minima* Osasco, 1898; *Phyllangia striata* Gümbel, 1861.

*Diploria crebriformis* (Michelotti, 1888) = *Favia daedala* Reuss,

1864; *Favia daedala* Reuss var. *hemisphaerica* Prever, 1922; *Favia minima* Osasco. Prever, 1922; *Symphyllia multisinuosa* DeAngelis, 1894; *Symphyllia intermedia* Prever, 1921; *Symphyllia canavarii* Prever, 1921; *Symphyllia brevisulcata* Prever, 1921; *Symphyllia bisinuosa* (Michelin). Prever, 1921; *Symphyllia ruvida* Prever, 1921; *Symphyllia irregularis* Prever, 1921; *Symphyllia isseli* Prever, 1921; *Favia irregularis* Prever, 1922; *Favia perrandii* Prever, 1922; *Favia cylindracea* Michelotti (pars). Prever, 1922; *Favia zuffardii* Prever, 1922; *Favia appennina* Prever, 1922; *Hydnophyllia dalpiazi* Prever, 1922 (pars). *Symphyllia vetusta* (Michelin). Osasco, 1898; *Phyllocoenia deperditia* Michelotti, 1861; *Latimaeandra dimorpha* Reuss, 1889; (pars), *Diploria intermedia* Sismonda and Michelotti, 1871; *Favia pulcherrima*, Michelotti, 1861; *Desmocladia septifera* Reuss, 1872; *Meandrina canavarii* Prever, 1921; *Hydnophyllia bellardii* (Milne Edwards and Haime). Prever, 1922 (pars).

*Goniastrea confertissima* (Reuss, 1868) = *Goniastrea cocchi* d'Achiardi. Sismonda and Michelotti, 1871; *Favia profunda* Reuss. Prever, 1922.

*Leptoria polygonalis* (Catullo, 1856) = *Leptoria ambigua* Prever, 1921; *Hydnophyllia hyerogliphica* Prever, 1922; *Hydnophyllia eocaenica* (Reuss). Prever, 1921 (pars); *Hydnophyllia crispata* (DeAngelis). Prever, 1922; *Hydnophyllia plana* Prever, 1922.

*Colpophyllia profunda* (Michelin, 1842). Growth forms with long, broad sinuous valleys include: *Meandrina labyrinthica* Michelotti, 1838; *Meandrina serpentinoides* Catullo, 1856; (?) *Symphyllia tiedemanni* Milne-Edwards, 1857; *Trochoseris distorta* (Michelin) (pars). Schauroth, 1865. *Ulophyllia flexuosa* d'Achiardi, 1867; *Coeloria* (?) *grandis* Reuss, 1869; *Plocophyllia oligocenica* Prever, 1922; *Meandrina fimbriata* Catullo, 1856; *Mycetophyllia interrupta* Reuss, 1864; *Pectina pseudomeandrites* d'Achiardi, 1866; *Hydnophyllia microlopha* (Reuss). Prever, 1921 (pars); *Hydnophyllia formosissima* (Catullo). Prever, 1922. *Ulophyllia irradians* Reuss, 1968 (pars); *Hydnophyllia scalaria* (Catullo). Reis, 1889; *Ulophyllia* ? *acutijuga* Reuss, 1868; *Trydacnophyllia undans* Prever, 1921; *Trydacnophyllia apennina* Prever, 1921; *Trydacnophyllia affinis* Prever, 1921; *Tridacnophyllia chichorium* Sismonda and Michelotti. Prever, 1921; *Mycetoseris pseudohydnophora* Reis. Prever, 1921; *Hydnophyllia daedalea* (Reuss). Reis, 1889; *Ulophyllia macrogyra* Reuss, 1868; *Latimaeandra daedalea* Reuss, 1868; (?) *Latimaeandra gastaldii* Haime in Bellardi, 1852; *Plerogyra crassisepta* DeAngelis, 1894; *Plerogyra deperdita* DeAngelis, 1894; *Ulophyllia distincta* Osasco, 1902; *Hydnophyllia microlopha* (Reuss). Prever, 1921 (pars); *Agaricia inflata* Catullo, 1856; *Mycetophyllia itaiica* d'Achiardi, 1867; *Ulophyllia laxa* DeAngelis, 1894; *Meandrina stellifera* Michelin, 1842; *Mycetophyllia interrupta* Reuss, 1864; *Meandrina cristata* Catullo, 1856; *Dimorphophyllia lobata* Reuss, 1864; *Hydnophyllia italica* Prever, 1922; *Hydnophyllia* cf. *grandis* (Reuss). Prever, 1922.

Growth forms with relatively short valleys include: *Pavonia dubia* (Catullo, 1856); *Meandrina filogranaeformis* Catullo, 1856; *Trochoseris distorta* (Michelin). Schauroth, 1865; *Meandrina stellata* Catullo, 1856; *Monticularia inaequalis* Gümbel, 1861; *Mycetophyllia multilamellosa* d'Achiardi, 1867; *Cyathoseris multisinuosa* Reuss, 1868; *Symphyllia confusa* Reuss, 1868; *Ulophyllia magnicostata* Sismonda and Michelotti, 1871; *Hydnophyllia tenera* (Reuss) (pars). Prever, 1922, *Latimaeandra tenera* Reuss, 1868.

Growth forms with relatively straight valleys include: *Meandrina scalaria* Catullo, 1856; *Hydnophora longicollis* Reuss, 1864; *Hydnophora maeandrinoides* (Michelin). Sismonda and Michelotti, 1871; *Ulophyllia irradians* Reuss, 1868 (pars); *Hydnophora anceps* Prever, 1921; *Latimaeandra dachiardii* Reuss, 1869; *Hydnophyllia eocaenica* (Reuss). Reis, 1889; *Hydnophyllia connectens* Reis, 1889.

Growth forms with narrow, closely appressed valleys include: *Meandrina vetusta* Michelin, 1842; *Meandrina phyrgia* Michelin, 1842; *Meandrina bellardii* Milne-Edwards and Haime (pars); Meandrina (?) *subcircularis* Catullo, 1856; *Hydnophyllia isseli* Prever 1921; *Symphyllia appennina* Prever, 1921; *Meandrina valleculosa* Gümbel, 1861; *Dendrogyra intermedia* Sismonda and Michelotti, 1871; *Hydnophora minoris* Osasco, 1898; *Hydno-*

*phora affinis* Sismonda and Michelotti, 1871; *Hydnophora elongata* Sismonda and Michelotti, 1871; *Hydnophora taramellii* Prever, 1921; *Symphyllia obliqua* Prever, 1921; *Hydnophora perrandii* Prever, 1921; *Symphyllia paronai* Prever, 1921; *Symphyllia crassa* Prever, 1921; *Coeloria* (?) *cerebriformis* Reuss, 1864;

"Hydnophorid" variants include: *Hydnophyllia mirabilis* Reis, 1889; *Hydnophyllia curvicollis* Reis, 1889; *Hydnophora magnifica* Osasco, 1898; *Hydnophyllia valleculosa* (Gümbel). Prever, 1922; *Hydnophora pulchra* Sismonda and Michelotti, 1871.

*Colpophyllia* questionable species: *Latimaeandra repanda* Michelotti, 1861; *Manicinia antiqua* Sismonda and Michelotti, 1871.

*Hydnophora* (?) *stavia* Sismonda and Michelotti, 1871.

*Montastrea irradians* (Milne-Edwards and Haime, 1848) = *Astrea radiata* Michelin, 1842; *Dendrophyllia inaequalis* Catullo, 1847; *Dendrophyllia maraschini* Catullo, 1856; *Sarcinula intermedia* Catullo, 1856; ? *Astrea lucasana* Defrance. Quenstedt, 1881.

*Montastrea deserta* (Catullo, 1856) = *Phyllocoenia distincta* Osasco, 1898; *Heliastraea appeninica* d'Achiardi, 1868; *Dimorphastraea bormidensis* Sismonda and Michelotti, 1871; *Heliastraea ambigua* Sismonda and Michelotti, 1871; *Heliastraea superficialis* Sismonda and Michelotti, 1871; *Confusastraea miocenica* Sismonda and Michelotti, 1871; *Plerastraea ornata* Sismonda and Michelotti, 1871; *Heliastraea eminens* Reuss, 1864; *Heliastraea subcoronata* Reuss, 1872; *Phyllangia grandis* Reuss, 1872. *Heliastraea beaudouini* (Haime). Reuss, 1869; *Thamnastraea olbiqua* Prever, 1921; *Confusastraea costulata* Osasco, 1902.

*Montastrea inaequalis* (Gümbel, 1861) = *Heliastraea dallagoi* Osasco, 1902; *Heliastraea inaequata* Gümbel, Reis, 1889; *Phyllocoenia lucasana* (Defrance). Oppenheim, 1909.

*Antiguastrea lucasiana* (Defrance, 1826) = *Astrea beaudouini* Haime in Bellardi, 1850; *Heliastraea inequata* Reis, 1889; *Sarcinula crispa* Catullo, 1856; *Sarcinula annulata* Catullo, 1856; *Heliastraea immersa* Reuss, 1868; *Heliastraea dallagoi* Osasco, 1902; *Heliastraea fontana* Oppenheim, 1899; *Heliastraea columnaris* Reuss, 1868; *Leptastraea anomala* Sismonda and Michelotti, 1871.

*Antiguastrea lucasiana* forma *michelottina* (Catullo, 1856) = *Astraea ingens* Catullo, 1856; *Isastraea affinis* Reuss, 1868; *Latimaeandra circumscripta* Reuss, 1868; *Isastraea elegans* Reuss, 1872; *Heliastraea columanaris* (Reuss) var. *tenuis* Osasco, 1902. *Astraea profundata* Catullo, 1856 is a probable synonym. *Heliastraea gemmans* (Astraea) Michelotti, 1861 may be a synonym.

*Antiguastrea lucasiana* forma *alveolaris* (Catullo, 1856) = *Astraea astroites* Goldfuss. Catullo, 1856; *Astraea montevialensis* Catullo, 1856; *Solenastraea columnaris* Reuss, 1868; *Solenastraea conferta* Reuss, 1868; *Astraea puritana* Catullo, 1856; *Astrangia princeps* Reuss, 1868; *Heliastraea fallax*, Sismonda and Michelotti, 1871; *Phyllocoenia ovalis* Gümbel, 1861; *Astraea brevissima* Catullo, 1856; *Stylina suessi* Reuss, 1868; *Stylina fasiculata* Reuss, 1868.

*Patallophyllia gnatae* Oppenheim, 1899 = *Trochocyathus sinuosus* (Brongniart). Reuss, 1869; *Trochosmilia bilobata* d'Achiardi, 1866.

*Cricocyathus annulatus* (Reuss, 1868) = *Leptaxis elliptica* Reuss, 1868.

*Agathiphyllia rochettina* (Michelin, 1842) = *Astrea burdigalensis* Milne-Edwards and Haime, 1850; *Astrea corsica* d'Orbigny, 1852; *Lobophyllia gregaria* Catullo, 1856; *Lobophyllia pseudorochettina* Catullo, 1856; *Agathiphyllia explanata* Reuss, 1864; *Agathiphyllia conglobata* Reuss, 1864; *Heliastraea guettardi* (Defrance). Reuss, 1869; *Thecosmilia* ? n.sp. Osasco, 1902; *Heliastraea maxima* Michelotti, 1861 (pars); *Brachyphyllia magna* d'Achiardi, 1867; *Heliastarea meneghinii* Reuss, 1869; *Brachyphyllia neglecta* Sismonda and Michelotti, 1871; *Brachyphyllia crassa* Osasco, 1898; *Lobophyllia succincta* Catullo, 1856; *Astraea affinis* Catullo, 1856; *Brachyphyllia umbellata* Reuss, 1869.

*Astrangia* (*Astrangia*) *striata* (Gümbel, 1861) = *Rhizangia hoernesi* Reuss, 1864; *Gombertangia felixi* Oppenheim, 1899. Possible synonyms; *Astrangia patula* Sismonda and Michelotti, 1871, and *Phyllangia propiniqua* Sismonda and Michelotti, 1871.

*Reussangia* (?) *oligocenica* (Osasco, 1898) = *Stylastraea pulchra* Osasco, 1898; *Cladangia minor* Osasco, 1898.

*Cladangia* ? questionable species: *Cladangia* ? *hybrida* d`Achiardi, 1867; *Cladangia mamillosa* d`Achiardi, 1868; *Cladangia proxima* Sismonda and Michelotti, 1871.

### Family OCULINIDAE

*Oculina* ? *solida* (Michelotti, 1861).

### Family MUSSIDAE

*Leptomussa variabilis* d`Achiardi, 1867 = *Caryophyllia cingulata* Catullo, 1856; *Caryophyllia globularis* Catullo, 1856; *Caryophyllia bithalamia* Catullo, 1856; *Caryophyllia biformis* Catullo, 1856; *Trochosmilia diversicostata* Reuss, 1869; *Coelosmilia elliptica* Reuss, 1868; *Leptomussa abbreviata* Reuss, 1869. *Lithophyllia debilis* is apparently based on juveniles of *L. variabilis*.

*Leptomussa* ? questionable species: *Leptomussa abnormis* Sismonda and Michelotti, 1871.

*Antillia* (?) *carcarensis* (Michelotti, 1847) = *Montlivaultia strozzi* Michelotti, 1861; *Montlivaultia poculum* Michelotti, 1861. This species may be *Antillophyllia*.

*Scolymia* (?) *detrita* (Michelin, 1842) = *Montlivaultia bormidensis* Milne-Edwards and Haime, 1857. *Trochoseris miocenica* Michelotti, 1861, is a possible synonym.

*Trochosmilia* questionable species: *Trochosmilia* ? *rhombica* d`Achiardi, 1866; *Trochosmilia granulosa* d`Achiardi, 1866; *Trochosmilia incerta* Sismonda and Michelotti, 1871.

*Ilariosmilia berica* (Catullo, 1847) = *Trochosmilia subcurvata* Reuss, 1864; *Lithophyllia brevis* Reuss, 1869; *Trochosmilia stipitata* Reuss, 1869.

### Family PECTINIIDAE

*Fungophyllia berica* (Catullo, 1856) = *Leptophyllia dilatata* Reuss, 1868; *Montlivaultia fungiformis* Osasco, 1902; *Trochoseris* ? *laevicostata* Osasco, 1902.

## Suborder CARYOPHYLLIINA

### Superfamily CARYOPHYLLIICAE

#### Family CARYOPHYLLIIDAE

*Lophosmilia elliptica* (Michelotti, 1861).

*Trochocyathus* ? *bourgueti* (Catullo, 1856).

*Ceratotrochus* (*Conotrochus*) *aequicostatus* (Schauroth, 1865)

*Parasmilia incurva* (Catullo, 1847) = *Coelosmilia vicentina* d`Achiardi, 1866; *Trochosmilia obliquacompressa* d`Achiardi, 1866; *Trochosmilia arguta* Reuss, 1868; *Circophyllia cylindroides* Reuss, 1869; *Parasmilia crassicostata* Reuss, 1868.

*Euphyllia contorta* (Catullo, 1847) = *Caryophyllia bisulcata* Catullo, 1847; *Caryophyllia pseudocernua* Catullo, 1847; *Turbinolia* ? *unisulcata* Catullo, 1847; *Lobophyllia caliculata* Catullo, 1856; *Lobophyllia pulchella* Catullo, 1856; *Dasyphyllia deformis* Reuss,

1868; *Plocophyllia constricta* Reuss, 1868; *Thecosmilia multilamellosa* d`Achiardi, 1868; *Plocophyllia caespitosa* Reuss, 1872; *Calamophyllia compressa* DeAngelis, 1894; *Calamophyllia planicostata* d`Achiardi, 1867; *Trochosmilia minuta* Reuss, 1868 appears to be based on a single juvenile corallum of *E. contorta*.

### Superfamily FLABELLICAE

#### Family FLABELLIDAE

*Flabellum laterocristatus* (Milne-Edwards and Haime, 1848). A probable synonym is *Flabellum dissitum* Michelotti, 1861.

*Flabellum appendiculatum* Brongniart, 1823) = *Trochocyathus ambiguus* Michelotti, 1861.

*Flabellum deperditum* Michelotti, 1861. A probable synonym is *Flabellum carcarense* Osasco, 1898.

*Conosmilia* ? *protensus* Sismonda and Michelotti, 1871. (This may be based on an isolated corallite of *Caulastrea pseudoflabellum*.

## Suborder DENDROPHYLLIINA

### Family DENDROPHYLLIIDAE

*Balanophyllia incerta* Sismonda and Michelotti, 1871 = *Trochocyathus guembeli* Reis, 1889; *Trochocyathus laterocristatus* Milne-Edwards and Haime, Reis, 1889; *Heteropsammia antiqua* DeAngelis, 1894; *Parasmilia cingulata* Catullo. Reis, 1889 (pars.)

*Desmopsammia subcylindria* (Phillipi, 1846) = *Balanophyllia praelonga* (Michelin). Roemer, 1863; *Balanophyllia calycina* Roemer, 1863; *Dendrophyllina* ? sp. Reuss, 1864; *Desmopsammia perlonga* Reis, 1889.

*Dendrophyllia rugosa* (Gumbel, 1861) = *Dendrophyllia nodosa* Reuss, 1864; *Dendrophyllia vicentina* Cenestrelli, 1908.

*Stichopsammia miocenica* (Sismonda and Michelotti, 1871).

## Subclass OCTOCORALLIA

### Order GORGONACEA

*Corallium* (?) *inaequale* DeAngelis, 1894.

*Isis* (?) *granifera* Osasco, 1898.

*Parisis brevis* (d`Achiardi, 1868).

### Order COENOTHECALIA

*Parapolytremacis bellardii* (Haime, 1852) = *Heliopora astraeoides* Gumbel, 1861; *Millepora mammillosa* Reuss, 1869; *Millepora globularis* Catullo, 1856.

## Class HYDROZOA

### Order MILLEPORINA

*Millepora* (?) *depauperata* Reuss, 1864 = *Millepora cylindrica* Reuss, 1868.

*Milleaster verrucosa* Reuss, 1868.

# REFERENCES

D`ACHIARDI, A., 1867, Corallari fossili del terreno nummulitico dell Alpi Venete. Parte 1: Soc. Ital. Sci. Nat. Milano, Mem., v. 2, no. 4, 54 p.

———, 1868A, Corallari fossili del terreno nummulitico dell Alpi Venete. Parte II: Soc. Ital. Sci. Nat. Milano, Mem., v. 4, no. 1, 31 p.

———, 1868B, Studio comparativo fra i coralli dei terreni terziarii del Piemonte e dell Alpi Venete: Univ. Pisa, Ann., v. 10, no. 2, p. 73–144.

ADEY, W. H., 1975, The algal ridges and coral reefs of St. Croix: Their structure and Holocene development: Atoll Res. Bull., No. 187, 67 p.

———, 1978, Coral reef morphogenesis: A multidimensional model: Science, v. 202, p. 831–837.

———, AND MACINTYRE, I. G., 1973, Crustose coralline algae: a re-evaluation in the geological sciences: Geol. Soc. America Bull., v. 84, p. 883–904.

ANGELIS D`OSSAT, G. DE, 1894, I corallari dei terreni Terziari dell Italia settentrionale. Collezione Michelotti. Museo Geologico della R. Università di Roma: R. Accad. Lincei, Roma, Mem., Cl. Sci. Fis., Mat., Nat., v. 1, p. 164–280.

ANSTEY, R. L., AND CHASE, T. L., 1974, Environments through time: Minneapolis, Burgess Pub. Co., 36 p.

BANDY, O. L., 1964, General correlation of foraminiferal structure with environment, *in* Imbrie, J., and Newell, N. D., Eds., Approaches to Paleoecology: New York, Academic Press, p. 75–90.

BARTA-CALMUS, S., 1973, Revision de collections de madréporaires provenant du Nummulitique du sud-est de la France, de l'Italie et de la Yougoslavie septentrionales [Thèse de Doc. d'etat es Sci. Nat.]: Paris, Univ., 694 p.

BAYER, F. M., 1956, Octocorallia, *in* Moore, R. C., Ed., Treatise on Invertebrate Paleontology: Geol. Soc. America, Univ. Kansas Press, Pt. F (Coelenterata), p. 166–231.

———, 1961, The shallow water octocarallia of the West Indian region: Studies on the fauna of Curaçao and other Caribbean Islands: The Hague, Martinus Nijhoff, v. 12, pt. 55, 373 p.

BEGGIATO, F. S., 1865, Antracoterio di Zovencedo e di Monteviale nel Vicentino: Mem. Soc. Ital. Sci. Nat., v. 1, 9 p.

BERGGREN, W. A., 1972, A Cenozoic time scale—some implications for regional geography and paleobiogeography: Lethaia, v. 5, no. 2, p. 195–215.

BERNARD, H. M., 1901, On the necessity for a provisional nomenclature for those forms of life which cannot be at once arranged in a natural system: Linnean Soc., 113th Session, Proc., p. 10–11.

———, 1902, The species problem in corals: Nature, v. 65, p. 560.

———, 1906, Catalogue of the madreporarian corals in the British Museum (Natural History). Vol. 6. The family Poritidae. 2—The genus *Porites*. 2—*Porites* of the Atlantic and West Indies, with the European fossil forms. The genus *Goniopora*, supplement to vol. IV: London, British Museum (Nat. Hist.), v. 6, 173 p.

BITTNER, A., 1883, Mittheilungen über das Alttertiär der Colli Berici: Verhandl. Kaiser. Geol. Reichsanst., p. 82–94.

BLOW, W. H., 1969, Late middle Eocene to Recent planktonic foraminiferal biostratigraphy: International Conf. Pl. Microfossils, 1st, Geneva, Proc., Leiden, E. J. Brill Pub. p. 199–421.

BOLLI, H. M., 1966, Zonation of Cretaceous to Pliocene marine sediments based on planktonic foraminifera: Bol. Inform. Assoc. Venezolana Geol., Min. y Petrol., v. 9, no. 1, p. 3–32.

BOSELLINI, A., CARRARO, F., CORSI, M., DE VECCHI, G. P., GATTO, G. O., MALARODA, R., STURANI, C., UNGARO, S., AND ZANETTIN, B., 1967, Note illustrative della Carta geologica d'Italia alla scala 1:100,000. Foglio 49, Verona: Roma, Serv. Geol. Italia, Nuova Tecnica Grafica, 61 p.

BRAGA, G., 1965, Briozoi dell Oligocene di Possagno (Trevignano Occidentale). II contributo alla conosenza dei Briozoi del Terziario Veneto: Bol. Soc. Paleont. Ital., v. 4, no. 2, p. 216–244.

BRAKEL, W. H., 1977, Corallite variation in *Porites* and the species problem in corals: International Coral Reef Symp., 3rd, Miami, Proc., v. 1, p. 457–462.

BRUNN, J., CHEVALIER, J.-P., AND MARIE, P., 1955, Quelquesformes nouvelles de Polypiers et de Foraminifères de l'Oligocene et du Miocene du NW de la Grèce: Soc. Géol. France, Bull., sér. 6, v. 5, nos. 1–3, p. 195–202.

CANESTRELLI, C., 1908, Revisione della fauna Oligocenica de Laverda nel Vicentino: Soc. Ligustica Sci. Nat. Geogr. Genova, Atti, v. 19, p. 27–29, p. 97–152.

CARY, L. R., 1918, The Gorgonaceae as a factor in the formation of coral reefs: Washington, D.C., Carnegie Institute, Pub. 213, p. 341–362.

———, 1931, Studies on the coral reef at Tutuila, American Samoa, with special reference to the Alcyonaria: Washington, D.C., Carnegie Institute, v. 27, no. 3, p. 53–98.

CATALÁ, R. L. A., 1964, Carnival sous la mer: Paris, R. Sicard Pub.

CATULLO, T., 1852, Cenni sopra il terreno di sedimento superiore delle Provincie Venet e descrizoine di alcuni polipaj fossili ch'esse racchuide: R. 1st. Veneto Sci., Mem., v. 4, p. 1–44.

———, 1856, Dei terreno di sedimento superiore della Venezie e dei fossili Bryozoari, Antozoari, e Spongiari: Padua, p. 1–88.

CHEVALIER, J.-P., 1955, Les polypiers Anthozoaires du Stampien de Gaas (Landes): Soc. Hist. Nat. Toulouse, Bull., v. 90, nos. 3–4, p. 375–410.

———, 1971, Les scléractiniaires de la Mélanesie Française. Première partie: Expédition Française sur les réciefs coralliens de la Nouvelle Caledonia, Edit. Found. Singer-Polignac, v. 5, 307 p.

———, 1975, Les scléractiniaires de la Mélanesie Française. Deuxième partie: Expédition Française sur les réciefs coralliens de la Nouvelle Caledonia, Edit. Found. Singer-Polignac, v. 7, 407 p.

———, 1977, Aperçu sur la faune corallienne récifale du Neogene: Corals Fossil Coral Reefs. Symp., 2nd, Paris, Trans., Mém. Bur. Rech. Géol. et Mim., Mém., v. 89, p. 359–366.

CITA, M. B., AND PICCOLI, G., 1964, Les stratotypes du Paléogene d'Italie: Mém. Bur. Rech. Géol. et Min., Mém., v. 28, p. 672–684.

COLETTI, F., PICCOLI, G., SAMBUGAR, B., AND VENDEMIATI DEI MEDICI, M. C., 1973, I molluschi fossili di Castelgomberto e il loro significato nella paleoecologia dell'Oligocene Veneto: Mem. 1st. Geol., Mineral. Univ. Padova, v. 28, 31 p.

COLLINS, A. C., 1958, Foraminifera: Great Barrier Reef Exped., 1928–29, Sci. Repts. v. 6, no. 6, p. 335–437.

CONNELL, J. H., 1973, Population ecology of reef building corals, *in* Jones, O. A., and Endean, R., Eds., Biology and Geology of Coral Reefs: New York, Academic Press, v. 2 (Biology 1), p. 205–245.

CUFFEY, R. J., 1973, Bryozoan distribution in the modern reefs of Eniwetok Atoll and the Bermuda Platform: Pacific Geol., v. 6, p. 25–50.

———, 1977, Bryozoan contributions to reefs and bioherms through geologic time, *in* Frost, S. H., Weiss, M. P., and Saunders, J. B., Eds., Reefs and Related Carbonates—Ecology and Sedimentology: Tulsa, Ok., Am. Assoc. Petroleum Geologists, Studies in Geology, no. 4, p. 181–194.

———, AND FONDA, S. S., 1977, Cryptic bryozoan species assemblages in modern coral reefs off Andros and Eleuthera, Bahamas: International Coral Reef Symp., 3rd., Miami, Proc., v. 1, p. 81–86.

————, AND KISSLING, D. L., 1973, Ecologic roles and paleoecologic implications of bryozoans on modern coral reefs in the Florida Keys (Abs.): Geol. Soc. America, Abs. with Programs, v. 5, no. 2, p. 152–153.

DAL PIAZ, G., 1930, I mammiferi dell'Oligocene veneto. *Trigonias ombonii*: Mem. 1st. Geol., Univ. Padova, v. 9, 63 p.

DAL PRA, A., 1965, Rocce vulcano-detritiche terziarie fra l'Astico e il Brenta nella regione di Thiene (Vicenza): Atti Mem. Accad. patavina Sci., Lett. Arti, Cl. Sci. Mat., Nat., v. 77, p. 201–215.

————, AND MEDIZZA, F., 1965, Manifestazioni vulcaniche paleiceniche nella zona di Laverda sulle colline tra Thiene e Bassano (Vicenza): Atti Accad. naz. Lincei, Rend. Cl. Sci. Fis., Mat., e Nat., pt. 8, v. 39, p. 106–112.

DAHL, A. L., PATTEN, B. C., SMITH, S. V., AND ZIEMAN, J. C., JR., 1974, A preliminary coral reef ecosystem model: Atoll Res. Bull., No. 172, p. 7–36.

DEFRANCE, M., 1826, Polypiers: Dictionnaire des Sci. Nat., t. 42, p. 377–398.

DIENI, I., AND PROTO DECIMA, F., 1964, *Cribrohantkenina* ed altri Hantkeninidae nell'Eocene superiore di Castelnuovo (Colli Euganei): Riv. Ital. Paleont., v. 70, no. 3, p. 555–592.

DINESIN, Z. D., 1977, The coral fauna of the Chagos Archipelego: International Coral Reef Symp., 3rd, Miami, Proc., v. 1, p. 155–161.

DODGE, R. E., ALLER, R. C., AND THOMSON, J., 1974, Coral growth related to resuspension of bottom sediments: Nature, v. 247, p. 574–577.

EBBS, N. K., 1966, The coral inhabiting polychaetes of the northern Florida reef tract, Part 1, Aphroditidae, Polynoidae, Amphinomidae, Eunicidae, Lysaretidae: Bull. Mar. Sci., v. 16, p. 485–555.

FABIANI, R., 1908, Paleontologia dei Colli Berici: Mem. Soc. Ital. Sci., v. 15, pt. 3A, p. 45–248.

————, 1915, Il Paleogene del Veneto: Mem. 1st. Geol. Univ. Padova, v. 3, 336 p.

FELIX, J. P., 1885, Kritische studien ueber die Tertiäre korallenfauna des Vicentins nebst beschreibung einiger neuen Arten: Deutsch. Geol. Gesell., Zeitschr., v. 37, p. 379–421.

————, 1916, Ueber *Hydnophyllia* und einige andere Korallen aus dem vicentinischen Tertiär: Naturforsch. Gesell. Leipzig, Sitzungsber., v. 43, p. 1–30.

————, 1925, Fossilum Catalogus 1: Animalia. Pt. 28, Anthozoa eocaenica et oligocaenica: Berlin, W. Junk, 296 p.

FRANCAVILLA, F., FRASCARI RITONDALE SPANO, F., AND ZECCHI, R., 1970, Alghe e marcroforaminiferi al limite Eocene-Oligocene presso Barbarano (Vicenza): Gior. Geol. (2a), v. 36, p. 653–678.

FROST, S. H., 1972, Evolution of Cenozoic Caribbean coral faunas: Conf. Geol. del Caribe, 6, Margarita, Venezuela, Mem., p. 461–464.

————, 1974, Oligocene barrier-reef lagoon biofacies, N. E. Italy (Abs.): Geol. Soc. America, Abstracts with programs, v. 6, no. 7, p. 1038.

————, 1974, Cenozoic reef systems of Caribbean—prospects for paleoecologic synthesis, *in* Frost, S. H., Weiss, M. P., and Saunders, J. B., Eds., Reefs and Related Carbonates—Ecology and Sedimentology: Tulsa, Ok., Am. Assoc. Petroleum Geologists, Studies in Geology, no. 4, p. 93–110.

————, 1977B, Oligocene reef coral biogeography, Caribbean and western Tethys: Corals Fossil Coral Reefs Symp., 2nd, Paris, Trans., Mem. B.R.G.M., v. 89, p. 342–352.

————, 1977C, Miocene to Holocene evolution of Caribbean Province reef building corals: International Coral Reef Symp., 3rd, Miami, Proc., v. 2, p. 353–359.

————, 1977D, Ecologic controls of Caribbean and Mediterranean Oligocene reef coral communities: International Coral Reef Symp., 3rd, Miami, Proc., v. 2, p. 367–373.

————, AND LANGENHEIM, R. L., 1974, Cenozoic Reef Biofacies. Tertiary Larger Foraminifera and Scleractinian Corals from Chiapas, Mexico: DeKalb, Northern Illinois Univ. Press, 388 p.

————, AND SCHAFERSMAN, S. D., 1978, Oligocene reef community succession, Damon Mound, Texas: Gulf Coast Assoc. Geol. Soc. Trans., v. 28, p. 143–160.

————, AND WEISS, M. P., 1979, Patch-reef communities and succession in the Oligocene of Antigua, West Indies: Geol. Soc. America Bull., v. 90, pt. 1, p. 612–616, pt. 11, p. 1094–1141, Doc. no. M90702.

GARDINIER, J. S., 1903, The fauna and geography of the Maldive and Laccadive Archipelagoes: Cambridge, Cambridge Univ. Press, v. 2, p. 755–790.

GEISTER, J., 1977, The influence of wave exposure on the ecological zonation of Caribbean coral reefs: International Coral Reef Symp., 3rd, Miami, Proc., v. 1, p. 23–29.

————, AND UNGARO, S., 1977, The Oligocene coral formations of the Colli Berici (Vicenza, northern Italy: Eclogae Geol. Helvetiae, v. 70, p. 811–823.

GINSBURG, R. N., AND SCHROEDER, J. H., 1973, Growth and fossilization of algae cup-reefs, Bermuda: Sedimentology, v. 20, p. 575–614.

GLYNN, P. W., 1973, Aspects of the ecology of coral reefs in the western Atlantic region, *in* Jones, D. A., and Endean, R., Eds., Biology and Geology of Coral Reefs: New York, London, Academic Press, v. 11 (Biology 1), p. 271–324.

GOREAU, T. F., 1959, The ecology of Jamaican coral reefs. 1. Species composition and zonation: Ecology, v. 40, p. 67–90.

————, AND GOREAU, N. I., 1973, The ecology of Jamaican coral reefs. 2. Geomorphology, zonation, and sedimentary phases: Bull. Mar. Sci., v. 23, no. 2, p. 399–464.

————, AND YONGE, C. M., 1968, Coral community on muddy sand: Nature, v. 217, p. 421–423.

GRAUS, R. R., CHAMBERLAIN, J. A., JR., AND BOKER, A. M., 1977, Structural modification of corals in relation to waves and currents, *in* Frost, S. H., Weiss, M. P., and Saunders, J. B., Eds., Reefs and Related Carbonates—Ecology and Sedimentology: Tulsa, Ok., Am. Assoc. Petroleum Geologists, Studies in Geology, No. 4, p. 135–153.

GRAVIER, M. C., 1911, Les récifs et les madréporaires de la Baie de Tadjourah: Annal. 1st. oceanogr., Monaco, v. 2, no. 3, 104 p.

GÜMBEL, C. W. VON, 1861, Geognostische beschreibung des Bayrischen Alpengebirges und seines Vorlandes: Gotha, v. 1, 600 p.

HARDENBOL, J., 1968, The "Priabonian" type section (a preliminary note): Mem. Bur. Rech. Géol. et Min., Mém., v. 58, p. 629–635.

HARTMAN, W. D., 1977, Sponges as reef builders and shapers, *in* Frost, S. H., Weiss, M. P., and Saunders, J. B., Eds., Reefs and Related Carbonates—Ecology and Sedimentology: Tulsa, Ok., Am. Assoc. Petroleum Geologists, Studies in Geology, No. 4, p. 127–134.

HERB, R., AND HEKEL, H., 1973, Biostratigraphy, variability, and facies relations of some upper Eocene *Nummulites* from northern Italy: Eclog. Geol. Helvetiae, v. 66, no. 2, p. 419–445.

HOFFMEISTER, J. E., 1925, Some corals from American Samoa and the Fiji Islands: Washington, D.C., Carnegie Institute, Pub. 343, Papers Dept. Marine Biol., v. 22, 90 p.

HUBBARD, J. A. E. B., 1973, Sediment-shifting experiments: A guide to functional behavior in colonial corals, *in* Boardman, R. S., Cheetham, A., and Oliver, W., Eds., Animal Colonies: Stroudsburg, Pa., Dowden, Hutchinson and Ross, p. 31–42.

———, AND POCOCK, Y. P., 1972, Sediment rejection by recent scleractinian corals: a key to paleoenvironmental reconstruction: Geol. Rundsch., v. 61, p. 598–626.

KINZIE, R. A., III, 1973, Coral reef project—papers in memory of Dr. Thomas F. Goreau. 5. The zonation of West Indian gorgonians: Bull. Mar. Sci., v. 23, no. 1, p. 93–155.

KRANZ, W., 1910–1915, Das Tertiar zwischen Castelgomberto, Montecchio Maggiore, Creazzo, und Monteviale im Vicentin: Neues Jahrb. Mineral., Geol., u Paleont., Beil-Bd., v. 29 (1910), p. 180–268; v. 32 (1911), p. 701–729; v. 38 (1915), p. 273–324.

LAND, L. S., 1976, Early dissolution of sponge spicules from reef sediments: Jour. Sed. Petrology, v. 46, p. 967–969.

LABOREL, J., 1970, Madréporaires et hydrocoralliares récifaux des côtes Brésiliennes. Systématique, écologie, répartition verticale et géographique: Campagne *Calypso* côtes Atl., Amerique Sud., fasc. 9, p. 171–229.

———, 1969, Les peuplements de madréporaires des côtes tropicales du Brésil-Ecologie: Ann. Univ. Abidjan, sér. E, II, fasc. 3, 260 p.

LANG, J. C., 1970, Interspecific aggression within the scleractinian reef corals [unpub. Ph.D. thesis]: New Haven, Conn., Yale Univ.

———, 1971, Interspecific aggression by scleractinian corals. I. The rediscovery of *Scolymia cubensis* (Milne Edwards and Haime): Bull. Mar. Sci., v. 21, no. 4, p. 952–959.

———, 1973, Coral reef project—papers in memory of Dr. Thomas F. Goreau. II. Interspecific aggression by scleractinian corals. 2. Why the race is not only to the swift: Bull. Mar. Sci., v. 23, no. 2, p. 260–279.

MACINTYRE, I. G., AND GLYNN, P. W., 1976, Evolution of modern Caribbean fringing reef, Galeta Point, Panama: Am. Assoc. Petroleum Geologists Bull., v. 60, p. 1054–1072.

MARASCHINI, P., 1824, Sulle formazioni della rocce del Vicentino: Saggio Geol. Minerva, 232 p.

MARSHALL, S. M., AND ORR, A. P., 1931, Sedimentation on Low Isles and its relation to coral growth: Great Barrier Reef Expedition, Scientific Repts., v. 1, p. 93–133.

MASPERONI BASTIANUTTI, C., 1964, Studio micropaleontologico della serie di terreni Priaboniano-Langhiani del Trevigiano occidentale: Mem 1st Geol., Mineral. Univ. Padova, v. 24, 65 p.

MATTHEWS, R. K., 1965, Genesis of Recent lime mud in southern British Honduras [unpub. Ph.D. thesis]: Houston, Texas, Rice Univ., 139 p.

MAXWELL, W. G. H., 1968, Atlas of the Great Barrier Reef: Amsterdam, Elsevier, 258 p.

MESOLELLA, K. J., SEALY, H. A., AND MATTHEWS, R. K., 1970, Facies geometrics within Pleistocene reefs of Barbados, West Indies: Am. Assoc. Petroleum Geologists Bull., v. 54, p. 1899–1917.

MICHELIN, J., 1840–1847, Iconographie zoophytologique. Description par localités et terrains des polypiers fossiles de France, et pays environments: Paris, 348 p.

MICHELOTTI, G., 1847, Description des fossiles des terrains miocenes de l'Italie septentrionale. Précis de la faune miocène de la Haute-Italie: Naturk, Holland. Maatsch, Wetensch. Haarlem, Verh., ser. 2, v. 3, 408 p.

———, 1861, Études sur le Miocène Inferieur de l'Italie Septentrionale: Naturk, Holland. Maatsch, Wetensch. Haarlem, Verh., ser. 2, v. 15, no. 2, 183 p.

MUNIER-CHALMAS, E., 1891, Étude du Tithonique, du Crétacé et du Tertiare du Vicentin: Thèse, Paris.

NEMENZO, F., 1955, On the scleractinian fauna of Puerto Galera Bay, Oriental Mindoro, and Laguimanoe Bay, Quezon: Bull. Nat., Appl. Sci., v. 15, p. 131–138.

NEUMANN, A. C., AND LAND, L. S., 1975, Lime mud deposition and calcareous algae in the Bight of Abaco, Bahamas: Jour. Sed. Petrology, v. 45, p. 763–786.

NEWMAN, W. A., AND LADD, H. S., 1974, Origin of coral-inhabiting balanids (Cirripedia, Thorocica): Verhandl, Naturforsch. Ges., Basel, v. 84, no. 1, p. 381–396.

ODUM, H. T., AND ODUM, E. P., 1955, Trophic structure and productivity of a windward coral reef community on Eniwetok Atoll: Ecol. Monogr., v. 25, no. 3, p. 291–320.

Oppenheim, P., 1900, Beiträge zur Kentniss des Oligocän und seiner Fauna in den venetianischen Voralpen: Zeitschr. Deutsch. Geol. Ges., v. 52, no. 2, p. 243–326.

———, 1901, Die Priabonaschichten und ihre fauna: Paleontographica, v. 47, 348 p.

———, 1909, Ueber schichtenfolge und fossilien von Laverda in der Marostica (Venetien): Paleontographica, v. 61, p. 36–55.

Opresko, D. M., 1973, Abundance and distribution of shallow water gorgonians in the area of Miami, Florida: Bull. Mar. Sci., v. 23, no. 3, p. 535–558.

Osasco, E., 1898, Di alcuni corallari oligocenici del Piemonte e della Ligiura: R. Accad. Sci. Torino., Atti., v. 33, p. 104–114.

———, 1902, Contribuzione allo studio dei coralli cenozoici del Veneto: Palaeontogr. Italica, v. 8, p. 99–120.

Pang, R. K., 1973, Coral reef project—papers in memory of Dr. Thomas F. Goreau. 9. The ecology of some Jamaican excavating sponges: Bull. Mar. Sci., v. 23, no. 1, p. 227–243.

Pfister, T., 1974, Palaeontologische untersuchungen an oligocaenen korallen der umgebung von San Luca (Provinz Vicenza-Nord-Italien) [unpub.]: Lizentiatsarbeit, Switzerland, Univ. Bern, 120 p.

———, 1977A, Systematische und paleoökologische untersuchungen an Oligozänenkorallen der umgebung von San Luca (Provinz Vicenza-Nord-Italien) [unpub. Ph.D. thesis]: Switzerland, Univ. Bern, 274 p.

———, 1977B, Das problem der variationsbreite van korallen am beispiel der oligozanen *Antiguastrea lucasiana* (Defrance): Eclog. Geol. Helvetiae., v. 70, no. 3, p. 825–843.

———, 1980, Systematische und paläoökologische untersuchungen an Oligozanen korallen der umgebung von San Luca (Provinz Vicenza-Nord-Italien): Schweizer, Palaont. Abhl. (in press).

Piccoli, G., 1958, Contributo alla conoscenza del vulcanismo terziario veneto: Atti Accad. Naz. Lincei, Rend. Cl. Sci. Fis., Mat., e Nat., pt. 8, v. 24, no. 5, p. 550–556.

———, 1965, Studio geologico del vulcanismo paleogenico veneto: Mem. Inst. Geol., Univ. Padova, v. 26, 98 p.

———, 1966, Studio geologico del vulcanismo paleogenico veneto: Mem. Inst. Geol., Univ. Padova, v. 26, 100 p.

———, 1967, Illustrazione della carta geologica del Marósticano occidentale fra Thiene e la valle del Torrente Laverda nel Vicentino: Mem. Inst. Geol., Univ. Padova, v. 26.

Pichon, M., 1974, Free living scleractinian coral communities in the coral reefs of Madagascar: International Coral Reef Symp., 2nd, Brisbane, Proc., v. 2, p. 173–181.

Porter, J. W., 1976, Autotrophy, heterotrophy, and resource partitioning in Caribbean reef building corals: Am. Naturalist, v. 110, p. 731–742.

Preston, E. M., and Preston, J. L., 1975, Ecological structure in a West Indian gorgonian fauna: Bull. Mar. Sci., v. 25, no. 2, p. 248–258.

Prever, P., 1921, 1922, I coralli oligocenici de Sassello nell Appennino Ligure. 1. Corallari a calici confluenti: Palaeontogr. Italica, v. 27 (1921), p. 53–100; v. 28 (1922), p. 1–40.

Proto Decima, F., and Sedea, R., 1970, Segnalazione di Oligocene marino nei Colli Euganei (Padova): Atti Accad. Naz. Lincei, Rend, Cl. Sci. Fis., Mat., e Nat., pt. 8, v. 48, no. 6, p. 646–653.

Reis, O., 1889, Korallen der Reiterschichten: Geognost. Jahresh., v. 2, p. 91–162.

Reiswig, H. M., Coral reef project—papers in memory of Dr. Thomas F. Goreau. 8. Population dynamics of three Jamaican Demospongiae: Bull. Mar. Sci., v. 23, no. 2, p. 191–226.

Reuss, A., 1864, Die fossilen Foraminifera, Anthozoen, und Bryozoen von Oberburg in Steiermark: K. Akad. Wiss., Wien, Math.–Naturwiss. Cl., Denkschr., v. 23, p. 1–36.

———, 1868, Paläeontologischen Studien ueber die älter Tertiärschichten der Alpen. 1. Die fossilen Anthozoen der Schichten von Castelgomberto: K. Akad. Wiss., Wien, Math.–Naturwiss. Cl., Denkschr., v. 28, p. 129–148.

———, 1869, Paläeontologischen Studien ueber die älter Tetiärschichten der Alpen. 2. Die fossilen Anthozoen und Bryozoen der Schichtengruppe von Crosara: K. Akad. Wiss., Wien, Math.–Naturwiss. Cl., Denkschr., v. 29, p. 215–298.

———, 1872, Paläeontologischen Studien ueber die älter Tertiärschichten der Alpen. 3. Die fossilen Anthozoen der Schichtengruppe von S. Giovanni Ilarione und von Ronca: K. Akad. Wiss., Wien, Math.–Naturwiss. Cl., Denkschr., v. 32, p. 1–60.

Ritondale Spano, Frascari, F., 1967, Osservazioni stratigrafiche e tettoniche sui dintorni di Calvene nel vicentiono: Gior. Geol., v. 34, no. 1, p. 307–340.

———, 1969, Serie paleogeniche nell area pedemontana a sud dell'Altopiano di Asiago (Vicenza): Mem. Bur. Rech. Geol. et Min., Mém., v. 69, p. 173–181.

Robertson, R., 1970, Review of the predators and parasites of stony corals with special reference to symbiotic prosobranch gastropods: Pacific Sci., v. 24, p. 43–54.

Rossi, D., and Semenza, E., 1958, Le scogliere oligoceniche dei Colli Berici: Ann. Univ. Ferrara (n.s.), Sez. 9, Sci. Geol. e Mineral., v. 3, no. 3, p. 49–70.

———, and ———, 1962, Recenti studi sull'Oligocene dei Colli Berici: Mem. Soc. Geol. Ital., v. 3, no. 3, p. 67–70.

Roveda, V., 1961, Contributo allo studio di alcuni macroforaminiferi di Priabona: Riv. Ital. Paleont. e Strat., v. 67, no. 2, p. 153–224.

Roy, K. J., and Smith, S. V., 1971, Sedimentation and coral reef development in turbid water: Fanning lagoon: Pacific Sci., v. 25, p. 235–248.

RUTZLER, K., AND MACINTYRE, I. G., 1978, Siliceous sponge spicules in coral reef sediments: Marine Biology, v. 49, p. 147–159.

SCHAFERSMAN, S. D., AND FROST, S. H., 1979, Tropical Cenozoic paleo-oceanography and correlated events in phylogeny and biogeography of scleractinian corals: Am. Assoc. Petroleum Geologists, Soc. Econ. Paleontologists Mineralogists Ann. Conv., Houston, Abs. with programs, p. 157.

SCHUHMACHER, H., 1977, Initial phases in reef development, studied at artificial reef types off Eilat, (Red Sea): Helgoländer Wiss. Meersunters, v. 30, p. 400–411.

SCHWEIGHAUSER, J., 1953, Mikropalaeontologische und stratigraphische untersuchungen im Palaeocaen und Eocaen des Vicentin (Nord-Italien): Schweiz. Palaont., Abbl., v. 70, p. 97.

SESTINI, N. FANTINI, 1960, La Fauna oligocenica dei dintorni di Ovada (Alessandria): Riv. Italiana Paleont. e Strat., v. 66, no. 3, p. 403–434.

SEUSS, E., 1868, Ueber die gliederung des vicentinischen Tertiärgebirges: Sitzber. Akad. Wiss., Math., Nat. Cl., v. 58, p. 265–279.

SHINN, E. A., 1972, Coral reef recovery in Florida and the Persian Gulf: Houston, Shell Oil Co. Pub., Environmental Conserv. Dept., 9 p.

SISMONDA, E., AND MICHELOTTI, G., 1871, Matériaux pour servir a la paléontologie du terrain tertiare de Piemont: R. Accad. Sci. Torino, Mem., ser. 2, v. 25, p. 257–362.

STEIGLITZ, R. D., 1972, Scanning electron microscopy of the fine fraction of Recent carbonate sediments from Bimini, Bahamas: Jour. Sed. Petrology, v. 42, p. 211–226.

——, 1973, Carbonate needles: additional organic sources: Geol. Soc. America Bull., v. 84, p. 927–930.

STOCKMAN, K. W., GINSBURG, R. N., AND SHINN, E. A., 1967, The production of lime mud by algae in south Florida: Jour. Sed. Petrology, v. 37, p. 633–648.

STODDART, D. R., 1969, Ecology and morphology of Recent reef corals: Bio. Rev., v. 44, p. 433–498.

THOMASSIN, B., 1971, Les biotopes de sables coralliens dérivant des appareils récifaux de la region Tuléar (S.W. de Madagascar), in Cochin, C. Mukundan, and Pillai, C. S. G., Eds., Corals Coral Reefs Symp., Mandapam Camp, 1969, Mar. Biol. Assn. India, p. 291–313.

UMBGROVE, J. H. F., 1928, De koraalriffen in de Baai van Batavia: Bandoeng Java, Wetensch. Meded. Dienst Mijnbouw, v. 7, p. 68.

UNGARO, S., 1969, Etude micropaléontogique et stratigraphique de l'Éocene supérieur (Priabonian) de Mossano (Colli Berici): Mém. Bur. Rech. Géol. et Min., Mém., no. 69, pt. 3, p. 267–280.

——, 1972, Studio microbiostratigrafica dei terreni sottostanti la scogliera di Lumignano (Colli Berici - Vicenza): Ann. Univ. Ferrara (n.s.), Sez. 9, Sci. Geol., Paleont., v. 5, pt. 3, p. 71–85.

VAIL, P. R, MITCHUM, R. M., JR., AND THOMPSON, S., III, 1977, Seismic stratigraphy and global changes of sea level, Part 3: Relative changes of sea level from coastal onlap, in Payton, C. E., Ed., Seismic Stratigraphy—applications of hydrocarbon exploration: Am. Assoc. Petroleum Geologists Mem. 26, p. 63–97.

VAUGHAN, T. W., 1907, Recent Madreporaria of the Hawaiian Islands and Laysan: U.S. Nat. Mus. Bull. 59, p. 1–222, p. 415–427.

——, 1919, Fossil corals from Central America, Cuba, and Puerto Rico, with an account of the American Tertiary, Pleistocene, and Recent coral reefs: U.S. Nat. Mus. Bull. 103, p. 189–524.

VENZO, S., 1938, La presenza del Cattiano a Molluschi nel Trevigniano e nel Bassanese. Serie terziaria e geomorfologia del Trevigniano occidentale: Boll. Soc. Geol. Ital., v. 57, no. 2, p. 179–206.

VERWEY, J., 1931, Coral reef studies, 2. The depth of coral reefs in relation to their oxygen consumption and the penetration of light in the water: Treubia, v. 13, no. 2, p. 169–198.

WALLACE, C. C., 1978, The coral genus Acropora (Scleractinia: Astrocoeniina: Acroporidae) in the Central and Southern Great Barrier Reef Province: Mem. Queensland. Mus., v. 18, no. 2, p. 273–319.

WALLACE, R. J., AND SCHAFERSMAN, S. D., 1977, Patch-reef ecology and sedimentology of Glovers Reef Atoll, Belize, in Frost, S. H., Weiss, M. P., and Saunders, J. B., Eds. Reefs and Related Carbonates/Ecology and Sedimentology: Am. Assoc. Petroleum Geologists, Studies in Geology, no. 4, p. 37–52.

WANTLAND, K. F., 1975, Distribution of Holocene benthonic foraminifera on the Belize Shelf, in Wantland, K. F., and Pusey, W. C., III, Eds., Belize Shelf-Carbonate Sediments, Clastic Sediments, and Ecology: Am. Assoc. Petroleum Geologists, Studies in Geology, No. 2, p. 332–399.

WARME, J. E., 1977, Carbonate borers—their role in reef ecology and preservation, in Frost, S. H., Weiss, M. P., and Saunders, J. B., Eds., Reefs and Related Carbonates—Ecology and Sedimentology: Am. Assoc. Petroleum Geologists, Studies in Geology, no. 4, p. 261–279.

WELLS, J. W., 1954, Recent corals of the Marshall Islands: U.S. Geol. Survey Professional Paper 260–261, p. 385–486.

——, 1957, Chapter 20, Coral Reefs, in Hedgepeth, J. W., Ed., Treatise on Marine Ecology and Paleoecology: Geol. Soc. America Mem. 67, p. 609–631.

——, 1969, Aspects of Pacific coral reefs: Micronesica, v. 5, no. 2, p. 317–322.

——, 1973A, Note on the scleractinian corals Scolymia lacera and S. cubensis in Jamaica: Bull. Mar. Sci., v. 21, no. 4, p. 960–963.

——, 1973B, Coral reef project–papers in memory of Dr. Thomas F. Goreau. 2. New and old scleractinian corals from Jamaica: Bull. Mar. Sci., v. 23, no. 1, p. 16–58.

WIJSMAN-BEST, M., 1972, Systematics and ecology of New Caledonian Faviinae (Coelenterata-Scleractinia): Proefsch., Bijdragen tot de Dierkunde, v. 42, no. 1, p. 71.

————, 1974A, Habitat-induced modification of reef corals (Faviidae) and its consequences for taxonomy: International Coral Reef Symp., 2nd, Brisbane, Proc., v. 2, p. 217–228.

————, 1974B, Biological results of the Snellius Expedition. 25. Faviidae collected by the Snellius Expedition. 1. The genus *Favia*: Zool. Meded., v. 48, no. 22, p. 249–260.

————, 1976, Biological results of the Snellius Expedition. 27. Faviidae collected by the Snellius Expedition. 2. The genera *Favites, Goniastrea, Platygyra, Oulophyllia, Leptoria, Hydnophora,* and *Caulastrea*: Zool. Meded., v. 50, no. 4, p. 45–61.

————, 1977, Indo-Pacific coral species belonging to the subfamily Montastreinae Vaoughan and Wells, 1943 (Seleractinia—Coelenterata). Part 1. The genera *Montastrea* and *Plesiastrea*: Zool. Meded., v. 52, no. 7, p. 81–97.

WRAY, J. L., 1977, Calcareous Algae: Amsterdam, Elsevier, 180 p.

YONGE, C. M., 1963, The biology of coral reefs: Advances in Marine Biology, v. 1, p. 209–260.

SEPM SPECIAL PUBLICATION NO. 30, P. 541–544, MAY 1981

# ABOUT THE AUTHORS

### LJUBO BABIĆ

Ljubo Babić was born in Zagreb, Yugoslavia, and received his education at the University of Zagreb, where he was awarded the Ph.D. degree in 1974. From 1962 until 1965 he worked at the Institute of Geological Research of Croatia, and from there he began teaching at the University of Zagreb. He became an Assistant Professor in 1974, and in 1979 he was promoted to Associate Professor on the Faculty of Science. He is a recipient of the "Milan Miličević" medal awarded in 1973 for the best geological paper by a young research geologist. Initially, his interests lay in the fields of regional geology and stratigraphy, but more recently his interests have changed to sedimentology and paleoenvironmental interpretation of Mesozoic and Tertiary facies trends in western Yugoslavia. In addition to academic duties he has served as leader of various stratigraphic research projects for INA-Naftaplin Company during the period 1967–1976. More recently (1977) he was instrumental in establishing a Sedimentology Section in the Croatian Geological Survey and was its first Director.

### RAINER BRADNER

Rainer Bradner was born in Kitzbühel, Austria (Tyrol), and received his Ph.D. degree from the University of Innsbruck in 1972. Since then, Bradner has served as Assistant Professor in the Geological Institute of the University of Innsbruck. His main field of interest is carbonate sedimentology; his recent research has dealt with Triassic tectonics and sedimentation of the Tethyan realm. Bradner is presently completing his "Habilitation" accreditation at the University of Innsbruck.

### TREVOR P. BURCHETTE

Trevor Burchette was born in London, England. He obtained his B.S. degree in Geology (1973) at University College, Cardiff, of the University of Wales, and researched for his Ph.D. degree (1977) at both Cardiff and the University of Newcastle-upon-Tyne. Following this, he was a Royal Society post-doctoral research fellow at the Technische Universität Braunschweig (West Germany). He is currently a sedimentologist with the British Petroleum Company, Ltd. in London. Burchette's research interests include the interpretation of Paleozoic and Mesozoic shallow water carbonate and mixed carbonate-terrigenous depositional environments, reefal and broad shelf depositional processes, and the paleoecology and diagenesis of limestones.

### STANKO BUSER

Stanko Buser was born in Boletina, Yugoslavia (Slovenia), and received his Ph.D. degree in 1966 from the University of Ljubljana, Yugoslavia. Since 1959 he has been associated as a regional geologist with the Geological Survey of Ljubljana, where he has been working on a comprehensive geological map of Slovenia. His special fields of interest have centered around Paleozoic and Mesozoic stratigraphy, paleoenvironments, and tectonics, and his publications reflect these areas of endeavor. He has also been investigating various mineral resources (copper, bauxite, and mercury).

### JOŽE ČAR

Jože Čar was born in Idrija, northwestern Yugoslavia (Slovenia), and received his B.S. degree in Geology in 1966 from Edvard Kardelj University in Ljubljana, Yugoslavia. Upon completion of his degree, he was hired by the Idrija Mercury Mine. Here his work dealt with a reconstruction of the Middle Triassic fault trench in which the Idrija mercury ore body is located. He has also been involved with both stratigraphic and sedimentological problems concerning the clastic "Langobard layers" of the Idrija region. Since 1977, he has been employed by the Institute for Karst Research of the Slovene Academy of Sciences in Postojna. Here his interests are mainly devoted to karst-hydrologic problems although he has continued to do research on Middle Triassic sedimentary rocks in western Slovenia.

### FEDERICO CARBONE

Federico Carbone was born in Reggio Calabria, Italy, and received his degree of Doctor in Geological Sciences in 1969. Subsequently, he was under contract to the Geological Survey of Italy, where he was engaged in research on carbonate sedimentology. He then became a staff scientist of the National Research Council at their "Center for the Geology of Central Italy," which is located at the Institute of Geology and Paleontology at the University of Rome. Carbone's research is devoted to carbonate sedimentology, especially facies analyses of Mesozoic carbonate shelfs.

### ERIK FLÜGEL

Erik Flügel was born in Fürstenfeld, Austria, and received his Ph.D. degree in geology and paleontology at the University of Graz, Austria, in 1957. From 1958 to 1962, he served as curator of paleontology at the Natural History Museum in Vienna. From 1962 until 1972, he was affiliated with the Technical University Darmstadt, West Germany. Since 1972, he has been with the Paleontological Institute of Erlangen-Nürnberg University, where he holds the chair of Paleontology. His main field of interest is facies analyses of carbonate rocks in which paleontological and sedimentological data are integrated. His work has dealt with fossil marine invertebrates, calcareous algae, and carbonate microfacies. His current research is devoted to the paleoecology of Triassic reefs in the Alpine-Mediterranean region and to Permian shelf and reef carbonates in the Southern Limestone Alps.

### STANLEY H. FROST

Stan Frost was born in northern Illinois and received his Ph.D. degree in 1966 from the University of Illinois at Champaign-Urbana. Other than a brief fling with Devonian carbonates of Nevada, his main interest centers on the paleoecology and sedimentology of Cenozoic reef buildups, their coral and large foram faunas, and the application of modern reefs as a standard of comparison in reconstructing the geologic complexities of their ancient counterparts. As Assistant, then Associate Professor of Geology at Northern Illinois University (DeKalb) from 1965 to 1977, he pursued research on Tertiary reefs of the Caribbean-western Atlantic and western Tethyan regions and on modern reefs of the Caribbean-western Atlantic and Indo-Pacific regions. Since 1977, he has been employed by Gulf Oil Company, Research and Development Section, in Houston, Texas and is presently studying Cretaceous and Tertiary reefs and carbonates on a worldwide basis.

### MARK M. LONGMAN

Mark W. Longman was born in Washington, D.C., and received his Ph.D. degree from the University of Texas at Austin in 1976. Since then, he has worked for Cities Service Company at their Exploration and Production Research Laboratory in Tulsa, Oklahoma. Much of his work has involved oil-producing Miocene reefs in the Philippine Islands, but he has also examined Tertiary reefs in the Mediterranean region and Indonesia. His work on Tertiary reefs with observations of sedimentologic facies distribution, combined with his work on depositional processes on modern reefs in the Caribbean and southwest Asia regions, has led to the paper in this volume. Other areas of Longman's research include carbonate petrology and paleoecology.

### JEAN-PIERRE MASSE

Jean-Pierre Masse was born in Cavaillon, France, and received his Ph.D. degree in 1976

from the University of Marseille. He is associated with the University, at the Centre National de la Recherche Scientifique. Masse has worked on Recent tropical carbonates in Madagascar and Sénégal, but his primary research interest is in Lower Cretaceous (Urgonian) limestones of Provence, southeastern France. Presently, he is working on Lower Cretaceous biosedimentological models for the peri-Mediterranean realm.

### BOJAN OGORELEC

Bojan Ogorelec was born in Klagenfurt, Austria, and received his B.S. degree in 1970 from the University of Ljubljana, Yugoslavia. Since 1971, he has been employed by the Geological Survey in Ljubljana as a sedimentologist. His research interests have centered around various aspects of both Ancient and Recent carbonates including petrology, depositional environments, geochemistry, and diagenesis. In 1974, Ogorelec studied his specialities at the Institute of Sedimentology, Heidelberg, West Germany.

### JEAN M. PHILIP

Jean M. Philip was born in Marseille, France, and obtained his State doctorate thesis in 1970. He presently teaches stratigraphy and paleontology at the University of Provence. His research interests center around Upper Cretaceous rudistid buildups in the Mediterranean region, including stratigraphy, paleoecology, sedimentology, and paleobiogeography.

### WERNER E. PILLER

Werner E. Piller was born in Vienna, Austria, and received his Ph.D. degree from the University of Vienna in 1975. He is an Assistant Professor of Geology at the Institute of Paleontology of the University of Vienna. His research interests center around the Upper Triassic carbonates of the Northern Limestone Alps of Austria, especially the paleoecology and taxonomy of foraminifers. Recently, he has been engaged in research on Miocene carbonates of the Vienna Basin and their contained corallinacean algae.

### LADISLAV PLACER

Ladislav Placer was born in Divača near Trieste, Yugoslavia (Slovenia), and obtained his B.S. degree from Edvard Kardelj University in Ljubljana, Yugoslavia, in 1969. His first employment was at the famous Idrija Mercury Mines; presently, he is employed at the Geological Survey in Ljubljana. His major areas of interest include structural geology and regional tectonics, and he is especially interested in the structural history that led to the formation of the Middle Triassic Idrija mercury ore body.

## WERNER RESCH

Werner Resch was born in Vorarlberg, Austria, and received his Ph.D. degree from the University of Innsbruck in 1963. His dissertation topic treated various problems relating to Subalpine molasse sequences. He has done post-graduate work on micropaleontology at the Austrian Geological Survey in Vienna and at the University of Kiel in West Germany. He is presently employed as an Assistant Professor at the University of Innsbruck, Austria. In 1978 he completed his "Habilitation" thesis on Triassic foraminifers.

## ROBERT RIDING

Robert Riding was born in Birmingham, England, and received his Ph.D. degree from the University of Sheffield. An early interest in reefs led to his doctoral work on carbonate mound complexes in the Carboniferous of the Cantabrian Mountains of northern Spain. This in turn stimulated an interest in calcareous algae which was further developed during a post-doctoral fellowship at McGill University, Montreal, Canada, where he did work on the Devonian reefs of western Canada. In 1972 he returned to England, to the University of Newcastle-upon-Tyne, and is presently on the staff of University College, Cardiff, Wales. Riding's specialization in calcareous algae has taken him to a range of geologic deposits, mainly Paleozoic, in several parts of North America and Europe, and the role of algae in the Silurian Gotland reefs is a strong interest currently. He firmly believes in the need to combine paleontological and sedimentological studies in order to interpret the inherent complexities of algal and reef carbonates. In 1974, he founded and continues to edit "The Reef Newsletter," and he has been a Visiting Professor at the Paleontological Institute of the University of Erlangen-Nürnberg, West Germany. He is presently Secretary of the Paleontological Association, United Kingdom.

## PRISKA SCHÄFER

Priska Schäfer was born in Wilhelmshaven, West Germany, and received her Ph.D. degree in 1978 from the Paleontological Institute of the University of Erlangen-Nürnberg, West Germany. Her dissertation topic dealt with the microfacies and paleoecology of Alpine Upper Triassic organic buildups. She is presently a scientific collaborator at the Paleontological Institute in Erlangen, and she is broadening her scientific investigations to include Triassic (Carnian-Norian) reef algae and corals of Greece and Sicily.

## BABA SENOWBARI-DARYAN

Baba Senowbari-Daryan was born in Tabriz, Iran. He left Iran in 1965 to pursue his education in West Germany and Austria. He received his Ph.D. degree in 1978 from the Institute of Paleontology, University of Erlangen-Nürnberg. His dissertation topic was concerned with paleontology and microfacies analyses of Triassic reef carbonates of the Northern Limestone Alps of Austria. He is presently a scientific collaborator at the Paleontological Institute at Erlangen, West Germany, where he is extending his work on Upper Triassic reefs into Sicily, Greece, and Yugoslavia.

## GIUSEPPE SIRNA

Giuseppe Sirna was born in Catania, Italy. In 1964 he was appointed an Assistant Professor of Paleontology at the University of Rome, and in 1975 he became a full Professor in Earth Sciences. His specialized areas of interest are Mesozoic pelecypods and gastropods. Presently, Sirna is Director of the Geological and Paleontological Institute at the University of Rome, Italy. He is also Vice President of the Italian Paleontological Association.

## DRAGOMIR SKABERNE

Dragomir Skaberne was born in Ljubljana, Yugoslavia (Slovenia), and received his B.S. degree from the University of Ljubljana in 1973. During the 1976–1977 academic year he was enrolled in a specialization course at the Institute of Sedimentology in Heidelberg, West Germany. Skaberne works as an assistant in the Department of Geology at the University of Ljubljana and is interested in a wide range of geological topics, including ore deposits, petrography, clastic sedimentation, depositional environments, geochemistry, and diagenesis.

## DENYS B. SMITH

Denys Smith was born in Wybunbury, England, and received his Doctor of Science degree at the University of Birmingham. Apart from a brief attachment to the New Mexico Bureau of Mines and Mineral Resources at Socorro, New Mexico, where he worked on the classic Late Permian (Guadalupian) shelf carbonates of the Carlsbad area, he has been employed throughout his career by the Geological Survey of Great Britain. His main interests are in the stratigraphy, sedimentology, and paleogeography of Permian sediments and evaporites in and around the Zechstein Basin. He is presently a District Geologist based in London.

## TORSTEN STEIGER

Torsten Steiger was born in West Germany and is presently studying at the University of Erlangen-Nürnberg. His area of specialization is microfacies analyses and paleontology of Upper Jurassic limestones in the Alpine region of the Tethyan Geosyncline. Since 1980, Steiger has been devoting considerable time to core studies dealing with the African Atlantic Continental Margin, as part of the Deep Sea Drilling Program.

## DRAGICA TURNŠEK

Dragica Turnšek was born in Šalamenci, Yugoslavia (Slovenia), and she received her Ph.D. degree in 1965 from the University of Ljubljana, Yugoslavia. Since 1954 she has been associated with the Institute of Paleontology of the Slovene Academy of Sciences and Arts in Ljubljana. In 1971 she was a Humboldt scholar at Stuttgart University in West Germany. Her research interests include paleontology and paleoecology of Mesozoic Cnidaria (corals, stromatoporoids, and chaetetids). Her publications focus on cnidarians from Yugoslavia, Spain, France, and Roumania. She is presently engaged in synthesizing the work relating to Triassic reef building corals in Yugoslavia.

## JOŽICA ZUPANIČ

Jožica Zupanič was born in Beograd, Yugoslavia, and was awarded a Ph.D. degree in 1974 from the University of Zagreb, Yugoslavia, where she is presently an Assistant Professor of Geology. Her main interest is sedimentology, including facies and environmental analyses. She is also interested in turbidite and pelagic sedimentation. For many years she has been involved in cooperative projects in stratigraphy and sedimentology for INA-Naftaplin Company.

# AUTHORS ADDRESSES

LJUBO BABIĆ
Department of Geology
University of Zagreb
YU-41 Zagreb
Yugoslavia

RAINER BRADNER
Institut für Geologie und Paläontologie der
    Universität Innsbruck
A-6020 Innsbruck
Universitätstrasse 4/2
Austria

TREVOR P. BURCHETTE*
2 The Close
Montreal Park, Sevenoaks
Kent TN13 2HE
United Kingdom

STANKO BUSER
Geological Survey
Parmova 33
61000 Ljubljana
Yugoslavia

JOŽE ČAR
Institute For Karst, SAZU
Titov trg 2
67230 Postojna
Yugoslavia

FEDERICO CARBONE
Istituto di Geologia
Piazzale del Scienza
Cita Universitaria
00100 Rome
Italy

ERIK FLÜGEL
Institut für Paläontologie
Universität Erlangen-Nürnberg
    Loewenichstrasse 28
D-8520 Erlangen
West Germany

STANLEY H. FROST
Gulf Research & Development Company
P.O. Box 36506
Houston, Texas 77036
U.S.A.

MARK W. LONGMAN
Cities Service Research Laboratory
Box 50408
Tulsa, Oklahoma 74110
U.S.A.

JEAN-PIERRE MASSE
Université de Marseille-Luminy
Laboratoire de Géologie
13000 Marseille
France

BOJAN OGORELEC
Geological Survey
Parmova 33
61000 Ljubljana
Yugoslavia

JEAN M. PHILIP
Université D'Aix-Marseille I
Laboratoire de Géologie Historique et
    de Paléontologie
Place Victor Hugo
13331 Marseille Cedex 3
France

WERNER PILLER
Paläontologisches Institut der Universität Wien
A-1010 Wien 1
Universitätstrasse 7
Austria

LADISLAV PLACER
Geological Survey
Parmova 33
6100 Ljubljana
Yugoslavia

ANTE POLŠAK
Department of Geology
University of Zagreb
YU-41 Zagreb
Yugoslavia

WERNER RESCH
Institut für Geologie und Paläontologie der
    Universität Innsbruck
A-6020 Innsbruck
Universitätstrasse 4/2
Austria

ROBERT RIDING
Department of Geology
University College
Cardiff CF1 1XL
United Kingdom

PRISKA SCHÄFER**
Institut für Paläontologie
Universität Erlangen-Nürnberg
    Loewenichstrasse 28
D-8520 Erlangen
West Germany

**BABA SENOWBARI-DARYAN**
Institut für Paläontologie
Universität Erlangen-Nürnberg
  Loewenichstrasse 28
D-8520 Erlangen
West Germany

**GIUSEPPE SIRNA**
Istituto di Geologia i Paleontologia
Cita Universitaria
00100 Rome
Italy

**DRAGOMIR SKABERNE**
Department of Geology
University of Ljubljana
Aškerčeva 20
61000 Ljubljana
Yugoslavia

**DENYS B. SMITH**
Institute of Geological Sciences
5, Princes Gate
London SW7 1QN
United Kingdom

**TORSTEN STEIGER**
Institut für Paläontologie
Universität Erlangen-Nürnberg
  Loewenichstrasse 28
D-8520 Erlangen
West Germany

**DRAGICA TURNŠEK**
Institute for Paleontology SAZU
Stari trg 3
61000 Ljubljana
Yugoslavia

**JOŽICA ZUPANIČ**
Department of Geology
University of Zagreb
YU-41 Zagreb
Yugoslavia

Present Addresses:

**TREVOR P. BURCHETTE**\*
British Petroleum Co., Ltd.
Britannic House, Moor Lane
London EC2Y 9BU
United Kingdom

**PRISKA SCHÄFER**\*\*
Geological Institute
University of Marburg
D-3550 Marburg
West Germany